St. Olaf College

AUG 12 1983

Science Library

Hua, Lo-Keng.

Hua Loo Keng

Introduction to Number Theory

Translated from the Chinese by Peter Shiu

With 14 Figures

Springer-Verlag
Berlin Heidelberg New York 1982

Hua Loo Keng

Institute of Mathematics
Academia Sinica
Beijing
The People's Republic of China

Peter Shiu

Department of Mathematics
University of Technology
Loughborough
Leicestershire LE 11 3 TU
United Kingdom

QA
241
H7513

ISBN 3-540-10818-1 Springer-Verlag Berlin Heidelberg New York
ISBN 0-387-10818-1 Springer-Verlag New York Heidelberg Berlin

Library of Congress Cataloging in Publication Data. Hua, Loo-Keng, 1910 –. Introduction to number theory. Translation of: Shu lun tao yin. Bibliography: p. Includes index. 1. Numbers, Theory of. I. Title. QA241.H7513. 512'.7. 82-645. ISBN 0-387-10818-1 (U.S.). AACR2

This work is subject to copyright. All rights are reserved, whether the whole or part of the material is concerned, specifically those of translation, reprinting, reuse of illustrations, broadcasting, reproduction by photocopying machine or similar means, and storage in data banks. Under § 54 of the German Copyright Law where copies are made for other than private use a fee is payable to "Verwertungsgesellschaft Wort", Munich.

© Springer-Verlag Berlin Heidelberg 1982
Printed in Germany

Typesetting: Buchdruckerei Dipl.-Ing. Schwarz' Erben KG, Zwettl. Printing and binding: Konrad Triltsch, Würzburg
2141/3140-5 4 3 2 1 0

Preface to the English Edition

The reasons for writing this book have already been given in the preface to the original edition and it suffices to append a few more points.

In the original edition I collected various recent results in number theory and put them in a text book suitable for teaching purposes. The book contains: The elementary proof of the prime number theorem due to Selberg and Erdős; Roth's theorem; A. O. Gelfond's solution to Hilbert's seventh problem; Siegel's theorem on the class number of binary quadratic forms; Linnik's proof of the Hilbert-Waring theorem; Selberg's sieve method and Schnirelman's theorem on the Goldbach problem; Vinogradov's result concerning least quadratic non-residues. It also contains some of my own results, for example, on the estimation of complete trigonometric sums, on least primitive roots, and on the Prouhet-Tarry problem. The reader can see that the book is much influenced by the work of Landau, Hardy, Mordell, Davenport, Vinogradov, Erdős and Mahler. In the quarter of a century between the two editions of the book there have been, of course, many new and exciting developments in number theory, and I am grateful to Professor Wang Yuan for incorporating many new results which will guide the reader to the literature concerning the latest developments.

It has been doubtful in the past whether number theory is a "useful" branch of mathematics. It is futile to get too involved in the argument but it may be relevant to point out some specific examples of applications. The fundamental principle behind the *Public Key Code* is the following: It is not difficult to construct a large prime number but it is not easy to factorize a large composite integer. For example, it only takes 45 seconds computing time to find the first prime exceeding 2^{200} (namely $2^{200} + 235$, a number with 61 digits), but the computing time required to factorize a product of two primes, each with 61 digits, exceeds 4 million million years. According to Fermat's theorem: if p is prime then $a^{p-1} \equiv 1 \pmod{p}$, and if n is composite then $a^{\phi(n)} \equiv 1 \pmod{n}$, $\phi(n) < n - 1$. The determination of whether n is prime by this method is quite fast and this is included in the book. Next the location of the zeros of the Riemann Zeta function is a problem in pure mathematics. However, an interesting problem emerged during calculations of these zeros: Can mathematicians always rely on the results obtained from computing machines, and if there are mistakes in the machines how do we find out? Generally speaking calculations by machines have to be accepted by faith. For this reason Rosser, Schoenfeld and Yohe were particularly careful when they used computers to calculate the zeros of the Riemann Zeta function. In their critical examination of the program they discovered that there were several logical errors in the machine itself. The machine has been in use for some years and no-one had found these errors until the three mathematicians wanted to scrutinize the results on a problem which has no practical applications. Apart from these there are applications from algebraic number theory and from the theory of rational approximations to real numbers which we need not mention.

Finally I must point out that this English edition owes its existence to Professor Heini Halberstam for suggesting it, to Dr. Peter Shiu for translating it and to Springer-Verlag for publishing it. I am particularly grateful to Peter Shiu for his excellent translation and to Springer-Verlag for their beautiful printing.

March 1981, Beijing Hua Loo-Keng

Preface to the Original Edition

This preface has been revised more than once. The reason is that, during the last fifteen years, the author's knowledge of mathematics has changed and the needs of the readers are different. Moreover the content of the book has been so expanded during this period that the old preface has become quite unsuitable.

Everything is still very clear in my memory. The plan for the book was conceived round about 1940 when I first lectured on number theory at Kwang Ming University. I had written some 85 thousand words (characters) for the first draft and I estimated that another 25 thousand words were needed to complete the manuscript. But where was I to publish the work? I therefore could not summon up the energy required to complete the project. Later when lecturing in America I made additions and revisions to the manuscripts, but these were made for my teaching requirements and not with a view to publishing the book.

The real effort required for the task was given after the liberation. Since our country has very few reference books there is need for a broad introductory text in number theory. It seems a little peculiar that, even though we have been busier after the liberation, with the help of comrades the project actually has progressed faster. The book has also increased in size with the addition of new chapters and the incorporation of recent results which are within its scope.

Apart from giving a broad introduction to number theory and some of its fundamental principles the author has also tried to emphasize several points to its readers.

First there is a close relationship between number theory and mathematics as a whole. In the history of mathematics we often see the various problems, methods and concepts in number theory having a significant influence on the progress of mathematics. On the other hand there are also frequent instances of applying the methods and results of the other branches of mathematics to solve concrete problems in number theory. However it is often not easy to see this relationship in many existing introductory books. Indeed many "self-contained" books for beginners in number theory give an erroneous impression to their readers that number theory is an isolated and independent branch of mathematics. In this book the author tries to highlight this relationship within the scope of elementary number theory. For example: the relationship between the prime number theorem and Fourier series (the limitation on the nature of the book does not allow us to describe the relationship between the prime number theorem and integral functions); the partition problem, the four squares problem and their relationship to modular functions, the theory of quadratic forms, modular transformations and their relationship to Lobachevskian geometry etc.

Secondly an important progression in mathematics is the development of abstract concepts from concrete examples. Specific concrete examples are often the basis of abstract notions and the methods employed on the examples are frequently the source of deep and powerful techniques in advanced mathematics. One cannot go very far by merely learning bare definitions and methods from abstract notions without knowing the source of the definitions in the concrete situation. Indeed such an approach may lead to insurmountable difficulties later in research situations. The history of mathematics is full of examples in which whole subjects were developed from methods employed to tackle practical problems, for example, in mechanics and in physics. As for mathematics itself the most fundamental notions are "numbers" and "shapes". From "shapes" we have geometric intuition and from "numbers" we have arithmetic operations which are rich sources for mathematics. In this book the author tries to bring out the concrete examples underlying the abstract notions hoping that the readers may remember them when they make further advances in mathematics. For example, in Chapter 4 and Chapter 14, concrete examples are given to illustrate abstract algebra; indeed the example on finite fields describes the situation of general finite fields.

Thirdly, for beginners engaging in research, a most difficult feature to grasp is that of quality – that is the depth of a problem. Sometimes authors work courageously and at length to arrive at results which they believe to be significant and which experts consider to be shallow. This can be explained by the analogy of playing chess. A master player can dispose of a beginner with ease no matter how hard the latter tries. The reason is that, even though the beginner may have planned a good number of moves ahead, by playing often the master has met many similar and deeper problems; he has read standard works on various aspects of the game so that he can recall many deeply analyzed positions. This is the same in mathematical research. We have to play often with the masters (that is, try to improve on the results of famous mathematicians); we must learn the standard works of the game (that is, the "well-known" results). If we continue like this our progress becomes inevitable. This book attempts to direct the reader to work in this way. Although the nature of the book excludes the very deep results in number theory the author introduces different methods with varying depths. For example, in the estimation of the partition function $p(n)$, the simplest of algebraic methods is used first to get a rough estimate, then using a slightly deeper method the asymptotic formula for $\log p(n)$ is obtained. It is also indicated how an asymptotic formula for $p(n)$ can be obtained by a Tauberian method and how an asymptotic expansion for $p(n)$ can be obtained using results in advanced modular function theory and methods in analytic number theory. It is then easy to judge the various levels of depth in the methods used by following the successive improvement of results.

The book is not written for a university course; its content far exceeds the syllabus for a single course in number theory. However lecturers can use it as a course text by taking Chapters 1 – 6 together with a suitable selection from the other chapters. Actually the book does not demand much previous knowledge in mathematics. Second year university students could understand most of the book, and those who know advanced calculus could understand the whole book apart from Sections 9.2, 12.14, 12.15 and 17.9 where some knowledge of complex functions theory is required. Those studying by themselves should not find any special difficulties either.

I am eternally grateful to the following comrades: Yue Min Yi, Wang Yuan, Wu Fang, Yan Shi Jian, Wei Dao Zheng, Xu Kong Shi and Ren Jian Hua. Since 1953, when I began my lectures, they have continually given me suggestions, and sometimes even offer to help with the revision. They have also assisted me throughout the stages of publication, particularly comrade Yue Min Yi. I would also like to thank Professor Zhang Yuan Da for his valuable suggestion on a method of preparing the manuscript for the typesetter.

Although we have collectively laboured over the book it must still contain many mistakes. I should be grateful if readers would inform me of these, whether they are misprints, errors in content, or other suggestions. There is much material that appears here for the first time in a book, as well as some unpublished research material, so that there must be plenty of room for improvement. Concerning this point we invite the readers for their valuable contributions.

September 1956, Beijing Hua Loo-Keng

Table of Contents

List of Frequently Used Symbols

$[\alpha]$ = the greatest integer not exceeding α.
$\{\alpha\} = \alpha - [\alpha]$ = the fractional part of α.
$\langle\alpha\rangle$ = the distance of α from the nearest integer, that is $\min(\alpha - [\alpha], [\alpha] + 1 - \alpha)$.
$(a, b, \ldots, c)$ = the greatest common divisor of $a, b, \ldots, c$.
$[a, b, \ldots, c]$ = the least common multiple of $a, b, \ldots, c$.
$a | b$ means a divides b.
$a \nmid b$ means a does not divide b.
$p^u \| a$ means $p^u | a$ and $p^{u+1} \nmid a$.
$a \equiv b \pmod{m}$ means $m | a - b$.
$a \not\equiv b \pmod{m}$ means $m \nmid a - b$.
$\prod_{d|m}$ and $\sum_{d|m}$ denote the product and the sum over the divisors d of m.
$\left(\frac{n}{p}\right)$ is Legendre's symbol; see §3.1.
$\left(\frac{n}{m}\right)$ is Jacobi's symbol; see §3.6.
$\left(\frac{d}{m}\right)$ where d is not a perfect square, $d \equiv 0$ or $1 \pmod 4$ and $m > 0$, is Kronecker's symbol; see §12.3.
$\operatorname{ind} n$ denotes the index of n; see §3.8.
$\partial^0 f$ denotes the degree of the polynomial $f(x)$.
$\ll$, O, o, $\sim$ see §5.1.
$\omega(n)$ denotes the number of distinct prime divisors of n.
$\Omega(n)$ denotes the total number of prime divisors of n.
$\max(a, b, \ldots, c)$ denotes the greatest number among $a, b, \ldots, c$.
$\min(a, b, \ldots, c)$ denotes the least number among $a, b, \ldots, c$.
$\Re s$ denotes the real part of the complex number s.
γ denotes Euler's constant.
$\{a, b, c\}$ represents the quadratic form $ax^2 + bxy + cy^2$; see §12.1.
(z_1, z_2, z_3, z_4) denotes the cross ratio of the four points z_1, z_2, z_3, z_4; see §13.3.
$A \overset{L}{=} B$ means that the matrices A and B are left associated.
$N(\mathfrak{M})$ denotes the norm of $\mathfrak{M}$; see §14.9.
$\{a_n\}$ denotes the sequence $a_1, a_2, \ldots$.
$\sim$ is an equivalence sign; see §12.1, §13.6, §14.5 and §16.12.

$[a_0, a_1, \ldots, a_N]$ or $a_0 + \frac{1}{a_1 +} \frac{1}{a_2 +} \cdots \frac{1}{+ a_N}$ denotes a finite continued fraction;
$p_n/q_n = [a_0, a_1, \ldots, a_n]$ is the n-th convergent of a continued fraction.
$S(\alpha) = \alpha^{(1)} + \alpha^{(2)} + \cdots + \alpha^{(n)}$ is the trace of α.
$N(\alpha) = \alpha^{(1)}\alpha^{(2)} \cdots \alpha^{(n)}$ is the norm of α.
$\Delta(\alpha_1, \ldots, \alpha_n)$ denotes the discriminant of $\alpha_1, \ldots, \alpha_n$; $\Delta = \Delta(R(\vartheta))$ denotes the discriminant of the integral basis for $R(\vartheta)$. See §16.3 and §16.4.
$\varphi(m)$ is Euler's function; see §2.3.
$\operatorname{li} x$ see §5.2.
$\pi(x)$ see §5.3.
$\mu(m)$ see §6.1.
$d(n)$ see §6.1.
$\sigma(n)$ see §6.1.
$\Lambda(n)$ see §6.1.
$\Lambda_1(n)$ see §6.1.
$\chi(n)$ see §7.2.
$p(n)$ see §8.2.
$\vartheta(n)$ see §9.1.
$\psi(n)$ see §9.1.
$g(k)$ see §18.1.
$G(k)$ see §18.1.
$v(k)$ see §18.5.
$N(k)$ see §18.6.
$M(k)$ see §18.6.
$\zeta(s) = \sum_{n=1}^{\infty} 1/n^s$ is the Riemann Zeta function.
$e(f(x)) = e^{2\pi i f(x)}$, $e_q(f(x)) = e^{2\pi i f(x)/q}$.
$S(a, \chi) = \sum_{n=1}^{m} \chi(n) e^{2\pi i a n/m}$ is a character sum.
$\tau(\chi) = S(1, \chi)$.
$S(n, m) = \sum_{x=0}^{m-1} e^{2\pi i n x^2/m}$, $(n, m) = 1$, is a Gauss sum.
$S(q, f(x)) = \sum_{x=0}^{q-1} e_q(f(x))$.

Chapter 1. The Factorization of Integers

Throughout this chapter the small Latin letters

$$a, b, \ldots, n, \ldots, p, \ldots, x, y, z$$

represent integers. The main purpose of the chapter is to prove the *Fundamental Theorem of Arithmetic* (Theorem 5.3) and its various applications.

1.1 Divisibility

We call the numbers $1, 2, 3, \ldots$ the *natural numbers* and

$$\ldots, -2, -1, 0, 1, 2, \ldots$$

the *integers*, so that the natural numbers are sometimes called the positive integers. It is clear that the sum, the difference, and the product of two integers are also integers. We say that the set of integers is "closed with respect to the three operations of addition, subtraction and multiplication".

Let α be a real number. We denote by $[\alpha]$ the greatest integer not exceeding α. For example

$$[3] = 3, \qquad [\sqrt{2}] = 1, \qquad [\pi] = 3, \qquad [-\pi] = -4.$$

If α is positive, then $[\alpha]$ is simply the integer part of α, and we always have

$$[\alpha] \leqslant \alpha < [\alpha] + 1.$$

We now take α to be a rational number a/b, $b > 0$. Then we have

$$0 \leqslant \frac{a}{b} - \left[\frac{a}{b}\right] < 1$$

or

$$0 \leqslant a - b\left[\frac{a}{b}\right] < b$$

giving

$$a = \left[\frac{a}{b}\right] b + r, \qquad 0 \leqslant r < b.$$

We have therefore proved:

Theorem 1.1. *Let a and b be any two integers with $b > 0$. Then there exist integers q and r satisfying*

$$a = qb + r, \qquad 0 \leqslant r < b. \quad \square$$

The number r in the theorem is called the (non-negative) remainder of a when divided by b.

Definition. If the remainder of a when divided by b is zero – that is, if there exists an integer c such that $a = bc$, then we say that a is a *multiple* of b. We also say that b *divides* a, and we write $b|a$, and we call b a *divisor* of a. Clearly we always have $1|a$, $b|0$ and, for any $a \neq 0$, $a|a$. If b does not divide a, then we write $b \nmid a$. Finally, if $a = bc$ and b is neither a nor 1, then we call b a *proper divisor* of a.

Concerning divisibility we have the following obvious theorems:

Theorem 1.2. *Suppose that $b \neq 0$, $c \neq 0$. Then*

1) *if $b|a$ and $c|b$, then $c|a$;*
2) *if $b|a$, then $bc|ac$;*
3) *if $c|d$ and $c|e$, then, for any m, n, $c|dm + en$.* $\square$

Theorem 1.3. *If b is a proper divisor of a, then $1 < |b| < |a|$.* $\square$

Exercise 1. If n is a positive integer, then

$$\left[\frac{[n\alpha]}{n}\right] = [\alpha].$$

Exercise 2. If n is a positive integer, then

$$[\alpha] + \left[\alpha + \frac{1}{n}\right] + \cdots + \left[\alpha + \frac{n-1}{n}\right] = [n\alpha].$$

Exercise 3. Prove the inequality

$$[2\alpha] + [2\beta] \geqslant [\alpha] + [\alpha + \beta] + [\beta].$$

1.2 Prime Numbers and Composite Numbers

We divide the natural numbers into three classes:

(i) 1, the only number with exactly one natural number divisor, namely 1 itself.

(ii) p, numbers with exactly two natural number divisors, namely 1 and p itself. In other words p is an integer greater than 1 with no proper divisors.

(iii) n, numbers with proper divisors (so that n has more than two divisors).

We call the numbers p in the second class the *prime numbers*, and the numbers n in the third class the *composite numbers*. We usually denote a prime number by the letter p. An integer is said to be *even* or *odd* according to whether it is divisible by 2 or not. Clearly even integers greater than 2 cannot be prime numbers.

Theorem 2.1. *Every integer greater than* 1 *is a product of prime numbers.*

Proof. Let $n > 1$. If n is prime, then there is nothing to prove. Suppose now that n is not prime and that q_1 is the least proper divisor. By Theorem 1.3, q_1 must be a prime number. Let $n = q_1 n_1$, $1 < n_1 < n$. If n_1 is prime, then the required result is proved; otherwise we let q_2 be the least prime divisor of n_1 giving

$$n = q_1 q_2 n_2, \qquad 1 < n_2 < n_1 < n.$$

Continuing the argument we have $n > n_1 > n_2 > \cdots > 1$, and the process must terminate before n steps so that eventually we have

$$n = q_1 q_2 \cdots q_s$$

where $q_1, \ldots, q_s$ are prime numbers. The theorem is proved. □

Example. $10725 = 3^1 \cdot 5^2 \cdot 11^1 \cdot 13^1$.

We can arrange the prime numbers in Theorem 2.1 as follows

$$n = p_1^{a_1} p_2^{a_2} \cdots p_k^{a_k}, \qquad a_1 > 0, a_2 > 0, \ldots, a_k > 0,$$
$$p_1 < p_2 < \cdots < p_k.$$

We call this the *standard factorization* of n. We shall see in section 5 that this standard factorization of a natural number is unique. This uniqueness theorem is known as "The Fundamental Theorem of Arithmetic".

1.3 Prime Numbers

The first few prime numbers are

$$2, 3, 5, 7, 11, 13, 17, 19, 23, 29, 31, 37, 41, 43, \ldots.$$

If N is not too large, it is not difficult to determine all the prime numbers not exceeding N. The method is known as the *sieve of Eratosthenes*. If $n \leqslant N$ and n is not prime, then n must be divisible by a prime not exceeding $\sqrt{N}$. We first list all the integers between 2 and N:

$$2, 3, 4, 5, \ldots, N.$$

We then successively remove the following:

(i) 4, 6, 8, 10, ..., that is even integers from 2^2 onwards;
(ii) 9, 15, 21, 27, ..., that is multiples of 3 from 3^2 onwards;
(iii) 25, 35, 55, 65, ..., that is multiples of 5 from 5^2 onwards;
... .

Continuing in this way we remove all those integers which are multiples of a prime not exceeding $\sqrt{N}$. The remaining numbers are all the prime numbers not exceeding N. All existing tables of prime numbers are built up this way with small modifications to the method. The most accurate table of prime numbers is by Lehmer: *List of prime numbers from* 1 *to* 10,006,721, Carnegie Institution, Washington 165 (1914). Lehmer also published a factor table: *Factor table for the first ten millions*, Carnegie Institution, Washington 105 (1909).

An example of a 39 digits prime number is

$$2^{127} - 1 = 1701, 41183, 46046, 92317, 31687, 30371, 58841, 05727,$$

and a 79 digits prime number is

$$180(2^{127} - 1)^2 + 1.$$

Up to the present (1981) the largest known prime is $2^{44497} - 1$, a number with 13395 digits. The following number

$$2^{257} - 1 = 231, 58417, 84746, 32390, 84714, 19700, 17375, 81570,$$
$$65399, 69331, 28112, 80789, 15168, 01582, 62592, 79871$$

is known to be composite, but its prime factorization is not known. These facts can be established with the aid of computing machines and special methods. We shall describe some of these methods later (see §3.9 and §16.15), but we cannot go into the details concerning the actual computations. A table of prime numbers up to 5000 is given at the end of Chapter 3.

1.4 Integral Modulus

By a *modulus* we mean a set of integers which is closed with respect to the operations of addition and subtraction. In other words, if m and n are integers in a modulus, then $m \pm n$ also belong to the modulus. The modulus containing only the integer 0 is called the zero modulus. The set of all integers forms a modulus, as does the set of integers which are multiples of a fixed integer k. We shall presently be concerned with integral moduli.

Theorem 4.1. 1) *The number* 0 *belongs to every modulus*;

2) *Let a, b belong to a modulus and m, n be any integers. Then $am + bn$ belongs to the modulus.*

Proof. 1) Take any a in the modulus. Then $0 = a - a$ belongs to the modulus.

2) If a is in the modulus, then $2a = a + a$, $3a = 2a + a, \ldots, ma$ are also in the modulus. Similarly nb belongs to the modulus and so the required result follows. □

Theorem 4.2. *Let a, b be any two integers. Then the set of numbers of the form $am + bn$ forms a modulus.*

Proof. This is trivial. □

Theorem 4.3. *Any non-zero modulus is the set of multiples of a fixed positive integer.*

Proof. Let d be the least positive integer in the modulus. We claim that every number in the modulus must be a multiple of d. For, suppose the contrary and let n be a number in the modulus which is not a multiple of d. Then, by Theorem 1.1, there are integers q and r such that

$$n = dq + r, \qquad 1 \leqslant r < d.$$

From the definition of a modulus, we see that $r = n - dq$ belongs to the modulus, and this contradicts the defining minimal property of d. Therefore every member of the modulus is a multiple of d. It is also clear that every multiple of d is in the modulus. The theorem is proved. □

Definition. Let a, b be any two integers and consider the modulus of the set of numbers of the form $am + bn$. If this is not the zero modulus, then the number d in the proof of Theorem 4.3 is called the *greatest common divisor* of a and b, and is denoted by (a, b).

Theorem 4.4. *The greatest common divisor (a, b) has the following properties:*

(i) *There exist integers x, y such that $ax + by = (a, b)$;*

(ii) *Given integers x, y we always have $(a, b) | ax + by$;*

(iii) *If $e|a$, $e|b$ then $e|(a, b)$.*

Proof. (i) and (ii) are immediate consequences of Theorem 4.3, and (iii) follows from (i). □

Definition. If $(a, b) = 1$, then we say that a and b are *coprime*.

Note: We introduced the well known method of successive divisions known as the *Euclidean algorithm* in the proof of Theorem 4.3. The detailed explanation of this method was also published in our country in the year 1247.

Example. We take $a = 323$, $b = 221$. From Euclid's algorithm we first have

$$323 = 221 \cdot 1 + 102.$$

Note that 102 belongs to the modulus of numbers $ax + by$. Next

$$221 = 102 \cdot 2 + 17,$$

so that 17 also belongs to the modulus. Since

$$102 = 17 \cdot 6$$

it follows that 17 is the least positive integer in the modulus, that is $17 = (323, 221)$.

This method can be used to determine the integers x, y in (i) of Theorem 4.4. In fact we have

$$\begin{aligned} 17 &= 221 - 2 \cdot 102 \\ &= 221 - 2(323 - 221) \\ &= 3 \cdot 221 - 2 \cdot 323 \end{aligned}$$

so that $x = -2$, $y = 3$.

This ancient method here is a fundamental pillar of elementary number theory.

1.5 The Fundamental Theorem of Arithmetic

Theorem 5.1. *Let p be a prime, and $p|ab$. Then either $p|a$ or $p|b$.*

Proof. If $p \nmid a$, then $(a, p) = 1$. By Theorem 4.4, there are integers x, y such that

$$xa + yp = 1,$$

and so

$$x \cdot ab + yb \cdot p = b.$$

But $p|ab$, so that $p|b$. □

Theorem 5.2. *If $c > 0$ and $(a, b) = d$, then $(ac, bc) = dc$.*

Proof. There are integers x, y such that

$$xa + yb = d,$$

or

$$xac + ybc = dc,$$

and so $(ac, bc)|dc$. On the other hand, from $d|a$ we deduce that $cd|ca$; similarly $cd|cb$. Thus $dc|(ac, bc)$. The required result follows. □

Theorem 5.3. *The standard factorization of a natural number n is unique. In other words, there is only one way of writing n as a product of prime numbers, apart from the ordering of the factors.*

Proof. From Theorem 5.1 we see that if $p|abc\ldots l$, then p must divide one of $a,b,c,\ldots,l$. In particular if $a,b,c,\ldots,l$ are all primes, then p must be one of $a,b,c,\ldots,l$.

Suppose now that

$$n = p_1^{a_1}p_2^{a_2}\cdots p_k^{a_k} = q_1^{b_1}q_2^{b_2}\cdots q_j^{b_j}$$

represent two standard factorization of n. We conclude from the above that each p must be a q, and each q must be a p. Therefore $k = j$. Also from

$$p_1 < p_2 < \cdots < p_k, \qquad q_1 < q_2 < \cdots < q_k$$

we conclude that $p_i = q_i$, $1 \leqslant i \leqslant k$.

If $a_i > b_i$, then on dividing n by $p_i^{b_i}$ we have

$$p_1^{a_1}\cdots p_i^{a_i-b_i}\cdots p_k^{a_k} = p_1^{b_1}\cdots p_{i-1}^{b_{i-1}}p_{i+1}^{b_{i+1}}\cdots p_k^{b_k},$$

which is impossible since only the left hand side is a multiple of p_i. Similarly we cannot have $a_i < b_i$. The theorem is proved. □

It is appropriate to insert here the explanation of excluding the number 1 from the definition of a prime number. If 1 is treated as a prime, then we shall have no unique factorization, since we can insert any power of 1 in the factorization.

Exercise 1. Prove that $\log_{10} 2$ and $\sqrt{2}$ are irrational numbers.

Exercise 2. Let

$$\log_{10}\frac{1025}{1024} = a, \qquad \log_{10}\frac{1024^2}{1023\cdot 1025} = b, \qquad \log_{10}\frac{81^2}{80\cdot 82} = c,$$

$$\log_{10}\frac{125^2}{124\cdot 126} = d, \qquad \log_{10}\frac{99^2}{98\cdot 100} = e.$$

Show that

$$196\log_{10} 2 = 59 + 5a + 8b - 3c - 8d + 4e.$$

Express $\log_{10} 3$ and $\log_{10} 41$ in terms of a, b, c, d, e. Determine $\log_{10} 2$ to ten decimal places and discuss the practical application of the method. (Given $\log_e 10 = 2.3025850930$.)

1.6 The Greatest Common Factor and the Least Common Multiple

Let $x_1,\ldots,x_n$ be any n numbers. We denote by $\min(x_1,\ldots,x_n)$ and $\max(x_1,\ldots,x_n)$ the least and the greatest numbers among $x_1,\ldots,x_n$ respectively.

The following theorem is clear.

Theorem 6.1. *Let a, b be two positive integers with prime divisors $p_1, \ldots, p_s$ so that we can write*

$$a = p_1^{a_1} \cdots p_s^{a_s}, \qquad a_\nu \geqslant 0,$$

$$b = p_1^{b_1} \cdots p_s^{b_s}, \qquad b_\nu \geqslant 0, \quad p_1 < p_2 < \cdots < p_s.$$

Then

$$(a, b) = p_1^{c_1} \cdots p_s^{c_s}$$

where $c_\nu = \min(a_\nu, b_\nu)$, $1 \leqslant \nu \leqslant s$. □

Definition. Let a, b be two positive integers. Integers which are divisible by both a and b are called *common multiples* of a and b. The least of all the positive common multiples is called the *least common multiple* of a and b. Since ab is certainly a positive common multiple, the least common multiple always exists.

Theorem 6.2. *Under the hypothesis of Theorem* 6.1, *the least common multiple of a, b is given by*

$$e = p_1^{e_1} \cdots p_s^{e_s}$$

where $e_\nu = \max(a_\nu, b_\nu)$, $1 \leqslant \nu \leqslant s$.

Proof. Clearly both a and b divide e. Moreover, if

$$e' = p_1^{m_1} \cdots p_s^{m_s}$$

is divisible by a, then $a_\nu \leqslant m_\nu$. Therefore, if e' is divisible by both a and b, then $a_\nu \leqslant m_\nu$ and $b_\nu \leqslant m_\nu$, and hence $\max(a_\nu, b_\nu) \leqslant m_\nu$. Therefore $e | e'$ and the theorem is proved. □

Theorem 6.3. *Any common multiple of a, b is a multiple of the least common multiple.* □

Theorem 6.4. *Let $[a, b]$ denote the least common multiple of a, b. Then*

$$[a, b](a, b) = ab.$$

Proof. Let

$$a = p_1^{a_1} \cdots p_s^{a_s}, \qquad b = p_1^{b_1} \cdots p_s^{b_s}, \qquad p_1 < p_2 < \cdots < p_s.$$

Then

$$ab = p_1^{a_1 + b_1} \cdots p_s^{a_s + b_s}.$$

Also

$$[a, b](a, b) = p_1^{\max(a_1, b_1) + \min(a_1, b_1)} \cdots p_s^{\max(a_s, b_s) + \min(a_s, b_s)}.$$

Since we always have

$$x + y = \max(x, y) + \min(x, y),$$

the theorem is proved. □

We now define inductively the greatest common factor and the least common multiple of n integers as follows: Let $a_1, \ldots, a_n$ be integers. The greatest common factor is the number

$$(a_1, \ldots, a_n) = ((a_1, \ldots, a_{n-1}), a_n),$$

and the least common multiple is the number

$$[a_1, \ldots, a_n] = [[a_1, \ldots, a_{n-1}], a_n].$$

Theorem 6.5. *Let*

$$a_1 = p_1^{e_{11}} \cdots p_s^{e_{1s}}, \quad \ldots, \quad a_n = p_1^{e_{n1}} \cdots p_s^{e_{ns}},$$

$$p_1 < p_2 < \cdots < p_s, \qquad e_{\mu\nu} \geqslant 0.$$

Then

$$(a_1, \ldots, a_n) = p_1^{e_1} \cdots p_s^{e_s}, \qquad e_\nu = \min(e_{1\nu}, \ldots, e_{n\nu}),$$

$$[a_1, \ldots, a_n] = p_1^{d_1} \cdots p_s^{d_s}, \qquad d_\nu = \max(e_{1\nu}, \ldots, e_{n\nu}). \quad \square$$

Exercise 1. Prove the following two equations

$$(a_1, \ldots, a_n) = ((a_1, \ldots, a_s), (a_{s+1}, \ldots, a_n)),$$

$$[b_1, \ldots, b_n] = [[b_1, \ldots, b_s], [b_{s+1}, \ldots, b_n]].$$

Exercise 2. Prove the following two equations

$$(a_1, \ldots, a_n) = \frac{a_1 a_2 \cdots a_n}{[a_2 \cdots a_n, a_1 a_3 \cdots a_n, \ldots, a_1 \cdots a_{n-1}]},$$

$$[a_1, \ldots, a_n] = \frac{a_1 a_2 \cdots a_n}{(a_2 \cdots a_n, a_1 a_3 \cdots a_n, \ldots, a_1 \cdots a_{n-1})}.$$

Exercise 3. Let $a_1, \ldots, a_n$ be n integers. Then $(a_1, \ldots, a_n)$ is the least positive integer belonging to the modulus of integers of the form $a_1 x_1 + \cdots + a_n x_n$.

Exercise 4. Find x, y, z such that

$$6x + 15y + 20z = 17.$$

Exercise 5. (Chinese publication (1372).)

There is a certain sum of money in yuens. On division by seventy-seven, there is a remainder of negative fifty. On division by seventy-eight, there is no remainder. How much money is there? (Answer 2106 yuens.)

1.7 The Inclusion-Exclusion Principle

Theorem 7.1. *Let there be N objects, and suppose that N_α of them possess the property α, N_β of them possess the property $\beta, \ldots, N_{\alpha\beta}$ of them possess both the properties α and $\beta, \ldots, N_{\alpha\beta\gamma}$ of them possess the three properties α, β and $\gamma, \ldots$. Then the number of objects which do not possess any of the properties $\alpha, \beta, \gamma, \ldots$ is given by*

$$\left.\begin{aligned} &N - N_\alpha - N_\beta - \cdots \\ &\quad + N_{\alpha\beta} + \cdots \\ &\quad - N_{\alpha\beta\gamma} - \cdots \\ &\quad + \cdots - \cdots . \end{aligned}\right\} \text{(A)}$$

Proof. Let P be an object which possesses k of the properties $\alpha, \beta, \ldots$. Then P occurs exactly once in the full set of N objects, k times in the enumeration of the $N_\alpha, N_\beta, \ldots$ objects,

$$\binom{k}{2} = \frac{1}{2}k(k-1)$$

times in the enumeration of the $N_{\alpha\beta}, \ldots$ objects,

$$\binom{k}{3} = \frac{1}{6}k(k-1)(k-2)$$

times in the enumeration of the $N_{\alpha\beta\gamma}, \ldots$ objects, $\ldots$. If $k \geqslant 1$, then the number of times P occurs in the enumeration (A) is

$$1 - \binom{k}{1} + \binom{k}{2} - \binom{k}{3} + \cdots = (1-1)^k = 0.$$

But if $k = 0$, then P is one of those objects which do not possess any of the properties $\alpha, \beta, \gamma, \ldots$, and it occurs exactly once in the enumeration (A). The theorem is proved. □

We now apply this principle as follows: For "property α" we mean "not exceeding a", $\ldots$.

Theorem 7.2. *Let $a, b, \ldots, k, l$ be non-negative numbers. Then we have*

$$\begin{aligned} \max(a, b, \ldots, k, l) = a + b + \cdots + k + l \\ - \min(a, b) - \cdots - \min(k, l) \\ + \min(a, b, c) + \cdots \\ - \cdots + \cdots \\ \pm \min(a, b, \ldots, k, l). \end{aligned}$$

Proof. We take the first N ($> \max(a, b, \ldots, k, l)$) positive integers. The number of integers without the properties $\alpha, \beta, \ldots$ is $N - \max(a, b, \ldots, k, l)$. The required result follows from Theorem 7.1. □

Theorem 7.1 can also be used to prove the following two theorems:

Theorem 7.3.

$$[a_1, \ldots, a_n] = a_1 \cdots a_n(a_1, a_2)^{-1} \cdots (a_{n-1}, a_n)^{-1}(a_1, a_2, a_3) \cdots (a_1, \ldots, a_n)^{(-1)^{n+1}}. \quad \square$$

Theorem 7.4.

$$(a_1, \ldots, a_n) = a_1 \cdots a_n[a_1, a_2]^{-1} \cdots [a_{n-1}, a_n]^{-1}[a_1, a_2, a_3] \cdots [a_1, \ldots, a_n]^{(-1)^{n+1}}. \quad \square$$

Note: Exercises 1 and 2 in 1.6, and Theorems 7.3 and 7.4 establish a "principle of duality" whereby ( ) and [ ] can be interchanged.

Exercise. Let $a, b, \ldots, k, l$ be positive integers. Determine the number of integers in $1, 2, \ldots, n$ which are coprime with $a, b, \ldots, k, l$.

1.8 Linear Indeterminate Equations

From Theorem 4.4 we have at once:

Theorem 8.1. *A necessary and sufficient condition for the equation*

$$ax + by = n$$

to have integer solutions in x, y *is that* $(a, b)|n$. $\square$

Theorem 8.2. *Let* $(a, b) = 1$, *and* x_0, y_0 *be a set of solutions to*

$$ax + by = n. \tag{1}$$

Then each set of solutions to (1) *are given by*

$$x = x_0 + bt, \qquad y = y_0 - at.$$

Moreover, given any integer t, *these are solutions to* (1).

Proof. From $ax + by = n$ and $ax_0 + by_0 = n$ we have $a(x - x_0) + b(y - y_0) = 0$. Since $(a, b) = 1$ we deduce that $a|y - y_0$. Let $y = y_0 - at$, so that $x = x_0 + bt$. The required result follows from substituting these into (1). $\square$

Theorem 8.3. *Let* $(a, b) = 1, a > 0, b > 0$. *Then every integer greater than* $ab - a - b$ *is representable as* $ax + by$ $(x \geqslant 0, y \geqslant 0)$. *Moreover,* $ab - a - b$ *is not representable as such.*

Proof. From Theorem 8.2 we know that the solutions to the equation $n = ax + by$ take the form

$$x = x_0 + bt, \qquad y = y_0 - at.$$

We now select t so that x and y are non-negative. We can choose t so that $0 \leqslant y_0 - at < a$, or $0 \leqslant y_0 - at \leqslant a - 1$. From the hypothesis, we have

$$(x_0 + bt)a = n - (y_0 - at)b > ab - a - b - (a - 1)b = -a$$

or

$$x_0 + bt > -1,$$

so that

$$x_0 + bt \geqslant 0.$$

Finally, suppose if possible that

$$ab - a - b = ax + by, \qquad x \geqslant 0, \quad y \geqslant 0.$$

Then we have

$$ab = (x + 1)a + (y + 1)b.$$

Since $(a, b) = 1$, it follows that $a|y + 1$, $b|x + 1$, so that $y + 1 \geqslant a$, $x + 1 \geqslant b$ and hence

$$ab = (x + 1)a + (y + 1)b \geqslant 2ab,$$

which is impossible. □

The above theorem can be interpreted as follows: If $a > 0$, $b > 0$, $(a, b) = 1$, then $ab - a - b$ is the largest integer not representable as $ax + by$ $(x \geqslant 0, y \geqslant 0)$. We can generalize this to the following problem: Let a, b, c be three positive integers satisfying $(a, b, c) = 1$. Determine the largest integer not representable as $ax + by + cz$ $(x \geqslant 0, y \geqslant 0, z \geqslant 0)$. This is an unsolved problem.

Exercise 1. Let $a > 0$, $b > 0$ and $(a, b) = 1$. Then the number of non-negative solutions to the equation $ax + by = n$ is equal to

$$\left[\frac{n}{ab}\right] \quad \text{or} \quad \left[\frac{n}{ab}\right] + 1.$$

(*Hint*: $[\alpha] - [\beta] = [\alpha - \beta]$ or $[\alpha - \beta] + 1$.)

Exercise 2. Let a, b, c be positive integers satisfying $(a, b) = (b, c) = (c, a) = 1$. Determine the largest integer not representable as

$$bcx + cay + abz, \qquad x \geqslant 0, \quad y \geqslant 0, \quad z \geqslant 0.$$

(*Answer*: $2abc - ab - bc - ca$.)

Exercise 3. Determine the number of solutions to

$$x + 2y + 3z = n, \qquad x \geqslant 0, \quad y \geqslant 0, \quad z \geqslant 0.$$

(*Hint*: The required number is the coefficient of x^n in the power series expansion for

$$\frac{1}{(1-x)(1-x^2)(1-x^3)}.$$

The power series can be obtained by the method of partial fractions. *Answer*:

$$\frac{(n+3)^2}{12} - \frac{7}{72} + \frac{(-1)^n}{8} + \frac{2}{9}\cos\frac{2n\pi}{3}.)$$

Exercise 4. (Ancient Chinese publication.) Cockerel one, five cents; chicken one, three cents; baby chicks three, one cent. One hundred cents are paid for one hundred birds. How many cockerels, chickens and baby chicks are there?

1.9 Perfect Numbers

Theorem 9.1. Let $\sigma(n)$ *denote the sum of the divisors of* n. *If* $n = p_1^{a_1} \cdots p_s^{a_s}$, *then*

$$\sigma(n) = \frac{p_1^{a_1+1} - 1}{p_1 - 1} \cdots \frac{p_s^{a_s+1} - 1}{p_s - 1}.$$

Proof. All the divisors of n are of the form

$$p_1^{x_1} \cdots p_s^{x_s}, \qquad 0 \leqslant x_1 \leqslant a_1, \ldots, 0 \leqslant x_s \leqslant a_s.$$

Therefore we have

$$\begin{aligned}\sigma(n) &= \sum_{x_1=0}^{a_1} \cdots \sum_{x_s=0}^{a_s} p_1^{x_1} \cdots p_s^{x_s} \\ &= \sum_{x_1=0}^{a_1} p_1^{x_1} \cdot \sum_{x_2=0}^{a_2} p_2^{x_2} \cdots \sum_{x_s=0}^{a_s} p_s^{x_s} \\ &= \frac{p_1^{a_1+1} - 1}{p_1 - 1} \cdots \frac{p_s^{a_s+1} - 1}{p_s - 1}. \quad \square\end{aligned}$$

An immediate consequence of this theorem is:

Theorem 9.2. *If* $(m, n) = 1$, *then* $\sigma(mn) = \sigma(m)\sigma(n)$. $\square$

Note: $\sigma(n)$ is called an arithmetic function. An arithmetic function possessing the property of Theorem 9.2 is called a *multiplicative function.*

Definition. A positive integer n is called a *perfect number* if $\sigma(n) = 2n$. Examples of perfect numbers are:

$$6 = 1 + 2 + 3, \qquad 28 = 1 + 2 + 4 + 7 + 14.$$

Theorem 9.3. *Let $p = 2^n - 1$ be prime. Then*

$$\tfrac{1}{2}p(p+1) = 2^{n-1}(2^n - 1)$$

is perfect. Moreover, every even perfect number is of this form.

Proof. 1) From Theorem 9.1 we have

$$\sigma(\tfrac{1}{2}p(p+1)) = \frac{2^n - 1}{2 - 1}\frac{p^2 - 1}{p - 1} = (2^n - 1)(p + 1) = p(p + 1).$$

2) Let a be any even perfect number. Set

$$a = 2^{n-1}u, \qquad u > 1, \quad 2 \nmid u.$$

Then, by Theorem 9.2,

$$2^n u = 2a = \sigma(a) = \frac{2^n - 1}{2 - 1}\sigma(u),$$

and so

$$\sigma(u) = \frac{2^n u}{2^n - 1} = u + \frac{u}{2^n - 1}.$$

But u and $u/(2^n - 1)$ are both divisors of u. Since $\sigma(u)$ is the sum of all the divisors of u, it follows that u has only two divisors, so that u is prime and $u/(2^n - 1) = 1$. The theorem is proved. □

Exercise 1. Verify that $\sigma(m) = \sigma(n) = m + n$ has the following three solutions:

m	284	17296	9363584
n	220	18416	9437056

Exercise 2. Prove that if a positive integer is the product of its proper divisors, then it must be a cube of a prime or a product of two distinct primes.

1.10 Mersenne Numbers and Fermat Numbers

Whether there exists an *odd* perfect number is a famous difficult problem. From the previous section we see that the determination of even perfect numbers is reduced to the determination of Mersenne primes, that is prime numbers of the form $2^n - 1$, since there is now a one-to-one correspondence between Mersenne primes and even perfect numbers. Whether there exist infinitely many Mersenne primes is another difficult unsolved problem is number theory.

Theorem 10.1. *If $n > 1$ and $a^n - 1$ is prime, then $a = 2$ and n is prime.*

Proof. If $a > 2$, then $(a-1)|(a^n-1)$ so that $a^n - 1$ cannot be prime. Again, if $a = 2$ and $n = kl$, where k is a proper divisor of n, then $(2^k - 1)|(2^n - 1)$ so that $2^n - 1$ cannot be prime. □

The problem of the primality of $2^n - 1$ is thus reduced to that of $2^p - 1$ where p is prime. We usually write

$$M_p = 2^p - 1$$

for a Mersenne prime. Up to the present (1981) M_p has been proved prime for

$$p = 2, 3, 5, 7, 13, 17, 19, 31, 61, 89, 107, 127, 521, 607, 1279, 2203, 2281,$$
$$3217, 4253, 4423, 9689, 9941, 11213, 19937, 21701, 23209, 44497$$

so that there are 27 perfect numbers known to us.

Similarly to the Mersenne numbers, there are the so-called Fermat numbers.

Theorem 10.2. *If $2^m + 1$ is prime, then $m = 2^n$.*

Proof. If $m = qr$, where q is an odd divisor of m, then we have

$$2^{qr} + 1 = (2^r)^q + 1 = (2^r + 1)(2^{r(q-1)} - \cdots + 1)$$

and $1 < 2^r + 1 < 2^{qr} + 1$, so that $2^m + 1$ cannot be prime. □

Let

$$F_n = 2^{2^n} + 1.$$

We call F_n a Fermat number, and the first five Fermat numbers

$$F_0 = 3, \qquad F_1 = 5, \qquad F_2 = 17, \qquad F_3 = 257, \qquad F_4 = 65537$$

are all primes. On this evidence Fermat conjectured that F_n is prime for all n. However, in 1732, Euler showed that

$$F_5 = 2^{2^5} + 1 = 641 \times 6700417$$

so that Fermat's conjecture is false.

Note: The divisibility of F_5 by 641 can be proved as follows: Let $a = 2^7$, $b = 5$ so that $a - b^3 = 3$, $1 + ab - b^4 = 1 + 3b = 2^4$. Therefore

$$2^{2^5} + 1 = (2a)^4 + 1 = (1 + ab - b^4)a^4 + 1 = (1 + ab)a^4 + 1 - a^4b^4,$$

and this must be divisible by $1 + ab = 2^4 + 5^4 = 641$.

It has been found that many Fermat numbers F_n are composite, but no Fermat prime has been found apart from the first five numbers. Therefore Fermat's conjecture has been a most unfortunate one, and indeed it is now conjectured that there are only finitely many Fermat primes.

There is an interesting geometry problem associated with F_n, namely that Gauss proved that if F_n is prime, then a regular polygon with F_n sides can be constructed using only straight edge and compass.

1.11 The Prime Power in a Factorial

Theorem 11.1. *Let p be a prime number. Then the (exact) power of p that divides $n!$ is given by*

$$\left[\frac{n}{p}\right]+\left[\frac{n}{p^2}\right]+\left[\frac{n}{p^3}\right]+\cdots.$$

(There are only finitely many non-zero terms in this series.)

Proof. From

$$\begin{aligned} n! = 1 \cdot 2 \cdots (p-1) \\ \cdot p \cdot (p+1) \cdots (2p) \cdots (p-1)p \cdots \\ \cdot p^2 \cdots \\ \cdots \end{aligned}$$

we see that there are $[\frac{n}{p}]$ multiples of p, $[\frac{n}{p^2}]$ multiples of p^2, and so on. The theorem follows. □

Theorem 11.2. *The number*

$$\binom{n}{r} = \frac{n!}{r!(n-r)!}$$

is an integer.

Proof. We use the fact that $[\alpha] - [\beta]$ is either $[\alpha - \beta]$ or $[\alpha - \beta] + 1$. From Theorem 11.1 we see that the power of p in $\binom{n}{r}$ is

$$\sum\left(\left[\frac{n}{p^m}\right]-\left[\frac{r}{p^m}\right]-\left[\frac{n-r}{p^m}\right]\right),$$

a non-negative integer. □

Example. If $n = 1000$, $p = 3$, then

$$\left[\frac{1000}{3}\right] = 333, \qquad \left[\frac{1000}{3^2}\right] = \left[\frac{333}{3}\right] = 111,$$

$$\left[\frac{1000}{3^3}\right] = 37, \quad \left[\frac{1000}{3^4}\right] = 12, \quad \left[\frac{1000}{3^5}\right] = 4, \quad \left[\frac{1000}{3^6}\right] = 1.$$

Therefore the exact power of 3 which divides 1000! is

$$333 + 111 + 37 + 12 + 4 + 1 = 498.$$

Exercise 1. Determine the exact power of 7 which divides 10000!.

Exercise 2. Determine the exact power of 5 which divides $\binom{1000}{500}$.

Exercise 3. Prove that if $r + s + \cdots + t = n$, then

$$\frac{n!}{r!s!\cdots t!}$$

is an integer. Prove further that if n is prime and $\max(r, s, \ldots, t) < n$, then the above number is a multiple of n.

1.12 Integral Valued Polynomials

Definition. By an *integral valued polynomial* we mean a polynomial $f(x)$ in the variable x which only takes integer values whenever x is an integer.

Example. Polynomials with integer coefficients are integral valued polynomials. The polynomial

$$\binom{x}{r} = \frac{x(x-1)\cdots(x-r+1)}{r!}$$

is an integral valued polynomial.

We shall write $\Delta f(x)$ for $f(x+1) - f(x)$.

Theorem 12.1.

$$\Delta\binom{x}{r} = \binom{x}{r-1}.$$

Proof.

$$\Delta\binom{x}{r} = \frac{(x+1)x\cdots(x-r+2)}{r!} - \frac{x(x-1)\cdots(x-r+1)}{r!}$$

$$= \frac{x\cdots(x-r+2)}{r!}((x+1)-(x-r+1)) = \binom{x}{r-1}. \quad \square$$

Theorem 12.2. *Every integral valued polynomial of degree k can be written as*

$$a_k\binom{x}{k} + a_{k-1}\binom{x}{k-1} + \cdots + a_1\binom{x}{1} + a_0,$$

where $a_k, \ldots, a_0$ are integers. Moreover, given any set of integers $a_k, \ldots, a_0$, the above is an integral valued polynomial.

Proof. Any polynomial $f(x)$ of degree k can be written as

$$f(x) = \alpha_k \binom{x}{k} + \alpha_{k-1} \binom{x}{k-1} + \cdots + \alpha_1 \binom{x}{1} + \alpha_0.$$

Now

$$\Delta f(x) = \alpha_k \binom{x}{k-1} + \alpha_{k-1} \binom{x}{k-2} + \cdots + \alpha_1.$$

Writing $\Delta^2 f(x)$ for $\Delta(\Delta f(x))$, and $\Delta^r f(x) = \Delta(\Delta^{r-1} f(x))$ we see that

$$f(0) = \alpha_0, \qquad (\Delta f(x))_{x=0} = \alpha_1, \quad \ldots, \quad (\Delta^r f(x))_{x=0} = \alpha_r, \ldots.$$

If $f(x)$ is integral valued, then so are $\Delta f(x)$, $\Delta^2 f(x), \ldots$. Therefore $f(0)$, $(\Delta f(x))_{x=0}, \ldots, (\Delta^r f(x))_{x=0}, \ldots$ are all integers; that is $\alpha_k, \ldots, \alpha_0$ are integers. The last part of the theorem is trivial. □

The same method can be used to prove:

Theorem 12.3. *Let $f(x)$ be an integral valued polynomial. Given any integer x, a necessary and sufficient condition for $f(x)$ to be a multiple of m is that*

$$m \mid (a_k, \ldots, a_0),$$

where $a_k, \ldots, a_0$ are integers given in Theorem 12.2. □

Theorem 12.4 (Fermat). *Let p be a prime number. Then, for any integer x, $x^p - x$ is a multiple of p.*

Proof. If $p = 2$, then the result follows at once from $x^2 - x = x(x-1)$. Assume therefore that $p > 2$, and let $f(x) = x^p - x$. Now $f(0) = 0$ and

$$\begin{aligned}\Delta f(x) &= (x+1)^p - x^p - (x+1) + x \\ &= \binom{p}{1} x^{p-1} + \binom{p}{2} x^{p-2} + \cdots + \binom{p}{p-1} x,\end{aligned}$$

where the coefficients (by Exercise 11.3) are all integers. With $x = 0$, we see that $f(1)$ is a multiple of p; with $x = 1$, we see that $f(2)$ is a multiple of p; and so on. Therefore $f(x)$ is always a multiple of p if $x \geqslant 0$. If x is a negative integer, we can deduce the result from

$$x^p - x = -[(-x)^p - (-x)].$$

The theorem is proved. □

Exercise 1. Generalize Theorems 12.2 and 12.3 to several variables.

Exercise 2. Prove that $n(n+1)(2n+1)$ is a multiple of 6.

Exercise 3. Prove that, as m and n run through the set of all positive integers,

$$m + \tfrac{1}{2}(m+n-1)(m+n-2)$$

also runs through the whole set of positive integers, and with no repetition.

Exercise 4. Prove that if a polynomial of degree k takes integer values for $k+1$ successive integers, then it must be an integral valued polynomial.

Exercise 5. If $f(-x) = -f(x)$, then we call $f(x)$ an odd polynomial. Prove that an odd integral valued polynomial can be written as

$$a_0 + a_1\frac{x}{1}\binom{x}{1} + a_2\frac{x}{2}\binom{x+1}{3} + \cdots + a_m\frac{x}{m}\binom{x+m-1}{2m-1},$$

where $a_1, \ldots, a_m$ are integers.

1.13 The Factorization of Polynomials

Theorem 13.1. *Let $g(x)$ and $h(x)$ be two polynomials with integer coefficients:*

$$g(x) = a_l x^l + \cdots + a_0, \qquad a_l \neq 0,$$
$$h(x) = b_m x^m + \cdots + b_0, \qquad b_m \neq 0,$$

and

$$g(x)h(x) = c_{l+m}x^{l+m} + \cdots + c_0.$$

Then

$$(a_l, \ldots, a_0)(b_m, \ldots, b_0) = (c_{l+m}, \ldots, c_0).$$

Proof. We may assume without loss that $(a_l, \ldots, a_0) = 1$, $(b_m, \ldots, b_0) = 1$. Suppose that $p|(c_{l+m}, \ldots, c_0)$ and

$$p|(a_l, \ldots, a_{u+1}), \qquad p \nmid a_u,$$
$$p|(b_m, \ldots, b_{v+1}), \qquad p \nmid b_v.$$

From the definition we have

$$c_{u+v} = \sum_{s+t=u+v} a_s b_t,$$

and apart from the term $a_u b_v$, each term is a multiple of p. Since $p \nmid a_u b_v$, it follows that $p \nmid c_{u+v}$, and so $p \nmid (c_{l+m}, \ldots, c_0)$, contradicting our assumption. Therefore no prime can divide $(c_{l+m}, \ldots, c_0)$. □

Definition. Let $f(x)$ be a polynomial with rational coefficients. Suppose that there are two non-constant polynomials $g(x)$ and $h(x)$ with rational coefficients such that $f(x) = g(x)h(x)$. Then $f(x)$ is said to be *reducible*. *Irreducible* means not reducible.

Example. $x^2 - 2$ and $x^2 + 1$ are irreducible polynomials, whereas $3x^2 + 8x + 4$ is reducible and the factorization is $(3x + 2)(x + 2)$.

Theorem 13.2 (Gauss). *Let $f(x)$ be a polynomial with integer coefficients. If $f(x) = g(x)h(x)$ where $g(x)$ and $h(x)$ are polynomials with rational coefficients, then there exists a rational number γ such that*

$$\gamma g(x), \qquad \frac{1}{\gamma} h(x)$$

have integer coefficients.

Proof. We may assume that the greatest common factor of the coefficients of $f(x)$ is 1. There are integers M, N such that

$$Mg(x) = a_l x^l + \cdots + a_0, \qquad a_i \text{ integer};$$

$$Nh(x) = b_m x^m + \cdots + b_0, \qquad b_i \text{ integer};$$

$$MNf(x) = c_{l+m} x^{l+m} + \cdots + c_0.$$

From our assumption and Theorem 13.1 we have

$$MN = (c_{l+m}, \ldots, c_0) = (a_l, \ldots, a_0)(b_m, \ldots, b_0).$$

Let

$$\gamma = \frac{M}{(a_l, \ldots, a_0)} = \frac{(b_m, \ldots, b_0)}{N},$$

and the required result follows. □

Theorem 13.3 (Eisenstein). *Let $f(x) = c_n x^n + \cdots + c_0$ be a polynomial with integer coefficients. If $p \nmid c_n$, $p | c_i$ $(0 \leqslant i < n)$ and $p^2 \nmid c_0$, then $f(x)$ is irreducible.*

Proof. Suppose, if possible, that $f(x)$ is reducible. By Theorem 13.2 we have that

$$f(x) = g(x)h(x),$$

$$g(x) = a_l x^l + \cdots + a_0, \qquad h(x) = b_m x^m + \cdots + b_0,$$

$$l + m = n, \qquad l > 0, \qquad m > 0,$$

where a_j and b_k are integers. From $c_0 = a_0 b_0$ and $p | c_0$ we see that either $p | a_0$ or $p | b_0$. Suppose that $p | a_0$. Then, from $p^2 \nmid a_0 b_0 = c_0$ we deduce that $p \nmid b_0$.

Next, the coefficients for $g(x)$ cannot all be a multiple of p, since otherwise $p | c_n$. We can therefore suppose that $p | (a_0, \ldots, a_{r-1})$, $p \nmid a_r$, $1 \leqslant r \leqslant l$. From $c_r =$

$a_r b_0 + \cdots + a_0 b_r$ we deduce that $p \nmid c_r$. But $r \leqslant l < n$ and so we have a contradiction. The theorem is proved. □

As a corollary we have:

Theorem 13.4. *$x^m - p$ is irreducible, so that $\sqrt[m]{p}$ is an irrational number.* □

Theorem 13.5. *The polynomial*

$$\frac{x^p - 1}{x - 1} = x^{p-1} + \cdots + x + 1$$

is irreducible.

Proof. Write $x = y + 1$ so that we have

$$\frac{1}{y}((y + 1)^p - 1) = y^{p-1} + py^{p-2} + \binom{p}{2} y^{p-3} + \cdots + p.$$

It is easy to see that each coefficient, apart from the first, is a multiple of p, and that the constant term is not a multiple of p^2. □

Exercise. Prove that the following polynomials are irreducible:

$$x^2 + 1, \qquad x^4 + 1, \qquad x^6 + x^3 + 1.$$

Notes

1.1. Up to the present there are 27 known Mersenne primes, namely $M_p = 2^p - 1$ where

$$p = 2, 3, 5, 7, 13, 17, 19, 31, 61, 89, 107, 127, 521, 607, 1279, 2203,$$
$$2281, 3217, 4253, 4423, 9689, 9941, 11213, 19937, 21701,$$
$$23209, 44497.$$

The twelfth Mersenne prime, namely M_{127}, was found by Lucas in 1876 and the remaining fifteen have been found since 1952 with the aid of electronic computers. Thus M_{44497} is the largest known prime with 13395 digits which was discovered in 1979 (see [54]).

1.2. It is known that any odd perfect number must
(i) exceed 10^{50} (see [26]),
(ii) have a prime factor exceeding 100110 (see [27]).

Chapter 2. Congruences

2.1 Definition

Let m be a natural number. If $a - b$ is a multiple of m, then we say that a and b are *congruent* mod m, and we write $a \equiv b \pmod{m}$. If a, b are not congruent mod m, then we write $a \not\equiv b \pmod{m}$.

Example. $31 \equiv -9 \pmod{10}$.

If a, b are integers, then we always have $a \equiv b \pmod{1}$.

The notion of congruence occurs frequently and even in our daily lives; for example we may consider the days of the week as a congruence problem with modulus 7. Again in the ancient calendar in our country we count the years with respect to the modulus 60. Indeed our country made some significant contribution to the theory of congruence. For example, the Chinese remainder theorem originates from ancient publications concerning solutions to problems such as the following:

There is a certain number. When divided by three this number has remainder two; when divided by five, it has remainder three; when divided by seven, it has remainder two. What is the number?

With our notation here, the number concerned is an integer x such that $x \equiv 2 \pmod{3}$, $x \equiv 3 \pmod{5}$, $x \equiv 2 \pmod{7}$. The problem is therefore a problem of the solutions to simultaneous congruences.

2.2 Fundamental Properties of Congruences

Theorem 2.1. (i) $a \equiv a \pmod{m}$ (*reflexive*); (ii) *If* $a \equiv b \pmod{m}$, *then* $b \equiv a \pmod{m}$ (*symmetric*); (iii) *If* $a \equiv b$, $b \equiv c \pmod{m}$, *then* $a \equiv c \pmod{m}$ (*transitive*). □

These three properties here show that being congruent is an equivalence relation. The set of integers can then be partitioned into equivalence classes so that integers in each class are congruent among themselves, and two integers from different classes are not congruent. We call these equivalence classes *residue classes*. It is clear that, for the modulus m, we have precisely m residue classes: the classes whose members have remainder $r = 0, 1, 2, \ldots, m - 1$ when divided by m.

If we select one member from each residue class, then the set of numbers formed is called a *complete residue system*.

Theorem 2.2. *If* $a \equiv b$, $a_1 \equiv b_1 \pmod{m}$, *then we have* $a + a_1 \equiv b + b_1$, $a - a_1 \equiv b - b_1$, $aa_1 \equiv bb_1 \pmod{m}$. □

Theorem 2.2 has the following interpretation: Let A, B be any two residue classes from which we select any representatives a, b. Denote by C the residue class which contains $a + b$ (or $a - b$ or ab). Then C depends on A, B but not on the representatives a, b. In other words, the sum of any two integers from A, B must belong to C. We can therefore define C to be the *sum* of the two classes A, B and we denote it by $C = A + B$. Similarly we can define $A - B$ and $A \cdot B$. We see from Theorem 2.2 that, with respect to residue classes mod m, the operations of addition, subtraction and multiplication are closed. We note that division is not always possible; for example $3 \cdot 2 \equiv 1 \cdot 2, 2 \equiv 2 \pmod{4}$, but $3 \not\equiv 1 \pmod{4}$. However we do have the following:

Theorem 2.3. *If* $ac \equiv bd$, $c \equiv d \pmod{m}$ *and* $(c, m) = 1$, *then* $a \equiv b \pmod{m}$.

Proof. From $(a - b)c + b(c - d) = ac - bd \equiv 0 \pmod{m}$, we have $m|(a - b)c$. But $(c, m) = 1$, so that $m|a - b$. □

We denote by O the residue class of all multiples of m. Then $A + O = A$ and $A \cdot O = O$. Again, if we let I be the residue class of integers with remainder 1 when divided by m, then $A \cdot I = A$. From our example and Theorem 2.3 we see that from $A \cdot B = A \cdot C$ we may *not* deduce that $B = C$; but if the members of A are coprime with m (Note: if A has one member which is coprime with m, then every member must also be coprime with m), then we have $B = C$.

If we take m to be a prime number, then apart from the class O, every class is coprime with m. Therefore, for a prime modulus, the operations of addition, subtraction, multiplication and division are closed, except that we cannot divide by the class O.

2.3 Reduced Residue System

As we said earlier, if a residue class A contains an element which is coprime with m, then every element of A is coprime with m, and we call A a class coprime with m. If A and m are coprime, then we can, by Theorem 2.3, define B/A. In particular, we write A^{-1} for I/A. For example:

A	0	1	2	3	4
A^{-1}	×	1	3	2	4

(mod 5)

A	0	1	2	3	4	5
A^{-1}	×	1	×	×	×	5

(mod 6)

A	0	1	2	3	4	5	6
A^{-1}	×	1	4	5	2	3	6

(mod 7)

The sign "×" in the table means "undefined".

Definition. We denote by $\varphi(m)$ the number of residue classes $(\bmod m)$ coprime with m. This function $\varphi(m)$ is called Euler's function. If we select one member of each residue class coprime with m:

$$a_1, \ldots, a_{\varphi(m)},$$

then we call this set of integers a *reduced residue system.*

Example. $\varphi(1) = 1, \quad \varphi(2) = 1, \quad \varphi(3) = 2, \quad \varphi(4) = 2.$

We may also describe $\varphi(m)$ as the number of positive integers not exceeding m and coprime with m. If $m = p$ is a prime, then $\varphi(p) = p - 1$.

Theorem 3.1. *Let $a_1, a_2, \ldots, a_{\varphi(m)}$ be a reduced residue system, and suppose that $(k, m) = 1$. Then $ka_1, ka_2, \ldots, ka_{\varphi(m)}$ is also a reduced residue system.*

Proof. Clearly we have $(ka_i, m) = 1$, so that each ka_i represents a residue class coprime with m. If $ka_i \equiv ka_j \pmod m$, then, since $(k, m) = 1$, we have $a_i \equiv a_j \pmod m$. Therefore the members ka_i represent distinct residue classes. The theorem is proved. □

Theorem 3.2 (Euler). *If $(k, m) = 1$, then $k^{\varphi(m)} \equiv 1 \pmod m$.*

Proof. From Theorem 3.1 we have

$$\prod_{v=1}^{\varphi(m)} (ka_v) \equiv \prod_{v=1}^{\varphi(m)} a_v \pmod m.$$

Since $(m, a_i) = 1$, it follows that $k^{\varphi(m)} \equiv 1 \pmod m$. □

Taking $m = p$ we have Fermat's theorem (Theorem 1.12.4).

Theorem 3.3. *Let p be a prime. Then, for all integers a, we have $a^p \equiv a \pmod p$.* □

2.4 The Divisibility of $2^{p-1} - 1$ by p^2

In 1828 Abel asked if there are primes p and integers a such that $a^{p-1} \equiv 1 \pmod{p^2}$? According to Jacobi: if $p \leqslant 37$, then the above has the solutions $p = 11, a = 3$ or 9; $p = 29$, $a = 14$; and $p = 37$, $a = 18$.

Recent research work on Fermat's last theorem has added some impetus to this problem. We have the following result concerning Fermat's last theorem: Let p be an odd prime. If there are integers x, y, z such that $x^p + y^p + z^p = 0$, $p \nmid xyz$, then

$$2^{p-1} \equiv 1 \pmod{p^2}, \tag{1}$$

and

$$3^{p-1} \equiv 1 \pmod{p^2}, \tag{2}$$

and more recently we know also that $n^{p-1} \equiv 1 \pmod{p^2}$ for $n = 2, 3, \ldots, 47$. We do not know if there exists a prime p such that both (1) and (2) hold.

Definition. If $a^{p-1} \equiv 1 \pmod{p^2}$, then we call a a *Fermat solution*.

It is clear that the product of two Fermat solutions is a Fermat solution, the product of a Fermat solution and a non-Fermat solution is a non-Fermat solution. In the prime factorization of a non-Fermat solution there must be a prime divisor which is a non-Fermat solution.

Theorem 4.1. *Let a, b be two Fermat solutions with respect to p. Then there does not exist q such that $qp = a \pm b$, $p \nmid q$.*

Proof. From the definition we have $a^p \equiv a$, $b^p \equiv b \pmod{p^2}$,

$$a^p \pm b^p \equiv a \pm b \pmod{p^2}. \tag{3}$$

If $qp = a \pm b$, $p \nmid q$, then $a^p = (\mp b + qp)^p \equiv \mp b^p \pmod{p^2}$ giving $a^p \pm b^p \equiv 0 \pmod{p^2}$. Substituting this into (3) yields $a \pm b = qp \equiv 0 \pmod{p^2}$, which is a contradiction. □

Theorem 4.2. 3 *is a Fermat solution with respect to* 11.

Proof. We have $3^5 = 243 \equiv 1 \pmod{11^2}$ so that $3^{10} \equiv 1 \pmod{11^2}$. □

Theorem 4.3. 2 *is a Fermat solution with respect to* 1093.

Proof. Let $p = 1093$. Then $3^7 = 2187 = 2p + 1$, so that

$$3^{14} \equiv 4p + 1 \pmod{p^2}); \tag{4}$$

also

$$2^{14} = 16384 = 15p - 11,$$

so that

$$\begin{aligned}
2^{28} &\equiv -330p + 121 \pmod{p^2}, \\
3^2 \cdot 2^{28} &\equiv -2970p + 1089 \pmod{p^2} \\
&\equiv -2969p - 4 \\
&\equiv 310p - 4 \pmod{p^2}, \\
3^2 \cdot 2^{28} \cdot 7 &\equiv 2170p - 28 \\
&\equiv -16p - 28 \pmod{p^2}.
\end{aligned}$$

Therefore

$$3^2 \cdot 2^{26} \cdot 7 \equiv -4p - 7 \pmod{p^2}.$$

From the binomial theorem we have

$$3^{14} \cdot 2^{182} \cdot 7^7 \equiv (-4p - 7)^7 \equiv -7 \cdot 4p \cdot 7^6 - 7^7 \pmod{p^2},$$

and hence

$$3^{14} \cdot 2^{182} \equiv -4p - 1 \pmod{p^2}. \tag{5}$$

From (4) and (5) we have

$$3^{14} \cdot 2^{182} \equiv -3^{14}, \qquad 2^{182} \equiv -1 \pmod{p^2}.$$

Therefore

$$2^{1092} \equiv 1 \pmod{p^2}. \quad \square$$

Theorem 4.4. 3 *is a non-Fermat solution with respect to* 1093.

Proof. If 3 were a Fermat solution, then so would 3^7 be one. Since -1 is clearly a Fermat solution, and $3^7 - 1 = 2p$, we obtain the required contradiction from Theorem 4.1. □

Theorem 4.5. *There exists no prime* $p < 100$ *which satisfies* (1) *and* (2) *simultaneously.*

Proof. Suppose that 2 and 3 are both Fermat solutions. Then 2^l, 3^m and 2^l3^m are all Fermat solutions, and of course 1 is also a Fermat solution. The theorem now follows from Theorem 4.1 and the following calculations:

$$\begin{array}{lllll}
2=3-1, & 3=2+1, & 5=2+3, & 7=2^2+3, & 11=2+3^2, \\
13=2^2+3^2, & 17=2^3+3^2, & 19=2^4+3, & 23=-2^2+3^3, & 29=2+3^3, \\
31=2^2+3^3, & 37=2^6-3^3, & 41=2^5+3^2, & 43=2^4+3^3, & 47=2^4\cdot 3-1, \\
53=2\cdot 3^3-1, & 59=2^5+3^3, & 61=2^6-3, & 67=2^6+3, & 71=2^3\cdot 3^2-1, \\
73=2^6+3^2, & 79=-2+3^4, & 83=2+3^4, & 89=2^3+3^4, & 97=2^4+3^4. \quad \square
\end{array}$$

Recently Lehmer has proved that if $p \leqslant 253{,}747{,}889$, then there must exist $m \leqslant 47$ such that $m^{p-1} \not\equiv 1 \pmod{p^2}$. This makes some contribution towards Fermat's last theorem.

2.5 The Function $\varphi(m)$

Theorem 5.1. *Let* $(m, m') = 1$, *and let* x *run over a complete residue system* $\bmod m$, *and* x' *run over a complete residue system* $\bmod m'$. *Then* $mx' + m'x$ *runs over a complete residue system* $\bmod mm'$.

Proof. Consider the mm' numbers $mx' + m'x$. If

$$mx' + m'x \equiv my' + m'y \pmod{mm'},$$

then

$$mx' \equiv my' \pmod{m'},$$

$$m'x \equiv m'y \pmod{m}.$$

From $(m, m') = 1$ we have $x' \equiv y' \pmod{m'}$, $x \equiv y \pmod{m}$. The theorem is proved. □

Theorem 5.2. *Let $(m, m') = 1$, and let x run over a reduced residue system* mod m, *and x' run over a reduced residue system* mod m. *Then $mx' + m'x$ runs over a reduced residue system* mod mm'.

Proof. 1) We first prove that $mx' + m'x$ is coprime with mm'. Suppose otherwise. Then there exists p such that $p|(mm', mx' + m'x)$. If $p|m$, then $p|m'x$. Since $(m, m') = 1$, it follows that $p\nmid m'$ and so $p|x$. Thus $p|(m, x)$ which is impossible.

2) We next prove that every integer a coprime with mm' must be congruent mod mm' to an integer of the form $mx' + m'x$, $(x, m) = (x', m') = 1$.

By Theorem 5.1 there are integers x, x' such that $a \equiv mx' + m'x \pmod{mm'}$. We now prove that $(x, m) = (x', m') = 1$. If $(x, m) = d \neq 1$, then $(a, m) = (mx' + m'x, m) = (m'x, m) = (x, m) = d \neq 1$, which contradicts the hypothesis. Similarly we must have $(x', m') = 1$.

3) We have already proved in Theorem 5.1 that the numbers $mx' + m'x$ are incongruent. Therefore the theorem is proved. □

We have in fact proved that $\varphi(m)$ is a multiplicative function – that is:

Theorem 5.3. *If $(m, m') = 1$, then $\varphi(mm') = \varphi(m)\varphi(m')$.* □

A multiplicative function is completely determined by the values it takes at the prime powers. Thus, if the standard factorization of m is given by

$$m = p_1^{l_1} \cdots p_s^{l_s}, \qquad p_1 < p_2 < \cdots < p_s$$

then, from Theorem 5.3, we have

$$\varphi(m) = \varphi(p_1^{l_1}) \cdots \varphi(p_s^{l_s}).$$

Theorem 5.4. *We have*

$$\varphi(p^l) = p^l\left(1 - \frac{1}{p}\right),$$

and

$$\varphi(m) = m \prod_{p|m} \left(1 - \frac{1}{p}\right),$$

where p runs over the distinct prime divisors of m.

Proof. Consider the integers in the interval $1 \leqslant n \leqslant p^l$. There are precisely p^{l-1} integers which are multiples of p and the others are coprime with p so that

$$\varphi(p^l) = p^l - p^{l-1} = p^l\left(1 - \frac{1}{p}\right).$$

The second equation in the theorem follows from this and the multiplicative property of the function. □

Example: $\varphi(300) = \varphi(2^2 \cdot 3 \cdot 5^2) = 2^2 \cdot 3 \cdot 5^2(1 - \frac{1}{2})(1 - \frac{1}{3})(1 - \frac{1}{5}) = 80$.

Exercise 1. Prove that $\sum_{d|m} \varphi(d) = m$, where in the sum, d runs over all the positive divisors of m.

Exercise 2. Let P be the product of the distinct prime divisors of (m, n). Prove that

$$\frac{\varphi(mn)}{\varphi(m)\varphi(n)} = \frac{P}{\varphi(P)}.$$

Exercise 3. Use Theorem 1.7.1 to prove Theorem 5.4.

2.6 Congruences

We first discuss the solubility of the congruence

$$ax + b \equiv 0 \pmod{m}, \tag{1}$$

and the number of incongruent solutions.

The congruence (1) is equivalent to the equation $ax + b = my$, where we seek integer solutions x, y. This indeterminate equation has already been discussed in §1.8, and we shall now advance one step further.

If $(a, m) = 1$, then we can choose x_0, y_0 according to Theorem 1.4.4 so that $ax_0 + my_0 = 1$. Thus $x = -bx_0$ is a solution to (1), and we now proceed to show that this solution is unique. If $ax' + b \equiv 0 \pmod{m}$ and $ax + b \equiv 0 \pmod{m}$, then $a(x - x') \equiv 0 \pmod{m}$. Since $(a, m) = 1$, we have $x \equiv x' \pmod{m}$. This proves that there is only one residue class whose members satisfy (1); in other words, there is only one solution x to (1) satisfying $0 \leqslant x < m$.

If $(a, m) = d > 1$, then d must divide b, or else there is no solution. We then have

$$\frac{a}{d}x + \frac{b}{d} \equiv 0 \quad \left(\text{mod}\, \frac{m}{d}\right), \qquad \left(\frac{a}{d}, \frac{m}{d}\right) = 1. \tag{2}$$

We have already proved that (2) has a unique solution x_1 satisfying $0 \leqslant x_1 < m/d$, and $x = x_1 + (m/d)t$ are all solutions to (2). Therefore

$$x_1, \quad x_1 + \frac{m}{d}, \quad x_1 + \frac{2m}{d}, \quad \ldots, \quad x_1 + (d-1)\frac{m}{d}$$

are all incongruent $(\bmod m)$ solutions to (1). We have therefore proved the following:

Theorem 6.1. *If $(a, m) | b$, then there are (a, m) incongruent* $(\bmod m)$ *solutions to* (1). *Otherwise* (1) *has no solution.* □

Theorem 6.2. *A necessary and sufficient condition for the congruence $ax_1 + \cdots + a_n x_n + b \equiv 0 \pmod m$ to have a solution $(x_1, \ldots, x_n)$ is that $(a_1, \ldots, a_n, m) | b$. If this condition is satisfied, then the number of incongruent* $(\bmod m)$ *solutions is $m^{n-1}(a_1, \ldots, a_n, m)$.*

Proof. The case $n = 1$ is settled by Theorem 6.1. We now proceed by induction. Let $(a_1, \ldots, a_n, m) = d$ and $(a_1, \ldots, a_{n-1}, m) = d_1$, so that $(d_1, a_n) = d$. From Theorem 6.1 we know that there are $d \cdot (m/d_1)$ solutions to

$$a_n x_n + b \equiv 0 \pmod{d_1}, \qquad 0 \leqslant x_n < m.$$

Corresponding to a solution x_n we set

$$\frac{a_n x_n + b}{d_1} = b_1.$$

From the induction hypothesis, the number of solutions to the congruence $a_1 x_1 + \cdots + a_{n-1} x_{n-1} + b_1 d_1 \equiv 0 \pmod m$ is $m^{n-2}(a_1, \ldots, a_{n-1}, m) = m^{n-2} d_1$. Therefore the total number of solutions is given by

$$\frac{md}{d_1} \cdot m^{n-2} d_1 = m^{n-1} d$$

as required. □

2.7 The Chinese Remainder Theorem

Theorem 7.1. *Let m be the least common multiple of m_1 and m_2. The condition for the solubility of the simultaneous congruences*

$$x \equiv a_1 \pmod{m_1},$$
$$x \equiv a_2 \pmod{m_2},$$

is

$$(m_1, m_2) | a_1 - a_2. \tag{1}$$

If (1) *holds, then the solution is unique* $\bmod m$.

Proof. 1) Let $(m_1, m_2) = d$. If the simultaneous congruences have a solution, then $x \equiv a_1$, $x \equiv a_2 \pmod{d}$ and hence $d | a_1 - a_2$.

2) If $d | a_1 - a_2$, then the solutions to $x \equiv a_1 \pmod{m_1}$ are given by $x = a_1 + m_1 y$. Substituting this into the second congruence gives $a_1 + m_1 y \equiv a_2 \pmod{m_2}$. From the proof of Theorem 6.1 this congruence has a unique solution $\bmod m_2/d$. Therefore the simultaneous congruences have a unique solution $x \bmod m$. □

Theorem 7.2. *If* $(m_i, m_j) = 1$ $(1 \leqslant i < j \leqslant n)$, *then the simultaneous congruences*

$$x \equiv a_i \pmod{m_i}, \qquad 1 \leqslant i \leqslant n$$

have a unique solution $\bmod m_1 \cdots m_n$.

Proof. Apply mathematical induction to Theorem 7.1. □

Let us now discuss the ancient method of solutions to this type of problem. We already stated the problem of "*What is the number?*" in §1. The solution to this problem was published as a song in 1593, and it goes as follows:

"*Three people walking together, 'tis rare that one be seventy,*
Five cherry blossom trees, twenty one branches bearing flowers,
Seven disciples reunite for the half-moon,
Take away (multiple of) one hundred and five and you shall know."

We recall that the problem was to solve the simultaneous congruences $x \equiv 2 \pmod 3$, $x \equiv 3 \pmod 5$, $x \equiv 2 \pmod 7$. The meaning of the song here is as follows: Multiply by 70 the remainder of x when divided by 3, multiply by 21 the remainder of x when divided by 5, multiply by 15 (the number of days in half a Chinese (synodic) month) the remainder of x when divided by 7. Add the three results together, and then subtract a suitable multiple of 105 and you shall have the required smallest solution. For our specific example, we have

$$2 \times 70 + 3 \times 21 + 2 \times 15 = 233$$

and on subtracting twice 105 we have the required solution 23.

How do we explain this ancient method of solution, and in particular where do 70, 21, 15 come from? The answer is as follows: 70 is a multiple of 5 and 7 which has remainder 1 when divided by 3. 21 is a multiple of 3 and 7 which has remainder 1 when divided by 5. 15 is a multiple of 3 and 5 which has remainder 1 when divided by 7. It follows that $70a + 21b + 15c$ must have remainders a, b and c when divided by 3, 5 and 7 respectively.

We may further investigate how they obtained 70, 21 and 15. They had to solve

$$x \equiv 0 \pmod{m_1}, \qquad x \equiv 0 \pmod{m_2}, \qquad x \equiv 1 \pmod{m_3}$$

where $(m_1, m_2) = (m_2, m_3) = (m_3, m_1) = 1$. That is, how did they find $x = m_1 m_2 y$

where y satisfies $m_1m_2y \equiv 1 \pmod{m_3}$? The answer is that they used their own version of the Euclidean algorithm to solve the indeterminate equation $m_1m_2y - m_3z = 1$.

The following exercises are all from ancient Chinese publications. Exercises 2, 3, 4 are dated 1275.

Exercise 1. Replace 3, 5, 7 by 3, 7, 11 and determine the three numbers which correspond to 70, 21, 15.

Exercise 2. Seven with remainder one, eight with remainder two, nine with remainder three. What is the number?

Exercise 3. Eleven with left over three, twelve with left over two, thirteen with left over one. What is the number?

Exercise 4. Two with left over one, five with left over two, seven with left over three, nine with left over four. What is the number?

Exercise 5. There is a number. It has no remainder when divided by five. It has a remainder ten when divided by seven hundred and fifteen. It has a remainder one hundred and forty when divided by two hundred and forty seven. It has a remainder two hundred and forty five when divided by three hundred and ninety one. It has a remainder one hundred and nine when divided by one hundred and eighty seven. May we ask what is the number? (Answer: Ten thousand and twenty.)

2.8 Higher Degree Congruences

Let m be a fixed natural number, and let $f(x) = a_nx^n + \cdots + a_0$ be a polynomial with integer coefficients. We now discuss the congruence

$$f(x) \equiv 0 \pmod{m}. \tag{1}$$

If x_0 is a solution, then $x_0 + mt$ is also a solution. This means that if x_0 satisfies (1), then each member of the residue class represented by x_0 also satisfies (1). Therefore, when we speak of the number of solutions to (1) we mean the number of incongruent solutions.

The number of solutions to a higher degree congruence is quite irregular. For example:

1. The congruence $x^3 - x = (x - 1)x(x + 1) \equiv 0 \pmod 6$ has six solutions.
2. The congruence $x^2 + 1 \equiv 0 \pmod 3$ has no solution.
3. The congruence $(x - 1)(x - p - 1) \equiv 0 \pmod{p^2}$ has p solutions, namely 1, $p + 1, 2p + 1, \ldots, (p - 1)p + 1$.

We see therefore that the solutions to higher degree congruences are difficult and complicated. The following theorem helps a little.

Theorem 8.1. *Let* $(m_1, m_2) = 1$. *Then the number of solutions to the congruence*

$$f(x) \equiv 0 \pmod{m_1 m_2} \tag{2}$$

is the product of the numbers of solutions to the congruences

$$f(x) \equiv 0 \pmod{m_1}, \tag{3}$$

$$f(x) \equiv 0 \pmod{m_2}. \tag{4}$$

If

$$m = m_1 m_2 = p_1^{l_1} \cdots p_s^{l_s} \qquad (p_1 < p_2 < \cdots < p_s)$$

is the standard prime factorization of m, *then the number of solutions to* (2) *is the product of the numbers of solutions to the* s *congruences*:

$$f(x) \equiv 0 \pmod{p_i^{l_i}}, \qquad 1 \leqslant i \leqslant s.$$

Proof. It is clear that each solution to (2) is also a solution to (3) and (4). Conversely, let c_1 and c_2 be solutions to (3) and (4) respectively, and let c be a solution of $c \equiv c_1 \pmod{m_1}$ and $c \equiv c_2 \pmod{m_2}$. The solution c exists and is unique mod m according to the Chinese remainder theorem. Moreover, this c satisfies (2) because $m_1 | f(c)$, $m_2 | f(c)$ so that $m | f(c)$. □

2.9 Higher Degree Congruences to a Prime Power Modulus

Theorem 9.1. *Let* p *be a prime number. The number of solutions (including repeated ones) to the congruence*

$$f(x) = a_n x^n + \cdots + a_0 \equiv 0 \pmod{p} \tag{1}$$

does not exceed n.

Proof. We can assume that $p \nmid a_n$. The theorem becomes trivial if (1) has no solutions. If a is a solution, then we can write

$$f(x) = (x - a) f_1(x) + r_1,$$

where we see that $p | r_1$ by substituting a for x. Therefore $f(x) \equiv (x - a) f_1(x) \pmod{p}$. If a is also a solution to $f_1(x) \equiv 0 \pmod{p}$, then we have similarly that $f_1(x) \equiv (x - a) f_2(x) \pmod{p}$, and in this case we call a a repeated solution to $f(x) \equiv 0 \pmod{p}$. If $f(x) \equiv (x - a)^h g_1(x) \pmod{p}$ where $g_1(a) \not\equiv 0 \pmod{p}$, then we call a a repeated solution of order h to $f(x) \equiv 0 \pmod{p}$. From our proof so far, we see that the degree of $g_1(x)$ is $n - h$.

Suppose now that b is another solution. Then

$$0 \equiv f(b) \equiv (b - a)^h g_1(b) \pmod{p}.$$

Since $p \nmid (b-a)$, it follows that $g_1(b) \equiv 0 \pmod{p}$. If b is a repeated solution of order k to $g_1(x) \equiv 0 \pmod{p}$, then we have, as before,

$$f(x) \equiv (x-a)^h (x-b)^k g_2(x) \pmod{p}.$$

Proceeding in this way we have

$$f(x) \equiv (x-a)^h (x-b)^k \cdots (x-c)^l g(x) \pmod{p},$$

where $g(x)$ is a polynomial of degree $n-h-k-\cdots-l$ and $g(x) \equiv 0 \pmod{p}$ has no solution.

The theorem is proved. □

Since $1, 2, \ldots, p-1$ are solutions to $x^{p-1} \equiv 1 \pmod{p}$ we see that

$$x^{p-1} - 1 \equiv (x-1)(x-2)\cdots(x-(p-1)) \pmod{p}. \tag{2}$$

Substituting $x = 0$ into this, and noting that $p-1$ is even if $p > 2$, we have:

Theorem 9.2 (Wilson). *If p is a prime, then $(p-1)! \equiv -1 \pmod{p}$.* □

Theorem 9.3. *Let $f'(x) = na_n x^{n-1} + \cdots + 2a_2 x + a_1$. If $f(x) \equiv 0$, $f'(x) \equiv 0 \pmod{p}$ have no common solution, then the two congruences $f(x) \equiv 0 \pmod{p^l}$ and $f(x) \equiv 0 \pmod{p}$ have the same number of solutions.*

Proof. We prove this by induction on l, the case $l = 1$ being trivial. Let x_1 be a solution to $f(x) \equiv 0 \pmod{p^{l-1}}$, so that

$$f(x_1 + p^{l-1}y) \equiv f(x_1) + p^{l-1}yf'(x_1) \pmod{p^l},$$

because $(x + p^{l-1}y)^n \equiv x^n + np^{l-1}yx^{n-1} \pmod{p^l}$. But $p \nmid f'(x_1)$ so that there exists a unique y such that

$$f(x_1 + p^{l-1}y) \equiv 0 \pmod{p^l}. \quad \square$$

Theorem 9.4. *The congruence $x^{p-1} \equiv 1 \pmod{p^l}$ has $p-1$ solutions.*

Proof. This is an immediate consequence of Theorem 9.3. □

2.10 Wolstenholme's Theorem

Theorem 10.1. *Let p be a prime number greater than 3, and denote by $\frac{1}{s}$ an integer s^* such that $ss^* \equiv 1 \pmod{p^2}$. Then we have*

$$1 + \frac{1}{2} + \frac{1}{3} + \cdots + \frac{1}{p-1} \equiv 0 \pmod{p^2}.$$

Proof. Let

$$(x-1)(x-2)\cdots(x-(p-1)) = x^{p-1} - s_1 x^{p-2} + \cdots + s_{p-1}, \tag{1}$$

so that

$$s_{p-1} = (p-1)!.$$

Since

$$(x-1)(x-2)\cdots(x-(p-1)) \equiv x^{p-1} - 1 \pmod{p}, \tag{2}$$

it follows that

$$p|(s_1, \ldots, s_{p-2}). \tag{3}$$

We set $x = p$ in (1). Then

$$(p-1)! = p^{p-1} - s_1 p^{p-2} + \cdots - s_{p-2}p + s_{p-1},$$

or

$$p^{p-2} - s_1 p^{p-3} + \cdots + s_{p-3}p - s_{p-2} = 0.$$

Since $p > 3$, we have, by (3), that

$$s_{p-2} \equiv 0 \pmod{p^2},$$

or

$$p^2 | (p-1)!\left(1 + \frac{1}{2} + \cdots + \frac{1}{p-1}\right),$$

or

$$1^* + 2^* + \cdots + (p-1)^* \equiv 0 \pmod{p^2},$$

as required. □

Chapter 3. Quadratic Residues

3.1 Definitions and Euler's Criteria

Definition 1. Let m be an integer greater than 1, and suppose that $(m, n) = 1$. If $x^2 \equiv n \pmod{m}$ is soluble, then we call n a *quadratic residue* mod m; otherwise we call n a *quadratic non-residue* mod m.

We can now divide the set of integers coprime with n into two classes: the class of quadratic residues and the class of quadratic non-residues.

Example. The numbers $1, 2, 4$ are quadratic residues and $3, 5, 6$ are quadratic non-residues mod 7.

Definition 2 (Legendre's symbol). Let p be an odd prime, and suppose that $p \nmid n$. We let

$$\left(\frac{n}{p}\right) = \begin{cases} 1 & \text{if } n \text{ is a quadratic residue mod } p, \\ -1 & \text{if } n \text{ is a quadratic non-residue mod } p. \end{cases}$$

If is easy to see that if $n \equiv n' \pmod{p}$ and $p \nmid n$, then

$$\left(\frac{n}{p}\right) = \left(\frac{n'}{p}\right).$$

Theorem 1.1. *Let $p > 2$. There are $\frac{1}{2}(p-1)$ quadratic residues, and $\frac{1}{2}(p-1)$ quadratic non-residues in any reduced residue system. Moreover $1^2, \ldots, (\frac{1}{2}(p-1))^2$ form a complete set of quadratic residues.*

Proof. If

$$x^2 \equiv n \pmod{p} \tag{1}$$

is soluble, then there are at most two solutions. From $(p-x)^2 \equiv (-x)^2 = x^2 \equiv n \pmod{p}$, we see that one of the roots of (1) must satisfy

$$1 \leqslant x \leqslant \tfrac{1}{2}(p-1). \tag{2}$$

That is, if (1) is soluble, there must be a solution satisfying (2). Also $1^2, 2^2, \ldots, (\frac{1}{2}(p-1))^2$ are incongruent numbers because $a^2 - b^2 = (a-b)(a+b)$ and neither of these factors, being smaller than p, is a multiple of p. The theorem is proved. □

Theorem 1.2 (Euler's Criterion). *Let p be an odd prime. Then we have*

$$n^{\frac{1}{2}(p-1)} \equiv \left(\frac{n}{p}\right) \pmod{p}.$$

Proof. 1) If $(\frac{n}{p}) = 1$, then there exists x such that $x^2 \equiv n \pmod{p}$, and so

$$n^{\frac{1}{2}(p-1)} \equiv x^{p-1} \equiv 1 \pmod{p}.$$

2) From Theorem 2.9.1 we know that there are at most $\frac{1}{2}(p-1)$ solutions to $n^{\frac{1}{2}(p-1)} \equiv 1 \pmod{p}$. Combining with 1) we see that this equation actually has $\frac{1}{2}(p-1)$ solutions, that is the quadratic residues mod p, and no other.

3) We have

$$p|(n^{p-1} - 1) = (n^{\frac{1}{2}(p-1)} - 1)(n^{\frac{1}{2}(p-1)} + 1).$$

Therefore, if $p \nmid (n^{\frac{1}{2}(p-1)} - 1)$, then

$$n^{\frac{1}{2}(p-1)} + 1 \equiv 0 \pmod{p}.$$

The theorem is proved. □

We have, as a consequence of this theorem:

Theorem 1.3. *If $p \nmid mn$, then* $\left(\frac{m}{p}\right)\left(\frac{n}{p}\right) = \left(\frac{mn}{p}\right)$. □

Thus, $(\frac{n}{p})$ is a multiplicative function of n. We also deduce:

Theorem 1.4. (i) *The product of two quadratic residues is a quadratic residue.*

(ii) *The product of two quadratic non-residues is a quadratic residue.*

(iii) *The product of a quadratic residue and a quadratic non-residue is a quadratic non-residue.* □

3.2 The Evaluation of Legendre's Symbol

From Theorem 1.3 we see that the evaluation of Legendre's symbol reduces to the evaluation of

$$\left(\frac{-1}{p}\right), \quad \left(\frac{2}{p}\right), \quad \left(\frac{q}{p}\right)$$

where q is an odd prime. For if

$$n = \pm 2^m \cdot q_1^{l_1} \cdots q_s^{l_s}, \qquad 2 < q_1 < \cdots < q_s,$$

then

$$\left(\frac{n}{p}\right)=\left(\frac{\pm 1}{p}\right)\left(\frac{2}{p}\right)^m\left(\frac{q_1}{p}\right)^{l_1}\cdots\left(\frac{q_s}{p}\right)^{l_s}.$$

Taking $n = -1$ in Theorem 1.2 we have

$$\left(\frac{-1}{p}\right)\equiv(-1)^{\frac{p-1}{2}}\pmod p,$$

and since both sides of the congruence must be ± 1, we have

Theorem 2.1. *If* $p > 2$, *then* $(\frac{-1}{p}) = (-1)^{\frac{1}{2}(p-1)}$. □

In other words, -1 is a quadratic residue or non-residue $\bmod p$, according to whether $p \equiv 1$ or $3 \pmod 4$. It follows from this that the odd prime divisors of $x^2 + 1$ must be congruent to $1 \pmod 4$.

Theorem 2.2 (Gauss's Lemma). Let $p > 2$, $p \nmid n$. *Denote by m the number of least positive residues of the* $\frac{1}{2}(p-1)$ *numbers* $n, 2n, \ldots, \frac{1}{2}(p-1)n \pmod p$ *which exceed* $p/2$. *Then*

$$\left(\frac{n}{p}\right)=(-1)^m.$$

Example 1. $p = 7, n = 10$. We have

$$10, 20, 30 \equiv 3, 6, 2 \pmod 7.$$

There is exactly one least positive residue which exceeds $\frac{7}{2}$. Therefore $m = 1$ and $(\frac{10}{7}) = -1$.

Example 2. $p = 11, n = 2$. We have the residues $2, 4, 6, 8, 10 \pmod{11}$, and there are three which exceed $\frac{11}{2}$. Therefore $(\frac{2}{11}) = -1$.

Proof of Theorem 2.2. Let $l = \frac{1}{2}(p-1) - m$, and let $a_1, \ldots, a_l$ be those residues which are less than $p/2$, and $b_1, \ldots, b_m$ be those residues which are greater than $p/2$. Then

$$\prod_{s=1}^{l} a_s \prod_{t=1}^{m} b_t \equiv \prod_{k=1}^{\frac{1}{2}(p-1)} kn = \left(\frac{p-1}{2}\right)!\,n^{\frac{p-1}{2}} \pmod p. \tag{1}$$

Since $1 \leqslant p - b_t \leqslant \frac{1}{2}(p-1)$ it follows that a_s and $p - b_t$ are $\frac{1}{2}(p-1)$ integers in the interval from 1 to $\frac{1}{2}(p-1)$. We now prove that they are distinct by proving $a_s \neq p - b_t$. Suppose, if possible, $a_s + b_t = p$. Then there are integers x, y such that

$$xn + yn \equiv 0 \pmod p, \quad 1 \leqslant x \leqslant \tfrac{1}{2}(p-1), \quad 1 \leqslant y \leqslant \tfrac{1}{2}(p-1)$$

or $x + y \equiv 0 \pmod p$, which is impossible. Therefore

$$\prod_{s=1}^{l} a_s \prod_{t=1}^{m} (p - b_t) = \left(\frac{p-1}{2}\right)!.$$

From (1) we see that the left hand side of this equation is

$$\equiv (-1)^m \prod_{s=1}^{l} a_s \prod_{t=1}^{m} b_t \equiv (-1)^m n^{\frac{1}{2}(p-1)} \left(\frac{p-1}{2}\right)! \pmod{p}.$$

Therefore

$$n^{\frac{1}{2}(p-1)} \equiv (-1)^m \pmod{p}.$$

From Euler's criterion we see that $(\frac{n}{p}) \equiv (-1)^m \pmod{p}$, and so $(\frac{n}{p}) = (-1)^m$. □

If we take $n = 2$ in Theorem 2.2, then

$$2,\ 2\cdot 2,\ 2\cdot 3, \ldots, \tfrac{1}{2}(p-1)\cdot 2$$

are already in the interval from 0 to p. We can now determine the number of integers k satisfying $\frac{p}{2} < 2k < p$, or $\frac{p}{4} < k < \frac{p}{2}$, which gives

$$m = \left[\frac{p}{2}\right] - \left[\frac{p}{4}\right].$$

Let $p = 8a + r$, $r = 1, 3, 5, 7$. Then

$$m = 2a + \left[\frac{r}{2}\right] - \left[\frac{r}{4}\right] \equiv 0, 1, 1, 0 \pmod{2}.$$

Therefore we have:

Theorem 2.3. *If $p > 2$, then $(\frac{2}{p}) = (-1)^{\frac{1}{8}(p^2-1)}$.* □

In other words 2 is a quadratic residue or non-residue mod p, according to whether $p \equiv \pm 1$ or $\pm 3 \pmod{8}$. It follows from this that every odd prime divisor of $x^2 - 2$ must be congruent to $\pm 1 \pmod{8}$.

Exercise. Let n be a positive integer such that $4n + 3$ and $8n + 7$ are primes. Prove that $2^{4n+3} - 1 = M_{4n+3}$ is composite. Use this to prove the following concerning Mersenne numbers:

$$23|M_{11}, \quad 47|M_{23}, \quad 167|M_{83}, \quad 263|M_{131},$$
$$359|M_{179}, \quad 383|M_{191}, \quad 479|M_{239}, \quad 503|M_{251}.$$

3.3 The Law of Quadratic Reciprocity

Theorem 3.1. *Let p, q be two distinct odd primes. Then*

$$\left(\frac{p}{q}\right)\left(\frac{q}{p}\right) = (-1)^{\frac{1}{2}(p-1)\frac{1}{2}(q-1)}.$$

In other words, if $p \equiv q \equiv 3 \pmod 4$, then exactly one of the two congruences $x^2 \equiv p \pmod q$, $x^2 \equiv q \pmod p$ is soluble. Otherwise the two congruences are either both soluble or both insoluble. This is the famous and important *Law of Quadratic Reciprocity* in elementary number theory which was discovered by Legendre and proved by Gauss, who named it "*the queen of number theory*". The later research work on algebraic number theory by Kummer, Eisenstein, Hilbert, Takagi, Artin, Furtwängler seem to justify the name.

Proof. We do not, for the moment, exclude the case $q = 2$, and we suppose that p, q are distinct primes. When $1 \leqslant k \leqslant \frac{1}{2}(p-1)$ we can write

$$kq = q_k p + r_k, \qquad q_k = \left[\frac{kq}{p}\right], \qquad 1 \leqslant r_k \leqslant p-1.$$

Let

$$a = \sum_{s=1}^{l} a_s, \qquad b = \sum_{t=1}^{m} b_t$$

where a_s and b_t are defined in the previous section. Then we have

$$\sum_{k=1}^{\frac{1}{2}(p-1)} r_k = a + b. \tag{1}$$

We saw in the proof of Gauss's lemma that a_s, $p - b_t$ are the same as $1, 2, \ldots, \frac{1}{2}(p-1)$. Therefore

$$\frac{p^2-1}{8} = 1 + 2 + \cdots + \frac{1}{2}(p-1) = a + mp - b, \tag{2}$$

and

$$\frac{p^2-1}{8}q = \sum_{k=1}^{\frac{1}{2}(p-1)} kq = p\sum_{k=1}^{\frac{1}{2}(p-1)} q_k + \sum_{k=1}^{\frac{1}{2}(p-1)} r_k = p\sum_{k=1}^{\frac{1}{2}(p-1)} q_k + a + b. \tag{3}$$

Subtracting (2) from (3), we have

$$\frac{p^2-1}{8}(q-1) = p\sum_{k=1}^{\frac{1}{2}(p-1)} q_k - mp + 2b,$$

or

$$\frac{p^2-1}{8}(q-1) \equiv \sum_{k=1}^{\frac{1}{2}(p-1)} q_k - m \pmod 2. \tag{4}$$

1) (Alternative proof of Theorem 2.3). We take $q = 2$ so that q_k are all 0, and hence

$$\frac{p^2-1}{8} \equiv -m \pmod 2.$$

2) Let $q > 2$. Then

$$m \equiv \sum_{k=1}^{\frac{1}{2}(p-1)} q_k \pmod 2.$$

Therefore

$$\left(\frac{q}{p}\right) = (-1)^m = (-1)^{\sum_{k=1}^{\frac{1}{2}(p-1)} q_k} = (-1)^{\sum_{k=1}^{\frac{1}{2}(p-1)} \left[\frac{kq}{p}\right]}.$$

Similarly we have

$$\left(\frac{p}{q}\right) = (-1)^{\sum_{l=1}^{\frac{1}{2}(q-1)} \left[\frac{lp}{q}\right]},$$

so that

$$\left(\frac{p}{q}\right)\left(\frac{q}{p}\right) = (-1)^{\sum_{k=1}^{\frac{1}{2}(p-1)} \left[\frac{kq}{p}\right] + \sum_{l=1}^{\frac{1}{2}(q-1)} \left[\frac{lp}{q}\right]}.$$

If we can prove that

$$\sum_{k=1}^{\frac{1}{2}(p-1)} \left[\frac{kq}{p}\right] + \sum_{l=1}^{\frac{1}{2}(q-1)} \left[\frac{lp}{q}\right] = \frac{p-1}{2}\frac{q-1}{2} \quad \text{or} \quad \equiv \frac{p-1}{2}\frac{q-1}{2} \pmod 2,$$

then the theorem will follow. It suffices therefore to prove the following lemma.

Lemma.

$$\sum_{k=1}^{\frac{1}{2}(p-1)} \left[\frac{kq}{p}\right] + \sum_{l=1}^{\frac{1}{2}(q-1)} \left[\frac{lp}{q}\right] = \frac{p-1}{2}\frac{q-1}{2}.$$

Proof. Consider the rectangle with vertices:

$$(0,0), (0, \tfrac{1}{2}q), (\tfrac{1}{2}p, 0), (\tfrac{1}{2}p, \tfrac{1}{2}q)$$

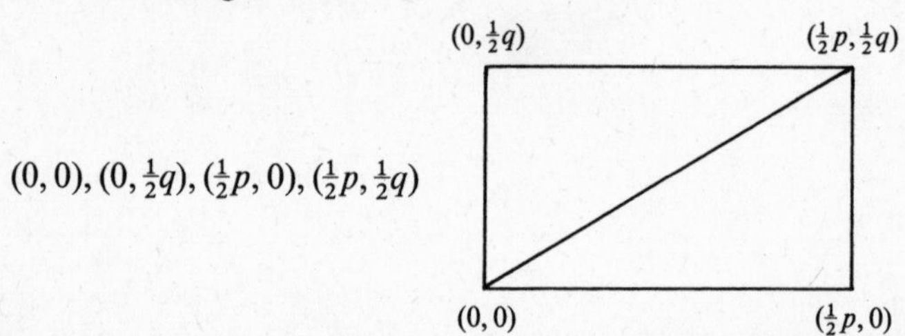

The diagonal from the origin does not pass through any lattice point (a point with integer coordinates). This is because if (x, y) is a lattice point on the diagonal, then $xq - yp = 0$ and so $p|x, q|y$, showing that (x, y) must lie outside the rectangle. The total number of lattice points in the rectangle is $\frac{1}{2}(p-1) \cdot \frac{1}{2}(q-1)$. The number of lattice points in the two triangular regions below and above the diagonal are respectively

$$\sum_{k=1}^{\frac{1}{2}(p-1)} \left[\frac{kq}{p}\right], \quad \sum_{l=1}^{\frac{1}{2}(q-1)} \left[\frac{lp}{q}\right].$$

The lemma is therefore proved. □

Example 1. Determine those primes $p > 3$ of which 3 is a quadratic residue.

From the law of quadratic reciprocity we have

$$\left(\frac{3}{p}\right) = \left(\frac{p}{3}\right)(-1)^{\frac{p-1}{2}}.$$

Now

$$\left(\frac{p}{3}\right)=\begin{cases}\left(\frac{1}{3}\right)=1, & \text{if } p\equiv 1 \pmod 3,\\ \left(\frac{-1}{3}\right)=-1, & \text{if } p\equiv 2 \pmod 3;\end{cases}$$

$$(-1)^{\frac{p-1}{2}}=\begin{cases}1, & \text{if } p\equiv 1 \pmod 4,\\ -1, & \text{if } p\equiv -1 \pmod 4.\end{cases}$$

It follows from the Chinese remainder theorem that

$$\left(\frac{3}{p}\right)=\begin{cases}1, & \text{if } p\equiv \pm 1 \pmod{12},\\ -1, & \text{if } p\equiv \pm 5 \pmod{12}.\end{cases}$$

Example 2. Determine those primes $p \neq 5$ of which 5 is a quadratic residue.
From the law of quadratic reciprocity we have $(\frac{5}{p})=(\frac{p}{5})$, and

$$\left(\frac{1}{5}\right)=1,\quad \left(\frac{2}{5}\right)=(-1)^{\frac{5^2-1}{8}}=-1,\quad \left(\frac{3}{5}\right)=\left(\frac{-2}{5}\right)=-1,\quad \left(\frac{4}{5}\right)=1,$$

so that

$$\left(\frac{5}{p}\right)=\begin{cases}1, & \text{if } p\equiv \pm 1 \pmod 5,\\ -1, & \text{if } p\equiv \pm 2 \pmod 5.\end{cases}$$

Example 3. Determine those primes p of which 10 is a quadratic residue.
From Example 2 and the Chinese remainder theorem we have

$$\left(\frac{10}{p}\right)=\begin{cases}+1, & \text{if } p\equiv \pm 1, \pm 3, \pm 9, \pm 13 \pmod{40},\\ -1, & \text{if } p\equiv \pm 7, \pm 11, \pm 17, \pm 19 \pmod{40}.\end{cases}$$

Example 4. Determine the solubility of $x^2 \equiv -1457 \pmod{2389}$.
Here $p = 2389$ is a prime. Since $-1457 = -31 \times 47$ it follows from

$$\left(\frac{-1}{p}\right)=1,\quad \left(\frac{31}{p}\right)=\left(\frac{p}{31}\right)=\left(\frac{2}{31}\right)=1,$$

$$\left(\frac{47}{p}\right)=\left(\frac{p}{47}\right)=\left(\frac{3}{47}\right)\left(\frac{13}{47}\right)=-\left(\frac{47}{3}\right)\left(\frac{47}{13}\right)$$

$$=-\left(\frac{2}{3}\right)\left(\frac{8}{13}\right)=-\left(\frac{2}{3}\right)\left(\frac{2}{13}\right)=-1,$$

that $(\frac{-1457}{2389})=-1$, so that the congruence is not soluble.

Exercise 1. Show that $\left(\frac{3}{73}\right)=1, \left(\frac{17}{7^3}\right)=-1.$

Exercise 2. Show that $\left(\frac{195}{1901}\right)=-1, \left(\frac{74}{101}\right)=-1, \left(\frac{365}{1847}\right)=1.$

Exercise 3. Show that

$$\text{if} \quad p \equiv \pm 1 \quad \text{or} \quad \pm 5 \pmod{24}, \quad \text{then} \left(\frac{6}{p}\right) = 1;$$

$$\text{if} \quad p \equiv \pm 7 \quad \text{or} \quad \pm 11 \pmod{24}, \quad \text{then} \left(\frac{6}{p}\right) = -1.$$

3.4 Practical Methods for the Solutions

Although the theory above is simple and beautiful, it is nevertheless rather negative. By this we mean the following. If, following our theory, the congruence is insoluble, then the problem is finished. However, if the congruence is soluble, we may further ask for the actual solutions to the congruence, and the method does not give us the solutions. In actual fact, when p is large, the determination of the solutions to $x^2 \equiv n \pmod{p}$ is no easy matter. However, if $p \equiv 3 \pmod 4$ or $p \equiv 5 \pmod 8$, then we have the following methods.

1) $p \equiv 3 \pmod 4$. Since $(\frac{n}{p}) = 1$, we have $n^{\frac{1}{2}(p-1)} \equiv 1 \pmod p$ and so $(n^{\frac{1}{4}(p+1)})^2 \equiv n \pmod{p}$. That is $n^{\frac{1}{4}(p+1)}$ is a solution to the congruence.

2) $p \equiv 5 \pmod 8$. We first determine the solution corresponding to $n = -1$. From Wilson's theorem we have

$$-1 \equiv (p-1)! \equiv 1 \cdot 2 \cdots \left(\frac{p-1}{2}\right) \cdot \left(p - \left(\frac{p-1}{2}\right)\right) \cdots (p-2)(p-1) \pmod{p}$$

$$\equiv \left(1 \cdot 2 \cdots \frac{1}{2}(p-1)\right)^2 \equiv \left(\left(\frac{1}{2}(p-1)\right)!\right)^2 \pmod{p}. \tag{1}$$

This gives us a solution. From $(\frac{n}{p}) = 1$, we have

$$n^{\frac{1}{2}(p-1)} - 1 \equiv 0 \pmod{p}.$$

Now n satisfies

$$n^{\frac{1}{4}(p-1)} \equiv 1 \pmod{p}$$

or

$$n^{\frac{1}{4}(p-1)} \equiv -1 \pmod{p}.$$

From the first congruence we have

$$n^{\frac{1}{4}(p+3)} \equiv (n^{\frac{1}{8}(p+3)})^2 \equiv n \pmod{p}.$$

From the second congruence we have

$$(n^{\frac{1}{8}(p+3)})^2 \equiv -n \pmod{p},$$

so that

$$\left(n^{\frac{1}{8}(p+3)}\left(\frac{p-1}{2}\right)!\right)^2 \equiv n \pmod{p}.$$

3) $p \equiv 1 \pmod 8$. This is a more difficult case. When p is not too large, we usually use the method of successive eliminations. The congruence $x^2 \equiv n \pmod p$ is equivalent to the indeterminate equation $x^2 = n + py$. We may assume that $0 < n < p$ and $0 < x < p/2$ so that $x^2 < p^2/4$, and hence $0 < y < p/4$, which disposes of a large number of trials. Next we let $e > 2, p \nmid e$, and we let $n_1, n_2, n_3, \ldots$ be the quadratic non-residues. Denote by $V_1, V_2, \ldots$ the solutions to

$$n + py \equiv n_1, \qquad n + py \equiv n_2, \ldots \pmod e.$$

If $y \equiv V_i \pmod e$, then $py + n$ is a quadratic non-residue mod e, and is therefore not a square. We may therefore discard those $y \equiv V_i \pmod e$. We may further discard more values of y by choosing different values of e until the number of trials is small enough not to be troublesome.

Example. Solve $x^2 \equiv 73 \pmod{127}$.

We try to solve $x^2 = 127y + 73$ where $1 \leqslant y \leqslant 31$. We take $e = 3, n_1 = 2$. From $73 + 127y \equiv 2 \pmod 3$, that is $y \equiv 1 \pmod 3$, we see that the remaining values for y are:

$$2, 3, 5, 6, 8, 9, 11, 12, 14, 15, 17, 18, 20, 21, 23, 24, 26, 27, 29, 30.$$

We next take $e = 5$, $n_1 = 2$, $n_2 = 3$. From $127y + 73 \equiv 2, 3 \pmod 5$, we have $V_1 \equiv 2$, $V_2 \equiv 0 \pmod 5$ and so the remaining values for y are now

$$3, 6, 8, 9, 11, 14, 18, 21, 23, 24, 26, 29.$$

We next take $e = 7$, $n_1 = 3$, $n_2 = 5$, $n_3 = 6$. From the congruences $127y + 73 \equiv 3, 5, 6 \pmod 7$, or $y + 3 \equiv 3, 5, 6 \pmod 7$ we have $y \equiv 0, 2, 3 \pmod 7$, so that we are left with only the six values

$$6, 8, 11, 18, 26, 29$$

for the trials. In fact $73 + 8 \times 127 = 1089 = 33^2$ so that $x \equiv \pm 33 \pmod{127}$ are the solutions.

Note. In this method, having taken e and e', there is no need to take ee'. Again, having taken an odd e, there is no need to take $2e$.

All we discuss here is related to the work of Gauss. We see therefore that this "Prince of mathematics" is not only a theoretician, but also an expert problem solver.

3.5 The Number of Roots of a Quadratic Congruence

Theorem 5.1. *Let $l > 0$, $p \nmid n$. If $p > 2$, then the congruence $x^2 \equiv n \pmod{p^l}$ has $1 + (\frac{n}{p})$ solutions. If $p = 2$, then we have the following three cases.*

1) $l = 1$. *There is one root.*
2) $l = 2$. *There are two or no roots depending on whether $n \equiv 1$ or $3 \pmod 4$.*
3) $l > 2$. *There are four or no roots depending on whether $n \equiv 1$ or $n \not\equiv 1 \pmod 8$.*

Proof. We first discuss the three cases associated with $p = 2$.

1) This is trivial.

2) The congruence $x^2 \equiv 1 \pmod 4$ has the solutions $\pm 1 \pmod 4$ and the congruence $x^2 \equiv 3 \pmod 4$ has no solution.

3) If $x^2 \equiv n \pmod{2^l}$ is soluble, then x must be odd, say $2k + 1$. Since

$$(2k+1)^2 = 4k(k+1) + 1 = 8 \cdot \frac{k(k+1)}{2} + 1 \equiv 1 \pmod 8,$$

it follows that the congruence is not soluble if $n \not\equiv 1 \pmod 8$.

Suppose now that $n \equiv 1 \pmod 8$. When $l = 3$, there are clearly the four roots $1, 3, 5, 7$. We now proceed by induction on l. Let a satisfy $a^2 \equiv n \pmod{2^{l-1}}$. Then

$$(a + 2^{l-2}b)^2 \equiv a^2 + 2^{l-1}b \pmod{2^l}.$$

We take $b = (n - a^2)/2^{l-1}$. Then $a + 2^{l-2}b$ is a solution with respect to mod 2^l. Therefore a solution to $x^2 \equiv n \pmod{2^l}$ certainly exists. Let x_1 be a solution, and let x_2 be any solution. Then $x_1^2 - x_2^2 \equiv (x_1 - x_2)(x_1 + x_2) \equiv 0 \pmod{2^l}$, and since $x_1 - x_2$, $x_1 + x_2$ are both even it follows that $\frac{1}{2}(x_1 - x_2) \cdot \frac{1}{2}(x_1 + x_2) \equiv 0 \pmod{2^{l-2}}$. But $\frac{1}{2}(x_1 - x_2)$ and $\frac{1}{2}(x_1 + x_2)$ must be of opposite parity, since otherwise their sum x_1 cannot be odd. Therefore we have either $x_1 \equiv x_2 \pmod{2^{l-1}}$ or $x_1 \equiv -x_2 \pmod{2^{l-1}}$, and this means that $x_2 = \pm x_1 + k2^{l-1}$ ($k = 0$ or 1). Hence there are at most four solutions to $x^2 \equiv n \pmod{2^l}$. Since $\pm x_1, \pm x_1 + 2^{l-1}$ are actually incongruent solutions we see that the congruence has exactly four solutions.

When $p > 2$ and $l = 1$, the result is trivial, and the remaining part of the theorem follows from Theorem 2.9.3. □

From the results of Chapter 2 we can determine the number of solutions to a quadratic congruence to any integer modulus m.

3.6 Jacobi's Symbol

Throughout this section m denotes a positive odd integer.

Definition. Let the standard factorization of m be $p_1 \cdots p_t$, where the p_r may be repeated. If $(n, m) = 1$ then we define the *Jacobi's symbol* by

$$\left(\frac{n}{m}\right) = \prod_{r=1}^{t} \left(\frac{n}{p_r}\right).$$

Examples. $\left(\frac{1}{m}\right) = 1$. If $(a, m) = 1$, then $\left(\frac{a^2}{m}\right) = 1$.

Note: If $\left(\frac{n}{m}\right) = 1$, it does *not* follow that $x^2 \equiv n \pmod{m}$ is soluble.

Theorem 6.1. *Let m and m' be positive odd integers.* (i) *If $n \equiv n' \pmod{m}$ and $(n, m) = 1$, then $\left(\frac{n}{m}\right) = \left(\frac{n'}{m}\right)$.* (ii) *If $(n, m) = (n, m') = 1$, then $\left(\frac{n}{m}\right)\left(\frac{n}{m'}\right) = \left(\frac{n}{mm'}\right)$.* (iii) *If $(n, m) = (n', m) = 1$, then $\left(\frac{n}{m}\right)\left(\frac{n'}{m}\right) = \left(\frac{nn'}{m}\right)$.* □

Theorem 6.2. $\left(\frac{-1}{m}\right) = (-1)^{\frac{1}{2}(m-1)}$.

Proof. It suffices to prove that

$$\sum_{i=1}^{t} \frac{p_i - 1}{2} \equiv \frac{\prod_{i=1}^{t} p_i - 1}{2} \pmod{2},$$

which certainly holds when $t = 1$. Given any two odd integers u, v we always have

$$\frac{u-1}{2} + \frac{v-1}{2} \equiv \frac{uv-1}{2} \pmod{2} \quad (\text{or } (u-1)(v-1) \equiv 0 \pmod{4}). \qquad (1)$$

It follows by induction that

$$\begin{aligned}\sum_{i=1}^{t} \frac{p_i - 1}{2} &\equiv \sum_{i=1}^{t-1} \frac{p_i - 1}{2} + \frac{p_t - 1}{2} \\ &\equiv \frac{\prod_{i=1}^{t-1} p_i - 1}{2} + \frac{p_t - 1}{2} \equiv \frac{\prod_{i=1}^{t} p_i - 1}{2} \pmod{2}. \quad \square\end{aligned}$$

Theorem 6.3. $\left(\frac{2}{m}\right) = (-1)^{\frac{1}{8}(m^2-1)}$.

Proof. This is similar to the above, except that we replace (1) by

$$\frac{u^2v^2 - 1}{8} \equiv \frac{u^2 - 1}{8} + \frac{v^2 - 1}{8} \pmod{2}. \quad \square$$

Theorem 6.4. *Let m, n be coprime positive odd integers. Then*

$$\left(\frac{m}{n}\right)\left(\frac{n}{m}\right) = (-1)^{\frac{n-1}{2}\frac{m-1}{2}}.$$

Proof. Let $m = \prod p, n = \prod q$. Then

$$\left(\frac{m}{n}\right)\left(\frac{n}{m}\right) = \left(\prod_p \prod_q \left(\frac{p}{q}\right)\right)\left(\prod_p \prod_q \left(\frac{q}{p}\right)\right) = \prod_p \prod_q \left(\frac{p}{q}\right)\left(\frac{q}{p}\right)$$

$$= \prod_p \prod_q (-1)^{\frac{p-1}{2}\frac{q-1}{2}} = (-1)^{\frac{n-1}{2}\frac{m-1}{2}}$$

where we have used (1). □

In using the Legendre's symbol we must always ensure that the denominator is a prime. In using Jacobi's symbol however, we can avoid the factorization process. For example:

$$\left(\frac{383}{443}\right) = -\left(\frac{443}{383}\right) = -\left(\frac{60}{383}\right) = -\left(\frac{2^2}{383}\right)\left(\frac{15}{383}\right) = -\left(\frac{15}{383}\right)$$

$$= \left(\frac{383}{15}\right) = \left(\frac{8}{15}\right) = \left(\frac{2}{15}\right) = 1.$$

If we delete the condition that m, m' are positive in Theorem 6.4, then we have:

Theorem 6.5. *Let m, n be coprime odd integers. If m, n are both negative, then*

$$\left(\frac{n}{|m|}\right)\left(\frac{m}{|n|}\right) = -(-1)^{\frac{m-1}{2}\frac{n-1}{2}}.$$

Otherwise, the required value is $(-1)^{\frac{1}{2}(m-1)\frac{1}{2}(n-1)}$. □

Example. Determine the solubility of $x^2 \equiv -286 \pmod{4272943}$.

Here $p = 4272943$ is a prime, and we have to evaluate $\left(\frac{-286}{p}\right)$. Since $\left(\frac{-1}{p}\right) = -1, \left(\frac{2}{p}\right) = 1$, we have

$$\left(\frac{-286}{p}\right) = \left(\frac{-1}{p}\right)\left(\frac{2}{p}\right)\left(\frac{143}{p}\right) = -\left(\frac{143}{p}\right).$$

We now determine $\left(\frac{143}{p}\right)$ as follows: We have

$$4272943 = 29880 \times 143 + 103^*,$$

$$143 = 2 \times 103 - 63,$$

$$103 = 2 \times 63 - 23,$$

$$63 = 2 \times 23 + 17^*,$$

$$23 = 2 \times 17 - 11,$$

$$17 = 2 \times 11 - 5^*,$$

$$11 = 2 \times 5 + 1$$

where each step with a * denotes a change of sign. Therefore

$$\left(\frac{143}{p}\right) = (-1)^3 = -1.$$

Thus $\left(\frac{-286}{p}\right) = 1$, and the congruence is soluble. Gauss determined the solutions as ± 1493445.

3.7 Two Terms Congruences

Let p be prime. We now discuss the congruence $x^k \equiv n \pmod{p}$.

Theorem 7.1. *The congruence*

$$x^k \equiv 1 \pmod{p} \tag{1}$$

has $(k, p-1)$ *roots.*

Proof. 1) Let $d = (k, p-1)$ and let s, t be integers such that $sk + t(p-1) = d$. We then have $x^d = (x^k)^s(x^{p-1})^t$, so that every root of (1) is a root of

$$x^d \equiv 1 \pmod{p}, \tag{2}$$

and conversely.

2) It suffices to prove that (2) has d roots. From Theorem 2.9.1 the number of roots for (2) certainly cannot exceed d. Also, there are $p-1$ roots to $x^{p-1} \equiv 1 \pmod{p}$. Again, by Theorem 2.9.1 the number of roots for

$$\frac{x^{p-1}-1}{x^d-1} = (x^d)^{\frac{p-1}{d}-1} + \cdots + x^d + 1 \equiv 0 \pmod{p}$$

does not exceed $p - 1 - d$, so that the number of roots for (2) must be at least d. The theorem is proved. □

Theorem 7.2. *Either the congruence* $x^k \equiv n \pmod{p}$, $p \nmid n$ *has no solution or it has* $(k, p-1)$ *solutions.*

Proof. If x_0 is a solution, then $(x_0^{-1}x)^k \equiv x^k x_0^{-k} \equiv 1 \pmod{p}$. The required result follows from Theorem 7.1. □

Theorem 7.3. *If x runs over a reduced set of residues* mod p, *then* x^k *take* $(p-1)/(k, p-1)$ *different values.*

Proof. From Theorem 7.2 we see that there are $(k, p-1)$ distinct residues whose k-th power have the same residue mod p. The $p-1$ residues are now partitioned into $(p-1)/(k, p-1)$ classes, and there is a one-to-one correspondence. □

Definition. Let h be an integer, and $(h, n) = 1$. The least positive integer l such that $h^l \equiv 1 \pmod{n}$ is called the *order* of $h \pmod{n}$.

Theorem 7.4. *If $h^m \equiv 1 \pmod{n}$, then $l | m$.*

Proof. Suppose the contrary. Then there are integers q, r such that $m = ql + r$, $0 < r < l$. Now $h^r \equiv h^m(h^l)^{-q} \equiv 1 \pmod{n}$ contradicts with the definition of l. □

Theorem 7.5. *Let $l | p - 1$, and denote by $\varphi(l)$ the number of incongruent integers with order l. Then $\varphi(l)$ is the Euler's function.*

Proof. We first establish certain properties of $\varphi(l)$.

1) If $(l_1, l_2) = 1$, then $\varphi(l_1 l_2) = \varphi(l_1)\varphi(l_2)$. Let h_1 and h_2 be integers with orders l_1 and l_2 respectively, and let l be the order of $h_1 h_2$. From $1 \equiv (h_1 h_2)^{ll_2} \equiv h_1^{ll_2} \pmod{p}$, and Theorem 7.4 we see that $l_1 | ll_2$. Since $(l_1, l_2) = 1$, we have $l_1 | l$, and similarly $l_2 | l$. Therefore $l = l_1 l_2$, that is the order of $h_1 h_2$ is $l_1 l_2$. Thus, given any h_1, h_2 with orders l_1, l_2, we can construct $h_1 h_2$ whose order is $l_1 l_2$. We now prove that if we do not have $h_1 \equiv h'_1, h_2 \equiv h'_2 \pmod{p}$, then $h_1 h_2 \not\equiv h'_1 h'_2 \pmod{p}$. For if $h_1 h_2 \equiv h'_1 h'_2 \pmod{p}$, then $h_1 h_1'^{-1} \equiv h'_2 h_2^{-1} \pmod{p}$. But the order of $h_1 h_1'^{-1}$ divides l_1 and the order of $h'_2 h_2^{-1}$ divides l_2, so that $h_1 h_1'^{-1} \equiv h'_2 h_2^{-1} \equiv 1 \pmod{p}$ which contradicts our assumption. Conversely, if h is an integer with order $l_1 l_2$ where $(l_1, l_2) = 1$, then $h_1 = h^{l_2}$, $h_2 = h^{l_1}$ are integers with orders l_1, l_2. Therefore $\varphi(l_1)\varphi(l_2) = \varphi(l_1 l_2)$.

2) If q is prime, then $\varphi(q^t) = q^t - q^{t-1}$. The number of roots of $x^{q^t} - 1 \equiv 0 \pmod{p}$ is q^t. If x satisfies this congruence and its order is not q^t, then it must satisfy $x^{q^{t-1}} - 1 \equiv 0 \pmod{p}$. But the number of roots of this congruence is q^{t-1}. Therefore $\varphi(q^t) = q^t - q^{t-1}$.

That $\varphi(l)$ is Euler's function follows from the two properties in 1) and 2). □

3.8 Primitive Roots and Indices

From Theorem 7.5 we see that there are $\varphi(p - 1)$ incongruent numbers with order $p - 1 \pmod{p}$.

Definition 1. A positive integer whose order is $p - 1$ is called a *primitive root* of p.

Let g be a primitive root of p. Then $g^0, g^1, \ldots, g^{p-2}$ are incongruent $\pmod{p}$.

Definition 2. Corresponding to each integer n not divisible by p, there exists a such that

$$n \equiv g^a \pmod{p}, \qquad 0 \leqslant a < p - 1.$$

We call a the *index* of $n \pmod{p}$ and we denote it by $\operatorname{ind}_g n$ or simply $\operatorname{ind} n$. If b is such that $n \equiv g^b \pmod{p}$, then $b \equiv \operatorname{ind} n \pmod{p - 1}$.

The function ind is similar to the logarithm function in that there are following properties:

1) $\operatorname{ind} nm \equiv \operatorname{ind} m + \operatorname{ind} n \pmod{p-1}, p \nmid mn$;
2) $\operatorname{ind} n^l \equiv l \operatorname{ind} n \pmod{p-1}, p \nmid n$.

Note: We do not define $\operatorname{ind} n$ when $p|n$; this is similar to not defining log 0.

Definition 3. Let $p \nmid n$. If the congruence

$$x^k \equiv n \pmod{p} \tag{1}$$

is soluble, then we call n a *k-th power residue* mod p; otherwise we call n a *k-th power non-residue.*

Theorem 8.1. *A necessary and sufficient condition for n to be a k-th power residue* mod p *is that* $(k, p-1)$ *divides* $\operatorname{ind} n$.

Proof. Let $a = \operatorname{ind} n$ and $y = \operatorname{ind} x$. Then (1) is equivalent to $ky \equiv a \pmod{p-1}$, and a necessary and sufficient condition for this to be soluble is that $(k, p-1)$ divides a. □

"*Base interchange formula*". It is clear that the index depends on the primitive root chosen. Let g_1 be another primitive root and $g_1 \equiv g^b \pmod{p}$. Then $n \equiv g_1^a \equiv (g^b)^a \pmod{p}$ or

$$\operatorname{ind}_g n \equiv ab \equiv (\operatorname{ind}_g g_1)(\operatorname{ind}_{g_1} n) \pmod{p-1}.$$

This is similar to the base interchange formula for the logarithm function.

We list the least primitive roots for all the primes up to 5000 at the end of this chapter.

3.9 The Structure of a Reduced Residue System

Let m be a natural number. We ask whether there exists g such that $g^0, g^1, g^2, \ldots, g^{\varphi(m)-1} \pmod{m}$ form a reduced residue system. If g exists, then we call it a primitive root of m.

Theorem 9.1. *A necessary and sufficient condition for m to have a primitive root is that $m = 2, 4, p^l$ or $2p^l$, where p is an odd prime.*

Proof. 1) Let the standard factorization of m be

$$m = p_1^{l_1} p_2^{l_2} \cdots p_s^{l_s}, \qquad p_1 < p_2 < \cdots < p_s.$$

From Euler's theorem, any integer a not divisible by p_i must satisfy

$$a^{\varphi(p_i^{l_i})} \equiv 1 \pmod{p_i^{l_i}}.$$

Let l be the least common multiple of $\varphi(p_1^{l_1}), \ldots, \varphi(p_s^{l_s})$ so that $a^l \equiv 1 \pmod{m}$. Therefore there can be no primitive root if $l < \varphi(m)$. If $p > 2$, then $\varphi(p^l)$ is even, so that m cannot have two distinct odd prime divisors. If m has a primitive root, then m must be of the form $2^l, p^l$ or $2^c p^l$. If $c \geqslant 2$, then $\varphi(2^c) = 2^{c-1}$ is also even, and so $2^c p^l$ cannot have primitive roots. Therefore m must be of the form $2^l, p^l$ or $2p^l$.

2) $m = 2^l$. If $l = 1$, then 1 is a primitive root. If $l = 2$, then 3 is a primitive root. Let $l \geqslant 3$. We prove by induction that for all odd a, we have

$$a^{2^{l-2}} \equiv 1 \pmod{2^l}.$$

This is easy, since if

$$a^{2^{l-3}} = 1 + 2^{l-1}\lambda,$$

then

$$a^{2^{l-2}} \equiv (1 + 2^{l-1}\lambda)^2 \equiv 1 \pmod{2^l}.$$

Therefore there is no primitive root for $m = 2^l$ $(l > 2)$.

3) $m = p^l$. The case $l = 1$ has already been settled in §8. Let g be a primitive root of p. If $g^{p-1} - 1 \not\equiv 0 \pmod{p^2}$, then we take $r = g$; if $g^{p-1} - 1 \equiv 0 \pmod{p^2}$, then we take $r = g + p$. We then have

$$r^{p-1} - 1 \equiv (g + p)^{p-1} - 1 \equiv -g^{p-2}p \not\equiv 0 \pmod{p^2}.$$

Therefore such an r is a primitive root of p^2. Let

$$r^{p-1} - 1 = kp, \quad p \nmid k.$$

Since

$$(1 + kp)^{p^s} \equiv 1 + kp^{s+1} \pmod{p^{s+2}}, \qquad s \geqslant 0,$$

we can prove as before that

$$(r^{p-1})^{p^s} \equiv 1 + kp^{s+1} \pmod{p^{s+2}}.$$

Hence

$$r^{p^{l-2}(p-1)} \equiv 1 + kp^{l-1} \pmod{p^l}, \qquad l \geqslant 2. \tag{1}$$

If the order of r is e, then $e|(p-1)p^{l-1} = \varphi(p^l)$. Since r is a primitive root of p, we see that $(p-1)|e$. We deduce from (1) that $e = \varphi(p^l)$; that is r is a primitive root of p^l.

4) $m = 2p^l$. We take g to be a primitive root of p^l. If g is odd, then g is also a primitive root of $2p^l$; if g is even, then $g + p^l$ is a primitive root of $2p^l$. □

Theorem 9.2. *Let $l > 2$. Then the order of 5 with respect to the modulus 2^l is 2^{l-2}.*

Proof. We first prove that, for $a \geqslant 3$,

$$5^{2^{a-3}} \equiv 1 + 2^{a-1} \pmod{2^a}.$$

This clearly holds when $a = 3$, and we now proceed by induction. We have

$$5^{2^{a-2}} = (5^{2^{a-3}})^2 \equiv (1 + 2^{a-1} + k2^a)^2 \equiv 1 + 2^a \pmod{2^{a+1}}.$$

Therefore $5^{2^{l-3}} \not\equiv 1 \pmod{2^l}$ and $5^{2^{l-2}} \equiv 1 \pmod{2^l}$. That is, the order of 5 is 2^{l-2} $\pmod{2^l}$. □

Theorem 9.3. *Let $l > 2$. Then, given any odd a, there exists b such that*

$$a \equiv (-1)^{\frac{a-1}{2}} 5^b \pmod{2^l}, \qquad b \geqslant 0.$$

Proof. If $a \equiv 1 \pmod 4$, then by Theorem 9.2, 5^b $(0 \leqslant b < 2^{l-2})$ gives 2^{l-2} distinct numbers mod 2^l; moreover they are all congruent 1 (mod 4). Therefore there must be an integer b such that $a \equiv 5^b \pmod{2^l}$.

If $a \equiv 3 \pmod 4$, then $-a \equiv 1 \pmod 4$, and the required result follows from the above. □

Theorem 9.4. *Let $m = 2^l \cdot p_1^{l_1} \cdots p_s^{l_s}$ (standard factorization) with $l \geqslant 0, l_1 > 0, \ldots, l_s > 0$. We define δ to be 0 or 1 or 2 according to whether $l = 0, 1$ or $l = 2$ or $l > 2$ respectively. Then the reduced residue system of m can be represented by the products of $s + \delta$ numbers.*

Proof. 1) Suppose that $m = m'm''$, $(m', m'') = 1$. Let $a_1, \ldots, a_{\varphi(m')}$ be a reduced residue system mod m', and that $a_i \equiv 1 \pmod{m''}$ (this is always possible). Let $b_1, \ldots, b_{\varphi(m'')}$ be a reduced residue system mod m'' and that $b_j \equiv 1 \pmod{m'}$. Then $a_i b_j$ represent a reduced residue system mod mm', and its number is $\varphi(m'm'')$. Also, if $a_i b_j \equiv a_s b_t \pmod{m'm''}$, then $a_i \equiv a_s \pmod{m'}$, $b_j \equiv b_t \pmod{m''}$.

2) From Theorems 9.1 and 9.3 we know that the reduced residue system mod m, where $m = p^l$ $(p > 2)$, is the product of a single number. If $m = 2^l$ where $l > 1$, then the reduced residue system is the product of δ numbers. Combining this with 1), the theorem is proved. □

This theorem points out an important principle. In group theory this result is known as the Fundamental Theorem of Abelian groups.

Exercise. Prove that if $k < p, n = kp^2 + 1$ and

$$2^k \not\equiv 1, \qquad 2^{n-1} \equiv 1 \pmod n,$$

then n is a prime number.

Hints: (i) First prove that n has a prime divisor congruent 1 (mod p). Let d be the least positive integer such that $2^d \equiv 1 \pmod n$. Deduce that $d \nmid k, d \mid n-1$ and $p \mid d$. Then obtain the conclusion from $p \mid d \mid \varphi(n)$.

(ii) Deduce from $n = kp^2 + 1 = (up + 1)(vp + 1)$ that n cannot be composite.

Note: Taking $p = 2^{127} - 1, k = 180$, Miller and Wheeler proved, with the aid of a computer, that $180(2^{127} - 1)^2 + 1$ is prime. (*Nature* **168** (1951), 838).

The least primitive roots for primes less than 5000. An asterisk indicates that 10 is a primitive root.

p	$p-1$	g	p	$p-1$	g	p	$p-1$	g
3	2	2	241	$2^{4}\cdot 3\cdot 5$	7	569	$2^{3}\cdot 71$	3
5	2^{2}	2	251	$2\cdot 5^{3}$	6	571*	$2\cdot 3\cdot 5\cdot 19$	3
7*	$2\cdot 3$	3	257*	2^{3}	3	577*	$2^{6}\cdot 3^{2}$	5
11	$2\cdot 5$	2	263*	$2\cdot 131$	5	587	$2\cdot 293$	2
13	$2^{2}\cdot 3$	2	269*	$2^{2}\cdot 67$	2	593*	$2^{4}\cdot 37$	3
17*	2^{4}	3	271	$2\cdot 3^{3}\cdot 5$	6	599	$2\cdot 13\cdot 23$	7
19*	$2\cdot 3^{2}$	2	277	$2^{2}\cdot 3\cdot 23$	5	601	$2^{3}\cdot 3\cdot 5^{2}$	7
23*	$2\cdot 11$	5	281	$2^{3}\cdot 5\cdot 7$	3	607	$2\cdot 3\cdot 101$	3
29*	$2^{2}\cdot 7$	2	283	$2\cdot 3\cdot 47$	3	613	$2^{2}\cdot 3^{2}\cdot 17$	2
31	$2\cdot 3\cdot 5$	3	293	$2^{2}\cdot 73$	2	617	$2^{3}\cdot 7\cdot 11$	3
37	$2^{2}\cdot 3^{2}$	2	307	$2\cdot 3^{2}\cdot 17$	5	619*	$2\cdot 3\cdot 103$	2
41	$2^{3}\cdot 5$	6	311	$2\cdot 5\cdot 31$	17	631	$2\cdot 3^{2}\cdot 5\cdot 7$	3
43	$2\cdot 3\cdot 7$	3	313*	$2^{3}\cdot 3\cdot 13$	10	641	$2^{7}\cdot 5$	3
47*	$2\cdot 23$	5	317	$2^{2}\cdot 79$	2	643	$2\cdot 3\cdot 107$	11
53	$2^{2}\cdot 13$	2	331	$2\cdot 3\cdot 5\cdot 11$	3	647*	$2\cdot 17\cdot 19$	5
59*	$2\cdot 29$	2	337*	$2^{4}\cdot 3\cdot 7$	10	653	$2^{2}\cdot 163$	2
61*	$2^{2}\cdot 3\cdot 5$	2	347	$2\cdot 173$	2	659*	$2\cdot 7\cdot 47$	2
67	$2\cdot 3\cdot 11$	2	349	$2^{2}\cdot 3\cdot 29$	2	661	$2^{2}\cdot 3\cdot 5\cdot 11$	2
71	$2\cdot 5\cdot 7$	7	353	$2^{5}\cdot 11$	3	673	$2^{5}\cdot 3\cdot 7$	5
73	$2^{3}\cdot 3^{2}$	5	359	$2\cdot 179$	7	677	$2^{2}\cdot 13^{2}$	2
79	$2\cdot 3\cdot 13$	3	367*	$2\cdot 3\cdot 61$	6	683	$2\cdot 11\cdot 31$	5
83	$2\cdot 41$	2	373	$2^{2}\cdot 3\cdot 31$	2	691	$2\cdot 3\cdot 5\cdot 23$	3
89	$2^{3}\cdot 11$	3	379*	$2\cdot 3^{3}\cdot 7$	2	701*	$2^{2}\cdot 5^{2}\cdot 7$	2
97*	$2^{5}\cdot 3$	5	383*	$2\cdot 191$	5	709*	$2^{2}\cdot 3\cdot 59$	2
101	$2^{2}\cdot 5^{2}$	2	389*	$2^{2}\cdot 97$	2	719	$2\cdot 359$	11
103	$2\cdot 3\cdot 17$	5	397	$2^{2}\cdot 3^{2}\cdot 11$	5	727*	$2\cdot 3\cdot 11^{2}$	5
107	$2\cdot 53$	2	401	$2^{4}\cdot 5^{2}$	3	733	$2^{2}\cdot 3\cdot 61$	6
109*	$2^{2}\cdot 3^{3}$	6	409	$2^{3}\cdot 3\cdot 17$	21	739	$2\cdot 3^{2}\cdot 41$	3
113*	$2^{4}\cdot 7$	3	419*	$2\cdot 11\cdot 19$	2	743*	$2\cdot 7\cdot 53$	5
127	$2\cdot 3^{2}\cdot 7$	3	421	$2^{2}\cdot 3\cdot 5\cdot 7$	2	751	$2\cdot 3\cdot 5^{3}$	3
131*	$2\cdot 5\cdot 13$	2	431	$2\cdot 5\cdot 43$	7	757	$2^{2}\cdot 3^{3}\cdot 7$	2
137	$2^{3}\cdot 17$	3	433*	$2^{4}\cdot 3^{3}$	5	761	$2^{2}\cdot 5\cdot 19$	6
139	$2\cdot 3\cdot 23$	2	439	$2\cdot 3\cdot 73$	15	769	$2^{8}\cdot 3$	11
149*	$2^{2}\cdot 37$	2	443	$2\cdot 13\cdot 17$	2	773	$2^{2}\cdot 193$	2
151	$2\cdot 3\cdot 5^{2}$	6	449	$2^{6}\cdot 7$	3	787	$2\cdot 3\cdot 131$	2
157	$2^{2}\cdot 3\cdot 13$	5	457	$2^{3}\cdot 3\cdot 19$	13	797	$2^{2}\cdot 199$	2
163	$2\cdot 3^{4}$	2	461*	$2^{2}\cdot 5\cdot 23$	2	809	$2^{3}\cdot 101$	3
167*	$2\cdot 83$	5	463	$2\cdot 3\cdot 7\cdot 11$	3	811*	$2\cdot 3^{4}\cdot 5$	3
173	$2^{2}\cdot 43$	2	467	$2\cdot 233$	2	821*	$2^{2}\cdot 5\cdot 41$	2
179*	$2\cdot 89$	2	479	$2\cdot 239$	13	823*	$2\cdot 3\cdot 137$	3
181*	$2^{2}\cdot 3^{2}\cdot 5$	2	487*	$2\cdot 3^{5}$	3	827	$2\cdot 7\cdot 59$	2
191	$2\cdot 5\cdot 19$	19	491*	$2\cdot 5\cdot 7^{2}$	2	829	$2^{2}\cdot 3^{2}\cdot 23$	2
193*	$2^{6}\cdot 3$	5	499*	$2\cdot 3\cdot 83$	7	839	$2\cdot 419$	11
197	$2^{2}\cdot 7^{2}$	2	503*	$2\cdot 251$	5	853	$2^{2}\cdot 3\cdot 71$	2
199	$2\cdot 3^{2}\cdot 11$	3	509*	$2^{2}\cdot 127$	2	857*	$2^{3}\cdot 107$	3
211	$2\cdot 3\cdot 5\cdot 7$	2	521	$2^{2}\cdot 5\cdot 13$	3	859	$2\cdot 3\cdot 11\cdot 13$	2
223*	$2\cdot 3\cdot 37$	3	523	$2\cdot 3^{2}\cdot 29$	2	863*	$2\cdot 431$	5
227	$2\cdot 113$	2	541*	$2^{2}\cdot 3^{3}\cdot 5$	2	877	$2^{2}\cdot 3\cdot 73$	2
229*	$2^{2}\cdot 3\cdot 19$	6	547	$2\cdot 3\cdot 7\cdot 13$	2	881	$2^{4}\cdot 5\cdot 11$	3
233*	$2^{3}\cdot 29$	3	557	$2^{2}\cdot 139$	2	883	$2\cdot 3^{2}\cdot 7^{2}$	2
239	$2\cdot 7\cdot 17$	7	563	$2\cdot 281$	2	887*	$2\cdot 443$	5

p	$p-1$	g	p	$p-1$	g	p	$p-1$	g
907	$2\cdot 3\cdot 151$	2	1283	$2\cdot 641$	2	1663*	$2\cdot 3\cdot 277$	3
911	$2\cdot 5\cdot 7\cdot 13$	17	1289	$2^3\cdot 7\cdot 23$	6	1667	$2\cdot 7^2\cdot 17$	2
919	$2\cdot 3^3\cdot 17$	7	1291*	$2\cdot 3\cdot 5\cdot 43$	2	1669	$2^2\cdot 3\cdot 139$	2
929	$2^5\cdot 29$	3	1297*	$2^4\cdot 3^4$	10	1693	$2^2\cdot 3^2\cdot 47$	2
937*	$2^3\cdot 3^2\cdot 13$	5	1301*	$2^2\cdot 5^2\cdot 13$	2	1697*	$2^5\cdot 53$	3
941*	$2^2\cdot 5\cdot 47$	2	1303*	$2\cdot 3\cdot 7\cdot 31$	6	1699	$2\cdot 3\cdot 283$	3
947	$2\cdot 11\cdot 43$	2	1307	$2\cdot 653$	2	1709*	$2^2\cdot 7\cdot 61$	3
953*	$2^3\cdot 7\cdot 17$	3	1319	$2\cdot 659$	13	1721	$2^3\cdot 5\cdot 43$	3
967	$2\cdot 3\cdot 7\cdot 23$	5	1321	$2^3\cdot 3\cdot 5\cdot 11$	13	1723	$2\cdot 3\cdot 7\cdot 41$	3
971*	$2\cdot 5\cdot 97$	6	1327*	$2\cdot 3\cdot 13\cdot 17$	3	1733	$2^2\cdot 433$	2
977*	$2^4\cdot 61$	3	1361	$2^4\cdot 5\cdot 17$	3	1741*	$2^2\cdot 3\cdot 5\cdot 29$	2
983*	$2\cdot 491$	5	1367*	$2\cdot 683$	5	1747	$2\cdot 3^2\cdot 97$	2
991	$2\cdot 3^2\cdot 5\cdot 11$	6	1373	$2^2\cdot 7^3$	2	1753	$2^2\cdot 3\cdot 73$	7
997	$2^2\cdot 3\cdot 83$	7	1381*	$2^2\cdot 3\cdot 5\cdot 23$	2	1759	$2\cdot 3\cdot 293$	6
1009	$2^4\cdot 3^2\cdot 7$	11	1399	$2\cdot 3\cdot 233$	13	1777*	$2^4\cdot 3\cdot 37$	5
1013	$2^2\cdot 11\cdot 23$	3	1409	$2^7\cdot 11$	3	1783*	$2\cdot 3^4\cdot 11$	10
1019*	$2\cdot 509$	2	1423	$2\cdot 3^2\cdot 79$	3	1787	$2\cdot 19\cdot 47$	2
1021*	$2^2\cdot 3\cdot 5\cdot 17$	10	1427	$2\cdot 23\cdot 31$	2	1789*	$2^2\cdot 3\cdot 149$	6
1031	$2\cdot 5\cdot 103$	14	1429*	$2^2\cdot 3\cdot 7\cdot 17$	6	1801	$2^3\cdot 3^2\cdot 5^2$	11
1033*	$2^3\cdot 3\cdot 43$	5	1433*	$2^3\cdot 179$	3	1811*	$2\cdot 5\cdot 181$	6
1039	$2\cdot 3\cdot 173$	3	1439	$2\cdot 719$	7	1823*	$2\cdot 911$	5
1049	$2^3\cdot 131$	3	1447*	$2\cdot 3\cdot 241$	3	1831	$2\cdot 3\cdot 5\cdot 61$	3
1051*	$2\cdot 3\cdot 5^2\cdot 7$	7	1451	$2\cdot 5^2\cdot 29$	2	1847*	$2\cdot 13\cdot 71$	5
1061	$2^2\cdot 5\cdot 53$	2	1453	$2^2\cdot 3\cdot 11^2$	2	1861*	$2^2\cdot 3\cdot 5\cdot 31$	2
1063*	$2\cdot 3^2\cdot 59$	3	1459	$2\cdot 3^6$	5	1867	$2\cdot 3\cdot 311$	2
1069*	$2^2\cdot 3\cdot 89$	6	1471	$2\cdot 3\cdot 5\cdot 7^2$	6	1871	$2\cdot 5\cdot 11\cdot 17$	14
1087*	$2\cdot 3\cdot 181$	3	1481	$2^3\cdot 5\cdot 37$	3	1873*	$2^4\cdot 3^2\cdot 13$	10
1091*	$2\cdot 5\cdot 109$	2	1483	$2\cdot 3\cdot 13\cdot 19$	2	1877	$2^2\cdot 7\cdot 67$	2
1093	$2^2\cdot 3\cdot 7\cdot 13$	5	1487*	$2\cdot 743$	5	1879	$2\cdot 3\cdot 313$	6
1097*	$2^3\cdot 137$	3	1489	$2^4\cdot 3\cdot 31$	14	1889	$2^5\cdot 59$	3
1103*	$2\cdot 19\cdot 29$	5	1493	$2^2\cdot 373$	2	1901	$2^2\cdot 3^2\cdot 19$	2
1109*	$2^2\cdot 277$	2	1499	$2\cdot 7\cdot 107$	2	1907	$2\cdot 953$	2
1117	$2^2\cdot 3^2\cdot 31$	2	1511	$2\cdot 5\cdot 151$	11	1913*	$2^3\cdot 239$	3
1123	$2\cdot 3\cdot 11\cdot 17$	2	1523	$2\cdot 761$	2	1931	$2\cdot 5\cdot 193$	2
1129	$2^3\cdot 3\cdot 47$	11	1531*	$2\cdot 3^2\cdot 5\cdot 17$	2	1933	$2^2\cdot 3\cdot 7\cdot 23$	5
1151	$2\cdot 5^2\cdot 23$	17	1543*	$2\cdot 3\cdot 257$	5	1949*	$2^2\cdot 487$	2
1153*	$2^7\cdot 3^2$	5	1549*	$2^2\cdot 3^2\cdot 43$	2	1951	$2\cdot 3\cdot 5^2\cdot 13$	3
1163	$2\cdot 7\cdot 83$	5	1553*	$2^4\cdot 97$	3	1973	$2^2\cdot 17\cdot 29$	2
1171*	$2\cdot 3^2\cdot 5\cdot 13$	2	1559	$2\cdot 19\cdot 41$	19	1979*	$2\cdot 23\cdot 43$	2
1181*	$2^2\cdot 5\cdot 59$	7	1567*	$2\cdot 3^3\cdot 29$	3	1987	$2\cdot 3\cdot 331$	2
1187	$2\cdot 593$	2	1571*	$2\cdot 5\cdot 157$	2	1993*	$2^2\cdot 3\cdot 83$	5
1193*	$2^2\cdot 149$	3	1579*	$2\cdot 3\cdot 263$	3	1997	$2^2\cdot 499$	2
1201	$2^4\cdot 3\cdot 5^2$	11	1583*	$2\cdot 7\cdot 113$	5	1999	$2\cdot 3^3\cdot 37$	3
1213*	$2^2\cdot 3\cdot 101$	2	1597	$2^2\cdot 3\cdot 7\cdot 19$	11	2003	$2\cdot 7\cdot 11\cdot 13$	5
1217*	$2^6\cdot 19$	3	1601	$2^6\cdot 5^2$	3	2011	$2\cdot 3\cdot 5\cdot 67$	3
1223*	$2\cdot 13\cdot 47$	5	1607*	$2\cdot 11\cdot 73$	5	2017*	$2^5\cdot 3^2\cdot 7$	5
1229*	$2^2\cdot 307$	2	1609	$2^3\cdot 3\cdot 67$	7	2027	$2\cdot 1013$	2
1231	$2\cdot 3\cdot 5\cdot 41$	3	1613	$2^2\cdot 13\cdot 31$	3	2029*	$2^2\cdot 3\cdot 13^2$	2
1237	$2^2\cdot 3\cdot 103$	2	1619*	$2\cdot 809$	2	2039	$2\cdot 1019$	7
1249	$2^5\cdot 3\cdot 13$	7	1621*	$2^2\cdot 3^4\cdot 5$	2	2053	$2^2\cdot 3^3\cdot 19$	2
1259*	$2\cdot 17\cdot 37$	2	1627	$2\cdot 3\cdot 271$	3	2063*	$2\cdot 1031$	5
1277	$2^2\cdot 11\cdot 29$	2	1637	$2^2\cdot 409$	2	2069*	$2^2\cdot 11\cdot 47$	2
1279	$2\cdot 3^2\cdot 71$	3	1657	$2^3\cdot 3^2\cdot 23$	11	2081	$2^5\cdot 5\cdot 13$	3

p	$p-1$	g	p	$p-1$	g	p	$p-1$	g
2083	$2\cdot 3\cdot 347$	2	2477	$2^2\cdot 619$	2	2903*	$2\cdot 1451$	5
2087	$2\cdot 7\cdot 149$	5	2503	$2\cdot 3^2\cdot 139$	3	2909*	$2^2\cdot 727$	2
2089	$2^3\cdot 3^2\cdot 29$	7	2521	$2^3\cdot 3^2\cdot 5\cdot 7$	17	2917	$2^2\cdot 3^6$	5
2099*	$2\cdot 1049$	2	2531	$2\cdot 5\cdot 11\cdot 23$	2	2927*	$2\cdot 7\cdot 11\cdot 19$	5
2111	$2\cdot 5\cdot 211$	7	2539*	$2\cdot 3^3\cdot 47$	2	2939*	$2\cdot 13\cdot 113$	2
2113*	$2^6\cdot 3\cdot 11$	5	2543*	$2\cdot 31\cdot 41$	5	2953	$2^3\cdot 3^3\cdot 41$	13
2129	$2^4\cdot 7\cdot 19$	3	2549*	$4\cdot 7^2\cdot 13$	2	2957	$2^2\cdot 739$	2
2131	$2\cdot 3\cdot 5\cdot 71$	2	2551	$2\cdot 3\cdot 5^3\cdot 17$	6	2963	$2\cdot 1481$	2
2137*	$2^3\cdot 3\cdot 89$	10	2557	$2^2\cdot 3^2\cdot 71$	2	2969	$2^3\cdot 7\cdot 53$	3
2141*	$2^2\cdot 5\cdot 107$	2	2579*	$2\cdot 1289$	2	2971*	$2\cdot 3^3\cdot 5\cdot 11$	10
2143*	$2\cdot 3^2\cdot 7\cdot 17$	3	2591	$2\cdot 5\cdot 7\cdot 37$	7	2999	$2\cdot 1499$	17
2153*	$2^3\cdot 269$	3	2593*	$2^5\cdot 3^4$	7	3001	$2^3\cdot 3\cdot 5^3$	14
2161	$2^4\cdot 3^3\cdot 5$	23	2609	$2^4\cdot 163$	3	3011*	$2\cdot 5\cdot 7\cdot 43$	2
2179*	$2\cdot 3^2\cdot 11^2$	7	2617*	$2^3\cdot 3\cdot 109$	5	3019*	$2\cdot 3\cdot 503$	2
2203	$2\cdot 3\cdot 367$	5	2621*	$2^2\cdot 5\cdot 131$	2	3023*	$2\cdot 1511$	5
2207*	$2\cdot 1103$	5	2633*	$2^3\cdot 7\cdot 47$	3	3037	$2^2\cdot 3\cdot 11\cdot 23$	2
2213	$2^2\cdot 7\cdot 79$	2	2647	$2\cdot 3^3\cdot 7^2$	3	3041	$2^5\cdot 5\cdot 19$	3
2221*	$2^2\cdot 3\cdot 5\cdot 37$	2	2657*	$2^5\cdot 83$	3	3049	$2^3\cdot 3\cdot 127$	11
2237	$2^2\cdot 13\cdot 43$	2	2659	$2\cdot 3\cdot 443$	2	3061	$2^2\cdot 3^2\cdot 5\cdot 17$	6
2239	$2\cdot 3\cdot 373$	3	2663*	$2\cdot 11^3$	5	3067	$2\cdot 3\cdot 7\cdot 73$	2
2243	$2\cdot 19\cdot 59$	2	2671	$2\cdot 3\cdot 5\cdot 89$	7	3079	$2\cdot 3^4\cdot 19$	6
2251*	$2\cdot 3^2\cdot 5^3$	7	2677	$2^2\cdot 3\cdot 223$	2	3083	$2\cdot 23\cdot 67$	2
2267	$2\cdot 11\cdot 103$	2	2683	$2\cdot 3^2\cdot 149$	2	3089	$2^4\cdot 193$	3
2269*	$2^2\cdot 3^4\cdot 7$	2	2687*	$2\cdot 17\cdot 79$	5	3109	$2^2\cdot 3\cdot 7\cdot 37$	6
2273*	$2^5\cdot 71$	3	2689	$2^7\cdot 3\cdot 7$	19	3119	$2\cdot 1559$	7
2281	$2^3\cdot 3\cdot 5\cdot 19$	7	2693	$2^2\cdot 673$	2	3121	$2^4\cdot 3\cdot 5\cdot 13$	7
2287	$2\cdot 3^2\cdot 127$	19	2699*	$2\cdot 19\cdot 71$	2	3137*	$2^6\cdot 7^2$	3
2293	$2^2\cdot 3\cdot 191$	2	2707	$2\cdot 3\cdot 11\cdot 41$	2	3163	$2\cdot 3\cdot 17\cdot 31$	3
2297*	$2^3\cdot 7\cdot 41$	5	2711	$2\cdot 5\cdot 271$	7	3167*	$2\cdot 1583$	5
2309*	$2^2\cdot 577$	2	2713*	$2^3\cdot 3\cdot 113$	5	3169	$2^2\cdot 3^2\cdot 11$	7
2311	$2\cdot 3\cdot 5\cdot 7\cdot 11$	3	2719	$2\cdot 3^2\cdot 151$	3	3181	$2^2\cdot 3\cdot 5\cdot 53$	7
2333	$2^2\cdot 11\cdot 53$	2	2729*	$2^3\cdot 11\cdot 31$	3	3187	$2\cdot 3^3\cdot 59$	2
2339*	$2\cdot 7\cdot 167$	2	2731	$2\cdot 3\cdot 5\cdot 7\cdot 13$	3	3191	$2\cdot 5\cdot 11\cdot 29$	11
2341*	$2^2\cdot 3^2\cdot 5\cdot 13$	7	2741*	$2^2\cdot 5\cdot 137$	2	3203	$2\cdot 1601$	2
2347	$2\cdot 3\cdot 17\cdot 23$	3	2749	$2^2\cdot 3\cdot 229$	6	3209	$2^3\cdot 401$	3
2351	$2\cdot 5^2\cdot 47$	13	2753*	$2^6\cdot 43$	3	3217	$2^4\cdot 3\cdot 67$	5
2357	$2^2\cdot 19\cdot 31$	2	2767*	$2\cdot 3\cdot 461$	3	3221*	$2^2\cdot 5\cdot 7\cdot 23$	10
2371*	$2\cdot 3\cdot 5\cdot 79$	2	2777*	$2^3\cdot 347$	3	3229	$2^2\cdot 3\cdot 269$	6
2377	$2^3\cdot 3^3\cdot 11$	5	2789*	$2^2\cdot 17\cdot 41$	2	3251*	$2\cdot 5^3\cdot 13$	6
2381	$2^2\cdot 5\cdot 7\cdot 17$	3	2791	$2\cdot 3^2\cdot 5\cdot 31$	6	3253	$2^2\cdot 3\cdot 271$	2
2383*	$2\cdot 3\cdot 397$	5	2797	$2^2\cdot 3\cdot 233$	2	3257*	$2^3\cdot 11\cdot 37$	3
2389*	$2^2\cdot 3\cdot 199$	2	2801	$2^4\cdot 5^2\cdot 7$	3	3259*	$2\cdot 3\cdot 181$	3
2393	$2^3\cdot 13\cdot 23$	3	2803	$2\cdot 3\cdot 467$	2	3271	$2\cdot 3\cdot 5\cdot 109$	3
2399	$2\cdot 11\cdot 109$	11	2819*	$2\cdot 1409$	2	3299*	$2\cdot 17\cdot 97$	2
2411*	$2\cdot 5\cdot 241$	6	2833*	$2^4\cdot 3\cdot 59$	5	3301*	$2^2\cdot 3\cdot 5^2\cdot 11$	6
2417*	$2^4\cdot 151$	3	2837	$2^2\cdot 709$	2	3307	$2\cdot 3\cdot 19\cdot 29$	2
2423*	$2\cdot 7\cdot 173$	5	2843	$2\cdot 7^2\cdot 29$	2	3313*	$2^4\cdot 3^2\cdot 23$	10
2437*	$2^2\cdot 3\cdot 7\cdot 29$	2	2851*	$2\cdot 3\cdot 5^2\cdot 19$	2	3319	$2\cdot 3\cdot 7\cdot 79$	6
2441	$2^3\cdot 5\cdot 61$	6	2857	$2^3\cdot 3\cdot 7\cdot 17$	11	3323	$2\cdot 11\cdot 151$	2
2447*	$2\cdot 1223$	5	2861*	$2^2\cdot 5\cdot 11\cdot 13$	2	3329	$2^8\cdot 13$	3
2459*	$2\cdot 1229$	2	2879	$2\cdot 1439$	7	3331*	$2\cdot 3^2\cdot 5\cdot 37$	3
2467	$2\cdot 3^2\cdot 137$	2	2887	$2\cdot 3\cdot 13\cdot 37$	5	3343*	$2\cdot 3\cdot 557$	5
2473*	$2^3\cdot 3\cdot 103$	5	2897*	$2^4\cdot 181$	3	3347	$2\cdot 7\cdot 239$	2

p	$p-1$	g	p	$p-1$	g	p	$p-1$	g
3359	$2\cdot 23\cdot 73$	11	3779*	$2\cdot 1889$	2	4229*	$2^2\cdot 7\cdot 151$	2
3361	$2^5\cdot 3\cdot 5\cdot 7$	22	3793	$2^4\cdot 3\cdot 79$	5	4231	$2\cdot 3^2\cdot 5\cdot 47$	3
3371*	$2\cdot 5\cdot 337$	2	3797	$2^2\cdot 13\cdot 73$	2	4241	$2^4\cdot 5\cdot 53$	3
3373	$2^2\cdot 3\cdot 281$	5	3803	$2\cdot 1901$	2	4243	$2\cdot 3\cdot 7\cdot 101$	2
3389*	$2^2\cdot 7\cdot 11^2$	3	3821*	$2^2\cdot 5\cdot 191$	3	4253	$2^2\cdot 1063$	2
3391	$2\cdot 3\cdot 5\cdot 113$	3	3823	$2\cdot 3\cdot 7^2\cdot 13$	3	4259*	$2\cdot 2129$	2
3407*	$2\cdot 13\cdot 131$	5	3833*	$2^3\cdot 479$	3	4261*	$2^2\cdot 3\cdot 5\cdot 71$	2
3413	$2^2\cdot 853$	2	3847*	$2\cdot 3\cdot 641$	5	4271	$2\cdot 5\cdot 7\cdot 61$	7
3433*	$2^3\cdot 3\cdot 11\cdot 13$	5	3851*	$2\cdot 5^2\cdot 7\cdot 11$	2	4273	$2^4\cdot 3\cdot 89$	5
3449	$2^3\cdot 431$	3	3853	$2^2\cdot 3^2\cdot 107$	2	4283	$2\cdot 2141$	2
3457	$2^7\cdot 3^3$	7	3863*	$2\cdot 1931$	5	4289	$2^6\cdot 67$	3
3461*	$2^2\cdot 5\cdot 173$	2	3877	$2^2\cdot 3\cdot 17\cdot 19$	2	4297	$2^3\cdot 3\cdot 179$	5
3463*	$2\cdot 3\cdot 577$	3	3881	$2^3\cdot 5\cdot 97$	13	4327*	$2\cdot 3\cdot 7\cdot 103$	3
3467	$2\cdot 1733$	2	3889	$2^4\cdot 3^5$	11	4337*	$2^4\cdot 271$	3
3469*	$2^2\cdot 3\cdot 17^2$	2	3907	$2\cdot 3^2\cdot 7\cdot 31$	2	4339*	$2\cdot 3^2\cdot 241$	10
3491	$2\cdot 5\cdot 349$	2	3911	$2\cdot 5\cdot 17\cdot 23$	13	4349*	$2^2\cdot 1087$	2
3499	$2\cdot 3\cdot 11\cdot 53$	2	3917	$2^2\cdot 11\cdot 89$	2	4357	$2^2\cdot 3^2\cdot 11^2$	2
3511	$2\cdot 3^3\cdot 5\cdot 13$	7	3919	$2\cdot 3\cdot 653$	3	4363	$2\cdot 3\cdot 727$	2
3517	$2^2\cdot 3\cdot 293$	2	3923	$2\cdot 37\cdot 53$	2	4373	$2^2\cdot 1093$	2
3527*	$2\cdot 41\cdot 43$	5	3929	$2^3\cdot 491$	3	4391	$2\cdot 5\cdot 439$	14
3529	$2^3\cdot 3^2\cdot 7^2$	17	3931	$2\cdot 3\cdot 5\cdot 131$	2	4397	$2^2\cdot 7\cdot 157$	2
3533	$2^2\cdot 883$	2	3943*	$2\cdot 3^3\cdot 73$	3	4409	$2^3\cdot 19\cdot 29$	3
3539*	$2\cdot 29\cdot 61$	2	3947	$2\cdot 1973$	2	4421*	$2^2\cdot 5\cdot 13\cdot 17$	3
3541	$2^2\cdot 3\cdot 5\cdot 59$	7	3967*	$2\cdot 3\cdot 661$	6	4423*	$2\cdot 3\cdot 11\cdot 67$	3
3547	$2\cdot 3^2\cdot 197$	2	3989*	$2^2\cdot 997$	2	4441	$2^3\cdot 3\cdot 5\cdot 37$	21
3557	$2^2\cdot 7\cdot 127$	2	4001	$2^5\cdot 5^3$	3	4447*	$2\cdot 3^2\cdot 13\cdot 19$	3
3559	$2\cdot 3\cdot 593$	3	4003	$2\cdot 3\cdot 23\cdot 29$	2	4451*	$2\cdot 5^2\cdot 89$	2
3571*	$2\cdot 3\cdot 5\cdot 7\cdot 17$	2	4007*	$2\cdot 2003$	5	4457*	$2^3\cdot 557$	3
3581*	$2^2\cdot 5\cdot 179$	2	4013	$2^2\cdot 17\cdot 59$	2	4463*	$2\cdot 23\cdot 97$	5
3583	$2\cdot 3^2\cdot 199$	3	4019*	$2\cdot 7^2\cdot 41$	2	4481	$2^7\cdot 5\cdot 7$	3
3593*	$2^3\cdot 449$	3	4021	$2^2\cdot 3\cdot 5\cdot 67$	2	4483	$2\cdot 3^3\cdot 83$	2
3607*	$2\cdot 3\cdot 601$	5	4027	$2\cdot 3\cdot 11\cdot 61$	3	4493	$2^2\cdot 1123$	2
3613	$2^2\cdot 3\cdot 7\cdot 43$	2	4049	$2^4\cdot 11\cdot 23$	3	4507	$2\cdot 3\cdot 751$	2
3617*	$2^5\cdot 113$	3	4051*	$2\cdot 3^4\cdot 5^2$	10	4513	$2^5\cdot 3\cdot 47$	7
3623*	$2\cdot 1811$	5	4057*	$2^3\cdot 3\cdot 13^2$	5	4517	$2^2\cdot 1129$	2
3631	$2\cdot 3\cdot 5\cdot 11^2$	21	4073*	$2^3\cdot 509$	3	4519	$2\cdot 3^2\cdot 251$	3
3637	$2^2\cdot 3^2\cdot 101$	2	4079	$2\cdot 2039$	11	4523	$2\cdot 7\cdot 17\cdot 19$	5
3643	$2\cdot 3\cdot 607$	2	4091*	$2\cdot 5\cdot 409$	2	4547	$2\cdot 2273$	2
3659*	$2\cdot 31\cdot 59$	2	4093	$2^2\cdot 3\cdot 11\cdot 31$	2	4549	$2^2\cdot 3\cdot 379$	6
3671	$2\cdot 5\cdot 367$	13	4099	$2\cdot 3\cdot 683$	2	4561	$2^4\cdot 3\cdot 5\cdot 19$	11
3673*	$2^3\cdot 3^3\cdot 17$	5	4111	$2\cdot 3\cdot 5\cdot 137$	17	4567*	$2\cdot 3\cdot 761$	3
3677	$2^2\cdot 919$	2	4127	$2\cdot 2063$	5	4583*	$2\cdot 29\cdot 79$	5
3691	$2\cdot 3^2\cdot 5\cdot 41$	2	4129	$2^5\cdot 3\cdot 43$	13	4591	$2\cdot 3^3\cdot 5\cdot 17$	11
3697	$2^4\cdot 3\cdot 7\cdot 11$	5	4133	$2^2\cdot 1033$	2	4597	$2^2\cdot 3\cdot 383$	5
3701*	$2^2\cdot 5^2\cdot 37$	2	4139*	$2\cdot 2069$	2	4603	$2\cdot 3\cdot 13\cdot 59$	2
3709*	$2^2\cdot 3^2\cdot 103$	2	4153*	$2^3\cdot 3\cdot 173$	5	4621	$2^2\cdot 3\cdot 5\cdot 7\cdot 11$	2
3719	$2\cdot 11\cdot 13^2$	7	4157	$2^2\cdot 1039$	2	4637	$2^2\cdot 19\cdot 61$	2
3727*	$2\cdot 3^4\cdot 23$	3	4159	$2\cdot 3^3\cdot 7\cdot 11$	3	4639	$2\cdot 3\cdot 773$	3
3733	$2^2\cdot 3\cdot 311$	2	4177*	$2^4\cdot 3^2\cdot 29$	5	4643	$2\cdot 11\cdot 211$	5
3739	$2\cdot 3\cdot 7\cdot 89$	7	4201	$2^3\cdot 3\cdot 5^2\cdot 7$	11	4649	$2^3\cdot 7\cdot 83$	3
3761	$2^4\cdot 5\cdot 47$	3	4211*	$2\cdot 5\cdot 421$	6	4651*	$2\cdot 3\cdot 5^2\cdot 31$	3
3767	$2\cdot 7\cdot 269$	5	4217*	$2^3\cdot 17\cdot 31$	3	4657	$2^4\cdot 3\cdot 97$	15
3769	$2^3\cdot 3\cdot 157$	7	4219*	$2\cdot 3\cdot 19\cdot 37$	2	4663	$2\cdot 3^2\cdot 7\cdot 37$	3

p	$p-1$	g	p	$p-1$	g	p	$p-1$	g
4673*	$2^6 \cdot 73$	3	4793*	$2^3 \cdot 599$	3	4931*	$2 \cdot 5 \cdot 17 \cdot 29$	6
4679	$2 \cdot 2339$	11	4799	$2 \cdot 2399$	7	4933	$2^2 \cdot 3^2 \cdot 137$	2
4691*	$2 \cdot 5 \cdot 7 \cdot 67$	2	4801	$2^6 \cdot 3 \cdot 5^2$	7	4937*	$2^3 \cdot 617$	3
4703*	$2 \cdot 2351$	5	4813	$2^2 \cdot 3 \cdot 401$	2	4943*	$2 \cdot 7 \cdot 358$	7
4721	$2^4 \cdot 5 \cdot 59$	6	4817*	$2^4 \cdot 7 \cdot 43$	3	4951	$2 \cdot 3^2 \cdot 5^2 \cdot 11$	6
4723	$2 \cdot 3 \cdot 787$	2	4831	$2 \cdot 3 \cdot 5 \cdot 7 \cdot 23$	3	4957	$2^2 \cdot 3 \cdot 7 \cdot 59$	2
4729	$2^3 \cdot 3 \cdot 197$	17	4861	$2^2 \cdot 3^5 \cdot 5$	11	4967*	$2 \cdot 13 \cdot 191$	5
4733	$2^2 \cdot 7 \cdot 13^2$	5	4871	$2 \cdot 5 \cdot 487$	11	4969	$2^3 \cdot 3^3 \cdot 23$	11
4751	$2 \cdot 5^3 \cdot 19$	19	4877	$2^2 \cdot 23 \cdot 53$	2	4973	$2^2 \cdot 11 \cdot 113$	2
4759	$2 \cdot 3 \cdot 13 \cdot 61$	3	4889	$2^3 \cdot 13 \cdot 47$	3	4987	$2 \cdot 3^2 \cdot 277$	2
4783*	$2 \cdot 3 \cdot 797$	6	4903	$2 \cdot 3 \cdot 19 \cdot 43$	3	4993	$2^7 \cdot 3 \cdot 13$	5
4787	$2 \cdot 2393$	2	4909	$2^2 \cdot 3 \cdot 409$	6	4999	$2 \cdot 3 \cdot 7^2 \cdot 17$	3
4789	$2^2 \cdot 3^2 \cdot 7 \cdot 19$	2	4919	$2 \cdot 2459$	13			

Chapter 4. Properties of Polynomials

4.1 The Division of Polynomials

We consider polynomials $f(x)$ with rational coefficients and we denote by $\partial^0 f$ the degree of the polynomial.

Definition 1.1. Let $f(x)$ and $g(x)$ be two polynomials with $g(x)$ not identically zero. If there is a polynomial $h(x)$ such that $f(x) = g(x)h(x)$, then we say that $g(x)$ *divides* $f(x)$, and we write $g(x)|f(x)$ or $g|f$. If $g(x)$ does not divide $f(x)$, then we write $g\nmid f$.

Clearly we have the following: (i) $f|f$; (ii) if $f|g$ and $g|f$, then f and g differ only by a constant divisor, and we call them *associated polynomials*; (iii) if $f|g$ and $g|h$, then $f|h$; (iv) if $f|g$, then $\partial^0 f \leqslant \partial^0 g$.

If $f|g$ and $g\nmid f$, then we call f a proper divisor of g and it is easy to see that, in this case, $\partial^0 f < \partial^0 g$.

Theorem 1.1. *Let $f(x)$ and $g(x)$ be any two polynomials with $g(x)$ not identically zero. Then there are two polynomials $q(x)$ and $r(x)$ such that $f = q \cdot g + r$, where either $r = 0$ or $\partial^0 r < \partial^0 g$.*

Proof. We prove this by induction on the degree of f. If $\partial^0 f < \partial^0 g$, then we can take $q = 0$, $r = f$. If $\partial^0 f \geqslant \partial^0 g$, we let

$$f = \alpha_n x^n + \cdots, \qquad \partial^0 f = n,$$

$$g = \beta_m x^m + \cdots, \qquad \partial^0 g = m,$$

so that

$$\partial^0(f - \alpha_n \beta_m^{-1} x^{n-m} g) < \partial^0 f.$$

From the induction hypothesis, there are two polynomials $h(x)$ and $r(x)$ such that

$$f - \alpha_n \beta_m^{-1} x^{n-m} g = hg + r$$

where either $r = 0$ or $\partial^0 r < \partial^0 g$. We now put

$$q(x) = h(x) + \alpha_n \beta_m^{-1} x^{n-m}$$

so that $f = qg + r$ as required. □

Definition 1.2. By an *ideal* we mean a set I of polynomials satisfying the following conditions:

(i) If $f, g \in I$, then $f + g \in I$;
(ii) If $f \in I$ and h is any polynomial, then $fh \in I$.

Example. The multiples of a fixed polynomial $f(x)$ forms an ideal.

Theorem 1.2. *Given any ideal I, there exists a polynomial $f \in I$ such that any polynomial in I is a multiple of f; that is I is the ideal of the set of multiples of f.*

Proof. Let f be a polynomial in I with the least degree. If g is a polynomial in I which is not a multiple of f, then, according to Theorem 1.1, there are polynomials $q(x)$ and $r(x)$ ($\neq 0$) such that

$$g = qf + r, \qquad \partial^0 r < \partial^0 f.$$

Since $f \in I$, it follows from (ii) that $qf \in I$, and hence from (i) that $g - qf \in I$, that is $r \in I$. But this contradicts the minimal degree property of f. The theorem is proved. □

Definition 1.3. Let f and g be two polynomials. Consider the set of polynomials of the form $mf + ng$ where m, n are polynomials. From Theorem 1.2 we see that this set is identical with the set of polynomial which are multiple of a polynomial d. We call this polynomial d the *greatest common divisor* of f and g, and we write $(f, g) = d$. For the sake of uniqueness we shall take the leading coefficient of (f, g) to be 1, that is a *monic* polynomial.

Theorem 1.3. *The greatest common divisor (f, g) has the following properties:*

(i) *There are two polynomials m, n such that $(f, g) = mf + ng$;*
(ii) *For every pair of polynomials m, n we have $(f, g) | mf + ng$;*
(iii) *If $l | f$ and $l | g$, then $l | (f, g)$.* □

Definition 1.4. If $(f, g) = 1$, then we say that f and g are *coprime.*

Theorem 1.4. *Let p be an irreducible polynomial. If $p | fg$, then either $p | f$ or $p | g$.*

Proof. If $p \nmid f$, then $(f, p) = 1$. Thus, from Theorem 1.3 there are polynomials m, n such that $mf + np = 1$ so that $mfg + ngp = g$. Since $p | fg$, it follows that $p | g$. □

4.2 The Unique Factorization Theorem

Theorem 2.1. *Any polynomial can be factorized into a product of irreducible polynomials. If associated polynomials are treated as identical, then, apart from the ordering of the factors, this factorization is unique.* □

The theorem can be proved by mathematical induction on the degree of the polynomial.

Theorem 2.2. *Let $f(x)$ and $g(x)$ be two polynomials with rational coefficients, and that $f(x)$ be irreducible. Suppose that $f(x) = 0$ and $g(x) = 0$ have a common root. Then $f(x)|g(x)$.*

Proof. Since f and g have a common zero, it follows that $(f, g) \neq 1$. Let $d(x)$ be the greatest common factor of $f(x)$ and $g(x)$. Then $d(x)$ and $f(x)$ are associated polynomials, because $f(x)$ is irreducible. Therefore $f(x)|g(x)$. □

From this theorem we deduce the following: If $f(x)$ is an irreducible polynomial of degree n, then the zeros

$$\vartheta^{(1)}, \vartheta^{(2)}, \ldots, \vartheta^{(n)}$$

are distinct. Moreover, if $\vartheta^{(i)}$ is a zero of another polynomial $g(x)$ with rational coefficients, then the other $n - 1$ numbers are also the zeros of $g(x)$.

Theorem 2.3. *Let f and g be monic polynomials:*

$$f = p_1^{a_1} \cdots p_s^{a_s}, \qquad a_v \geqslant 0,$$

$$g = p_1^{b_1} \cdots p_s^{b_s}, \qquad b_v \geqslant 0,$$

where p_v are distinct irreducible monic polynomials. Then

$$(f, g) = p_1^{c_1} \cdots p_s^{c_s}$$

where $c_v = \min(a_v, b_v)$. □

Definition 2.1. Let f and g be two polynomials. Polynomials which are divisible by both f and g are called *common multiples* of f and g. Those common multiples which have the least degree are called the *least common multiples*, and we denote by $[f, g]$ the monic least common multiple.

Theorem 2.4. *Under the same hypothesis as Theorem 2.3 we have*

$$[f, g] = p_1^{d_1} \cdots p_s^{d_s}$$

where $d_v = \max(a_v, b_v)$. □

From this we deduce:

Theorem 2.5. *A least common multiple divides every common multiple.* □

Theorem 2.6. *Let f, g be monic polynomials. Then $fg = [f, g](f, g)$.* □

4.3 Congruences

Let $m(x)$ be a polynomial. If $m(x)|f(x) - g(x)$, then we say that $f(x)$ is congruent to $g(x)$ modulo $m(x)$ and we write

$$f(x) \equiv g(x) \pmod{m(x)}.$$

With respect to any modulus $m(x)$ we have: (i) $f \equiv f \pmod{m}$; (ii) if $f \equiv g \pmod{m}$, then $g \equiv f \pmod{m}$; (iii) if $f \equiv g$, $g \equiv h \pmod{m}$, then $f \equiv h \pmod{m}$; (iv) if $f \equiv g$, $f_1 \equiv g_1 \pmod{m}$, then $f \pm f_1 \equiv g \pm g_1$, $ff_1 \equiv gg_1 \pmod{m}$.

Being congruent is an equivalence relation which partitions the set polynomials into equivalence classes. From (iv) we see that addition and multiplication can be defined on these classes. We denote by 0 the class whose members are divisible by $m(x)$. If $m(x)$ is irreducible we can even define division on the set of equivalence classes (except by 0, of course). Specifically, if $f(x)$ is not a multiple of $m(x)$, then there are polynomials $a(x)$, $b(x)$ such that $a(x)f(x) + b(x)m(x) = 1$ which means that there is a polynomial $a(x)$ such that $a(x)f(x) \equiv 1 \pmod{m(x)}$.

We state this as a theorem.

Theorem 3.1. *Let $m(x)$ be irreducible. Then any non-zero equivalence class has a reciprocal. That is, if A is a non-zero equivalence class, then there exists a class B such that for any polynomials $f(x)$ and $g(x)$ in A and B respectively we have $f(x)g(x) \equiv 1$ $\pmod{m(x)}$.* □

We now give an example to illustrate the ideas in this section. Let $m(x) = x^2 + 1$, an irreducible polynomial. Each equivalence class contains a unique polynomial $ax + b$ which we may take as the representative. The addition and subtraction of classes is given by $ax + b \pm (a_1x + b_1) = (a \pm a_1)x + (b \pm b_1)$. Multiplication is given by $(ax + b)(a_1x + b_1) = aa_1x^2 + (ab_1 + a_1b)x + bb_1 \equiv (ab_1 + a_1b)x + bb_1 - aa_1 \pmod{x^2 + 1}$. Using the ordered pair (a, b) to denote the class containing $ax + b$ we then have

$$(a, b) \pm (a_1, b_1) = (a \pm a_1, b \pm b_1),$$
$$(a, b)(a_1, b_1) = (ab_1 + ba_1, bb_1 - aa_1).$$

From

$$(ax + b)(-ax + b) \equiv a^2 + b^2 \pmod{x^2 + 1},$$

we see that the inverse of (a, b) is $\left(-\frac{a}{a^2 + b^2}, \frac{b}{a^2 + b^2}\right)$. In other words we have the arithmetic of the complex number $ai + b$.

Extending the idea here, if $m(x)$ is a monic polynomial of degree n, then each equivalence class possesses a unique polynomial with degree less than n, say

$$\alpha_1 x^{n-1} + \alpha_2 x^{n-2} + \cdots + \alpha_{n-1} x + \alpha_n$$

and the arithmetic of the congruence modulo $m(x)$ becomes the arithmetic of these polynomials. The sum of two such polynomials is obtained by adding the corresponding coefficients, and the product is the ordinary product polynomial reduced modulo $m(x)$.

Exercise 1. Let α_1, α_2, α_3 be distinct. Determine a quadratic polynomial $f(x)$ satisfying $f(\alpha_1) = \beta_1$, $f(\alpha_2) = \beta_2$, $f(\alpha_3) = \beta_3$.

Answer: The Lagrange interpolation formula

$$f(x) = \beta_1 \frac{(x - \alpha_2)(x - \alpha_3)}{(\alpha_1 - \alpha_2)(\alpha_1 - \alpha_3)} + \beta_2 \frac{(x - \alpha_3)(x - \alpha_1)}{(\alpha_2 - \alpha_3)(\alpha_2 - \alpha_1)} + \beta_3 \frac{(x - \alpha_1)(x - \alpha_2)}{(\alpha_8 - \alpha_1)(\alpha_3 - \alpha_2)}.$$

Exercise 2. Let $m_1(x)$ and $m_2(x)$ be two non-associated irreducible polynomials. Let $f_1(x)$ and $f_2(x)$ be two given polynomials. Prove that there exists a polynomial $f(x)$ such that $f(x) \equiv f_i(x) \pmod{m_i(x)}$, $i = 1, 2$.

4.4 Integer Coefficients Polynomials

It is clear that the set of integer coefficients polynomials is closed with respect to addition, subtraction and multiplication. A set of integer coefficients polynomials is called an ideal if (i) $f + g$ belongs to the set whenever f and g belong to the set, (ii) fg belongs to the set whenever f belongs to the set, and g is any integer coefficients polynomial.

Theorem 4.1. (Hilbert) *Every ideal A possesses a finite number of polynomials $f_1, \ldots, f_n$ with the following property: Every polynomial $f \in A$ is representable as $f = g_1 f_1 + \cdots + g_n f_n$ where $g_1, \ldots, g_n$ are integer coefficients polynomials.*

Proof. 1) Denote by B the set of leading coefficients of members of A. We claim that B forms an integral modulus. To see this, we observe that if a, $b \in B$, where $f(x) = ax^n + \cdots$, $g(x) = bx^m + \cdots$, then by (ii) we know that $f(x)x^m$, $g(x)x^n \in A$ so that

$$f(x)x^m \pm g(x)x^n = (a \pm b)x^{m+n} + \cdots$$

are in A. Therefore $a \pm b \in B$ which proves our claim. From Theorem 1.4.3 members of B are multiples of an integer d. Let the corresponding polynomial with leading coefficient d be

$$f_1 = dx^l + d_1 x^{l-1} + \cdots + d_{l-1}x + d_l.$$

2) Let $f \in A$. Then there are two polynomials $q(x)$ and $r(x)$ such that $f(x) = q(x)f_1(x) + r(x)$ where $\partial^0 r < \partial^0 f_1$ or $r = 0$. This is certainly so if the degree of f is less than that of f_1. If $f(x) = ax^n + \cdots + a_n$ $(n \geqslant l)$, then by 1) we see that $d|a$, and

$$f(x) - \frac{a}{d} x^{n-l} f_1(x)$$

is a polynomial with degree at most $n - 1$. If the degree here is greater than or equal to l, then its leading coefficient is again divisible by d. Continuing the argument we see that our claim is valid.

3) If every member of A has degree at least l, then the theorem is proved. Otherwise we let d' be the greatest common divisor of the leading coefficients of members of A whose degree are less than l, and we let

$$f_2 = d'x^{l'} + d'_1 x^{l'-1} + \cdots \quad (d|d')$$

be the corresponding polynomial in A. From the above, we see that members of A whose degree lies between l' and l can be written as $f(x) = q(x)f_2(x) + r(x)$ where $\partial^0 r < \partial^2 f_2$ or $r = 0$. Continuing this argument the theorem is proved. □

4.5 Polynomial Congruences with a Prime Modulus

In this section all the polynomials have integer coefficients and p is a fixed prime number.

Definition 5.1. If the corresponding coefficients of two polynomials $f(x)$ and $g(x)$ differ by multiples of p, then we say that $f(x)$ and $g(x)$ are *congruent modulo p*, and we write $f(x) \equiv g(x) \pmod{p}$. By the degree $\partial^0 f$ of $f(x)$ modulo p we mean the highest degree of $f(x)$ whose coefficient is not a multiple of p.

For example $7x^2 + 16x + 9 \equiv 2x + 2 \pmod 7$, and $\partial^0(7x^2 + 16x + 9) = 1 \pmod 7$. But with respect to the modulus 3, $\partial^2(7x^2 + 16x + 9) = 2$. Clearly we have (i) $f(x) \equiv f(x) \pmod{p}$; (ii) if $f \equiv g \pmod{p}$, then $g \equiv f \pmod{p}$; (iii) if $f \equiv g$, $g \equiv h \pmod{p}$, then $f \equiv h \pmod{p}$; (iv) if $f \equiv g$, $f_1 \equiv g_1 \pmod{p}$, then $f \pm f_1 \equiv g \pm g_1$ and $ff_1 \equiv gg_1 \pmod{p}$. We note particularly that

$$(f(x))^p \equiv f(x^p) \pmod{p}.$$

Definition 5.2. Let $f(x)$ and $g(x)$ be polynomials with $g(x)$ not identically zero mod p. If there is a polynomial $h(x)$ such that $f(x) \equiv h(x)g(x) \pmod{p}$, then we say that $g(x)$ *divides* $f(x)$ modulo p. We call $g(x)$ a *divisor* of $f(x)$ modulo p, and we write $g(x)|f(x) \pmod{p}$.

Example. From $x^5 + 3x^4 - 4x^3 + 2 \equiv (2x^2 - 3)(3x^3 - x^2 + 1) \pmod 5$ we see that $2x^2 - 3|x^5 + 3x^4 - 4x^2 + 2 \pmod 5$.

We have the following: (i) $f(x)|f(x) \pmod{p}$; (ii) if $f(x)|g(x)$ and $g(x)|f(x) \pmod{p}$, then $f(x)$ and $g(x)$ differs only by a constant factor; that is, there exists an integer a such that $f(x) \equiv ag(x) \pmod{p}$. In this case we say that $f(x)$ and $g(x)$ are *associated modulo p*. It is easy to see that every polynomial has $p - 1$ associates modulo p. Moreover, there is a unique monic associated polynomial. (iii) If $f|g$, $g|h$ (mod p), then $f|h$ (mod p). (iv) Let $f(x)$ and $g(x)$ be two polynomials with $g(x)$ not identically zero modulo p. Then there are two polynomials $q(x)$ and $r(x)$ such that $f(x) \equiv q(x)g(x) + r(x) \pmod{p}$, where either $\partial^0 r < \partial^0 g$, or $r(x) \equiv 0 \pmod{p}$.

Definition 5.3. If a polynomial $f(x)$ cannot be factorized into a product of two polynomials with smaller degrees mod p, then we say that $f(x)$ is an *irreducible polynomial* mod p, or that $f(x)$ is *prime* mod p.

Example. We take $p = 3$. There are three non-associated linear polynomials, namely x, $x + 1$, $x + 2$, which are irreducible. There are nine non-associated quadratic polynomials, namely x^2, $x^2 + x$, $x^2 + 2x$, $x^2 + 1$, $x^2 + x + 1$, $x^2 + 2x + 1$, $x^2 + 2$, $x^2 + x + 2$, $x^2 + 2x + 2$. Of these there are 6 $(= (x + a)(x + b))$ which are reducible, and the three irreducible ones are $x^2 + 1$, $x^2 + x + 2$, $x^2 + 2x + 2$.

We note that if a polynomial is irreducible mod p, then it is irreducible and from this we deduce that $x^2 + 2x + 2$ has no rational zeros. The determination of the number of irreducible polynomials mod p of degree n is an interesting problem which we shall solve in §9.

Theorem 5.1. *Any polynomial can be written as a product of irreducible polynomials* mod p, *and this product representation is unique apart from associates and ordering of the factors.* □

We can define, similarly to §1, the greatest common divisor and the least common multiple. If we denote by (f, g) the monic greatest common divisor, then we have

Theorem 5.2. *Given polynomials $f(x)$ and $g(x)$, there are polynomials $m(x)$ and $n(x)$ such that* $m(x)f(x) + n(x)g(x) \equiv (f(x), g(x)) \pmod{p}$. □

4.6 On Several Theorems Concerning Factorizations

Definition 6.1. Let $f(x) = a_n x^n + a_{n-1} x^{n-1} + \cdots$ be a polynomial. The polynomial $n a_n x^{n-1} + (n-1) a_{n-1} x^{n-2} + \cdots$ is called the *derivative* of $f(x)$ and is denoted by $f'(x)$.

Clearly we have $(f(x) + g(x))' = f'(x) + g'(x)$, and it is not difficult to prove that $(f(x)g(x))' = f'(x)g(x) + g'(x)f(x)$.

Definition 6.2. If a polynomial $f(x)$ is divisible by the square of a non-constant polynomial mod p, then we say that $f(x)$ has *repeated factors* mod p.

For example, $x^5 + x^4 - x^3 - x^2 + x + 1$ has the repeated factors $(x^2 + 1)^2$ modulo 3.

Theorem 6.1. *A necessary and sufficient condition for $f(x)$ to have repeated factors is that the degree of $(f(x), f'(x))$ is at least* 1. □

Theorem 6.2. *If $p \nmid n$, then $x^n - 1$ has no repeated factors* mod p. □

Theorem 6.3. *Let $(m, n) = d$. Then $(x^m - 1, x^n - 1) = x^d - 1$.* □

Theorem 6.4. *Let $(m, n) = d$. Then*

$$(x^{p^m-1} - 1, x^{p^n-1} - 1) = x^{p^d-1} - 1. \quad \square$$

4.7 Double Moduli Congruences

Definition 7.1. Let p be a prime number and $\varphi(x)$ be a polynomial. If $f_1(x) - f_2(x)$ is a multiple of $\varphi(x) \bmod p$, then we say that f_1 and f_2 are congruent to the double moduli p, $\varphi(x)$ and we write

$$f_1(x) \equiv f_2(x) \pmod{\text{d}\, p, \varphi(x)}.$$

For example, $x^5 + 3x^4 + x^2 + 4x + 3 \equiv 0$ (modd 5, $2x^2 - 3$).

Double moduli congruences have the following properties:

1) $f(x) \equiv f(x)$ (modd $p, \varphi(x)$);

2) If $f \equiv g$ (modd p, φ), then $g \equiv f$ (modd p, φ);

3) If $f \equiv g$ and $g \equiv h$ (modd p, φ), then $f \equiv h$ (modd p, φ);

4) If $f \equiv g$ and $f_1 \equiv g_1$ (modd p, φ), then $f \pm f_1 \equiv g \pm g_1$ and $ff_1 \equiv gg_1$ (modd p, φ);

5) Suppose that the degree of $\varphi(x)$ (mod p) is n. Then every polynomial is congruent to one of the following polynomials

$$a_1 + a_2 x + \cdots + a_n x^{n-1}, \qquad 0 \leqslant a_i \leqslant p - 1. \tag{1}$$

It is clear that there are p^n polynomials in (1), no two of them are congruent (modd p, $\varphi(x)$), and any polynomial must be congruent to one of them (modd p, $\varphi(x)$).

Definition 7.2. We call the p^n polynomials in (1) a *complete residue system* (modd p, $\varphi(x)$). By discarding those polynomials which are not coprime with $\varphi(x)$ we have a *reduced residue system* (modd p, $\varphi(x)$).

Theorem 7.1. *Let $(g(x), \varphi(x)) = 1$. Then, as $f(x)$ runs through a complete (or reduced) residue system* (modd p, $\varphi(x)$), *so does $f(x)g(x)$.*

Proof. If $g(x)f_1(x) \equiv g(x)f_2(x)$ (modd p, $\varphi(x)$), then from $(g(x), \varphi(x)) = 1$ we deduce that $f_1(x) \equiv f_2(x)$ (modd p, $\varphi(x)$). The required result follows easily from this. □

4.8 Generalization of Fermat's Theorem

Let p be a prime number, and $\varphi(x)$ be an irreducible polynomial of degree $n \bmod p$.

Theorem 8.1. *Let $f(x)$ be a polynomial not divisible by $\varphi(x) \bmod p$. Then*

$$(f(x))^{p^n-1} \equiv 1 \pmod{p, \varphi(x)}. \tag{1}$$

Given any polynomial $f(x)$, we have

$$(f(x))^{p^n} \equiv f(x) \pmod{p, \varphi(x)}, \tag{2}$$

and in particular, we have

$$x^{p^n} \equiv x \pmod{p, \varphi(x)}.$$

Proof. Let $f_1(x), \ldots, f_{p^n-1}(x) \pmod{p, \varphi(x)}$ be a reduced residue system modd p, $\varphi(x)$. Then $ff_1, \ldots, ff_{p^n-1}$ is also a reduced residue system. Therefore

$$\prod_{i=1}^{p^n-1} f_i(x) \equiv \prod_{i=1}^{p^n-1} (f(x)f_i(x)) \pmod{p, \varphi(x)},$$

or

$$((f(x))^{p^n-1} - 1) \prod_{i=1}^{p^n-1} f_i(x) \equiv 0 \pmod{p, \varphi(x)},$$

and hence

$$(f(x))^{p^n-1} \equiv 1 \pmod{p, \varphi(x)}. \quad \square$$

This theorem is a generalization of Fermat's theorem in Chapter 1. We note that (2) is a special case of (1), but we observe that (1) can also be deduced from (2), since

$$(f(x))^{p^n} \equiv f(x^{p^n}) \equiv f(x) \pmod{p, \varphi(x)}.$$

Exercise. Generalize Euler's theorem in Chapter 2.

Theorem 8.2. *Any irreducible polynomial of degree n must divide $x^{p^n-1} - 1$* $(\bmod\, p)$. $\square$

Theorem 8.3. *The number of roots of $f(X) \equiv 0 \pmod{p, \varphi(x)}$ does not exceed the degree of $f(X)$.*

Proof. Let $g(x)$ be a root of the congruence, and let

$$f(X) = a_n X^n + a_{n-1} X^{n-1} + \cdots,$$

so that

$$\begin{aligned} f(X) - f(g(x)) &= a_n(X^n - (g(x))^n) + a_{n-1}(X^{n-1} - (g(x))^{n-1}) + \cdots \\ &= (X - g(x))h(X). \end{aligned}$$

If $g_1(x)$ is another root distinct from $g(x)$, then $h(g_1(x)) \equiv 0 \pmod{p, \varphi(x)}$, and the required result follows. □

Theorem 8.4. *x^{p^n-1} is not divisible by any irreducible polynomial of degree greater than n,* $\bmod p$.

Proof. Let $\psi(x)$ be an irreducible polynomial with degree $m > n$, $\bmod p$, and suppose, if possible, that $x^{p^n} \equiv x \pmod{p, \psi(x)}$. There are p^m incongruent polynomials $f(x) \bmod p, \psi(x)$. From $(f(x))^p \equiv f(x^p) \pmod{p}$ we deduce that $(f(x))^{p^n} \equiv f(x^{p^n}) \equiv f(x) \pmod{p, \psi(x)}$. This means that the number of roots of $X^{p^n} \equiv X \pmod{p, \psi(x)}$, being p^m, exceeds p^n. This is impossible by Theorem 8.3 so that the theorem is proved. □

Theorem 8.5. *Let $\psi(x)$ be an irreducible polynomial of degree l,* $\bmod p$. *If $\psi(x) | x^{p^n} - x$* $\pmod{p}$, *then $l | n$.*

Proof. From Theorem 8.2 and the hypothesis, we have

$$\psi(x) | (x^{p^n-1} - 1, x^{p^l-1} - 1) \pmod{p},$$

and from Theorem 6.3,

$$\psi(x) | x^{p^d-1} - 1 \pmod{p}, \qquad d = (n, l).$$

Moreover, from Theorem 8.4 we see that $l \leqslant d = (n, l)$ so that $l = d$, and hence $l | n$. □

Exercise. Let $\psi(x)$ and $\varphi(x)$ be irreducible polynomials $\bmod p$. Then a necessary and sufficient condition for the solubility of $\psi(X) \equiv 0 \pmod{p, \varphi(x)}$ is that $\partial^0 \psi | \partial^0 \varphi$. Prove further that if it is soluble then it can be factorized into a product of linear factors.

4.9 Irreducible Polynomials mod p

Theorem 9.1. *The product of all the irreducible polynomials of degree n* $\pmod{p}$, *is equal to*

$$\frac{x^{p^n} - x}{\prod_{q_1} (x^{p^{n/q_1}} - x)} \frac{\prod_{q_1, q_2} (x^{p^{n/q_1 q_2}} - x)}{\prod_{q_1, q_2, q_3} (x^{p^{n/q_1 q_2 q_3}} - x)} \cdots \pmod{p},$$

where $q_1, q_2, \ldots$ run over the distinct prime divisors of n.

Proof. By Theorem 6.1 the polynomial $x^{p^n} - x$ has not repeated factors, so that it can be factorized into a product of various distinct irreducible polynomials of the form

$$\psi(x) = x^d + a_1 x^{d-1} + \cdots,$$

where $\psi(x) | x^{p^d} - x$, $d | n$.

We now apply the inclusion-exclusion principle of §1.7. We already know that $x^{p^n} - x$ is a product of various irreducible polynomials of degree m where $m|n$. We exclude all those polynomials whose degrees divide n/q_1; but those polynomials whose degrees possess n/q_1q_2 as divisors have been excluded twice, so that we have to re-include them, and so on. □

Theorem 9.2. *The total number of irreducible polynomials of degree n* (mod p), *is equal to*

$$\frac{1}{n}\left(p^n - \sum_{q_1} p^{n/q_1} + \sum_{q_1, q_2} p^{n/q_1q_2} - \sum p^{n/q_1q_2q_3} + \cdots\right).$$

Here the sums are over the distinct prime divisors q_i of n.

Proof. The degree of the polynomial in Theorem 9.1 is

$$N = p^n - \sum_{q_1} p^{n/q_1} + \cdots, \tag{1}$$

and each of its factor has degree n, so that the result follows. □

Let $n = q_1^{l_1} \cdots q_s^{l_s}$, where q_i are the distinct prime divisors of n. Now

$$N \equiv (-1)^s p^{n/q_1 \cdots q_s} \pmod{p^{n/q_1 \cdots q_s + 1}}.$$

Therefore $N > 0$, so that we have:

Theorem 9.3. *There always exists an irreducible polynomial of degree n* (mod p). □

4.10 Primitive Roots

The content of this section is very similar to §3.8, and we shall therefore omit the details.

Let $(f(x), \varphi(x)) = 1$. Suppose that there exists a polynomial $g(x)$ such that $(g(x))^m \equiv f(x)$ (modd $p, \varphi(x)$). Then we call $f(x)$ an m-th *residue* modd p, $\varphi(x)$.

A polynomial $f(x)$ is, or is not, a quadratic residue according to whether

$$(f(x))^{\frac{1}{2}(p^n - 1)} \equiv 1 \quad (\text{modd } p, \varphi(x)),$$

or

$$(f(x))^{\frac{1}{2}(p^n - 1)} \equiv -1 \quad (\text{modd } p, \varphi(x)).$$

Definition. The least positive integer l satisfying $(f(x))^l \equiv 1 \pmod{p, \varphi(x)}$ is called the *order* of $f(x)$.

As before, it can be proved that l divides $p^n - 1$, and that there are precisely $\varphi(l)$ polynomials having order l. There are therefore $\varphi(p^n - 1)$ polynomials with order $p^n - 1$, and these polynomials are called the primitive roots (modd $p, \varphi(x)$). If $f(x)$ is a primitive root, then $(f(x))^v$, $v = 1, 2, \ldots, p^n - 1$ represent all the non-zero incongruent polynomials, modd $p, \varphi(x)$.

It is not difficult to prove that the product $\prod_v (X - f_v(x))$, where f_v runs over all the primitive roots, is equal to

$$\frac{X^{p^n-1} - 1}{\prod_q (X^{(p^n-1)/q} - 1)} \frac{\prod_{q,q_1} (X^{(p^n-1)/qq_1} - 1)}{\prod_{q,q_1,q_2} (X^{(p^n-1)/qq_1q_2} - 1)} \cdots, \tag{1}$$

where q_i runs over all the distinct prime divisors of $p^n - 1$.

Exercise. Prove that the product of all the non-zero incongruent polynomials is congruent to $-1 \pmod{p, \varphi(x)}$.

4.11 Summary

We may summarize the discussions of this chapter in the language of modern algebra or abstract algebra.

We have a set of objects which we denote by R. The number of objects in R may be finite or infinite.

1. If we can define the operations of addition and subtraction in R and that these operations are closed in R, then we call R an *integral modulus*. For example: The set of even integers forms an integral modulus; the set of polynomials with even integer coefficients forms an integral modulus.

An integral modulus is also known as an *Abelian group*.

2. If we can define the operations of addition, subtraction and multiplication which are closed in R, then we call R a *ring*. For example: The set of integers forms a ring; the set of integer coefficients polynomials forms a ring.

3. By an *ideal* E, we mean a subset of a ring R which satisfies the following conditions:

i) If $a, b \in E$, then $a - b \in E$;

ii) If $a \in E$ and $r \in R$, then $ar \in E$.

For example: The subset of even integers forms an ideal in the ring of integers. In the ring of integer coefficient polynomials, we may form the ideal of polynomials having the form $f(x)(x^2 + 1) + 2g(x)x$, where f and g run over all integer coefficient polynomials.

4. If in R we can define the operations of addition, subtraction, multiplication and division (except by 0), and that these operations are closed in R, then we call R a *field*.

For example: The set of rational numbers forms a field. The residue classes modulo a fixed irreducible polynomial forms a field, which is known as an algebraic extension field in modern algebra. Next, take a prime number p and an irreducible polynomial $\varphi(x)$ of degree n. The residue classes with respect to the double modulus p and $\varphi(x)$ forms a field with p^n elements.

Students who master the various concrete examples discussed in this chapter will find it easier to learn the abstract concepts of modern algebra.

Chapter 5. The Distribution of Prime Numbers

In this chapter we give some basic results concerning the distribution of prime numbers. The reader will only require some knowledge of the calculus—this chapter is a first introduction to analytic number theory and we shall omit all the deeper investigations.

5.1 Order of Infinity

In the discussion of the distribution of prime numbers we must understand the notion of the comparison of the order of growth between two functions. We often use the symbols

$$\ll, \quad O, \quad o, \quad \sim,$$

the meanings of which we shall now give.

Let n be a positive integer which tends to infinity (or x a continuous variable which tends to infinity). Let $\varphi(n)$ (or $\varphi(x)$) be a positive function of n (or x), and let $f(n)$ (or $f(x)$) be any function. If there exists A which does not depend on n (or x) such that

$$|f| \leqslant A\varphi,$$

then we write

$$f \ll \varphi.$$

If $f - g \ll \varphi$, then it is more convenient to write

$$f = g + O(\varphi).$$

Also $f = o(\varphi)$ and $f \sim \varphi$ mean

$$\lim_{n\to\infty} \frac{f(n)}{\varphi(n)} = 0 \quad \text{and 1 respectively}$$

$$\left(\text{or} \quad \lim_{x\to\infty} \frac{f(x)}{\varphi(x)} = 0 \quad \text{and 1 respectively}\right).$$

We have the following examples:

$$\sin x \ll 1, \qquad x + \frac{1}{x} \ll x \ll x + \frac{1}{x},$$

$$x + \frac{1}{x} = o(x^2), \qquad x + \sin x \sim x,$$

$$x + \sin x = x + O(1).$$

Naturally "x tending to infinity" may be replaced by "x tending to l" where l is a finite number. For example, as $x \to 0$, we have

$$x^2 = O(x), \qquad \sin x \sim x, \qquad 1 + x \sim 1.$$

However, unless otherwise stated, we shall assume that the variable is tending to infinity.

It is easy to verify the following properties: (i) $\varphi \ll \varphi$; (ii) if $f \ll \varphi$ and $\varphi \ll \psi$, then $f \ll \psi$; (iii) if $f \ll \varphi$ and $g \ll \psi$, then $f + g \ll \varphi + \psi$ and $fg \ll \varphi\psi$. The properties (ii) and (iii) still hold if we replace $\ll$ by $o(\ )$. We also have (iv) $\varphi \sim \varphi$; (v) if $\psi \sim \varphi$, then $\varphi \sim \psi$; (vi) if $\varphi \sim \psi$, and $\psi \sim \chi$, then $\varphi \sim \chi$; (vii) if $\psi \sim \varphi$ and $\psi_1 \sim \varphi_1$, then $\psi\psi_1 \sim \varphi\varphi_1$.

5.2 The Logarithm Function

The logarithm function $\log x$ frequently enters in the discussion of the distribution of prime numbers. We assume the reader already knows the definition of $\log x$ and we shall recall the following simple properties. Since

$$e^x = 1 + x + \cdots + \frac{x^n}{n!} + \frac{x^{n+1}}{(n+1)!} + \cdots,$$

it follows that for positive x and for all n

$$x^{-n}e^x > \frac{x}{(n+1)!}.$$

Since, for any fixed n, the right-hand side tends to infinity as $x \to \infty$, it follows that e^x grows faster than any fixed power of x. We can therefore write $x^n = o(e^x)$. If α is positive, then $x^\alpha = O(x^{[\alpha]+1}) = o(e^x)$.

Since $\log x$ is the inverse function of e^x, on substituting $\log y$ for x in the above, we see that $(\log y)^\alpha = o(y)$, or

$$\log x = o(x^\delta), \qquad \delta > 0.$$

In other words $\log x$ grows slower than any fixed positive power of x. It is easy to see that $\log\log x$ is even smaller than $\log x$.

Theorem 2.1.

$$\sum_{n=1}^{x} \frac{1}{n} \sim \log x.$$

Proof. The result follows at once from

$$\log x = \int_1^x \frac{dt}{t} \leqslant \sum_{n=1}^{x} \frac{1}{n} \leqslant 1 + \int_1^x \frac{dt}{t} = 1 + \log x. \quad \square$$

Theorem 2.2. *Let*

$$\operatorname{li} x = \lim_{\eta \to 0} \left(\int_0^{1-\eta} + \int_{1+\eta}^{x} \right) \frac{dt}{\log t}.$$

Then

$$\operatorname{li} x \sim \frac{x}{\log x}.$$

Proof. We have

$$\lim_{x\to\infty} \frac{\operatorname{li} x}{\dfrac{x}{\log x}} = \lim_{x\to\infty} \frac{(\operatorname{li} x)'}{\left(\dfrac{x}{\log x}\right)'}$$

$$= \lim_{x\to\infty} \frac{\dfrac{1}{\log x}}{\dfrac{1}{\log x} - \dfrac{1}{\log^2 x}}$$

$$= 1. \quad \square$$

5.3 Introduction

The distribution of prime numbers is the most interesting branch of number theory. The various conjectures and theorems are mostly the result of empirical observations. We now consider several problems and the ancient conjectures associated with them.

(i) Let $\pi(x)$ denote the number of primes not exceeding x. Then we have the following table which suggests:

1) There are infinitely many primes; that is $\pi(x) \to \infty$.

2) However, there are relatively few primes comparing with all the integers. That is, almost all numbers are not primes in the sense that

x	$\pi(x)$	$\frac{x}{\log x}$	$\operatorname{li} x$	$\frac{\pi(x)}{\operatorname{li} x}$	$\frac{\pi(x)}{x}$
1 000	168	145	178	0.94...	0.1680
10 000	1 229	1 086	1 246	0.98...	0.1229
50 000	5 133	4 621	5 167	0.993...	0.1026
100 000	9 592	8 686	9 630	0.996...	0.0959
500 000	41 538	38 103	41 606	0.9983...	0.0830
1 000 000	78 498	72 382	78 628	0.9983...	0.0785
2 000 000	148 933	137 848	149 055	0.9991...	0.0745
5 000 000	348 513	324 149	348 638	0.9996...	0.0697
10 000 000	664 579	620 417	664 918	0.9994...	0.0665
20 000 000	1 270 607	1 189 676	1 270 905	0.9997...	0.0635
90 000 000	5 216 954	4 913 897	5 217 810	0.99983...	0.0580
100 000 000	5 761 455	5 428 613	5 762 209	0.99986...	0.0576
1 000 000 000	50 847 478	48 254 630	50 849 235	0.99996...	0.0508

$$\frac{\pi(x)}{x} \to 0.$$

3) The number of primes not exceeding x is asymptotically $\operatorname{li} x$; that is

$$\pi(x) \sim \operatorname{li} x \sim \frac{x}{\log x}.$$

We note that 3) implies 1) and 2).
4) The best approximation to $\pi(x)$ is $\operatorname{li} x$.
5) $\pi(x) < \operatorname{li} x$.
In this chapter our deepest result is Chebyshev's theorem which states that

$$\frac{x}{\log x} \ll \pi(x) \ll \frac{x}{\log x}.$$

This result implies 1) and 2). The statement 3) is the famous *prime number theorem* which we shall prove in Chapter 9. The problem raised in 4) belongs to a difficult branch of analytic number theory and its discussion is outside the scope of this book. Finally, despite the convincing evidence from the table, 5) is actually false; this was proved by Littlewood.

(ii) We know that

$$5, 13, 17, 29, \ldots, 10006721,$$

are all primes congruent 1 mod 4. A natural question is whether there are infinitely many such primes. Associated with this problem Dirichlet's theorem gives the following general answer: Let a, b be coprime integers. Then there are infinitely many primes of the form $an + b$. In this chapter we shall only discuss particular examples of this theorem, the proof of which is given in Chapter 9.

(iii) We have

$$6 = 3 + 3, \qquad 8 = 3 + 5, \qquad 10 = 5 + 5, \qquad 12 = 5 + 7,$$

$$14 = 7 + 7, \qquad 16 = 3 + 13, \qquad 18 = 5 + 13, \qquad 20 = 7 + 13,$$

$$22 = 3 + 19, \qquad 24 = 5 + 19, \qquad \ldots .$$

This suggests the following: Every even integer greater than 4 must be the sum of two odd prime numbers. This is the famous *Goldbach's problem.* If this problem is settled to be true, then we can deduce that every odd integer greater than 7 must be the sum of three odd primes. This is because if n is an odd integer greater than 7, then $n - 3$ is an even integer greater than 4 so that $n - 3 = p_1 + p_2$ or $n = 3 + p_1 + p_2$.

The unsolved Goldbach's problem is extremely difficult. I. M. Vinogradov proved that every sufficiently large odd integer is the sum of three primes. The author proved that "almost all" even numbers are the sum of two primes. V. Brun proved that every sufficiently large even integer is the sum of two numbers each having at most 9 prime factors. (See Notes.)

(iv) We also note that

$$3, 5;\ 5, 7;\ 11, 13;\ 17, 19;\ 29, 31; \ldots ;$$

$$10016957,\ 10016959; \ldots ; 10^9 + 7,\ 10^9 + 9; \ldots$$

are all pairs of primes having difference 2; we call such pairs *prime twins.* More specifically we know that there are 1224 pairs less than 100,000 and 8164 pairs less than 1,000,000. At present (1957) the largest pair known to us is

$$1000\,000\,009\,649,\ 1000\,000\,009\,651.$$

From the evidence here it is natural to conjecture that there are infinitely many pairs of prime twins. This too is a famous unsolved problem. (See Notes.) In the theory of numbers we shall always have more unsolved problems than solved ones. For example we also have

$$5, 7, 11;\ 11, 13, 17;\ 17, 19, 23; \ldots ; 101, 103, 107;$$

$$\ldots ;\ 10\,014\,491,\ 10\,014\,493,\ 10\,041\,497; \ldots$$

all primes. It is conjectured that there are infinitely many primes p such that $p + 2$, $p + 6$ are also primes. Advancing even further:

(v) We can verify that $n^2 - n + 17$ is always prime when $0 \leqslant n \leqslant 16$, and that $n^2 - n + 41$ is always prime when $0 \leqslant n \leqslant 40$.

We now suggest the following interesting problem: Let N be any given number. Can we always find a prime p such that

$$n^2 - n + p$$

is always prime when $0 \leqslant n \leqslant N$? This too is an unsolved problem and, in the author's view, it is even more difficult than (iii) and (iv). If this problem is solved affirmatively, then (iv) can also be settled. Let us see why. In order for the polynomial $n^2 - n + p$ to (successively) take prime values, n must be restricted to the integers from 0 to $p-1$. We now construct a sequence of polynomials $n^2 - n + p_i$ with the following property: When $0 \leqslant n \leqslant p_{i-1}$, the number $n^2 - n + p_i$ is always prime. We note that if (v) is solved, then this construction is certainly possible. Now taking $n = 1$ and 2 will give p_i, $p_i + 2$ both primes, and taking $n = 1, 2, 3$ will give $p_i, p_i + 2, p_i + 6$ all primes. This shows that (iv) follows as a consequence of (v).

(vi) Another difficult unsolved problem is whether there are infinitely many primes of the form $n^2 + 1$. We know that

$$2, 5, 17, 37, \ldots, 65537, \ldots$$

are all primes of this form and it is conjectured that there are infinitely many such primes. (See Notes.)

(vii) Let p_n denote the n-th prime. We may ask about the distribution of the values $p_n - p_{n-1}$. From (iv) we see that $p_n - p_{n-1}$ may be as small as 2, but what about its maximum value – that is, an order estimate for $p_n - p_{n-1}$ as $n \to \infty$.

(viii) The so-called Bertrand's postulate states that there always exists a prime in any interval from n to $2n$. This is comparatively easy and we shall prove it in §7. A more delicate conjecture is that "there always exists a prime in any interval from n^2 to $(n+1)^2$." This is a difficult unsolved problem.

5.4 The Number of Primes is Infinite

Theorem 4.1. *The number of primes is infinite; that is $\pi(x) \to \infty$ as $x \to \infty$.*

Proof. Let $2, 3, \ldots, p$ be all the primes not exceeding p and let

$$q = 2 \cdot 3 \cdot \cdots \cdot p + 1.$$

Then q is not a multiple of $2, 3, \ldots, p$ and hence either q is prime or q is divisible by a prime between p and q. Therefore there always exists a prime greater than p, and so it follows that the number of primes is not finite. □

This method can be generalized to give the following:

Theorem 4.2. *Let $f(x)$ be any polynomial with integer coefficients. Then the numbers*

$$f(1), f(2), f(3), \ldots$$

contain infinitely many distinct prime divisors.

Proof. Let

$$f(x) = a_0 x^n + a_1 x^{n-1} + \cdots + a_n, \qquad n \geqslant 1.$$

If $a_n = 0$, then our sequence of numbers contains all the primes as divisors. We assume therefore that $a_n \neq 0$.

Suppose that the sequence of numbers has only finitely many prime divisors $p_1, p_2, \ldots, p_v$. We consider $f(p_1 \cdots p_v a_n y)$, a polynomial in y with all the coefficients a multiple of a_n. Let

$$f(p_1 \cdots p_v a_n y) = a_n g(y)$$

where

$$g(y) = 1 + A_1 y + A_2 y^2 + \cdots + A_n y^n$$

is a polynomial with integer coefficients such that $p_1, \ldots, p_v$ divide $A_1, A_2, \ldots, A_n$. If there exists an integer y_0 such that $g(y_0) \neq \pm 1$, then $g(y_0)$ must contain a prime divisor distinct from $p_1, \ldots, p_v$, and so the theorem follows at once. But $g(y) = \pm 1$ has at most $2n$ solutions so that the theorem is proved. □

A different method of proof of Theorem 4.1 was given by Euler. This method, which we give below, opens the door for analytic number theory.

Theorem 4.3. *The series $\sum_p 1/p$ is divergent; here the summation is over all primes p. Therefore the number of primes is infinite.*

We first prove:

Theorem 4.4 (Euler's identity). *Let $f(n)$ be defined for all positive integers n, and $f(n)$ not identically zero. Suppose that*

$$f(nn') = f(n)f(n') \qquad \textit{whenever} \qquad (n, n') = 1.$$

Then we have the following identity

$$\sum_{n=1}^{\infty} f(n) = \prod_p (1 + f(p) + f(p^2) + \cdots);$$

the condition for the validity of this identity is either

$$\text{(i)} \qquad \sum_{n=1}^{\infty} |f(n)| \qquad \textit{converges}$$

or

$$\text{(ii)} \qquad \prod_p (1 + |f(p)| + |f(p^2)| + \cdots) \qquad \textit{converges.}$$

Moreover, if $f(nn') = f(n)f(n')$ for all n, n', then, subject to the same convergence conditions, we have that

$$\sum_{n=1}^{\infty} f(n) = \prod_p \frac{1}{1 - f(p)}.$$

Proof. We have, for all n, $f(1)f(n) = f(n)$, and there exists n such that $f(n) \neq 0$. Therefore $f(1) = 1$.

1) Suppose that the series

$$\sum_{n=1}^{\infty} |f(n)| \tag{1}$$

converges to the sum $\bar{S}$. Now consider

$$P(x) = \prod_{p \leqslant x} (1 + f(p) + f(p^2) + \cdots).$$

For any p, the series $\sum_{n=1}^{\infty} |f(p^n)|$ is part of the series (1) so that it must also converge. This means that $P(x)$ is a finite product of absolutely convergent series. Therefore

$$P(x) = \sum{}' f(n)$$

where the summation is over all integers n having prime factors $\leqslant x$. Let

$$S = \sum_{n=1}^{\infty} f(n)$$

so that

$$|S - P(x)| \leqslant \sum_{n > x} |f(n)|.$$

When $x \to \infty$, $|S - P(x)| \to 0$ so that $P(x) \to S$.

Using this result on $|f(n)|$ we see that

$$\prod_{p} (1 + |f(p)| + |f(p^2)| + \cdots)$$

converges to $\bar{S}$.

2) Suppose that

$$\prod_{p} (1 + |f(p)| + |f(p^2)| + \cdots)$$

converges to $\bar{P}$. Then

$$\begin{aligned}\bar{P}(x) &= \prod_{p \leqslant x} (1 + |f(p)| + |f(p^2)| + \cdots) \\ &= \sum{}' |f(n)| \geqslant \sum_{n \leqslant x} |f(n)|.\end{aligned}$$

Therefore

$$\sum_{n=1}^{\infty} |f(n)|$$

converges. From our result in 1) we see that the first part of the theorem is proved. The last part follows from

$$\begin{aligned}1 + f(p) + f(p^2) + \cdots &= 1 + f(p) + (f(p))^2 + \cdots \\ &= \frac{1}{1 - f(p)}. \quad \square\end{aligned}$$

Proof of Theorem 4.3. We put $f(n) = 1/n$ in the above theorem. If $\sum 1/p$ converges, then we deduce that

$$\prod\left(1 - \frac{1}{p}\right) \quad \text{and} \quad \prod\left(1 - \frac{1}{p}\right)^{-1}$$

converge, and we infer from the theorem that

$$\sum_{n=1}^{\infty} \frac{1}{n}$$

also converges, which is impossible. Therefore Theorem 4.3 is proved. □

From $0 < 1 - \frac{1}{p} < 1$ we deduce:

Theorem 4.5. $\prod(1 - \frac{1}{p})$ *diverges to zero.* □

Exercise 1. Prove that there are infinitely many primes of the form $6n - 1$.

Exercise 2. Prove that there are infinitely many primes of the form $4n - 1$.

Exercise 3. Prove that

$$\prod_p \frac{p^2}{p^2 - 1} = \frac{\pi^2}{6}.$$

(Note that $\sum_{n=1}^{\infty} \frac{1}{n^2} = \frac{\pi^2}{6}$.)

5.5 Almost all Integers are Composite

Theorem 5.1.

$$\lim_{n \to \infty} \frac{\pi(n)}{n} = 0.$$

That is, the ratio of the number of primes in $1, 2, \ldots, n$ *and* n *tends to zero as* n *tends to infinity, or almost all integers are composite.*

Proof. We prove the slightly more general result

$$\lim_{x \to \infty} \frac{\pi(x)}{x} = 0,$$

where x tends to infinity through all real numbers.

We first observe the following useful and simple fact. The number of integers not exceeding x that are divisible by a is $[x/a]$. Here $[\xi]$ denotes the integer part of ξ.

Denote by $\bar{\omega}(x,r)$ the number of positive integers not exceeding x and not divisible by the first r primes $2, 3, 5, \ldots, p_r$. Then, from Theorem 1.7.1 we have

$$\bar{\omega}(x,r) = [x] - \sum_{1 \leqslant i \leqslant r} \left[\frac{x}{p_i}\right] + \sum_{1 \leqslant i < j \leqslant r} \left[\frac{x}{p_i p_j}\right] - \cdots$$

(it is not difficult to give a direct proof of this).

Clearly then

$$\pi(x) \leqslant \bar{\omega}(x,r) + r,$$

so that

$$\begin{aligned} \pi(x) &< x - \sum \frac{x}{p_i} + \sum \frac{x}{p_i p_j} - \cdots + r + 2^r \\ &= x \prod_{i=1}^{r} \left(1 - \frac{1}{p_i}\right) + r + 2^r \\ &< x \prod_{i=1}^{r} \left(1 - \frac{1}{p_i}\right) + 2^{r+1}. \end{aligned}$$

From Theorem 4.5 we know that, as $r \to \infty$,

$$\prod_{i=1}^{r} \left(1 - \frac{1}{p_i}\right) \to 0.$$

Let $\varepsilon > 0$. We can take $r = r(\varepsilon)$ so that

$$\pi(x) < \tfrac{1}{2}\varepsilon x + 2^{r+1},$$

and therefore, for sufficiently large x,

$$\pi(x) < \varepsilon x.$$

The theorem is proved. □

5.6 Chebyshev's Theorem

The theorem in this section is an important result in elementary number theory and one should try to give the most elementary argument to prove it.

Theorem 6.1. *When $n \geqslant 2$ we have*

$$\frac{1}{8} \leqslant \pi(n) \frac{H(n)}{n} < 6,$$

where

$$H(n) = \sum_{v=2}^{n} \frac{1}{v}.$$

That is $\pi(n)$ is about the same as the reciprocal of the average of $(\frac{1}{2}, \frac{1}{3}, \frac{1}{4}, \ldots, \frac{1}{n})$.

We shall require the following two lemmas:

Lemma 1. *When $k \geqslant 0$ we have*

$$\pi(2^{k+1}) \leqslant 2^k.$$

Proof. When $x > 9$ we have, by considering even and odd numbers, that $\pi(x) \leqslant x/2$. Also

$$\pi(2) = 1 = 2^0, \qquad \pi(4) = 2 = 2^1, \qquad \pi(8) = 4 = 2^2. \quad \square$$

Lemma 2. *When $l > 0$ we have*

$$\tfrac{1}{2}l \leqslant H(2^l) \leqslant l.$$

Proof.

$$H(2^l) = \frac{1}{2} + \left(\frac{1}{3} + \frac{1}{4}\right) + \left(\frac{1}{5} + \frac{1}{6} + \frac{1}{7} + \frac{1}{8}\right) + \cdots$$

$$+ \left(\frac{1}{2^{l-1}+1} + \cdots + \frac{1}{2^l}\right) \geqslant \frac{1}{2} + \left(\frac{1}{4} + \frac{1}{4}\right) + \left(\frac{1}{8} + \frac{1}{8} + \frac{1}{8} + \frac{1}{8}\right)$$

$$+ \cdots + \left(\frac{1}{2^l} + \cdots + \frac{1}{2^l}\right) = \frac{1}{2}l.$$

$$H(2^l) = \left(\frac{1}{2} + \frac{1}{3}\right) + \left(\frac{1}{4} + \frac{1}{5} + \frac{1}{6} + \frac{1}{7}\right) + \cdots + \frac{1}{2^l} \leqslant \left(\frac{1}{2} + \frac{1}{2}\right)$$

$$+ \left(\frac{1}{4} + \frac{1}{4} + \frac{1}{4} + \frac{1}{4}\right) + \cdots + \left(\frac{1}{2^{l-1}} + \cdots + \frac{1}{2^{l-1}}\right) + \frac{1}{2^l} \leqslant l. \quad \square$$

Proof of Theorem 6.1. We first prove that

$$\prod_{n<p\leqslant 2n} p \,\Bigg|\, \binom{2n}{n} = \frac{(2n)!}{n!\,n!} \,\Bigg|\, \prod_{p^r \leqslant 2n < p^{r+1}} p^r. \tag{1}$$

(i) Any prime in the interval from n to $2n$ must divide $(2n)!$ but not $n!$, so that the left hand side of the formula holds.

(ii) The power of p in $\binom{2n}{n}$ is

$$\sum_{m=1}^{r} \left(\left[\frac{2n}{p^m}\right] - 2\left[\frac{n}{p^m}\right]\right) \leqslant r,$$

since each term in the sum is at most 1. This proves the right hand side of the formula (1).

From (1) we now have

$$n^{\pi(2n)-\pi(n)} < \prod_{n<p\leqslant 2n} p \leqslant \binom{2n}{n} \leqslant \prod_{p^r \leqslant 2n < p^{r+1}} p^r \leqslant (2n)^{\pi(2n)}, \qquad n \geqslant 1. \tag{2}$$

Since

$$\binom{2n}{n} = \frac{2n(2n-1)\cdots(n+1)}{n(n-1)\cdots 1}$$
$$= 2\left(2 + \frac{1}{n-1}\right)\cdots\left(2 + \frac{v}{n-v}\right)\cdots\left(2 + \frac{n-1}{1}\right) \geqslant 2^n$$

and

$$\binom{2n}{n} \leqslant (1+1)^{2n} = 2^{2n},$$

we deduce from (2) that

$$n^{\pi(2n)-\pi(n)} < 2^{2n}, \qquad 2^n \leqslant (2n)^{\pi(2n)}, \qquad n \geqslant 1. \tag{3}$$

Let $n = 2^k$, $k = 0, 1, 2, \ldots,$ so that we have

$$2^{k(\pi(2^{k+1})-\pi(2^k))} < 2^{2^{k+1}}, \qquad 2^{2^k} \leqslant 2^{(k+1)\pi(2^{k+1})}, \qquad k \geqslant 0,$$

or

$$k(\pi(2^{k+1}) - \pi(2^k)) < 2^{k+1}, \qquad 2^k \leqslant (k+1)\pi(2^{k+1}). \tag{4}$$

From Lemma 1 we have

$$(k+1)\pi(2^{k+1}) - k\pi(2^k) < 2^{k+1} + \pi(2^{k+1}) \leqslant 3 \cdot 2^k, \qquad k \geqslant 0.$$

Taking $k = 0, 1, \ldots, k$ and adding the corresponding results we have

$$(k+1)\pi(2^{k+1}) < 3(2^0 + 2^1 + \cdots + 2^k) < 3 \cdot 2^{k+1}, \qquad k \geqslant 0. \tag{5}$$

From (4) and (5) we have

$$\frac{1}{2}\frac{2^{k+1}}{k+1} \leqslant \pi(2^{k+1}) < 3\frac{2^{k+1}}{k+1}, \qquad k \geqslant 0. \tag{6}$$

Let n be an integer greater than 1 and choose k so that

$$2^{k+1} \leqslant n < 2^{k+2}, \qquad k \geqslant 0.$$

From Lemma 2 we have

$$\pi(n) \leqslant \pi(2^{k+2}) < 3\frac{2^{k+2}}{k+2} \leqslant 6\frac{2^{k+1}}{H(2^{k+2})} \leqslant 6\frac{n}{H(n)}, \tag{7}$$

and

$$\pi(n) \geqslant \pi(2^{k+1}) \geqslant \frac{1}{2}\frac{2^{k+1}}{k+1} = \frac{1}{8}\frac{2^{k+2}}{\frac{1}{2}(k+1)}$$
$$\geqslant \frac{1}{8}\frac{2^{k+2}}{H(2^{k+1})} \geqslant \frac{1}{8}\frac{n}{H(n)}. \tag{8}$$

This holds for all $n \geqslant 2$. Therefore

$$\frac{1}{8} \leqslant \pi(n)\frac{H(n)}{n} < 6. \quad \square$$

Theorem 6.2.

$$\frac{1}{8} \leqslant \frac{\pi(n)}{\dfrac{n}{\log n}} \leqslant 12, \qquad n \geqslant 2.$$

Proof. When $n \geqslant 2$, we have

$$\log\frac{n}{2} = \int_2^n \frac{dt}{t} < \frac{1}{2} + \frac{1}{3} + \cdots + \frac{1}{n} < \int_1^n \frac{dt}{t} = \log n.$$

When $n \geqslant 4$, we have

$$\log\frac{n}{2} \geqslant \frac{1}{2}\log n.$$

Also

$$\tfrac{1}{2}\log 3 \leqslant \tfrac{1}{2} + \tfrac{1}{3}, \qquad \tfrac{1}{2}\log 2 \leqslant \tfrac{1}{2}$$

so that the required result follows from the previous theorem. $\square$

We note, of course, that Theorem 4.1 and Theorem 5.1 are consequences of this theorem.

5.7 Bertrand's Postulate

Bertrand's postulate was first proved by Chebyshev.

Theorem 7.1. *Given any real $x \geqslant 1$, there exists a prime in the interval x to $2x$.*

Proof. 1) We begin by giving a good estimate for the binomial coefficient

$$\binom{2n}{n} = \frac{(2n)!}{n!n!},$$

namely, for $n \geqslant 5$, we have that

$$\frac{1}{2n}2^{2n} < \binom{2n}{n} < \frac{1}{4}2^{2n}. \tag{1}$$

The left hand side inequality follows from

$$(2n)\binom{2n}{n} = \frac{2}{1}\cdot\frac{3}{1}\cdot\frac{4}{2}\cdot\frac{5}{2}\cdot \cdots \cdot\frac{2n-1}{n-1}\cdot\frac{2n-1}{n-1}\cdot\frac{2n}{n}\cdot\frac{2n}{n} > 2^{2n},$$

and we shall use induction for the right hand side inequality in (1). When $n = 5$, we have

$$\binom{2n}{n} = 252 < 256 = \frac{1}{4} \cdot 2^{10}.$$

Since

$$\binom{2(n+1)}{n+1} = \frac{(2n)!(2n+1)(2n+2)}{(n!)^2(n+1)(n+1)} < 4\binom{2n}{n},$$

the inductive argument is complete.

2) Let $b \geqslant 10$. We denote by $\{\xi\}$ the least integer $\geqslant \xi$, and we set

$$a_1 = \left\{\frac{b}{2}\right\}, \quad a_2 = \left\{\frac{b}{2^2}\right\}, \quad \ldots, \quad a_k = \left\{\frac{b}{2^k}\right\}, \quad \ldots.$$

We then have $a_1 \geqslant a_2 \geqslant \cdots \geqslant a_k \geqslant \cdots$, and

$$a_k < \frac{b}{2^k} + 1 = 2\frac{b}{2^{k+1}} + 1 \leqslant 2a_{k+1} + 1.$$

Since both the outsides are integers we have

$$a_k \leqslant 2a_{k+1}. \tag{2}$$

Let m be the greatest integer such that $a_m \geqslant 5$, so that $a_{m+1} < 5$, and hence, by (2), $a_m < 10$. Since $2a_1 \geqslant b$, the m intervals

$$a_m < \eta < 2a_m, \qquad a_{m-1} < \eta \leqslant 2a_{m-1}, \quad \ldots, \quad a_1 < \eta \leqslant 2a_1$$

covers the whole interval $10 < \eta \leqslant b$. Therefore

$$\prod_{10<p\leqslant b} p \leqslant \prod_{a_1<p\leqslant 2a_1} p \prod_{a_2<p\leqslant 2a_2} p \cdots \prod_{a_m<p\leqslant 2a_m} p.$$

From

$$\prod_{n<p\leqslant 2n} p < \binom{2n}{n} < 2^{2(n-1)},$$

we have

$$\prod_{10<p\leqslant b} p \leqslant 2^{2(a_1-1+a_2-1+\cdots+a_m-1)}$$

$$< 2^{2\left(\frac{b}{2}+\frac{b}{2^2}+\cdots+\frac{b}{2^m}\right)} < 2^{2b}. \tag{3}$$

3) We already proved earlier that the power of p in $\binom{2n}{n}$ does not exceed r, where r is the greatest integer satisfying $p^r \leqslant 2n$. It follows that if $p > \sqrt{2n}$, then p^2 does not divide $\binom{2n}{n}$.

We further observe that, when $n \geqslant 3$, the primes p satisfying $\frac{2}{3}n < p \leqslant n$ cannot divide $\binom{2n}{n}$. This is because $3p > 2n$, so that only p and $2p$, and not other multiples of p, may occur among the divisors of $(2n)!$, whereas p^2 clearly is a divisor of $(n!)^2$. Therefore such a prime p cannot divide $\binom{2n}{n}$. (This is the most important point in this proof.)

Collecting our results we have

$$\binom{2n}{n} \leqslant \prod_{p \leqslant \sqrt{2n}} p^r \prod_{\sqrt{2n} < p \leqslant \frac{2}{3}n} p \prod_{n < p \leqslant 2n} p$$
$$\leqslant \prod_{p \leqslant \sqrt{2n}} (2n) \prod_{\sqrt{2n} < p \leqslant \frac{2}{3}n} p \prod_{n < p \leqslant 2n} p.$$

From (1) and (3) we see that, for $n \geqslant 50$ (so that $\sqrt{2n} \geqslant 10$),

$$2^{2n} < (2n)^{\sqrt{2n}+1} \prod_{\sqrt{2n} < p \leqslant \frac{2}{3}n} p \prod_{n < p \leqslant 2n} p$$
$$< (2n)^{\sqrt{2n}+1} 2^{\frac{4}{3}n} \prod_{n < p \leqslant 2n} p. \tag{4}$$

If there is no prime number between n and $2n$, then

$$2^{2n} < (2n)^{\sqrt{2n}+1} 2^{\frac{4}{3}n},$$

or

$$2^{\frac{2}{3}n} < (2n)^{\sqrt{2n}+1}. \tag{5}$$

But this is clearly impossible if n is sufficiently large. We now determine an explicit bound for the validity of this inequality. We use $n \leqslant 2^{n-1}$ (this can be proved by induction) to give

$$2n = (\sqrt[6]{2n})^6 < ([\sqrt[6]{2n}] + 1)^6 \leqslant 2^{6[\sqrt[6]{2n}]} \leqslant 2^{6\sqrt[6]{2n}}. \tag{6}$$

From (5) we have (using $n \geqslant 50$) that

$$2^{2n} < (2n)^{3(1+\sqrt{2n})} < 2^{\sqrt[6]{2n}(18+18\sqrt{2n})} < 2^{\sqrt[6]{2n} \times 20\sqrt{2n}} = 2^{20(2n)^{\frac{2}{3}}},$$

that is $(2n)^{\frac{1}{3}} < 20$ or $n < \frac{1}{2} \cdot 20^3 = 4000$. Thus (5) can hold only if $n < 4000$ and we have therefore proved that, if $n \geqslant 4000$, there is always a prime p satisfying $n < p \leqslant 2n$.

4) We complete the proof by observing that

$$2, 3, 5, 7, 13, 23, 43, 83, 163, 317, 631, 1259, 2503, 4001 \tag{7}$$

is a chain of prime numbers, each one being smaller than twice its predecessor. Now, given any n $(1 \leqslant n < 4000)$ we can select the smallest prime p in (7) which exceeds n, and we denote by p' its predecessor. Then we have

$$p' \leqslant n < p \leqslant 2p' \leqslant 2n.$$

The proof of the theorem is complete. □

Theorem 7.2. *There exist two positive constants* α *and* β *such that*

$$\alpha \frac{n}{\log n} < \pi(2n) - \pi(n) < \beta \frac{n}{\log n}, \qquad n \geqslant 2.$$

Proof. The right hand side inequality in the theorem follows at once from Theorem 6.2. We now prove the remaining inequality.

The theorem is trivial if $n < 4000$. Suppose then that $n \geqslant 4000$. From (4) and (6) we have that

$$\begin{aligned} \prod_{n<p\leqslant 2n} p &> 2^{2n-\frac{4}{3}n}(2n)^{-(1+\sqrt{2n})} \\ &> 2^{\frac{1}{3}(2n-\sqrt[6]{2n}(18+18\sqrt{2n}))} \\ &> 2^{\frac{1}{3}(2n-19(2n)^{\frac{2}{3}})} \\ &\geqslant 2^{\frac{2}{3}n(1-19/20)} = 2^{\frac{1}{30}n}. \end{aligned}$$

From

$$\prod_{n<p\leqslant 2n} p < (2n)^{\pi(2n)-\pi(n)},$$

we have

$$\pi(2n) - \pi(n) > \frac{\log 2}{30} \cdot \frac{n}{\log 2n},$$

and the theorem is proved. □

Note: Although Theorem 7.1 settles Bertrand's postulate, it is not a very sharp result. Deep analytic methods can be used to give much better results concerning the gaps between successive primes, but these are beyond the scope of this book.

Exercise. Use differential calculus to determine the bound for the validity of (5).

5.8 Estimation of a Sum by an Integral

Theorem 8.1. *Let* $f(x)$ *be increasing and non-negative for* $x \geqslant a$. *Then, for* $\xi \geqslant a$, *we have*

$$\left| \sum_{a\leqslant n\leqslant \xi} f(n) - \int_a^{\xi} f(x)\,dx \right| \leqslant f(\xi).$$

Proof. We set $b = [\xi]$. Then

$$\int_a^b f(x)\,dx = \sum_{i=a}^{b-1} \int_i^{i+1} f(x)\,dx \begin{cases} \geqslant \sum_{i=a}^{b-1} f(i) \\ \leqslant \sum_{i=a}^{b-1} f(i+1), \end{cases}$$

or

$$f(a) + \cdots + f(b-1) \leqslant \int_a^b f(x)\,dx \leqslant f(a+1) + \cdots + f(b);$$

also

$$0 \leqslant \int_b^\xi f(x)\,dx \leqslant f(\xi),$$

and so the theorem follows. □

Example 1. Let $\lambda \geqslant 0$, $f(x) = x^\lambda$. Then

$$\left| \sum_{a \leqslant n \leqslant \xi} n^\lambda - \frac{\xi^{\lambda+1} - a^{\lambda+1}}{\lambda+1} \right| \leqslant \xi^\lambda.$$

From Example 1, we have, for $\lambda \geqslant 0$,

$$\sum_{1 \leqslant n \leqslant \xi} n^\lambda = \frac{\xi^{\lambda+1}}{\lambda+1} + O(\xi^\lambda). \tag{1}$$

This implies that

$$\sum_{1 \leqslant n \leqslant \xi} n^\lambda = O(\xi^{\lambda+1}).$$

Example 2. Let $f(x) = \log x$, $\xi \geqslant 1$ and $T(\xi) = \sum_{n \leqslant \xi} \log n$. Then we have

$$\left| T(\xi) - \int_1^\xi \log x\,dx \right| \leqslant \log \xi,$$

or

$$|T(\xi) - \xi \log \xi + \xi - 1| \leqslant \log \xi. \tag{2}$$

In particular, if ξ is an integer n, then

$$n \log n - n + 1 - \log n \leqslant \log n! \leqslant n \log n - n + 1 + \log n,$$

or

$$n^{n-1}e^{-n+1} \leqslant n! \leqslant n^{n+1}e^{-n+1}. \tag{3}$$

Exercise 1. Let ξ be an integer. Determine one further dominating term in (1); that is, find c so that the following holds for $\lambda \geqslant 1$:

$$\sum_{1 \leqslant n \leqslant \xi} n^{\lambda} = \frac{\xi^{\lambda+1}}{\lambda+1} + c\xi^{\lambda} + O(\xi^{\lambda-1}).$$

Exercise 2. Use Theorem 8.1 to study the sum

$$\sum_{3 \leqslant n \leqslant \xi} \log\log n.$$

Concerning decreasing functions we have:

Theorem 8.2. *Let $f(x)$ be decreasing and non-negative for $x \geqslant a$. Then the limit*

$$\lim_{N \to \infty} \left(\sum_{n=a}^{N} f(n) - \int_a^N f(x)\,dx \right) = \alpha \tag{4}$$

exists, and that $0 \leqslant \alpha \leqslant f(a)$. Moreover, if $f(x) \to 0$ as $x \to \infty$, then for $\xi \geqslant a+1$, we have,

$$\left| \sum_{a \leqslant n \leqslant \xi} f(n) - \int_a^{\xi} f(v)\,dv - \alpha \right| \leqslant f(\xi - 1). \tag{5}$$

Proof. Let

$$g(\xi) = \sum_{a \leqslant n \leqslant \xi} f(n) - \int_a^{\xi} f(x)\,dx.$$

Then

$$\begin{aligned} g(n) - g(n+1) &= -f(n+1) + \int_n^{n+1} f(x)\,dx \\ &\geqslant -f(n+1) + f(n+1) = 0. \end{aligned}$$

Also

$$\begin{aligned} g(N) &= \sum_{n=a}^{N-1} \left(f(n) - \int_n^{n+1} f(x)\,dx \right) + f(N) \\ &\geqslant \sum_{n=a}^{N-1} (f(n) - f(n)) + f(N) = f(N) \geqslant 0, \end{aligned}$$

so that $g(n)$ is a decreasing function, and that

$$0 \leqslant g(n) \leqslant g(a) = f(a).$$

Therefore $g(n)$ has a limit which we denote by α, so that $0 \leqslant \alpha \leqslant f(a)$.

Suppose now that $f(x) \to 0$ as $x \to \infty$. Then

$$\begin{aligned}
g(\xi) - \alpha &= \sum_{a \leqslant n \leqslant \xi} f(n) - \int_a^\xi f(x)\,dx - \lim_{N\to\infty}\left(\sum_{n=a}^{N} f(n) - \int_a^N f(x)\,dx\right)\\
&= \sum_{n=a}^{[\xi]} f(n) - \int_a^{[\xi]} f(x)\,dx - \int_{[\xi]}^{\xi} f(x)\,dx - \lim_{N\to\infty}\left(\sum_{n=a}^{N} f(n) - \int_a^N f(x)\,dx\right)\\
&= -\int_{[\xi]}^{\xi} f(x)\,dx - \lim_{N\to\infty}\left(\sum_{n=[\xi]+1}^{N} f(n) - \int_{[\xi]}^{N} f(x)\,dx\right)\\
&= -\int_{[\xi]}^{\xi} f(x)\,dx + \lim_{N\to\infty}\sum_{n=[\xi]+1}^{N}\int_{n-1}^{n}(f(x) - f(n))\,dx
\end{aligned}$$

$$\begin{cases}
\leqslant \lim\limits_{N\to\infty}\sum\limits_{n=[\xi]+1}^{N}\int_{n-1}^{n}(f(n-1) - f(n))\,dx = f([\xi]) \leqslant f(\xi - 1)\\
\geqslant -\int_{[\xi]}^{\xi} f(x)\,dx \geqslant -(\xi - [\xi])f([\xi]) \geqslant -f(\xi - 1),
\end{cases}$$

and so the theorem is proved. □

Example 3. We take $a = 1$, $f(x) = 1/x$. Then the number α is known as Euler's constant, and is usually denoted by γ. Therefore $0 \leqslant \gamma \leqslant 1$, and

$$\sum_{1 \leqslant n \leqslant \xi} \frac{1}{n} = \log \xi + \gamma + O\left(\frac{1}{\xi}\right). \tag{6}$$

Example 4. Let $0 < \sigma \neq 1$, $f(x) = x^{-\sigma}$. Then there is a constant $\alpha = \alpha(a, \sigma)$ which depends on a and σ, such that when $a \geqslant 1$ we have

$$\left|\sum_{a \leqslant n \leqslant \xi} \frac{1}{n^\sigma} - \frac{\xi^{1-\sigma} - a^{1-\sigma}}{1-\sigma} - a\right| \leqslant \frac{1}{(\xi - 1)^\sigma}. \tag{7}$$

From this we deduce the following: If $\sigma > 1$, then the series

$$\sum_{n=1}^{\infty} \frac{1}{n^\sigma}$$

converges, and when $\xi \geqslant 1$ we have

$$\sum_{n \geqslant \xi} \frac{1}{n^\sigma} = \frac{1}{(\sigma - 1)\xi^{\sigma - 1}} + O\left(\frac{1}{\xi^\sigma}\right). \tag{8}$$

The four results (1), (3), (6), (8) are used very frequently and the reader is advised to remember them.

Exercise 1. Prove that, for $\xi \geqslant 2$,

$$\sum_{1 \leqslant n \leqslant \xi} \frac{\log n}{n} = \frac{1}{2}\log^2 \xi + c_1 + O\left(\frac{\log \xi}{\xi}\right).$$

Exercise 2. Prove that, for $\xi \geqslant 2$,

$$\sum_{2 \leqslant n \leqslant \xi} \frac{1}{n \log n} = \log\log \xi + c_2 + O\left(\frac{1}{\xi \log \xi}\right).$$

5.9 Consequences of Chebyshev's Theorem

The letters $c_1, c_2, \ldots$ used in this section represent absolute constants.

Theorem 9.1. *There exists a constant c_1 such that, for $\xi \geqslant 1$,*

$$\left| \sum_{p \leqslant \xi} \frac{\log p}{p} - \log \xi \right| < c_1.$$

Here $\sum_{p \leqslant \xi}$ represents the summation of all primes p not exceeding ξ.

Proof. 1) We assume first that $\xi = x$ is an integer. From Theorem 1.11.1, we have

$$T(x) = \log x! = \log \prod_{p \leqslant x} p^{\left[\frac{x}{p}\right] + \left[\frac{x}{p^2}\right] + \cdots} = \sum_{p \leqslant x} \left(\left[\frac{x}{p}\right] + \left[\frac{x}{p^2}\right] + \cdots \right) \log p.$$

From

$$\frac{x}{p} - 1 < \left[\frac{x}{p}\right] + \left[\frac{x}{p^2}\right] + \cdots \leqslant \frac{x}{p} + \frac{x}{p^2} + \cdots \leqslant \frac{x}{p} + \frac{x}{p(p-1)},$$

we have

$$\sum_{p \leqslant x} \frac{x \log p}{p} - \sum_{p \leqslant x} \log p < T(x) \leqslant x \left(\sum_{p \leqslant x} \frac{\log p}{p} + \sum_{p \leqslant x} \frac{\log p}{p(p-1)} \right). \tag{1}$$

From Theorem 6.2, we have

$$\sum_{p \leqslant x} \log p \leqslant \log x \cdot \pi(x) \leqslant c_2 x.$$

We also have

$$\sum_{p\leqslant x}\frac{\log p}{p(p-1)}\leqslant\sum_{2\leqslant n\leqslant x+1}\frac{\log n}{(n-1)^2}\leqslant\sum_{n=1}^{\infty}\frac{\log(n+1)}{n^2}=c_3,$$

so that we now have, from (1), that

$$\left|T(x)-x\sum_{p\leqslant x}\frac{\log p}{p}\right|\leqslant c_4x.$$

From Example 8.2 we have $|T(x)-x\log x|<c_5x$. But

$$\left|x\sum_{p\leqslant x}\frac{\log p}{p}-x\log x\right|\leqslant\left|T(x)-x\sum_{p\leqslant x}\frac{\log p}{p}\right|+\left|T(x)-x\log x\right|$$
$$<c_4x+c_5x=c_6x,$$

so that

$$\left|\sum_{p\leqslant x}\frac{\log p}{p}-\log x\right|<c_6.$$

2) Let ξ be real. Then

$$\sum_{p\leqslant\xi}\frac{\log p}{p}=\sum_{p\leqslant[\xi]}\frac{\log p}{p}.$$

From our earlier result we have

$$\left|\sum_{p\leqslant\xi}\frac{\log p}{p}-\log[\xi]\right|<c_6.$$

But

$$|\log[\xi]-\log\xi|=\int_{[\xi]}^{\xi}d(\log t)=\int_{[\xi]}^{\xi}\frac{dt}{t}\leqslant\int_{[\xi]}^{\xi}dt\leqslant 1,$$

so that

$$\left|\sum_{p\leqslant\xi}\frac{\log p}{p}-\log\xi\right|<c_6+1=c_1.$$

The theorem is proved. □

Theorem 9.2. *There exists a constant* c_7 *such that, for* $\xi\geqslant 2$,

$$\sum_{p\leqslant\xi}\frac{1}{p}=\log\log\xi+c_7+O\left(\frac{1}{\log\xi}\right).$$

Proof. Let

$$S(n)=\sum_{p\leqslant n}\frac{\log p}{p},$$

so that, by Theorem 9.1,

$$S(n) = \log n + r_n, \qquad r_n = O(1).$$

Therefore

$$\sum_{p \leqslant \xi} \frac{1}{p} = \sum_{p \leqslant \xi} \frac{\log p}{p} \cdot \frac{1}{\log p} = \sum_{2 \leqslant n \leqslant \xi} \frac{S(n) - S(n-1)}{\log n}$$

$$= \sum_{2 \leqslant n \leqslant \xi} \frac{\log n - \log (n-1)}{\log n} + \sum_{2 \leqslant n \leqslant \xi} \frac{r_n - r_{n-1}}{\log n} = \sum\nolimits_1 + \sum\nolimits_2. \tag{2}$$

Now the function

$$f(x) = -\frac{\log\left(1 - \dfrac{1}{x}\right)}{\log x}, \qquad x \geqslant 2,$$

is decreasing, and $f(x) \to 0$ as $x \to \infty$. Therefore, by Theorem 8.2, we have

$$\sum\nolimits_1 = -\sum_{2 \leqslant n \leqslant \xi} \frac{\log\left(1 - \dfrac{1}{n}\right)}{\log n} = -\int_2^{\xi} \frac{\log\left(1 - \dfrac{1}{x}\right)}{\log x} dx + c_8 + O(f(\xi)).$$

Since

$$f(x) = \frac{1}{x \log x} + O\left(\frac{1}{x^2 \log x}\right),$$

the integral

$$\int_2^{\infty} \frac{-\log\left(1 - \dfrac{1}{x}\right) - \dfrac{1}{x}}{\log x} dx$$

converges to c_9, so that

$$\sum\nolimits_1 = \int_2^{\xi} \frac{dx}{x \log x} + c_8 + \int_2^{\xi} \frac{-\log\left(1 - \dfrac{1}{x}\right) - \dfrac{1}{x}}{\log x} dx + O\left(\frac{1}{\xi \log \xi}\right)$$

$$= \log\log \xi - \log\log 2 + c_8 + c_9 + \int_{\xi}^{\infty} \frac{\log\left(1 - \dfrac{1}{x}\right) + \dfrac{1}{x}}{\log x} dx + O\left(\frac{1}{\xi \log \xi}\right)$$

$$= \log\log \xi + c_{10} + O\left(\frac{1}{\xi \log \xi}\right), \tag{3}$$

where we have used

$$\int_{\xi}^{\infty} \frac{\log\left(1 - \frac{1}{x}\right) + \frac{1}{x}}{\log x} = O\left(\int_{\xi}^{\infty} \frac{dx}{x^2 \log x}\right) = O\left(\frac{1}{\log \xi} \int_{\xi}^{\infty} \frac{dx}{x^2}\right) = O\left(\frac{1}{\xi \log \xi}\right).$$

Next, from the convergence of the positive terms series

$$\sum_{n=2}^{\infty} \left(\frac{1}{\log n} - \frac{1}{\log(n+1)}\right)$$

and $r_n = O(1)$ we deduce that the series

$$\sum_{n=2}^{\infty} r_n \left(\frac{1}{\log n} - \frac{1}{\log(n+1)}\right)$$

converges to c_{11}. Also

$$\sum_{n>\xi} r_n \left(\frac{1}{\log n} - \frac{1}{\log(n+1)}\right) = O\left(\sum_{n>\xi} \left|\frac{1}{\log n} - \frac{1}{\log(n+1)}\right|\right)$$
$$= O\left(\sum_{n>\xi} \frac{1}{n \log^2 n}\right) = O\left(\frac{1}{\log \xi}\right).$$

Therefore

$$\Sigma_2 = \sum_{2 \leqslant n \leqslant \xi} r_n \left(\frac{1}{\log n} - \frac{1}{\log(n+1)}\right) + O\left(\frac{r_\xi}{\log \xi}\right)$$
$$= \sum_{n=2}^{\infty} r_n \left(\frac{1}{\log n} - \frac{1}{\log(n+1)}\right) - \sum_{n>\xi} r_n \left(\frac{1}{\log n} - \frac{1}{\log(n+1)}\right) + O\left(\frac{1}{\log \xi}\right)$$
$$= c_{11} + O\left(\frac{1}{\log \xi}\right). \tag{4}$$

From (2), (3) and (4) we arrive at

$$\sum_{p \leqslant \xi} \frac{1}{p} = \log\log \xi + c_{10} + c_{11} + O\left(\frac{1}{\log \xi}\right)$$
$$= \log\log \xi + c_7 + O\left(\frac{1}{\log \xi}\right).$$

The theorem is proved. □

Theorem 9.3. *There exists a constant c_{12} such that, for $\xi \geqslant 2$,*

$$\prod_{p \leqslant \xi} \left(1 - \frac{1}{p}\right) = \frac{c_{12}}{\log \xi} + O\left(\frac{1}{\log^2 \xi}\right).$$

Proof. Since

$$\sum_{p>\xi}\left(\log\left(1-\frac{1}{p}\right)+\frac{1}{p}\right)=O\left(\sum_{p>\xi}\frac{1}{p^2}\right)=O\left(\sum_{n>\xi}\frac{1}{n^2}\right)=O\left(\frac{1}{\xi}\right),$$

it follows from the previous theorem that

$$\begin{aligned}\log\prod_{p\leqslant\xi}\left(1-\frac{1}{p}\right)&=\sum_{p\leqslant\xi}\log\left(1-\frac{1}{p}\right)=-\sum_{p\leqslant\xi}\frac{1}{p}+\sum_{p\leqslant\xi}\left[\log\left(1-\frac{1}{p}\right)+\frac{1}{p}\right]\\&=-\log\log\xi-c_7+O\left(\frac{1}{\log\xi}\right)+\sum_{p>2}\left(\log\left(1-\frac{1}{p}\right)+\frac{1}{p}\right)\\&\quad-\sum_{p>\xi}\left(\log\left(1-\frac{1}{p}\right)+\frac{1}{p}\right)=-\log\log\xi+c_{13}+O\left(\frac{1}{\log\xi}\right),\end{aligned}$$

where

$$c_{13}=-c_7+\sum_{p>2}\left(\log\left(1-\frac{1}{p}\right)+\frac{1}{p}\right).$$

Therefore

$$\begin{aligned}\sum_{p\leqslant\xi}\left(1-\frac{1}{p}\right)&=e^{-\log\log\xi+c_{13}+O\left(\frac{1}{\log\xi}\right)}=\frac{e^{c_{13}}}{\log\xi}\cdot c^{O\left(\frac{1}{\log\xi}\right)}\\&=\frac{c_{12}}{\log\xi}\left(1+O\left(\frac{1}{\log\xi}\right)\right)(c_{12}=e^{c_{13}}),\end{aligned}$$

where we have used

$$e^{O\left(\frac{1}{\log\xi}\right)}=1+O\left(\frac{1}{\log\xi}\right).$$

The theorem is proved. □

Theorems 9.2 and 9.3 are quantitative elaborations of Theorems 4.3 and 4.5.

Exercise 1. Let p_n denote the n-th prime. Prove that there are constants c_1, c_2 such that

$$c_1 n\log n < p_n < c_2 n\log n, \qquad n\geqslant 2.$$

Exercise 2. Prove that there exists a positive constant c such that

$$\varphi(n)>\frac{cn}{\log\log n}, \qquad n\geqslant 3.$$

Exercise 3. Prove that the infinite series

$$\sum_p\frac{1}{p(\log\log p)^h}$$

converges or diverges according to whether $h > 1$ or $h \leqslant 1$. Here $\sum_p$ represents the summation over all the prime numbers.

5.10 The Number of Prime Factors of n

Let n be a positive integer. We denote by $\omega(n)$ the number of distinct prime factors of n and by $\Omega(n)$ the total number of prime factors of n. That is, if $n = p_1^{a_1} \cdots p_s^{a_s}$, then

$$\omega(n) = s, \qquad \Omega(n) = a_1 + \cdots + a_s. \tag{1}$$

If n is a prime, then $\omega(n) = \Omega(n) = 1$; but as n tends to infinity through powers of 2, then

$$\Omega(n) = \frac{\log n}{\log 2} \to \infty;$$

and if $n = p_1 p_2 \cdots p_s$ is the product of the first s primes, then as $n \to \infty$, $\omega(n) = s \to \infty$. Thus the behaviours of $\omega(n)$ and $\Omega(n)$ are rather irregular and there is certainly no asymptotic formula for them. However, we do have the following:

Theorem 10.1. *There are positive constants c_1, c_2 such that*

$$\sum_{n \leqslant x} \omega(n) = x \log\log x + c_1 + o(x), \tag{2}$$

$$\sum_{n \leqslant x} \Omega(n) = x \log\log x + c_2 + o(x). \tag{3}$$

Proof. 1) We have

$$\sum_{n \leqslant x} \omega(n) = \sum_{n \leqslant x} \sum_{p \mid n} 1 = \sum_{p \leqslant x} \left[\frac{x}{p}\right] = \sum_{p \leqslant x} \frac{x}{p} + O(\pi(x))$$

and so (2) follows from Theorem 9.2 and Theorem 6.2.

2) We have

$$\sum_{n \leqslant x} \Omega(n) = \sum_{n \leqslant x} \sum_{p^m \mid n} 1 = \sum_{p^m \leqslant x} \left[\frac{x}{p^m}\right] = \sum_{p \leqslant x} \left[\frac{x}{p}\right] + \sum_{\substack{p^m \leqslant x \\ m \geqslant 2}} \left[\frac{x}{p^m}\right],$$

and, by Theorem 6.2,

$$\sum_{\substack{p^m \leqslant x \\ m \geqslant 2}} 1 \leqslant \sum_{p^2 \leqslant x} 1 + \sum_{p^3 \leqslant x} 1 + \cdots + \sum_{p^{\left[\frac{\log x}{\log 2}\right]} \leqslant x} 1 \leqslant \frac{\log x}{\log 2} \sum_{p^2 \leqslant x} 1 = \frac{\log x}{\log 2} \pi(\sqrt{x}) = o(x).$$

Therefore

$$\sum_{n \leqslant x} \Omega(n) = \sum_{n \leqslant x} \omega(n) + \sum_{\substack{p^m \leqslant x \\ m \geqslant 2}} \frac{x}{p^m} + o(x).$$

But the series

$$\sum_{m=2}^{\infty} \sum_{p} \frac{1}{p^m} = \sum_{p} \left(\frac{1}{p^2} + \frac{1}{p^3} + \cdots \right) = \sum_{p} \frac{1}{p(p-1)} = c$$

converges, so that

$$\sum_{n \leqslant x} \Omega(n) = \sum_{n \leqslant x} \omega(n) + x(c + o(1)) + o(x) = x \log\log x + c_2 x + o(x). \quad \square$$

Theorem 10.2 (Hardy-Ramanujan). *Let $\varepsilon > 0$, and let $f(n)$ denote either $\omega(n)$ or $\Omega(n)$. Then the number of positive integers $n \leqslant x$ satisfying*

$$|f(n) - \log\log n| > (\log\log n)^{\frac{1}{2}+\varepsilon} \tag{4}$$

is $o(x)$, as $x \to \infty$.

Proof (Turan). Since $\log\log x - 1 < \log\log n \leqslant \log\log x$ when $x^{1/e} < n \leqslant x$, and the number of positive integers $n \leqslant x^{1/e}$ is $[x^{1/e}] = o(x)$, it suffices to prove that the number of positive integers $n \leqslant x$ satisfying

$$|f(n) - \log\log x| > (\log\log x)^{\frac{1}{2}+\varepsilon} \tag{5}$$

is $o(x)$ as $x \to \infty$.

Next, from $\Omega(n) \geqslant \omega(n)$, and by (2) and (3)

$$\sum_{n \leqslant x} (\Omega(n) - \omega(n)) = O(x)$$

so that the number of positive integers $n \leqslant x$ satisfying $\Omega(n) - \omega(n) > (\log\log x)^{\frac{1}{2}}$ is

$$O\left(\frac{x}{(\log\log x)^{\frac{1}{2}}} \right) = o(x)..$$

Therefore we need only consider the case $f(n) = \omega(n)$.

We consider a pair p, q of distinct prime divisors of n (p, q and q, p are treated as two different pairs). Each p may take $\omega(n)$ values and for each fixed p, q may take $\omega(n) - 1$ values. Therefore we have

$$\omega(n)(\omega(n) - 1) = \sum_{\substack{pq \mid n \\ p \neq q}} 1 = \sum_{pq \mid n} 1 - \sum_{p^2 \mid n} 1.$$

Summing over $n = 1, 2, \ldots, [x]$ we have

$$\sum_{n \leqslant x} \omega^2(n) - \sum_{n \leqslant x} \omega(n) = \sum_{n \leqslant x} \left(\sum_{pq \mid n} 1 - \sum_{p^2 \mid n} 1 \right) = \sum_{pq \leqslant x} \left[\frac{x}{pq} \right] - \sum_{p^2 \leqslant x} \left[\frac{x}{p^2} \right]. \tag{6}$$

Since

$$\sum_{p^2 \leqslant x} \left[\frac{x}{p^2} \right] \leqslant \sum_{p^2 \leqslant x} \frac{x}{p^2} \leqslant x \sum_{p} \frac{1}{p^2} = O(x)$$

and

$$\sum_{pq \leqslant x} \left[\frac{x}{pq}\right] = x \sum_{pq \leqslant x} \frac{1}{pq} + O(x),$$

it follows from (2) and (6) that

$$\sum_{n \leqslant x} \omega^2(n) = x \sum_{pq \leqslant x} \frac{1}{pq} + O(x \log\log x). \tag{7}$$

Now

$$\left(\sum_{p \leqslant \sqrt{x}} \frac{1}{p}\right)^2 \leqslant \sum_{pq \leqslant x} \frac{1}{pq} \leqslant \left(\sum_{p \leqslant x} \frac{1}{p}\right)^2,$$

and $\sum_{p \leqslant \xi} 1/p = \log\log \xi + O(1)$, so that both the outsides in the above are

$$(\log\log x + O(1))^2 = (\log\log x)^2 + O(\log\log x).$$

It now follows from (7) that

$$\sum_{n \leqslant x} \omega^2(n) = x(\log\log x)^2 + O(x \log\log x), \tag{8}$$

and so

$$\begin{aligned}\sum_{n \leqslant x} (\omega(n) - \log\log x)^2 &= \sum_{n \leqslant x} \omega^2(n) - 2\log\log x \sum_{n \leqslant x} \omega(n) + [x](\log\log x)^2 \\ &= x(\log\log x)^2 + O(x \log\log x) - 2\log\log x(x \log\log x + O(x)) \\ &\quad + (x + O(1))(\log\log x)^2 = O(x \log\log x).\end{aligned} \tag{9}$$

Given any $\delta > 0$, if there are δx positive integers $n \leqslant x$ such that (5) holds, then

$$\sum_{n \leqslant x} (\omega(n) - \log\log x)^2 \geqslant \delta x(\log\log x)^{1+2\varepsilon}, \tag{10}$$

which contradicts with (9). Therefore the number of positive integers $n \leqslant x$ such that (5) holds is $o(x)$, and the theorem is proved. □

From this we see that

$$\omega(n) \sim \log\log n \qquad \text{and} \qquad \Omega(n) \sim \log\log n$$

for *almost all n*.

5.11 A Prime Representing Function

Theorem 11.1 (Miller). *There exists a fixed number α such that if*

$$\alpha = \alpha_0, \qquad 2^{\alpha_0} = \alpha_1, \ldots, \qquad 2^{\alpha_n} = \alpha_{n+1}, \ldots,$$

then $[\alpha_n]$ is always prime.

Proof. We construct a sequence of primes $\{p_n\}$ by induction: Take $p_1 = 3$. By Theorem 7.1 there exists a prime p_{n+1} satisfying

$$2^{p_n} < p_{n+1} < p_{n+1} + 1 \leqslant 2^{p_n+1}.$$

If $p_{n+1} + 1 = 2^{p_n+1}$, then $p_{n+1} = 2^{p_n+1} - 1$ cannot be prime (because it has the divisor $2^{\frac{1}{2}(p_n+1)} - 1$). Therefore

$$2^{p_n} < p_{n+1} < p_{n+1} + 1 < 2^{p_n+1}.$$

Using logarithm base 2 we define

$$\log^{(n)} x = \log^{(n-1)}(\log x).$$

Consider the sequences

$$u_n = \log^{(n)} p_n, \qquad v_n = \log^{(n)}(p_n + 1).$$

From $p_n < \log p_{n+1} < \log(p_{n+1} + 1) < p_n + 1$, we see that $u_n < u_{n+1} < v_{n+1} < v_n$, so that u_n and v_n are monotonic sequences. Therefore there exists α such that $\lim_{n\to\infty} u_n = \alpha$, and $u_n < \alpha < v_n$. That is, $p_n < \alpha_n < p_n + 1$ and so $[\alpha_n] = p_n$. □

Exercise 1. Prove that there does not exist a non-constant polynomial $f(n)$ with integer coefficients which takes prime values for all n.

Exercise 2. Let $P(x_1, x_2, \ldots, x_k)$ be a polynomial with integer coefficients. Let $f(n) = P(n, 2^n, 3^n, \ldots, k^n)$. Prove that if $f(n) \to \infty$, as $n \to \infty$, then $f(n)$ represents infinitely many composite numbers.

5.12 On Primes in an Arithmetic Progression

We saw in the exercises in §5 that there are infinitely many primes of the form $4n - 1$ and $6n - 1$. This suggests the following: If a and b are coprime integers, then there are infinitely many primes of the form $an + b$. This is the famous Dirichlet's theorem which we shall prove in Chapter 9. Here we study the following special situation.

We assume that a, b are positive and that b is fixed. We observe that if, given any a, there is always a prime of the form $an + b$ $(n > 0)$, then Dirichlet's theorem follows. For if there exists n such that $an + b = p_1$ $(> b)$ is prime, and (replacing a by ap_1) there exists n such that $ap_1 n + b = p_2$ $(> p_1)$ is prime, and so on, then there are infinitely many primes of the form $an + b$.

Theorem 12.1. *Let $k > 1$. Then there are infinitely many primes of the form $kn + 1$.*

From what we said earlier it suffices to prove that there always exists a prime of the form $kn + 1$.

The roots of the equation $x^k = 1$ are given by

$$e^{2\pi i a/k}, \qquad a = 0, 1, \ldots, k - 1.$$

Let

$$F_n(x) = \prod_{(a,n)=1} (x - e^{2\pi i a/n}),$$

where the product is over a reduced set of residues $a \bmod n$. Clearly we have

$$x^k - 1 = \prod_{n|k} F_n(x)$$

where the product is over the divisors n of k, since each root on the left hand side must occur on the right hand side, and conversely without any repetition. Let

$$x^k - 1 = F_k(x)G_k(x),$$

where $G_k(x)$ is the least common multiple of the various polynomials $x^n - 1$ $(n|k, n < k)$, and its leading coefficient is 1. Therefore $G_k(x)$ is an integer coefficient polynomial, and by Theorem 1.13.2 we see that $F_k(x)$ is also an integer coefficient polynomial.

If x is an integer not equal to ± 1, then

$$F_k(x)G_k(x) \neq 0,$$

that is, $F_k(x)$ and $G_k(x)$ are non-zero integers.

Lemma 1. *Let n be a proper divisor of k. Then for all integers $x \neq \pm 1$, we have*

$$\left(x^n - 1, \frac{x^k - 1}{x^n - 1}\right) \Big| k.$$

Proof. Let $x^n - 1 = y, k = nd$. Then

$$\frac{x^k - 1}{x^n - 1} = \frac{(y+1)^d - 1}{y} = y^{d-1} + \binom{d}{1} y^{d-2} + \cdots + \binom{d}{2} y + d$$

$$\equiv d \pmod{y}. \quad \square$$

Lemma 2. *Let x be an integer not equal to ± 1. Then each common prime divisor of $F_k(x)$ and $G_k(x)$ must be a divisor of k.*

Proof. Let $p|(F_k(x), G_k(x))$. From

$$p|G_k(x) = \prod_{\substack{n|k \\ n<k}} F_n(x)$$

we see that there must be an integer n such that

$$p|F_n(x), \qquad (n|k, n < k),$$

so that

$$p|x^n - 1.$$

Again, from $p|F_k(x)$, we have

$$p\left|\frac{x^k - 1}{x^n - 1}\right..$$

Therefore

$$p\left|\left(x^n - 1, \frac{x^k - 1}{x^n - 1}\right)\right.$$

and the required result follows from Lemma 1. □

Proof of Theorem 12.1. Let $x = ky$. Then

$$F_k(x)G_k(x) = x^k - 1 \equiv -1 \pmod{k}.$$

We can select y such that $F_k(x) \neq \pm 1$; this is possible because the equation $F_k(x) = \pm 1$ has only finitely many solutions. There must be a prime divisor p in $F_k(x)$, and, by Lemma 2, p does not divide $G_k(x)$. In other words, for each proper divisor n of k,

$$x^n \not\equiv 1 \pmod{p}. \tag{1}$$

But

$$x^k \equiv 1 \pmod{p}.$$

We now prove that $k \mid p - 1$. Suppose otherwise. Then there are integers s and t such that

$$(k, p - 1) = sk + t(p - 1).$$

That is, corresponding to $n = (k, p - 1)$, we have

$$x^n \equiv (x^k)^s(x^{p-1})^t \equiv 1 \pmod{p},$$

which contradicts (1). Therefore $p \equiv 1 \pmod{k}$; that is there exists a prime of the form $kn + 1$. As we already observed, this proves the theorem. □

Exercise. Prove that there are infinitely many primes of the form $8n + 5$.

Hint: Consider $q = 3^2 \cdot 5^2 \cdot 7^2 \cdot \cdots \cdot p^2 + 2^2$, and prove that each prime p of the form $x^2 + y^2$ must be congruent $1 \pmod 4$.

Notes

5.1. There has been much progress towards the Goldbach problem in recent years using sieve methods. Perhaps the most exciting is the following result of J. R. Chen [19], [20].

Let n be a sufficiently large even integer and denote by $P_n(1,2)$ the number of primes $p \leqslant n$ such that either $n-p$ is a prime or a product of two primes. Then

$$P_n(1,2) > 0.67 \prod_{\substack{p|n \\ p>2}} \frac{p-1}{p-2} \prod_{p>2} \left(1 - \frac{1}{(p-1)^2}\right) \frac{n}{\log^2 n}.$$

It follows, of course, from this that every sufficiently large even integer is a sum of a prime and an integer having at most two prime factors. The proof of Chen's theorem is given in the book "Sieve Methods" by H. Halberstam and H. E. Richert [28] where there is also a comprehensive bibliography.

5.2. Concerning the prime twins problem J. R. Chen [20] also proved that there are infinitely many primes p such that $p+2$ is either a prime or has two prime factors.

5.3. H. Iwaniec (unpublished) has proved that there are infinitely many integers n such that n^2+1 is either a prime or has two prime factors.

5.4. The principle of the "large sieve" was invented by Yu. Linnik and A. Renyi, and was substantially developed by K. F. Roth [50] and E. Bombieri [9] (see also the books by H. L. Montgomery [44] and E. Bombieri [10]). From his result Bombieri deduced the following theorem on the average value of $\pi(x;k,l)$: Given any $A>0$, there exists $B=B(A)>0$ such that

$$\sum_{k \leqslant x^{\frac{1}{2}}/\log^B x} \max_{(l,k)=1} \left| \pi(x;k,l) - \frac{\operatorname{li} x}{\varphi(k)} \right| = O\left(\frac{x}{\log^A x}\right).$$

(A. I. Vinogradov [59] independently proved a slightly weaker result).

5.5. There has also been much recent work on the distribution of $d_n = p_{n+1} - p_n$ where p_n is the n-th prime number. For example, H. L. Montgomery [44] proved that $d_n = O(p_n^{\frac{3}{5}+\varepsilon})$, where ε is any positive number, with the implied constant depending on ε; M. N. Huxley [31] improved this to $d_n = O(p_n^{\frac{7}{12}+\varepsilon})$ and very recently this has been improved to $d_n = O(p_n^{\frac{11}{20}+\varepsilon})$ by H. Halberstam, D. R. Heath-Brown and H. E. Richert.

We observe that $d_n = 2$ whenever p_n, p_{n+1} are prime twins. Concerning unconditional lower bounds for d_n, E. Bombieri and H. Davenport [11] proved that

$$E = \inf_{n\to\infty}\inf \frac{d_n}{\log p_n} \leqslant \frac{1}{8}(2+\sqrt{3}) = 0.46650\ldots,$$

and this has been improved to $E \leqslant \frac{1}{4}(\frac{\pi}{4}+1) = 0.4463\ldots$ (see [32]).

5.6. Besides the problems on the distribution of primes mentioned in the text there is also the problem of the least prime in an arithmetic progression, that is the estimate of the least prime $P(k,l)$ in the arithmetic progression $kn+l$ $(n=1,2,\ldots)$ where k,l are coprime positive integers. S. Chowla has conjectured that $P(k,l) = O(k^{1+\varepsilon})$ and Yu. Linnik was the first to prove that there is an absolute constant c such that $P(k,l) = O(k^c)$. Later C. T. Pan [45] gave a computable estimate for the value of c, and the present best estimate gives $c < 15$ which is due to J. R. Chen (unpublished).

5.7. In 1922 G. H. Hardy and J. E. Littlewood conjectured that every sufficiently large integer is the sum of two squares and a prime. This was proved by Yu. Linnik [40] using rather complicated methods. However there is now a simpler proof, based on E. Bombieri's mean value theorem for $\pi(x; k, l)$, of this conjecture (see P. D. T. A. Elliot and H. Halberstam [23]).

Many of the problems mentioned in these notes are also discussed in the author's book [30].

Chapter 6. Arithmetic Functions

6.1 Examples of Arithmetic Functions

Definition 1. By an arithmetic function $f(n)$ we mean a function whose domain is the set of positive integers.

Examples. Any sequence a_n is an arithmetic function. Specifically we can have $n!$, $\sin n$, $d(n) = \sum_{d|n} 1$ or $r(n)$ where $r(n)$ is the number of solutions to the equation $n = x^2 + y^2$.

Definition 2. Let $f(n)$ be an arithmetic function such that if $(a, b) = 1$, then

$$f(a, b) = f(a)f(b). \tag{1}$$

Then we call $f(n)$ a *multiplicative function.* If (1) holds regardless of the condition $(a, b) = 1$, then we say that $f(n)$ is *completely multiplicative.*

From this definition we see that if $f(n)$ is a multiplicative function and if $p_1, \ldots, p_r$ are distinct prime numbers, then

$$f(p_1^{a_1} \cdots p_r^{a_r}) = f(p_1^{a_1}) \cdots f(p_r^{a_r}),$$

so that $f(n)$ is determined by the values it takes at the prime powers. Moreover, if $f(n)$ is completely multiplicative, then

$$f(p_1^{a_1} \cdots p_r^{a_r}) = (f(p_1))^{a_1} \cdots (f(p_r))^{a_r},$$

so that $f(n)$ is determined by the values it takes at the primes. It is clear that the product of two multiplicative functions is multiplicative and the product of two completely multiplicative functions is completely multiplicative.

Example 1. The function

$$\Delta(n) = \begin{cases} 1 & \text{if} \quad n = 1, \\ 0 & \text{if} \quad n \neq 1, \end{cases}$$

is completely multiplicative.

Example 2. The function $E_\lambda(n) = n^\lambda$ is completely multiplicative.

Example 3. The Möbius function is defined by:

$$\mu(n) = \begin{cases} 1 & \text{if } n = 1, \\ (-1)^r & \text{if } n \text{ is the product of } r \text{ distinct primes,} \\ 0 & \text{if } n \text{ is divisible by a prime square.} \end{cases}$$

It is easy to see that

$$\mu(1) = 1, \quad \mu(2) = -1, \quad \mu(3) = -1, \quad \mu(4) = 0, \quad \mu(5) = -1, \quad \mu(6) = 1,$$
$$\mu(7) = -1, \quad \mu(8) = 0, \quad \mu(9) = 0, \quad \mu(10) = 1, \quad \mu(11) = -1, \ldots.$$

Here $\mu(n)$ is multiplicative, but not completely multiplicative.

Example 4. The number of positive integers not exceeding n and coprime with n is denoted by $\varphi(n)$, and it is called Euler's function. This function is also multiplicative, but not completely multiplicative.

Example 5. The divisor function $d(n) = \sum_{d|n} 1$ is also multiplicative, but not completely multiplicative. More generally, the function $\sigma_\lambda(n) = \sum_{d|n} d^\lambda$ is multiplicative. We note that $\sigma_0(n) = d(n)$.

Example 6. Von Mangoldt's function is defined by:

$$\Lambda(n) = \begin{cases} \log p, & \text{if } p \text{ is the only prime factor of } n, \\ 0, & \text{otherwise.} \end{cases}$$

We have

$$\Lambda(1) = 0, \quad \Lambda(2) = \log 2, \quad \Lambda(3) = \log 3, \quad \Lambda(4) = \log 2, \quad \Lambda(5) = \log 5,$$
$$\Lambda(6) = 0, \quad \Lambda(7) = \log 7, \quad \Lambda(8) = \log 2, \quad \Lambda(9) = \log 3, \quad \Lambda(10) = 0, \ldots$$

and we see that $\Lambda(n)$ is *not* multiplicative.

Example 7. We define

$$\Lambda_1(n) = \begin{cases} \dfrac{1}{m}, & \text{if } n \text{ is the } m\text{-th power of a prime,} \\ 0, & \text{otherwise.} \end{cases}$$

We have

$$\Lambda_1(1) = 0, \quad \Lambda_1(2) = 1, \quad \Lambda_1(3) = 1, \quad \Lambda_1(4) = \tfrac{1}{2}, \quad \Lambda_1(5) = 1, \quad \Lambda_1(6) = 0,$$
$$\Lambda_1(7) = 1, \quad \Lambda_1(8) = \tfrac{1}{3}, \quad \Lambda_1(9) = \tfrac{1}{2}, \quad \Lambda_1(10) = 0, \ldots,$$

and that $\Lambda_1(n)$ is *not* multiplicative.

Example 8. Let p be a fixed prime number. If $p^a \| n$, we define $V_p(n) = p^{-a}$. This function is completely multiplicative and it is not difficult to prove that $V_p(n+m) \leqslant \max(V_p(n), V_p(m))$.

Example 9. Let $r(n)$ denote the number of solutions to the equation $n = x^2 + y^2$. We shall prove in §7 that $\frac{1}{4}r(n)$ is a multiplicative function. However, from $r(3) = 0$, $r(9) = 4$ we see that it is not completely multiplicative.

6.2 Properties of Multiplicative Functions

Theorem 2.1. *Let $f(n)$ be a multiplicative function which is not identically zero. Then $f(1) = 1$.*

Proof. Let $f(a) \neq 0$. From $f(a) = f(a)f(1)$ we deduce that $f(1) = 1$. □

Theorem 2.2. *Let $g(n)$ and $h(n)$ be multiplicative functions. Then the function*

$$f(n) = \sum_{d|n} g(d)h\left(\frac{n}{d}\right) = \sum_{d|n} g\left(\frac{n}{d}\right)h(d) \tag{1}$$

is also multiplicative.

Proof. The second equation in (1) follows from the substitution $d' = n/d$. Suppose that $(a, b) = 1$. Then

$$f(a,b) = \sum_{d|ab} g(d)h\left(\frac{ab}{d}\right).$$

Let $u = (a, d)$, $v = (b, d)$ so that $uv = d$ and hence

$$\begin{aligned} f(ab) &= \sum_{u|a} \sum_{v|b} g(uv)h\left(\frac{ab}{uv}\right) \\ &= \sum_{u|a} g(u)h\left(\frac{a}{u}\right) \sum_{v|b} g(v)h\left(\frac{b}{v}\right) \\ &= f(a)f(b). \quad \square \end{aligned}$$

Theorem 2.3. *Let $f(n)$ be a multiplicative function which is not identically zero. Then*

$$\sum_{d|n} \mu(d)f(d) = \prod_{p|n} (1 - f(p)), \tag{2}$$

where p runs through the prime divisors of n.

Proof. We put $g(n) = \mu(n)f(n)$, $h(n) = 1$ in Theorem 2.2, so that the left hand side of (2) is a multiplicative function. It is clear that the right hand side of (2) is also multiplicative. It follows that we only need to prove (2) when $n = 1$ and $n = p^l$, and these two cases can be verified easily. □

Theorem 2.4. *Let $f(n)$ be multiplicative. Then*

$$f((m,n))f([m,n]) = f(m)f(n),$$

where $[m,n]$ is the least common multiple of m and n.

Proof. Let

$$m = p_1^{l_1} \cdots p_s^{l_s}, \qquad l_v \geqslant 0,$$

$$n = p_1^{r_1} \cdots p_s^{r_s}, \qquad r_v \geqslant 0.$$

Then

$$f(m) = f(p_1^{l_1}) \cdots f(p_s^{l_s}),$$

$$f(n) = f(p_1^{r_1}) \cdots f(p_s^{r_s}),$$

$$f((m,n)) = f(p_1^{\min(l_1,r_1)}) \cdots f(p_s^{\min(l_s,r_s)}),$$

$$f\left(\frac{mn}{(m,n)}\right) = f(p_1^{\max(l_1,r_1)}) \cdots f(p_s^{\max(l_s,r_s)}).$$

Since

$$f(p^l)f(p^r) = f(p^{\max(l,r)})f(p^{\min(l,r)}),$$

the theorem follows. □

6.3 The Möbius Inversion Formula

Theorem 3.1. *Let $n > 0$. We have*

$$\sum_{d|n} \mu(d) = \sum_{d|n} \mu(n/d) = \Delta(n) = \begin{cases} 1, & \text{if } n = 1, \\ 0, & \text{if } n \neq 1. \end{cases}$$

Proof. This follows from taking $f(d) = 1$ in Theorem 2.3. □

Theorem 3.2. *Let $0 < \eta_0 \leqslant \eta_1$ and let $h(k)$ be a completely multiplicative function which is not identically zero. If for any η satisfying $\eta_0 \leqslant \eta \leqslant \eta_1$ we have*

$$g(\eta) = \sum_{1 \leqslant k \leqslant \eta_1/\eta} f(k\eta)h(k), \tag{1}$$

then

$$f(\eta) = \sum_{1 \leqslant k \leqslant \eta_1/\eta} \mu(k)g(k\eta)h(k); \tag{2}$$

the converse also holds.

Proof. From (1) we have

$$\sum_{1\leqslant k\leqslant \eta_1/\eta} \mu(k)g(k\eta)h(k) = \sum_{1\leqslant k\leqslant \eta_1/\eta} \mu(k)h(k) \sum_{1\leqslant m\leqslant \eta_1/k\eta} f(mk\eta)h(m).$$

Let $mk = r$. From Theorem 3.1 we have

$$\begin{aligned}\sum_{1\leqslant k\leqslant \eta_1/\eta} \mu(k)g(k\eta)h(k) &= \sum_{1\leqslant k\leqslant \eta_1/\eta} \mu(k) \sum_{\substack{1\leqslant k\leqslant \eta_1/\eta \\ k\mid r}} f(r\eta)h(k)h\left(\frac{r}{k}\right) \\ &= \sum_{1\leqslant r\leqslant \eta_1/\eta} f(r\eta)h(r) \sum_{\substack{1\leqslant k\leqslant \eta_1/\eta \\ k\mid r}} \mu(k) \\ &= \sum_{1\leqslant r\leqslant \eta_1/\eta} f(r\eta)h(r) \sum_{k\mid r} \mu(k) \\ &= \sum_{1\leqslant r\leqslant \eta_1/\eta} f(r\eta)h(r)\Delta(r) = f(\eta)h(1) = f(\eta)\end{aligned}$$

which proves (2).

Suppose instead that (2) holds. Then

$$\begin{aligned}\sum_{1\leqslant k\leqslant \eta_1/\eta} f(k\eta)h(k) &= \sum_{1\leqslant k\leqslant \eta_1/\eta} h(k) \sum_{1\leqslant m\leqslant \eta_1/k\eta} \mu(m)g(mk\eta)h(m) \\ &= \sum_{1\leqslant k\leqslant \eta_1/\eta} \sum_{\substack{1\leqslant k\leqslant \eta_1/\eta \\ k\mid r}} \mu(r/k)g(r\eta)h(k)h(r/k) \\ &= \sum_{1\leqslant r\leqslant \eta_1/\eta} g(r\eta)h(r) \sum_{\substack{1\leqslant k\leqslant \eta_1/\eta \\ k\mid r}} \mu(r/k) \\ &= \sum_{1\leqslant r\leqslant \eta_1/\eta} g(r\eta)h(r)\Delta(r) = g(\eta)\end{aligned}$$

which proves (1). □

We can extend this theorem as follows:

Theorem 3.3. *Let $\xi_0 \geqslant 1$ and let $H(k)$ be a completely multiplicative function which is not identically zero. If for all real ξ satisfying $1 \leqslant \xi \leqslant \xi_0$ we have*

$$G(\xi) = \sum_{1\leqslant k\leqslant \xi} F(\xi/k)H(k), \tag{3}$$

then we have, for such ξ,

$$F(\xi) = \sum_{1\leqslant k\leqslant \xi} \mu(k)G(\xi/k)H(k); \tag{4}$$

the converse also holds.

Proof. Let $f(\eta) = F(1/\eta)$ and $g(\eta) = G(1/\eta)$. Then from (3) and (4) we have

$$g(\eta) = G(1/\eta) = \sum_{1\leqslant k\leqslant 1/\eta} F\left(\frac{1}{\eta k}\right)H(k) = \sum_{1\leqslant k\leqslant 1/\eta} f(\eta k)H(k),$$

$$f(\eta) = F(1/\eta) = \sum_{1 \leqslant k \leqslant 1/\eta} \mu(k) G\left(\frac{1}{\eta k}\right) H(k) = \sum_{1 \leqslant k \leqslant 1/\eta} \mu(k) g(\eta k) H(k).$$

These are just formulae (1) and (2) with $\eta_1 = 1 \geqslant 1/\xi_0 = \eta_0$. □

We now apply this to the following:

Theorem 3.4. *When* $\xi \geqslant 1$ *we have*

$$\left| \sum_{1 \leqslant k \leqslant \xi} \frac{\mu(k)}{k} \right| \leqslant 1. \tag{5}$$

Proof. In (3) we set $F(\xi) = H(k) = 1$ so that $G(\xi) = [\xi]$. From (4) we have

$$1 = \sum_{1 \leqslant k \leqslant \xi} \mu(k) \left[\frac{\xi}{k}\right]. \tag{6}$$

If $1 \leqslant \xi < 2$, then (5) clearly holds. Suppose now that $\xi \geqslant 2$, and let $x = [\xi]$. Then

$$\begin{aligned} \left| x \sum_{k=1}^{x} \frac{\mu(k)}{k} - 1 \right| &= \left| \sum_{k=1}^{x} \mu(k) \left(\frac{x}{k} - \left[\frac{x}{k}\right]\right) \right| \\ &= \left| \sum_{k=2}^{x} \mu(k) \left(\frac{x}{k} - \left[\frac{x}{k}\right]\right) \right| \leqslant \sum_{k=2}^{x} 1 = x - 1. \end{aligned}$$

Therefore

$$x \left| \sum_{k=1}^{x} \frac{\mu(k)}{k} \right| \leqslant 1 + (x - 1) = x,$$

and the required result follows. □

6.4 The Möbius Transformation

Another consequence of Theorem 3.3 is the following:

Theorem 4.1. *Let $h(k)$ be a completely multiplicative function which is not identically zero, and let n_0 be a positive integer. If for all n satisfying $1 \leqslant n \leqslant n_0$, we have*

$$g(n) = \sum_{d \mid n} f(d) h\left(\frac{n}{d}\right), \tag{1}$$

then, for such n, we have

$$f(n) = \sum_{d \mid n} \mu(d) g\left(\frac{n}{d}\right) h(d); \tag{2}$$

the converse also holds.

Proof. We define $F(\xi)$ by setting $F(\xi) = f(\xi)$ when ξ is an integer and $F(\xi) = 0$ if ξ is not an integer, and we define $G(\xi)$ similarly. We can rewrite (1) and (2) as

$$G(n) = g(n) = \sum_{d|n} f(d)h\left(\frac{n}{d}\right) = \sum_{k|n} f\left(\frac{n}{k}\right)h(k) = \sum_{1 \leqslant k \leqslant n} F\left(\frac{n}{k}\right)h(k)$$

and

$$F(n) = f(n) = \sum_{d|n} \mu(d)g\left(\frac{n}{d}\right)h(d) = \sum_{d|n} \mu(d)G\left(\frac{n}{d}\right)h(d)$$

$$= \sum_{1 \leqslant d \leqslant n} \mu(d)G\left(\frac{n}{d}\right)h(d).$$

From the definition of $F(\xi)$ and $G(\xi)$ these two formulae can also be written as

$$G(\xi) = \sum_{1 \leqslant k \leqslant \xi} F\left(\frac{\xi}{k}\right)h(k),$$

$$F(\xi) = \sum_{1 \leqslant k \leqslant \xi} \mu(k)G\left(\frac{\xi}{k}\right)h(k).$$

Here ξ satisfies $1 \leqslant \xi \leqslant n_0$. Conversely (1) and (2) can be deduced from these formulae. The theorem now follows from Theorem 3.3 with $\xi_0 = n_0$. □

Definition. If

$$g(n) = \sum_{d|n} f(d) = \sum_{d|n} f\left(\frac{n}{d}\right),$$

then we call $g(n)$ the *Möbius transform* of $f(n)$. We also call $f(n)$ the *inverse Möbius transform* of $g(n)$.

From Theorem 4.1 we have

$$f(n) = \sum_{d|n} \mu(d)g\left(\frac{n}{d}\right) = \sum_{d|n} \mu\left(\frac{n}{d}\right)g(d).$$

From Theorem 2.2 we see that the Möbius transform, and the inverse Möbius transform, of a multiplicative function is multiplicative.

Example 1. From Theorem 3.1 we see that $\Delta(n)$ is the Möbius transform of $\mu(n)$.

Example 2. From $\sigma_\lambda(n) = \sum_{d|n} d^\lambda$, we see that $\sigma_\lambda(n)$ is the Möbius transform of the multiplicative function $E_\lambda(n) = n^\lambda$, and therefore $\sigma_\lambda(n)$ is a multiplicative function. Since

$$\sigma_\lambda(p^l) = \sum_{m=0}^{l} p^{m\lambda} = \frac{p^{\lambda(l+1)} - 1}{p^\lambda - 1} \qquad (\lambda \neq 0),$$

we deduce that if $n = \prod_v p_v^{l_v}$, then

$$\sigma_\lambda(n) = \prod_v \frac{p_v^{\lambda(l_v+1)} - 1}{p_v^\lambda - 1}.$$

In particular, when $\lambda = 0$, we have

$$d(n) = \sigma_0(n) = \prod_v (l_v + 1),$$

which we already proved in an earlier exercise.

Example 3. The function $E_0(n) = 1$ is the Möbius transform of $\Delta(n)$.

Example 4. Let n be fixed and let the integers $1, 2, \ldots, a, \ldots, n$ be partitioned into distinct classes according to the value of the greatest common divisor (n, a). If $d = (n, a)$, then we can write $n = dk$ and $1 = (k, a/d)$. Now the number of integers a satisfying $1 = (k, a/d)$ is precisely $\varphi(n/d)$ and we therefore deduce that

$$n = \sum_{d|n} \varphi\left(\frac{n}{d}\right) = \sum_{d|n} \varphi(d).$$

In other words, the function $E_1(n) = n$ is the Möbius transform of $\varphi(n)$. From this we deduce a result in §5, Chapter 2, namely that $\varphi(n)$ is multiplicative. Moreover, from the Möbius inversion formula we have:

Theorem 4.2.

$$\varphi(n) = n \sum_{d|n} \frac{\mu(d)}{d}. \quad \square$$

Example 5. More generally we denote by $\varphi_\lambda(n)$ the inverse Möbius transform of $E_\lambda(n)$ so that $\varphi_1(n) = \varphi(n)$. Then $\varphi_\lambda(n)$ is a multiplicative function. Also, when $n = \prod_v p_v^{l_v}$, we have

$$\varphi_\lambda(n) = n^\lambda \sum_{d|n} \frac{\mu(d)}{d^\lambda} = n^\lambda \prod_{p|n} \left(1 - \frac{1}{p^\lambda}\right).$$

We leave the verification for this to the reader.

Example 6. Consider a prime modulus p. Let the polynomial $x^{p^n} - x$ be factorized into a product of irreducible factors. If m is the degree of one of its factors, then we know that $m|n$. Conversely any irreducible polynomial of degree m must be one of its factors. Denote by Φ_n the number of irreducible polynomials (mod p) of degree n. Then, concerning the degree of a polynomial, we have the formula

$$p^n = \sum_{m|n} m\Phi_m.$$

That is, the function p^n is the Möbius transform of $n\Phi_n$. From the inversion formula, we have

$$n\Phi_n = \sum_{m|n} \mu(m) p^{n/m},$$

which gives another proof of Theorem 4.9.2.

Example 7. We seek the Möbius transform of $\Lambda(n)$. Let $n = p_1^{l_1} \cdots p_v^{l_v}$ be the standard factorization of n. Then

$$\begin{aligned}
\sum_{d|n} \Lambda(d) &= \sum_{s_1=0}^{l_1} \cdots \sum_{s_r=0}^{l_r} \Lambda(p_1^{s_1} \cdots p_r^{s_r}) \\
&= \sum_{s_1=1}^{l_1} \Lambda(p_1^{s_1}) + \cdots + \sum_{s_r=1}^{l_r} \Lambda(p_r^{s_r}) \\
&= \sum_{s_1=1}^{l_r} \log p_1 + \cdots + \sum_{s_r=1}^{l_r} \log p_r \\
&= l_1 \log p_1 + \cdots + l_r \log p_r \\
&= \log n,
\end{aligned}$$

that is $\log n$ is the Möbius transform of $\Lambda(n)$.

Example 8. Since $\Lambda(n)$ is the inverse Möbius transform of $\log n$, it follows that

$$\begin{aligned}
\Lambda(n) = \sum_{d|n} \mu(d) \log n/d &= \log n \sum_{d|n} \mu(d) - \sum_{d|n} \mu(d) \log d \\
&= \Delta(n) \log n - \sum_{d|n} \mu(d) \log d.
\end{aligned}$$

Since $\Delta(n) \log n$ is always zero, it follows that $\Lambda(n)$ is the Möbius transform of $-\mu(n) \log n$.

Collecting our results we have the following table, where $g(n)$ represents the Möbius transform of $f(n)$.

$f(n)$	$\mu(n)$	$\Delta(n)$	$\varphi_\lambda(n)$	$E_\lambda(n)$	$-\mu(n) \log n$	$\Lambda(n)$
$g(n)$	$\Delta(n)$	$E_0(n)$	$E_\lambda(n)$	$\sigma_\lambda(n)$	$\Lambda(n)$	$\log n$

Exercise 1. Let $g(n)$ and $g_1(n)$ be the Möbius transforms of $f(n)$ and $f_1(n)$ respectively. Prove that

$$\sum_{d|n} g(d) f_1\left(\frac{n}{d}\right) = \sum_{d|n} f(d) g_1\left(\frac{n}{d}\right).$$

Exercise 2. Evaluate the inverse Möbius transform of $g(n)g_1(n)$.

Exercise 3. The Möbius transform of the Möbius transform of $f(n)$ is given by

$$\sum_{a|n} f(a) d\left(\frac{n}{a}\right).$$

Exercise 4. Use the method of Example 6 to prove formula (1) of §10, Chapter 4.

6.5 The Divisor Function

Theorem 5.1. *We have, for all positive integers m, n,*

$$d(m, n) \leqslant d(m)d(n).$$

Proof. If p is a prime, then

$$d(p^a \cdot p^b) = d(p^{a+b}) = a + b + 1 \leqslant (a + 1)(b + 1) = d(p^a)d(p^b).$$

Since $d(n)$ is a multiplicative function, the result follows. □

Theorem 5.2. *Let* $\varepsilon > 0$. *Then*

$$d(n) = O(n^\varepsilon). \tag{1}$$

Here the O-constant depends on ε.

Proof. Let $n = \prod_{p|n} p^a$ be the standard factorization of n. We have

$$p^{a\varepsilon} \geqslant 2^{a\varepsilon} = e^{a\varepsilon \log 2} \geqslant a\varepsilon \log 2 \geqslant \tfrac{1}{2}(a + 1)\varepsilon \log 2.$$

If $p^\varepsilon \geqslant 2$, then $p^{a\varepsilon} \geqslant 2^a \geqslant a + 1$. Therefore

$$\frac{d(n)}{n^\varepsilon} = \prod_{p|n} \frac{a+1}{p^{a\varepsilon}} = \prod_{\substack{p|n \\ p^\varepsilon < 2}} \frac{a+1}{p^{a\varepsilon}} \prod_{\substack{p|n \\ p^\varepsilon \geqslant 2}} \frac{a+1}{p^{a\varepsilon}}$$

$$\leqslant \prod_{\substack{p|n \\ p^\varepsilon < 2}} \frac{a+1}{\frac{1}{2}(a+1)\varepsilon \log 2} \prod_{\substack{p|n \\ p^\varepsilon \geqslant 2}} \frac{a+1}{a+1} \leqslant \prod_{p^\varepsilon < 2} \frac{2}{\varepsilon \log 2},$$

and the required result follows. □

Theorem 5.3. *Let q be a non-negative integer and* $\xi \geqslant 2$. *Then*

$$\sum_{1 \leqslant n \leqslant \xi} (d(n))^q = O(\xi(\log \xi)^{2^q - 1}), \tag{2}$$

$$\sum_{1 \leqslant n \leqslant \xi} \frac{(d(n))^q}{n} = O((\log \xi)^{2^q}). \tag{3}$$

Proof. We first prove (3) by induction on q. We know that the result holds when $q = 0$, and we now assume that it holds when q is replaced by $q - 1$. Then

$$\sum_{1 \leqslant n \leqslant \xi} \frac{(d(n))^q}{n} = \sum_{1 \leqslant n \leqslant \xi} \frac{(d(n))^{q-1}}{n} \sum_{u|n} 1$$

$$= \sum_{1 \leqslant u \leqslant \xi} \sum_{\substack{1 \leqslant n \leqslant \xi \\ u|n}} \frac{(d(n))^{q-1}}{n}.$$

Let $n = uv$ and using $d(uv) \leqslant d(u)d(v)$ we see that

$$\sum_{1 \leqslant n \leqslant \xi} \frac{(d(n))^q}{n} \leqslant \sum_{1 \leqslant u \leqslant \xi} \frac{(d(u))^{q-1}}{u} \sum_{1 \leqslant v \leqslant \xi/u} \frac{(d(v))^{q-1}}{v}$$
$$= O((\log \xi)^{2^q}).$$

To prove (2) we again use induction on q:

$$\begin{aligned}\sum_{1 \leqslant n \leqslant \xi} (d(n))^q &= \sum_{1 \leqslant n \leqslant \xi} (d(n))^{q-1} \sum_{u \mid n} 1 \\ &= \sum_{1 \leqslant u \leqslant \xi} \sum_{\substack{1 \leqslant n \leqslant \xi \\ u \mid n}} (d(n))^{q-1} \\ &\leqslant \sum_{1 \leqslant u \leqslant \xi} (d(u))^{q-1} \sum_{1 \leqslant v \leqslant \xi/u} (d(v))^{q-1} \\ &\leqslant \xi \sum_{1 \leqslant u \leqslant \xi} \frac{(d(u))^{q-1}}{u} O((\log \xi)^{2^{q-1}-1}) \\ &= O(\xi(\log \xi)^{2^q - 1}). \quad \square\end{aligned}$$

This theorem can be made much sharper. We give only a very important special case as an example.

Theorem 5.4. *If* $\xi \geqslant 1$, *then*

$$\sum_{1 \leqslant n \leqslant \xi} d(n) = \xi \log \xi + (2\gamma - 1)\xi + O(\sqrt{\xi}),$$

where γ *is Euler's constant.*

Proof. We have

$$\sum_{1 \leqslant n \leqslant \xi} d(n) = \sum_{1 \leqslant n \leqslant \xi} \sum_{u \mid n} 1 = \sum_{1 \leqslant uv \leqslant \xi} 1.$$

In other words $\sum_{1 \leqslant n \leqslant \xi} d(n)$ is the number of lattice points in the first quadrant which lie below the rectangular hyperbola $uv = \xi$. By a lattice point we mean a point with integer coordinates.

By erecting two perpendiculars to the axes passing through the point $(\sqrt{\xi}, \sqrt{\xi})$ the region concerned is divided into a square together with two regions each having the same number of lattice points inside. That is

$$\begin{aligned}\sum_{1 \leqslant uv \leqslant \xi} 1 &= [\sqrt{\xi}]^2 + 2 \sum_{u=1}^{[\sqrt{\xi}]} \sum_{[\sqrt{\xi}] < v \leqslant \xi/u} 1 \\ &= -[\sqrt{\xi}]^2 + 2 \sum_{u=1}^{[\sqrt{\xi}]} \left[\frac{\xi}{u}\right] \\ &= -\xi + O(\sqrt{\xi}) + 2 \sum_{u=1}^{\sqrt{\xi}} \frac{\xi}{u} + O(\sqrt{\xi}).\end{aligned}$$

Since

$$\sum_{u=1}^{\sqrt{\xi}} \frac{1}{u} = \frac{1}{2}\log \xi + \gamma + O\left(\frac{1}{\sqrt{\xi}}\right),$$

it follows that

$$\sum_{1 \leqslant n \leqslant \xi} d(n) = \xi \log \xi + (2\gamma - 1)\xi + O(\sqrt{\xi}).$$

Exercise 1. Prove that, for $\xi \geqslant 2$,

$$\sum_{1 \leqslant n \leqslant \xi} \frac{d(n)}{n} = \frac{1}{2}\log^2 \xi + 2\gamma \log \xi + c + O(\xi^{-\frac{1}{2}} \log \xi).$$

Exercise 2. Prove that, for any positive ε, we have

$$\sigma(n) = O(n^{1+\varepsilon}).$$

Exercise 3. Prove that, for $\xi \geqslant 2$,

$$\sum_{1 \leqslant n \leqslant \xi} \sigma(n) = \frac{1}{12}\pi^2\xi^2 + O(\xi \log \xi).$$

(The reader may use the result $\sum_{n=1}^{\infty} 1/n^2 = \pi^2/6$, a formula which will be proved in Exercise 8.7.1.)

6.6 Two Theorems Related to Asymptotic Densities

Definition 1. Let there be a set of positive integers, and denote by $N(x)$ the number of elements in the set not exceeding x. Suppose that

$$\lim_{x \to \infty} \frac{N(x)}{x} = \alpha.$$

Then we say that the set has asymptotic density α:

Examples. The set of odd positive integers has asymptotic density $\frac{1}{2}$. The set of all perfect squares has asymptotic density 0.

In this section we shall use the result

$$\sum_{n=1}^{\infty} \frac{\mu(n)}{n^2} = \frac{6}{\pi^2}, \tag{1}$$

the proof of which is given in Exercise 8.7.1.

Definition 2. A positive integer which is not divisible by any prime square is called a *square-free number*.

The set of square-free numbers has asymptotic density $6/\pi^2$. More precisely we have

Theorem 6.1. *Let $Q(x)$ denote the number of square free numbers not exceeding x. Then, as $x \to \infty$,*

$$Q(x) = \frac{6x}{\pi^2} + O(\sqrt{x}). \tag{2}$$

Proof. We partition the set of positive integers not exceeding x into subsets according to their largest square divisor q^2. The number of positive integers not exceeding x having largest square divisor q^2 is $Q(x/q^2)$ so that

$$[x] = \sum_{q=1}^{[\sqrt{x}]} Q\left(\frac{x}{q^2}\right).$$

Let $x = y^2$. Then

$$[y^2] = \sum_{q=1}^{[y]} Q\left(\left(\frac{y}{q}\right)^2\right).$$

From Theorem 3.3 we have

$$\begin{aligned} Q(y^2) &= \sum_{1 \leqslant k \leqslant y} \mu(k)\left[\frac{y^2}{k^2}\right] \\ &= y^2 \sum_{1 \leqslant k \leqslant y} \frac{\mu(k)}{k^2} + \sum_{1 \leqslant k \leqslant y} O(1) \\ &= \frac{6}{\pi^2} y^2 + y^2 O\left(\sum_{k > y} \frac{1}{k^2}\right) + O(y) \\ &= \frac{6}{\pi^2} y^2 + O(y), \end{aligned}$$

where we have used formula (5.8.8). The required result follows. □

We can restate Theorem 6.1 as:

Theorem 6.2. *If $x \geqslant 1$, then*

$$\sum_{n \leqslant x} |\mu(n)| = \frac{6x}{\pi^2} + O(\sqrt{x}). \quad \square \tag{3}$$

The number of pairs of integers x, y satisfying $1 \leqslant x \leqslant y \leqslant n$ is equal to $n(n+1)/2$. Let us denote by $\Phi(n)$ the number of those pairs satisfying $(x, y) = 1$. We can prove that

$$\lim_{n \to \infty} \frac{\Phi(n)}{\frac{1}{2}n(n+1)} = \frac{6}{\pi^2}.$$

We can interpret this result by saying that the probability that two given integers are coprime is $6/\pi^2$. Here we prove a sharper theorem.

Theorem 6.3.

$$\Phi(n) = \sum_{m \leqslant n} \varphi(n) = \frac{3n^2}{\pi^2} + O(n \log n).$$

Proof. We have

$$\begin{aligned}
\Phi(n) &= \sum_{m=1}^{n} m \sum_{d|m} \frac{\mu(d)}{d} = \sum_{dd' \leqslant n} d'\mu(d) \\
&= \sum_{d=1}^{n} \mu(d) \sum_{d'=1}^{[n/d]} d' = \frac{1}{2} \sum_{d=1}^{n} \mu(d) \left(\left[\frac{n}{d}\right]^2 + \left[\frac{n}{d}\right] \right) \\
&= \frac{1}{2} \sum_{d=1}^{n} \mu(d) \left(\frac{n^2}{d^2} + O\left(\frac{n}{d}\right) \right) \\
&= \frac{1}{2} n^2 \sum_{d=1}^{n} \frac{\mu(d)}{d^2} + O\left(n \sum_{d=1}^{n} \frac{1}{d} \right) \\
&= \frac{1}{2} n^2 \sum_{d=1}^{\infty} \frac{\mu(d)}{d^2} + O\left(n^2 \sum_{n+1}^{\infty} \frac{1}{d^2} \right) + O(n \log n) \\
&= \frac{3n^2}{\pi^2} + O(n) + O(n \log n) \\
&= \frac{3n^2}{\pi^2} + O(n \log n)
\end{aligned}$$

as required. □

6.7 The Representation of Integers as a Sum of Two Squares

We first introduce the function

$$\chi(n) = \begin{cases} 0, & \text{if } 2|n, \\ (-1)^{\frac{1}{2}(n-1)}, & \text{if } 2\nmid n. \end{cases}$$

It is easy to verify that $\chi(n)$ is multiplicative. We write

$$\delta(n) = \sum_{d|n} \chi(d),$$

the Möbius transform of $\chi(n)$, so that $\delta(n)$ is also multiplicative. If $n = \prod_{p|n} p^l$ is the standard factorization of n, then

$$\delta(n) = \prod_{p|n} (1 + \chi(p) + \chi(p^2) + \cdots + \chi(p^l)).$$

Using the function $\chi(n)$ we can restate Theorem 3.5.1 as follows:

Theorem 7.1. *Let $V(n)$ denote the number of solutions to the congruence $x^2 \equiv -1 \pmod{n}$. Then*

$$V(n) = \begin{cases} 0, & \text{if } 4|n, \\ \prod_{p|n} (1 + \chi(p)), & \text{if } 4\nmid n. \end{cases}$$

In the product here p runs through all the distinct prime divisors of n. □

It is not difficult to deduce this theorem from Theorem 3.5.1 and Theorem 2.8.1. The main aim of this section is to prove:

Theorem 7.2. *Let $r(n)$ denote the number of solutions to the equation $n = x^2 + y^2$ in integers x, y. Then*

$$r(n) = 4\delta(n).$$

We shall require two auxiliary results for the proof of this theorem.

Theorem 7.3. *We have the identity*

$$(x_1^2 + y_1^2)(x_2^2 + y_2^2) = (x_1x_2 + y_1y_2)^2 + (x_1y_2 - y_1x_2)^2.$$

Proof. Direct multiplication gives the result at once. □

Exercise 1. Prove the identity:

$$\begin{aligned}(x_1^2 + x_2^2 + x_3^2 + x_4^2)&(y_1^2 + y_2^2 + y_3^2 + y_4^2)\\ &= (x_1y_1 + x_2y_2 + x_3y_3 + x_4y_4)^2 + (x_1y_2 - x_2y_1 + x_3y_4 - x_4y_3)^2\\ &\quad + (x_1y_3 - x_3y_1 + x_4y_2 - x_2y_4)^2 + (x_1y_4 - x_4y_1 + x_2y_3 - x_3y_2)^2.\end{aligned}$$

Exercise 2. Prove the identity:

$$\begin{aligned}(x_1^2 + x_2^2 + x_3^2 + x_4^2 &+ x_5^2 + x_6^2 + x_7^2 + x_8^2)\\ &\times (y_1^2 + y_2^2 + y_3^2 + y_4^2 + y_5^2 + y_6^2 + y_7^2 + y_8^2)\\ &= (x_1y_1 + x_2y_2 + x_3y_3 + x_4y_4 + x_5y_5 + x_6y_6 + x_7y_7 + x_8y_8)^2\\ &\quad + (x_1y_2 - x_2y_1 - x_3y_4 + x_4y_3 - x_5y_6 + x_6y_5 - x_7y_8 + x_8y_7)^2\\ &\quad + (x_1y_3 + x_2y_4 - x_3y_1 - x_4y_2 + x_5y_7 - x_6y_8 - x_7y_5 + x_8y_6)^2\\ &\quad + (x_1y_4 - x_2y_3 + x_3y_2 - x_4y_1 - x_5y_8 - x_6y_7 + x_7y_6 + x_8y_5)^2\\ &\quad + (x_1y_5 + x_2y_6 - x_3y_7 + x_4y_8 - x_5y_1 - x_6y_2 + x_7y_3 - x_8y_4)^2\\ &\quad + (x_1y_6 - x_2y_5 + x_3y_8 + x_4y_7 + x_5y_2 - x_6y_1 - x_7y_4 - x_8y_3)^2\\ &\quad + (x_1y_7 + x_2y_8 + x_3y_5 - x_4y_6 - x_5y_3 + x_6y_4 - x_7y_1 - x_8y_2)^2\\ &\quad + (x_1y_8 - x_2y_7 - x_3y_6 - x_4y_5 + x_5y_4 + x_6y_3 + x_7y_2 - x_8y_1)^2.\end{aligned}$$

Theorem 7.4. *Let $n > 1$ be such that the congruence*

$$l^2 \equiv -1 \pmod{n} \tag{1}$$

has a solution. Then there exists a unique pair of integers x, y satisfying

$$x^2 + y^2 = n, \quad x > 0, \quad y > 0, \quad (x, y) = 1, \quad y \equiv lx \pmod{n}. \tag{2}$$

Proof. Clearly if (2) is soluble, then so is (1). A necessary condition for (1) to be soluble is that n is representable as

$$n = 2^a p_1^{l_1} \cdots p_s^{l_s}, \qquad a = 0 \text{ or } 1,$$

and p_i $(i = 1, 2, \ldots, s)$ is a prime $\equiv 1 \pmod 4$. We now use induction to prove the theorem.

1) We consider first the case $n = p^\lambda$. If $\lambda = 1$, then from $l^2 + 1 \equiv 0 \pmod p$ we see that when $(x, p) = 1$, we have $x^2 l^2 + x^2 \equiv 0 \pmod p$. We shall presently choose y and x so that $x^2 l^2 \equiv y^2 \pmod p$, and $x^2 < p$, $y^2 < p$. Let x and y take the values $0, 1, \ldots, [\sqrt{p}]$ and consider the various differences $xl - y$. Since there are $([\sqrt{p}] + 1)^2 > p$ such differences, there must be two which are congruent mod p. Let $x_1 l - y_1 \equiv x_2 l - y_2 \pmod p$, or $(x_1 - x_2) l \equiv y_1 - y_2 \pmod p$, and we can assume that $x_1 - x_2 > 0$ so that $x_1 - x_2 < \sqrt{p}$, $|y_1 - y_2| < \sqrt{p}$ and this then gives our desired x and y. For this pair x, y we have $x^2 + y^2 = tp$, and it is easy to see that $t = 1$, $(x, y) = 1$.

The congruence $y \equiv mx \pmod p$ is soluble, and from $x^2(1 + m^2) \equiv 0 \pmod p$ we see that $m \equiv \pm l$. If $m = l$, then we take the pair (x, y), while if $m = -l$, then we take the pair (y, x).

Now assume that $p \neq 2$ and that the theorem holds for $n = p^\lambda$. Let $(-l)^2 \equiv -1 \pmod{p^{\lambda+1}}$ so that there exist u, v such that

$$p^\lambda = u^2 + v^2, \quad u > 0, \quad v > 0, \quad (u, v) = 1, \quad v \equiv -lu \pmod{p^\lambda}.$$

When $n = p^{\lambda+1}$, we have

$$p^{\lambda+1} = (xu + yv)^2 + (xv - yu)^2 = X^2 + Y^2 \quad (X > 0,\ Y > 0).$$

First we have $(X, Y) = 1$, since otherwise $p | (X, Y)$, but

$$X \equiv xu + yv \equiv xu - l^2 xu \equiv xu(1 - l^2) \not\equiv 0 \pmod p,$$

which is impossible.

Next, because $(X, p) = 1$, the congruence $Xm \equiv Y \pmod{p^{\lambda+1}}$ is soluble. Thus $X^2 + Y^2 m^2 \equiv 0 \pmod{p^{\lambda+1}}$ or $1 + m^2 \equiv 0 \pmod{p^{\lambda+1}}$. From Theorem 2.9.3 this congruence has only two solutions, so that $m = \pm l$. The desired result follows from the discussion in the case $\lambda = 1$.

2) Let $n = ab$, $a > 1$, $b > 1$, $(a, b) = 1$, and suppose that

$$l^2 \equiv -1 \pmod{n},$$

$$u^2 + v^2 = a, \quad u > 0, \quad v > 0, \quad (u, v) = 1, \quad v \equiv lu \pmod{a},$$

$$x^2 + y^2 = b, \quad x > 0, \quad y > 0, \quad (x, y) = 1, \quad y \equiv lx \pmod{b}.$$

From Theorem 7.3 we have

$$n = ab = (xv + yu)^2 + (xu - yv)^2 = X^2 + Y^2.$$

(If $xu - yv > 0$, then let $xu - yv = Y$; otherwise we let $xu - yv = -Y$.)

We now prove the following:

(i) $(X, Y) = 1$. Let $p|(X, Y)$. Then

$$xv + yu = ps,$$

$$xu - yv = pt,$$

or

$$x(u^2 + v^2) = p(sv + tu),$$

$$y(u^2 + v^2) = p(su - tv).$$

Since $(x, y) = 1$, we must have $p|(u^2 + v^2)$, that is $p|a$. Similarly $p|b$. But this contradicts $(a, b) = 1$.

(ii) $X \equiv lY \pmod{n}$. From our assumption we have

$$xv + yu \equiv lxu - lyv \equiv l(xu - yv) \pmod{a},$$

$$xv + yu \equiv -lyv + lxu \equiv l(xu - yv) \pmod{b}.$$

Since $(a, b) = 1$, it follows that $X \equiv lY \pmod{n}$.

3) Uniqueness. Suppose that there are two pairs (X, Y), (X', Y') both satisfying the conditions. Then

$$n^2 = (XX' + YY')^2 + (XY' - YX')^2.$$

But

$$XX' + YY' \equiv XX'(1 + l^2) \equiv 0 \pmod{n},$$

so that

$$XX' + YY' = n, \quad XY' - YX' = 0.$$

From $XY' - YX' = 0$, we have

$$\frac{X}{X'} = \frac{Y}{Y'} = C,$$

so that $X^2 + Y^2 = C^2(X'^2 + Y'^2)$ giving $C = \pm 1$. Also from $X > 0$, $X' > 0$ we see that $C = 1$. The proof of our theorem is complete. □

Proof of Theorem 7.2. From Theorem 7.1 and Theorem 7.4 we see that the number of solutions to $x^2 + y^2 = n$, $(x, y) = 1$ is $4V(n)$. We now consider the equation $x^2 + y^2 = n$, and we partition the various solutions into sets according to $(x, y) = d$. The number of solutions satisfying $(x, y) = d$ is equal to the number of solutions satisfying

$$\left(\frac{x}{d}\right)^2 + \left(\frac{y}{d}\right)^2 = \frac{n}{d^2},$$

that is $4V(n/d^2)$. Therefore

$$r(n) = 4 \sum_{d^2|n} V\left(\frac{n}{d^2}\right) = 4 \sum_{d|n} V\left(\frac{n}{d}\right) \lambda(d),$$

where $\lambda(d) = 1$ or 0 according to whether d is a square or not. Since $V(n)$ and $\lambda(n)$ are both multiplicative it follows that $r(n)/4$ is multiplicative.

Since $\delta(n)$ is also multiplicative the theorem will follow if we show that $r(n) = 4\delta(n)$ when $n = p^l$. Now, if $2|m$, then

$$\frac{r(p^m)}{4} = V(p^m) + V(p^{m-2}) + \cdots + V(p^2) + V(1)$$

$$= \begin{cases} 0 + \cdots + 0 + 1 = 1, & \text{if } p = 2, \\ 0 + \cdots + 0 + 1 = 1, & \text{if } p \equiv 3 \pmod 4, \\ 2 + \cdots + 2 + 1 = & \\ \quad = \dfrac{m}{2} \cdot 2 + 1 = m + 1, & \text{if } p \equiv 1 \pmod 4, \end{cases}$$

and if $2 \nmid m$, then

$$\frac{r(p^m)}{4} = V(p^m) + \cdots + V(p)$$

$$= \begin{cases} 1, & \text{if } p = 2, \\ 0, & \text{if } p \equiv 3 \pmod 4, \\ m + 1, & \text{if } p \equiv 1 \pmod 4. \end{cases}$$

On the other hand we have

$$\delta(p^m) = 1 + \chi(p) + \cdots + \chi(p^m)$$

$$= \begin{cases} 1 + 0 + 0 + \cdots + 0 = 1, & \text{if } p = 2, \\ 1 - 1 + \cdots + 1 = 1, & \text{if } p \equiv 3 \pmod 4, \ 2|m, \\ 1 - 1 + \cdots - 1 = 0, & \text{if } p \equiv 3 \pmod 4, \ 2\nmid m, \\ 1 + 1 + \cdots + 1 = m + 1, & \text{if } p \equiv 1 \pmod 4. \end{cases}$$

The theorem is proved. □

Theorem 7.5. *Denote by A and B the number of divisors of n which are congruent* 1 *and* 3 (mod 4) *respectively. Then* $r(n) = 4(A - B)$.

Proof. This is an immediate consequence of Theorem 7.2. □

Theorem 7.6. *Let* $\varepsilon > 0$. *Then*

$$r(n) = O(n^{\varepsilon}).$$

Proof. Since $r(n) \leqslant 4d(n)$, the required result follows from Theorem 5.2. □

6.8 The Methods of Partial Summation and Integration

Theorem 8.1 (Abel). *Let* $a \leqslant b$ *and let n vary in* $a \leqslant n \leqslant b$. *Let* γ_n *and* ε_n *be complex numbers and*

$$s_n = \sum_{a \leqslant m \leqslant n} \gamma_m.$$

Then

$$\left|\sum_{n=a}^{b} \gamma_n \varepsilon_n\right| \leqslant \max_{a \leqslant n \leqslant b} |s_n| \left(\sum_{a \leqslant m \leqslant b-1} |\varepsilon_m - \varepsilon_{m+1}| + |\varepsilon_b| \right). \tag{1}$$

Proof. Let $S_{a-1} = 0$. Then

$$\begin{aligned}\sum_{n=a}^{b} \gamma_n \varepsilon_n &= \sum_{n=a}^{b} (s_n - s_{n-1}) \varepsilon_n \\ &= \sum_{n=a}^{b} s_n \varepsilon_n - \sum_{n=a}^{b-1} s_n \varepsilon_{n+1} \\ &= \sum_{n=a}^{b-1} s_n (\varepsilon_n - \varepsilon_{n+1}) + s_b \varepsilon_b,\end{aligned}$$

so that

$$\begin{aligned}\left|\sum_{n=a}^{b} \gamma_n \varepsilon_n\right| &\leqslant \sum_{n=a}^{b-1} |s_n| \, |\varepsilon_n - \varepsilon_{n+1}| + |s_b| \, |\varepsilon_b| \\ &\leqslant \max_{a \leqslant n \leqslant b} |s_n| \left(\sum_{a \leqslant n \leqslant b-1} |\varepsilon_n - \varepsilon_{n+1}| + |\varepsilon_b| \right). \quad \square\end{aligned}$$

Theorem 8.2. *In the previous theorem if* ε_n *is a positive decreasing sequence, then*

$$\left|\sum_{n=a}^{b} \gamma_n \varepsilon_n\right| \leqslant \max_{a \leqslant n \leqslant b} |s_n| \varepsilon_a. \quad \square \tag{2}$$

We now apply this to the following:

Theorem 8.3. *If $s > 0$, then*

$$\left|\sum_{n\geqslant a}\frac{\chi(n)}{n^s}\right| \leqslant \frac{1}{a^s},$$

so that the series $\sum_{n=1}^{\infty}\chi(n)/n^s$ converges when $s > 0$.

Proof. We have

$$\chi(a) + \chi(a+1) + \chi(a+2) + \chi(a+3) = 0,$$

so that

$$\left|\sum_{a\leqslant m\leqslant b}\chi(m)\right| \leqslant 1.$$

From Theorem 8.2 we deduce that

$$\left|\sum_{n=a}^{b}\frac{\chi(n)}{n^s}\right| \leqslant \frac{1}{a^s}.$$

Since the right hand side is independent of b, the theorem follows. □

Note: In the next section we shall require

$$\sum_{n=1}^{\infty}\frac{\chi(n)}{n} = 1 - \frac{1}{3} + \frac{1}{5} - \frac{1}{7} + \cdots = \frac{\pi}{4}.$$

This can be proved using the series expansion for $\tan^{-1} x$ in ordinary calculus.

Analogous to Theorems 8.1 and 8.2 we have:

Theorem 8.4. *Let $\xi \leqslant \eta$ and let x vary in $\xi \leqslant x \leqslant \eta$. Suppose that $f(x)$ and $g(x)$ are continuous and $g(x)$ is differentiable. Let*

$$f_1(x) = \int_{\xi}^{x} f(t)\,dt.$$

Then

$$\left|\int_{\xi}^{\eta} f(x)g(x)\,dx\right| \leqslant \max_{\xi\leqslant x\leqslant\eta}|f_1(x)|\left(\int_{\xi}^{\eta}|g'(x)|\,dx + |g(\eta)|\right).$$

Moreover, if $g'(x) \leqslant 0$ and $g(x) > 0$, then

$$\left|\int_{\xi}^{\eta} f(x)g(x)\,dx\right| \leqslant g(\xi)\max_{\xi\leqslant x\leqslant\eta}|f_1(x)|.$$

Proof. From integration by parts we have

$$\int_{\xi}^{\eta} f(x)g(x)\,dx = \int_{\xi}^{\eta} g(x)\,df_1(x)$$

$$= g(\eta)f_1(\eta) - \int_{\xi}^{\eta} f_1(x)g'(x)\,dx,$$

and hence

$$\left|\int_{\xi}^{\eta} f(x)g(x)\,dx\right| \leqslant \max_{\xi \leqslant x \leqslant \eta} |f_1(x)| \left(|g(\eta)| + \int_{\xi}^{\eta} |g'(x)|\,dx\right).$$

The last part of the theorem is also clear. □

Example. Let $a > 0$. Prove that

$$\left|\int_{a}^{\infty} \cos x^2\,dx\right| = \left|\int_{a^2}^{\infty} \frac{\cos y\,dy}{2y^{1/2}}\right| \leqslant \frac{1}{2a} \max_{a^2 \leqslant \eta} \left|\int_{a^2}^{\eta} \cos y\,dy\right| \leqslant \frac{\pi}{a}.$$

6.9 The Circle Problem

Theorem 9.1.

$$\sum_{1 \leqslant n \leqslant x} r(n) = \pi x + O(\sqrt{x}).$$

Proof. From Theorem 7.2 we have

$$\sum_{1 \leqslant n \leqslant x} r(n) = 4 \sum_{1 \leqslant n \leqslant x} \sum_{d \mid n} \chi(d)$$

$$= 4 \sum_{1 \leqslant d \leqslant x} \chi(d) \sum_{1 \leqslant n \leqslant x} 1$$

$$= 4 \sum_{1 \leqslant d \leqslant x} \chi(d) \left[\frac{x}{d}\right].$$

Here we divide the sum into two parts. From Theorem 8.3 we have

$$\sum\nolimits_1 = 4 \sum_{1 \leqslant d \leqslant \sqrt{x}} \chi(d) \left[\frac{x}{d}\right]$$

$$= 4x \sum_{1 \leqslant d \leqslant \sqrt{x}} \frac{\chi(d)}{d} + O(\sqrt{x})$$

$$= 4x \sum_{d=1}^{\infty} \frac{\chi(d)}{d} + O(\sqrt{x})$$

$$= \pi x + O(\sqrt{x});$$

the other part is

$$\sum\nolimits_2 = 4 \sum_{\sqrt{x} \leqslant d \leqslant x} \chi(d)\left[\frac{x}{d}\right],$$

and from Theorem 8.2 we have

$$\sum\nolimits_2 = O(\sqrt{x}).$$

The theorem is proved. □

Another proof of the theorem is the following: Clearly $\sum_{0 \leqslant n \leqslant x} r(n)$ is the number of pairs of integers u, v satisfying $u^2 + v^2 \leqslant x$. In other words the sum is the number of lattice points inside the circle centre at the origin with radius $\sqrt{x}$. This circle has area πx. We partition the plane into unit squares with orthogonal lines passing through the lattice points. To each point (u, v) in our circle we assign the square whose four corners have the coordinates (u, v), $(u + 1, v)$, $(u, v + 1)$, $(u + 1, v + 1)$. These squares must lie inside the circle $u^2 + v^2 = (\sqrt{x} + \sqrt{2})^2$ and they include the circle $u^2 + v^2 = (\sqrt{x} - \sqrt{2})^2$. Therefore

$$\pi(\sqrt{x} - \sqrt{2})^2 \leqslant \sum_{n \leqslant r} r(n) \leqslant \pi(\sqrt{x} + \sqrt{2})^2,$$

and the required result follows at once. We observe that this second proof can be used as a proof for

$$1 - \frac{1}{3} + \frac{1}{5} - \frac{1}{7} + \cdots = \frac{\pi}{4}.$$

Concerning the problem of the number of lattice points inside a closed curve, the Czech mathematician M. V. Jarnik proved the following:

Theorem 9.2. *Let $l \geqslant 1$ be the length of a rectifiable simple closed curve and let A be the area of the region bounded by the curve. If N is the number of lattice points inside the curve, then*

$$|A - N| < l.$$

Proof (Steinhaus). We first prove the following two simple lemmas.

Lemma 1. *Let C be a rectifiable curve inside a unit square with the two end points on the boundary of the square. If C crosses the two diagonals of the square, then its length must be at least* 1.

Proof. If the two end points are on the opposite sides of the square, then the result follows at once. Suppose next that the two end points are on two adjacent sides of

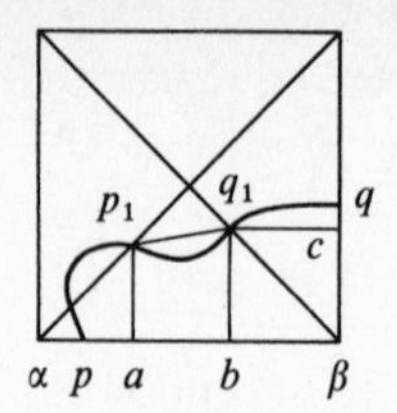

the square as shown in the diagram. It is easy to see that

$$l \geqslant \overline{ap_1} + \overline{p_1q_1} + \overline{q_1c} \geqslant \overline{\alpha a} + \overline{ab} + \overline{b\beta} = \overline{\alpha\beta} = 1.$$

A similar argument applies when the two end points are on the same side of the square.

Lemma 2. *Let C be a rectifiable curve inside a unit square with the two end points on the boundary of the square so that the square is partitioned into two regions. Suppose that C does not pass through the centre of the square, and denote by Δ the region which does not contain the centre. Then the area of Δ must be less than the length of C.*

Proof. We consider separately the cases shown in the following diagrams:

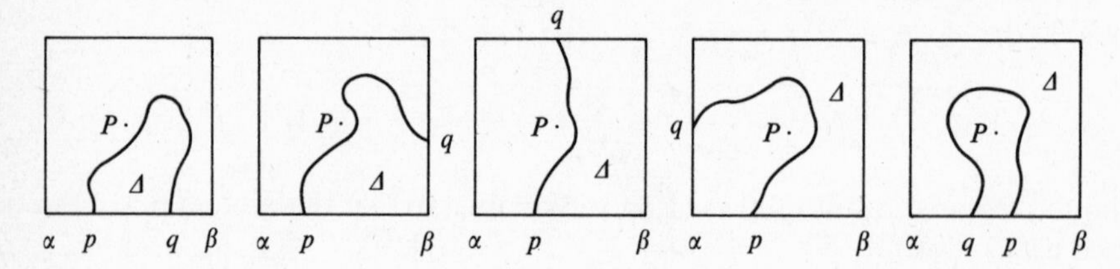

Let A be the area of the region Δ and l be the length of C. In the first two cases it is easy to see that every point of C is of distance at most l from the base line $\alpha\beta$ so that Δ must lie inside a rectangle with sides 1 and l and hence $A < l$. In the remaining three cases we see from Lemma 1 that $l \geqslant 1$ and so $A < 1 \leqslant l$.

We can now proceed to prove the theorem. Denote by I the region inside the curve. We form a net of unit squares in the plane with the lines

$$x = m + \tfrac{1}{2}, \qquad y = n + \tfrac{1}{2} \qquad (m, n = 0, \pm 1, \pm 2, \ldots).$$

Let $Q_1, Q_2, \ldots, Q_k$ be those squares which contain part of the boundary of I, let C_i be the part of the curve in Q_i, let Ω_i be the intersection of Q_i and I, and define

$$N_i = \begin{cases} 1, & \text{if } \Omega_i \text{ contains a lattice point,} \\ 0, & \text{otherwise.} \end{cases}$$

We let A_i be the area of Ω_i, l_i the length of C_i, so that our theorem will follow if we can prove that $|A_i - N_i| < l_i$.

Now the case when the whole of I lies inside a Q follows at once since $l \geqslant 1$. We can assume therefore that C_i is made up of a number of sections of the curve and Q_i is partitioned into regions $D_i^{(s)}$.

If the lattice point does not lie in any $D_i^{(s)}$ so that it lies on C_i, then $N_i = 0$, $0 < A_i < 1$ and $l_i \geqslant 1$ so that our required result follows. If the lattice point lies inside a $D_i^{(s)}$ we denote by $A_i^{(s)}$ the area of $D_i^{(s)}$. If $D_i^{(s)}$ is not in I, then $N_i = 0$, $A_i \leqslant 1 - A_i^{(s)}$; if $D_i^{(s)}$ is in I, then $N_i = 1$, $1 - A_i \leqslant 1 - A_i^{(s)}$ and, from Lemma 2, we have $1 - A_i^{(s)} < l_i$. The theorem is proved. □

It is clear that Theorem 9.1 is an immediate consequence of Theorem 9.2.

Exercise 1. Find the asymptotic formula for the number of lattice points inside an ellipse centre at the origin.

Exercise 2. Prove that the number of lattice points inside the sphere $u^2 + v^2 + w^2 \leqslant x$ is given by

$$\tfrac{4}{3}\pi x^{3/2} + O(x).$$

Exercise 3. Generalize the previous exercise to a sphere in n-dimensions.

Exercise 4. Determine the order of $\sum_{n \leqslant x} r^2(n)$.

Exercise 5. The number of lattice points inside the circle $u^2 + v^2 \leqslant x$ with coprime coordinates is given by

$$\frac{6}{\pi}x + O(\sqrt{x}\log x).$$

6.10 Farey Sequence and Its Applications

Farey sequence was discovered well over a hundred years ago, but its significance in number theory is revealed only in modern times.

Definition 1. By the *Farey sequence* of order n we mean the fractions in the interval from 0 to 1, whose denominators are $\leqslant n$, arranged in ascending order of magnitude. That is, they are numbers of the form

$$\frac{a}{b}, \qquad (a, b) = 1, \qquad 0 \leqslant a \leqslant b \leqslant n$$

arranged into an increasing sequence. We denote by $\mathfrak{F}_n$ the Farey sequence of order n.

Example: $\mathfrak{F}_7$ is the sequence

$$\tfrac{0}{1}, \tfrac{1}{7}, \tfrac{1}{6}, \tfrac{1}{5}, \tfrac{1}{4}, \tfrac{2}{7}, \tfrac{1}{3}, \tfrac{2}{5}, \tfrac{3}{7}, \tfrac{1}{2}, \tfrac{4}{7}, \tfrac{3}{5}, \tfrac{2}{3}, \tfrac{5}{7}, \tfrac{3}{4}, \tfrac{4}{5}, \tfrac{5}{6}, \tfrac{6}{7}, 1.$$

The total number of fractions in $\mathfrak{F}_n$ is $1 + \sum_{m=1}^{n} \varphi(m)$. These fractions divide the interval $0 \leqslant x \leqslant 1$ into $\sum_{m=1}^{n} \varphi(m)$ parts, and $\mathfrak{F}_{n+1}$ is obtained from adding the

$\varphi(n+1)$ numbers

$$\frac{a}{n+1}, \qquad (a, n+1) = 1, \qquad 0 < a \leqslant n.$$

Theorem 10.1. *Let ξ be an irrational number, $0 < \xi < 1$. Let a_m/b_m, a'_m/b'_m be two successive Farey fractions of order n satisfying*

$$\frac{a_m}{b_m} < \xi < \frac{a'_m}{b'_m}.$$

Then (i) *a_m/b_m is an increasing function of n, while a'_m/b'_m is a decreasing function of n, and*

$$\lim_{n\to\infty} \frac{a_m}{b_m} = \xi = \lim_{n\to\infty} \frac{a'_m}{b'_m};$$

(ii) *b_m and b'_m are increasing and unbounded functions of n.*

Proof. We note that every rational number in the interval $[0, 1]$ is a term in a Farey sequence. The theorem follows once from the definition of a Farey sequence of order n. □

Theorem 10.2. *Let a/b, a'/b' be two successive terms in $\mathfrak{F}_n$. Then $b + b' \geqslant n + 1$. If $a/b < a'/b'$, then $ba' - ab' = 1$.*

Proof. Since $(a, b) = 1$, there are integers x, y such that

$$bx - ay = 1, \qquad n - b < y \leqslant n. \tag{1}$$

It follows at once that

$$y > 0, \qquad (x, y) = 1, \qquad \frac{x}{y} = \frac{a}{b} + \frac{1}{by} > \frac{a}{b}.$$

It suffices to prove that $x/y = a'/b'$. This is because we can then deduce that $x = a'$, $y = b'$, $ba' - ab' = 1$ and $b + b' > n$. Suppose that $x/y \neq a'/b'$. Then

$$\frac{a}{b} < \frac{a'}{b'} < \frac{x}{y}.$$

From this we deduce that

$$\frac{x}{y} - \frac{a}{b} = \frac{x}{y} - \frac{a'}{b'} + \frac{a'}{b'} - \frac{a}{b} \geqslant \frac{1}{b'y} + \frac{1}{b'b} = \frac{b+y}{ybb'} > \frac{n}{ybb'} \geqslant \frac{1}{by}.$$

But we have, from (1),

$$\frac{x}{y} - \frac{a}{b} = \frac{1}{by},$$

giving a contradiction. The theorem is proved. □

Theorem 10.3. *Suppose that $a/b < a''/b'' < a'/b'$ are three successive Farey fractions. Then*

$$\frac{a''}{b''} = \frac{a+a'}{b+b'}.$$

Proof. From Theorem 10.2 we have $a''b - b''a = 1$ and $a'b'' - b'a'' = 1$, and so, on subtraction, $a''(b+b') - b''(a+a') = 0$. The required result follows. □

Definition 2. Let a/b and a'/b' be two successive Farey fractions. Then we call $(a+a')/(b+b')$ the *mediant* of the two fractions.

Theorem 10.4. *The mediant lies between the two fractions a/b and a'/b', and the distance from them are*

$$\frac{1}{b(b+b')} \quad \text{and} \quad \frac{1}{b'(b+b')}$$

respectively.

Proof. We assume that $a/b < a'/b'$. Then

$$\frac{a'}{b'} - \frac{a+a'}{b+b'} = \frac{ba'-ab'}{b'(b+b')} = \frac{1}{b'(b+b')} > 0,$$

$$\frac{a+a'}{b+b'} - \frac{a}{b} = \frac{a'b-ab'}{b(b+b')} = \frac{1}{b(b'+b)} > 0. \quad \square$$

Theorem 10.5. *Let ξ be a real number, $0 < \xi < 1$. Then there always exists a/b in $\mathfrak{F}_n$ such that*

$$\left|\xi - \frac{a}{b}\right| < \frac{1}{bn}, \qquad 0 < b \leqslant n.$$

Proof. We partition the interval $(0, 1)$ into subintervals by the points in $\mathfrak{F}_n$ together with their mediants. Now ξ must be in one of these subintervals one of whose end point is a/b while the other is $(a+a')/(b+b')$. Therefore we have

$$\left|\xi - \frac{a}{b}\right| \leqslant \left|\frac{a+a'}{b+b'} - \frac{a}{b}\right| = \frac{1}{b(b+b')} \leqslant \frac{1}{b(n+1)} < \frac{1}{bn}.$$

The theorem is proved. □

Theorem 10.6. *Let ξ and η be any two real numbers, $\eta \geqslant 1$. There always exists a rational number a/b such that*

$$\left|\xi - \frac{a}{b}\right| < \frac{1}{b\eta}, \qquad 0 < b \leqslant \eta.$$

Proof. We can assume without loss that $0 < \xi < 1$ and the required result follows at once from Theorem 10.5. □

Theorem 10.7. *Let ξ be any real number. There always exists a rational number a/b such that*

$$\left|\xi - \frac{a}{b}\right| < \frac{1}{b^2}. \tag{2}$$

If ξ is irrational, then there are infinitely many such a/b satisfying this inequality.

Proof. Clearly we need only examine the case when ξ is irrational, $0 < \xi < 1$. Let a_n/b_n, a'_n/b'_n be two successive terms in $\mathfrak{F}_n$ satisfying

$$\frac{a_n}{b_n} < \xi < \frac{a'_n}{b'_n}.$$

From the proof of Theorem 10.5 we see that one of these must satisfy the inequality (2). Our theorem now follows from Theorem 10.1. □

Theorem 10.8. *Let ξ be any irrational number. Then there exist infinitely many rational numbers a/b such that*

$$\left|\xi - \frac{a}{b}\right| < \frac{1}{\sqrt{5}\,b^2}. \tag{3}$$

Proof. We can assume without loss that $0 < \xi < 1$. Let a/b and a'/b' be two successive Farey fractions of order n satisfying $a/b < \xi < a'/b'$. Let $\omega = b'/b$ and we consider separately the following two cases.

1) Suppose that $\omega > (1 + \sqrt{5})/2$ or $\omega < (\sqrt{5} - 1)/2$. Then, from Theorem 10.2, we have

$$\frac{a'}{b'} - \frac{a}{b} = \frac{1}{bb'} = \frac{1}{b^2\omega}.$$

Since

$$\begin{aligned}\frac{1}{\omega} - \frac{1}{\sqrt{5}}\left(1 + \frac{1}{\omega^2}\right) &= -\frac{1}{\sqrt{5}\,\omega^2}(\omega^2 - \sqrt{5}\,\omega + 1)\\ &= -\frac{1}{\sqrt{5}\,\omega^2}\left(\omega - \frac{1}{2}(\sqrt{5} + 1)\right)\left(\omega - \frac{1}{2}(\sqrt{5} - 1)\right) < 0,\end{aligned}$$

we have

$$\frac{a'}{b'} - \frac{a}{b} < \frac{1}{\sqrt{5}\,b^2}\left(1 + \frac{1}{\omega^2}\right) = \frac{1}{\sqrt{5}}\left(\frac{1}{b^2} + \frac{1}{b'^2}\right),$$

$$\frac{a}{b} + \frac{1}{\sqrt{5}}\frac{1}{b^2} > \frac{a'}{b'} - \frac{1}{\sqrt{5}}\frac{1}{b'^2}.$$

Therefore the two intervals

$$\left(\frac{a}{b}, \frac{a}{b} + \frac{1}{\sqrt{5}\,b^2}\right) \quad \text{and} \quad \left(\frac{a'}{b'} - \frac{1}{\sqrt{5}\,b'^2}, \frac{a'}{b'}\right)$$

overlap, and so one of them must contain ξ giving

$$\left|\xi - \frac{a}{b}\right| < \frac{1}{\sqrt{5}\,b^2}, \quad \text{or} \quad \left|\xi - \frac{a'}{b'}\right| < \frac{1}{\sqrt{5}\,b'^2}. \tag{4}$$

2) Suppose that $(\sqrt{5} - 1)/2 < \omega < (1 + \sqrt{5})/2$. Then

$$b + b' > \tfrac{1}{2}(\sqrt{5} + 1)b, \qquad b + b' < \tfrac{1}{2}(\sqrt{5} + 1)b'.$$

Therefore we can deal with the intervals

$$\left(\frac{a}{b}, \frac{a + a'}{b + b'}\right) \quad \text{and} \quad \left(\frac{a + a'}{b + b'}, \frac{a'}{b'}\right)$$

with the method in 1). That is, there are three possibilities; apart from the two situations in (4) we also have

$$\left|\xi - \frac{a + a'}{b + b'}\right| < \frac{1}{\sqrt{5}\,(b + b')^2}.$$

Therefore, given any n, there always exist a, b such that (3) holds. Since ξ is irrational, b and b' tend to infinity with n according to Theorem 10.1, and so our theorem is proved. □

Exercise. Prove that the denominators of two successive Farey fractions are different.

6.11 Vinogradov's Method of Estimating Sums of Fractional Parts

Let $\{\alpha\}$ be the fractional part of α; that is $\{\alpha\} = \alpha - [\alpha]$. The purpose of this section is to study sums of the form

$$\sum_{A \leqslant x < B} \{f(x)\}.$$

We shall apply the results in the next section.

Theorem 11.1. *Let $m > 0$, $(a, m) = 1$, $h \geqslant 0$ and c be real. Suppose that*

$$c \leqslant \psi(x) \leqslant c + h, \qquad \text{for} \qquad x = 0, \ldots, m,$$

and let

$$S = \sum_{x=0}^{m-1} \left\{\frac{ax + \psi(x)}{m}\right\}.$$

Then

$$|S - \tfrac{1}{2}m| \leqslant h + \tfrac{1}{2}.$$

Proof. Clearly we have

$$\left|S - \frac{1}{2}m\right| \leqslant \sum_{x=0}^{m-1} \left|\left\{\frac{ax + \psi(x)}{m}\right\} - \frac{1}{2}\right| \leqslant \frac{1}{2}m.$$

The theorem therefore follows at once if $m \leqslant 2h + 1$. Suppose now that $m > 2h + 1$. Let r be the least positive residue of $ax + [c] \bmod m$. We then have

$$S = \sum_{r=0}^{m-1} \left\{\frac{r + \Phi(r)}{m}\right\}, \tag{1}$$

where

$$\Phi(r) = \psi(x) - [c].$$

Hence

$$\{c\} \leqslant \Phi(r) \leqslant \{c\} + h. \tag{2}$$

If $0 \leqslant r < m - [h + \{c\}]$, then

$$0 \leqslant \{c\} \leqslant r + \Phi(r) \leqslant m - [h + \{c\}] - 1 + \{c\} + h < m,$$

or

$$0 \leqslant \frac{r + \Phi(r)}{m} < 1;$$

therefore

$$\left\{\frac{r + \Phi(r)}{m}\right\} = \frac{r + \Phi(r)}{m},$$

or

$$\frac{r}{m} + \frac{\{c\}}{m} \leqslant \left\{\frac{r + \Phi(r)}{m}\right\} \leqslant \frac{r}{m} + \frac{\{c\} + h}{m}. \tag{3}$$

If $m - [h + \{c\}] \leqslant r < m$, let $r = m - s$. Then for $s = 1, 2, \ldots, [h + \{c\}]$, we have

$$\left\{\frac{r + \Phi(r)}{m}\right\} = \left\{1 + \frac{\Phi(r) - s}{m}\right\}.$$

If $\Phi(r) - s \geqslant 0$, then from $\Phi(r) - s \leqslant h + \{c\} - 1 < m$ we see that

$$\frac{\{c\} - s}{m} \leqslant \left\{\frac{r + \Phi(r)}{m}\right\} = \frac{\Phi(r) - s}{m} \leqslant \frac{h + \{c\} - s}{m}; \tag{4}$$

and if $\Phi(r) - s < 0$, then from $0 < m + \{c\} - s \leqslant r + \Phi(r) < m$ we see that

$$\frac{r + \{c\}}{m} \leqslant \left\{\frac{r + \Phi(r)}{m}\right\} = \frac{r + \Phi(r)}{m} \leqslant \frac{r + h + \{c\}}{m}. \tag{5}$$

From (4) and (5) we have

$$-1 + \frac{r}{m} + \frac{\{c\}}{m} \leqslant \left\{\frac{r + \Phi(r)}{m}\right\} \leqslant \frac{r}{m} + \frac{h + \{c\}}{m}. \tag{6}$$

Now from (1), (3) and (6) we arrive at

$$\{c\} - (h + \{c\}) \leqslant S - \sum_{r=0}^{m-1} \frac{r}{m} \leqslant h + \{c\},$$

and hence

$$-h \leqslant S - \tfrac{1}{2}(m-1) \leqslant h + 1.$$

The theorem is proved. □

Theorem 11.2. *Let m be an integer, $A > 2, 1 \leqslant m \leqslant A^{1/3}, (a, m) = 1, k \geqslant 1$. Suppose that*

$$S = \sum_{x=M}^{M+m-1} \{f(x)\},$$

where $f(x)$ has a continuous second derivative in $M \leqslant x \leqslant M + m - 1$ and satisfies

$$f'(M) = \frac{a}{m} + \frac{\vartheta}{m^2}, \qquad (a, m) = 1, \qquad |\vartheta| < 1,$$

$$\frac{1}{A} \leqslant |f''(x)| \leqslant \frac{k}{A}.$$

Then

$$|S - \tfrac{1}{2}m| \leqslant \tfrac{1}{2}(k + 5).$$

Proof. From the mean value theorem of differential calculus we have

$$f(M + y) = f(M) + yf'(M) + \frac{y^2}{2} f''(M + \vartheta' y), \qquad |\vartheta'| < 1.$$

In Theorem 11.1 we take

$$\psi(y) = m\left(f(M) + \frac{\vartheta}{m^2} y + \frac{1}{2} y^2 f''(M + \vartheta' y)\right).$$

From the continuity of $f''(x)$ and from $|f''(x)| > 1/A$ we see that $f''(x)$ does not change sign. We can therefore assume without loss that $f''(x) > 0$. Then we have

$$m\left(f(M) - \frac{m}{m^2}\right) < \psi(y) < m\left(f(M) + \frac{m}{m^2} + \frac{1}{2}\frac{m^2}{A} k\right),$$

or

$$mf(M) - 1 < \psi(y) < mf(M) + 1 + \tfrac{1}{2}k.$$

The result follows from taking $c = mf(M) - 1$ and $h = 2 + k/2$ in Theorem 11.1. □

Theorem 11.3. *Let $k \geqslant 1$ and let $f(x)$ have a continuous second derivative in $M \leqslant x \leqslant M + m$, and*

$$\frac{1}{A} \leqslant |f''(x)| \leqslant \frac{k}{A}.$$

Then

$$S = \sum_{x=M}^{M+m-1} \{f(x)\} = \frac{1}{2}m + O(\Delta),$$

where

$$\Delta = (k^2 m \log A + kA)A^{-\frac{1}{3}}.$$

Proof. We take $\tau = A^{1/3}$, $M = M_1$. We see from Theorem 10.6 that there exist a_1, m, ϑ_1 such that

$$f'(M_1) = \frac{a_1}{m_1} + \frac{\vartheta_1}{m_1 \tau}, \qquad 0 < m_1 \leqslant \tau, \qquad (a_1, m_1) = 1, \qquad |\vartheta_1| < 1. \tag{7}$$

From Theorem 11.2 we have

$$\sum_{x=M_1}^{M_1+m_1-1} \{f(x)\} = \frac{1}{2}m_1 + \frac{\vartheta'_1}{2}(k+5), \qquad |\vartheta'_1| \leqslant 1.$$

We next take $M_2 = M_1 + m_1$ and again from Theorem 10.6 there exist a_2, m_2, ϑ_2 such that

$$f'(M_2) = \frac{a_2}{m_2} + \frac{\vartheta_2}{m_2 \tau}, \qquad 0 < m_2 \leqslant \tau, \qquad (a_2, m_2) = 1, \qquad |\vartheta_2| < 1,$$

and

$$\sum_{x=M_2}^{M_2+m_2-1} \{f(x)\} = \frac{1}{2}m_2 + \frac{\vartheta'_2}{2}(k+5), \qquad |\vartheta'_2| \leqslant 1.$$

Continuing this way, if after s steps we have

$$0 \leqslant M + m - 1 - M_{s+1} < \tau,$$

then

$$|S - \tfrac{1}{2}(m_1 + \cdots + m_s) - \tfrac{1}{2}(M + m - M_{s+1})|$$
$$\leqslant \frac{s}{2}(k+5) + \frac{1}{2}(M + m - M_{s+1}),$$

or (since $M_{s+1} = M + m_1 + \cdots + m_s$)

$$|S - \tfrac{1}{2}m| < \tfrac{1}{2}s(k+5) + \tfrac{1}{2}(\tau + 1). \tag{8}$$

We now have to estimate s. Suppose that $0 < q < \tau$, $(p, q) = 1$. If p, q are given, we can estimate how many $m_1, \ldots, m_s$ are equal to q. From $|f''(x)| > 1/A$ and its continuity we know that $f''(x)$ does not change sign. It follows that the set of values x satisfying

$$\frac{p}{q} - \frac{1}{q\tau} \leqslant f'(x) \leqslant \frac{p}{q} + \frac{1}{q\tau} \tag{9}$$

forms an interval. Let x_1, x_2 be any two points in the interval, so that

$$-\frac{2}{q\tau} < f'(x_1) - f'(x_2) < \frac{2}{q\tau}.$$

Hence

$$\left| \int_{x_1}^{x_2} f''(t)\, dt \right| < \frac{2}{q\tau},$$

and so

$$\frac{1}{A}|x_2 - x_1| < \frac{2}{q\tau}.$$

This shows that the length of the interval of values x which satisfies (9) is at most $2A/q\tau$. It follows that the number of m_i which are equal to q is at most $2A/q^2\tau + 1$.

Next, for fixed q, we estimate the number of values p which satisfy (9). Suppose that $p_1 > p_2$ and

$$\frac{p_1}{q} - \frac{1}{q\tau} \leqslant f'(x_1) \leqslant \frac{p_1}{q} + \frac{1}{q\tau},$$

$$\frac{p_2}{q} - \frac{1}{q\tau} \leqslant f'(x_2) \leqslant \frac{p_2}{q} + \frac{1}{q\tau}.$$

Then

$$\left| \int_{x_2}^{x_1} f''(t)\, dt \right| = |f'(x_1) - f'(x_2)| \geqslant \frac{p_1 - p_2}{q} - \frac{2}{q\tau},$$

and so

$$\frac{mk}{A} \geqslant |x_1 - x_2| \cdot \frac{k}{A} \geqslant \frac{p_1 - p_2}{q} - \frac{2}{q\tau},$$

and hence

$$p_1 - p_2 + 1 \leqslant \frac{kmq}{A} + \frac{2}{\tau} + 1.$$

This shows that the number of p is at most

$$\frac{kmq}{A} + \frac{2}{\tau} + 1.$$

Collecting our results we see that if we write $f'(M_i)$ as in formula (7), then the number of fractions a_i/m_i whose denominator m_i is q is

$$\leqslant \left(\frac{2A}{q^2\tau}+1\right)\left(\frac{kmq}{A}+\frac{2}{\tau}+1\right)$$
$$=\frac{km}{\tau}\left(\frac{2}{q}+\frac{q}{\tau^2}\right)+\left(\frac{2A}{q^2\tau}+1\right)\left(1+\frac{2}{\tau}\right).$$

Summing over $q = 1, 2, \ldots, [\tau]$ we see that

$$s \leqslant \frac{km}{\tau}\left(2\log\tau + 2 + \frac{\tau^2+\tau}{2\tau^2}\right) + O\left(\frac{A}{\tau}\right)$$
$$= O\left(\frac{km}{\tau}\log A + \frac{A}{\tau}\right).$$

The theorem follows from substituting this into (8). □

6.12 Application of Vinogradov's Theorem to Lattice Point Problems

We already proved in Theorem 9.1 that the number $R(x)$ of lattice points inside the circle $u^2 + v^2 \leqslant x$ satisfies $R(x) = \pi x + O(\sqrt{x})$. In this section we shall prove the following sharper result.

Theorem 12.1 (Sierpinski). *Let* $x \geqslant 2$. *Then*

$$R(x) = \pi x + O(x^{\frac{1}{3}}\log x).$$

This result is not the best known. Using more complicated analytic tools the author proved in 1942 that, for $\varepsilon > 0$,

$$R(x) = \pi x + O(x^{\frac{13}{40}+\varepsilon}).$$

(See Note 6.1.) A famous problem in number theory is the conjecture that

$$R(x) = \pi x + O(x^{\frac{1}{4}+\varepsilon}).$$

We require the following result for the proof of Theorem 12.1.

Theorem 12.2. *Let* $f(x)$ *have a continuous second derivative in the interval* $Q \leqslant x \leqslant R$, *and let*

$$\sigma(x) = \int_0^x \left(\frac{1}{2} - \{t\}\right)dt.$$

Then

$$\sum_{Q<x\leqslant R} f(x) = \int_Q^R f(x)\,dx + (\tfrac{1}{2} - \{R\})f(R) - (\tfrac{1}{2} - \{Q\})f(Q) - \sigma(R)f'(R)$$

$$+ \sigma(Q)f'(Q) + \int_Q^R \sigma(x)f''(x)\,dx.$$

Proof. Let x_1 be an integer, $Q \leqslant \alpha < \beta \leqslant R$, $x_1 < \alpha < \beta < x_1 + 1$. From integration by parts we have

$$-\int_\alpha^\beta f(x)\,dx = \int_\alpha^\beta f(x)\frac{d}{dx}\left(\frac{1}{2} - \{x\}\right)dx$$

$$= (\tfrac{1}{2} - \{\beta\})f(\beta) - (\tfrac{1}{2} - \{\alpha\})f(\alpha) - \sigma(\beta)f'(\beta) + \sigma(\alpha)f'(\alpha)$$

$$+ \int_\alpha^\beta \sigma(x)f''(x)\,dx. \tag{1}$$

Letting $\alpha \to x_1$, $\beta \to x_1 + 1$ we have

$$-\int_{x_1}^{x_1+1} f(x)\,dx = -\tfrac{1}{2}f(x_1+1) - \tfrac{1}{2}f(x_1) + \int_{x_1}^{x_1+1} \sigma(x)f''(x)\,dx.$$

From this it follows that

$$-\int_{[Q]+1}^{[R]} f(x)\,dx = -\sum_{[Q]+1\leqslant x\leqslant [R]} f(x) + \tfrac{1}{2}f([Q]+1) + \tfrac{1}{2}f([R])$$

$$+ \int_{[Q]+1}^{[R]} \sigma(x)f''(x)\,dx. \tag{2}$$

If in (1) we let $\alpha = Q$, $\beta \to [Q] + 1$, then

$$-\int_Q^{[Q]+1} f(x)\,dx = \frac{-1}{2}f([Q]+1) - \left(\frac{1}{2} - \{Q\}\right)f(Q) + \sigma(Q)f'(Q)$$

$$+ \int_Q^{[Q]+1} \sigma(x)f''(x)\,dx. \tag{3}$$

Similarly we have

$$-\int_{[R]}^{R} f(x)\,dx = (\tfrac{1}{2} - \{R\})f(R) - \tfrac{1}{2}f([R]) - \sigma(R)f'(R)$$

$$+ \int_{[R]}^{R} \sigma(x)f''(x)\,dx. \tag{4}$$

The required formula is obtained by adding (2), (3) and (4). □

Proof of Theorem 12.1. By considering the diagram associated with the circle problem it is easy to see that

$$R(x) = 1 + 4[\sqrt{x}] + 8 \sum_{0<u\leqslant\sqrt{\frac{x}{2}}} [\sqrt{x-u^2}] - 4\left[\sqrt{\frac{x}{2}}\right]^2. \tag{5}$$

Clearly we have

$$\sum_{0<u\leqslant\sqrt{\frac{x}{2}}} [\sqrt{x-u^2}] = \sum_{0<u\leqslant\sqrt{\frac{x}{2}}} \sqrt{x-u^2} - \sum_{0<u\leqslant\sqrt{\frac{x}{2}}} \{\sqrt{x-u^2}\}$$

$$= \sum\nolimits_1 - \sum\nolimits_2.$$

Let us estimate $\sum_1$. Take $f(u) = \sqrt{x-u^2}$ so that from Theorem 12.2 we have

$$\sum\nolimits_1 = \int_0^{\sqrt{\frac{x}{2}}} \sqrt{x-u^2}\,du + \left(\frac{1}{2} - \left\{\sqrt{\frac{x}{2}}\right\}\right)\sqrt{\frac{x}{2}} - \frac{1}{2}\sqrt{x} + \sigma\left(\sqrt{\frac{x}{2}}\right)$$

$$- x\int_0^{\sqrt{\frac{x}{2}}} \frac{\sigma(u)\,du}{(x-u^2)^{3/2}} = \frac{\pi}{8}x + \frac{x}{4} + \left(\frac{1}{2} - \left\{\sqrt{\frac{x}{2}}\right\}\right)\sqrt{\frac{x}{2}} - \frac{1}{2}\sqrt{x} + O(1).$$

From Theorem 11.3 we have

$$\sum\nolimits_2 = \frac{1}{2}\sqrt{\frac{x}{2}} + O(x^{\frac{1}{3}}\log x).$$

The theorem follows from substituting these estimates into (5). □

A similar problem to the circle problem is the Dirichlet divisor problem. We already proved in Theorem 5.4 that

$$\sum_{1\leqslant n\leqslant\xi} d(n) = \xi\log\xi + (2\gamma - 1)\xi + O(\xi^{\frac{1}{2}+\varepsilon}).$$

Here we prove:

Theorem 12.3 (Voronoi). *If* $\xi \geqslant 2$, *then*

$$\sum_{1 \leqslant n \leqslant \xi} d(n) = \xi \log \xi + (2\gamma - 1)\xi + O(\xi^{\frac{1}{3}} \log^2 \xi).$$

With reference to this problem Yin has improved the result by replacing the error term with $O(\xi^{\frac{12}{37}+\varepsilon})$. Again the conjecture is that it should be $O(x^{\frac{1}{4}+\varepsilon})$.

Proof. From the proof of Theorem 5.4 we have

$$\sum_{1 \leqslant n \leqslant \xi} d(n) = 2 \sum_{1 \leqslant u \leqslant \sqrt{\xi}} \left[\frac{\xi}{u}\right] - [\sqrt{\xi}]^2. \tag{6}$$

We take $f(u) = 1/u$ and from Theorem 12.2 we have

$$\sum_{1 \leqslant u \leqslant \sqrt{\xi}} \frac{1}{u} = \lim_{\varepsilon \to 0} \sum_{1-\varepsilon < u \leqslant \sqrt{\xi}} \frac{1}{u} = \int_1^{\sqrt{\xi}} \frac{du}{u} + \left(\frac{1}{2} - \{\sqrt{\xi}\}\right)\xi^{-\frac{1}{2}}$$

$$+ \frac{1}{2} + \sigma(\sqrt{\xi})\xi^{-1} + 2\int_1^{\sqrt{\xi}} \sigma(x)x^{-3}\, dx.$$

We note that

$$\int_1^{\infty} \sigma(x)x^{-3}\, dx = \frac{1}{2}\int_1^{\infty} \left(\frac{1}{2} - \{x\}\right)x^{-2}\, dx$$

$$= \frac{1}{4} - \frac{1}{2}\sum_{n=1}^{\infty} \int_0^1 \frac{x}{(n+x)^2}\, dx$$

$$= \frac{1}{4} - \frac{1}{2}\sum_{n=1}^{\infty} \left\{\log(n+1) - \log n - \frac{1}{n+1}\right\}$$

$$= -\frac{1}{4} + \frac{1}{2}\gamma,$$

and so we have

$$2\sum_{1 \leqslant u \leqslant \sqrt{\xi}} \frac{\xi}{u} = \xi \log \xi + 2\left(\frac{1}{2} - \{\sqrt{\xi}\}\right)\xi^{\frac{1}{2}} + 2\gamma\xi + O(1). \tag{7}$$

We now estimate

$$S = \sum_{1 \leqslant u \leqslant \sqrt{\xi}} \left\{\frac{\xi}{u}\right\}.$$

We take t_0 so that $[\sqrt{\xi}]2^{-t_0} \geqslant 2\xi^{\frac{1}{3}} \geqslant [\sqrt{\xi}]2^{-t_0-1}$. Then clearly we have

$$S = \sum_{t=0}^{t_0} \sum_{[\sqrt{\xi}]2^{-t-1} \leqslant u \leqslant [\sqrt{\xi}]2^{-t}} \left\{\frac{\xi}{u}\right\} + O(\xi^{\frac{1}{3}}).$$

From Theorem 11.3 (replacing m by $[\sqrt{\xi}]2^{-t-1}$, and A by $[\sqrt{\xi}]^3\xi^{-1}2^{-(3t+1)}$), we have

$$\sum_{[\sqrt{\xi}]2^{-t-1} \leqslant u \leqslant [\sqrt{\xi}]2^{-t}} \left\{\frac{\xi}{u}\right\} = \frac{1}{2^{t+2}}[\sqrt{\xi}] + O(\xi^{\frac{1}{3}} \log \xi).$$

Therefore

$$S = \tfrac{1}{2}[\sqrt{\xi}] + O(\xi^{\frac{1}{3}} \log^2 \xi). \tag{8}$$

Noting that $[\sqrt{\xi}]^2 = \xi - 2\{\sqrt{\xi}\}\xi^{\frac{1}{2}} + O(1)$, we see that the theorem follows from (6), (7) and (8). □

6.13 Ω-Results

A number of famous problems in number theory are concerned with the accuracy of various asymptotic formulae; that is the problem of reducing the size of the error term in the formula. These results are generally called O-results, and our Theorem 12.1 and Theorem 12.3 are such examples. On the other hand we may also estimate how large the error term must be; that is we can prove that some error terms *cannot* be smaller than a certain order. These types of results are called Ω-results.

In §12 we mentioned that the O-term in Theorem 12.1 is conjectured to be $O(x^{\frac{1}{4}+\varepsilon})$. Here we prove that if $\varepsilon > 0$, then the formula

$$R(x) = \pi x + O(x^{\frac{1}{4}-\varepsilon})$$

does *not* hold. Actually we shall prove a very general result.

In this section K, K_1, K_2, K_3 represent absolute constants. At various places we may use the same symbol to denote different constants, but this should not cause any confusion.

Theorem 13.1 (Erdös-Fuchs). *Let $c > 0$ and let $a_1, a_2, \ldots$ be integers satisfying $0 \leqslant a_1 \leqslant a_2 \leqslant \cdots$. Let $f(n)$ denote the number of solutions to the equation $a_i + a_j = n$, and*

$$r(x) = \sum_{n \leqslant x} f(n)$$

so that $r(x)$ is the number of pairs of integers a_i, a_j satisfying $a_i + a_j \leqslant x$. Then the formula

$$r(x) = cx + o(x^{\frac{1}{4}} \log^{-\frac{1}{2}} x)$$

cannot hold.

We shall first deal with the following auxiliary results.

Theorem 13.2. *Let a_n be real numbers such that*

$$\psi(\vartheta) = \sum_{n=-\infty}^{\infty} a_n e^{in\vartheta}$$

converges uniformly, and that $\sum_{n=-\infty}^{\infty} a_n^2$ converges. Then

$$\frac{1}{2\pi}\int_{-\pi}^{\pi} |\psi(\vartheta)|^2\, d\vartheta = \sum_{n=-\infty}^{\infty} a_n^2.$$

Proof. Clearly we have

$$|\psi(\vartheta)|^2 = \sum_{n=-\infty}^{\infty}\ \sum_{m=-\infty}^{\infty} a_n a_m e^{i(n-m)\vartheta}.$$

The required result follows from integrating term by term over $-\pi$ to π. □

Theorem 13.3. *Let $b_n \geqslant 0$ and let $\varphi(z) = \sum_{n=1}^{\infty} b_n z^n$ be convergent for $|z| < 1$. If $0 < \alpha < \pi$, $z = re^{i\vartheta}$ $(0 < r < 1)$, then we have*

$$\frac{1}{2\alpha}\int_{-\alpha}^{\alpha} |\varphi(z)|^2\, d\vartheta \geqslant \frac{1}{6\pi}\int_{-\pi}^{\pi} |\varphi(z)|^2\, d\vartheta.$$

Proof. We introduce the function

$$q(\vartheta) = \begin{cases} 1 - \left|\dfrac{\vartheta}{\alpha}\right|, & \text{when } |\vartheta| \leqslant \alpha, \\ 0, & \text{when } \alpha < |\vartheta| \leqslant \pi. \end{cases}$$

Then we have

$$\int_{-\alpha}^{\alpha} |\varphi(z)|^2\, d\vartheta \geqslant \int_{-\pi}^{\pi} |q(\vartheta)|^2 |\varphi(z)|^2\, d\vartheta = \sum_{m,n=1}^{\infty} b_n b_m r^{n+m} \int_{-\pi}^{\pi} |q(\vartheta)|^2 e^{i(n-m)\vartheta}\, d\vartheta.$$

When $m \neq n$, we have

$$\int_{-\pi}^{\pi} |q(\vartheta)|^2 e^{i(n-m)\vartheta}\, d\vartheta = 2\int_{0}^{a} \left(1 - \frac{\vartheta}{\alpha}\right)^2 \cos(n-m)\vartheta\, d\vartheta$$

$$= \frac{4}{\alpha(n-m)^2}\left(1 - \frac{\sin(n-m)\alpha}{\alpha(n-m)}\right) \geqslant 0,$$

while when $m = n$,

$$\int_{-\pi}^{\pi} |q(\vartheta)|^2 \, d\vartheta = \frac{2\alpha}{3},$$

and therefore we have

$$\int_{-\alpha}^{\alpha} |\varphi(z)|^2 \, d\vartheta \geqslant \frac{2\alpha}{3} \sum_{n=1}^{\infty} b_n^2 r^{2n} = \frac{\alpha}{3\pi} \int_{-\pi}^{\pi} |\varphi(z)|^2 \, d\vartheta. \quad \square$$

Theorem 13.4. *Suppose that* $|z| < 1$ *and let*

$$(1 - z)^{-r} = \sum_{n=0}^{\infty} \gamma_n z^n.$$

Then there exist constants c, C such that

$$0 < c < \frac{\gamma_n}{n^{r-1}} < C < \infty.$$

Proof. From the binomial theorem we have

$$\gamma_n = \frac{r(r+1) \cdots (r+n-1)}{1 \cdot 2 \cdot \cdots \cdot n}.$$

Since

$$\begin{aligned}\int_{v-\frac{1}{2}}^{v+\frac{1}{2}} \log t \, dt &= \int_{0}^{\frac{1}{2}} \{\log(v+t) + \log(v-t)\} \, dt \\ &= \int_{0}^{\frac{1}{2}} \left\{\log v^2 + \log\left(1 - \frac{t^2}{v^2}\right)\right\} dt \\ &= \log v + O\left(\frac{1}{v^2}\right),\end{aligned}$$

it follows that

$$\begin{aligned}\sum_{l=1}^{n} \log(r+l-1) &= \sum_{l=1}^{n} \int_{r+l-\frac{3}{2}}^{r+l-\frac{1}{2}} \log t \, dt + O\left(\sum_{l=1}^{n} \frac{1}{(r+l-1)^2}\right) \\ &= \int_{r-\frac{1}{2}}^{r-\frac{1}{2}+n} \log t \, dt + O(1) \\ &= (r - \tfrac{1}{2} + n) \log(r - \tfrac{1}{2} + n) - (r - \tfrac{1}{2} + n) + O(1) \\ &= (r - \tfrac{1}{2} + n) \log n - n + O(1)\end{aligned}$$

and

$$\log n! = \sum_{l=1}^{n} \log l = (\tfrac{1}{2} + n)\log n - n + O(1),$$

$$\log \gamma_n = (r - 1)\log n + O(1),$$

and the theorem follows. □

Theorem 13.5. *If* $b_n = o(n^{\frac{1}{2}} \log^{-1} n)$, *then when* $0 < r < 1$ *we have*

$$\sum_{n=0}^{\infty} b_n r^n = o\left((1 - r)^{-\frac{3}{2}} \log^{-1} \frac{1}{1 - r}\right).$$

Proof. From the hypothesis we have

$$\sum_{n=0}^{\infty} b_n r^n \leqslant K \sum_{n \leqslant (1-r)^{-\frac{1}{2}}} n^{\frac{1}{2}} r^n + \varepsilon_1(r) \log^{-1} \frac{1}{1 - r} \sum_{n > (1-r)^{-\frac{1}{2}}} n^{\frac{1}{2}} r^n,$$

where $\varepsilon_1(r) \to 0$ as $r \to 1$. In the first sum there are at most $(1 - r)^{-\frac{1}{2}}$ terms, each of which is at most $(1 - r)^{-\frac{1}{4}}$, so that the sum is at most $(1 - r)^{-\frac{3}{4}}$. From Theorem 13.4 the second sum is

$$\leqslant \varepsilon_1(r) \log^{-1} \frac{1}{1 - r} \sum_{n=1}^{\infty} n^{\frac{1}{2}} r^n$$

$$\leqslant \varepsilon(r) \log^{-1} \frac{1}{1 - r} (1 - r)^{-\frac{3}{2}}.$$

Together we have

$$\sum_{n=1}^{\infty} b_n r^n \leqslant K(1 - r)^{-\frac{3}{4}} + \varepsilon(r) \log^{-1} \frac{1}{1 - r} (1 - r)^{-\frac{3}{2}}$$

$$= o\left(\log^{-1} \frac{1}{1 - r} (1 - r)^{-\frac{3}{2}}\right). \qquad \square$$

Theorem 13.6. *Let* $f(x)$ *and* $g(x)$ *be two continuous real functions in the interval* (a, b). *Then*

$$\left| \int_a^b f(x) g(x)\, dx \right| \leqslant \left(\int_a^b f^2(x)\, dx \int_a^b g^2(x)\, dx \right)^{\frac{1}{2}}.$$

Proof. Let λ be any real number and consider

$$\lambda^2 \int_a^b f^2(x)\, dx + 2\lambda \int_a^b f(x) g(x)\, dx + \int_a^b g^2(x)\, dx$$

$$= \int_a^b (\lambda f(x) + g(x))^2\, dx \geqslant 0.$$

The discriminant of the quadratic expression cannot be positive and so the theorem follows. □

Proof of Theorem 13.1. Suppose that $\frac{1}{2} < r < 1$, $z = re^{i\vartheta}$, $1 - r < \alpha < \pi/2$. Let

$$g(z) = \sum_{k=1}^{\infty} z^{a_k},$$

so that we have at once

$$g^2(z) = \sum_{n=0}^{\infty} f(n)z^n$$

and

$$(1 - z)^{-1}g^2(z) = \sum_{n=0}^{\infty} r(n)z^n.$$

If formula (1) holds, then

$$\begin{aligned}(1 - z)^{-1}g^2(z) &= c \sum_{n=0}^{\infty} nz^n + h(z)\\ &= cz(1 - z)^{-2} + h(z),\end{aligned} \tag{2}$$

where

$$h(z) = \sum_{n=0}^{\infty} v_n z^n, \qquad v_n = o(n^{\frac{1}{4}} \log^{-\frac{1}{2}} n).$$

We shall now derive a contradiction. From (2) we have

$$\begin{aligned}\int_{-\alpha}^{\alpha} |g(z)|^2\, d\vartheta &= \int_{-\alpha}^{\alpha} |cz(1 - z)^{-1} + (1 - z)h(z)|\, d\vartheta\\ &\leqslant c \int_{-\pi}^{\pi} |1 - z|^{-1}\, d\vartheta + \int_{-\alpha}^{\alpha} |1 - z|\,|h(z)|\, d\vartheta,\end{aligned} \tag{3}$$

and from Theorem 13.2 and Theorem 13.4 we have

$$\int_{-\pi}^{\pi} |1 - z|^{-1}\, d\vartheta = \int_{-\pi}^{\pi} |(1 - z)^{-\frac{1}{2}}|^2\, d\vartheta < K \sum_{n=1}^{\infty} \frac{r^{2n}}{n} < K \log \frac{1}{1 - r}.$$

From Theorems 13.6 and 13.5 we have

$$\begin{aligned}\int_{-\alpha}^{\alpha} |1 - z|\,|h(z)|\, d\vartheta &\leqslant \sqrt{\int_{-\alpha}^{\alpha} |1 - z|^2\, d\vartheta \int_{-\alpha}^{\alpha} |h(z)|^2\, d\vartheta}\\ &\leqslant \sqrt{(2\alpha(1 + r^2) - 4r \sin \alpha) \int_{-\pi}^{\pi} |h(z)|^2\, d\vartheta}\end{aligned}$$

$$\leqslant \left\{(2\alpha(1-r)^2 + 4r(\alpha - \sin\alpha))\varepsilon(r)(1-r)^{-\frac{3}{2}}\log^{-1}\frac{1}{1-r}\right\}^{\frac{1}{2}}$$

$$\leqslant \varepsilon(r)\,\alpha^{\frac{3}{2}}(1-r)^{-\frac{3}{4}}\log^{-\frac{1}{2}}\frac{1}{1-r},$$

where $\varepsilon(r) \to 0$ as $r \to 1$. Therefore, from (3) we arrive at

$$\int_{-\alpha}^{\alpha} |g(z)|^2\, d\vartheta \leqslant K_1 \log\frac{1}{1-r} + \varepsilon(r)a^{\frac{3}{2}}(1-r)^{-\frac{3}{4}}\log^{-\frac{1}{2}}\frac{1}{1-r}. \tag{4}$$

On the other hand, from Theorem 13.3, we have

$$\int_{-\alpha}^{\alpha} |g(z)|^2\, d\vartheta \geqslant \frac{\alpha}{3\pi}\int_{-\pi}^{\pi} |g(z)|^2\, d\vartheta = \frac{a}{3\pi}\sum_{k=1}^{\infty} r^{2a_k} = \frac{\alpha}{3\pi}g(r^2).$$

From (2) and Theorem 13.4 we have

$$\begin{aligned} g^2(r^2) &= cr^2(1-r^2)^{-1} + (1-r^2)h(r^2) \\ &= cr^2(1-r^2)^{-1} + (1-r^2)O(\textstyle\sum n^{-\frac{1}{4}}r^{2n}) \\ &> K(1-r)^{-1} - O((1-r)^{1-\frac{5}{4}}) \\ &> K(1-r)^{-1}. \end{aligned}$$

Therefore

$$\int_{-\alpha}^{\alpha} |g(z)|^2\, d\vartheta > K_2\alpha(1-r)^{-\frac{1}{2}}. \tag{5}$$

We take $K_2\varepsilon^{-\frac{2}{3}} > 1 + K_1$ and let $\alpha = \varepsilon^{-\frac{2}{3}}(1-r)^{\frac{1}{2}}\log(1/1-r)$. Then from (4) and (5), we arrive at

$$K_2\varepsilon^{-\frac{2}{3}} < K_1 + 1$$

which is a contradiction. Our theorem is proved. □

6.14 Dirichlet Series

A Dirichlet series is a series of the form

$$F(s) = \sum_{n=1}^{\infty}\frac{f(n)}{n^s}.$$

Here we call $F(s)$ the generating function of $f(n)$. This book does not discuss the fundamental properties of Dirichlet series. Instead we only deal with the various formulae and their transformations. We do not even discuss the region of convergence for the series.

If $f(n)$ is a multiplicative function, then

$$F(s) = \prod_p \left(1 + \frac{f(p)}{p^s} + \frac{f(p^2)}{p^{2s}} + \cdots\right),$$

where p runs over all the primes. Also if $f(n)$ is completely multiplicative, then

$$F(s) = \prod_p \left(1 - \frac{f(p)}{p^s}\right)^{-1}.$$

If

$$G(s) = \sum_{n=1}^{\infty} \frac{g(n)}{n^s},$$

then

$$\begin{aligned} F(s)G(s) &= \sum_{l=1}^{\infty} \frac{f(l)}{l^s} \sum_{m=1}^{\infty} \frac{g(m)}{m^s} \\ &= \sum_{n=1}^{\infty} \frac{1}{n^s} \sum_{d|n} f(d) g\left(\frac{n}{d}\right). \end{aligned}$$

Therefore $F(s)F(s)$ is the generating function of $\sum_{d|n} f(d)g(n/d)$. We can use this to derive Theorem 4.2.

Let

$$\zeta(s) = \sum_{n=1}^{\infty} \frac{1}{n^s}.$$

This is the famous *Riemann zeta function* in analytic number theory. We have the product formula

$$\zeta(s) = \prod_p \left(1 - \frac{1}{p^s}\right)^{-1}. \tag{1}$$

Therefore

$$\begin{aligned} \frac{1}{\zeta(s)} &= \prod_p \left(1 - \frac{1}{p^s}\right) = \prod_p \left(1 + \frac{\mu(p)}{p^s} + \frac{\mu(p^2)}{p^{2s}} + \cdots\right) \\ &= \sum_{n=1}^{\infty} \frac{\mu(n)}{n^s}. \end{aligned} \tag{2}$$

If $g(n)$ is the Möbius transformation of $f(n)$, then their generating functions $G(s)$ and $F(s)$ are related by $G(s) = \zeta(s)F(s)$. The inverse Möbius transform theorem then becomes $F(s) = G(s)/\zeta(s)$.

We also have

$$\sum_{n=1}^{\infty} \frac{d(n)}{n^s} = \zeta^2(s), \tag{3}$$

and

$$\sum_{n=1}^{\infty} \frac{|\mu(n)|}{n^s} = \prod_p \left(1 + \frac{1}{p^s}\right) = \frac{\prod_p \left(1 - \frac{1}{p^{2s}}\right)}{\prod_p \left(1 - \frac{1}{p^s}\right)} = \frac{\zeta(s)}{\zeta(2s)}. \tag{4}$$

Taking the logarithmic derivative of (1) we have

$$\begin{aligned}\frac{\zeta'(s)}{\zeta(s)} &= -\sum_p \frac{\log p}{p^s}\left(1 - \frac{1}{p^s}\right)^{-1} \\ &= -\sum_p \log p \sum_{m=1}^{\infty} \frac{1}{p^{ms}} \\ &= -\sum_{n=2}^{\infty} \frac{\Lambda(n)}{n^s}. \end{aligned} \tag{5}$$

Since

$$\zeta'(s) = -\sum_{n=2}^{\infty} \frac{\log n}{n^s}, \tag{6}$$

these two formulae give a new proof of the Möbius transform relationship between $\log n$ and $\Lambda(n)$. Now

$$\begin{aligned}\log \zeta(s) &= -\sum_p \log\left(1 - \frac{1}{p^s}\right) \\ &= \sum_p \sum_{m=1}^{\infty} \frac{1}{mp^{sm}} = \sum_{n=1}^{\infty} \frac{\Lambda_1(n)}{n^s}. \end{aligned} \tag{7}$$

Also

$$\zeta''(s) = \sum_{n=1}^{\infty} \frac{\log^2 n}{n^s}.$$

From

$$\sum_{n=1}^{\infty} \frac{\Lambda(n) \log n}{n^s} = \left(\frac{\zeta'(s)}{\zeta(s)}\right)'$$

and

$$\sum_{n=1}^{\infty} \frac{1}{n^s}\left(\sum_{d|n} \Lambda(d)\Lambda\left(\frac{n}{d}\right)\right) = \left(\frac{\zeta'(s)}{\zeta(s)}\right)^2,$$

using

$$\frac{\zeta''(s)}{\zeta(s)} = \frac{d}{ds}\frac{\zeta'(s)}{\zeta(s)} + \left(\frac{\zeta'(s)}{\zeta(s)}\right)^2 \tag{8}$$

we arrive at

$$\sum_{d|n} \mu(d) \log^2 \frac{n}{d} = \sum_{d|n} \Lambda(d) \Lambda\left(\frac{n}{d}\right) + \Lambda(n) \log n.$$

The results in §8 can also be expressed as follows. Let

$$L(s) = \sum_{n=1}^{\infty} \frac{\chi(n)}{n^s}.$$

Then we have

$$\sum_{n=1}^{\infty} \frac{r(n)}{n^s} = 4L(s)\zeta(s). \tag{9}$$

In the study of analytic number theory we study the analytic properties of $F(s)$ and use these properties to derive results concerning the function $f(n)$.

Exercise 1. Discuss the region of convergence for the series in (1)–(9).

Exercise 2. Establish the following:

$$\frac{\zeta^3(s)}{\zeta(2s)} = \sum_{n=1}^{\infty} \frac{d(n^2)}{n^s}, \qquad (s > 1).$$

$$\frac{\zeta^4(s)}{\zeta(2s)} = \sum_{n=1}^{\infty} \frac{(d(n))^2}{n^s}, \qquad (s > 1).$$

$$\frac{\zeta(s-1)}{\zeta(s)} = \sum_{n=1}^{\infty} \frac{\varphi(n)}{n^s}, \qquad (s > 2).$$

$$\zeta(s)\zeta(s-a) = \sum_{n=1}^{\infty} \frac{\sigma_a(n)}{n^s}, \qquad s > \max(1, a+1).$$

$$\frac{\zeta(s)\zeta(s-a)\zeta(s-b)\zeta(s-a-b)}{\zeta(2s-a-b)} = \sum_{n=1}^{\infty} \frac{\sigma_a(n)\sigma_b(n)}{n^s},$$

$$s > \max(1, a+1, b+1, a+b+1).$$

6.15 Lambert Series

Definition. We call

$$F(x) = \sum_{n=1}^{\infty} f(n) \frac{x^n}{1-x^n} \tag{1}$$

a *Lambert series*. Here $F(x)$ is the generating function of $f(n)$.

Expanding (1) into a power series we have

$$F(x) = \sum_{n=1}^{\infty} f(n) \sum_{m=1}^{\infty} x^{mn}$$

$$= \sum_{n=1}^{\infty} g(n)x^n,$$

where $g(n) = \sum_{d|n} f(d)$.

Thus if $g(n)$ is the Möbius transform of $f(n)$, then $g(n)$ is the coefficient of the power series whose sum is the Lambert series generating function of $f(n)$.

We now take $g(n) = \Delta(n)$, giving

$$x = \sum_{n=1}^{\infty} \frac{\mu(n)x^n}{1 - x^n}. \tag{2}$$

Again, if we take $g(n) = n$, then

$$\sum_{n=1}^{\infty} nx^n = \frac{x}{(1 - x)^2},$$

so that

$$\sum_{n=1}^{\infty} \frac{\varphi(n)x^n}{1 - x^n} = \frac{x}{(1 - x)^2}. \tag{3}$$

A similar method gives

$$\sum_{n=1}^{\infty} d(n)x^n = \frac{x}{1 - x} + \frac{x^2}{1 - x^2} + \frac{x^3}{1 - x^3} + \cdots \tag{4}$$

and

$$\sum_{n=1}^{\infty} r(n)x^n = 4\left(\frac{x}{1 - x} - \frac{x^3}{1 - x^3} + \frac{x^5}{1 - x^5} - \cdots\right). \tag{5}$$

Notes

6.1. The present best result on the circle problem is

$$R(x) = \pi x + O(x^{\frac{12}{37}+\varepsilon})$$

by J. R. Chen [17]. There is a similar result for the Dirichlet's divisor problem (see Yin [65] and G. A. Kolesnik [34]).

6.2. Concerning the Ω-result with respect to the divisor problem, H. E. Richert [49] has proved the following: Let α_n $(n = 1, 2, 3, \ldots)$ be a complex sequence such that

$$\sum_{n \leqslant x} \alpha_n = x + O(x^{\frac{1}{4}-\delta})$$

holds for some $\delta > 0$. Then, given any $\varepsilon > 0$ and any constant c, the following asymptotic formula cannot hold:

$$\sum_{mn \leqslant x} \alpha_m \alpha_n = x \log x + cx + O(x^{\frac{1}{4}-\varepsilon}).$$

Chapter 7. Trigonometric Sums and Characters

7.1 Representation of Residue Classes

Let m be a positive integer. We have seen that the set of integers can be partitioned into residue classes

$$A_0, A_1, \dots, A_{m-1}$$

where A_s is the set of integers congruent to $s \bmod m$. We can define the operation of addition on these residue classes by

$$A_s + A_t = A_u, \qquad u = \begin{cases} s+t & \text{if } s+t<m, \\ s+t-m & \text{if } s+t \geqslant m. \end{cases}$$

This definition satisfies properties associated with groups. Within the theory of groups there is a representation theory whereby more abstract objects are given concrete representations, and this theory has very useful applications (for example, in electronics). In this section we discuss the method of representing residue classes which form an additive group.

To replace the more abstract notion of a residue class we assign to each A_u a complex number ξ_u, bearing in mind that the representation should have the property that if

$$A_u + A_v = A_w \tag{1}$$

then

$$\xi_u \xi_v = \xi_w. \tag{2}$$

An immediate candidate for such a representation is

$$\xi_u = e^{2\pi i u/m}.$$

The advantages of this representation are: (i) integers belonging to the same residue class are assigned the same number; that is, if $u = v + km$, then

$$\xi_u = e^{2\pi i(v+km)/m} = e^{2\pi i v/m} = \xi_v;$$

(ii) if $u + v \equiv w \pmod m$, then

$$\xi_u \xi_v = \xi_w.$$

After giving this representation the abstract notion of adding residue classes becomes the concrete one of multiplication of complex numbers. Thus it is possible that some results on congruences can be obtained from the results in trigonometric sums. This is the underlying reason for the important place occupied by research in trigonometric sums in the theory of numbers.

Let a be any integer. Then

$$\xi_u^a = e^{2\pi iau/m}$$

also possesses the properties (i) and (ii), and so there are m different representations. We now prove that there are no other representations.

Let η_u be any complex number with the above properties. Then from $mu \equiv 0 \pmod m$ we have $\eta_u^m = \eta_0$. But $\eta_0^2 = \eta_0$ so that if $\eta_0 \neq 0$, then $\eta_0 = 1$ and we see that η_u must be an m-th root of unity. If we let $\eta_1 = e^{2\pi ia/m}$, then

$$\eta_u = \eta_1^u = e^{2\pi iau/m}.$$

If $\eta_0 = 0$, then $\eta_u = 0$; that is, all the representations are zero which we exclude from our discussion.

Theorem 1.1. *We have, according to whether m divides n or not,*

$$\frac{1}{m}\sum_{a=0}^{m-1} \xi_n^a = 1 \quad \text{or} \quad 0,$$

that is

$$\frac{1}{m}\sum_{a=0}^{m-1} e^{2\pi ian/m} = 1 \quad \text{or} \quad 0,$$

Proof. If $m|n$, then the theorem is obvious. If $m\nmid n$, then

$$\sum_{a=0}^{m-1} \xi_n^a = \frac{1-\xi_n^m}{1-\xi_n} = 0. \quad \square$$

From this theorem we see that the number of solutions to the congruence

$$f(x_1, \ldots, x_n) \equiv N \pmod m, \qquad 0 \leqslant x_v \leqslant m-1$$

can be represented by

$$\frac{1}{m}\sum_{x_1=0}^{m-1} \cdots \sum_{x_n=0}^{m-1} \sum_{a=0}^{m-1} e^{2\pi ia(f(x_1,\ldots,x_n)-N)/m}.$$

After giving this representation the problem of congruence is now given an analytic interpretation.

For the system of integers we have:

Theorem 1.2. *We have, according to whether n is 0 or not,*

$$\int_0^1 e^{2\pi i n x}\,dx = 1 \quad \text{or} \quad 0. \quad \square$$

From this theorem we see that the number of sets of integer solutions to the equation

$$f(x_1, \ldots, x_n) = N, \qquad a_r \leqslant x_r \leqslant b_r$$

is equal to

$$\sum_{a_1 \leqslant x_1 \leqslant b_1} \cdots \sum_{a_n \leqslant x_n \leqslant b_n} \int_0^1 e^{2\pi i (f(x_1, \ldots, x_n) - N)\alpha}\,d\alpha.$$

Example 1. Fermat's problem is to prove: when $k \geqslant 3$,

$$\int_0^1 \left(\sum_{x=1}^{N} e^{2\pi i x^k \alpha}\right)^2 \left(\sum_{x=1}^{N} e^{-2\pi i x^k \alpha}\right) d\alpha = 0.$$

Example 2. Goldbach's problem is to prove:

$$\int_0^1 \left(\sum_{p \leqslant 2N} e^{2\pi i p \alpha}\right)^2 e^{-4\pi i N \alpha}\,d\alpha > 0.$$

Of course, in these two examples, the new representations do not assist the solutions of the problems.

Exercise 1. Let $(m, n) = 1$,

$$S = \sum_{x=0}^{m-1} \sum_{y=0}^{m-1} \xi(x)\eta(y) e^{2\pi i x y n/m},$$

$$\sum_{x=0}^{m-1} |\xi(x)|^2 = X_0, \qquad \sum_{y=0}^{m-1} |\eta(x)|^2 = Y_0.$$

Show that

$$|S| \leqslant \sqrt{X_0 Y_0 m}.$$

7.2 Character Functions

We already know that multiplication is closed within a reduced residue system. That is, if

$$A_{a_1}, A_{a_2}, \ldots, A_{a_{\varphi(m)}}$$

are the residue classes mod m corresponding to $(a_u, m) = 1$, then

$$A_{a_u} A_{a_v}$$

is still one of these classes. We now ask if there is also a representation for these classes.

Definition. By a *character* $\chi(n)$ mod m we mean a function on n, defined when $(n, m) = 1$, and $\chi(n)$ satisfies the following:

1) $\chi(1) \neq 0$;
2) If $a \equiv b \pmod{m}$, then $\chi(a) = \chi(b)$;
3) $\chi(ab) = \chi(a)\chi(b)$.

Sometimes it is convenient to add: if $(n, m) > 1$, then

$$\chi(n) = 0.$$

Example. $\chi(n) = 1$ is clearly a character. We call this the *principal character* and we denote it by χ_0.

We can deduce from the definition that $\chi(1) = 1$, $\bar{\chi}(n)$ is also a character, and that the product of two characters is a character.

As an example we first take $m = p$, a prime number. Take g to be a primitive root mod p. Then the function

$$\chi_a(n) = e^{2\pi i a \operatorname{ind} n/(p-1)}$$

is a representative because it has the following properties:

1) $\chi_a(1) = 1 \neq 0$;
2) if $n \equiv n' \pmod{p}$, then

$$\operatorname{ind} n \equiv \operatorname{ind} n' \pmod{p-1},$$

so that $\chi_a(n) = \chi_a(n')$;

3)
$$\begin{aligned}\chi_a(nn') &= e^{2\pi i a \operatorname{ind}(nn')/(p-1)} \\ &= e^{2\pi i a(\operatorname{ind} n + \operatorname{ind} n')/(p-1)} \\ &= \chi_a(n)\chi_a(n').\end{aligned}$$

More specifically, when p is an odd prime we take $a = (p-1)/2$ so that

$$\chi_{\frac{1}{2}(p-1)}(n) = e^{\pi i \operatorname{ind} n} = \left(\frac{n}{p}\right).$$

That is the Legendre symbol is a character. From the above we see that there are $p - 1$ characters mod p and it is not difficult to prove that there are only $p - 1$ distinct characters.

We now generalize our discussion to the following:

1) $m = p^l$ where p is an odd prime.

From Theorem 3.9.1 there exists a primitive root mod p^l, so that if $p \nmid n$ we can define ind n, that is

$$n \equiv g^{\operatorname{ind} n} \pmod{p^l}.$$

From this we can obtain $\varphi(p^l)$ characters:

$$\chi_a(n) = e^{2\pi i a \operatorname{ind} n/\varphi(p^l)}, \qquad 1 \leqslant a \leqslant \varphi(p^l).$$

Clearly $\chi_a(1) = 1$, and there exists a character

$$\chi_1(n) = e^{2\pi i \operatorname{ind} n/\varphi(p^l)}$$

with the property: if $n \not\equiv 1 \pmod{p^l}$, then $\chi_1(n) \neq 1$.

2) $m = 2^l$.

2.1) $l = 1$. There is only the principal character.

2.2) $l = 2$. Besides the principal character there is the character

$$\chi(1) = 1, \qquad \chi(3) = -1.$$

2.3) $l > 2$. By Theorem 3.9.3, when n is an odd prime, there is an integer b such that

$$n \equiv (-1)^{\frac{1}{2}(n-1)} 5^b \pmod{2^l}, \qquad b \geqslant 0.$$

We now define

$$\chi_{a,c}(n) = (-1)^{\frac{1}{2}(n-1)a} e^{2\pi i c b/2^{l-2}}.$$

Here a may take two distinct values mod 2 and c may take 2^{l-2} distinct values mod 2^{l-2}, so that there are $\varphi(2^l) = 2^{l-1}$ characters. Also

$$\chi_{1,1}(n) = (-1)^{\frac{1}{2}(n-1)} e^{2\pi i b/2^{l-2}}$$

has the following property: if $\chi_{1,1}(n) = 1$, then $n \equiv 1 \pmod{2^l}$ or $n \equiv -5^{2^{l-3}}$ $\pmod{2^l}$. When $n \equiv -5^{2^{l-3}} \pmod{2^l}$, we have $\chi_{0,1}(n) = -1 \neq 1$. That is, if $n \not\equiv 1$ $\pmod{2^l}$ then we can select a character $\chi_{a,c}(n) \neq 1$.

3) The general case. Let

$$m = p_1^{l_1} \cdots p_s^{l_s}, \qquad l_\nu > 0,$$

be the standard factorization for m. Let a character mod $p_\nu^{l_\nu}$ be

$$\chi^{(\nu)}(n),$$

so that

$$\chi(n) = \prod_{\nu=1}^{s} \chi^{(\nu)}(n) \tag{1}$$

is a character mod m. There are thus $\varphi(m)$ characters mod m.

Conversely, if the modulus of a character $\chi(n)$ is

$$k = k_1 \cdots k_v,$$

where k_i are pairwise coprime, then there exist characters $\chi_i(n) \bmod k_i$ $(i = 1, \ldots, v)$ such that

$$\chi(n) = \chi_1(n) \cdots \chi_v(n).$$

In order to understand this we need only prove the case $v = 2$. From the Chinese remainder theorem, given any n, we can find n_1 and n_2 such that

$$n_1 \equiv n \pmod{k_1}, \qquad n_1 \equiv 1 \pmod{k_2},$$
$$n_2 \equiv 1 \pmod{k_1}, \qquad n_2 \equiv n \pmod{k_2}.$$

We define

$$\chi_1(n) = \chi(n_1), \qquad \chi_2(n) = \chi(n_2)$$

and it is not difficult to prove that $\chi_1(n)$ is a character $\bmod k_1$ and $\chi_2(n)$ is a character $\bmod k_2$. From the definition of n_1 and n_2, we have

$$n_1 n_2 \equiv n \pmod{k_1}, \qquad n_1 n_2 \equiv n \pmod{k_2},$$

so that

$$n_1 n_2 \equiv n \pmod{k}.$$

Therefore

$$\chi(n) = \chi(n_1 n_2) = \chi(n_1)\chi(n_2) = \chi_1(n)\chi_2(n).$$

Theorem 2.1. *The $\varphi(m)$ characters so constructed are all distinct.*

Proof. Suppose that

$$\prod_{v=1}^{s} \chi^{(v)}(n) = \prod_{v=1}^{s} \chi_1^{(v)}(n).$$

From the fact that $\chi^{(v)}(n)/\chi_1^{(v)}(n)$ is also a character $\bmod p_v^{l_v}$ it suffices to prove that if

$$\prod_{v=1}^{s} \chi^{(v)}(n)$$

is the principal character, then $\chi^{(v)}(n)$ is the principal character $\bmod p_v^{l_v}$. Take

$$n \equiv 1 \pmod{p_v^{l_v}}, \qquad 1 \leqslant v \leqslant s-1,$$
$$n \equiv a \pmod{p_s^{l_s}},$$

and we see that for all a $(p_s \nmid a)$,

$$\chi^{(s)}(a) = 1,$$

that is $\chi^{(s)}$ is the principal character $\bmod p_s^{l_s}$. The theorem is proved. □

Theorem 2.2. *If $n \not\equiv 1 \pmod{m}$, then we can select, from among the $\varphi(m)$ characters, a $\chi(n)$ such that $\chi(n) \neq 1$.*

Proof. From the hypothesis there must exist a prime p_ν such that $n \not\equiv 1 \pmod{p_\nu^{l_\nu}}$, and from earlier there exists $\chi^{(\nu)}(n) \neq 1$. If $\mu \neq \nu$ we take $\chi^{(\mu)}$ to be the principal character, and now

$$\chi(n) = \prod_{\nu=1}^{s} \chi^{(\nu)}(n)$$

is the required character. □

Theorem 2.3.

$$\sum_n \chi(n) = \begin{cases} \varphi(m), & \text{if } \chi = \chi_0, \\ 0, & \text{if } \chi \neq \chi_0, \end{cases}$$

where the sum is over a complete set of residues mod m.

Proof. The theorem is obvious if $\chi = \chi_0$. When $\chi \neq \chi_0$, there must be an integer a such that $(a, m) = 1$, and $\chi(a) \neq 1$. From

$$\chi(a) \sum_n \chi(n) = \sum_n \chi(an) = \sum_n \chi(n),$$

or

$$(\chi(a) - 1) \sum_n \chi(n) = 0,$$

the theorem follows. □

Theorem 2.4. *Let c denote the total number of characters* mod m. *Then*

$$\sum_\chi \chi(n) = \begin{cases} c, & \text{if } n \equiv 1 \pmod{m}, \\ 0, & \text{if } n \not\equiv 1 \pmod{m}, \end{cases}$$

where the sum is over all the characters.

Proof. From $n^{\varphi(m)} \equiv 1 \pmod{m}$ we deduce that

$$(\chi(n))^{\varphi(m)} = 1,$$

so that the number of characters c is finite.

If $n \equiv 1 \pmod{m}$, then the theorem is obvious. If $n \not\equiv 1 \pmod{m}$, from Theorem 2 there is a character $X(a)$ such that $X(n) \neq 1$. From

$$X(n) \sum_\chi \chi(n) = \sum_\chi X(n)\chi(n) = \sum_\chi \chi(n),$$

we have

$$(X(n) - 1) \sum_{\chi} \chi(n) = 0,$$

and the theorem is proved. □

Theorem 2.5. *The total number of characters is* $\varphi(m)$.

In other words, the characters that we constructed are precisely all the characters.

Proof. From Theorems 2.3 and 2.4, we have

$$\sum_{n,\chi} \chi(n) = \begin{cases} \sum_{n} \sum_{\chi} \chi(n) = c, \\ \sum_{\chi} \sum_{n} \chi(n) = \varphi(m). \quad \square \end{cases}$$

Definition. We call (1) the *standard factorization of a character*. More specifically we let

$$\chi_1(n, 2^l) = (-1)^{(n-1)/2}, \qquad \chi_2(n, 2^l) = e^{2\pi i b/2^{l-2}}, \qquad \chi(n, p^l) = e^{2\pi i \operatorname{ind} n/\varphi(p^l)}$$

(the definition of b is given in Theorem 3.9.3). Let $m = 2^a \prod_{p_\nu} p_\nu^{l_\nu}$ be the standard factorization of m. Then any character $\chi(n) \bmod m$ has the factorization:

$$\chi(n) = \begin{cases} \prod_{p_\nu} (\chi(n, p_\nu^{l_\nu}))^{c_\nu}, & \text{if } a = 0, 1, \\ (\chi_1(n, 2^l))^{c_0} \prod_{p_\nu} (\chi(n, p_\nu^{l_\nu}))^{c_\nu}, & \text{if } a = 2, \\ (\chi_1(n, 2^l))^{c_\nu} (\chi_2(n, 2^l))^{c_0'} \prod_{p_\nu} (\chi(n, p_\nu^{l_\nu}))^{c_\nu}, & \text{if } a \geqslant 3, \end{cases}$$

$$(c_0 = 0, 1, \qquad 0 \leqslant c_0' < 2^{l-2}, \qquad 0 \leqslant c_\nu < \varphi(p_\nu^{l_\nu})).$$

Exercise 1. If $\chi \neq \chi_0$, then for any two positive integers u and v ($v \geqslant u$) we have

$$\left| \sum_{n=u}^{v} \chi(n) \right| \leqslant \frac{\varphi(m)}{2}.$$

Exercise 2. If $(l, m) = 1$, then

$$\sum_{\chi} \frac{\chi(n)}{\chi(l)} = \begin{cases} \varphi(m), & \text{when } n \equiv l \pmod{m}, \\ 0, & \text{when } n \not\equiv l \pmod{m}. \end{cases}$$

7.3 Types of Characters

Definition. A character $\chi(n) \bmod m$ is said to be *primitive* if, for every divisor M of m, $0 < M < m$, there exists an integer a satisfying

$$a \equiv 1 \pmod{M}, \qquad (a, m) = 1, \qquad \chi(a) \neq 1.$$

A character which is not primitive is called an *improper character*.

Example 1. If $m > 1$, then the principal character mod m is improper, since 1 is a divisor of m.

Example 2. If $m = p$, then any non-principal character mod p is primitive.

Example 3. If $m = p^l$ ($l > 1$) and p is an odd prime, then a necessary and sufficient condition for the character

$$\chi_a(n) = e^{2\pi i a \operatorname{ind} n/\varphi(m)}$$

to be improper is that $p|a$. Thus, every improper character mod p^l induces a character mod p^{l-1}.

Example 4. $m = 2^l$.

If $l = 1$, then there is only the principal character. If $l = 2$, then the non-principal character

$$\chi(1) = 1, \qquad \chi(3) = -1$$

is primitive. When $l \geqslant 3$, if

$$\chi_{a,c}(n) = (-1)^{(n-1)a/2} e^{2\pi i c b/2^{l-2}}$$

is an improper character, then

$$\chi_{a,c}(n) = \chi_{a,c}(n + 2^{l-1})$$

and the converse also holds. That is

$$\begin{aligned}(-1)^{\frac{n-1}{2}a} e^{2\pi i c b/2^{l-2}} &= (-1)^{\frac{1}{2}a(n-1+2^{l-1})} e^{2\pi i c b'/2^{l-2}} \\ &= (-1)^{\frac{1}{2}a(n-1)} e^{2\pi i c b'/2^{l-2}},\end{aligned}$$

or

$$c(b - b') \equiv 0 \pmod{2^{l-2}},$$

where the definition of b' is given by

$$n + 2^{l-1} \equiv (-1)^{\frac{n-1}{2}} 5^{b'} \pmod{2^l}.$$

From

$$\begin{aligned}n + 2^{l-1} &\equiv n + n2^{l-1} \pmod{2^l} \\ &\equiv n(1 + 2^{l-1}) \pmod{2^l} \\ &\equiv n5^{2^{l-3}} \pmod{2^l},\end{aligned}$$

we have

$$b' \equiv b + 2^{l-3} \pmod{2^{l-2}}.$$

That is, a necessary and sufficient condition for $\chi_{a,c}(n)$ to be primitive is that $2 \nmid c$.

Let us take a more specific example. When $l = 3$,

$$\chi_{a,c}(n) = (-1)^{\frac{n-1}{2}a + cb},$$

where $b = 0, 1, 1, 0$ when $n = 1, 3, 5, 7$. If $c = 1$, then

$$\chi_{a,1}(1) = 1, \qquad \chi_{a,1}(3) = -(-1)^a,$$
$$\chi_{a,1}(5) = -1, \qquad \chi_{a,1}(7) = (-1)^a$$

are primitive characters, and we can simply write them as

$$\chi_{0,1}(n) = \left(\frac{2}{n}\right) \quad \text{and} \quad \chi_{1,1}(n) = \left(\frac{-2}{n}\right).$$

When $c = 0$, $a = 1$,

$$\chi_{1,0}(1) = 1, \qquad \chi_{1,0}(3) = -1,$$
$$\chi_{1,0}(5) = 1, \qquad \chi_{1,0}(7) = -1$$

is an improper character, that is $\chi_{1,0} = (-1/n)$.

In the character representation in §2 we have

$$\chi(n) = \prod_{\nu} \chi^{(\nu)}(n).$$

If one of the characters $\chi^{(\nu)}(n)$ is improper, then $\chi(n)$ itself is also improper. Conversely, if $\chi(n)$ is an improper character, then at least one of the characters $\chi^{(\nu)}(n)$ is improper.

We next investigate the situation under which there is a real valued primitive character. If a character is real, then each of its factor characters is also real. When p is an odd prime, in

$$(\chi(n, p^l))^{c_\nu} = e^{2\pi i c_\nu \operatorname{ind} n/\varphi(p^l)}$$

the value c_ν must be a multiple of $\varphi(p^l)/2$. If this character is also primitive, then from Example 3, l must be equal to 1.

Suppose that

$$(\chi_2(n, 2^l))^{c_0'} = e^{2\pi i c_0' b/2^{l-2}}$$

is a real character. Then we must have

$$2^{l-3} | c_0'.$$

If this character is also primitive, then from Example 4, we must have $l \leqslant 3$. Therefore there can be no real primitive character if $l > 3$. There cannot be any real primitive character either if $l = 1$. For if $m = 2m'$, $2 \nmid m'$, then from

$$n \equiv n' \pmod{m'}, \qquad (n, m) = 1, \qquad (n', m) = 1$$

we deduce that $n \equiv n' \pmod{m}$ giving $\chi(n) = \chi(n')$ so that $\chi(n)$ is improper. Summarizing, the possibility for the existence of real primitive character occurs when

$$m = 2^a p_1 p_2 \cdots p_s$$

where p_i are distinct odd primes and $a = 0, 2, 3$. Moreover, if the character is primitive, then $c_v = \varphi(p)/2$ or

$$(\chi(n, p))^{\frac{1}{2}(p-1)} = e^{\pi i \operatorname{ind} n} = \left(\frac{n}{p}\right).$$

Thus, if $a = 0$, then the real primitive character is the Jacobi symbol

$$\left(\frac{n}{m}\right), \qquad (n, m) = 1.$$

If $a = 2$, then the real primitive character is

$$(-1)^{\frac{n-1}{2}}\left(\frac{n}{m/4}\right), \qquad (n, m) = 1,$$

and if $a = 3$, then there are two types of real primitive character:

$$(-1)^{\frac{3}{8}(n^2-1)}\left(\frac{n}{m/8}\right), \qquad (n, m) = 1,$$

$$(-1)^{\frac{n-1}{2}+\frac{n^2-1}{8}}\left(\frac{n}{m/8}\right) = (-1)^{\frac{3}{8}((n-2)^2-9)}\left(\frac{n}{m/8}\right), \qquad (n, m) = 1.$$

7.4 Character Sums

Let

$$S(a, \chi) = \sum_{n=1}^{m} \chi(n) e^{2\pi i a n/m}.$$

Theorem 4.1. *Let $(m_1, m_2) = 1$ and let χ be factorized into*

$$\chi(n) = \chi_1(n)\chi_2(n)$$

where $\chi_1(n)$ is a character $\bmod m_1$ *and $\chi_2(n)$ is a character* $\bmod m_2$. *Then*

$$S(a, \chi) = \chi_1(m_2)\chi_2(m_1)S(a, \chi_1)S(a, \chi_2).$$

Proof. Let $n = m_1 n_2 + m_2 n_1$. Then as n_1, n_2 run over the complete sets of residues $\bmod m_1$, $\bmod m_2$ respectively, n runs over the complete set of residues $\bmod m_1 m_2$.

Therefore

$$S(a,\chi) = \chi_1(m_2)\chi_2(m_1) \sum_{n_1=1}^{m_1} \sum_{n_2=1}^{m_2} \chi_1(n_1)\chi_2(n_2)e^{2\pi i a(m_1 n_2 + m_2 n_1)/m_1 m_2}$$

$$= \chi_1(m_2)\chi_2(m_1)S(a,\chi_1)S(a,\chi_2). \quad \square$$

Thus the study of character sums mod m is reduced to that of character sums to a prime power modulus.

Theorem 4.2. *Let $m = p^l$. If $p|a$ and χ is a primitive character, or if $p \nmid a$ and χ is an improper character (but we exclude the case $l = 1$, $\chi = \chi_0$), then*

$$S(a,\chi) = 0.$$

Proof. We make the substitution

$$n = x(1 + p^{l-1}y).$$

When $1 \leqslant x \leqslant p^{l-1}$, $p \nmid x$ and $1 \leqslant y \leqslant p$, the number n runs over the reduced residue system mod p^l, and conversely. Therefore

$$S(a,\chi) = \sum_{\substack{x=1 \\ p \nmid x}}^{p^{l-1}} \chi(x)e^{2\pi i a x/p^l} \sum_{y=1}^{p} \chi(1 + p^{l-1}y)e^{2\pi i a x y/p}.$$

If $\chi(n)$ is improper, then $\chi(1 + p^{l-1}y) = 1$, so that

$$S(a,\chi) = \begin{cases} 0, & \text{if } p \nmid a, \\ p \sum_{x=1}^{p^{l-1}} \chi(x)e^{2\pi i a x/p^l}, & \text{if } p|a. \end{cases}$$

If $\chi(n)$ is primitive, then there exists u such that $\chi(1 + p^{l-1}u) \neq 1$; now $p|a$ and so from

$$\chi(1 + p^{l-1}u) \sum_{y=1}^{p} \chi(1 + p^{l-1}y) = \sum_{y=1}^{p} \chi(1 + p^{l-1}(y + u))$$

$$= \sum_{y=1}^{p} \chi(1 + p^{l-1}y),$$

we have

$$\sum_{y=1}^{p} \chi(1 + p^{l-1}y) = 0.$$

Therefore $S(a,\chi) = 0$ also. $\square$

We shall write

$$\tau(\chi) = S(1,\chi).$$

If $(a, m) = 1$, then

$$\chi(a)S(a, \chi) = \sum_{n=1}^{m} \chi(an)e^{2\pi ian/m}$$
$$= S(1, \chi).$$

Theorem 4.3. *Let*

$$C_q(n) = \sum_{(a, q)=1} e^{2\pi ian/q},$$

where a runs over a reduced set of residues mod q. *Then*

1) $c_q(n)$ *is a multiplicative function of q; that is if* $(q_1, q_2) = 1$, *then*

$$C_{q_1}(n)C_{q_2}(n) = C_{q_1q_2}(n);$$

2)
$$C_{p^l}(n) = \begin{cases} p^l - p^{l-1}, & \text{if } p^l|n, \\ -p^{l-1}, & \text{if } p^l \nmid n, \quad p^{l-1}|n, \\ 0, & \text{if } p^{l-1} \nmid n; \end{cases}$$

3)
$$C_q(1) = \mu(q).$$

Proof. 1) can be proved by the substitution $a = q_1a_2 + q_2a_1$ with the familiar method described earlier.

2) follows from

$$C_{p^l}(n) = \sum_{a=1}^{p^l} e^{2\pi ian/p^l} - \sum_{a=1}^{p^{l-1}} e^{2\pi ian/p^{l-1}}.$$

3) follows from 1) and 2). □

Theorem 4.4. *If* $\chi(n)$ *is a primitive character, then*

$$|\tau(\chi)|^2 = m.$$

Proof. First consider the case $m = p^l$. We have easily

$$\begin{aligned}|\tau(\chi)|^2 &= \tau(\chi)\bar{\tau}(\chi) \\ &= \sum_{n=1}^{p^l} \chi(n)e^{2\pi in/p^l} \sum_{q=1}^{p^l} \bar{\chi}(q)e^{-2\pi iq/p^l} \\ &= \sum_{n=1}^{p^l} \chi(n)e^{2\pi in/p^l} \sum_{q=1}^{p^l} \bar{\chi}(nq)e^{-2\pi inq/p^l} \\ &= \sum_{q=1}^{p^l} \bar{\chi}(q) \sum_{\substack{n=1 \\ p \nmid n}}^{p^l} e^{2\pi i(1-q)n/p^l}.\end{aligned}$$

If $p^{l-1} \nmid (q-1)$, then from Theorem 4.3, the inner sum on the right hand side in the above is 0. We need therefore only examine the situation when $p^{l-1}|(q-1)$, that is

$q = 1 + p^{l-1}u$, $0 \leqslant u \leqslant p-1$. But now clearly

$$|\tau(\chi)|^2 = p^l - p^{l-1} - \sum_{u=1}^{p-1} \bar{\chi}(1 + p^{l-1}u)p^{l-1}$$

$$= p^l - p^{l-1} \sum_{u=1}^{p} \bar{\chi}(1 + p^{l-1}u).$$

Now if $\chi(n)$ is primitive, then there exists v such that $\chi(1 + p^{l-1}v) \neq 0, 1$ so that $\bar{\chi}(1 + p^{l-1}v) \neq 0, 1$. From

$$\bar{\chi}(1 + p^{l-1}v) \sum_{u=1}^{p} \bar{\chi}(1 + p^{l-1}u) = \sum_{u=1}^{p} \bar{\chi}(1 + p^{l-1}(u + v)) = \sum_{u=1}^{p} \bar{\chi}(1 + p^{l-1}u),$$

we have

$$\sum_{u=1}^{p} \bar{\chi}(1 + p^{l-1}u) = 0.$$

Therefore the case $m = p^l$ is proved, and the general case follows at once from Theorem 4.1. □

We see therefore that

$$\tau(\chi) = \varepsilon\sqrt{m}, \qquad |\varepsilon| = 1.$$

However, the determination of ε is no easy matter. For real primitive characters we know much more and in the next section we shall determine ε when χ is a real primitive character.

Theorem 4.5. *Let χ be a real primitive character. Then, for odd m, we have*

$$\tau(\chi) = \begin{cases} \pm\sqrt{m} & \text{if } m \equiv 1 \pmod 4, \\ \pm i\sqrt{m} & \text{if } m \equiv 3 \pmod 4. \end{cases}$$

Proof. This is similar to the proof of Theorem 4.4. If $m = p$, then

$$(\tau(\chi))^2 = \sum_{q=1}^{p} \chi(q) \sum_{n=1}^{p-1} e^{2\pi i(1+q)n/p} = \chi(-1)p.$$

We already have

$$\chi(-1) = \left(\frac{-1}{p}\right) = (-1)^{\frac{p-1}{2}},$$

so that the theorem follows. □

7.5 Gauss Sums

The trigonometric sum

$$S(n, m) = \sum_{x=0}^{m-1} e^{2\pi i x^2 n/m}, \qquad (n, m) = 1$$

is the famous Gauss sum. In this formula the summation can be taken over any complete set of residues mod m.

Theorem 5.1. *If* $(m, m') = 1$, *then*

$$S(n, mm') = S(nm', m)S(nm, m').$$

Proof. Let $x = my + m'z$. Then

$$\begin{aligned} S(n, mm') &= \sum_{x=1}^{mm'} e^{2\pi i x^2 n/mm'} \\ &= \sum_{y=1}^{m'} \sum_{z=1}^{m} e^{2\pi i n(my+m'z)^2/mm'} \\ &= \sum_{y=1}^{m'} e^{2\pi i m n y^2/m'} \sum_{z=1}^{m} e^{2\pi i m' n z^2/m} \end{aligned}$$

and hence the result. □

We see that in order to evaluate a Gauss sum we need only deal with the case $m = p^l$.

Theorem 5.2. *Let*

$$\delta = \begin{cases} 1, & \text{when } p \text{ is an odd prime,} \\ 2, & \text{when } p = 2. \end{cases}$$

Then, for $l \geqslant 2\delta$, *we have*

$$S(n, p^l) = pS(n, p^{l-2}).$$

Proof. Let $x = y + p^{l-\delta}z$. Then, from $2(l - \delta) \geqslant l$, we have

$$\begin{aligned} S(n, p^l) &= \sum_{y=1}^{p^{l-\delta}} \sum_{z=1}^{p^\delta} e^{2\pi i(y+p^{l-\delta}z)^2 n/p^l} \\ &= \sum_{y=1}^{p^{l-\delta}} e^{2\pi i y^2 n/p^l} \cdot \sum_{z=1}^{p^\delta} e^{4\pi i y z n/p^\delta} \\ &= p^\delta \sum_{\substack{y=1 \\ p \mid y}}^{p^{l-\delta}} e^{2\pi i y^2 n/p^l} \\ &= p^\delta \sum_{x=1}^{p^{l-\delta-1}} e^{2\pi i x^2 n/p^{l-2}}. \end{aligned}$$

When $p > 2$, this is what is required. When $p = 2$, then from

$$p \sum_{x=1}^{p^{l-3}} e^{2\pi i x^2 n/p^{l-2}} = \sum_{x=1}^{p^{l-2}} e^{2\pi i x^2 n/p^{l-2}},$$

the result also follows. □

From this theorem we see that the crucial points in the evaluation of a Gauss sum rest on the determination of

$$S(n,2), \qquad S(n,4), \qquad S(n,8)$$

and

$$S(n,p), \qquad p \text{ an odd prime.}$$

Theorem 5.3. *If* $2 \nmid n$, *then*

$$S(n,2) = 0,$$
$$S(n,4) = 2(1+i^n),$$
$$S(n,8) = 4e^{\frac{\pi i}{4}n}.$$

Proof. Clearly we have

$$S(n,2) = 1 + e^{\frac{2\pi i}{2}n} = 1 - 1 = 0,$$

$$S(n,4) = 1 + e^{\frac{2\pi i}{4}n} + e^{\frac{2\pi i}{4}4n} + e^{\frac{2\pi i}{4}9n}$$
$$= 1 + i^n + 1 + i^n = 2(1+i^n),$$

$$S(n,8) = 2(1 + e^{\frac{2\pi i}{8}n} + e^{\frac{2\pi i}{8}4n} + e^{\frac{2\pi i}{8}9n})$$
$$= 4e^{\frac{\pi i}{4}n}. \qquad \square$$

Theorem 5.4. *If p is an odd prime, then*

$$S(n,p) = \left(\frac{n}{p}\right) S(1,p) = \left(\frac{n}{p}\right) \tau(\chi).$$

Here

$$\chi(a) = \left(\frac{a}{p}\right).$$

Proof. The number of solutions to the congruence

$$x^2 \equiv u \pmod{p}$$

is

$$1 + \left(\frac{u}{p}\right),$$

and therefore

$$\sum_{x=1}^{p} e^{2\pi i x^2 n/p} = \sum_{u=1}^{p} \left(1 + \left(\frac{u}{p}\right)\right) e^{2\pi i u n/p} = \sum_{u=1}^{p} \left(\frac{u}{p}\right) e^{2\pi i u n/p}$$
$$= \left(\frac{n}{p}\right) \sum_{v=1}^{p} \left(\frac{v}{p}\right) e^{2\pi i v/p},$$

which is the required result. $\square$

Theorem 5.5.

$$S(1,p) = \begin{cases} \sqrt{p}, & \textit{if} \quad p \equiv 1 \pmod 4, \\ i\sqrt{p}, & \textit{if} \quad p \equiv 3 \pmod 4. \end{cases}$$

Proof. From the above theorem and Theorem 4.5 we have

$$S(1,p) = \begin{cases} \pm\sqrt{p}, & \text{if} \quad p \equiv 1 \pmod 4, \\ \pm i\sqrt{p}, & \text{if} \quad p \equiv 3 \pmod 4, \end{cases}$$

which, combining into a single formula, gives

$$\tfrac{1}{2}(1 + i^p)(1 - i)S(1,p) = \pm\sqrt{p}.$$

If we can prove that

$$\Re\{\tfrac{1}{2}(1 + i^p)(1 - i)S(1,p)\} > -\sqrt{p},$$

where $\Re\{x\}$ represents the real part of x, then the theorem will follow. Now it is easy to see that

$$S(1,p) - 1 = \sum_{x=1}^{p-1} e^{2\pi i x^2/p} = \sum_{x=1}^{\frac{1}{2}(p-1)} (e^{2\pi i x^2/p} + e^{2\pi i (p-x)^2/p})$$

$$= 2\sum_{x=1}^{\frac{1}{2}(p-1)} e^{2\pi i x^2/p}. \tag{1}$$

Let $f(x)$ be any function. Then

$$\sum_{x=1}^{\frac{1}{2}(p-1)} f(x) + \sum_{x=1}^{\frac{1}{2}(p-1)} f\left(\frac{p}{2} - x\right) = \sum_{x=1}^{p-1} f\left(\frac{x}{2}\right).$$

This formula clearly holds because the first term on the left hand side is merely the sum of those terms on the right hand side when x is even, and the second term is the sum on the right hand side when x is odd. We take $f(x) = e^{2\pi i x^2/p}$ and note that $f(\frac{p}{2} - x) = i^p e^{2\pi i x^2/p}$. Then, from (1), we have

$$\tfrac{1}{2}(1 + i^p)(S(1,p) - 1) = \sum_{x=1}^{p-1} e^{2\pi i x^2/4p} = W + Z, \tag{2}$$

where

$$W = \sum_{x \leqslant \sqrt{p}} e^{2\pi i x^2/4p}, \qquad Z = \sum_{\sqrt{p} < x \leqslant p-1} e^{2\pi i x^2/4p}. \tag{3}$$

From (2) we have

$$\tfrac{1}{2}(1 + i^p)(1 - i)S(1,p) - \tfrac{1}{2}(1 + i^p)(1 - i) = (1 - i)(W + Z).$$

Since $\Re\{\frac{1}{2}(1 + i^p)(1 - i)\}$ is 1 or 0, it follows that

$$\Re\{\tfrac{1}{2}(1 + i^p)(1 - i)S(1,p)\} \geqslant \Re\{(1 - i)(W + Z)\} \geqslant \Re(1 - i)W - \sqrt{2}|Z|. \tag{4}$$

From $\cos x + \sin x \geqslant 1$ when $0 \leqslant x \leqslant \pi/2$, we deduce that

$$\Re\{(1-i)W\} = \sum_{x \leqslant \sqrt{p}} \left(\cos\frac{\pi x^2}{2p} + \sin\frac{\pi x^2}{2p}\right) \geqslant [\sqrt{p}] \geqslant \frac{1}{2}\sqrt{p}. \tag{5}$$

On the other hand, if we write in Z,

$$v_x = e^{2\pi i x(x+1)/4p}, \qquad w_x = \operatorname{cosec}\frac{\pi x}{2p}, \qquad q = [\sqrt{p}],$$

then

$$(v_x - v_{x-1})w_x = 2ie^{2\pi i x^2/4p}. \tag{6}$$

Therefore, from (3) and (6) we have

$$2iZ = \sum_{x=q+1}^{p-1} (v_x - v_{x-1})w_x,$$

that is

$$\begin{aligned} 2|Z| &= \left|\sum_{x=q+1}^{p-1} v_x(w_x - w_{x+1}) + v_{p-1}w_p - v_q w_{q+1}\right| \\ &\leqslant \sum_{x=q+1}^{p-1} (w_x - w_{x+1}) + w_p + w_{q+1} = 2w_{q+1} \\ &\leqslant \frac{2p}{q+1} \leqslant 2\sqrt{p} \end{aligned} \tag{7}$$

(because w_x is decreasing). From (4), (5) and (7) we finally have

$$\Re\{\tfrac{1}{2}(1+i^p)(1-i)S(1,p)\} \geqslant (\tfrac{1}{2} - \sqrt{2})\sqrt{p} > -\sqrt{p}.$$

The theorem is therefore proved. □

Summarizing we have the following result:

Theorem 5.6. *If m is odd, then*

$$S(n,m) = \begin{cases} \left(\dfrac{n}{m}\right)\sqrt{m}, & \text{if } m \equiv 1 \pmod 4, \\ i\left(\dfrac{n}{m}\right)\sqrt{m}, & \text{if } m \equiv 3 \pmod 4. \end{cases}$$

Proof. We use induction on the number of distinct prime divisors of m. If $m = p^l$, then we have by Theorems 5.2 and 5.4, that

$$S(n,p^l) = \begin{cases} p^{\frac{l}{2}}, & \text{if } 2|l, \\ p^{\frac{1}{2}(l-1)}S(n,p) = \left(\dfrac{n}{p}\right)p^{\frac{1}{2}(l-1)}S(1,p) \end{cases}$$

$$=\begin{cases}\left(\frac{n}{p}\right)p^{\frac{l}{2}}, & \text{if } 2\nmid l, \quad p\equiv 1 \pmod 4,\\ i\left(\frac{n}{p}\right)p^{\frac{l}{2}}, & \text{if } 2\nmid l, \quad p\equiv 3 \pmod 4.\end{cases}$$

Moreover, from Theorem 5.1 and the induction hypothesis, we have

$$\begin{aligned}S(n,mm') &= S(nm',m)S(nm,m')\\ &= \left(\frac{nm'}{m}\right)\left(\frac{nm}{m'}\right)i^{\left(\frac{m-1}{2}\right)^2}\sqrt{m}\cdot i^{\left(\frac{m'-1}{2}\right)^2}\sqrt{m'}\\ &= \left(\frac{n}{mm'}\right)\left(\frac{m'}{m}\right)\left(\frac{m}{m'}\right)i^{\left(\frac{m-1}{2}\right)^2+\left(\frac{m'-1}{2}\right)^2}\sqrt{mm'}\\ &= \left(\frac{n}{mm'}\right)(-1)^{\frac{m-1}{2}\frac{m'-1}{2}}i^{\left(\frac{m-1}{2}\right)+\left(\frac{m'-1}{2}\right)^2}\sqrt{mm'}\\ &= \left(\frac{n}{mm'}\right)\sqrt{mm'}\,i^{\left(\frac{m+m'}{2}-1\right)^2}\\ &= \begin{cases}\left(\frac{n}{mm'}\right)\sqrt{mm'}, & \text{if } mm'\equiv 1 \pmod 4,\\ i\left(\frac{n}{mm'}\right)\sqrt{mm'}, & \text{if } mm'\equiv 3 \pmod 4.\end{cases}\end{aligned}$$

(Here we have used the law of quadratic reciprocity.) □

Theorem 5.7.

$$S(n,2^l)=\begin{cases}0, & \textit{if } l=1\\ (1+i^n)2^{\frac{l}{2}}, & \textit{if } l \textit{ is even}\\ 2^{\frac{l+1}{2}}e^{\frac{\pi i}{4}n}, & \textit{if } l>1 \textit{ and odd.}\end{cases}$$

Proof. From Theorem 5.3 we see that the result holds when $l = 1, 2, 3$. When $l > 3$ the result follows from Theorems 5.2 and 5.3. □

Theorem 5.8. *Let $\chi(n)$ be a real primitive character. Then*

$$\tau(\chi)=\begin{cases}\sqrt{m}, & \textit{if } \chi(-1)=1,\\ i\sqrt{m}, & \textit{if } \chi(-1)=-1.\end{cases}$$

Proof. From §3 we know that m can be written as $m = 2^a m'$, where $a = 0, 2, 3$ and m' is a product of distinct primes; moreover

1) if $a = 0$, then

$$\chi(n)=\left(\frac{n}{m}\right), \qquad (n,m)=1;$$

2) if $a = 2$, then

$$\chi(n) = (-1)^{\frac{n-1}{2}}\left(\frac{n}{m'}\right), \qquad (n, m) = 1;$$

3) if $a = 3$, then

$$\chi(n) = (-1)^{\frac{1}{8}(n^2-1)}\left(\frac{n}{m'}\right) \quad \text{or} \quad (-1)^{\frac{1}{2}(n-1)+\frac{1}{8}(n^2-1)}\left(\frac{n}{m'}\right), \qquad (n, m) = 1.$$

Here $\left(\frac{n}{m}\right)$ and $\left(\frac{n}{m'}\right)$ are Jacobi's symbols. We now consider the three separate cases.

1) $a = 0$. Let $m = p_1 \cdots p_s$ and we use induction on s. When $s = 1$ the result follows from Theorems 5.4 and 5.5. Let $s > 1$ and put $m = p_1 m'$. Then, from Theorem 4.1 we have

$$\tau(\chi) = \chi_1(m')\chi_2(p_1)\tau(\chi_1)\tau(\chi_2),$$

where χ_1, χ_2 have the moduli p_1, m' respectively, and $\chi(n) = \chi_1(n)\chi_2(n)$. Therefore, from Theorem 3.6.4 and the induction hypothesis, we have

$$\begin{aligned}
\tau(\chi) &= \left(\frac{m'}{p_1}\right)\left(\frac{p_1}{m'}\right) \cdot \begin{Bmatrix} \sqrt{p_1} \\ i\sqrt{p_1} \end{Bmatrix} \cdot \begin{Bmatrix} \sqrt{m'} \\ i\sqrt{m'} \end{Bmatrix} \\
&= (-1)^{\frac{p_1-1}{2}\frac{m'-1}{2}} \cdot \begin{Bmatrix} \sqrt{p_1} \\ i\sqrt{p_1} \end{Bmatrix} \cdot \begin{Bmatrix} \sqrt{m'} \\ i\sqrt{m'} \end{Bmatrix} \\
&= \begin{cases} \sqrt{p_1 m'} = \sqrt{m}, & \text{if } m \equiv 1 \pmod 4 \quad \text{or} \quad \chi(-1) = 1, \\ i\sqrt{p_1 m'} = i\sqrt{m}, & \text{if } m \equiv 3 \pmod 4 \quad \text{or} \quad \chi(-1) = -1. \end{cases}
\end{aligned}$$

2) $a = 2$. Let $m = 2^2 m'$. If $m' = 1$, then $\chi(1) = 1$, $\chi(3) = -1$ so that

$$\tau(\chi) = \sum_{n=1}^{4} \chi(n) e^{2\pi i n/4} = e^{2\pi i/4} - e^{6\pi i/4} = 2i.$$

If $m' > 1$, then from Theorem 4.1 and 1)

$$\begin{aligned}
\tau(\chi) &= (-1)^{\frac{m'-1}{2}}\left(\frac{4}{m'}\right) 2i \\
&\quad \cdot \begin{cases} \sqrt{m'} = i\sqrt{m}, & \text{if } m' \equiv 1 \pmod 4 \quad \text{or} \quad \chi(-1) = -1, \\ i\sqrt{m'} = \sqrt{m}, & \text{if } m' \equiv 3 \pmod 4 \quad \text{or} \quad \chi(-1) = 1. \end{cases}
\end{aligned}$$

3) $a = 3$. Let $m = 2^3 m'$. When $m' = 1$, we have

$$\tau(\chi) = \sum_{n=1}^{8} \chi(n) e^{2\pi i n/8} = \begin{cases} e^{2\pi i/8} - e^{6\pi i/8} - e^{10\pi i/8} + e^{14\pi i/8} = \sqrt{8}, & \text{if } \chi(-1) = 1, \\ e^{2\pi i/8} + e^{6\pi i/8} - e^{10\pi i/8} - e^{14\pi i/8} = i\sqrt{8}, & \text{if } \chi(-1) = -1. \end{cases}$$

Suppose that $m' > 1$. If $\chi(n) = (-1)^{\frac{1}{8}(n^2-1)}\left(\frac{n}{m'}\right)$, then

$$\tau(\chi) = (-1)^{\frac{1}{8}(m'^2-1)}\left(\frac{8}{m'}\right)\sqrt{8}$$

$$\cdot\begin{cases}\sqrt{m'} = \sqrt{m}, & \text{if} \quad m' \equiv 1 \pmod 4 \quad \text{or} \quad \chi(-1) = 1,\\ i\sqrt{m'} = i\sqrt{m}, & \text{if} \quad m' \equiv 3 \pmod 4 \quad \text{or} \quad \chi(-1) = -1.\end{cases}$$

If $\chi(n) = (-1)^{\frac{1}{2}(n-1)+\frac{1}{8}(n^2-1)}\left(\frac{n}{m'}\right)$, then

$$\tau(\chi) = (-1)^{\frac{1}{2}(m'-1)+\frac{1}{8}(m'^2-1)}\left(\frac{8}{m'}\right)i\sqrt{8}$$

$$\cdot\begin{cases}\sqrt{m'} = i\sqrt{m}, & \text{if} \quad m' \equiv 1 \pmod 4 \quad \text{or} \quad \chi(-1) = -1,\\ i\sqrt{m'} = \sqrt{m}, & \text{if} \quad m' \equiv 3 \pmod 4 \quad \text{or} \quad \chi(-1) = 1.\end{cases}$$

Collecting 1), 2) and 3) the theorem is proved. □

7.6 Character Sums and Trigonometric Sums

We have seen in the previous section the relationship between Gauss sums and character sums. We now proceed to establish certain relationships between trigonometric sums and character sums.

Theorem 6.1. *Let p be a prime, and $d|p-1$. Then a necessary and sufficient condition for an integer x to be a d-th power non-residue* mod p *is that*

$$\frac{1}{d}\sum_{a=1}^{d} e^{2\pi i a \operatorname{ind} x/d} = 0;$$

otherwise the formula is equal to 1.

Proof. By Theorem 3.8.1 whether x is a d-th power residue or not depends on whether $d|\operatorname{ind} x$ or $d\nmid\operatorname{ind} x$. Using trigonometric sums this means that

$$\frac{1}{d}\sum_{a=1}^{d} e^{2\pi i a \operatorname{ind} x/d} = \begin{cases}1, & \text{if } x \text{ is a } d\text{-th power residue mod } p,\\ 0, & \text{if } x \text{ is a } d\text{-th power non-residue mod } p.\end{cases}$$ □

Theorem 6.2. *Let p be a prime, $p\nmid a$, $(p-1, k) = d$. Then*

$$\sum_{x=1}^{p} e^{2\pi i a x^k/p} = \sum_{b=1}^{d-1} S(a, \chi^b),$$

where

$$\chi(u) = e^{2\pi i \operatorname{ind} u/d}.$$

Proof. The congruence $x^k \equiv u \pmod p$ has either no root, or $d = (p-1, k)$ roots. Therefore, from Theorem 6.1 we have

$$\sum_{x=1}^{p} e^{2\pi i a x^k/p} = 1 + \sum_{u=1}^{p-1} e^{2\pi i a u/p} \sum_{b=1}^{d} e^{2\pi i b \operatorname{ind} u/d}$$

$$= 1 + \sum_{b=1}^{d} \sum_{u=1}^{p-1} e^{2\pi i a u/p} \chi^b(u)$$

$$= 1 + \sum_{u=1}^{p-1} e^{2\pi i a u/p} + \sum_{b=1}^{d-1} \sum_{u=1}^{p-1} e^{2\pi i a u/p} \chi^b(u)$$

$$= \sum_{b=1}^{d-1} S(a, \chi^b). \quad \square$$

From Theorem 4.5 we see that $|S(a, \chi^b)| \leqslant \sqrt{p}$ so that we have:

Theorem 6.3. *Let $d = (k, p-1)$. Then*

$$\left| \sum_{x=1}^{p} e^{2\pi i a x^k/p} \right| \leqslant (d-1)\sqrt{p}. \quad \square$$

Exercise. Study the trigonometric sum

$$\sum_{x=0}^{m-1} e^{2\pi i x^k n/m}, \qquad (n, m) = 1$$

by following Theorems 5.1 and 5.2.

7.7 From Complete Sums to Incomplete Sums

Theorem 7.1. *Let $g(x)$ be periodic with period q, and*

$$g(x) = \begin{cases} 1, & \text{if } 0 \leqslant x < m, \\ 0, & \text{if } m \leqslant x < q. \end{cases}$$

Then $g(x)$ is representable as

$$g(x) = \frac{m}{q} + \frac{1}{q} \sum_{n=1}^{q-1} e^{2\pi i n x/q} (1 - e^{-2\pi i n m/q})/(1 - e^{-2\pi i n/q}).$$

Proof. Clearly

$$g(x) = \frac{1}{q} \sum_{n=0}^{q-1} e^{2\pi i n x/q} \sum_{t=0}^{m-1} e^{-2\pi i n t/q}$$

$$= \frac{m}{q} + \frac{1}{q} \sum_{n=1}^{q-1} e^{2\pi i n x/q} \frac{1 - e^{-2\pi i n m/q}}{1 - e^{-2\pi i n/q}}. \quad \square$$

Theorem 7.2. *Let α be a real number and*

$$S = \sum_{q' < n \leqslant q''} e^{2\pi i n \alpha}.$$

Then

$$|S| \leqslant \min\left(q'' - q', \frac{1}{2\langle\alpha\rangle}\right),$$

where

$$\langle\alpha\rangle = \min(\alpha - [\alpha], [\alpha] + 1 - \alpha).$$

Proof. Clearly we have

$$|S| \leqslant q'' - q'.$$

If $\alpha \neq [\alpha]$, we let $Q = q'' - q'$ so that

$$|S| = \left|\sum_{n=0}^{Q-1} e^{2\pi i n\alpha}\right| = \left|\frac{1 - e^{2\pi i Q\alpha}}{1 - e^{2\pi i\alpha}}\right|$$

$$\leqslant \frac{2}{|1 - e^{2\pi i\alpha}|} = \frac{1}{|\sin \pi\alpha|}$$

$$\leqslant \frac{1}{2\langle\alpha\rangle}$$

(when $0 \leqslant \xi \leqslant \frac{1}{2}$, $\sin \pi\xi \geqslant 2\xi$, so that $|\sin \pi\xi| \geqslant 2\langle\xi\rangle$).

Theorem 7.3. *If* $2 \nmid q$, *then*

$$\left|\sum_{x=0}^{m-1} e^{2\pi i x^2/q} - \frac{m}{q}\sum_{x=0}^{q-1} e^{2\pi i x^2/q}\right| \leqslant \sqrt{q}\log q.$$

Proof. Clearly we can assume that $m \leqslant q$. From Theorem 7.1 we have

$$\sum_{x=0}^{m-1} e^{2\pi i x^2/q} = \sum_{x=0}^{q-1} e^{2\pi i x^2/q} g(x)$$

$$= \frac{m}{q}\sum_{x=0}^{q-1} e^{2\pi i x^2/q} + \frac{1}{q}\sum_{n=1}^{q-1}\sum_{x=0}^{q-1} e^{2\pi i(x^2+nx)/q}\frac{1 - e^{-2\pi i nm/q}}{1 - e^{-2\pi i n/q}}.$$

From the formula for a Gauss sum we have

$$\left|\sum_{x=0}^{q-1} e^{2\pi i(x^2+nx)/q}\right| = \left|\sum_{x=0}^{q-1} e^{2\pi i(x+\frac{1}{2}n)^2/q}\right|^* \leqslant \sqrt{q},$$

so that

$$\left|\sum_{x=0}^{q-1} e^{2\pi i x^2/q} - \frac{m}{q}\sum_{x=0}^{q-1} e^{2\pi i x^2/q}\right|$$

$$\leqslant \frac{1}{\sqrt{q}}\sum_{n=1}^{q-1}\frac{1}{2\left\langle\frac{n}{q}\right\rangle}$$

* Here $\frac{1}{2}$ represents the solution to the congruence $2x \equiv 1 \pmod q$.

$$\leqslant \frac{1}{\sqrt{q}} \sum_{n=1}^{\frac{1}{2}(q-1)} \frac{q}{n} = \sqrt{q} \sum_{n=1}^{\frac{1}{2}(q-1)} \frac{1}{n}$$

$$< \sqrt{q} \sum_{n=1}^{\frac{1}{2}(q-1)} \left(-\log\left(1 - \frac{1}{2n}\right) + \log\left(1 + \frac{1}{2n}\right)\right)$$

$$= \sqrt{q} \sum_{n=1}^{\frac{1}{2}(q-1)} (-\log(2n-1) + \log(2n+1))$$

$$= \sqrt{q} \log q. \quad \square$$

Theorem 7.4 (Pólya). *Let p be an odd prime, $1 \leqslant m \leqslant p$, and χ be a non-principal character* mod p. *Then*

$$\left| \sum_{x=0}^{m-1} \chi(x) \right| < \sqrt{p} \log p.$$

Proof. From Theorem 7.1 we have

$$\sum_{x=0}^{m-1} \chi(x) = \sum_{x=0}^{p-1} \chi(x) g(x)$$

$$= \frac{m}{p} \sum_{x=0}^{p-1} \chi(x) + \frac{1}{p} \sum_{x=0}^{p-1} \chi(x) \sum_{n=1}^{p-1} e^{2\pi i n x/p} \frac{1 - e^{-2\pi i n m/p}}{1 - e^{-2\pi i n/p}}.$$

From Theorem 2.3, Theorem 4.4 and Theorem 7.2 we have

$$\left| \sum_{x=0}^{m-1} \chi(x) \right| \leqslant \frac{1}{p} \sum_{n=1}^{p-1} \left| \frac{1 - e^{-2\pi i n m/p}}{1 - e^{-2\pi i n/p}} \right| \left| \sum_{x=0}^{p-1} \chi(x) e^{2\pi i n x/p} \right|$$

$$\leqslant \frac{1}{\sqrt{p}} \sum_{n=1}^{p-1} \frac{1}{2\left\langle \frac{n}{p} \right\rangle} < \sqrt{p} \log p. \quad \square$$

This theorem has the following application:

Theorem 7.5. *Let p be an odd prime and $d \mid (p-1)$. Then there is always a d-th power non-residue* mod p *which is less than $\sqrt{p} \log p$.*

Proof. Let R represent a d-th power residue not exceeding m. Then

$$R = \sum_{x=1}^{m} \frac{1}{d} \sum_{a=1}^{d} e^{2\pi i a \operatorname{ind} x/d} = \frac{1}{d} \sum_{a=1}^{d} \sum_{x=1}^{m} e^{2\pi i a \operatorname{ind} x/d}$$

$$= \frac{m}{d} + \frac{1}{d} \sum_{a=1}^{d-1} \sum_{x=1}^{m} (\chi(x))^a,$$

where $\chi(x) = e^{2\pi i \operatorname{ind} x/d}$. From Theorem 7.4, we have

$$\left| R - \frac{m}{d} \right| < \frac{d-1}{d} \sqrt{p} \log p, \tag{1}$$

and so

$$R < \frac{m}{d} + \frac{d-1}{d}\sqrt{p}\log p.$$

Now if $m = \sqrt{p}\log p$, then

$$R < \frac{m}{d} + \frac{d-1}{d}m = m,$$

so that a d-th power non-residue less than $\sqrt{p}\log p$ exists. □

In particular there must be a quadratic non-residue less than $\sqrt{p}\log p$. The determination of the smallest exponent δ such that the least quadratic residue satisfies $O(p^\delta)$ is a famous difficult problem. The result of Vinogradov is:

Theorem 7.6. *For sufficiently large p the least quadratic non-residue does not exceed*

$$p^{\frac{1}{2\sqrt{e}}}\log^2 p \ll p^{\frac{1}{3.2}}.$$

Proof. Let

$$T = [p^{\frac{1}{2\sqrt{e}}}\log^2 p], \qquad m = \sqrt{p}\log^2 p,$$

and suppose that $1, 2, \ldots, T$ are all quadratic residues. Since every quadratic non-residue must have a prime divisor which is also a quadratic non-residue, it follows that every quadratic non-residue not exceeding m must have a prime divisor q satisfying $T < q \leqslant m$. Therefore, denoting by N the number of quadratic non-residues not exceeding m, we have

$$N \leqslant \sum_{T<q\leqslant m}\left[\frac{m}{q}\right] < m\sum_{T<q\leqslant m}\frac{1}{q},$$

and hence, by Theorem 5.9.2,

$$\begin{aligned} N &< m\log\frac{\log m}{\log T} + O\left(\frac{m}{\log T}\right) \\ &= m\left(\frac{1}{2} + \log\frac{1+\dfrac{4\log\log p}{\log p}}{1+\dfrac{4\sqrt{e}\log\log p}{\log p}}\right) + O\left(\frac{m}{\log T}\right) \\ &= m\left(\frac{1}{2} - \frac{4(\sqrt{e}-1)\log\log p}{\log p}\right) + O\left(\frac{m}{\log T}\right). \end{aligned}$$

From (1) we have

$$N = \frac{m}{2} + O(\sqrt{p}\log p) = \frac{m}{2} + O\left(\frac{m}{\log p}\right)$$

so that

$$\frac{m}{2}+O\left(\frac{m}{\log p}\right)<m\left(\frac{1}{2}-\frac{4(\sqrt{e}-1)\log\log p}{\log p}\right)+O\left(\frac{m}{\log p}\right),$$

that is

$$\log\log p=O(1),$$

which is impossible if p is sufficiently large. The theorem is therefore proved. □

7.8 Applications of the Character Sum $\sum_{x=1}^{p}\left(\frac{x^2+ax+b}{p}\right)$

Theorem 8.1. *The number of integers a such that a and a + 1 are both quadratic residues* mod p *is*

$$\frac{1}{4}\left(p-4-\left(\frac{-1}{p}\right)\right).$$

Before we prove this theorem we have to evaluate a sum first.

Theorem 8.2. *Let* $p>2$, $a^2-4b\not\equiv 0 \pmod p$. *Then*

$$\sum_{x=1}^{p}\left(\frac{x^2+ax+b}{p}\right)=-1,$$

where in the formula the value 0 *is given to those terms in which* $p|x^2+ax+b$.

Proof. We can assume that $a=0$, since otherwise we can use the substitution $y=x+a/2$.

Now suppose that $p\nmid b$. From Euler's criterion we have

$$\sum_{x=1}^{p}\left(\frac{x^2+b}{p}\right)\equiv\sum_{x=1}^{p}(x^2+b)^{\frac{1}{2}(p-1)} \pmod p. \tag{1}$$

Let g be a primitive root of p. If $0<c<p-1$, then

$$\sum_{x=1}^{p}x^c\equiv\sum_{v=0}^{p-2}g^{cv}=\frac{1-g^{c(p-1)}}{1-g^c}\equiv 0 \pmod p.$$

Substituting this into (1) yields

$$\sum_{x=1}^{p}\left(\frac{x^2+b}{p}\right)\equiv\sum_{x=1}^{p}x^{p-1}\equiv\sum_{x=1}^{p-1}1$$
$$\equiv -1 \pmod p.$$

Clearly

$$\left|\sum_{x=1}^{p}\left(\frac{x^2+b}{p}\right)\right|\leqslant p,$$

so that

$$\sum_{x=1}^{p}\left(\frac{x^2+b}{p}\right) = -1 \text{ or } p-1.$$

Since

$$\sum_{x=1}^{p}\left(\frac{x^2+b}{p}\right) = \left(\frac{b}{p}\right) + 2\sum_{x=1}^{\frac{1}{2}(p-1)}\left(\frac{x^2+b}{p}\right)$$
$$\equiv 1 \pmod 2,$$

we have

$$\sum_{x=1}^{p}\left(\frac{x^2+b}{p}\right) = -1. \quad \square$$

Proof of Theorem 8.1. The number of integers a with the property stated in the theorem can be represented by

$$\begin{aligned}
&\frac{1}{4}\sum_{a=1}^{p-2}\left(1+\left(\frac{a}{p}\right)\right)\left(1+\left(\frac{a+1}{p}\right)\right)\\
&=\frac{1}{4}\sum_{a=1}^{p-2}\left(1+\left(\frac{a}{p}\right)+\left(\frac{a+1}{p}\right)+\left(\frac{a(a+1)}{p}\right)\right)\\
&=\frac{1}{4}\left(p-2-\left(\frac{-1}{p}\right)-\left(\frac{1}{p}\right)-1\right)\\
&=\frac{1}{4}\left(p-4-\left(\frac{-1}{p}\right)\right)
\end{aligned}$$

(because $\sum_{a=1}^{p}\left(\frac{a}{p}\right)=0$). $\square$

From Theorem 8.1 we deduce at once:

Theorem 8.3. *If $p \geqslant 2$, then there must be a pair of consecutive integers which are both quadratic residues.* $\square$

Similarly we can prove:

Theorem 8.4. *The number of integers a such that a and $a+1$ are both quadratic non-residues* mod p *is*

$$\frac{1}{4}\left(p-2+\left(\frac{-1}{p}\right)\right),$$

so that, if $p \geqslant 5$, then there must be a pair of consecutive integers which are both quadratic non-residues. $\square$

Theorem 8.5. *There are $\frac{1}{2}(p-1)$ integers a such that a and $a+1$ are neither both quadratic residues nor both quadratic non-residues.*

Proof. The theorem follows at once from

$$\sum_{a=1}^{p-2}\left(1-\left(\frac{a}{p}\right)\left(\frac{a+1}{p}\right)\right)=p-1. \quad \square$$

Note: The problem concerning three consecutive quadratic residues involves the study of the character sum

$$\sum_{x=1}^{p}\left(\frac{x(x+1)(x+2)}{p}\right)$$

which is outside the scope of this book. However, we have the following application of character sums involving cubic polynomials.

Theorem 8.6 (Jacobsthal). *Let p be a prime $\equiv 1 \pmod 4$. Then a solution to the equation*

$$p = X^2 + Y^2$$

in integers X, Y is given by $2X = S(r)$, $2Y = S(u)$ where

$$\left(\frac{r}{p}\right)=1, \quad \left(\frac{u}{p}\right)=-1$$

and

$$S(k) = \sum_{x=1}^{p-1} = \left(\frac{x(x^2+k)}{p}\right).$$

Proof. Since

$$\begin{aligned}S(k) &= \sum_{x=1}^{\frac{1}{2}(p-1)}\left(\frac{x(x^2+k)}{p}\right)+\sum_{y=1}^{\frac{1}{2}(p-1)}\left(\frac{(p-y)((p-y)^2+k)}{p}\right)\\ &= 2\sum_{x=1}^{\frac{1}{2}(p-1)}\left(\frac{x(x^2+k)}{p}\right),\end{aligned}$$

we see that X and Y are actually integers. Also, if $p \nmid t$, then

$$\left(\frac{t}{p}\right)^3 S(k) = \sum_{x=1}^{p-1}\left(\frac{tx((tx)^2+t^2k)}{p}\right)=\sum_{x=1}^{p-1}\left(\frac{x(x^2+t^2k)}{p}\right)=S(t^2k).$$

Now consider

$$\begin{aligned}\frac{p-1}{2}((S(r))^2+(S(u))^2) &= \sum_{t=1}^{\frac{1}{2}(p-1)}(S(rt^2))^2+\sum_{t=1}^{\frac{1}{2}(p-1)}(S(ut^2))^2\\ &= \sum_{k=1}^{p-1}(S(k))^2 = \sum_{x=1}^{p-1}\sum_{y=1}^{p-1}\sum_{k=1}^{p-1}\left(\frac{xy(x^2+k)(y^2+k)}{p}\right).\end{aligned}$$

From Theorem 8.2 we see that the innermost sum here is

$$=\begin{cases}-2\left(\dfrac{xy}{p}\right), & \text{if } x \not\equiv \pm y \pmod p,\\ p-2, & \text{if } x \equiv \pm y \pmod p.\end{cases}$$

Therefore

$$\sum_{k=1}^{p-1} (S(k))^2 = 2(p-1)(p-2) - 2\sum \sum_{\substack{x \not\equiv \pm y \\ (\bmod p)}} \left(\frac{xy}{p}\right)$$

$$= 2p(p-1) - 2\sum_{x=1}^{p-1}\sum_{y=1}^{p-1}\left(\frac{xy}{p}\right) = 2p(p-1).$$

Collecting our results we have

$$(S(r))^2 + (S(u))^2 = 4p. \quad \square$$

7.9 The Problem of the Distribution of Primitive Roots

Theorem 9.1. *Let p be an odd prime and $p \nmid n$. If n is not a primitive root* $\bmod p$, *then*

$$\sum_{k|p-1} \frac{\mu(k)}{\varphi(k)} \sum_{\substack{a=1 \\ (a,k)=1}}^{k} e^{2\pi i a \operatorname{ind} n/k} = 0. \tag{1}$$

Proof. The inner sum on the left hand side of (1) is a multiplicative function of k, as are the functions $\mu(k)$ and $\varphi(k)$. Therefore the left hand side of (1) is equal to

$$\prod_{q|p-1}\left(1 + \frac{\mu(q)}{\varphi(q)} \sum_{\substack{a=1 \\ (a,q)=1}}^{q} e^{2\pi i a \operatorname{ind} n/q}\right),$$

where q runs over the prime divisors of $p-1$.

If n is not a primitive root, then $(\operatorname{ind} n, p-1) > 1$, and so there exists a prime divisor q of $p-1$ which divides $\operatorname{ind} n$. For this prime number we have

$$1 + \frac{\mu(q)}{\varphi(q)} \sum_{\substack{a=1 \\ (a,q)=1}}^{q} e^{2\pi i a \operatorname{ind} n/q} = 1 + \frac{-1}{q-1} \cdot (q-1) = 0.$$

The theorem is proved. $\square$

Theorem 9.2. *Let p be an odd prime, $1 \leqslant A < p$. If $\chi(n)$ is a non-principal character* $\bmod p$, *then*

$$\frac{1}{A+1}\left|\sum_{a=0}^{A}\sum_{n=-a}^{a} \chi(n)\right| \leqslant p^{\frac{1}{2}} - \frac{A+1}{p^{\frac{1}{2}}}. \tag{2}$$

Proof. We already have

$$|\tau(\chi)| = \left|\sum_{h=1}^{p-1} \chi(h) e^{2\pi i h/p}\right| = p^{\frac{1}{2}}.$$

If $p \nmid n$, then

$$\begin{aligned}\chi(n)\tau(\bar{\chi}) &= \chi(n)\sum_{h=1}^{p-1}\bar{\chi}(h)e^{2\pi ih/p}\\ &= \chi(n)\sum_{h=1}^{p-1}\bar{\chi}(nh)e^{2\pi inh/p}\\ &= \sum_{h=1}^{p-1}\bar{\chi}(h)e^{2\pi inh/p}.\end{aligned}$$

If we multiply the left hand side of (2) by $\tau(\bar{\chi})$, then

$$\begin{aligned}\frac{\sqrt{p}}{A+1}\left|\sum_{a=0}^{A}\sum_{n=-a}^{a}\chi(n)\right| &= \frac{1}{A+1}\left|\sum_{a=0}^{A}\sum_{n=-a}^{a}\chi(n)\tau(\bar{\chi})\right|\\ &= \frac{1}{A+1}\left|\sum_{a=0}^{A}\sum_{n=-a}^{a}\sum_{h=1}^{p-1}\bar{\chi}(h)e^{2\pi inh/p}\right|\\ &= \frac{1}{A+1}\left|\sum_{h=1}^{p-1}\bar{\chi}(h)\left(\frac{\sin(A+1)\pi h/p}{\sin \pi h/p}\right)^2\right|, \qquad (3)\end{aligned}$$

where we have used the formula

$$\sum_{a=0}^{A}\sum_{n=-a}^{a} e^{2\pi inh/p} = \left(\frac{\sin(A+1)\pi h/p}{\sin \pi h/p}\right)^2, \qquad (4)$$

the proof of which is not difficult.

From (3) and (4) we arrive at

$$\begin{aligned}\frac{\sqrt{p}}{A+1}\left|\sum_{a=0}^{A}\sum_{n=-a}^{a}\chi(n)\right| &\leqslant \frac{1}{A+1}\sum_{h=1}^{p-1}\left(\frac{\sin(A+1)\pi h/p}{\sin \pi h/p}\right)^2\\ &= \frac{1}{A+1}\sum_{h=1}^{p-1}\sum_{a=0}^{A}\sum_{n=-a}^{a} e^{2\pi inh/p}\\ &= \frac{1}{A+1}\sum_{a=0}^{A}\sum_{n=-a}^{a}\left(\sum_{h=1}^{p} e^{2\pi inh/p} - 1\right)\\ &= p-(A+1). \quad \square\end{aligned}$$

Theorem 9.3. *Let $h(p)$ denote the primitive root* mod p *with the least absolute value. Then*

$$|h(p)| < 2^m p^{\frac{1}{2}},$$

where m is the number of distinct prime divisors of $p-1$.

Proof. Let $p > 2$. From Theorem 9.1 we have

$$0 = \sum_{k|p-1}\frac{\mu(k)}{\varphi(k)}\sum_{\substack{u=1\\(u,k)=1}}^{k}\sum_{a=0}^{|h(p)|-1}{\sum_{n=-a}^{a}}' e^{2\pi iu \operatorname{ind} n/k},$$

where $\sum'$ means that we omit the term $n = 0$. On the right hand side of this equation the term $k = 1$ is equal to

$$\sum_{a=0}^{|h(p)|-1} \sum_{n=-a}^{a}{}' 1 = \sum_{a=0}^{|h(p)|-1} 2a = |h(p)|^2 - |h(p)|.$$

For those terms in which $k \neq 1$ we use Theorem 9.2, taking $A = |h(p)| - 1$, so that

$$\left| \sum_{a=0}^{|h(p)|-1} \sum_{n=-a}^{a}{}' \chi(n) \right| \leqslant |h(p)|p^{\frac{1}{2}} - \frac{|h(p)|^2}{p^{\frac{1}{2}}},$$

where

$$\chi(n) = e^{2\pi i u \operatorname{ind} n/k}.$$

Therefore

$$\begin{aligned} |h(p)|^2 - |h(p)| &\leqslant \left(|h(p)|p^{\frac{1}{2}} - \frac{|h(p)|^2}{p^{\frac{1}{2}}} \right) \sum_{k|p-1} \frac{|\mu(k)|}{\varphi(k)} \varphi(k) \\ &= 2^m \left(|h(p)|p^{\frac{1}{2}} - \frac{|h(p)|^2}{p^{\frac{1}{2}}} \right). \end{aligned}$$

That is

$$|h(p)| \leqslant \frac{2^m p^{\frac{1}{2}} + 1}{1 + 2^m/p^{\frac{1}{2}}} < 2^m p^{\frac{1}{2}}. \quad \square$$

From Theorem 9.3 we immediately deduce:

Theorem 9.4. *If* $p \equiv 1 \pmod 4$, *then we have the primitive root*

$$g(p) = |h(p)| < 2^m p^{\frac{1}{2}}.$$

Proof. We have to prove that $|h(p)|$ is a primitive root. Suppose otherwise, so that $-|h(p)|$ is now a primitive root. But

$$|h(p)|^l \equiv 1 \pmod p, \qquad l < p - 1,$$

and hence

$$(h(p))^{2l} \equiv 1 \pmod p.$$

From $-|h(p)|$ being a primitive root we see that $2l = p - 1$ and whence

$$|h(p)|^{\frac{p-1}{2}} \equiv 1 \pmod p.$$

This means that $|h(p)|$ is a quadratic residue, and since -1 is a quadratic residue, it follows that $-|h(p)|$ is also a quadratic residue. This contradicts $-|h(p)|$ being a primitive root. $\square$

Theorem 9.5. *The least positive primitive root* mod p *satisfies*

$$g(p) < 2^{m+1} p^{\frac{1}{2}}.$$

Proof. We take $A = [(g(p)) - 1)/2]$. Then

$$0 = \sum_{k|p-1} \frac{\mu(k)}{\varphi(k)} \sum_{\substack{u=1 \\ (u,k)=1}}^{k} \sum_{a=0}^{A} \sum_{n=A+1-a}^{A+1+a} e^{2\pi i u \operatorname{ind} n/k},$$

and here the term $k = 1$ on the right hand side is equal to

$$\sum_{a=0}^{A} \sum_{n=A+1-a}^{A+1+a} 1 = \sum_{a=0}^{A} (2a+1) = (A+1)^2;$$

while for the terms which correspond to $k \neq 1$, Theorem 9.2 gives

$$\left| \sum_{a=0}^{A} \sum_{n=A+1-a}^{A+1+a} e^{2\pi i u \operatorname{ind} n/k} \right| \leqslant (A+1)p^{\frac{1}{2}} - \frac{1}{p^{\frac{1}{2}}}(A+1)^2.$$

Therefore, similarly to the proof of Theorem 9.4, we have

$$(A+1)^2 \leqslant 2^m \left((A+1)p^{\frac{1}{2}} - \frac{1}{p^{\frac{1}{2}}}(A+1)^2 \right),$$

$$\frac{1}{2}(g(p) - 1) < A + 1 \leqslant \frac{2^m p^{\frac{1}{2}}}{1 + 2^m/p^{\frac{1}{2}}},$$

which gives

$$g(p) \leqslant \frac{2^{m+1} p^{\frac{1}{2}}}{1 + 2^m/p^{\frac{1}{2}}} + 1 < 2^{m+1} p^{\frac{1}{2}}. \quad \square$$

7.10 Trigonometric Sums Involving Polynomials

The main purpose of this section is to prove the following:

Theorem 10.1. *Let $f(x)$ denote a polynomial with integer coefficients,*

$$f(x) = a_k x^k + \cdots + a_1 x + a_0.$$

If $(a_k, \ldots, a_0, q) = 1$, then

$$S(q, f(x)) = \sum_{x=1}^{q} e^{2\pi i f(x)/q} = O(q^{1 - \frac{1}{k} + \varepsilon}),$$

where ε is any positive number, and the constant involved in the O-symbol depends only on k and ε.

Since $|e^{2\pi i a_0/q}| = 1$, we can always assume, without loss of generality, that $f(0) = 0$. We now divide the proof of the theorem into several stages.

Theorem 10.2. *If $(q_1, q_2) = 1$, then*

$$S(q_1 q_2, f(x)) = S(q_1, f(q_2 x)/q_2) S(q_2, f(q_1 x)/q_1).$$

Proof. Let $x = q_1 y + q_2 z$, so that as y and z run over the complete sets of residues mod q_2 and mod q_1 respectively, x runs over a complete set of residues mod $q_1 q_2$. Clearly we have

$$e^{2\pi i f(q_1 y + q_2 z)/q_1 q_2} = e^{2\pi i f(q_1 y)/q_1 q_2} \cdot e^{2\pi i f(q_2 z)/q_1 q_2},$$

so that

$$\begin{aligned} S(q_1 q_2, f(x)) &= \sum_{x=1}^{q_1 q_2} e^{2\pi i f(x)/q_1 q_2} \\ &= \sum_{y=1}^{q_2} \sum_{z=1}^{q_1} e^{2\pi i f(q_1 y)/q_1 q_2} \cdot e^{2\pi i f(q_2 z)/q_1 q_2} \\ &= S(q_1, f(q_2 x)/q_2) S(q_2, f(q_1 x)/q_1). \quad \square \end{aligned}$$

From this theorem we see that our discussion should centre on the case $q = p^l$.

Lemma 1. *Let $f(x)$ be a polynomial with integer coefficients* mod p *and let α be a root of*

$$f(x) \equiv 0 \pmod p$$

with multiplicity m. Let $p^u \| f(px + \alpha)$ and $g(x) = p^{-u} f(px + \alpha)$. Then the congruence*

$$g(x) \equiv 0 \pmod p$$

has at most m roots.

Proof. We can assume, without loss of generality, that $\alpha = 0$. We then have

$$f(x) = x^m f_1(x) + p f_2(x)$$

where $f_1(x)$ and $f_2(x)$ are polynomials with integer coefficients, $f_1(0) \not\equiv 0 \pmod p$, and the degree of $f_2(x)$ is less than m. Thus

$$f(px) = p^m x^m f_1(px) + p f_2(px).$$

Since p^{m+1} does not divide $p^m f_1(0)$, the coefficient of x^m, it follows that $u \leqslant m$. Also the degree of $p^{-u} f(px)$ is at most m (mod p), so that the lemma follows. $\square$

Lemma 2. *Let $f(x) = a_k x^k + \cdots + a_1 x$ be a polynomial with integer coefficients, $p \nmid (a_k, \ldots, a_1)$ and $p^t \| (k a_k, \ldots, 2a_2, a_1)$. Suppose that μ is a root of*

$$f'(x) \equiv 0 \pmod{p^{t+1}}, \qquad 0 \leqslant x < p,$$

* We use the symbol $p^u \| a$ to represent $p^u | a$ and $p^{u+1} \nmid a$. We also write $p^u \| S(x)$ if p^u divides all the coefficients of $S(x)$ and p^{u+1} does not.

and that $p^\sigma \| (f(\mu + px) - f(\mu))$. *Then*

$$1 \leqslant \sigma \leqslant k.$$

Proof. Suppose that $\sigma \geqslant k + 1$. Then, by hypothesis,

$$p^\sigma \left| \frac{p^h}{h!} f^{(h)}(\mu), \qquad 1 \leqslant h \leqslant k. \right.$$

That is, given any h $(1 \leqslant h \leqslant k)$ we always have

$$p^{k+1} \left| \frac{p^h}{h!} f^{(h)}(\mu), \right.$$

and so

$$p \left| \frac{1}{h!} f^{(h)}(\mu). \right.$$

It follows that $p|a_k$, $p|a_{k-1}, \ldots, p|a_1$. This contradicts the hypothesis $p \nmid (a_k, \ldots, a_1)$. □

Fundamental Lemma. *If* $p \nmid (a_k, \ldots, a_1)$, *then*

$$|S(p^l, f(x))| < C(k) p^{l\left(1 - \frac{1}{k}\right)}.$$

Proof. We use mathematical induction to prove this lemma. We first prove the case $l = 1$ (Mordell). We can clearly assume that $p > k$. Denote by N the number of solutions to the set of congruences

$$x_1^h + \cdots + x_k^h \equiv y_1^h + \cdots + y_k^h \pmod p, \quad 1 \leqslant x, \quad y \leqslant p, \quad h = 1, 2, \ldots, k. \tag{1}$$

We shall simplify the notations by writing $\sum_x$ for $\sum_{x=1}^p$, and $e_p(f(x))$ for $e^{2\pi i f(x)/p}$. Then, from Theorem 1.1, we have

$$\begin{aligned} &\sum_{a_k} \cdots \sum_{a_1} \left| \sum_x e_p(a_k x^k + \cdots + a_1 x) \right|^{2k} \\ &\quad = \sum_{x_1} \cdots \sum_{x_k} \sum_{y_1} \cdots \sum_{y_k} \sum_{a_k} \cdots \sum_{a_1} e_p(a_k(x_1^k + \cdots + x_k^k - y_1^k - \cdots - y_k^k) + \cdots \\ &\quad\quad + a_1(x_1 + \cdots + x_k - y_1 - \cdots - y_k)) = p^k N. \end{aligned}$$

Applying the theory of symmetric functions we have, from (1), that

$$(x - x_1) \cdots (x - x_k) \equiv (x - y_1) \cdots (x - y_k) \pmod p.$$

Therefore $x_1, \ldots, x_k$ and $y_1, \ldots, y_k$ only differ by ordering mod p, so that

$$N \leqslant k! p^k,$$

that is

$$\sum_{a_k} \cdots \sum_{a_1} \left| \sum_x e_p(a_k x^k + \cdots + a_1 x) \right|^{2k} \leqslant k! p^{2k}. \tag{2}$$

For any λ ($\not\equiv 0 \pmod p$) and any μ, we have

$$|S(p, f(x))| = |S(p, f(\lambda x + \mu) - f(\mu))|.$$

All the sums of this form occur on the left hand side of (2). We now take it that any two polynomials with the same coefficients reduced mod p are equal mod p. We shall then determine the number of distinct polynomials $f(\lambda x + \mu) - f(\mu)$ ($\lambda = 1, \ldots, p-1$; $\mu = 0, 1, \ldots, p-1$). We can assume, without loss of generality, that $p \nmid a_k$. If $f(\lambda x + \mu) - f(\mu)$ is identical to $f(x)$ mod p, then

$$a_k \lambda^k \equiv a_k, \qquad k a_k \lambda^{k-1} \mu + a_{k-1} \lambda^{k-1} \equiv a_{k-1} \pmod p.$$

By Theorem 2.9.1 the congruence $\lambda^k \equiv 1 \pmod p$ has at most k roots, and each fixed λ determines μ uniquely. Therefore the number of polynomials of the form $f(\lambda x + \mu) - f(\mu)$ which are identical with $f(x)$ mod p is at most k. In other words there are at least $p(p-1)/k$ distinct polynomials $f(\lambda x + \mu) - f(\mu)$. Therefore

$$\frac{p(p-1)}{k} |S(p, f(x))|^{2k} \leqslant k! p^{2k},$$

that is

$$|S(p, f(x))| \leqslant \left(\frac{k \cdot k!}{p(p-1)} \right)^{\frac{1}{2k}} p \leqslant (2k \cdot k!)^{\frac{1}{2k}} p^{1 - \frac{1}{k}}.$$

Now suppose that $l > 1$, $p^t \| (k a_k, \ldots, 2a_2, a_1)$, and that $\mu_1, \ldots, \mu_r$ are distinct roots of

$$f'(x) \equiv 0 \pmod{p^{t+1}}, \qquad 0 \leqslant x < p,$$

with multiplicities $m_1, \ldots, m_r$ respectively. Let $m_1 + \cdots + m_r = m$, and it is easy to see that $m \leqslant k-1$. We now prove that

$$|S(p^l, f(x))| \leqslant k^2 \max(1, m) p^{\left(1 - \frac{1}{k}\right) l}.$$

From hypothesis, $p \nmid (a_k, \ldots, a_1)$, $p^t | (k a_k, \ldots, 2a_2, a_1)$ so that necessarily $p^t \leqslant k$.

1) $l < 2(t+1)$. Since $l > 1$ it follows that $t \geqslant 1$ and hence

$$|S(p^l, f(x))| \leqslant p^l \leqslant p^{l\left(1 - \frac{1}{k}\right)} \cdot p^{(2t+1)\frac{1}{k}} \leqslant p^{l\left(1 - \frac{1}{k}\right)} k^{\left(2 + \frac{1}{t}\right)\frac{1}{k}} \leqslant k^2 p^{l\left(1 - \frac{1}{k}\right)},$$

so that the theorem is established.

2) $l \geqslant 2(t+1)$. We write

$$S(p^l, f(x)) = \sum_{v=1}^{p} \sum_{\substack{0 \leqslant x \leqslant p^l - 1 \\ x \equiv v \pmod p}} e_{p^l}(f(x)) = \sum_{v=1}^{p} S_v.$$

If v is not one of the μ_i, then we let

$$x = y + p^{l-t-1}z, \qquad 0 \leqslant y < p^{l-t-1}, \qquad 0 \leqslant z < p^{t+1}.$$

From $f'(y) \not\equiv 0 \pmod{p^{l+1}}$ and Theorem 1.1, we have

$$S_v = \sum_{\substack{0 \leqslant x < p^l \\ x \equiv v \pmod p}} e_{p^l}(f(x)) = \sum_{\substack{0 \leqslant y < p^{l-t-1} \\ y \equiv v \pmod p}} \sum_{0 \leqslant z < p^{t+1}} e_{p^l}(f(y) - p^{l-t-1}f'(y)z)$$

$$= \sum_{\substack{0 \leqslant y < p^{l-t-1} \\ y \equiv v \pmod p}} e_{p^l}(f(y)) \sum_{0 \leqslant z < p^{l+1}} e_{p^{r+1}}(zf'(y)) = 0. \tag{3}$$

If $v = \mu_i$, then, with σ_i defined by Lemma 2, we have

$$S_{\mu_i} = \sum_{\substack{x=1 \\ x \equiv \mu_i \pmod p}}^{p^l} e_{p^l}(f(x)) = \sum_{v=1}^{p^{l-1}} e_{p^l}(f(\mu_i + py))$$

$$= e_{p^l}(f(\mu_i)) \sum_{y=1}^{p^{l-1}} e_{p^{l-\sigma_i}}(p^{-\sigma_i}(f(\mu_i + py) - f(\mu_i))).$$

Let $g_i = p^{-\sigma_i}(f(\mu_i + py) - f(\mu_i))$. Then, by Lemma 2,

$$|S_{\mu_i}| = p^{\sigma_i - 1}|S(p^{l-\sigma_i}, g_i(x))| \leqslant p^{\sigma_i\left(1 - \frac{1}{k}\right)}|S(p^{l-\sigma_i}, g_i(x))|. \tag{4}$$

From (3) and (4) we have

$$|S(p^l, f(x))| \leqslant \sum_{i=1}^{r} p^{\sigma_i\left(1 - \frac{1}{k}\right)}|S(p^{l-\sigma_i}, g_i(x))|.$$

If $l \geqslant \max(\sigma_1, \ldots, \sigma_r)$, then, by the induction hypothesis, Lemma 1, and the formula above, we have

$$|S(p^l, f(x))| \leqslant \sum_{i=1}^{r} m_i p^{\sigma_i\left(1 - \frac{1}{k}\right)} k^2 p^{(l-\sigma_i)\left(1 - \frac{1}{k}\right)} < mk^2 p^{l\left(1 - \frac{1}{k}\right)}.$$

If $l < \max(\sigma_1, \ldots, \sigma_r)$, then $l < k$ and

$$|S(p^l, f(x))| \leqslant \sum_{i=1}^{r} p^{\sigma_i - 1} p^{l-\sigma_i} \leqslant kp^{l\left(1 - \frac{1}{k}\right)}.$$

The proof of the fundamental lemma is now complete. □

Proof of Theorem 10.1. Let $q = p_1^{l_1} \cdots p_s^{l_s}$, where $p_1, \ldots, p_s$ are distinct prime numbers. From Theorem 10.2 we have

$$S(q, f(x)) = \prod_{p|q} S\left(p^l, \frac{f(qx/p^l)}{q/p^l}\right),$$

and so from the fundamental lemma,

$$|S(q, f(x))| \leqslant C_1^s q^{1 - \frac{1}{k}}.$$

We can assume that $C_1 > 1$ and so from Theorem 6.6.2

$$C_1^s = (2^s)^{\log C_1/\log 2} \leqslant C_2(k, \varepsilon)q^{\varepsilon}.$$

The theorem is proved. □

Notes

7.1. As a corollary of his proof of the Riemann hypothesis for functions in finite algebraic fields, A. Weil deduced the following result on complete trigonometric sums with $q = p$, a prime number. When $(a_k, \ldots, a_1) = 1$ and $f(x) = a_k x^k + \cdots + a_1 x$ we have

$$\left| \sum_{x=1}^{p} e^{2\pi i f(x)/p} \right| \leqslant (k-1)\sqrt{p}.$$

See the author's book [30].

7.2. Applying A. Weil's proof of the analogue of the Riemann hypothesis in finite algebraic fields, D. A. Burgess [12] has improved on G. Pólya's estimate for character sums. He proved the following: Let ε, δ be any two positive numbers and let p be a large prime number. Then, for any integers N, H with $H > p^{1/4+\delta}$, we have

$$\left| \sum_{n=N+1}^{N+H} \left(\frac{n}{p} \right) \right| < \varepsilon H,$$

where $(\frac{n}{p})$ is the Kronecker symbol. He also used this to give an estimate for $n_2(p)$, the least quadratic non-residue mod p, namely $n_2(p) = O(p^{\frac{1}{4}\sqrt{e}+\varepsilon})$. Burgess's method can be generalized and extended to give estimates for the least primitive root $h(p)$ and the least d-th power non-residue $n_d(p)$: $h(p) = O(p^{\frac{1}{4}+\varepsilon})$ (see D. A. Burgess [13] and Y. Wang [62]), $n_d(p) = O(p^{1/A+\varepsilon})$, $A = 4e^{1-1/d}$ $(d \geqslant 2)$; $n_d(p) = O(p^B)$, $B = (\log\log d + 2)/4\log d$ $(d > e^{33})$ (see Y. Wang [63]).

Chapter 8. On Several Arithmetic Problems Associated with the Elliptic Modular Function

8.1 Introduction

The following four important functions frequently occur in the theory of elliptic modular functions:

$$q_0 = \prod_{n=1}^{\infty} (1 - q^{2n}),$$

$$q_1 = \prod_{n=1}^{\infty} (1 + q^{2n}),$$

$$q_2 = \prod_{n=1}^{\infty} (1 + q^{2n-1}),$$

$$q_3 = \prod_{n=1}^{\infty} (1 - q^{2n-1}).$$

Following the tradition in the theory of elliptic modular functions we use q to represent the variable, which can be real or complex and which satisfies $|q| < 1$. The four infinite products then clearly converge.

We do not give any deep discussion on the properties of the elliptic modular function in this chapter. Indeed we do not even define an elliptic modular function and instead we shall study the following associated arithmetic problems: the partition of integers, the sum of four squares, and the transformation of power series related to q_0, q_1, q_2, q_3. The problems of convergence arising in the chapter are very simple and any reader familiar with advanced calculus can easily supply the details. (In §8 we also use n-dimensional multiple integration). We shall therefore omit all discussions on convergence in this chapter.

The following is the first and simplest relationship between q_1, q_2, q_3.

Theorem 1.1. *If* $|q| < 1$, *then*

$$q_1 q_2 q_3 = 1.$$

Proof. We have

$$q_2 q_3 = \prod_{n=1}^{\infty} (1 - q^{2(2n-1)}).$$

We rearrange the terms in q_1 by taking out all the powers of 2 from $2n$ giving

$$q_1 = \prod_{n=1}^{\infty} (1 + q^{2(2n-1)}) \prod_{n=1}^{\infty} (1 + q^{4(2n-1)}) \prod_{n=1}^{\infty} (1 + q^{8(2n-1)}) \cdots.$$

From this we see that

$$\begin{aligned} q_1q_2q_3 &= \prod_{n=1}^{\infty} (1 - q^{2(2n-1)}) \prod_{n=1}^{\infty} (1 + q^{2(2n-1)}) \prod_{n=1}^{\infty} (1 + q^{4(2n-1)}) \prod_{n=1}^{\infty} (1 + q^{8(2n-1)}) \cdots \\ &= \prod_{n=1}^{\infty} (1 - q^{4(2n-1)}) \prod_{n=1}^{\infty} (1 + q^{4(2n-1)}) \prod_{n=1}^{\infty} (1 + q^{8(2n-1)}) \cdots \\ &= \prod_{n=1}^{\infty} (1 - q^{8(2n-1)}) \prod_{n=1}^{\infty} (1 + q^{8(2n-1)}) \cdots = \cdots = 1. \quad \square \end{aligned}$$

The theorem can also be proved from the equation

$$q_0q_1q_2q_3 = \prod_{n=1}^{\infty} (1 - q^n) \prod_{n=1}^{\infty} (1 + q^n) = \prod_{n=1}^{\infty} (1 - q^{2n}) = q_0.$$

8.2 The Partition of Integers

Let n be a positive integer. Any collection of positive integers whose sum is equal to n is said to form a *partition* of n. For example:

$$\begin{aligned} 5 &= 4 + 1 = 3 + 2 = 3 + 1 + 1 = 2 + 2 + 1 \\ &= 2 + 1 + 1 + 1 = 1 + 1 + 1 + 1 + 1, \end{aligned}$$

so that there are 7 partitions of 5.

We denote by $p(n)$ the number of partitions of n, so that in the above example we have $p(5) = 7$.

If we restrict to those partitions of n in which each term in the partition does not exceed r, then we denote by $p_r(n)$ the number of such partitions. For example, $p_3(5) = 5$.

Theorem 2.1. *If* $|q| < 1$, *then*

$$1 + \sum_{n=1}^{\infty} p_r(n)q^n = \frac{1}{(1 - q)(1 - q^2) \cdots (1 - q^r)}.$$

Proof. The right hand side of the equation above is equal to

$$\begin{aligned} &(1 + q + q^2 + q^3 + \cdots + q^{x_1} + \cdots) \\ &\times (1 + q^2 + (q^2)^2 + (q^2)^3 + \cdots + (q^2)^{x_2} + \cdots) \\ &\times (1 + q^3 + (q^3)^2 + (q^3)^3 + \cdots + (q^3)^{x_3} + \cdots) \\ &\times \cdots \\ &\times (1 + q^r + (q^r)^2 + (q^r)^3 + \cdots + (q^r)^{x_r} + \cdots), \end{aligned}$$

and the coefficient of q^n is the number of non-negative integers solutions to

$$x_1 + 2x_2 + 3x_3 + \cdots + rx_r = n$$

which is $p_r(n)$. □

We can prove similarly:

Theorem 2.2. *If* $|q| < 1$, *then*

$$\frac{1}{q_0 q_3} = \frac{1}{(1-q)(1-q^2)(1-q^3)\cdots} = 1 + \sum_{n=1}^{\infty} p(n)q^n. \quad \square$$

Theorem 2.3. *Let $q(n)$ be the number of partitions of n into odd integers. Then*

$$\frac{1}{q_3} = \frac{1}{(1-q)(1-q^3)(1-q^5)\cdots} = 1 + \sum_{n=1}^{\infty} q(n)q^n. \quad \square$$

Theorem 2.4. *The coefficient of q^n in the expansion of $q_1 q_2$ is the number of partitions of n into unequal parts.* □

The reader should have no difficulty with the proofs of the above three theorems. From Theorem 1.1 together with the results of Theorems 2.3 and 2.4 we have

Theorem 2.5. *The number of partitions of n into unequal parts is equal to the number of partitions of n into odd parts.* □

8.3 Jacobi's Identity

Theorem 3.1. *If* $|q| < 1$, $z \neq 0$, *then*

$$\prod_{n=1}^{\infty} ((1-q^{2n})(1+q^{2n-1}z)(1+q^{2n-1}z^{-1})) = 1 + \sum_{n=1}^{\infty} q^{n^2}(z^n + z^{-n})$$

$$= \sum_{n=-\infty}^{\infty} q^{n^2} z^n. \tag{1}$$

Proof. The two series are clearly equal. Let

$$\varphi_m(z) = \prod_{n=1}^{m} \{(1+q^{2n-1}z)(1+q^{2n-1}z^{-1})\}$$

$$= X_0 + X_1(z+z^{-1}) + X_2(z^2+z^{-2}) + \cdots + X_m(z^m+z^{-m}), \tag{2}$$

where $X_0, X_1, \ldots, X_m$ are independent of z. The coefficient of z^m is clearly

$$X_m = q^{1+3+\cdots+(2m-1)} = q^{m^2}. \tag{3}$$

Also

$$\varphi_m(q^2z) = \prod_{n=1}^{m} \{(1 + q^{2n+1}z)(1 + q^{2n-3}z^{-1})\}$$

$$= \frac{1 + q^{-1}z^{-1}}{1 + qz} \cdot \frac{1 + q^{2m+1}z}{1 + q^{2m-1}z^{-1}} \varphi_m(z)$$

$$= \frac{1 + q^{2m+1}z}{qz + q^{2m}} \varphi_m(z);$$

that is

$$(qz + q^{2m})\varphi_m(q^2z) = (1 + q^{2m+1}z)\varphi_m(z).$$

Substituting (2) into here and comparing the coefficient of z^{1-n} we see that

$$X_n = \frac{q^{2n-1}(1 - q^{2m-2n+2})}{1 - q^{2m+2n}} X_{n-1},$$

or

$$X_n = q^{n^2} \frac{(1 - q^{2m-2n+2})(1 - q^{2m-2n+4}) \cdots (1 - q^{2m})}{(1 - q^{2m+2n})(1 - q^{2m+2n-2}) \cdots (1 - q^{2m+2})} X_0.$$

From (3) we have

$$X_0 = \frac{(1 - q^{4m})(1 - q^{4m-2}) \cdots (1 - q^{2m+2})}{(1 - q^2)(1 - q^4) \cdots (1 - q^{2m})},$$

so that when $0 \leqslant n \leqslant m - 1$,

$$X_n = \frac{q^{n^2}}{(1 - q^2)(1 - q^4) \cdots (1 - q^{2m})} X'_n,$$

where

$$X'_n = \frac{(1 - q^{2m-2n+2})(1 - q^{2m-2n+4}) \cdots (1 - q^{2m})}{(1 - q^{2m+2n})(1 - q^{2m+2n-2}) \cdots (1 - q^{2m+2})} (1 - q^{2m+2}) \cdots (1 - q^{4m})$$

$$= (1 - q^{2m-2n+2}) \cdots (1 - q^{2m})(1 - q^{2m+2n+2}) \cdots (1 - q^{4m}). \tag{4}$$

It follows that (2) can be written as

$$(1 - q^2)(1 - q^4) \cdots (1 - q^{2m})\varphi_m(z) = X'_0 + \sum_{n=1}^{m} q^{n^2}(z^n + z^{-n})X'_n. \tag{5}$$

As $m \to \infty$, $X'_n \to 1$ so that the identity in the theorem follows. However we still have to justify the process of taking the limit of the individual terms in the series. Let

$$u_{0,m} = X_0,$$

$$u_{n,m} = \begin{cases} \dfrac{q^{n^2}}{(1 - q^2)(1 - q^4) \cdots (1 - q^{2m})} X'_n(z^n + z^{-n}), & \text{if } 1 \leqslant n \leqslant m, \\ 0, & \text{if } n > m, \end{cases}$$

so that

$$\varphi_m(z) = \sum_{n=0}^{\infty} u_{n,m}. \tag{6}$$

As $m \to \infty$, the term $u_{n,m} \to u_n$ where

$$u_0 = \frac{1}{(1-q^2)(1-q^4)\cdots}, \qquad u_n = \frac{q^{n^2}(z^n + z^{-n})}{(1-q^2)(1-q^4)\cdots} \qquad (n > 0).$$

We have

$$|X_n'| < \prod_{k=1}^{\infty} (1 + |q|^{2k}) = K_1 \qquad \text{(say)}$$

and

$$\left|\frac{1}{(1-q^2)(1-q^4)\cdots(1-q^{2m})}\right| < \prod_{k=1}^{\infty} \frac{1}{1-|q|^{2k}} = K_2 \qquad \text{(say)},$$

so that

$$|u_{n,m}| \leqslant K_1 K_2 |q|^{n^2}(|z|^n + |z|^{-n}) = v_n.$$

Now v_n is independent of m and as $n \to \infty$,

$$\begin{aligned} \frac{v_{n+1}}{v_n} &= |q|^{2n+1}\left(\frac{|z|^{n+1} + |z|^{-(n+1)}}{|z|^n + |z|^{-n}}\right) \\ &< |q|^{2n+1}(|z| + |z|^{-1}) \to 0, \end{aligned}$$

so that $\sum v_n$ converges. This shows that the series (6) is uniformly convergent and therefore

$$\varphi_m(z) \to \sum_0^{\infty} u_n.$$

This completes the justification of taking the limit term by term. □

There are a number of interesting special examples included in Theorem 3.1. Taking $z = \pm 1$ and $z = q$ separately we have:

Theorem 3.2. *When* $|q| < 1$,

$$q_0 q_2^2 = \sum_{n=-\infty}^{\infty} q^{n^2}$$

and

$$q_0 q_3^2 = \sum_{n=-\infty}^{\infty} (-1)^n q^{n^2},$$

$$q_0 q_1^2 = \sum_{n=0}^{\infty} q^{n^2+n}. \quad □$$

Replacing q by $-q^{\frac{3}{2}}$ and taking $z = q^{\frac{1}{2}}$ we have

$$\prod_{n=1}^{\infty} ((1 - q^{3n})(1 - q^{3n-1})(1 - q^{3n-2})) = \sum_{n=-\infty}^{\infty} (-q^{\frac{3}{2}})^{n^2}(q^{\frac{n}{2}})$$

$$= \sum_{n=-\infty}^{\infty} (-1)^n q^{\frac{1}{2}(3n^2+n)}$$

and we deduce at once Euler's identity:

Theorem 3.3. *If* $|q| < 1$, *then*

$$q_0 q_3 = (1 - q)(1 - q^2)(1 - q^3) \cdots$$

$$= \sum_{n=-\infty}^{\infty} (-1)^n q^{\frac{1}{2}n(3n+1)} = 1 + \sum_{n=1}^{\infty} (-1)^n (q^{\frac{1}{2}n(3n-1)} + q^{\frac{1}{2}n(3n+1)})$$

$$= 1 - q - q^2 + q^5 + q^7 - q^{12} - q^{15} + \cdots. \quad \square$$

Again, replacing q by $q^{\frac{1}{2}}$ and z by $q^{\frac{1}{2}}$, we have

$$\prod_{n=1}^{\infty} (1 - q^n)(1 + q^n)(1 + q^{n-1}) = \sum_{n=-\infty}^{\infty} q^{\frac{1}{2}(n^2+n)},$$

giving:

Theorem 3.4. *If* $|q| < 1$, *then*

$$q_0 q_1 q_2 = \sum_{n=0}^{\infty} q^{\frac{1}{2}n(n+1)}. \quad \square$$

Note: The exponent $\frac{1}{2}n(n + 1)$ is commonly called a *triangular number*. From Theorem 1.1, we can restate Theorem 3.4 as:

Theorem 3.5. *If* $|q| < 1$, *then*

$$\frac{q_0}{q_3} = \frac{(1 - q^2)(1 - q^4) \cdots}{(1 - q)(1 - q^3) \cdots} = \sum_{n=0}^{\infty} q^{\frac{1}{2}n(n+1)}. \quad \square$$

We now prove:

Theorem 3.6. *If* $|q| < 1$, *then*

$$(q_0 q_3)^3 = ((1 - q)(1 - q^2)(1 - q^3) \cdots)^3$$

$$= \sum_{n=-\infty}^{\infty} (-1)^n n q^{\frac{1}{2}n(n+1)}$$

$$= 1 - 3q + 5q^3 - 7q^6 + \cdots.$$

Proof. We replace q and z in Theorem 3.1 by $q^{\frac{1}{2}}$ and $q^{\frac{1}{2}}\zeta$ respectively, giving

$$\prod_{n=1}^{\infty}((1-q^n)(1+q^n\zeta)(1+q^{n-1}\zeta^{-1})) = \sum_{n=-\infty}^{\infty} q^{\frac{1}{2}n(n+1)}\zeta^n,$$

or

$$\frac{\zeta+1}{\zeta}\prod_{n=1}^{\infty}((1-q^n)(1+q^n\zeta)(1+q^n\zeta^{-1})) = \sum_{n=-\infty}^{\infty} q^{\frac{1}{2}n(n+1)}\zeta^n.$$

We now study the situation when $\zeta \to -1$. Clearly

$$\lim_{\zeta\to-1}\prod_{n=1}^{\infty}((1-q^n)(1+q^n\zeta)(1+q^n\zeta^{-1})) = \left(\prod_{n=1}^{\infty}(1-q^n)\right)^3.$$

From

$$\begin{aligned}\sum_{n=-\infty}^{\infty}(-1)^n q^{\frac{1}{2}n(n+1)} &= \sum_{n=0}^{\infty}(-1)^n q^{\frac{1}{2}n(n+1)} + \sum_{n=-\infty}^{-1}(-1)^n q^{\frac{1}{2}n(n+1)} \\ &= \sum_{n=0}^{\infty}(-1)^n q^{\frac{1}{2}n(n+1)} + \sum_{m=0}^{\infty}(-1)^{m+1} q^{\frac{1}{2}m(m+1)} = 0,\end{aligned}$$

we have

$$\begin{aligned}\frac{\zeta}{\zeta+1}\sum_{n=-\infty}^{\infty} q^{\frac{1}{2}n(n+1)}\zeta^n &= \frac{\zeta}{\zeta+1}\sum_{n=-\infty}^{\infty} q^{\frac{1}{2}n(n+1)}(\zeta^n-(-1)^n) \\ &= \sum_{n=-\infty}^{\infty} q^{\frac{1}{2}n(n+1)}\frac{\zeta(\zeta^n-(-1)^n)}{\zeta+1}.\end{aligned}$$

Now

$$\lim_{\zeta\to-1}\frac{(\zeta^n-(-1)^n)}{\zeta+1} = n(-1)^{n-1},$$

so that

$$\lim_{\zeta\to-1}\frac{\zeta}{\zeta+1}\sum_{n=-\infty}^{\infty} q^{\frac{1}{2}n(n+1)}\zeta^n = \sum_{n=-\infty}^{\infty}(-1)^n n q^{\frac{1}{2}n(n+1)}.$$

The theorem therefore follows. (We have taken term by term limits twice, which is allowed since the series can be proved to be uniformly convergent.) □

Exercise 1. Prove that when $|q| < 1$,

$$\prod_{n=0}^{\infty}((1-q^{5n+1})(1-q^{5n+4})(1-q^{5n+5})) = \sum_{n=-\infty}^{\infty}(-1)^n q^{\frac{1}{2}n(5n+3)},$$

$$\prod_{n=0}^{\infty}((1-q^{5n+2})(1-q^{5n+3})(1-q^{5n+5})) = \sum_{n=-\infty}^{\infty}(-1)^n q^{\frac{1}{2}n(5n+1)}.$$

Exercise 2. Prove

$$q(1-q^{24})(1-q^{2\cdot 24})(1-q^{3\cdot 24})\cdots = q^{1^2}-q^{5^2}-q^{7^2}+q^{11^2}+q^{13^2}-q^{17^2}-\cdots,$$
$$q((1-q^{8})(1-q^{2\cdot 8})(1-q^{3\cdot 8})\cdots)^3 = q^{1^2}-3q^{3^2}+5q^{5^2}-7q^{7^2}+\cdots.$$

8.4 Methods of Representing Partitions

Theorem 4.1. *If* $|q| < 1$, *then*

$$(1+aq)(1+aq^3)(1+aq^5)\cdots = 1+\frac{aq}{1-q^2}+\frac{a^2q^4}{(1-q^2)(1-q^4)}+\cdots$$
$$+\frac{a^mq^{m^2}}{(1-q^2)\cdots(1-q^{2m})}+\cdots.$$

Proof. Let $F(a)$ represent the left hand side of the equation above, and let

$$F(a) = 1 + c_1a + c_2a^2 + \cdots.$$

From

$$(1+aq)F(aq^2) = (1+aq)(1+aq^3)(1+aq^5)\cdots$$
$$= F(a),$$

by comparing the coefficients of a^n, we see that

$$c_1 = q + c_1q^2, \qquad c_2 = c_1q^3 + c_2q^4, \ldots$$
$$c_m = c_{m-1}q^{2m-1} + c_mq^{2m}, \ldots$$

so that

$$c_m = \frac{q^{2m-1}}{1-q^{2m}}c_{m-1} = \frac{q^{1+3+\cdots+(2m-1)}}{(1-q^2)(1-q^4)\cdots(1-q^{2m})}$$
$$= \frac{q^{m^2}}{(1-q^2)(1-q^4)\cdots(1-q^{2m})}.$$

This proves the theorem. □

On taking separately $a = 1$ and $a = q$ in this theorem we deduce the following two theorems:

Theorem 4.2. *If* $|q| < 1$, *then*

$$q_2 = (1+q)(1+q^3)(1+q^5)\cdots$$
$$= 1+\frac{q}{1-q^2}+\frac{q^4}{(1-q^2)(1-q^4)}+\cdots$$
$$+\frac{q^{m^2}}{(1-q^2)(1-q^4)\cdots(1-q^{2m})}+\cdots. \quad □$$

Theorem 4.3. *If* $|q| < 1$, *then*

$$q_1 = (1+q^2)(1+q^4)(1+q^6)\cdots$$
$$= 1 + \frac{q^2}{1-q^2} + \frac{q^6}{(1-q^2)(1-q^4)} + \cdots$$
$$+ \frac{q^{m^2+m}}{(1-q^2)(1-q^4)\cdots(1-q^{2m})} + \cdots. \quad \square$$

Replacing q by $q^{\frac{1}{2}}$ in Theorem 4.3 we have:

Theorem 4.4. *If* $|q| < 1$, *then*

$$(1+q)(1+q^2)(1+q^3)\cdots = 1 + \frac{q}{1-q} + \frac{q^3}{(1-q)(1-q^2)} + \cdots$$
$$+ \frac{q^{\frac{1}{2}m(m+1)}}{(1-q)(1-q^2)\cdots(1-q^m)} + \cdots. \quad \square$$

Theorem 4.5. *If* $|q| < 1$, *then*

$$\frac{1}{(1-aq)(1-aq^2)(1-aq^3)\cdots} = 1 + \frac{aq}{1-q} + \frac{a^2q^2}{(1-q)(1-q^2)}$$
$$+ \frac{a^3q^3}{(1-q)(1-q^2)(1-q^3)} + \cdots.$$

Proof. Denote the left hand side of the above equation by $F(a)$. Then

$$F(aq) = \frac{1}{(1-aq^2)(1-aq^3)\cdots} = (1-aq)F(a).$$

Substituting the expansion

$$F(a) = 1 + \sum_{m=1}^{\infty} c_m a^m$$

into this equation we have $c_m q^m = c_m - c_{m-1}q$, or

$$c_m = \frac{q}{1-q^m} c_{m-1}.$$

Therefore

$$c_m = \frac{q^m}{(1-q)(1-q^2)\cdots(1-q^m)}. \quad \square$$

A special case of this theorem is:

Theorem 4.6. *If* $|q| < 1$, *then*

$$\frac{1}{q_0 q_3} = 1 + \frac{q}{1-q} + \frac{q^2}{(1-q)(1-q^2)}$$
$$+ \frac{q^3}{(1-q)(1-q^2)(1-q^3)} + \cdots. \quad \square$$

If we replace q and a by q^2 and q^{-1} respectively in Theorem 4.5, then we have:

Theorem 4.7. *If* $|q| < 1$, *then*

$$\frac{1}{q_3} = 1 + \frac{q}{1-q^2} + \frac{q^2}{(1-q^2)(1-q^4)} + \frac{q^3}{(1-q^2)(1-q^4)(1-q^6)} + \cdots. \quad \square$$

8.5 Graphical Method for Partitions

Let a partition of n be

$$n = a_1 + a_2 + a_3 + \cdots + a_s,$$

where the a_i are arranged in descending order of magnitude, that is

$$a_1 \geqslant a_2 \geqslant a_3 \geqslant \cdots \geqslant a_s.$$

We construct a diagram where there are a_1 points in the first row, a_2 points in the second row, etc. The points are equally spaced apart in each row and the first points of the rows form the first column. Such a diagram is called the *graph* of the partition. For example

$$\begin{array}{lllllll} \cdot & \cdot & \cdot & \cdot & \cdot & \cdot & \cdot \\ \cdot & \cdot & \cdot & \cdot & & & \\ \cdot & \cdot & \cdot & & & & \\ \cdot & \cdot & \cdot & & & & \\ \cdot & & & & & & \end{array}$$

is the graph of the partition

$$18 = 7 + 4 + 3 + 3 + 1.$$

Clearly we can also read a graph vertically as columns rather than rows, giving another partition which is known as the conjugate partition. For the above graph the conjugate partition is

$$18 = 5 + 4 + 4 + 2 + 1 + 1 + 1.$$

By reading graphs vertically or horizontally we have the following theorem:

Theorem 5.1. *The number of partitions of n into parts not exceeding m is equal to the number of partitions of n not exceeding m parts.* $\square$

Graphical methods can be used to prove more complicated theorems. For example:

Alternative proof of Theorem 4.2. Clearly the coefficient of q^n in the expansion

$$(1+q)(1+q^3)(1+q^5)\cdots$$

is the number $r(n)$ of partitions of n into unequal odd parts. For example:

$$15 = 11 + 3 + 1 = 9 + 5 + 1 = 7 + 5 + 3.$$

We now reconstruct the graph for the partition $15 = 11 + 3 + 1$ as follows:

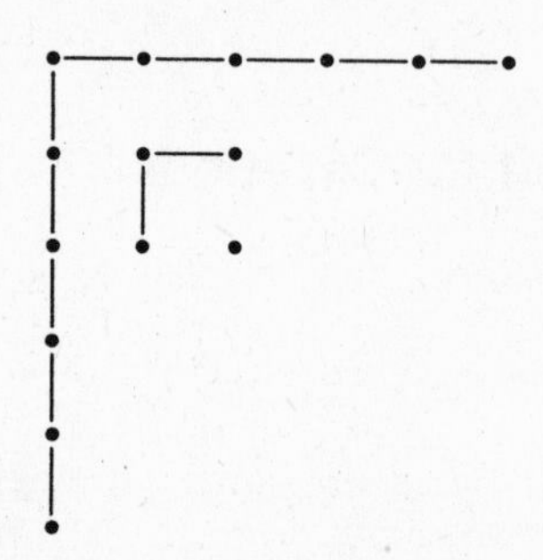

Since each part in the partition is odd and the parts are unequal, the resulting diagram is still a graph for a partition. But this graph has a special property: it reads the same horizontally and vertically. Such a graph is called a *self-conjugate graph* and its corresponding partition a *self-conjugate partition*. Therefore each partition into unequal odd parts corresponds to a self-conjugate partition, and conversely.

Therefore $r(n)$ is the number of self-conjugate partitions of n. Denote by t the number of points on the side of the largest square in a self-conjugate graph ($t = 3$ in the diagram). Then, corresponding to each fixed t, the number of self-conjugate partitions is equal to the number of partitions of $(n - t^2)/2$ not exceeding t parts. This is the same as the coefficient of q^n in the expansion

$$\frac{q^{t^2}}{(1-q^2)(1-q^4)\cdots(1-q^{2t})}.$$

We have therefore

$$(1+q)(1+q^3)(1+q^5)\cdots = \sum_{t=0}^{\infty} \frac{q^{t^2}}{(1-q^2)(1-q^4)\cdots(1-q^{2t})},$$

where the term $t = 0$ in the series is 1. This is Theorem 4.2. □

Exercise 1. Prove

$$\frac{1}{(1-q)(1-q^2)(1-q^3)\cdots} = 1 + \frac{q}{(1-q)^2} + \frac{q^4}{(1-q)^2(1-q^2)^2} + \frac{q^9}{(1-q)^2(1-q^2)^2(1-q^3)^2} + \cdots.$$

Exercise 2. Use the graphical method to prove Theorem 4.4.

Suggestion: Shift each row by one unit successively in the graph for the partition with unequal parts, for example

$$19 = 7 + 5 + 4 + 2 + 1$$

Now read the partition to the right of the line.

Another application of the graphical method is to give a proof of Theorem 3.3. This theorem clearly can be restated as follows:

Theorem 5.2. *Denote by $E(n)$ the number of partitions of n into an even number of unequal parts* (even partitions), *and by $U(n)$ the number of partitions of n into an odd number of unequal parts* (odd partitions). *Then*

$$E(n) - U(n) = \begin{cases} 0, & \text{if} \quad n \neq \frac{1}{2}k(3k \pm 1), \\ (-1)^k, & \text{if} \quad n = \frac{1}{2}k(3k \pm 1). \end{cases}$$

Proof (Franklin). In a graph of a partition of n we take the point in the extreme top right hand corner and draw a 45° line towards the bottom left hand corner, the end point of this line being a point of the graph. We denote this line by σ. We also denote by β the bottom line joining all the points in the last row.

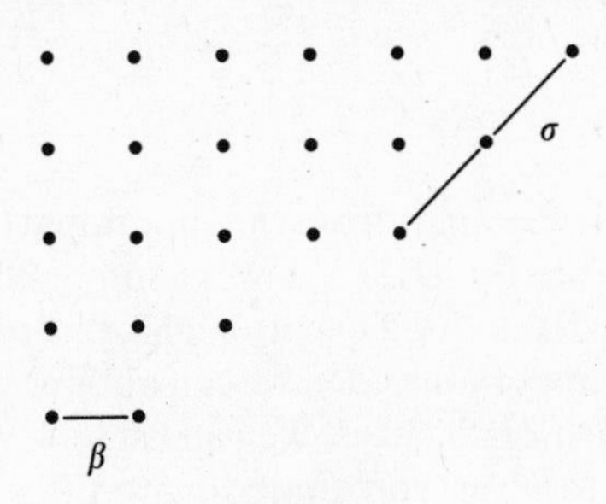

Figure 1

We can move the line β to the top right hand corner of the diagram so that it is just to the right of, and parallel to, σ (we use O to indicate this operation). We can also move σ to below, and parallel to, β (we use Ω to indicate this operation). Corresponding to these operations O or Ω we may obtain another graph for a partition of n, but it is possible that the resulting diagram cannot be a graph for a partition (the graphs of partitions are drawn in descending order for rows). In Figure 1, after the operation O we obtain Figure 2, whereas after the operation Ω we obtain Figure 3, and according to our rules, Figure 2 is a graph for a partition whereas Figure 3 is not.

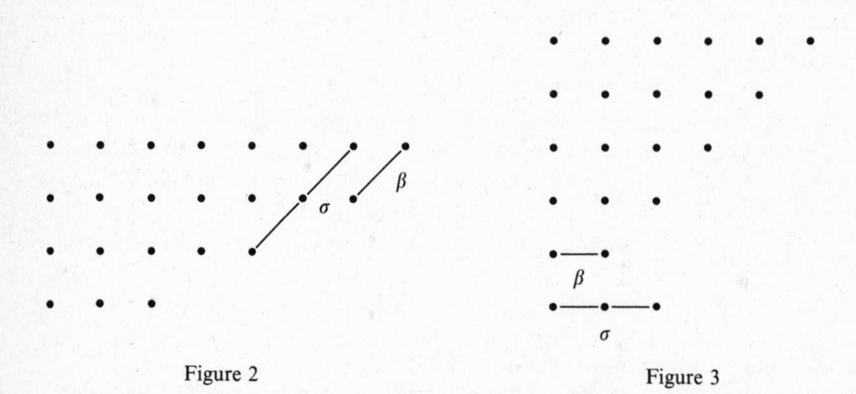

Figure 2 Figure 3

We now discuss the three separate cases:

1) $\beta < \sigma$. From Figure 1 we see that O is always possible while Ω is impossible.

2) $\beta > \sigma$. Here O is always impossible, and Ω is possible unless β and σ meet, and $\beta = \sigma + 1$ (Figure 4). In the exceptional situation we have a partition with two equal parts, contrary to what we require.

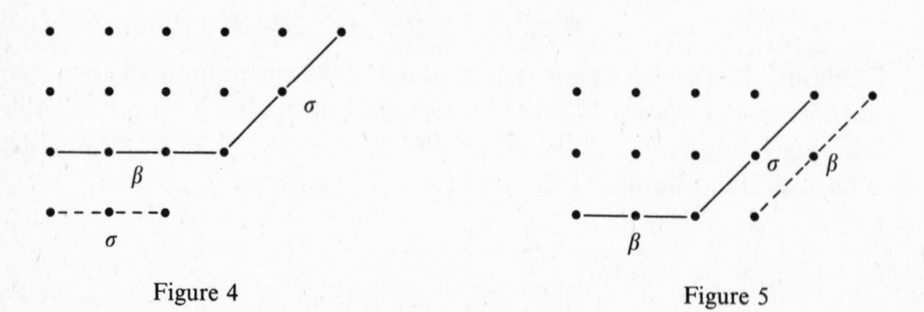

Figure 4 Figure 5

3) $\beta = \sigma$. Here O is possible apart from the situation when β and σ meet (Figure 5) which becomes impossible for O. Ω is always impossible.

From the above we see that if, for a partition of n, one of the operations O and Ω is possible and the other is impossible, then we can obtain an even (or odd) partition from an odd (or even) partition. That is, we can establish a bijection between the even and odd partitions. However, corresponding to Figure 4 and Figure 5, such a bijection cannot be established. In the first case, n must be of the form

$$n = (k + 1) + (k + 2) + \cdots + 2k = \tfrac{1}{2}(3k^2 + k)$$

while in the second case

$$n = k + (k + 1) + \cdots + (2k - 1) = \tfrac{1}{2}(3k^2 - k).$$

In either case we clearly have

$$E(n) - U(n) = (-1)^k. \quad \square$$

8.6 Estimates for $p(n)$

In this section we first use the simplest algebraic method to give the roughest estimates for $p(n)$ and we then use a slightly deeper method to determine an asymptotic formula for $\log p(n)$. However the still deeper method of applying Tauberian theory to determine the asymptotic formula for $p(n)$, and the even deeper method of applying results in modular function theory and analytic number theory to obtain the expansion for $p(n)$ is outside the scope of this book. Following the successive improvements of the results that can be obtained it is easy to judge the various levels of depth in the methods used.

Theorem 6.1. *If* $n > 1$, *then*

$$2^{[\sqrt{n}]} < p(n) < n^{3[\sqrt{n}]}.$$

Proof. 1) We first prove the inequality on the left hand side. From the integers $1, 2, \ldots, [\sqrt{n}]$ we select any r of them $a_1, a_2, \ldots, a_r$ and form the partition

$$n = a_1 + \cdots + a_r + (n - a_1 - \cdots - a_r). \tag{1}$$

Since

$$a_1 + \cdots + a_r \leqslant 1 + 2 + \cdots + [\sqrt{n}] \leqslant [\sqrt{n}]^2 \leqslant n$$

we see that (1) is a partition of n. The total number of ways of selecting these partitions is

$$1 + \binom{[\sqrt{n}]}{1} + \binom{[\sqrt{n}]}{2} + \cdots + \binom{[\sqrt{n}]}{r} + \cdots = (1+1)^{[\sqrt{n}]} = 2^{[\sqrt{n}]}$$

so that the left hand side inequality in the theorem follows.

2) We next prove the right hand inequality. Consider the graph of a partition of n:

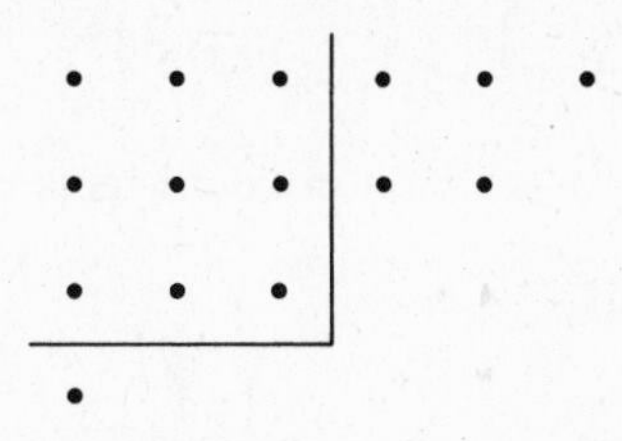

Let r be the side of the largest square in the top left hand corner in the diagram. The top right hand corner of the diagram has at most $n - r^2$ points forming a partition with at most r rows, and a similar interpretation can be given to the bottom left hand corner. If r is fixed, there are at most n^r partitions for the top right hand corner and similarly for the bottom left hand corner. Since clearly $r \leqslant [\sqrt{n}]$ we have

$$p(n) \leqslant \sum_{r=1}^{[\sqrt{n}]} n^{2r} < \sqrt{n}\, n^{2[\sqrt{n}]} < n^{3[\sqrt{n}]}. \quad \square$$

Theorem 6.2.

$$\lim_{n\to\infty} \frac{\log p(n)}{n^{\frac{1}{2}}} = \pi\sqrt{\frac{2}{3}}.$$

We shall require some preparation for the proof of this theorem.

Theorem 6.3.

$$np(n) = \sum_{lk \leqslant n} lp(n - lk).$$

Proof. Suppose that $|q| < 1$ and let

$$f(q) = \frac{1}{(1-q)(1-q^2)(1-q^3)\cdots} = 1 + \sum_{l=1}^{\infty} p(l)q^l.$$

Taking the logarithmic derivative of the product formula for $f(q)$ we have

$$\begin{aligned}\frac{f'(q)}{f(q)} &= \sum_{l=1}^{\infty} \frac{lq^{l-1}}{1-q^l}\\ &= \frac{1}{q}\sum_{l=1}^{\infty} l(q^l + q^{2l} + q^{3l} + \cdots)\\ &= \frac{1}{q}\sum_{l=1}^{\infty}\sum_{k=1}^{\infty} lq^{lk}.\end{aligned}$$

Differentiating $f(q)$ from the series expansion we have

$$\begin{aligned}\sum_{n=1}^{\infty} np(n)q^n &= qf'(q) = f(q)\sum_{l=1}^{\infty}\sum_{k=1}^{\infty} lq^{lk}\\ &= \left(1 + \sum_{\nu=1}^{\infty} p(\nu)q^{\nu}\right)\sum_{l=1}^{\infty}\sum_{k=1}^{\infty} lq^{lk}.\end{aligned}$$

The theorem follows from comparison of coefficients. $\square$

Theorem 6.4. *If* $n > \nu > 0$, *then*

$$\frac{1}{2}\frac{\nu}{\sqrt{n}} < n^{\frac{1}{2}} - (n-\nu)^{\frac{1}{2}} < \frac{1}{2}\frac{\nu}{\sqrt{n}} + \frac{\nu^2}{2n^{\frac{3}{2}}}.$$

Proof. This follows from the inequality

$$1 - \frac{x}{2} - \frac{x^2}{2} < (1 - x)^{\frac{1}{2}} < 1 - \frac{x}{2}, \qquad 0 < x < 1. \quad \square$$

Theorem 6.5. *If* $0 < x < 1$, *then*

$$\frac{1}{x^2} - c_1 < \frac{e^{-x}}{(1 - e^{-x})^2} < \frac{1}{x^2}.$$

Here c_1 (and later $c_2, c_3, \ldots$) represents a positive constant.

Proof. From

$$e^{\frac{1}{2}x} - e^{-\frac{1}{2}x} = x + \frac{2}{3!}\left(\frac{1}{2}x\right)^3 + \frac{2}{5!}\left(\frac{1}{2}x\right)^5 + \cdots > x$$

the inequality on the right hand side follows at once. Since

$$\frac{1}{e^{\frac{1}{2}x} - e^{-\frac{1}{2}x}} = \frac{1}{x}(1 + O(x^2)),$$

we have

$$\frac{1}{x^2} = \frac{e^{-x}}{(1 - e^{-x})^2} + O(1),$$

which establishes the inequality for the left hand side in the theorem. $\square$

Theorem 6.6. *Let* α *be positive. Then*

$$\frac{\pi^2 n}{6\alpha^2} - c_2\sqrt{n} < \sum_{k=1}^{\infty} \sum_{l=1}^{\infty} l e^{-\alpha l k n^{-\frac{1}{2}}} < \frac{\pi^2 n}{6\alpha^2}.$$

To be accurate here, c_2 depends on α.

Proof. From $\sum_{l=1}^{\infty} l x^l = x/(1 - x)^2$, we see that the double sum is equal to

$$\sum_{k=1}^{\infty} \frac{e^{-\alpha k n^{-\frac{1}{2}}}}{(1 - e^{-\alpha k n^{-\frac{1}{2}}})^2}. \tag{2}$$

From the inequality on the right hand side in Theorem 6.5 we see that this sum here is less than

$$\sum_{k=1}^{\infty} \frac{1}{(\alpha k n^{-\frac{1}{2}})^2} = \frac{n}{\alpha^2} \sum_{k=1}^{\infty} \frac{1}{k^2} = \frac{\pi^2 n}{6\alpha^2}.$$

Separating the sum (2) into two sums

$$\sum_{k=1}^{[\sqrt{n}]} + \sum_{k=[\sqrt{n}]+1}^{\infty} = \sum\nolimits_1 + \sum\nolimits_2,$$

and applying the left hand inequality of Theorem 6.5, we have

$$\sum_1 > \sum_{k=1}^{[\sqrt{n}]} \frac{1}{(\alpha k n^{-\frac{1}{2}})^2} + O(\sqrt{n})$$
$$= \frac{n}{\alpha^2} \sum_{k=1}^{[\sqrt{n}]} \frac{1}{k^2} + O(\sqrt{n})$$
$$= \frac{\pi^2 n}{6\alpha^2} + O\left(n \sum_{k > \sqrt{n}} \frac{1}{k^2}\right) + O(\sqrt{n})$$
$$= \frac{\pi^2 n}{6\alpha^2} + O(\sqrt{n}).$$

Applying the right hand inequality of Theorem 6.5, we have

$$\sum_2 = O\left(n \sum_{k > \sqrt{n}} \frac{1}{k^2}\right) = O(\sqrt{n}).$$

Collecting these results, our theorem is proved. □

Proof of Theorem 6.2. Let $c = \pi\sqrt{\frac{2}{3}}$.

1) We first prove

$$p(n) < e^{cn^{\frac{1}{2}}}. \tag{3}$$

When $n = 1$, (3) clearly holds. We now use induction on n. From Theorem 6.3 and the induction hypothesis we see that

$$np(n) < \sum_{lk \leqslant n} l e^{c(n-lk)^{\frac{1}{2}}}$$
$$< \sum_{lk \leqslant n} l e^{cn^{\frac{1}{2}} - \frac{c}{2} lkn^{-\frac{1}{2}}} \quad \text{(using Theorem 6.4)}$$
$$< e^{cn^{\frac{1}{2}}} \sum_{k=1}^{\infty} \sum_{l=1}^{\infty} l e^{-clk/(2n^{\frac{1}{2}})}$$
$$< e^{cn^{\frac{1}{2}}} \frac{\pi^2 n}{6(c/2)^2} = ne^{cn^{\frac{1}{2}}} \text{ (using Theorem 6.6)}$$

which proves (3).

2) We next prove: Given any positive ε there exists A $(= A(\varepsilon))$ such that

$$p(n) > \frac{1}{A} e^{(c-\varepsilon)n^{\frac{1}{2}}}.$$

We use induction on n, but the choice of A will not be made clear until later. From Theorems 6.3 and 6.4 together with the induction hypothesis we see that

$$np(n) > \frac{1}{A} e^{(c-\varepsilon)n^{\frac{1}{2}}} \sum\sum_{lk \leqslant n} l e^{-\frac{1}{2}(c-\varepsilon)(lkn^{-\frac{1}{2}} + l^2k^2n^{-\frac{3}{2}})}. \tag{4}$$

Since $e^{-x} \geqslant 1 - x$, the double sum is

$$\geqslant \sum_{lk \leqslant n} le^{-\frac{1}{2}(c-\varepsilon)lkn^{-\frac{1}{2}}}\left(1 - \frac{1}{2}(c-\varepsilon)\frac{l^2k^2}{n^{\frac{3}{2}}}\right)$$

$$= \sum_{lk \leqslant n} le^{-\frac{1}{2}(c-\varepsilon)lkn^{-\frac{1}{2}}} - \frac{(c-\varepsilon)}{2n^{\frac{3}{2}}} \sum_{lk \leqslant n} k^2l^3e^{-\frac{1}{2}(c-\varepsilon)lkn^{-\frac{1}{2}}}$$

$$= \sum\nolimits_1 - \frac{c-\varepsilon}{2n^{\frac{3}{2}}}\sum\nolimits_2 \qquad \text{(say)}. \tag{5}$$

For any positive t we always have $e^{-x} = O(x^{-t})$, so that

$$\sum_{lk > n} le^{-\frac{1}{2}(c-\varepsilon)lkn^{-\frac{1}{2}}} = O\left(n^{\frac{t}{2}} \sum_{lk>n} l^{1-\frac{t}{4}}k^{-\frac{1}{4}t}(lk)^{-\frac{3}{4}t}\right)$$

$$= O\left(n^{-\frac{1}{4}t} \sum_{l=1}^{\infty}\sum_{k=1}^{\infty} l^{1-\frac{1}{4}t}k^{-\frac{1}{4}t}\right)$$

$$= O(n^{-\frac{1}{4}t}), \qquad \text{if} \quad t > 8. \tag{6}$$

From this and Theorem 6.6 we have

$$\sum\nolimits_1 > \frac{2\pi^2 n}{3(c-\varepsilon)^2} - c_3\sqrt{n}$$

$$= \frac{2\pi^2 n}{3c^2} + \frac{2\pi^2 n}{3}\left(\frac{1}{(c-\varepsilon)^2} - \frac{1}{c^2}\right) - c_3\sqrt{n}$$

$$> (1 + 2\varepsilon c^{-1})n - c_3\sqrt{n} \tag{7}$$

(using $\dfrac{1}{(c-\varepsilon)^2} - \dfrac{1}{c^2} = 2\displaystyle\int_{c-\varepsilon}^{c} x^{-3}\,dx > 2\varepsilon c^{-3}$).

On the other hand, by the binomial theorem and Theorem 6.5,

$$\sum\nolimits_2 = \sum_{lk \leqslant n} k^2l^3e^{-\frac{1}{2}(c-\varepsilon)lkn^{-\frac{1}{2}}}$$

$$\leqslant \sum_{k=1}^{n} k^2 \sum_{l=1}^{\infty} l^3e^{-\frac{1}{2}(c-\varepsilon)lkn^{-\frac{1}{2}}}$$

$$\leqslant 12\sum_{k=1}^{n} k^2 \frac{e^{-\frac{1}{2}(c-\varepsilon)kn^{-\frac{1}{2}}}}{(1-e^{-\frac{1}{2}(c-\varepsilon)kn^{-\frac{1}{2}}})^4}$$

$$= O\left(n\sum_{k=1}^{n} \frac{1}{(1-e^{-\frac{1}{2}(c-\varepsilon)kn^{-\frac{1}{2}}})^2}\right). \tag{8}$$

We divide the sum in the bracket into two parts:

$$\sum_{k=1}^{n} = \sum_{k \leqslant \sqrt{n}} + \sum_{\sqrt{n} < k \leqslant n}.$$

In the first part $\frac{1}{2}(c-\varepsilon)kn^{\frac{1}{2}} < \frac{1}{2}c$, and when $x < \frac{1}{2}c$,

$$1 - e^{-x} = \int_0^x e^{-t}\,dt > e^{-\frac{1}{2}c}x,$$

which gives

$$\sum_{k \leqslant \sqrt{n}} \frac{1}{(1 - e^{-\frac{1}{2}(c-\varepsilon)kn^{-\frac{1}{2}}})^2} = O\left(n \sum_{k \leqslant \sqrt{n}} \frac{1}{k^2}\right) = O(n).$$

In the second part $\frac{1}{2}(c-\varepsilon)kn^{-\frac{1}{2}} \geqslant \frac{1}{2}(c-\varepsilon)$ and

$$1 - e^{-\frac{1}{2}(c-\varepsilon)kn^{-\frac{1}{2}}} > 1 - e^{-\frac{1}{2}(c-\varepsilon)},$$

so that

$$\sum_{\sqrt{n} < k \leqslant n} \frac{1}{(1 - e^{-\frac{1}{2}(c-\varepsilon)kn^{-\frac{1}{2}}})^2} = O\left(\sum_{k \leqslant n} 1\right) = O(n).$$

From this and (8) we see that

$$\sum\nolimits_2 = O(n^2). \tag{9}$$

Collecting (4), (5), (7), (9) we have

$$np(n) > \frac{1}{A} e^{(c-\varepsilon)n^{\frac{1}{2}}}((1 + 2\varepsilon c^{-1})n - c_4\sqrt{n}).$$

When

$$n > \left(\frac{c_4}{2\varepsilon c^{-1}}\right)^2$$

we have

$$p(n) > \frac{1}{A} e^{(c-\varepsilon)n^{\frac{1}{2}}}. \tag{10}$$

When $n \leqslant c_4^2(2\varepsilon c^{-1})^{-2}$ we take A large enough so that (10) holds. The theorem is proved. □

8.7 The Problem of Sums of Squares

Let $r_s(n)$ denote the number of sets of integer solutions $(x_1, \ldots, x_s)$ to the equation

$$x_1^2 + \cdots + x_s^2 = n.$$

From Theorem 6.7.5 we already have

$$r_2(n) = \sum_{u|n} (-1)^{\frac{1}{2}(u-1)},$$

where u runs over the odd divisors of n. This theorem is clearly equivalent to the following:

Theorem 7.1. *If* $|q| < 1$, *then*

$$q_0^2 q_2^4 = \left(\sum_{n=-\infty}^{\infty} q^{n^2} \right)^2$$

$$= 1 + 4\left(\frac{q}{1-q} - \frac{q^3}{1-q^3} + \frac{q^5}{1-q^5} - \cdots \right). \quad \square \tag{1}$$

We now prove:

Theorem 7.2. *If* $|q| < 1$, *then*

$$q_0^4 q_2^8 = \left(\sum_{n=-\infty}^{\infty} q^{n^2} \right)^4 = 1 + 8 \sum{}' \frac{mq^m}{1-q^m},$$

where $\sum{}'$ *means summation over all integers not divisible by* 4. *In other words*

$$r_4(n) = 8 \sum_{m|n}{}' m,$$

where m *runs over the divisors of* n *not divisible by* 4.

Before we prove this theorem we shall need some preparation. Let

$$u_r = \frac{q}{1-q^r},$$

so that

$$\frac{q^r}{(1-q^r)^2} = u_r(1+u_r). \tag{2}$$

Theorem 7.3.

$$\sum_{m=1}^{\infty} u_m(1+u_m) = \sum_{n=1}^{\infty} nu_n.$$

Proof. From formula (2) we have

$$\sum_{m=1}^{\infty} u_m(1+u_m) = \sum_{m=1}^{\infty} \frac{q^m}{(1-q^m)^2} = \sum_{m=1}^{\infty} \sum_{n=1}^{\infty} nq^{mn} = \sum_{n=1}^{\infty} nu_n. \quad \square$$

Theorem 7.4.

$$\sum_{m=1}^{\infty} (-1)^{m-1} u_{2m}(1+u_{2m}) = \sum_{n=1}^{\infty} (2n-1) u_{4n-2}.$$

Proof. From formula (2) we have

$$\begin{aligned}
&\sum_{m=1}^{\infty}(-1)^{m-1}u_{2m}(1+u_{2m})\\
&\quad=\sum_{m=1}^{\infty}(-1)^{m-1}\frac{q^{2m}}{(1-q^{2m})^2}=\sum_{m=1}^{\infty}(-1)^{m-1}\sum_{r=1}^{\infty}rq^{2mr}\\
&\quad=\sum_{r=1}^{\infty}r\sum_{m=1}^{\infty}(-1)^{m-1}q^{2mr}=\sum_{r=1}^{\infty}\frac{rq^{2r}}{1+q^{2r}}\\
&\quad=\sum_{r=1}^{\infty}\left(\frac{rq^{2r}}{1-q^{2r}}-\frac{2rq^{4r}}{1-q^{4r}}\right)=\sum_{n=1}^{\infty}\frac{(2n-1)q^{4n-2}}{1-q^{4n-2}}. \quad \square
\end{aligned}$$

Theorem 7.5. *Let ϑ be real and not an even multiple of π. Then*

$$\begin{aligned}
&(\tfrac{1}{4}\cot\tfrac{1}{2}\vartheta+u_1\sin\vartheta+u_2\sin 2\vartheta+\cdots)^2\\
&\quad=\left(\frac{1}{4}\cot\frac{1}{2}\vartheta\right)^2+C_0+\sum_{k=1}^{\infty}C_k\cos k\vartheta,
\end{aligned}\tag{3}$$

where

$$C_0=\frac{1}{2}\sum_{n=1}^{\infty}nu_n,$$

$$C_k=u_k(1+u_k-\tfrac{1}{2}k),\qquad k\geqslant 1.$$

Proof. The left hand side of formula (3) is equal to

$$\left(\frac{1}{4}\cot\frac{1}{2}\vartheta\right)^2+\frac{1}{2}\sum_{n=1}^{\infty}u_n\cot\frac{1}{2}\vartheta\sin n\vartheta+\sum_{m=1}^{\infty}\sum_{n=1}^{\infty}u_mu_n\sin m\vartheta\sin n\vartheta.$$

Now

$$\tfrac{1}{2}\cot\tfrac{1}{2}\vartheta\sin n\vartheta=\tfrac{1}{2}+\cos\vartheta+\cdots+\cos(n-1)\vartheta+\tfrac{1}{2}\cos n\vartheta,$$
$$2\sin m\vartheta\sin n\vartheta=\cos(m-n)\vartheta-\cos(m+n)\vartheta,$$

so that the formula is equal to

$$\begin{aligned}
&\left(\frac{1}{4}\cot\frac{1}{2}\vartheta\right)^2+\sum_{n=1}^{\infty}u_n\left(\frac{1}{2}+\cos\vartheta+\cdots+\cos(n-1)\vartheta+\frac{1}{2}\cos n\vartheta\right)\\
&\qquad+\frac{1}{2}\sum_{m=1}^{\infty}\sum_{n=1}^{\infty}u_mu_n(\cos(m-n)\vartheta-\cos(m+n)\vartheta).
\end{aligned}$$

From this we have

$$C_0=\frac{1}{2}\sum_{n=1}^{\infty}(u_n+u_n^2),$$

$$C_k=\frac{1}{2}u_k+\sum_{n=k+1}^{\infty}u_n+\frac{1}{2}\sum_{m-n=k}u_mu_n+\frac{1}{2}\sum_{n-m=k}u_mu_n-\frac{1}{2}\sum_{n+m=k}u_mu_n,$$

where $m\geqslant 1$, $n\geqslant 1$.

From Theorem 7.3 we see that

$$C_0 = \frac{1}{2}\sum_{n=1}^{\infty} nu_n,$$

and

$$C_k = \frac{1}{2}u_k + \sum_{l=1}^{\infty} u_{k+l} + \sum_{l=1}^{\infty} u_l u_{k+l} - \frac{1}{2}\sum_{l=1}^{k-1} u_l u_{k-l}.$$

Now

$$u_l u_{k-l} = u_k(1 + u_l + u_{k-l})$$

and

$$u_{k+l} + u_l u_{k+l} = u_k(u_l - u_{k+l}),$$

so that

$$\begin{aligned} C_k &= u_k\left(\frac{1}{2} + \sum_{l=1}^{\infty}(u_l - u_{k+l}) - \frac{1}{2}\sum_{l=1}^{k-1}(1 + u_l + u_{k-l})\right) \\ &= u_k\left(\frac{1}{2} + u_1 + \cdots + u_k - \frac{1}{2}(k-1) - (u_1 + \cdots + u_{k-1})\right) \\ &= u_k\left(1 + u_k - \frac{1}{2}k\right). \quad \square \end{aligned}$$

Theorem 7.6.

$$\left(\frac{1}{4} + \sum_{n=0}^{\infty} u_{4n+1} - \sum_{n=0}^{\infty} u_{4n+3}\right)^2 = \frac{1}{16} + \frac{1}{2}\sum_{\substack{m=1 \\ 4\nmid m}}^{\infty} mu_m.$$

Proof. In Theorem 7.5 we take $\vartheta = \pi/2$ giving

$$\begin{aligned} &\left(\frac{1}{4} + \sum_{n=0}^{\infty} u_{4n+1} - \sum_{n=0}^{\infty} u_{4n+3}\right)^2 \\ &\quad = \frac{1}{16} + \frac{1}{2}\sum_{n=1}^{\infty} nu_n + \sum_{m=1}^{\infty}(-1)^m C_{2m} \\ &\quad = \frac{1}{16} + \frac{1}{2}\sum_{n=1}^{\infty} nu_n + \sum_{m=1}^{\infty}(-1)^m u_{2m}(1 + u_{2m} - m) \\ &\quad = \frac{1}{16} + \frac{1}{2}\sum_{m=1}^{\infty}(2m-1)u_{2m-1} + \sum_{m=1}^{\infty}(-1)^m u_{2m}(1 + u_{2m}) \\ &\qquad + 2\sum_{m=1}^{\infty}(2m-1)u_{4m-2} \\ &\quad = \frac{1}{16} + \frac{1}{2}\sum_{m=1}^{\infty}(2m-1)u_{2m-1} + \sum_{m=1}^{\infty}(2m-1)u_{4m-2} \quad \text{(by Theorem 4)} \\ &\quad = \frac{1}{16} + \frac{1}{2}\sum_{\substack{n=1 \\ 4\nmid n}}^{\infty} nu_n. \quad \square \end{aligned}$$

Theorem 7.2 now follows easily from Theorem 7.1 and Theorem 7.6. From Theorem 7.2 we deduce at once:

Theorem 7.7. $r_4(n)/8$ *is a multiplicative function.* □

Theorem 7.8 (Lagrange). *Every positive integer is the sum of four squares.* □

Apart from these we also have the following application:

Theorem 7.9 (Jacobi). $q_2^8 - q_3^8 = 16qq_1^8$.

If we substitute the representation formulae in §1 into this identity then we have

$$\left(\prod_{n=1}^{\infty}(1+q^{2n-1})\right)^8 - \left(\prod_{n=1}^{\infty}(1-q^{2n-1})\right)^8 = 16q\left(\prod_{n=1}^{\infty}(1+q^{2n})\right)^8.$$

(Jacobi called this result "*Aequartro identica ratis abstrura*".)

Proof. We multiply both sides of the identity by q_0^4. Then from

$$(q_0q_2^2)^4 = \left(\sum_{n=-\infty}^{\infty} q^{n^2}\right)^4 = \sum_{n=0}^{\infty} r_4(n)q^n,$$

$$(q_0q_3^2)^4 = \sum_{n=0}^{\infty} r_4(n)(-1)^n q^n$$

and

$$(2q_0q_1^2)^4 = \left(\sum_{n=-\infty}^{\infty} q^{n(n+1)}\right)^4,$$

we see that our required identity is equivalent to

$$q\left(\sum_{n=-\infty}^{\infty} q^{n(n+1)}\right)^4 = 2\sum_{\substack{n=0 \\ 2\nmid n}}^{\infty} r_4(n)q^n.$$

Let $s_4(n)$ denote the number of solutions to

$$x_1(x_1+1) + \cdots + x_4(x_4+1) + 1 = n, \tag{4}$$

where n must be odd. Thus our theorem has the following arithmetical interpretation: if n is odd, then $s_4(n)$ is equal to $2r_4(n)$.

We multiply equation (4) by 4 and from completing squares we have

$$(2x_1+1)^2 + \cdots + (2x_4+1)^2 = 4n.$$

The $r_4(4n)$ solutions to the Diophantine equation

$$y_1^2 + y_2^2 + y_3^2 + y_4^2 = 4n$$

have only two types: (i) y_1, y_2, y_3, y_4 all odd, (ii) y_1, y_2, y_3, y_4 all even. From this it follows that

$$s_4(n) = r_4(4n) - r_4(n).$$

From Theorem 7.2 we have

$$r_4(4n) = 8 \sum_{m|2n} m = 8 \sum_{m|n} (m + 2m) = 3\left(8 \sum_{m|n} m\right) = 3r_4(n),$$

and hence

$$s_4(n) = 2r_4(n).$$

The theorem is proved. □

Exercise 1. Use the following method to prove that

$$1 + \frac{1}{2^2} + \frac{1}{3^2} + \frac{1}{4^2} + \cdots = \frac{\pi^2}{6}.$$

Obtain the asymptotic formula

$$A(x) = \frac{\pi^2}{2} x^2 + O(x^{\frac{3}{2}})$$

for the number $A(x)$ of lattice points inside the four dimensional sphere

$$u^2 + v^2 + w^2 + z^2 \leqslant x.$$

Find another representation for $A(x)$ with Theorem 7.2 and compare the results.

Note. From this exercise and (6.14.2) we deduce at once that

$$\sum_{n=1}^{\infty} \frac{\mu(n)}{n^2} = \frac{6}{\pi^2}.$$

Exercise 2. Show that

$$\left(\frac{1}{6} + \frac{x}{1-x} - \frac{x^2}{1-x^2} + \frac{x^4}{1-x^4} - \frac{x^5}{1-x^5} + \cdots\right)^2$$

$$= \frac{1}{36} + \frac{1}{3}\left(\frac{x}{1-x} + \frac{2x^2}{1-x^2} + \frac{4x^4}{1-x^4} + \frac{5x^5}{1-x^5} + \cdots\right).$$

Exercise 3. Use the identity

$$(1 - \cos n\vartheta)\cot^2 \tfrac{1}{2}\vartheta = (2n-1) + 4(n-1)\cos\vartheta + 4(n-2)\cos 2\vartheta + \cdots$$
$$+ 4\cos(n-1)\vartheta + \cos n\vartheta$$

to prove that

$$\left\{\frac{1}{8}\cot^2\frac{1}{2}\vartheta + \frac{1}{12} + \frac{x}{1-x}(1-\cos\vartheta) + \frac{2x^2}{1-x^2}(1-\cos 2\vartheta)\right.$$
$$\left. + \frac{3x^3}{1-x^3}(1-\cos 3\vartheta) + \cdots\right\}^2$$
$$= \left(\frac{1}{8}\cot^2\frac{1}{2}\vartheta + \frac{1}{12}\right)^2 + \frac{1}{12}\left\{\frac{1^3x}{1-x}(5+\cos\vartheta) + \frac{2^3x^2}{1-x^2}(5+\cos 2\vartheta)\right.$$
$$\left. + \frac{3^3x^3}{1-x^3}(5+\cos 3\vartheta) + \cdots\right\}.$$

8.8 Density

Let $r_s(n, q)$ denote the number of solutions to

$$x_1^2 + \cdots + x_s^2 \equiv n \pmod q. \tag{1}$$

Consider the substitution

$$x_1^2 + \cdots + x_s^2 = y.$$

There can be q^s values on the left hand side and q values on the right hand side. This means that corresponding to one value of y there are, on average, q^{s-1} solutions. We now consider the ratio between the number of solutions and the average number

$$\Delta_q(n) = \frac{r_s(n, q)}{q^{s-1}}.$$

Let

$$\partial_p(n) = \lim_{l\to\infty} \Delta_{p^l}(n);$$

we call this the *p-density* of the congruence (1).

We also define

$$\partial_0(n) = \lim_{\delta\to 0}\frac{1}{2\delta}\int\cdots\int_{n-\delta\leqslant x_1^2+\cdots+x_s^2\leqslant n+\delta} dx_1\cdots dx_s,$$

which we call the *real density* of the congruence (1).

We now calculate the values of the various densities.

Theorem 8.1. *When s is even the real density is equal to*

$$\frac{\pi^{\frac{s}{2}}}{\left(\frac{s}{2}-1\right)!}n^{\frac{s}{2}-1}. \tag{2}$$

Proof. We have, with polar coordinates,

$$\iint\limits_{1-x^2-y^2>0} (1-x^2-y^2)^{a-1}\,dx\,dy = \int_0^{2\pi} d\vartheta \int_0^1 (1-\rho^2)^{a-1}\rho\,d\rho = \frac{\pi}{a}.$$

We next use induction to prove the result:

$$V_s = \int\limits_{1-x_1^2-\cdots-x_s^2>0}\cdots\int dx_1\cdots dx_s = \frac{\pi^{\frac{s}{2}}}{\left(\frac{s}{2}\right)!}.$$

Let

$$x_\nu = y_{\nu-2}\sqrt{1-x_1^2-x_2^2} \qquad (\nu = 3,\ldots,s).$$

Then

$$V_s = \iint\limits_{1-x_1^2-x_2^2>0} (1-x_1^2-x_2^2)^{\frac{s-2}{2}}\,dx_1\,dx_2 \int\limits_{1-y_1^2-\cdots-y_{s-2}^2>0}\cdots\int dy_1\cdots dy_{s-2}$$

$$= \frac{\pi}{\frac{s}{2}} V_{s-2} = \frac{\pi^{s/2}}{\left(\frac{s}{2}\right)!}.$$

We then have

$$\partial_0(n) = \lim_{\delta\to 0}\frac{1}{2\delta}\left(\int\limits_{x_1^2+\cdots+x_s^2\leqslant n+\delta}\cdots\int dx_1\cdots dx_s - \int\limits_{x_1^2+\cdots+x_s^2\leqslant n-\delta}\cdots\int dx_1\cdots dx_s\right)$$

$$= \frac{\pi^{s/2}}{\left(\frac{s}{2}\right)!}\lim_{\delta\to 0}\frac{(n+\delta)^{s/2}-(n-\delta)^{s/2}}{2\delta} = \frac{\pi^{s/2}}{\left(\frac{s}{2}-1\right)!}n^{\frac{s}{2}-1}. \quad \square$$

In order to determine the p-density we shall require the following preparation. Let

$$A_{p^l}(n) = \sum_{\substack{a=1\\ p\nmid a}}^{p^l} \frac{1}{p^{sl}}\left(\sum_{x=1}^{p^l} e^{2\pi iax^2/p^l}\right)^s e^{-2\pi ian/p^l}.$$

Theorem 8.2.

$$\sum_{m=0}^{l} A_{p^m}(n) = \Delta_{p^l}(n).$$

Proof.

$$\sum_{m=0}^{l} A_{p^m}(n) = \sum_{m=0}^{l}\sum_{\substack{a=1\\ p\nmid a}}^{p^m}\frac{1}{p^{sm}}\left(\sum_{x=1}^{p^m} e^{2\pi iax^2/p^m}\right)^s e^{-2\pi ian/p^m}$$

$$= \sum_{m=0}^{l} \sum_{\substack{a=1 \\ p^{l-m}\|a}}^{p^l} \frac{1}{p^{sl}} \left(\sum_{x=1}^{p^l} e^{2\pi i a x^2/p^l} \right)^s e^{-2\pi i a n/p^l}$$

$$= \sum_{a=1}^{p^l} \frac{1}{p^{sl}} \left(\sum_{x=1}^{p^l} e^{2\pi i a x^2/p^l} \right)^s e^{-2\pi i a n/p^l}$$

$$= \frac{1}{p^{(s-1)l}} \cdot \frac{1}{p^l} \sum_{a=1}^{p^l} \left(\sum_{x=1}^{p^l} e^{2\pi i a x^2/p^l} \right)^s e^{-2\pi i a n/p^l}$$

$$= \frac{r_s(n, p^l)}{p^{(s-1)l}} = \Delta_{p^l}(n). \quad \square$$

Theorem 8.3. *Let $s = 4r$ and p be an odd prime. Then*

$$A_{p^l}(n) = p^{-2rl} C_{p^l}(n).$$

Proof. From Theorem 7.5.6 we know that if $p \nmid a$, then

$$\left(\sum_{x=1}^{p^l} e^{2\pi i a x^2/p^l} \right)^{4r} = p^{2rl},$$

so that

$$A_{p^l}(n) = p^{-2rl} \sum_{\substack{a=1 \\ p \nmid a}}^{p^l} e^{-2\pi i a n/p^l}.$$

On replacing a by $-a$ the theorem follows. $\square$

Theorem 8.4. *Let $s = 4r$. Then*

$$A_2(n) = 0, \qquad A_{2^l}(n) = (-1)^r 2^{-2r(l-1)} C_{2^l}(n).$$

Proof. From Theorem 7.5.3 we have $A_2(n) = 0$. Also, from Theorem 7.5.7 we see that if $2 \nmid a$, then

$$\sum_{x=1}^{2^l} e^{2\pi i a x^2/2^l} = \begin{cases} 2^{\frac{1}{2}l}(1 + i^a), & \text{if } 2 \mid l, \\ 2^{\frac{1}{2}(l+1)} e^{\frac{\pi i}{4} a}, & \text{if } 2 \nmid l. \end{cases}$$

From $(1 + i^a)^4 = -4$, $(e^{\frac{\pi i}{4} a})^4 = -1$, we have

$$\left(\sum_{x=1}^{2^l} e^{2\pi i a x^2/2^l} \right)^{4r} = (-1)^r 2^{2r(l+1)},$$

and Theorem 8.4 follows. $\square$

Theorem 8.5. *Let $s = 4r$, $p \neq 2$, $p^\tau \| n$. Then*

$$\partial_p(n) = (1 - p^{-2r}) \sum_{l=0}^{\tau} p^{-(2r-1)l} = (1 - p^{-2r})(p^\tau)^{-(2r-1)} \sigma_{2r-1}(p^\tau),$$

where

$$\sigma_t(n) = \sum_{d|n} d^t.$$

Proof. From Theorem 8.3 and Theorem 7.4.4 we have

$$\begin{aligned}
\partial_p(n) &= \sum_{l=0}^{\infty} A_{p^l}(n) = 1 + \sum_{l=1}^{\infty} p^{-2rl} C_{p^l}(n) \\
&= 1 + \sum_{l=1}^{\tau} p^{-2rl}(p^l - p^{l-1}) - p^{-2r(\tau+1)} p^{\tau} \\
&= \sum_{l=0}^{\tau} p^{-2rl+l} - \sum_{l=1}^{\tau+1} p^{-2rl+l-1} \\
&= \sum_{l=0}^{\tau} p^{-(2r-1)l}(1 - p^{-2r}). \quad \square
\end{aligned}$$

Theorem 8.6. *Suppose that* $s = 4r$ *and let* $2^{\tau} \| n$. *Then*

$$\partial_2(n) = \begin{cases} 1, & \text{if } \tau = 0, \\ (1 - 2^{2-2r} + 2^{(1-2r)(\tau+1)}(2^{2r} - 1))(1 - 2^{1-2r})^{-1}, & \text{if } \tau > 0,\ 2 \nmid r, \\ (1 - 2^{(1-2r)(\tau+1)}(2^{2r} - 1))(1 - 2^{1-2r})^{-1}, & \text{if } \tau > 0,\ 2 | r. \quad \square \end{cases}$$

The reader can supply the proof which is similar to Theorem 8.5.

Definition. Let

$$\mathfrak{S}_s(n) = \prod_p \partial_p(n)$$

and

$$\delta_s(n) = \partial_0(n) \mathfrak{S}_s(n) = \partial_0(n) \prod_p \partial_p(n).$$

Theorem 8.7. *If* $s = 4$, *then*

$$\delta_4(n) = r_4(n) = 8 \sum_{\substack{d|n \\ 4 \nmid d}} d.$$

Proof. Let $n = 2^{\tau} n'$, $2 \nmid n'$. Then from

$$\frac{1}{\zeta(s)} = \prod_p (1 - p^{-s})$$

and Theorem 8.5 we have

$$\prod_{p>2} \partial_p(n) = \frac{4}{3} \frac{1}{\zeta(2)} n'^{-1} \sigma(n') = \frac{8}{\pi^2} n'^{-1} \sigma(n').$$

Also, from Theorem 8.1, we have

$$\partial_0(n) = \pi^2 n,$$

so that

$$\partial_0(n) \prod_{p>2} \partial_p(n) = 2^{\tau+3}\sigma(n').$$

If n is odd, then the theorem is proved. If n is even, then, from Theorem 8.6, we have

$$\partial_2(n) = 3 \cdot 2^{-\tau}.$$

The theorem is proved. □

Theorem 8.8. *If* $s = 8$, *then*

$$\delta_8(n) = 16(-1)^n \sum_{d|n} (-1)^d d^3.$$

Proof. Let $n = 2^\tau n'$, $2 \nmid n'$. Then

$$\prod_{p>2} \partial_p(n) = \frac{16}{15} \frac{1}{\zeta(4)} n'^{-3}\sigma_3(n') = \frac{96}{\pi^4} n'^{-3}\sigma_3(n').$$

Also, from Theorem 8.1, we have

$$\partial_0(n) = \frac{\pi^4}{6} n^3,$$

so that

$$\partial_0(n) \prod_{p>2} \partial_p(n) = 16 \cdot 2^{3\tau}\sigma_3(n').$$

Also

$$\partial_2(n) = (1 - 2^{-3(\tau+1)} \cdot 15)(1 - \tfrac{1}{8})^{-1};$$

hence

$$\delta_8(n) = 16 \cdot \tfrac{8}{7}(2^{3\tau} - \tfrac{15}{8})\sigma_3(n').$$

When n is even

$$\begin{aligned}
\sum_{d|n} (-1)^d d^3 &= -\sigma_3(n') + 2^3\sigma_3(n') + 2^{3\cdot 2}\sigma_3(n') + \cdots + 2^{3\cdot\tau}\sigma_3(n') \\
&= -2\sigma_3(n') + \frac{2^{3(\tau+1)} - 1}{2^3 - 1}\sigma_3(n') \\
&= \left(-2 + \frac{2^{3(\tau+1)} - 1}{2^3 - 1}\right)\sigma_3(n') = \frac{8}{7}\left(2^{3\tau} - \frac{15}{8}\right)\sigma_3(n').
\end{aligned}$$

The theorem is proved. □

Exercise 1. Prove the following: Let $s = 2r$. If r is even, then

$$(1 - 2^{-r})\zeta(r)\mathfrak{S}_s(2^\tau n')$$

$$= \begin{cases} n'^{1-r}\sigma_{r-1}(n') & \text{if } \tau=0, \\ (1-2^{2-r}+2^{(1-r)(\tau+1)}(2^r-1))(1-2^{1-r})^{-1}n'^{1-r}\sigma_{r-1}(n'), & \text{if } \tau>0, 2\|r, \\ (1-2^{(1-r)(\tau+1)}(2^r-1))(1-2^{1-r})^{-1}n'^{1-r}\sigma_{r-1}(n'), & \text{if } \tau>0, 4|r. \end{cases}$$

If r is odd, then

$$L(r)\mathfrak{S}_s(2^\tau n') = \left(\left(\frac{-1}{n'}\right) + \left(\frac{-1}{r}\right)2^{(1-r)(\tau+1)}\right)n'^{1-r}\rho_{r-1}(n'),$$

where

$$L(r) = \sum_{n=1}^{\infty} \frac{\chi(n)}{n^r},$$

and $\chi(n) = 0, 1, 0, -1$ when $n \equiv 0, 1, 2, 3 \pmod 4$. Also

$$\rho_t(n) = \sum_{q|n} \left(\frac{-1}{q}\right) q^t.$$

Exercise 2. Prove that $\delta_2(n) = 2r_2(n)$.

Exercise 3. Prove that

$$\delta_6(n) = 16 \sum_{d|n} \chi\left(\frac{n}{d}\right) d^2 - 4 \sum_{d|n} \chi(d) d^2.$$

8.9 A Summary of the Problem of Sums of Squares

In the previous section we proved that $r_4(n) = \delta_4(n)$, but is this a mere coincidence? Actually we can prove that, for $3 \leqslant s \leqslant 8$, we have

$$r_s(n) = \delta_s(n),$$

and that this is no longer true if $s > 8$.

Up to the present $r_s(n)$ has been explicitly evaluated for $s \leqslant 24$. For example:

$$r_3(n) = \frac{16}{\pi} n^{\frac{1}{2}} \chi_2(n) K(-4n) \prod_{p^2|n} \left(1 + \frac{1}{p} + \cdots + \frac{1}{p^{\tau-1}} + \frac{1}{p^\tau}\left(1 - \left(\frac{-p^{-2\tau}n}{p}\right)\frac{1}{p}\right)^{-1}\right),$$

where the definition of τ is $p^{2\tau}|n$, $p^{2(\tau+1)} \nmid n$,

$$K(-4n) = \sum_{m=1}^{\infty} \left(\frac{-4n}{m}\right)\frac{1}{m},$$

and

$$\chi_2(n) = \begin{cases} 0, & \text{if } 4^{-a}n \equiv 7 \pmod 8, \\ 2^{-a}, & \text{if } 4^{-a}n \equiv 3 \pmod 8, \\ 3 \cdot 2^{-1-a}, & \text{if } 4^{-a}n \equiv 1, 2, 5, 6 \pmod 8, \end{cases}$$

and here the definition of a is $4^a | n$, $4^{a+1} \nmid n$.

$$r_{24}(n) = \tfrac{16}{691}\sigma_{11}^*(n) + \tfrac{128}{691}((-1)^{n-1}259\tau(n) - 512\tau(\tfrac{1}{2}n)),$$

where

$$\sigma_{11}^*(n) = \sum_{d|n}(-1)^d d^{11},$$

and $\tau(n)$ is the coefficient in the power series expansion

$$q((1-q)(1-q^2)\cdots)^{24} = \sum_{n=1}^{\infty} \tau(n)q^n$$

and if $n/2$ is not an integer, then $\tau(n/2) = 0$.

From Theorem 3.6 we have

$$((1-q)(1-q^2)(1-q^3)\cdots)^3 = \sum_{n=0}^{\infty}(-1)^n(2n+1)q^{\frac{1}{2}n(n+1)},$$

so that

$$\tau(n) = \sum_{\frac{1}{2}x_1(x_1+1)+\cdots+\frac{1}{2}x_8(\tau_8+1)=n-1} ((-1)^{x_1}(2x_1+1)+\cdots+(-1)^{x_8}(2x_8+1))$$

$$= \sum_{\substack{y_1^2+\cdots+y_8^2=8n \\ 2\nmid y_1\cdots y_8}} \sum_{i=1}^{8} (-1)^{\frac{1}{2}(y_i-1)} y_i.$$

The following table records the mathematicians who did the evaluations:

s	$r_s(n)$
2, 4, 6, 8	Jacobi, 1828
3	Dirichlet
5, 7	Eisenstein, Smith, Minkowski
10, 12	Liouville, 1864, 1866
14, 16, 18	Glaisher, 1907
20, 22, 24	Ramanujan, 1916
9, 11, 13	Lomadze, 1949
15, 17, 19	
21, 23	

Chapter 9. The Prime Number Theorem

9.1 Introduction

The main aim of this chapter is to prove the following formula:

$$\pi(x) \sim \frac{x}{\log x}. \tag{1}$$

Here $\pi(x)$ denotes the number of primes not exceeding x, and the formula (1) is the famous *prime number theorem*. In this chapter we shall give two proofs. The first proof makes use of some rather deep analytic tools (the reader needs to know a little advanced calculus and complex function theory) but is relatively straight-forward, the fundamental idea being due to N. Wiener. Although the other proof does not require much analytic knowledge and can indeed be classified as an elementary proof, it is more difficult to understand. This proof is due to Erdős and Selberg. One of the difficult problems in the long history of prime number theory is the search for an "elementary proof" of the prime number theorem and success came in 1949.

In the following sections we do not give a direct proof of the formula (1). Instead we prove two formulae, each of which is equivalent to (1).

Suppose that $x > 0$. Let

$$\vartheta(x) = \sum_{p \leqslant x} \log p, \tag{2}$$

$$\psi(x) = \sum_{n \leqslant x} \Lambda(n) = \sum_{p^m \leqslant x} \log p. \tag{3}$$

In formula (3) $\Lambda(n)$ is the von Mangoldt function of Example 6 in §6.1. $\vartheta(x)$ and $\psi(x)$ are called Chebyshev's functions. It is easy to see that

$$\psi(x) = \vartheta(x) + \vartheta(x^{\frac{1}{2}}) + \vartheta(x^{\frac{1}{3}}) + \cdots \tag{4}$$

and

$$\psi(x) = \sum_{p \leqslant x} \left[\frac{\log x}{\log p}\right] \log p, \tag{5}$$

where $[\frac{\log x}{\log p}]$ denotes the integer part of $\frac{\log x}{\log p}$.

Theorem 1.1. *We have*

$$\varlimsup_{x\to\infty} \frac{\pi(x)}{x(\log x)^{-1}} = \varlimsup_{x\to\infty} \frac{\vartheta(x)}{x} = \varlimsup_{x\to\infty} \frac{\psi(x)}{x} \tag{6}$$

and

$$\underline{\lim}_{x\to\infty} \frac{\pi(x)}{x(\log x)^{-1}} = \underline{\lim}_{x\to\infty} \frac{\vartheta(x)}{x} = \underline{\lim}_{x\to\infty} \frac{\psi(x)}{x}. \tag{7}$$

Proof. From (4) and (5) we derive easily

$$\vartheta(x) \leqslant \psi(x) \leqslant \sum_{p\leqslant x} \frac{\log x}{\log p} \log p = \pi(x)\log x,$$

so that

$$\overline{\lim}_{x\to\infty} \frac{\vartheta(x)}{x} \leqslant \overline{\lim}_{x\to\infty} \frac{\psi(x)}{x} \leqslant \overline{\lim}_{x\to\infty} \frac{\pi(x)}{x(\log x)^{-1}}.$$

Now let $0 < \alpha < 1$, $x > 1$. Then

$$\vartheta(x) \geqslant \sum_{x^\alpha < p \leqslant x} \log p \geqslant \{\pi(x) - \pi(x^\alpha)\}\log x^\alpha \geqslant \alpha\{\pi(x) - x^\alpha\}\log x.$$

Since $\lim_{x\to\infty} \frac{\log x}{x^{1-\alpha}} = 0$, it follows that

$$\overline{\lim}_{x\to\infty} \frac{\vartheta(x)}{x} \geqslant \alpha\, \overline{\lim}_{x\to\infty} \frac{\pi(x)}{x(\log x)^{-1}}$$

holds for any positive α less than 1. Therefore

$$\overline{\lim}_{x\to\infty} \frac{\vartheta(x)}{x} \geqslant \overline{\lim}_{x\to\infty} \frac{\pi(x)}{x(\log x)^{-1}}.$$

Collecting our results we have

$$\overline{\lim}_{x\to\infty} \frac{\pi(x)}{x(\log x)^{-1}} = \overline{\lim}_{x\to\infty} \frac{\vartheta(x)}{x} = \overline{\lim}_{x\to\infty} \frac{\psi(x)}{x}.$$

Formula (7) can be proved similarly. □

From Theorem 1.1 and Theorem 5.6.2 we have

Theorem 1.2. *There exist constants* $c_i > 0$ $(i = 1, 2, 3, 4)$ *such that for* $x \geqslant 2$,

$$c_1 x \leqslant \vartheta(x) \leqslant c_2 x, \tag{8}$$

and

$$c_3 x \leqslant \psi(x) \leqslant c_4 x. \quad \square \tag{9}$$

Also from Theorem 1.1 we see at once that in order to prove formula (1) we need only prove that

$$\psi(x) \sim x \tag{10}$$

or

$$\vartheta(x) \sim x. \tag{11}$$

Before we prove formula (10) we need some preparation.

9.2 The Riemann ζ-Function

From now on we write $s = \sigma + it$ for a complex number with σ and t real. The function defined by the series

$$\zeta(s) = \sum_{n=1}^{\infty} \frac{1}{n^s} \qquad (\sigma > 1) \tag{1}$$

is called the *Riemann ζ-function.*

Let $a > 1$. When $\sigma \geqslant a$, because

$$\left| \sum_{n=N}^{\infty} \frac{1}{n^s} \right| \leqslant \sum_{n=N}^{\infty} \frac{1}{n^\sigma} \leqslant \sum_{n=N}^{\infty} \frac{1}{n^a},$$

we see that the series for $\zeta(s)$ is uniformly convergent. Since a is any real number greater than 1, it follows that $\zeta(s)$ is an analytic function in the half plane $\sigma > 1$.

Theorem 2.1. *Let*

$$h(s) = \zeta(s) - \frac{1}{s-1}.$$

Then $h(s)$ is analytic in the half plane $\sigma > 0$, and

$$|h(s)| \leqslant \frac{|s|}{\sigma} \qquad (\sigma > 0).$$

Proof. Let

$$f_n(s) = n^{-s} - \int_n^{n+1} u^{-s}\, du,$$

so that

$$\zeta(s) = \sum_{n=1}^{\infty} n^{-s} = \sum_{n=1}^{\infty} f_n(s) + \int_1^{\infty} u^{-s}\, du = \sum_{n=1}^{\infty} f_n(s) + \frac{1}{s-1} \qquad (\sigma > 1). \tag{2}$$

Since

$$|n^{-s} - u^{-s}| = \left| \int_n^u s v^{-s-1}\, dv \right| \leqslant |s| \int_n^{n+1} v^{-\sigma-1}\, dv \qquad (n \leqslant u \leqslant n+1),$$

we have

$$|f_n(s)| = \left| \int_n^{n+1} (n^{-s} - u^{-s})\, du \right| \leqslant |s| \int_n^{n+1} v^{-\sigma-1}\, dv.$$

Suppose that $0 < a \leqslant \sigma \leqslant b$, $-T \leqslant t \leqslant T$. Then

$$\left| \sum_{n=N}^{\infty} f_n(s) \right| \leqslant \sum_{n=N}^{\infty} |f_n(s)| \leqslant |s| \int_N^{\infty} v^{-\sigma-1}\, dv = \frac{|s|}{\sigma} N^{-\sigma}$$

$$\leqslant \frac{\sqrt{b^2 + T^2}}{a} N^{-a},$$

so that the series $\sum_{n=1}^{\infty} f_n(s)$ is uniformly convergent in $0 < a \leqslant \sigma \leqslant b$, $-T \leqslant t \leqslant T$. Since a can be arbitrarily near 0, and b, T can be arbitrarily large it follows that $h(s) = \sum_{n=1}^{\infty} f_n(s)$ is analytic in the half plane $\sigma > 0$. From this we see that (2) can be used as an analytic continuation for $\zeta(s)$ into the half plane $\sigma > 0$, and $s = 1$ is the only simple pole with residue 1.

From (2) we derive at once

$$\left| \zeta(s) - \frac{1}{s-1} \right| = \left| \sum_{n=1}^{\infty} f_n(s) \right| \leqslant |s| \int_1^{\infty} v^{-\sigma-1}\, dv = \frac{|s|}{\sigma} \qquad (\sigma > 0).$$

The theorem is proved. □

Theorem 2.2. *In the half plane* $\sigma \geqslant 1$, $\zeta(s) \neq 0$.

Proof. When $\sigma > 1$ the series $\sum_{n=1}^{\infty} (1/n^s)$ converges absolutely so that from Theorem 5.4.4

$$\zeta(s) = \sum_{n=1}^{\infty} \frac{1}{n^s} = \prod_p (1 - p^{-s})^{-1}; \tag{3}$$

here the product is over all primes p. Since each factor in the product is non-zero and the product converges absolutely, it follows that $\zeta(s) \neq 0$ when $\sigma > 1$.

Since $\zeta(s)$ has a pole at $s = 1$ we are left to prove: when $t \neq 0$

$$\zeta(1 + it) \neq 0.$$

Now consider the function

$$\varphi_\varepsilon(t) = |\zeta(1 + \varepsilon)|^3 |\zeta(1 + \varepsilon + it)|^4 |\zeta(1 + \varepsilon + 2it)|, \qquad (\varepsilon > 0,\ t \neq 0).$$

From (3) we know that

$$\varphi_\varepsilon(t) = \prod_p a_p,$$

where

$$a_p = \left|1 - \frac{1}{p^{1+\varepsilon}}\right|^{-3} \cdot \left|1 - \frac{1}{p^{1+\varepsilon+it}}\right|^{-4} \cdot \left|1 - \frac{1}{p^{1+\varepsilon+2it}}\right|^{-1},$$

so that

$$\begin{aligned}\log a_p &= -3\log\left(1 - \frac{1}{p^{1+\varepsilon}}\right) - 4R\log\left(1 - \frac{1}{p^{1+\varepsilon+it}}\right) - R\log\left(1 - \frac{1}{p^{1+\varepsilon+2it}}\right)\\ &= \sum_{m=1}^{\infty} \frac{1}{m} p^{-(1+\varepsilon)m}(3 + 4\cos(mt\log p) + \cos(2mt\log p)).\end{aligned}$$

From $3 + 4\cos\vartheta + \cos 2\vartheta = 2(1 + \cos\vartheta)^2 \geqslant 0$, we have $\log a_p \geqslant 0$, that is

$$|\varphi_\varepsilon(t)| \geqslant 1. \tag{4}$$

Suppose that $\zeta(1 + it) = 0$. Then

$$\zeta(1 + \varepsilon + it) = \int_1^{1+\varepsilon} \zeta'(\sigma + it)\, d\sigma = O(\varepsilon).$$

From Theorem 2.1, we have

$$\varepsilon\zeta(1 + \varepsilon) = O(1)$$

so that, for any small ε, we have

$$\varphi_\varepsilon(t) = O(\varepsilon),$$

and this contradicts (4). □

Theorem 2.3. *Let*

$$\frac{\zeta'(s)}{\zeta(s)} + \frac{1}{s-1} = g(s).$$

When $\sigma \geqslant 1$, $g(s)$ has a continuous first derivative.

Proof. Differentiating the function $h(s)$ in Theorem 2.1 we have

$$\zeta'(s) = -\frac{1}{(s-1)^2} + h'(s).$$

Here $h'(s)$ is infinitely differentiable in $\sigma > 0$. Also from Theorem 2.2, we see that

$$\frac{1}{\zeta(s)} = \frac{s-1}{1 + (s-1)h(s)}$$

is regular in the half plane $\sigma \geqslant 1$, so that $1 + (s-1)h(s) \neq 0$ in the same half plane.

Therefore

$$\frac{\zeta'(s)}{\zeta(s)} = \frac{-\left(\dfrac{1}{(s-1)^2} - h'(s)\right)(s-1)}{1 + (s-1)h(s)} = -\frac{1}{s-1} + g(s),$$

and here $g(s)$ has the required property stated in the theorem. □

9.3 Several Lemmas

Theorem 3.1. *If $f(x)$ has a continuous first derivative, then*

$$\int_a^b f(x)e^{ixt}\,dx = O\left(\frac{1}{t}\right). \tag{1}$$

Proof. From integration by parts we have

$$\int_a^b f(x)e^{ixt}\,dx = \frac{1}{it}\left\{[f(x)e^{ixt}]_a^b - \int_a^b f'(x)e^{ixt}\,dx\right\} = O\left(\frac{1}{t}\right).$$

Theorem 3.2.

$$\int_{-\infty}^{\infty} \frac{\sin x}{x}\,dx = \pi. \tag{2}$$

Proof. Let

$$J = \int_0^\infty e^{-kx}\frac{\sin \alpha x}{x}\,dx \qquad (1 \leqslant \alpha \leqslant 2,\ 0 \leqslant k \leqslant 1).$$

Fix $k > 0$ so that the integrand is now a continuous function of α and x, and the partial derivative with respect to α is $e^{-kx}\cos \alpha x$, which is also continuous. From the convergence of the integral

$$\int_0^\infty e^{-kx}\,dx$$

we see that the integral

$$\int_0^\infty e^{-kx}\cos \alpha x\,dx$$

converges uniformly in $1 \leqslant \alpha \leqslant 2$. We can therefore differentiate J under the integral sign giving

$$\frac{dJ}{d\alpha} = \int_0^\infty e^{-kx} \cos \alpha x \, dx = \frac{k}{\alpha^2 + k^2}.$$

Here the right hand side is obtained from integrating by parts twice. From integration formulae we have

$$J = \tan^{-1} \frac{\alpha}{k} \qquad (1 \leqslant \alpha \leqslant 2, 0 < k \leqslant 1).$$

With α fixed, when $0 \leqslant k \leqslant 1$, J is uniformly convergent so that J is continuous for $0 \leqslant k \leqslant 1$. Therefore

$$\lim_{k \to 0+} J = \int_0^\infty \frac{\sin \alpha x}{x} dx = \lim_{k \to 0+} \tan^{-1} \frac{\alpha}{k} = \frac{\pi}{2}.$$

Taking in particular $\alpha = 1$, we have

$$\int_{-\infty}^\infty \frac{\sin x}{x} dx = 2 \int_0^\infty \frac{\sin x}{x} dx = \pi. \quad \square$$

Theorem 3.3. *Let $a < 0 < b$. If $f(x)$ has a continuous second derivative, then*

$$\lim_{\omega \to \infty} \frac{1}{\pi} \int_a^b f(x) \frac{\sin \omega x}{x} dx = f(0). \tag{3}$$

Proof. We consider

$$\int_a^b (f(x) - f(0)) \frac{\sin \omega x}{x} dx.$$

At the point 0, $(f(x) - f(0))/x$ has a continuous first derivative so that from Theorem 3.1 we have

$$\lim_{\omega \to \infty} \int_a^b (f(x) - f(0)) \frac{\sin \omega x}{x} dx = 0,$$

that is

$$\lim_{\omega \to \infty} \frac{1}{\pi} \int_a^b f(x) \frac{\sin \omega x}{x} dx = f(0) \lim_{\omega \to \infty} \frac{1}{\pi} \int_a^b \frac{\sin \omega x}{x} dx$$

$$= f(0)\frac{1}{\pi}\lim_{\omega\to\infty}\int_{c\omega}^{b\omega}\frac{\sin x}{x}\,dx$$

$$= f(0)\frac{1}{\pi}\int_{-\infty}^{\infty}\frac{\sin x}{x}\,dx,$$

and the result follows from Theorem 3.2. □

Theorem 3.4. *Let* $\lambda > 0$, *and*

$$K_\lambda(x) = \begin{cases} 1 - \dfrac{|x|}{2\lambda}, & \text{if } |x| \leqslant 2\lambda, \\ 0, & \text{if } |x| > 2\lambda. \end{cases}$$

Then

$$\frac{1}{\sqrt{2\pi}}\int_{-\infty}^{\infty} K_\lambda(t)e^{ixt}\,dt = k_\lambda(x), \tag{4}$$

where

$$k_\lambda(x) = \begin{cases} \dfrac{2\lambda}{\sqrt{2\pi}}\left(\dfrac{\sin\lambda x}{\lambda x}\right)^2, & \text{if } x \neq 0, \\ \dfrac{2\lambda}{\sqrt{2\pi}}, & \text{if } x = 0. \end{cases}$$

Proof. It is easy to see that

$$k_\lambda(x) = \frac{2}{\sqrt{2\pi}}\int_0^{2\lambda}\left(1 - \frac{t}{2\lambda}\right)\cos xt\,dt. \tag{5}$$

If $x = 0$, then clearly

$$k_\lambda(x) = \frac{1}{\sqrt{2\pi}}2\lambda.$$

If $x \neq 0$, then integration by parts gives the required result at once. □

Theorem 3.5. *We have*

$$K_\lambda(x) = \frac{1}{\sqrt{2\pi}}\int_{-\infty}^{\infty} k_\lambda(t)e^{ixt}\,dt. \tag{6}$$

In particular, with $\lambda = 1$, $x = 0$, *we have*

$$\frac{1}{\pi}\int_{-\infty}^{\infty}\frac{\sin^2 x}{x^2}\,dx = 1. \tag{7}$$

Proof. We first consider the integral

$$I(\omega) = \frac{1}{\sqrt{2\pi}}\int_{-\omega}^{\omega} k_\lambda(t)e^{ixt}\,dt = \frac{2}{\sqrt{2\pi}}\int_0^{\omega} k_\lambda(t)\cos xt\,dt.$$

From (5) we have

$$\begin{aligned}
I(\omega) &= \frac{2}{\pi}\int_0^{\omega}\int_0^{2\lambda}\left(1-\frac{u}{2\lambda}\right)\cos ut\cos xt\,du\,dt\\
&= \frac{1}{\pi}\int_0^{2\lambda}\left(1-\frac{u}{2\lambda}\right)du\int_0^{\omega}(\cos(u+x)t+\cos(u-x)t)\,dt\\
&= \frac{1}{\pi}\int_0^{2\lambda}\left(1-\frac{u}{2\lambda}\right)\left(\frac{\sin(u+x)\omega}{u+x}+\frac{\sin(u-x)\omega}{u-x}\right)du.
\end{aligned}$$

If $x > 2\lambda$ we have $\lim_{\omega\to\infty} I(\omega) = 0$ from Theorem 3.1; if $0 < x < 2\lambda$ we see from Theorem 3.1 and Theorem 3.3 that in the above formula the limit of the first term is 0 and the limit of the second term is $1 - x/2\lambda$. Since the integral in (6) is a continuous function of x, we see that $K_\lambda(2\lambda) = 0$, $K_\lambda(0) = 1$. The theorem is proved. □

Theorem 3.6. *Let* $f(t) \geqslant 0$ $(0 \leqslant t \leqslant \infty)$, *and for any* $T > 0$, *the interval* $0 \leqslant t \leqslant T$ *can be divided into a finite number of sections in each of which* $f(t)$ *is continuous. Suppose further that, for any* $\varepsilon > 0$, *the integral*

$$\int_0^{\infty} e^{-\varepsilon t}f(t)\,dt$$

converges. Then

$$\lim_{\varepsilon\to 0}\int_0^{\infty} e^{-\varepsilon t}f(t)\,dt = \int_0^{\infty} f(t)\,dt. \tag{8}$$

Proof. Since $f(t) \geqslant 0$, $\int_0^T f(t)\,dt$ increases with respect to T so that $\int_0^{\infty} f(t)\,dt$ exists either as a finite number or ∞. Now

$$\int_0^\infty e^{-\varepsilon t} f(t)\, dt \leqslant \int_0^\infty f(t)\, dt,$$

so that

$$\overline{\lim_{\varepsilon \to 0}} \int_0^\infty e^{-\varepsilon t} f(t)\, dt \leqslant \int_0^\infty f(t)\, dt.$$

On the other hand

$$\int_0^\infty e^{-\varepsilon t} f(t)\, dt \geqslant \int_0^T e^{-\varepsilon t} f(t)\, dt \geqslant e^{-\varepsilon T} \int_0^T f(t)\, dt,$$

so that

$$\underline{\lim_{\varepsilon \to 0}} \int_0^\infty e^{-\varepsilon t} f(t)\, dt \geqslant \int_0^T f(t)\, dt.$$

Letting $T \to \infty$ we have

$$\underline{\lim_{\varepsilon \to 0}} \int_0^\infty e^{-\varepsilon t} f(t)\, dt \geqslant \int_0^\infty f(t)\, dt,$$

and the theorem is proved. □

9.4 A Tauberian Theorem

Definition. If $f(x)$ is defined in $-\infty < x < \infty$ and satisfies

$$\underline{\lim_{\substack{y - x \to 0 \\ x \to \infty}}} \{f(y) - f(x)\} \geqslant 0 \qquad (y > x), \tag{1}$$

then we say that $f(x)$ is a *slowly decreasing function.*

Theorem 4.1. *Let $f(x)$ be a slowly decreasing function satisfying $|f(x)| < M$ $(-\infty < x < \infty)$. If*

$$\lim_{x \to \infty} \frac{1}{\sqrt{2\pi}} \int_{-\infty}^{\infty} k_\lambda(x - t) f(t)\, dt = l,$$

holds for every $\lambda > 0$, then $f(x) \to l$ $(x \to \infty)$.

Proof. From Theorem 3.5 we have

$$\frac{1}{\sqrt{2\pi}}\int_{-\infty}^{\infty} k_\lambda(x-t)\,dt = \frac{1}{\pi}\int_{-\infty}^{\infty}\frac{\sin^2 u}{u^2}\,du = 1,$$

so that, without loss of generality we can suppose that $l = 0$.

If $f(x) \not\to 0$, then there exists $\delta > 0$ and a sequence $(x_n)(x_n \to \infty)$ such that $f(x_n) < -\delta$ $(n = 1, 2, \ldots)$ or $f(x_n) > \delta$. Assume without loss that $f(x_n) > \delta$ $(n = 1, 2, \ldots)$. (The case $f(x_n) < -\delta$ can be proved in the same way.)

Since $f(x)$ is slowly decreasing, there exists $x_0 = x_0(\delta)$ and $\eta = \eta(\delta)$ such that

$$f(y) - f(x) \geqslant -\frac{\delta}{2} \qquad (x \geqslant x_0,\ 0 \leqslant y - x \leqslant 2\eta)$$

holds. Take a particular x in (x_n). Then

$$f(y) > \frac{\delta}{2} \qquad (x_0 \leqslant x \leqslant y \leqslant x + 2\eta,\ x \in \{x_n\}). \tag{2}$$

From (2), when $x \geqslant x_0$ and x in (x_n), we have

$$\begin{aligned}
&\frac{1}{\sqrt{2\pi}}\int_{-\infty}^{\infty} k_\lambda(x+\eta-t)f(t)\,dt \\
&\geqslant \frac{\delta}{2\sqrt{2\pi}}\int_{x}^{x+2\eta} k_\lambda(x+\eta-t)\,dt - \frac{M}{\sqrt{2\pi}}\int_{-\infty}^{x} k_\lambda(x+\eta-t)\,dt \\
&\quad - \frac{M}{\sqrt{2\pi}}\int_{x+2\eta}^{\infty} k_\lambda(x+\eta-t)\,dt \\
&= \frac{\delta}{2\sqrt{2\pi}}\int_{x-\eta}^{x+\eta} k_\lambda(x-u)\,du - \frac{M}{\sqrt{2\pi}}\int_{-\infty}^{x-\eta} k_\lambda(x-u)\,du - \frac{M}{\sqrt{2\pi}}\int_{x+\eta}^{\infty} k_\lambda(x-u)\,du \\
&= \frac{\delta}{\sqrt{2\pi}}\int_{0}^{\eta} k_\lambda(v)\,dv - \frac{2M}{\sqrt{2\pi}}\int_{\eta}^{\infty} k_\lambda(v)\,dv \\
&= \frac{\delta}{\pi}\int_{0}^{\lambda\eta}\frac{\sin^2 w}{w^2}\,dw - \frac{2M}{\pi}\int_{\lambda\eta}^{\infty}\frac{\sin^2 w}{w^2}\,dw \\
&\to \frac{\delta}{2}(\lambda \to \infty),
\end{aligned}$$

so that there exists a suitably large λ_0 such that

$$\frac{1}{\sqrt{2\pi}} \int_{-\infty}^{\infty} k_{\lambda_0}(x + \eta - t) f(t)\, dt > \frac{\delta}{4} \qquad (x \geqslant x_0,\ x \in \{x_n\}).$$

Let x increase without bound in (x_n) so that

$$\varliminf_{\substack{x \to \infty \\ x \in \{x_n\}}} \frac{1}{\sqrt{2\pi}} \int_{-\infty}^{\infty} k_{\lambda_0}(x + \eta - t) f(t)\, dt \geqslant \frac{\delta}{4},$$

which contradicts our supposition. Therefore $f(x) \to 0$ and the theorem is proved. □

Theorem 4.2 (Ikehara). *Let $h(t)$ be non-decreasing in $0 \leqslant t < \infty$, and suppose that for any finite T, $h(t)$ has only a finite number of discontinuities in $0 \leqslant t \leqslant T$. Suppose also that the integral*

$$f(s) = \int_0^{\infty} e^{-st} h(t)\, dt \qquad (\sigma > 1) \tag{3}$$

converges, and that given any finite $a > 0$, there exists a constant A such that

$$\lim_{\sigma \to 1} \left(f(s) - \frac{A}{s - 1} \right) = g(t) \tag{4}$$

uniformly in $|t| \leqslant a$, where $g(t)$ has a continuous derivative. Then

$$\lim_{t \to \infty} e^{-t} h(t) = A. \tag{5}$$

Proof. Let

$$a(t) = \begin{cases} e^{-t} h(t) & (t \geqslant 0), \\ 0 & (t < 0); \end{cases} \qquad A(t) = \begin{cases} A & (t \geqslant 0) \\ 0 & (t < 0). \end{cases}$$

We now prove the following: 1) For any $\lambda > 0$, the integral

$$I_\lambda(x) = \frac{1}{\sqrt{2\pi}} \int_{-\infty}^{\infty} k_\lambda(x - t)(a(t) - A(t))\, dt \tag{6}$$

exists; 2)

$$\lim_{x \to \infty} I_\lambda(x) = 0 \tag{7}$$

and 3) $a(t) - A(t)$ is a bounded slowly decreasing function. The theorem will then follow from these three points and Theorem 4.1.

Consider the integral

$$I_{\lambda,\varepsilon}(x) = \frac{1}{\sqrt{2\pi}} \int_{-\infty}^{\infty} k_\lambda(x-t)(a(t) - A(t))e^{-\varepsilon t}\,dt.$$

From our hypothesis this integral exists for any $\varepsilon > 0$, $\lambda > 0$. From Theorem 3.4 and the uniform convergence of

$$\int_{-\infty}^{\infty} (a(t) - A(t))e^{-(\varepsilon + iy)t}\,dt$$

in $|y| \leqslant 2\lambda$, it follows that

$$I_{\lambda,\varepsilon}(x) = \frac{1}{2\pi} \int_{-\infty}^{\infty} (a(t) - A(t))e^{-\varepsilon t}\,dt \int_{-2\lambda}^{2\lambda} K_\lambda(y)e^{i(x-t)y}\,dy$$

$$= \frac{1}{2\pi} \int_{-2\lambda}^{2\lambda} K_\lambda(y)e^{ixy}\,dy \int_{-\infty}^{\infty} (a(t) - A(t))e^{-(\varepsilon+iy)t}\,dt$$

$$= \frac{1}{2\pi} \int_{-2\lambda}^{2\lambda} K_\lambda(y)e^{ixy}\left(f(1 + \varepsilon + iy) - \frac{A}{\varepsilon + iy}\right)dy.$$

From (4) we have

$$\lim_{\varepsilon \to 0} I_{\lambda,\varepsilon}(x) = \frac{1}{2\pi} \int_{-2\lambda}^{2\lambda} g(y)K_\lambda(y)e^{ixy}\,dy. \tag{8}$$

From Theorem 3.1 we have

$$\lim_{x \to \infty} \lim_{\varepsilon \to 0} I_{\lambda,\varepsilon}(x) = 0. \tag{9}$$

On the other hand, from Theorem 3.6, we have

$$\lim_{\varepsilon \to 0} I_{\lambda,\varepsilon}(x) = \lim_{\varepsilon \to 0} \frac{1}{\sqrt{2\pi}}\left(\int_0^{\infty} k_\lambda(x-t)a(t)e^{-\varepsilon t}\,dt - A\int_0^{\infty} k_\lambda(x-t)e^{-\varepsilon t}\,dt\right)$$

$$= \frac{1}{\sqrt{2\pi}} \int_0^{\infty} k_\lambda(x-t)a(t)\,dt - \frac{A}{\sqrt{2\pi}} \int_0^{\infty} k_\lambda(x-t)\,dt$$

$$= \frac{1}{\sqrt{2\pi}} \int_{-\infty}^{\infty} k_\lambda(x-t)(a(t) - A(t))\,dt = I_\lambda(x),$$

and so from (8) we see that $I_\lambda(x)$ exists. This proves 1), and now 2) follows from (9).

Finally we prove 3). From the definition of $A(t)$ we see that it suffices to prove that $a(t)$ is a bounded slowly decreasing function.

From (7) we have

$$\lim_{x\to\infty} \frac{1}{\sqrt{2\pi}} \int_{-\infty}^{\infty} k_\lambda(x-t)a(t)\,dt = \lim_{x\to\infty} \frac{1}{\sqrt{2\pi}} \int_{-\infty}^{\infty} k_\lambda(x-t)A(t)\,dt$$

$$= \frac{A}{\sqrt{2\pi}} \lim_{x\to\infty} \int_{-\infty}^{\lambda x} \sqrt{\frac{2}{\pi}} \left(\frac{\sin u}{u}\right)^2 du$$

$$= \frac{A}{\pi} \int_{-\infty}^{\infty} \left(\frac{\sin u}{u}\right)^2 du = A,$$

so that there exists x_0 such that, when $x \geqslant x_0$,

$$\frac{1}{\sqrt{2\pi}} \int_{-\infty}^{\infty} k_\lambda(x-t)a(t)\,dt < A + 1;$$

that is

$$\int_{-\infty}^{\infty} \left(\frac{\sin t}{t}\right)^2 a\left(x - \frac{t}{\lambda}\right) dt < \pi(A+1) \qquad (x \geqslant x_0).$$

Since the integrand is non-negative, substituting $x + 2/\sqrt{\lambda}$ for x, we have

$$\int_{-\sqrt{\lambda}}^{\sqrt{\lambda}} \left(\frac{\sin t}{t}\right)^2 a\left(x + \frac{2}{\sqrt{\lambda}} - \frac{t}{\lambda}\right) dt < \pi(A+1) \qquad (x \geqslant x_0).$$

From our hypothesis $e^t a(t)$ is an increasing function of t so that

$$a(x)e^{-3/\sqrt{\lambda}} \int_{-\sqrt{\lambda}}^{\sqrt{\lambda}} \left(\frac{\sin t}{t}\right)^2 dt < \pi(A+1) \qquad (x \geqslant x_0).$$

Letting $\lambda \to \infty$ we have at once

$$a(x) \leqslant A + 1 \qquad (x \geqslant x_0).$$

When $x < x_0$, $h(x)$ is bounded and this implies that $a(x)$ is bounded in $-\infty < x < \infty$.

Now let $\delta > 0$. We have

$$a(x+\delta) - a(x) = e^{-x}\{e^{-\delta}h(x+\delta) - h(x)\} \geqslant e^{-x}h(x)(e^{-\delta} - 1),$$

so that

$$\varliminf_{\substack{x \to \infty \\ \delta \to 0}} \{a(x+\delta) - a(x)\} \geqslant 0.$$

This means that $a(x)$ is slowly decreasing. The theorem is proved. □

9.5 The Prime Number Theorem

In this section we apply Ikehara's theorem to prove the prime number theorem. We do not give a direct proof of the prime number theorem; instead we prove the equivalent theorem (see §1):

Theorem 5.1. $\psi(x) \sim x$.

Proof. From the definition of $\psi(x)$ we see that $\psi(x)$ is a non-negative increasing function with only finitely many discontinuities in the interval $0 \leqslant t \leqslant T$.

When $\sigma > 1$ we have, from Theorem 1.2 and formula (6.14.5), that

$$\begin{aligned}
\int_0^\infty e^{-st}\psi(e^t)\,dt &= \int_1^\infty u^{-(1+s)}\psi(u)\,du \\
&= \sum_{n=1}^\infty \int_n^{n+1} u^{-(1+s)}\psi(u)\,du = \sum_{n=1}^\infty \sum_{m \leqslant n} \Lambda(m) \int_n^{n+1} u^{-(s+1)}\,du \\
&= \frac{1}{s}\sum_{n=1}^\infty (n^{-s} - (n+1)^{-s}) \sum_{m \leqslant n} \Lambda(m) \\
&= \frac{1}{s}\lim_{N\to\infty} \sum_{n=1}^N (n^{-s} - (n+1)^{-s}) \sum_{m \leqslant n} \Lambda(m) \\
&= \frac{1}{s}\lim_{N\to\infty} \left\{ \sum_{n=1}^N \Lambda(n)n^{-s} - \left(\sum_{m \leqslant N} \Lambda(m) \right)(N+1)^{-s} \right\} \\
&= \frac{1}{s}\sum_{n=1}^\infty \frac{\Lambda(n)}{n^s} = -\frac{1}{s}\cdot\frac{\zeta'(s)}{\zeta(s)} \qquad (\sigma > 1).
\end{aligned}$$

From Theorem 2.3 we see that the function

$$-\frac{1}{s}\frac{\zeta'(s)}{\zeta(s)} - \frac{1}{s-1} = -\frac{1}{s}\left(\frac{\zeta'(s)}{\zeta(s)} + \frac{1}{s-1}\right) - \frac{1}{s}$$

has a continuous derivative in $\sigma \geqslant 1$, so that for any $a > 0$ the function is uniformly continuous in $1 \leqslant \sigma \leqslant 2$, $|t| \leqslant a$, and therefore there is a continuously differentiable function $g(t)$ satisfying

$$\lim_{\sigma \to 1} \left(-\frac{1}{s} \frac{\zeta'(s)}{\zeta(s)} - \frac{1}{s-1} \right) = g(t)$$

in $|t| \leqslant a$ uniformly. From Theorem 4.2 we see that

$$\lim_{t \to \infty} e^{-t} \psi(e^t) = 1.$$

Let $e^t = x$. Then

$$\lim_{x \to \infty} \frac{\psi(x)}{x} = 1,$$

which proves the theorem. □

Exercise 1. Let p_n be the n-th prime number. Prove that the prime number theorem is equivalent to

$$\lim_{n \to \infty} \frac{p_n}{n \log n} = 1.$$

Exercise 2. Use the prime number theorem to deduce that

$$M(x) = \sum_{n \leqslant x} \mu(n) = o(x).$$

Exercise 3. Use the prime number theorem to deduce that

$$\sum_{n=1}^{\infty} \frac{\mu(n)}{n} = 0.$$

Exercise 4. Let $n = p_1^{\alpha_1} \cdots p_k^{\alpha_k}$ and define

$$\omega(n) = k, \qquad \Omega(n) = \alpha_1 + \alpha_2 + \cdots + \alpha_k.$$

Let

$$\pi_k(x) = \sum_{\substack{n \leqslant x \\ \omega(n) = \Omega(n) = k}} 1, \qquad \tau_k(x) = \sum_{\substack{n \leqslant x \\ \Omega(n) = k}} 1,$$

$$\vartheta_k(x) = \sum_{p_1 \cdots p_k \leqslant x} \log(p_1 \cdots p_k), \qquad \prod_k(x) = \sum_{p_1 \cdots p_k \leqslant x} 1.$$

(*Note*: Here the sum is over all primes $p_1, \ldots, p_k$ satisfying $p_1 \cdots p_k \leqslant x$; the same set of primes $p_1, \ldots, p_k$ with a different ordering is treated differently.) Prove:

$$\prod_k(x) \sim \frac{kx(\log\log x)^{k-1}}{\log x} \qquad (k \geqslant 2),$$

$$\vartheta_k(x) \sim kx(\log\log x)^{k-1} \qquad (k \geqslant 2),$$

$$\pi_k(x) \sim \tau_k(x) \sim \frac{x(\log\log x)^{k-1}}{(k-1)!\log x} \qquad (k \geqslant 2).$$

9.6 Selberg's Asymptotic Formula

Throughout §6 – 8 we use the letters q and r to represent prime numbers.

Theorem 6.1 (Selberg). *Let* $x \geqslant 1$. *Then*

$$\vartheta(x)\log x + \sum_{p\leqslant x} \vartheta\left(\frac{x}{p}\right)\log p = 2x\log x + O(x) \tag{1}$$

and

$$\sum_{p\leqslant x}\log^2 p + \sum_{pq\leqslant x}\log p\log q = 2x\log x + O(x). \tag{2}$$

We first prove the following:

Lemma. *Let $F(x)$ and $G(x)$ be two functions defined for $x \geqslant 1$ and satisfying*

$$G(x) = \sum_{1\leqslant n\leqslant x} F\left(\frac{x}{n}\right)\log x.$$

Then

$$\sum_{n\leqslant x}\mu(n)G\left(\frac{x}{n}\right) = F(x)\log x + \sum_{n\leqslant x} F\left(\frac{x}{n}\right)\Lambda(n).$$

Proof. We have, from §6.4, $\Lambda(n) = \sum_{d|n}\mu(d)\log\frac{n}{d}$ so that

$$\begin{aligned}\sum_{n\leqslant x}\mu(n)G\left(\frac{x}{n}\right) &= \sum_{n\leqslant x}\mu(n)\sum_{m\leqslant\frac{x}{n}} F\left(\frac{x}{mn}\right)\log\frac{x}{n}\\ &= \sum_{l\leqslant x} F\left(\frac{x}{l}\right)\sum_{n|l}\mu(n)\left(\log\frac{x}{l}+\log\frac{l}{n}\right)\\ &= \sum_{l\leqslant x} F\left(\frac{x}{l}\right)\log\frac{x}{l}\cdot\sum_{n|l}\mu(n) + \sum_{l\leqslant x} F\left(\frac{x}{l}\right)\Lambda(l)\\ &= F(x)\log x + \sum_{l\leqslant x} F\left(\frac{x}{l}\right)\Lambda(l). \quad \square\end{aligned}$$

Proof of Theorem 6.1. Let γ be Euler's constant. From §5.8 we have

$$\sum_{n\leqslant x}\frac{1}{n} = \log x + \gamma + O\left(\frac{1}{x}\right).$$

Also

$$\begin{aligned}\sum_{n\leqslant x}\psi\left(\frac{x}{n}\right) &= \sum_{mn\leqslant x}\Lambda(m) = \sum_{n\leqslant x}\sum_{d|n}\Lambda(d)\\ &= \sum_{n\leqslant x}\log n = \int_1^x \log t\,dt + O(\log x) = x\log x - x + O(\log x).\end{aligned}$$

We apply the lemma with

$$F(x) = \psi(x) - x + \gamma + 1$$

so that

$$G(x) = \log x \sum_{1 \leqslant n \leqslant x} \psi\left(\frac{x}{n}\right) - x \log x \sum_{n \leqslant x} \frac{1}{n} + (\gamma + 1)x \log x + O(\log x)$$

$$= O(\log^2 x) = O(\sqrt{x}).$$

From the lemma we have

$$F(x) \log x + \sum_{n \leqslant x} F\left(\frac{x}{n}\right) \Lambda(n) = O\left(\sum_{n \leqslant x} \sqrt{\frac{x}{n}}\right) = O(x). \tag{4}$$

From Theorem 5.9.1 we have

$$\sum_{n \leqslant x} \frac{\Lambda(n)}{n} = \log x + O(1). \tag{5}$$

Therefore, from (3), (4), (5) and Theorem 1.2 we have

$$\psi(x) \log x + \sum_{n \leqslant x} \psi\left(\frac{x}{n}\right) \Lambda(n)$$

$$= x \log x + x \sum_{n \leqslant x} \frac{\Lambda(n)}{n} - (\gamma + 1) \log x - (\gamma + 1) \sum_{n \leqslant x} \Lambda(n) + O(x)$$

$$= 2x \log x + O(x). \tag{6}$$

From Theorem 1.2 we have

$$\sum_{n \leqslant x} \psi\left(\frac{x}{n}\right) \Lambda(n) - \sum_{p \leqslant x} \vartheta\left(\frac{x}{p}\right) \log p = \sum_{mn \leqslant x} \Lambda(m)\Lambda(n) - \sum_{pq \leqslant x} \log p \log q$$

$$= O\left(\sum_{\substack{p^\alpha q^\beta \leqslant x \\ \alpha \geqslant 2, \beta \geqslant 1}} \log p \log q\right) = O\left(\sum_{\substack{p^\alpha \leqslant x \\ \alpha \geqslant 2}} \log p \sum_{\substack{q^\beta \leqslant x/p^\alpha \\ \beta \geqslant 1}} \log q\right)$$

$$= O\left(\sum_{\substack{p^\alpha \leqslant x \\ \alpha \geqslant 2}} \log p \, \psi\left(\frac{x}{p^\alpha}\right)\right) = O\left(x \sum_{p \leqslant \sqrt{x}} \sum_{\alpha \geqslant 2} \frac{\log p}{p^\alpha}\right)$$

$$= O\left(x \sum_{p \leqslant \sqrt{x}} \frac{\log p}{p(p-1)}\right) = O(x) \tag{7}$$

and

$$\psi(x) = \vartheta(x) + \vartheta(x^{\frac{1}{2}}) + \cdots + \vartheta\left(x^{\left[\frac{\log 2}{\log x}\right]}\right)$$

$$= \vartheta(x) + O(\log x \cdot \vartheta(x^{\frac{1}{2}})) = \vartheta(x) + O(x^{\frac{1}{2}} \log x). \tag{8}$$

Formula (1) now follows from (6), (7) and (8).

Also from

$$\vartheta(x)\log x - \sum_{p\leqslant x}\log^2 p = \sum_{p\leqslant x}\log p\log\frac{x}{p} = \sum_{p\leqslant x}\log p\left(\sum_{n\leqslant\frac{x}{p}}\frac{1}{n}+O(1)\right)$$

$$= \sum_{n\leqslant x}\frac{1}{n}\sum_{p\leqslant\frac{x}{n}}\log p + O(\vartheta(x))$$

$$= O\left(x\sum_{n\leqslant x}\frac{1}{n^2}\right) + O(x) = O(x),$$

formula (2) follows at once. □

9.7 Elementary Proof of the Prime Number Theorem

Let

$$R(x) = \vartheta(x) - x. \tag{1}$$

We know from Theorem 1.1 that the prime number theorem is equivalent to

$$\lim_{x\to\infty}\frac{R(x)}{x} = 0. \tag{2}$$

Before we prove (2) we first establish the following lemmas.

Lemma 1. *If* $x \geqslant 3$, *then*

$$\sum_{pq\leqslant x}\frac{\log p\log q}{pq} = \frac{1}{2}\log^2 x + O(\log x),$$

$$\sum_{pq\leqslant x}\frac{\log p\log q}{pq\log pq} = \log x + O(\log\log x),$$

$$\sum_{p\leqslant x}\frac{\log p}{p\log\frac{2x}{p}} = O(\log\log x).$$

Proof. Let $A(n) = \sum_{p\leqslant n}\log p/p$. From Theorem 5.9.1 we have $A(n) = \log n + r_n$ where $r_n = O(1)$. Therefore

$$\sum_{pq\leqslant x}\frac{\log p\log q}{pq} = \sum_{p\leqslant x}\frac{\log p}{p}\sum_{q\leqslant\frac{x}{p}}\frac{\log q}{q} = \sum_{p\leqslant x}\frac{\log p}{p}\log\frac{x}{p} + O(\log x)$$

$$= \sum_{n\leqslant x}(A(n) - A(n-1))\log\frac{x}{n} + O(\log x)$$

$$= \sum_{n \leqslant x-1} A(n)\left\{\log\frac{x}{n} - \log\frac{x}{n+1}\right\} + O(\log x)$$

$$= \sum_{n \leqslant x} \log n \cdot \log\left(1 + \frac{1}{n}\right) + O\left(\sum_{n \leqslant x} \log\left(1 + \frac{1}{n}\right)\right) + O(\log x)$$

$$= \frac{1}{2}\log^2 x + O(\log x).$$

Using the same method we have, by partial summations,

$$\sum_{pq \leqslant x} \frac{\log p \log q}{pq \log pq} = \log x + O(\log\log x).$$

Also from

$$\sum_{n \leqslant x} \frac{1}{n \log\frac{2x}{n}} = \frac{1}{\log x}\sum_{n \leqslant x}\frac{1}{n} + \sum_{n \leqslant x}\frac{1}{n}\left(\frac{1}{\log\frac{2x}{n}} - \frac{1}{\log x}\right)$$

$$= \sum_{n \leqslant x}\frac{1}{n}\int_{\frac{2x}{n}}^{x}\frac{du}{u\log^2 u} + O(1)$$

$$= \int_2^x \frac{\sum_{\frac{2x}{u} \leqslant n \leqslant x}\frac{1}{n}}{u\log^2 u}\,du + O(1) = \int_2^x \frac{du}{u\log u} + O(1) = O(\log\log x),$$

we have

$$\sum_{p \leqslant x}\frac{\log p}{p\log\frac{2x}{p}} = \sum_{n \leqslant x}(A(n) - A(n-1))\frac{1}{\log\frac{2x}{n}}$$

$$= \sum_{n \leqslant x}\{\log n - \log(n-1)\}\frac{1}{\log\frac{2x}{n}} + O\left(\sum_{n \leqslant x} r_n\left|\frac{1}{\log\frac{2x}{n}} - \frac{1}{\log\frac{2x}{n+1}}\right|\right)$$

$$= O\left(\sum_{n \leqslant x}\frac{1}{n\log\frac{2x}{n}}\right) = O(\log\log x).$$

The lemma is proved. □

Lemma 2.

$$\vartheta(x) + \sum_{pq \leqslant x}\frac{\log p \log q}{\log pq} = 2x + O\left(\frac{x}{\log x}\right) \qquad (x \geqslant 2).$$

Proof. Let

$$B(n) = \sum_{pq \leqslant n} \log p \log q, \qquad C(n) = \sum_{p \leqslant n} \log^2 p.$$

Then we have

$$\begin{aligned}
\vartheta(x) + \sum_{pq \leqslant x} \frac{\log p \log q}{\log pq}
&= \sum_{n \leqslant x} \frac{C(n) - C(n-1)}{\log n} + \sum_{n \leqslant x} \frac{B(n) - B(n-1)}{\log n} \\
&= \frac{C([x])}{\log[x]} + \frac{B([x])}{\log[x]} + \sum_{n \leqslant x-1} \{C(n) + B(n)\} \left\{ \frac{1}{\log n} - \frac{1}{\log(n+1)} \right\} \\
&= 2x + O\left(\frac{x}{\log x}\right) + \sum_{n \leqslant x-1} (2n \log n + O(n)) \frac{\log\left(1 + \frac{1}{n}\right)}{\log n \log(n+1)} \\
&= 2x + O\left(\frac{x}{\log x}\right),
\end{aligned}$$

and the lemma is proved. □

Lemma 3.

$$R(x) \log x = \sum_{pq \leqslant x} \frac{\log p \log q}{\log pq} R\left(\frac{x}{pq}\right) + O(x \log\log x) \qquad (x \geqslant 3).$$

Proof. From Lemma 1 and Lemma 2 we have

$$\begin{aligned}
\sum_{p \leqslant x} \vartheta\left(\frac{x}{p}\right) \log p &= 2x \sum_{p \leqslant x} \frac{\log p}{p} - \sum_{p \leqslant x} \log p \sum_{qr \leqslant \frac{x}{p}} \frac{\log q \log r}{\log qr} \\
&\quad + O\left(x \sum_{p \leqslant x} \frac{\log p}{p \log \frac{2x}{p}} \right) \\
&= 2x \log x - \sum_{qr \leqslant x} \frac{\log q \log r}{\log qr} \vartheta\left(\frac{x}{qr}\right) + O(x \log\log x).
\end{aligned}$$

Substituting this into Selberg's asymptotic formula (that is, formula (6.1)), we have

$$\vartheta(x) \log x = \sum_{pq \leqslant x} \frac{\log p \log q}{\log pq} \vartheta\left(\frac{x}{pq}\right) + O(x \log\log x).$$

The result follows from substituting (1) into this and applying Lemma 1. □

Lemma 4.

$$|R(x)| \leqslant \frac{1}{\log x} \sum_{n \leqslant x} \left| R\left(\frac{x}{n}\right) \right| + O\left(\frac{x \log\log x}{\log x}\right) \qquad (x \geqslant 3).$$

Proof. Substituting (1) into formula (6.1) we have

$$R(x)\log x = -\sum_{p\leqslant x} R\left(\frac{x}{p}\right)\log p + O(x),$$

so that from Lemma 3 we see that

$$2|R(x)|\log x \leqslant \sum_{p\leqslant x}\left|R\left(\frac{x}{p}\right)\right|\log p + \sum_{pq\leqslant x}\frac{\log p\log q}{\log pq}\left|R\left(\frac{x}{pq}\right)\right| + O(x\log\log x).$$

From Lemma 2 and partial summations, and noting that $||a| - |b|| \leqslant |a - b|$, we see that

$$\begin{aligned}2|R(x)|\log x &\leqslant \sum_{n\leqslant x-1}\left(\sum_{p\leqslant n}\log p + \sum_{pq\leqslant n}\frac{\log p\log q}{\log pq}\right)\left(\left|R\left(\frac{x}{n}\right)\right| - \left|R\left(\frac{x}{n+1}\right)\right|\right)\\ &\quad + O\left(\sum_{p\leqslant x}\log p + \sum_{pq\leqslant x}\frac{\log p\log q}{\log pq}\right) + O(x\log\log x)\\ &\leqslant 2\sum_{n\leqslant x-1} n\left(\left|R\left(\frac{x}{n}\right)\right| - \left|R\left(\frac{x}{n+1}\right)\right|\right)\\ &\quad + O\left(\sum_{n\leqslant x-1}\frac{n}{\log 2n}\left|\left|R\left(\frac{x}{n}\right)\right| - \left|R\left(\frac{x}{n+1}\right)\right|\right|\right) + O(x\log\log x)\\ &\leqslant 2\sum_{n\leqslant x}\left|R\left(\frac{x}{n}\right)\right| + O\left(\sum_{n\leqslant x-1}\frac{n}{\log 2n}\left(\vartheta\left(\frac{x}{n}\right) - \vartheta\left(\frac{x}{n+1}\right)\right)\right)\\ &\quad + O\left(x\sum_{n\leqslant x-1}\frac{n}{\log 2n}\left(\frac{1}{n} - \frac{1}{n+1}\right)\right) + O(x\log\log x).\end{aligned}$$

From Theorem 1.2 we have

$$\begin{aligned}&\sum_{n\leqslant x-1}\frac{n}{\log 2n}\left(\vartheta\left(\frac{x}{n}\right) - \vartheta\left(\frac{x}{n+1}\right)\right)\\ &\qquad = \sum_{2\leqslant n\leqslant x-1}\vartheta\left(\frac{x}{n}\right)\left(\frac{n}{\log 2n} - \frac{n-1}{\log 2(n-1)}\right) + O(x)\\ &\qquad = O\left(x\sum_{n\leqslant x}\frac{1}{n\log n}\right) = O(x\log\log x),\end{aligned}$$

so that

$$2|R(x)|\log x \leqslant 2\sum_{n\leqslant x}\left|R\left(\frac{x}{n}\right)\right| + O(x\log\log x),$$

and the required result follows. □

Lemma 5. *If* $x > 1$, *then*

$$\sum_{n\leqslant x}\frac{\vartheta(n)}{n^2} = \log x + O(1),$$

and

$$\sum_{n \leqslant x} \vartheta\left(\frac{x}{n}\right) = x \log x + O(x).$$

Proof. Since

$$\sum_{p \leqslant n \leqslant x} \frac{1}{n^2} = \sum_{n \geqslant p} \frac{1}{n^2} - \sum_{n > x} \frac{1}{n^2} = \frac{1}{p} + O\left(\frac{1}{p^2}\right) + O\left(\frac{1}{x}\right),$$

we have

$$\begin{aligned} \sum_{n \leqslant x} \frac{\vartheta(n)}{n^2} &= \sum_{n \leqslant x} \frac{1}{n^2} \sum_{p \leqslant n} \log p = \sum_{p \leqslant x} \log p \sum_{p \leqslant n \leqslant x} \frac{1}{n^2} \\ &= \sum_{p \leqslant x} \log p \left(\frac{1}{p} + O\left(\frac{1}{p^2}\right) + O\left(\frac{1}{x}\right)\right) = \log x + O(1) \end{aligned}$$

and

$$\begin{aligned} \sum_{n \leqslant x} \vartheta\left(\frac{x}{n}\right) &= \sum_{n \leqslant x} \sum_{p \leqslant \frac{x}{n}} \log p = \sum_{p \leqslant x} \log p \sum_{n \leqslant \frac{x}{p}} 1 \\ &= \sum_{p \leqslant x} \log p \cdot \left(\frac{x}{p} + O(1)\right) = x \log x + O(x). \quad \square \end{aligned}$$

Lemma 6.

$$\sum_{n \leqslant x} \frac{\log n}{n} R(n) = - \sum_{n \leqslant x} \frac{1}{n} R(n) R\left(\frac{x}{n}\right) + O(x).$$

Proof. From Selberg's formula (that is (6.2)) and partial summations we have

$$\sum_{p \leqslant x} \log^2 p \log \frac{x}{p} + \sum_{pq \leqslant x} \log p \log q \log \frac{x}{pq} = 2x \log x + O(x).$$

Substituting

$$\log \frac{x}{p} = \sum_{p \leqslant n \leqslant x} \frac{1}{n} + O\left(\frac{1}{p}\right), \qquad \log \frac{x}{pq} = \sum_{p \leqslant n \leqslant \frac{x}{q}} \frac{1}{n} + O\left(\frac{1}{p}\right),$$

into the above formula and interchanging the summations we have

$$\sum_{n \leqslant x} \frac{1}{n} \sum_{p \leqslant n} \log^2 p + \sum_{n \leqslant x} \frac{1}{n} \sum_{p \leqslant n} \log p \sum_{q \leqslant \frac{x}{n}} \log q = 2x \log x + O(x);$$

that is

$$\sum_{n \leqslant x} \frac{\log n}{n} \vartheta(n) + \sum_{n \leqslant x} \frac{1}{n} \vartheta(n) \vartheta\left(\frac{x}{n}\right) = 2x \log x + O(x).$$

The required result follows from substituting (1) into this formula and then apply Lemma 5. $\square$

Lemma 7. *Let $0 < \sigma < 1$ and suppose that there exists x_0 such that, for $x > x_0$,*

$$|R(x)| < \sigma x. \tag{3}$$

Then there exists x_σ such that, when $x > x_\sigma$, the interval $((1-\sigma)^{16}x, x)$ contains a subinterval $(y, e^\delta y)$ with the property that

$$\left|\frac{R(z)}{z}\right| < \frac{\sigma + \sigma^2}{2},$$

when $y \leqslant z \leqslant e^\delta y$. Here $\delta = \sigma(1-\sigma)/32$.

Proof. From Lemma 6 we have

$$\begin{aligned}\left|\sum_{n \leqslant x} \frac{\log n}{n} R(n)\right| &\leqslant \left|\sum_{x_0 \leqslant n \leqslant \frac{x}{x_0}} \frac{1}{n} R(n) R\left(\frac{x}{n}\right)\right| + \left|\sum_{n < x_0} \frac{1}{n} R(n) R\left(\frac{x}{n}\right)\right| \\ &\quad + \left|\sum_{\frac{x}{x_0} < n \leqslant x} \frac{1}{n} R(n) R\left(\frac{x}{n}\right)\right| + O(x) \\ &\leqslant \sigma^2 x \sum_{x_0 \leqslant n \leqslant \frac{x}{x_0}} \frac{1}{n} + O(x) = \sigma^2 x \log x + O(x),\end{aligned}$$

so that when $x > x_1$,

$$\left|\sum_{x' \leqslant n \leqslant x} \frac{\log n}{n} R(n)\right| < \sigma^2(x + x')\log x + O(x),$$

where $x' = (1-\sigma)^{16}x$.

Suppose that $R(n)$ does not change sign in $(x' \leqslant n \leqslant x)$. Then there must exist y $(x' \leqslant y \leqslant x)$ so that

$$\left|\frac{R(y)}{y}\right| \sum_{x' \leqslant n \leqslant x} \log n < \sigma^2(x + x')\log x + O(x).$$

From

$$(1-\sigma)^{16} < \frac{1-\sigma}{1+15\sigma},$$

we see that

$$\begin{aligned}\left|\frac{R(y)}{y}\right| &< \sigma^2 \frac{x + x'}{x - x'} + O\left(\frac{1}{\log x}\right) < \frac{\sigma(1+7\sigma)}{8} + O\left(\frac{1}{\log x}\right) \\ &< \frac{\sigma(1+3\sigma)}{4} \qquad (x > x_1).\end{aligned} \tag{4}$$

But if $R(n)$ changes sign in (x', x), then clearly there exists y $(x' \leqslant y \leqslant x)$ such that $|R(y)| = O(\log y)$ so that (4) still holds.

When $1 < y < y'$ we have, by Lemma 2,

$$\sum_{y<p\leqslant y'} \log p \leqslant 2(y' - y) + O\left(\frac{y'}{\log y'}\right).$$

From (1) we have

$$|R(y') - R(y)| < (y' - y) + O\left(\frac{y'}{\log y'}\right). \tag{5}$$

Let $x' \leqslant y_1, y_2 \leqslant x_1$, and y_1 satisfying (4) and

$$e^{-\delta} \leqslant \frac{y_2}{y_1} \leqslant e^{\delta}.$$

From (4) and (5) we see that

$$\begin{aligned} \left|\frac{R(y_2)}{y_2}\right| &< \left|\frac{R(y_1)}{y_1}\right| \cdot \frac{y_1}{y_2} + \left|1 - \frac{y_1}{y_2}\right| + O\left(\frac{1}{\log x}\right) \\ &< \frac{\sigma(1 + 3\sigma)}{4} \cdot e^{\delta} + (e^{\delta} - 1) + O\left(\frac{1}{\log x}\right). \end{aligned}$$

Since $e^{\delta} < 1/(1 - \delta)$ $(0 < \delta < 1)$ we have

$$\begin{aligned} \left|\frac{R(y_2)}{y_2}\right| &< \frac{\sigma(1 + 3\sigma)}{4} \cdot \frac{1}{1 - \delta} + \left(\frac{1}{1 - \delta} - 1\right) + O\left(\frac{1}{\log x}\right) \\ &< \frac{\sigma(3 + 5\sigma)}{8} + O\left(\frac{1}{\log x}\right) < \frac{\sigma + \sigma^2}{4} \qquad (x > x_\sigma). \end{aligned}$$

When $y_1 \leqslant ((1 + 7\sigma)/(1 + 15\sigma))x$ we have $e^{\delta}y_1 < x$, so that we can take $y = y_1$. When $y_1 > ((1 + 7\sigma)/(1 + 15\sigma))x$, then

$$e^{-\delta}y_1 > \frac{1 - \sigma}{1 + 15\sigma}x > x_1$$

so that we can take $y = e^{-\delta}y_1$. The lemma is proved. □

Proof of the prime number theorem. We already know that there exist $c > 0$ and x_0' such that, for $x > x_0$,

$$\vartheta(x) > cx \tag{6}$$

(this is Theorem 1.2). From Selberg's formula, we have

$$\begin{aligned} \vartheta(x) &= 2x - \frac{1}{\log x} \sum_{p\leqslant x} \vartheta\left(\frac{x}{p}\right) \log p + O\left(\frac{x}{\log x}\right) \\ &= 2x - \frac{1}{\log x} \sum_{p\leqslant \frac{x}{x_0'}} \vartheta\left(\frac{x}{p}\right) \log p \end{aligned}$$

$$-\frac{1}{\log x}\sum_{\frac{x}{x_0'}<p\leqslant x}\vartheta\left(\frac{x}{p}\right)\log p+O\left(\frac{x}{\log x}\right)$$

$$\leqslant 2x-\frac{cx\log x}{\log x}+O\left(\frac{1}{\log x}\sum_{\frac{x}{x_0'}<p\leqslant x}\log p\right)+O\left(\frac{x}{\log x}\right)$$

$$=(2-c)x+O\left(\frac{x}{\log x}\right)<\left(2-\frac{c}{2}\right)x \qquad (x>x_0,\ c>0).$$

From (1) we have

$$|R(x)|<\sigma_0(x) \qquad \left(x>x_0,\ \sigma_0=\left|1-\frac{c}{2}\right|,\ 0<\sigma_0<1\right).$$

Let

$$\zeta=(1-\sigma_0)^{-16}, \qquad \delta=\frac{\sigma_0(1-\sigma_0)}{32}.$$

From Lemma 7 there exists $x_{\sigma_0}>x_0$ such that when $x>x_{\sigma_0}$, any interval (ζ^{v-1},ζ^v) $(\zeta\leqslant\zeta^v\leqslant x/x_{\sigma_0})$ will contain a subinterval $(y_v,e^\delta y_v)$ so that when $y_v\leqslant n\leqslant e^\delta y_v$,

$$\left|\frac{n}{x}R\left(\frac{x}{n}\right)\right|<\frac{\sigma_0+\sigma_0^2}{2}.$$

From Lemma 4 we have

$$|R(x)|<\frac{1}{\log x}\sum_{n\leqslant\frac{x}{x_{\sigma_0}}}\left|R\left(\frac{x}{n}\right)\right|+\frac{1}{\log x}\sum_{\frac{x}{x_{\sigma_0}}<n\leqslant x}\left|R\left(\frac{x}{n}\right)\right|+O\left(\frac{x}{\sqrt{\log x}}\right)$$

$$<\frac{\sigma_0 x}{\log x}\sum_{\substack{1\leqslant n\leqslant\frac{x}{x_{\sigma_0}}\\ n\notin(y_v,e^\delta y_v)}}\frac{1}{n}+\frac{\sigma_0+\sigma_0^2}{2}\frac{x}{\log x}\sum_{\zeta^v\leqslant\frac{x}{x_{\sigma_0}}}\sum_{y_v\leqslant n\leqslant e^\delta y}\frac{1}{n}+O\left(\frac{x}{\sqrt{\log x}}\right)$$

$$<\frac{\sigma_0 x}{\log x}\sum_{n\leqslant\frac{x}{x_{\sigma_0}}}\frac{1}{n}-\frac{(\sigma_0-\sigma_0^2)}{2}\frac{x}{\log x}\sum_{\zeta^v\leqslant\frac{x}{x_{\sigma_0}}}\sum_{y_v\leqslant n\leqslant e^\delta y_v}\frac{1}{n}+O\left(\frac{x}{\sqrt{\log x}}\right)$$

$$<\sigma_0 x-\frac{(\sigma_0-\sigma_0^2)}{2}\frac{x}{\log x}\sum_{\zeta^v\leqslant\frac{x}{x_{\sigma_0}}}\left(\delta+O\left(\frac{1}{\zeta^v}\right)\right)+O\left(\frac{x}{\sqrt{\log x}}\right)$$

$$<\sigma_0 x-\frac{(\sigma_0-\sigma_0^2)}{2}\frac{x}{\log x}\frac{\delta\log x}{\log\zeta}+O\left(\frac{x}{\sqrt{\log x}}\right)$$

$$< \sigma_0\left(1 - \frac{(1-\sigma_0)^2\sigma_0}{1024\log\dfrac{1}{1-\sigma_0}}\right)x + O\left(\frac{x}{\sqrt{\log x}}\right)$$

$$< \sigma_0\left(1 - \frac{(1-\sigma_0)^3}{1024}\right)x + O\left(\frac{x}{\sqrt{\log x}}\right)$$

$$< \sigma_0\left(1 - \frac{(1-\sigma_0)^3}{2000}\right)x = \sigma_1 x \qquad (x > x_{\sigma_1} > x_{\sigma_0}),$$

where $\sigma_1 < \sigma_0$. Repeating the above we arrive at

$$|R(x)| < \sigma_n x \qquad (x > x_{\sigma_n}),$$

where

$$\sigma_n = \sigma_{n-1}\left(1 - \frac{(1-\sigma_{n-1})^3}{2000}\right) \leqslant \sigma_{n-1}\left(1 - \frac{(1-\sigma_0)^3}{2000}\right) \leqslant \cdots$$

$$\leqslant \sigma_0\left(1 - \frac{(1-\sigma_0)^3}{2000}\right)^n,$$

so that

$$\lim_{n\to\infty} \sigma_n = 0$$

and the required result is proved. □

9.8 Dirichlet's Theorem

Theorem 8.1 (Dirichlet). *Let $k > 0$, $l > 0$, $(k, l) = 1$. Then there are infinitely many primes of the form $kn + l$.*

In this section we prove a stronger version of Theorem 8.1, namely:

Theorem 8.2. *Let $k > 0$, $l > 0$, $(k, l) = 1$. Then*

$$\sum_{\substack{p \leqslant x \\ p \equiv l (\bmod k)}} \frac{\log p}{p} = \frac{1}{\varphi(k)} \log x + O(1).$$

Here the condition of summation is over all primes not exceeding x which are of the form $kn + l$, and the constant implied by the O-symbol depends on k. We shall require the following lemmas for the proof of Theorem 8.2.

If χ is a non-principal character, we write

$$L(\chi) = \sum_{n=1}^{\infty} \frac{\chi(n)}{n}, \qquad L_1(\chi) = \sum_{n=1}^{\infty} \frac{\chi(n)\log n}{n}. \tag{1}$$

Lemma 1. *Let χ be a non-principal real character. Then $L(\chi) \neq 0$.*

Proof. Let

$$F(n) = \sum_{d|n} \chi(d).$$

Since

$$F(p^l) = \begin{cases} 1 + 1 + \cdots + 1 = l + 1 & \text{if} \quad \chi(p) = 1, \\ 1 - 1 + \cdots + 1 = 1 & \text{if} \quad \chi(p) = -1, \quad l \text{ even}, \\ 1 - 1 + \cdots - 1 = 0 & \text{if} \quad \chi(p) = -1, \quad l \text{ odd}, \end{cases}$$

and $F(n)$ is multiplicative, we have

$$F(n) \geqslant \begin{cases} 1, & \text{if } n \text{ is a perfect square}, \\ 0, & \text{otherwise}, \end{cases}$$

so that

$$G(x) = \sum_{n \leqslant x} \frac{F(n)}{n^{\frac{1}{2}}} \geqslant \sum_{1 \leqslant m \leqslant \sqrt{x}} \frac{1}{m} \to \infty.$$

On the other hand, when χ is a non-principal character, we have

$$\sum_{x \leqslant n \leqslant y} \frac{\chi(n)}{n^{\delta}} = O(x^{-\delta}), \qquad \sum_{x \leqslant n \leqslant y} \frac{\chi(n) \log n}{n^{\delta}} = O\left(\frac{\log x}{x^{\delta}}\right) \quad (\delta > 0, x > 1). \tag{2}$$

(This can be proved from Exercise 7.2.1 and Theorem 6.8.2.) From formula (5.8.4) we now have

$$\begin{aligned} G(x) &= \sum_{n \leqslant x} \frac{1}{n^{\frac{1}{2}}} \sum_{d|n} \chi(d) = \sum_{dd' \leqslant x} \frac{\chi(d)}{d^{\frac{1}{2}} d'^{\frac{1}{2}}} \\ &= \sum_{d' \leqslant \sqrt{x}} \frac{1}{d'^{\frac{1}{2}}} \sum_{\sqrt{x} < d \leqslant \frac{x}{d'}} \frac{\chi(d)}{d^{\frac{1}{2}}} + \sum_{d \leqslant \sqrt{x}} \frac{\chi(d)}{d^{\frac{1}{2}}} \sum_{d' \leqslant \frac{x}{d}} \frac{1}{d'^{\frac{1}{2}}} \\ &= \sum_{d' \leqslant \sqrt{x}} \frac{1}{d'^{\frac{1}{2}}} \{O(x^{-\frac{1}{4}})\} + \sum_{d \leqslant \sqrt{\tau}} \frac{\chi(d)}{d^{\frac{1}{2}}} \left\{2\sqrt{\frac{x}{d}} + c_1 + O\left(\sqrt{\frac{d}{x}}\right)\right\} \\ &= 2\sqrt{x} \sum_{d \leqslant \sqrt{x}} \frac{\chi(d)}{d} + O(1) \\ &= 2\sqrt{x}\, L(\chi) + O(1). \end{aligned}$$

If $L(\chi) = 0$, then $G(x) = O(1)$ which is impossible. The lemma is proved. □

Lemma 2.

$$L_1(\chi) \sum_{n \leqslant x} \frac{\mu(n)\chi(n)}{n} = \begin{cases} O(1), & \textit{if} \quad L(\chi) \neq 0, \\ -\log x + O(1), & \textit{if} \quad L(\chi) = 0. \end{cases}$$

Proof. In Theorem 6.3.3 we let $H(n) = \chi(n)$, $F(n) = n$. Then from

$$G(x) = \sum_{1 \leqslant n \leqslant x} F\left(\frac{x}{n}\right) H(n) = x \sum_{1 \leqslant n \leqslant x} \frac{\chi(n)}{n} = xL(\chi) + O(1),$$

we have

$$x = F(x) = \sum_{1 \leqslant n \leqslant x} \mu(n) G\left(\frac{x}{n}\right) H(n) = xL(\chi) \sum_{1 \leqslant n \leqslant x} \frac{\chi(n)\mu(n)}{n} + O(x),$$

giving

$$L(\chi) \sum_{n \leqslant x} \frac{\mu(n)\chi(n)}{n} = O(1).$$

If $L(\chi) \neq 0$, then

$$\sum_{n \leqslant x} \frac{\mu(n)\chi(n)}{n} = O(1)$$

and the result follows at once. If $L(\chi) = 0$, then in Theorem 6.3.3 we let $F(x) = x \log x$, $H(n) = \chi(n)$ so that

$$\begin{aligned} G(x) &= \sum_{n \leqslant x} F\left(\frac{x}{n}\right) H(n) = x \sum_{n \leqslant x} \frac{\chi(n)}{n} \log \frac{x}{n} \\ &= L(\chi) x \log x - L_1(\chi) x + O(\log x) \\ &= -L_1(\chi) x + O(\log x). \end{aligned}$$

From Example 5.8.2 we have

$$\sum_{n \leqslant x} \log \frac{x}{n} = O(x)$$

so that

$$\begin{aligned} x \log x &= \sum_{1 \leqslant n \leqslant x} \mu(n) G\left(\frac{x}{n}\right) H(n) \\ &= \sum_{n \leqslant x} \mu(n)\chi(n) \left\{ -L_1(\chi) \frac{x}{n} + O\left(\log \frac{x}{n}\right) \right\} \\ &= -L_1(\chi) x \sum_{n \leqslant x} \frac{\mu(n)\chi(n)}{n} + O(x). \end{aligned}$$

The proof of the lemma is complete. $\square$

Lemma 3.

$$\sum_{p \leqslant x} \frac{\chi(p) \log p}{p} = \begin{cases} O(1), & \text{if } L(\chi) \neq 0, \\ -\log x + O(1), & \text{if } L(\chi) = 0. \end{cases}$$

Proof.

$$
\begin{aligned}
\sum_{p\leqslant x}\frac{\chi(p)\log p}{p} &= \sum_{n\leqslant x}\frac{\chi(n)\Lambda(n)}{n} + O(1) \\
&= \sum_{n\leqslant x}\frac{\chi(n)}{n}\sum_{d\mid n}\mu(d)\log\frac{n}{d} + O(1) \\
&= \sum_{dd'\leqslant x}\frac{\chi(d)\chi(d')}{dd'}\mu(d)\log d' + O(1) \\
&= \sum_{d\leqslant x}\frac{\mu(d)\chi(d)}{d}\sum_{d'\leqslant \frac{x}{d}}\frac{\chi(d')\log d'}{d'} + O(1) \\
&= \sum_{d\leqslant x}\frac{\mu(d)\chi(d)}{d}\left\{L_1(\chi) + O\left(\frac{\log\frac{x}{d}}{x/d}\right)\right\} + O(1) \\
&= L_1(\chi)\sum_{d\leqslant x}\frac{\mu(d)\chi(d)}{d} + O(1),
\end{aligned}
$$

and the required result follows from Lemma 2. □

Lemma 4. *Suppose that χ is a non-principal character. Then $L(\chi) \neq 0$.*

Proof. Let N be the number of non-principal character $\chi \bmod k$ such that $L(\chi) = 0$, and let $\sum_{(\chi)}$ represent the sum over all characters mod k. Then, from Lemma 3 together with Theorem 7.2.4 and Theorem 7.2.5, we have

$$
\begin{aligned}
\varphi(k)\sum_{\substack{p\leqslant x\\ p\equiv 1(\mathrm{mod}\, k)}}\frac{\log p}{p} &= \sum_{(\chi)}\sum_{p\leqslant x}\frac{\chi(p)\log p}{p} \\
&= \sum_{\substack{p\leqslant x\\ p\nmid k}}\frac{\log p}{p} + \sum_{\substack{\chi\neq\chi_0\\ (\chi)}}\sum_{p\leqslant x}\frac{\chi(p)\log p}{p} \\
&= (1-N)\log x + O(1).
\end{aligned}
$$

But since

$$
\varphi(k)\sum_{\substack{p\leqslant x\\ p\equiv l(\mathrm{mod}\, k)}}\frac{\log p}{p} \geqslant 0
$$

we must have $0 \leqslant N \leqslant 1$. Now if χ is a complex character then we must have $L(\bar{\chi}) \neq 0$ also so that $N \geqslant 2$. But if χ is a real character we know from Lemma 1 that $L(\chi) \neq 0$, so that $N = 0$. The lemma is proved. □

Proof of Theorem 2. From Lemma 3 and Lemma 4 we have

$$
\sum_{p\leqslant x}\frac{\chi(p)\log p}{p} = O(1),
$$

so that, by Exercise 7.2.2,

$$\varphi(k) \sum_{\substack{p \leqslant x \\ p \equiv l (\operatorname{mod} k)}} \frac{\log p}{p} = \sum_{(\chi)} \bar{\chi}(l) \sum_{p \leqslant x} \frac{\chi(p) \log p}{p}$$

$$= \sum_{\substack{p \leqslant x \\ p \nmid k}} \frac{\log p}{p} + \sum_{\substack{(\chi) \\ \chi \neq \chi_0}} \bar{\chi}(l) \sum_{p \leqslant x} \frac{\chi(p) \log p}{p}$$

$$= \log x + O(1),$$

and the desired result follows. □

Exercise. Suppose that $(k, l) = 1$, $l \leqslant k$. Prove that

$$\lim_{x \to \infty} \frac{\pi(x; k, l)}{\dfrac{x}{\varphi(k) \log x}} = 1.$$

Suggestions: 1) Let

$$\vartheta_l(x) = \sum_{\substack{p \leqslant x \\ p \equiv l (\operatorname{mod} k)}} \log p, \qquad \psi_l(x) = \sum_{\substack{n \leqslant x \\ n \equiv l (\operatorname{mod} k)}} \Lambda(n),$$

so that

$$\varlimsup_{x \to \infty} \frac{\pi(x; k, l)}{\dfrac{x}{\varphi(k) \log x}} = \varlimsup_{x \to \infty} \frac{\vartheta_l(x)}{\dfrac{x}{\varphi(k)}} = \varlimsup_{x \to \infty} \frac{\psi_l(x)}{\dfrac{x}{\varphi(k)}};$$

$$\varliminf_{x \to \infty} \frac{\pi(x; k, l)}{\dfrac{x}{\varphi(k) \log x}} = \varliminf_{x \to \infty} \frac{\vartheta_l(x)}{\dfrac{x}{\varphi(k)}} = \varliminf_{x \to \infty} \frac{\psi_l(x)}{\dfrac{x}{\varphi(k)}}.$$

2) Prove:

$$\sum_{d \mid n} \mu(d) \log^2 \frac{n}{d} = \Lambda(n) \log n + \sum_{d \mid n} \Lambda(d) \Lambda\left(\frac{n}{d}\right).$$

Summing with respect to n over $1 \leqslant n \leqslant x$, $n \equiv l \ (\operatorname{mod} k)$ we have

$$\sum_{\substack{p \leqslant x \\ p \equiv l (\operatorname{mod} k)}} \log^2 p + \sum_{\substack{pq \leqslant x \\ pq \equiv l (\operatorname{mod} k)}} \log p \log q = \frac{2}{\varphi(k)} x \log x + O(x)$$

and

$$\vartheta_i(x) \log x + \sum_{p \leqslant x} \log p \, \vartheta_{l\bar{p}}\left(\frac{x}{p}\right) = \frac{2}{\varphi(k)} x \log x + O(x),$$

where $\bar{p}$ satisfies $p\bar{p} \equiv 1 \ (\operatorname{mod} k)$.

3) Let

$$\vartheta_l(x) = \frac{x}{\varphi(k)} + R_l(x),$$

so that

$$R_l(x)\log x = \sum_{pq \leqslant x} \frac{\log p \log q}{\log pq} R_{l\bar{p}\bar{q}}\left(\frac{x}{pq}\right) + O(x \log\log x).$$

4) $$|R_l(x)| \leqslant \frac{1}{\varphi(k)\log x} \sum_{\substack{1 \leqslant a \leqslant k \\ (a,k)=1}} \sum_{n \leqslant x} \left| R_a\left(\frac{x}{n}\right)\right| + O\left(\frac{x \log\log x}{\log x}\right).$$

5) $$\sum_{n \leqslant x} \frac{\log n}{n} R_l(n) = -\sum_{n \leqslant x} \frac{1}{n} \sum_{\alpha\beta \equiv l (\bmod k)} R_\alpha(n) R_\beta\left(\frac{x}{n}\right) + O(x).$$

6) If $0 < \sigma < 1$ and there exists x_0 such that when $x > x_0$ we have

$$|R_l(x)| < \frac{\sigma x}{\varphi(k)},$$

then there exists x_σ such that, when $x > x_\sigma$, the interval $((1-\sigma)^{16}x, x)$ contains a subinterval $(y, e^\delta y)$ $(\delta = \sigma(1-\sigma)/32)$ such that when $y \leqslant z \leqslant e^\delta y$ we have

$$\left|\frac{R_l(z)}{z}\right| < \frac{1}{\varphi(k)} \cdot \frac{\sigma + \sigma^2}{2}.$$

7) First use Theorem 8.2 to show that σ_0 and x_0 exist such that $0 < \sigma_0 < 1$ and when $x > x_0$

$$|R_l(x)| < \frac{\sigma_0}{\varphi(k)} x.$$

Then use this together with 4) and 6) to prove that

$$\lim_{x \to \infty} \frac{R_l(x)}{x} = 0.$$

Notes

9.1. The present best result on the error term of the prime number theorem, namely

$$\pi(x) = \operatorname{li} x + O(xe^{-c(\log x)^{\frac{3}{5}}}), \qquad c \text{ a positive constant},$$

is due to I. M. Vinogradov and H. M. Korobov and is based on estimates on trigonometric sums.

9.2. In recent years a number of mathematicians have obtained error term estimates in Selberg's elementary proof of the prime number theorem. For example:

$$\pi(x) = \operatorname{li} x + O\left(\frac{x}{\log^A x}\right)$$

where A is any constant however large and the O-constant depending on A (see E. Bombieri [8] and E. Wirsing [64]). An even better estimate is given by H. Diamond and G. J. Steinig [21].

Chapter 10. Continued Fractions and Approximation Methods

10.1 Simple Continued Fractions

By a *finite continued fraction* we mean an expression

$$a_0 + \cfrac{1}{a_1 + \cfrac{1}{a_2 + \cfrac{1}{a_3 + \ddots + \cfrac{1}{a_N}}}}.$$

We shall see that, as $N \to \infty$, the expression here tends to a definite number; we call the infinite continued fraction a *continued fraction.* It is convenient to denote the above expression by

$$a_0 + \frac{1}{a_1 +} \frac{1}{a_2 +} \cdots \frac{1}{+ a_N} \qquad \text{or} \qquad [a_0, a_1, a_2, \ldots, a_N].$$

It is easy to see that

$$[a_0] = \frac{a_0}{1}, \qquad [a_0, a_1] = \frac{a_0 a_1 + 1}{a_1}, \qquad [a_0, a_1, a_2] = \frac{a_2 a_1 a_0 + a_2 + a_0}{a_2 a_1 + 1}.$$

In general, we let $[a_0, a_1, \ldots, a_n] = p_n/q_n$, $0 \leqslant n \leqslant N$, where p_n, q_n are polynomials in $a_0, a_1, \ldots, a_n$. These polynomials are linear in any one a, and the denominator q_n is independent of a_0. We call p_n/q_n the n-th *convergent* of $[a_0, a_1, \ldots, a_N]$.

Theorem 1.1. *The convergents satisfy the following*:

$$\begin{aligned} &p_0 = a_0, \qquad p_1 = a_1 a_0 + 1, \qquad p_n = a_n p_{n-1} + p_{n-2} \qquad (2 \leqslant n \leqslant N), \\ &q_0 = 1, \qquad q_1 = a_1, \qquad q_n = a_n q_{n-1} + q_{n-2} \qquad (2 \leqslant n \leqslant N). \quad \square \end{aligned}$$

Theorem 1.2. *The convergents satisfy the following*:

$$p_n q_{n-1} - p_{n-1} q_n = (-1)^{n-1} \qquad (n \geqslant 1), \tag{1}$$

or

$$\frac{p_n}{q_n} - \frac{p_{n-1}}{q_{n-1}} = \frac{(-1)^{n-1}}{q_n q_{n-1}},$$

and

$$p_n q_{n-2} - p_{n-2} q_n = (-1)^n a_n \qquad (n \geqslant 2). \quad \square \tag{2}$$

Definition. Let a_0 be an integer, and $a_1, a_2, \ldots$ be positive integers. Then

$$a_0 + \frac{1}{a_1 +} \frac{1}{a_2 +} \cdots$$

is called a *simple continued fraction*. We shall only deal with simple continued fractions in this chapter.

From Theorems 1.1 and 1.2 we deduce at once:

Theorem 1.3. (i) *If* $n > 1$, *then* $q_n \geqslant q_{n-1} + 1$, *so that* $q_n \geqslant n$.

(ii)
$$\frac{p_{2n+1}}{q_{2n+1}} < \frac{p_{2n-1}}{q_{2n-1}}, \qquad \frac{p_{2n}}{q_{2n}} > \frac{p_{2n-2}}{q_{2n-2}}.$$

(iii) *Every convergent of a simple continued fraction is a reduced fraction.* $\square$

Let α be a real number. We take $a_0 = [\alpha]$ and we let $\alpha_1' = 1/(\alpha - [\alpha])$. We then take $a_1 = [\alpha_1]$ and we let $\alpha_2' = 1/(\alpha_1' - [\alpha_1'])$. We continue in this way by taking $a_n = [\alpha_n']$ and defining $\alpha_{n+1}' = 1/(\alpha_n' - [\alpha_n'])$. It is clear that if this process terminates, then α must be a rational number. Conversely, if α is a rational number p/q where $(p, q) = 1$, then $a_0 = [p/q]$ and

$$\frac{1}{\alpha_1'} = \frac{p}{q} - \left[\frac{p}{q}\right], \qquad 0 \leqslant \frac{1}{\alpha_1'} < 1,$$

or

$$p - \left[\frac{p}{q}\right] q = \frac{q}{\alpha_1'} (= r_1), \qquad 0 \leqslant r_1 < q$$

and similarly

$$q - r_1 \left[\frac{q}{r_1}\right] = \frac{r_1}{\alpha_2'} (= r_2), \qquad 0 \leqslant r_2 < r_1.$$

We see therefore that if α is rational, then the evaluation of its continued fraction is similar to the Euclidean algorithm, so that we have:

Theorem 1.4. *Every rational number is representable as a finite continued fraction.* $\square$

An immediate problem is the uniqueness of this representation. From $a + 1/1 = a + 1$ we see at once that there is no uniqueness. In other words, if $a_n > 1$, then $[a_0, \ldots, a_n] = [a_0, \ldots, a_{n-1}, a_n - 1, 1]$; if $a_n = 1$, then $[a_0, \ldots, a_n] = [a_0, \ldots, a_{n-1} + 1]$. Therefore each rational number has two representations, one with n odd, and the other with n even. If α is irrational, then the above method gives an infinite sequence $a_0, a_1, a_2, \ldots, a_n, \ldots$. For example, we have

$$\pi = [3, 7, 15, 1, 292, 1, 1, 1, 21, 31, 14, 2, 1, 2, 2, 2, \\ 2, 1, 84, 2, 1, 1, 15, 3, 13, 1, 4, 2, 6, 6, 1, \ldots].$$

Theorem 1.5. *Let* $\alpha_n = [a_0, a_1, \ldots, a_n]$. *Then* $\lim \alpha_n$ *exists.*

Proof. We have $\alpha_n = p_n/q_n$, and by Theorem 1.3 (ii), $\alpha_{2n+1} < \alpha_{2n-1}$, $\alpha_{2n} > \alpha_{2n-2}$. Next from Theorem 1.2 (1), $\alpha_1 \geqslant \alpha_{2n+1} \geqslant \alpha_{2n} \geqslant \alpha_2$, so that $\lim \alpha_{2n}$ and $\lim \alpha_{2n+1}$ exist. Finally, from Theorem 1.2 and Theorem 1.3 (i), we have $|\alpha_{2n} - \alpha_{2n-1}| = 1/q_{2n}q_{2n-1} \leqslant 1/2n(2n-1)$ so that $\lim \alpha_{2n} = \lim \alpha_{2n-1}$. □

Exercise. Prove that

$$p_n = \begin{vmatrix} a_0 & -1 & 0 & 0 & \cdots & 0 & 0 & 0 \\ 1 & a_1 & -1 & 0 & \cdots & 0 & 0 & 0 \\ 0 & 1 & a_2 & -1 & \cdots & 0 & 0 & 0 \\ \cdots & \cdots & \cdots & \cdots & \cdots & \cdots & \cdots & \cdots \\ 0 & 0 & 0 & 0 & \cdots & 1 & a_{n-1} & -1 \\ 0 & 0 & 0 & 0 & \cdots & 0 & 1 & a_n \end{vmatrix},$$

and that q_n is the determinant above with the first row and first column omitted.

Exercise 2. The sequence $(u_n) = (1, 1, 2, 3, 5, 8, 13, \ldots)$, where $u_1 = u_2 = 1$, $u_{i+1} = u_{i-1} + u_i$ $(i > 1)$, is called the *Fibonacci sequence*. Prove that (i) u_{n+2}/u_{n+1} is the n-th convergent of $(1 + \sqrt{5})/2$; (ii) in the continued fraction $[a_0, a_1, \ldots]$, if $a_i = 2$ $(i > 0)$ and $a_n = 1$ $(n \neq i)$, then for $m > i$ we have

$$\frac{p_m}{q_m} = \frac{u_{i+1}u_{m-i+3} + u_i u_{m-i+1}}{u_i u_{m-i+3} + u_{i-1}u_{m-i+1}}.$$

Exercise 3. A synodic month is the period of time between two new moons, and is 29.5306 days. When projected onto the star sphere, the path of the moon intersects the ecliptic (the path of the sun) at the ascending and the descending nodes. A draconic month is the period of time for the moon to return to the same node, and is 27.2123 days. Show that solar and lunar eclipses occur in cycles with a period of 18 years 10 days.

10.2 The Uniqueness of a Continued Fraction Expansion

Definition. We call $\alpha_n' = [a_n, a_{n+1}, \ldots]$ the $(n+1)$-th *complete quotient* of $[a_0, a_1, \ldots, a_n, \ldots]$.

Theorem 2.1. *We have*

$$\alpha = \alpha_0', \qquad \alpha = \frac{\alpha_1' a_0 + 1}{\alpha_1'}, \qquad \alpha = \frac{\alpha_n' p_{n-1} + p_{n-2}}{\alpha_n' q_{n-1} + q_{n-2}}, \qquad n \geqslant 2.$$

If α is rational, then this holds up to $n = N$.

Proof. Use mathematical induction. □

Theorem 2.2. *We always have $a_n = [\alpha_n']$, except when α is rational and $a_N = 1$ in which case we have $a_{N-1} = [\alpha_{N-1}'] - 1$. Therefore there are only two representations to a rational number.*

Proof. We have $\alpha_n' = a_n + 1/\alpha_{n+1}'$. If α is irrational or if α is rational and $n \neq N - 1$, then $\alpha_{n+1}' > 1$ so that $a_n < \alpha_n' < a_n + 1$, as required. If α is rational and $n = N - 1$, $\alpha_{n+1} = 1$, then $a_n = [\alpha_n'] - 1$. □

Theorem 2.3. *The representation of an irrational number by a simple continued fraction is unique.*

Proof. Suppose that $\alpha = [a_0, a_1, a_2, \ldots] = [b_0, b_1, b_2, \ldots]$. Certainly we have $a_0 = [\alpha] = b_0$, and similarly $a_1 = b_1$. Suppose now that $a_k = b_k$ for $k < n$, and we have to prove that $a_n = b_n$. From $\alpha = [a_0, \ldots, a_{n-1}, \alpha_n'] = [a_0, \ldots, a_{n-1}, \beta_n']$, we have

$$\alpha = \frac{\alpha_n' p_{n-1} + p_{n-2}}{\alpha_n' q_{n-1} + q_{n-2}} = \frac{\beta_n' p_{n-1} + p_{n-2}}{\beta_n' q_{n-1} + q_{n-2}},$$

so that $(\alpha_n' - \beta_n')(p_{n-1}q_{n-2} - p_{n-2}q_{n-1}) = 0$. From Theorem 1.2 we deduce that $\alpha_n' = \beta_n'$ and therefore $a_n = [\alpha_n'] = [\beta_n'] = b_n$. □

Theorem 2.4. *We have*

$$q_n\alpha - p_n = \frac{(-1)^n \delta_n}{q_{n+1}}, \qquad 0 < \delta_n < 1,$$

and δ_n/q_{n+1} is a decreasing function of n. (If α is rational, then this holds only for $1 \leqslant n \leqslant N - 2$, and $\delta_{N-1} = 1$.)

Proof. We have

$$\alpha = \frac{\alpha_{n+1}' p_n + p_{n-1}}{\alpha_{n+1}' q_n + q_{n-1}},$$

so that

$$\alpha - \frac{p_n}{q_n} = \frac{\alpha_{n+1}' p_n + p_{n-1}}{\alpha_{n+1}' q_n + q_{n-1}} - \frac{p_n}{q_n}$$

$$= \frac{-(p_n q_{n-1} - q_n p_{n-1})}{q_n(\alpha_{n+1}' q_n + q_{n+1})} = \frac{(-1)^n}{q_n(\alpha_{n+1}' q_n + q_{n-1})},$$

and hence

$$\delta_n = \frac{q_{n+1}}{\alpha'_{n+1} q_n + q_{n-1}} = \frac{a_{n+1} q_n + q_{n-1}}{\alpha'_{n+1} q_n + q_{n-1}}.$$

From this we see that $0 < \delta_n < 1$ except when $\alpha_{n+1} = \alpha'_{n+1}$. Also, from $\alpha'_n = 1 + 1/\alpha'_{n+1}$ we have

$$\frac{\delta_n}{q_{n+1}} = \frac{1}{\alpha'_{n+1} q_n + q_{n-1}} \geqslant \frac{1}{(a_{n+1} + 1) q_n + q_{n-1}} = \frac{1}{q_{n+1} + q_n}$$

$$\geqslant \frac{1}{a_{n-2} q_{n+1} + q_n} = \frac{1}{q_{n+2}} \geqslant \frac{\delta_{n+1}}{q_{n+2}}.$$

In the last inequality, equality sign holds only when $\alpha_{n+1} = \alpha'_{n+1}$, that is when α is rational and $n = N - 1$. □

From this theorem we deduce:

Theorem 2.5. *If α is irrational, then* $\lim p_n/q_n = \alpha$. □

Theorem 2.6. *We have*

$$\left| \alpha - \frac{p_n}{q_n} \right| \leqslant \frac{1}{q_n q_{n-1}} < \frac{1}{q_n^2},$$

with the equality sign only when α is rational and $n = N - 1$. □

10.3 The Best Approximation

Let α be a real number. Among the rational numbers with denominators not exceeding N, there is one which is closest to α, and we call it the *best rational approximation* to α. We now prove that the convergents p_n/q_n are the best rational approximations to α.

Theorem 3.1. *Suppose that $n \geqslant 1$, $0 < q \leqslant q_n$ and $p/q \neq p_n/q_n$. Then* $|p_n/q_n - \alpha| < |p/q - \alpha|$.

Proof. It suffices to prove that $|p_n - q_n\alpha| < |p - q\alpha|$.

(i) If $\alpha = [\alpha] + \frac{1}{2}$, then $p_1/q_1 = \alpha$ and the result follows at once.

(ii) If $\alpha < [\alpha] + \frac{1}{2}$, then the result holds when $n = 0$, and if $\alpha > [\alpha] + \frac{1}{2}$, then the result holds when $n = 1$. We now assume as induction hypothesis that the result holds for $n - 1$, and proceed to prove by induction.

If $q \leqslant q_{n-1}$, then from the induction hypothesis $|p_{n-1} - q_{n-1}\alpha| < |p - q\alpha|$, so that we may assume that $q_n \geqslant q > q_{n-1}$. If $q = q_n$, then

$$\left| \frac{p_n}{q_n} - \frac{p}{q} \right| \geqslant \frac{1}{q_n}, \qquad p \neq p_n.$$

Also

$$\left|\frac{p_n}{q_n} - \alpha\right| \leqslant \frac{1}{q_n q_{n+1}} \leqslant \frac{1}{2q_n}.$$

If $q_{n+1} = 2$, then $n = 1$, and $a_1 = a_2 = 1$, giving

$$\alpha = a_0 + \frac{1}{1+}\frac{1}{1+}\frac{1}{a_3+}\cdots,$$

which shows that $a_0 + \frac{1}{2} < \alpha < a_0 + 1$, and our required result clearly holds. We may therefore assume that $q_{n+1} > 2$, that is

$$\left|\frac{p_n}{q_n} - \alpha\right| \leqslant \frac{1}{q_n q_{n+1}} < \frac{1}{2q_n},$$

and so

$$\left|\frac{p}{q} - \alpha\right| \geqslant \left|\frac{p}{q} - \frac{p_n}{q_n}\right| - \left|\frac{p_n}{q_n} - \alpha\right| \geqslant \frac{1}{q_n} - \left|\frac{p_n}{q_n} - \alpha\right| > \left|\frac{p_n}{q_n} - \alpha\right|.$$

We may now assume that $q_n > q > q_{n-1}$. Let us write $up_n + vp_{n-1} = p$, $uq_n + vq_{n-1} = q$, so that $u(p_n q_{n-1} - p_{n-1} q_n) = pq_{n-1} - qp_{n-1}$. From Theorem 1.2 we have $u = \pm(pq_{n-1} - qp_{n-1})$, and similarly $v = \pm(pq_n - qp_n)$. The numbers u, v cannot be zero, and in fact from $q_n > q = uq_n + vq_{n-1}$ we see that they are of opposite signs. Now from Theorem 2.4, $p_n - q_n\alpha$ and $p_{n-1} - q_{n-1}\alpha$ have opposite signs, and therefore $u(p_n - q_n\alpha)$ and $v(p_{n-1} - q_{n-1}\alpha)$ have the same sign. Finally from $p - q\alpha = u(p_n - q_n\alpha) + v(p_{n-1} - q_{n-1}\alpha)$ we see that $|p - q\alpha| > |p_{n-1} - q_{n-1}\alpha| > |p_n - q_n\alpha|$. □

Example. From $\pi = [3, 7, 15, 1, 292, 1, 1, \ldots]$ we obtain the convergents

$$\frac{3}{1}, \frac{22}{7}, \frac{333}{106}, \frac{355}{113}, \frac{103993}{33102}, \frac{104348}{33215}, \ldots.$$

In the year 500 A. D. Chao Jung-Tze obtained both the crude estimate 22/7 and the good estimate 355/113 (this is more than a thousand years earlier than the earliest European record due to Otto). More interesting still the two estimates of Chao belong to the family of best approximations to π; in other words there is no fraction with denominator less than 113 which is closer to π than 355/113 is.

From Theorem 2.6 we have

$$\left|\pi - \frac{355}{113}\right| < \frac{1}{113 \times 33102} < \frac{1}{10^6}.$$

In fact $355/113 = 3.1415929\ldots$ whereas $\pi = 3.1415926\ldots$.

10.4 Hurwitz's Theorem

Theorem 4.1. *Of any two consecutive convergents to α, at least one of them satisfies the inequality $|\alpha - p/q| < 1/2q^2$.*

Proof. By Theorem 2.4, α lies between p_n/q_n and p_{n+1}/q_{n+1} so that we have

$$\left|\frac{p_{n+1}}{q_{n+1}} - \frac{p_n}{q_n}\right| = \left|\frac{p_n}{q_n} - \alpha\right| + \left|\frac{p_{n+1}}{q_{n+1}} - \alpha\right|.$$

If the theorem is false, then

$$\frac{1}{q_n q_{n+1}} = \left|\frac{p_{n+1}}{q_{n+1}} - \frac{p_n}{q_n}\right| \geqslant \frac{1}{2q_n^2} + \frac{1}{2q_{n+1}^2},$$

giving $(q_{n+1} - q_n)^2 \leqslant 0$, which is impossible if $n > 0$. □

It follows from this theorem that, if α is irrational, then there are infinitely many rational numbers p/q such that $|\alpha - p/q| < 1/2q^2$.

Theorem 4.2 (Hurwitz). *Of any three consecutive convergents to* α, *at least one of them satisfies the inequality*

$$\left|\alpha - \frac{p}{q}\right| < \frac{1}{\sqrt{5}\,q^2}.$$

Proof. Let $\beta_{n+1} = q_{n-1}/q_n$, so that, by Theorem 2.1,

$$\left|\frac{p_n}{q_n} - \alpha\right| = \frac{1}{q_n(\alpha'_{n+1}q_n + q_{n-1})} = \frac{1}{q_n^2(\alpha'_{n+1} + \beta_{n+1})}.$$

We proceed to prove that

$$\alpha'_i + \beta_i \leqslant \sqrt{5} \tag{1}$$

cannot hold for three consecutive values $i = n - 1, n, n + 1$. Suppose that (1) holds for $i = n - 1$ and $i = n$. From $\alpha'_{n-1} = a_{n-1} + 1/\alpha'_n$ and

$$\frac{1}{\beta_n} = \frac{q_{n-1}}{q_{n-2}} = \frac{a_{n-1}q_{n-2} + q_{n-3}}{q_{n-2}} = a_{n-1} + \beta_{n-1},$$

we have

$$\frac{1}{\alpha'_n} + \frac{1}{\beta_n} = \alpha'_{n-1} + \beta_{n-1} \leqslant \sqrt{5},$$

giving

$$1 = \frac{1}{\alpha'_n}\alpha'_n \leqslant \left(\sqrt{5} - \frac{1}{\beta_n}\right)(\sqrt{5} - \beta_n).$$

Thus $\beta_n + 1/\beta_n \leqslant \sqrt{5}$. Since β_n is a rational number we must have strict inequality, and since $\beta_n < 1$, we deduce easily that $\beta_n > \frac{1}{2}(\sqrt{5} - 1)$.

Similarly, if (1) holds for $i = n$ and $i = n + 1$, then $\beta_{n+1} > \frac{1}{2}(\sqrt{5} - 1)$, and we arrive at

$$a_n = \frac{1}{\beta_{n+1}} - \beta_n < \sqrt{5} - \beta_{n+1} - \beta_n < \sqrt{5} - (\sqrt{5} - 1) = 1,$$

which is impossible. The theorem is proved. □

As a corollary to this, we have

Theorem 4.3. *Given any irrational number* α, *there are infinitely many convergents satisfying the inequality*

$$\left|\frac{p}{q} - \alpha\right| < \frac{1}{\sqrt{5}\,q^2}. \quad \square$$

Theorem 4.4. *The number* $\sqrt{5}$ *in the above theorem is best possible in the sense that if* $A > \sqrt{5}$, *then there exists a real number* α *such that* $|\alpha - p/q| < 1/Aq^2$ *has only finitely many solutions.*

Proof. In fact we set $\alpha = \frac{1}{2}(\sqrt{5} - 1)$. Let us assume that

$$\frac{1}{2}(\sqrt{5} - 1) = \frac{p}{q} + \frac{\delta}{q^2}, \qquad |\delta| < \frac{1}{A} < \frac{1}{\sqrt{5}}.$$

Then

$$\frac{\delta}{q} - \frac{1}{2}\sqrt{5}\,q = -\frac{1}{2}q - p,$$

and on squaring

$$\frac{\delta^2}{q^2} - \sqrt{5}\,\delta = \left(\frac{1}{2}q + p\right)^2 - \frac{5}{4}q^2 = pq - q^2 + p^2.$$

For sufficiently large q, the left hand side here is less than 1 in magnitude, so that $pq - q^2 + p^2 = 0$ or $(2p + q)^2 = 5q^2$, which is impossible. □

10.5 The Equivalence of Real Numbers

Definition 5.1. Two real numbers ξ and η are said to be *equivalent* if there are integers a, b, c, d such that

$$\xi = \frac{a\eta + b}{c\eta + d}, \qquad ad - bc = \pm 1. \tag{1}$$

The relationship between ξ and η here is called a *modular transformation.*

Example 1. $\xi = a + \eta$ and $\xi = 1/\eta$ are modular transformations.

Example 2. $\xi = [a, \eta] = a + 1/\eta$ is also a modular transformation.

Example 3. We may view $\alpha = [a_0, a_1, \ldots, a_n, \alpha_n']$ as n successive modular transformations of Example 2. The resulting modular transformation is given by

$$\alpha = \frac{p_{n-1}\alpha_n' + p_{n-2}}{q_{n-1}\alpha_n' + q_{n-2}}.$$

The following properties belong to equivalence: (i) Every real number is equivalent to itself; (ii) if ξ is equivalent to η, then η is equivalent to ξ; (iii) if ξ is equivalent to η, and η is equivalent to ζ, then ξ is equivalent to ζ. To see (iii) we set $\xi = (a\eta + b)/(c\eta + d)$ and $\eta = (a_1\zeta + b_1)/(c_1\zeta + d_1)$. Then

$$\xi = \{(aa_1 + bc_1)\zeta + (ab_1 + bd_1)\}/\{(ca_1 + dc_1)\zeta + (cb_1 + dd_1)\},$$

where

$$\begin{aligned}(aa_1 + bc_1)(cb_1 + dd_1) - (ab_1 + bd_1)(ca_1 + dc_1)\\ = (ad - bc)(a_1d_1 - b_1c_1) = \pm 1.\end{aligned}$$

Definition 5.2. We call this last transformation above the *product* of the two previous transformations.

It is easy to see that every rational number is equivalent to 0, so that we have

Theorem 5.1. *Any two rational numbers are equivalent.* □

Theorem 5.2. *A necessary and sufficient condition for the modular transformation* (1) *to be representable as a continued fraction* $\xi = [a_0, a_1, \ldots, a_{k-1}, \eta]$ $(k \geqslant 2)$ *is that* $c > d > 0$.

Proof. From $\xi = [a_0, a_1, \ldots, a_{k-1}, \eta]$, we have $\xi = (p_{k-1}\eta + p_{k-2})/(q_{k-1}\eta + q_{k-2})$ and we see that the condition $c > d > 0$ is necessary. The sufficiency of the condition can be proved by induction on d. □

Theorem 5.3. *A necessary and sufficient condition for two irrational numbers* ξ *and* η *to be equivalent is that* $\xi = [a_0, a_1, \ldots, a_m, c_0, c_1, \ldots]$ *and* $\eta = [b_0, b_1, \ldots, b_n, c_0, c_1, \ldots]$. *In other words their continued fractions expansions are eventually identical.*

Proof. 1) Let $\omega = [c_0, c_1, \ldots]$. Then

$$\xi = [a_0, a_1, \ldots, a_m, \omega] = \frac{\omega p_m + p_{m-1}}{\omega q_m + q_{m-1}}, \qquad p_m q_{m-1} - q_m p_{m-1} = \pm 1.$$

Thus ω and ξ are equivalent. Similarly ω and η are equivalent, and hence ξ and η are equivalent.

2) Let ξ and η be equivalent, and $\eta = (a\xi + b)/(c\xi + d)$, $ad - bc = \pm 1$. We may assume that $c\xi + d > 0$. We expand ξ into continued fractions:

$$\begin{aligned}\xi &= [a_0, \ldots, a_k, a_{k+1}, \ldots] = [a_0, \ldots, a_{k-1}, \alpha_k']\\ &= (\alpha_k' p_{k-1} + p_{k-2})(\alpha_k' q_{k-1} + q_{k-2})^{-1}.\end{aligned}$$

It follows that $\eta = (P\alpha_k' + R)(Q\alpha_k' + S)^{-1}$, where $P = ap_{k-1} + bq_{k-1}$, $R = ap_{k-2} + bq_{k-2}$, $Q = cp_{k-1} + dq_{k-1}$, $S = cp_{k-2} + dq_{k-2}$; P, Q, R, S are integers satisfying $PS - QR = \pm 1$.

From Theorem 2.4 we have

$$p_{k-1} = \xi q_{k-1} + \frac{\delta}{q_{k-1}}, \qquad p_{k-2} = \xi q_{k-2} + \frac{\delta'}{q_{k-2}}, \qquad |\delta| < 1, \quad |\delta'| < 1,$$

so that

$$Q = (c\xi + d)q_{k-1} + \frac{c\delta}{q_{k-1}}, \qquad S = (c\xi + d)q_{k-2} + \frac{c\delta'}{q_{k-2}}.$$

From $c\xi + d > 0$, $q_{k-2} \geqslant k - 2$ and by Theorem 1.3, $q_{k-1} \geqslant q_{k-2} + 1$ we see that $Q > S > 0$ when k is sufficiently large. It follows from Theorem 5.2 that $\eta = [b_0, \ldots, b_n, \alpha_k']$ and the necessity of the condition is established. □

Denote by $M(\alpha)$ the greatest number such that, for any $\varepsilon > 0$, the inequality

$$\left|\alpha - \frac{p_i}{q_i}\right| \leqslant \frac{1}{(M(\alpha) - \varepsilon)q_i^2}$$

has infinitely many solutions. For example $M((\sqrt{5} - 1)/2) = \sqrt{5}$. Let

$$\alpha - \frac{p_i}{q_i} = \frac{1}{\lambda_i q_i^2};$$

then

$$\lambda_i = (-1)^i\left(\alpha_{i+1}' + \frac{q_{i-1}}{q_i}\right), \qquad \alpha_{i+1}' = [a_{i+1}, a_{i+2}, \ldots],$$

and

$$\frac{q_{i-1}}{q_i} = \frac{1}{q_i/q_{i-1}} = \frac{1}{a_i +} \frac{q_{i-2}}{q_{i-1}} = \frac{1}{a_i +} \frac{1}{a_{i-1} +} \frac{q_{i-3}}{q_{i-2}} = \cdots$$

$$= [0, a_i, a_{i-1}, \ldots, a_2, a_1].$$

Therefore

$$M(\alpha) = \overline{\lim_{i\to\infty}}\, \lambda_i = \overline{\lim_{i\to\infty}}\, ([a_{i+1}, a_{i+2}, \ldots] + [0, a_i, a_{i-1}, \ldots, a_2, a_1]).$$

If α and β are equivalent, then $a_i = b_i$ for all large i. We have therefore proved the following

Theorem 5.4. *If α and β are equivalent, then $M(\alpha) = M(\beta)$.* □

We see therefore that if $A > \sqrt{5}$ and if α is equivalent to $(\sqrt{5} - 1)/2$, then the inequality $|\alpha - p/q| < 1/Aq^2$ has only finitely many solutions. We may now ask for the value of $M(\alpha)$ when α is not equivalent to $(\sqrt{5} - 1)/2$. We have the following result: If α is not equivalent to $(\sqrt{5}-1)/2$, then $M(\alpha) \geqslant \sqrt{8}$. Specifically, for such α, the inequality $|\alpha - p/q| < 1/\sqrt{8}\,q^2$ has infinitely many solutions. Also, if α is equivalent to $1+\sqrt{2}$, then $M(\alpha) = \sqrt{8}$. For the general situation, we need the following:

Definition 5.3. By a *Markoff number* we mean a positive integer u such that there are integers v, w satisfying $u^2 + v^2 + w^2 = 3uvw$. The first eleven Markoff numbers are $1, 2, 5, 13, 29, 34, 89, 169, 194, 233, 433$. (We shall prove in the next chapter that the number of Markoff numbers is infinite.)

It can be proved that if

$$\alpha_u = \frac{1}{2u}\left(\sqrt{9u^2 - 4} + u + \frac{2v}{w}\right)$$

where u, v, w are related by the definition of the Markoff number u, then $M(\alpha_u) = \sqrt{9u^2 - 4}/u$. Furthermore if α is not equivalent to α_u for $1 \leqslant u \leqslant v$, then the inequality

$$\left|\alpha - \frac{p}{q}\right| < \frac{1}{M(\alpha_v)q^2}$$

has infinitely many solutions.

It follows from this that if α is not a root of a quadratic equation with rational coefficients, then $M(\alpha) \geqslant \sqrt{9u^2 - 4}/u$ for all u, and hence $M(\alpha) \geqslant 3$. Finally, if $0 < m_1 < m_2 < \cdots$ and

$$\alpha = [2, 2, \underbrace{1, 1, \ldots, 1}_{m_1}, 2, 2, \underbrace{1, \ldots, 1}_{m_2}, 2, 2, \underbrace{1, 1, \ldots, 1}_{m_3}, \ldots]$$

then $M(\alpha) = 3$. The proofs of these facts are outside the scope of this book.

10.6 Periodic Continued Fractions

Definition. A continued fraction is said to be *periodic* if there exist k and L such that $a_l = a_{l+k}$ for $l \geqslant L$. We call k the *period* and we write

$$[a_0, \ldots, a_{L-1}, \dot{a}_L, \ldots, \dot{a}_{L+k-1}].$$

For example, we have $\sqrt{2} = [1, \dot{2}]$, $\sqrt{3} = [1, \dot{1}, \dot{2}]$, $\sqrt{5} = [2, \dot{4}]$ and $\sqrt{7} = [2, \dot{1}, 1, 1, \dot{4}]$. In fact we have the following well known result.

Theorem 6.1. *A necessary and sufficient condition for a number to have a periodic continued fractions expansion is that it is a root of an irreducible quadratic with rational coefficients.* □

10.7 Legendre's Criterion

We saw that if p/q is a convergent of α, then $|\alpha - p/q| < 1/q^2$. The converse of this is not true. We shall now determine a necessary and sufficient condition for p/q to be a convergent of α. Let

$$\alpha - \frac{p}{q} = \frac{\varepsilon\vartheta}{q^2}, \qquad \varepsilon = \pm 1, \qquad 0 < \vartheta < 1,$$

and let

$$\frac{p}{q} = [a_0, \ldots, a_{n-1}] = \frac{p_{n-1}}{q_{n-1}}.$$

We can choose n so that $(-1)^{n-1} = \varepsilon$, and we can now write

$$\alpha - \frac{p_{n-1}}{q_{n-1}} = \frac{\varepsilon\vartheta}{q_{n-1}^2}.$$

We define β by

$$\alpha = \frac{p_{n-1}\beta + p_{n-2}}{q_{n-1}\beta + q_{n-2}},$$

so that

$$\frac{\varepsilon\vartheta}{q_{n-1}^2} = \frac{p_{n-1}\beta + p_{n-2}}{q_{n-1}\beta + q_{n-2}} - \frac{p_{n-1}}{q_{n-1}} = \frac{(-1)^{n-1}}{q_{n-1}(q_{n-1}\beta + q_{n-2})}.$$

Therefore

$$\vartheta = \frac{q_{n-1}}{q_{n-1}\beta + q_{n-2}}.$$

Solving this for β we have $\beta = (q_{n-1} - \vartheta q_{n-2})/\vartheta q_{n-1}$, and since $0 < \vartheta < 1$ we see that $\beta > 0$.

Now $\alpha = [a_0, \ldots, a_{n-1}, \beta]$. If $\beta \geqslant 1$, then $\beta = \alpha_n'$ $(= [a_n, a_{n+1}, \ldots])$, and this means that p/q is a convergent of α. If $0 < \beta < 1$, then $[a_{n-1} + 1/\beta] = a_{n-1} + c$, $c > 0$ so that $\alpha = [a_0, \ldots, a_{n-2}, a_{n-1}, + c, \ldots]$ and we see that $[a_0, \ldots, a_{n-1}]$ is not a convergent. Therefore the required necessary and sufficient condition is that $\beta \geqslant 1$; in other words we have:

Theorem 7.1 (Legendre). *Let* $\varepsilon\vartheta = q^2\alpha - pq$, $\varepsilon = \pm 1$, $0 < \vartheta < 1$, *and let* $p/q = [a_0, \ldots, a_{n-1}]$, $(-1)^{n-1} = \varepsilon$. *Then, a necessary and sufficient condition for* p/q *to be a convergent of* α *is that*

$$\vartheta \leqslant \frac{q_{n-1}}{q_{n-1} + q_{n-2}}. \quad \square$$

Since the right hand side of the above inequality exceeds $\frac{1}{2}$ we deduce at once the following

Theorem 7.2. *If a rational number p/q satisfies $|\alpha - p/q| < 1/2q^2$, then it is a convergent of α.* □

Theorem 7.3. *Let p, q be positive integers satisfying $|p^2 - \alpha^2 q^2| < \alpha$. Then p/q is a convergent of α.*

Proof. Let $\alpha^2 q^2 - p^2 = \varepsilon\delta\alpha$, $\varepsilon = \pm 1$, $0 \leqslant \delta < 1$. Then $\alpha q - p = \varepsilon\delta\alpha/(\alpha q + p)$, so that

$$\vartheta = \varepsilon q(\alpha q - p) = \frac{\delta\alpha q}{\alpha q + p} = \frac{\delta\alpha q_{n-1}}{\alpha q_{n-1} + p_{n-1}}, \qquad (-1)^{n-1} = \varepsilon.$$

From Theorem 7.1 we see that it suffices to prove that

$$\frac{\delta\alpha q_{n-1}}{\alpha q_{n-1} + p_{n-1}} < \frac{q_{n-1}}{q_{n-1} + q_{n-2}},$$

or that $\delta\alpha(q_{n-1} + q_{n-2}) < \alpha q_{n-1} + p_{n-1}$. Now this inequality clearly holds when $n = 2$ so that it suffices to establish $\alpha q_{n-1} - p_{n-1} < \alpha(q_{n-1} - q_{n-2})$ for $n > 2$. But $\alpha q_{n-1} - p_{n-1} = \varepsilon\delta\alpha/(\alpha q_{n-1} + p_{n-1})$, and by Theorem 1.3 we have

$$q_{n-1} - q_{n-2} \geqslant 1 > \frac{1}{\alpha q_{n-1} + p_{n-1}}.$$

The theorem is proved. □

10.8 Quadratic Indeterminate Equations

In this and the next sections d denotes a positve integer which is not a perfect square. We consider the equation

$$x^2 - dy^2 = l, \qquad 0 < |l| < \sqrt{d}$$

in the integer unknowns x and y.

Theorem 8.1. *In the continued fractions expansion of $\sqrt{d}$, the numbers α'_n must take the form*

$$\frac{\sqrt{d} + P_n}{Q_n}, \qquad P_n^2 \equiv d \pmod{Q_n},$$

where P_n and Q_n are integers.

Proof. We use induction on n. First, $\sqrt{d} - [\sqrt{d}] = 1/\alpha'_1$ so that the required result holds by setting $P_1 = [\sqrt{d}]$ and $Q_1 = d - [\sqrt{d}]^2$. We now assume as induction hypothesis that $\alpha'_n = (\sqrt{d} + P_n)/Q_n$. Since $\alpha'_n = a_n + 1/\alpha'_{n+1}$ we have to find two integers P_{n+1}, Q_{n+1} such that

$$\frac{\sqrt{d}+P_n}{Q_n} = a_n + \frac{Q_{n+1}}{\sqrt{d}+P_{n+1}},$$

and

$$d - P_{n+1}^2 \equiv 0 \pmod{Q_{n+1}}. \tag{1}$$

This means that we have to find P_{n+1}, Q_{n+1} so that

$$d + P_nP_{n+1} = a_nQ_nP_{n+1} + Q_nQ_{n+1}, \tag{2}$$

$$P_n + P_{n+1} = a_nQ_n \tag{3}$$

and (1) hold. On subtracting P_{n+1} times equation (3) from (2) we have

$$d - P_{n+1}^2 = Q_nQ_{n+1}. \tag{4}$$

If (4) holds, then (1) also holds; and (2) also follows from (3) and (4). It remains therefore to find P_{n+1}, Q_{n+1} so that (3) and (4) are satisfied.

We solve (3) for P_{n+1}. From $P_n^2 \equiv P_{n+1}^2 \pmod{Q_n}$ we see that $d - P_{n+1}^2 \equiv 0 \pmod{Q_n}$ so that there exists Q_{n+1} satisfying (4). The theorem is proved. □

Theorem 8.2. *The equation $x^2 - dy^2 = (-1)^nQ_n$ is always soluble. If $l \neq (-1)^nQ_n$ and $|l| < \sqrt{d}$, then the equation $x^2 - dy^2 = l$ has no solution.*

Proof. We have

$$\sqrt{d} = \frac{p_{n-1}\alpha_n' + p_{n-2}}{q_{n-1}\alpha_n' + q_{n-2}} = \frac{p_{n-1}(\sqrt{d}+P_n) + p_{n-2}Q_n}{q_{n-1}(\sqrt{d}+P_n) + q_{n-2}Q_n},$$

and since $\sqrt{d}$ is irrational, we have, on clearing the denominators,

$$p_{n-1} = q_{n-1}P_n + q_{n-2}Q_n, \qquad dq_{n-1} = p_{n-1}P_n + p_{n-2}Q_n.$$

On subtracting q_{n-1} times the second equation from p_{n-1} times the first equation we have

$$p_{n-1}^2 - dq_{n-1}^2 = (p_{n-1}q_{n-2} - p_{n-2}q_{n-1})Q_n = (-1)^nQ_n.$$

The last part of the theorem follows from Theorem 7.3. □

Theorem 8.3. *Let k be the period of the continued fraction expansion for $\sqrt{d}$. Let $n > L$ and $p_{n-1}^2 - dq_{n-1}^2 = (-1)^nQ_n$. Then $p_{n-1+lk}^2 - dq_{n-1+lk}^2 = (-1)^{n+lk}Q_n$.*

Proof. This follows at once from

$$\frac{\sqrt{d}+P_n}{Q_n} = \frac{\sqrt{d}+P_{n+lk}}{Q_{n+lk}}. \quad \square$$

10.9 Pell's Equation

We shall now consider Pell's equation

$$x^2 - dy^2 = \pm 1. \tag{1}$$

From Theorem 8.3, there exists Q such that the equation $x^2 - dy^2 = Q$ has infinitely many solutions. If we partition these solutions into Q^2 classes mod $|Q|$, then there must be one class with at least two solutions. That is, there are integers x_1, y_1 and x_2, y_2 such that

$$x_1^2 - dy_1^2 = x_2^2 - dy_2^2 = Q, \qquad x_1 > 0, \quad y_1 > 0, \quad x_2 > 0, \quad y_2 > 0,$$

and

$$x_1 \equiv x_2 \pmod{|Q|}, \qquad y_1 \equiv y_2 \pmod{|Q|}, \qquad x_1 \neq x_2.$$

We now show that

$$x = \frac{x_1x_2 - dy_1y_2}{Q}, \qquad y = \frac{x_1y_2 - x_2y_1}{Q}$$

are solutions to Pell's equation (1). First, we have

$$x_1x_2 - dy_1y_2 \equiv x_1^2 - dy_1^2 = Q \equiv 0 \pmod{|Q|},$$
$$x_1y_2 - x_2y_1 \equiv x_1y_1 - x_1y_1 \equiv 0 \pmod{|Q|},$$

so that x, y are integers. Secondly we have

$$Q^2(x^2 - dy^2) = (x_1x_2 - dy_1y_2)^2 - d(x_1y_2 - x_2y_1)^2$$
$$= (x_1^2 - dy_1^2)(x_2^2 - dy_2^2) = Q^2,$$

so that x, y are solutions to (1). Finally, they are not the trivial solutions $x = \pm 1$, $y = 0$, because $y = 0$ implies $x_1y_2 - x_2y_1 = 0$, and from $(x_1, y_1) = (x_2, y_2) = 1$ we deduce that $x_1 = x_2$, $y_1 = y_2$ contrary to our assumption. We have therefore proved

Theorem 9.1. *The Pell's equation $x^2 - dy^2 = 1$ has a non-trivial solution.* □

From Theorem 7.3 we see that $x/y = p_{n-1}/q_{n-1}$ must be a convergent of $\sqrt{d}$, and from Theorem 8.2 we know that there exists n such that $(-1)^n Q_n = 1$.

Theorem 9.2. *Let n be the least positive integer satisfying $(-1)^n Q_n = 1$. Then all the solutions to the equation $x^2 - dy^2 = 1$ are given by*

$$x + \sqrt{d}y = \pm(p_{n-1} + \sqrt{d}q_{n-1})^l, \qquad l \gtreqless 0.$$

Proof. Let $\varepsilon = p_{n-1} + \sqrt{d}\,q_{n-1} > 1$. Because $\pm 1/(x + \sqrt{d}y) = \pm(x - \sqrt{d}y)$, it suffices to show that all positive solutions to $x^2 - dy^2 = 1$ are given by $x + y\sqrt{d} = \varepsilon^m$ $(m > 0)$.

Let (x, y) be such a solution, so that $x + y\sqrt{d} > 1$. We may choose m so that $\varepsilon^m \leqslant x + y\sqrt{d} < \varepsilon^{m+1}$ or $1 \leqslant \varepsilon^{-m}(x + y\sqrt{d}) < \varepsilon$. Let

$$\varepsilon^{-m}(x + y\sqrt{d}) = (x_0 - y_0\sqrt{d})(x + y\sqrt{d}) = X + Y\sqrt{d},$$

and we shall prove that $X + Y\sqrt{d} = 1$. Since $\sqrt{d}$ is irrational, it follows that

$$(x_0 + y_0\sqrt{d})(x - y\sqrt{d}) = X - Y\sqrt{d}.$$

On multiplying the equations together we have $X^2 - dY^2 = 1$. Suppose now that $1 < X + \sqrt{d}\,Y < \varepsilon$. Then

$$0 < \varepsilon^{-1} < (X + \sqrt{d}\,Y)^{-1} = X - \sqrt{d}\,Y < 1.$$

We deduce easily that

$$2X = (X + \sqrt{d}\,Y) + (X - \sqrt{d}\,Y) > 1 + \varepsilon^{-1} > 0,$$

$$2\sqrt{d}\,Y = (X + \sqrt{d}\,Y) - (X - \sqrt{d}\,Y) > 1 - 1 = 0.$$

It follows from these that

$$X^2 - dY^2 = 1, \qquad X > 0, \qquad Y > 0$$

and

$$1 < X + \sqrt{d}\,Y < p_{n-1} + \sqrt{d}\,q_{n-1}.$$

Now $x = \sqrt{1 + dy^2}$ increases with y, so that $x + \sqrt{d}y$ increases as y increases. We deduce from the above that $Y < q_{n-1}$ and $X < p_{n-1}$, so that X/Y is a convergent with denominator less than q_{n-1}. This is impossible; therefore $X + \sqrt{d}\,Y = 1$. □

We see from the above that the equation $x^2 - dy^2 = 1$ is always soluble, but the equation $x^2 - dy^2 = -1$ may have no solution. For example, since $x^2 \equiv 0, 1 \pmod 4$ so that $x^2 - 3y^2 \equiv x^2 + y^2 \equiv 0, 1, 2 \pmod 4$, we see that the equation $x^2 - 3y^2 = -1$ is insoluble. In fact this example shows that $x^2 - dy^2 = -1$ is insoluble whenever $d \equiv 3 \pmod 4$.

However if x_0, y_0 satisfy $x_0^2 - dy_0^2 = -1$, then, by defining x_1, y_1 with $x_1 + \sqrt{d}y_1 = (x_0 + \sqrt{d}y_0)^2$ we see that $x_1^2 - dy_1^2 = 1$. It is not difficult to prove that if $x^2 - dy^2 = -1$ is soluble, then all the solutions to $x^2 - dy^2 = \pm 1$ are given by $\pm(p_{n-1} + \sqrt{d}\,q_{n-1})^l$ where n is the least positive integer satisfying $(-1)^n Q_n = -1$.

10.10 Chebyshev's Theorem and Khintchin's Theorem

Let ϑ be an irrational number. According to Theorem 2.4 there are infinitely many integers x, y satisfying

$$|x\vartheta - y| < \frac{1}{x}, \qquad y = [x\vartheta], \qquad (x, y) = 1. \tag{1}$$

It follows at once from this that, if $\varepsilon > 0$, then there exists an integer x such that $x\vartheta$ differs from $[x\vartheta]$ by less than ε. In other words the number 0 is a limit point of the point set

$$x\vartheta - [x\vartheta], \qquad x = 1, 2, 3, \ldots. \tag{2}$$

An immediate problem arising from this is the determination of the set of limit points of the point set (2). For this Chebyshev has proved that each point in the interval $(0, 1)$ is a limit point of the point set (2). In fact he proved the following stronger result.

Theorem 10.1. *Let ϑ be any irrational number and β be any real number. Then there are infinitely many integers x, y satisfying*

$$|\vartheta x - y - \beta| < \frac{3}{x}. \tag{3}$$

Proof. By Theorem 2.4 there are infinitely many integers $p, q > 0$ such that

$$\vartheta = \frac{p}{q} + \frac{\delta}{q^2}, \qquad |\delta| < 1, \qquad (p, q) = 1. \tag{4}$$

For fixed q and β we may choose an integer t such that $|q\beta - t| \leqslant \frac{1}{2}$ so that

$$\beta = \frac{t}{q} + \frac{\delta'}{2q}, \qquad |\delta'| \leqslant 1. \tag{5}$$

Since p, q are coprime, there exist integers x, y such that

$$\frac{q}{2} \leqslant x < \frac{3}{2}q, \qquad px - qy = t. \tag{6}$$

From (4) and (5) we have

$$\begin{aligned} |\vartheta x - y - \beta| &= \left| \frac{xp}{q} + \frac{x\delta}{q^2} - y - \frac{t}{q} - \frac{\delta'}{2q} \right| \\ &= \left| \frac{x\delta}{q^2} - \frac{\delta'}{2q} \right| < \frac{x}{q^2} + \frac{1}{2q}. \end{aligned}$$

Since $q > 2x/3$, we have $|\vartheta x - y - \beta| < 9/4x + 3/4x = 3/x$. Since q can be arbitrarily large, and $x \geqslant q/2$ by (6), the theorem is proved. □

According to this theorem there exists a constant c such that given any irrational ϑ and any real β the inequality

$$|\vartheta x - y - \beta| < \frac{c}{x} \tag{7}$$

has infinitely many integer solutions in $x > 0$ and y. In (3) we have $c = 3$, and we see from Theorem 4.4 that c must be at least $1/\sqrt{5}$. Khintchin has proved the following

Theorem 10.2. *Let ϑ be irrational, β be real and $\varepsilon > 0$. Then the inequality*

$$|x\vartheta - y - \beta| < \frac{1+\varepsilon}{\sqrt{5}\,x} \tag{8}$$

has infinitely many solutions in integers $x > 0$, and y.

Proof. By Theorem 4.3 there are infinitely many coprime pair of integers p, q such that $\vartheta = p/q + \delta/q^2$, where $0 < |\delta| < 1/\sqrt{5}$. We may assume that $\delta > 0$ since otherwise we can replace ϑ, β by $-\vartheta$, $-\beta$. Let ξ_1, ξ_2 be real numbers satisfying $\xi_2 - \xi_1 \geqslant 1$, and we shall specify them later. We can choose x, y such that

$$px - qy = [q\beta], \qquad \xi_1 q \leqslant x < \xi_2 q. \tag{9}$$

Then we have

$$\begin{aligned} |x\vartheta - y - \beta| &= \left|\frac{p}{q}x + \frac{\delta x}{q^2} - y - \frac{[q\beta]}{q} - \frac{\tau}{q}\right| \\ &= \frac{1}{q}\left|\frac{x\delta}{q} - \tau\right| = \frac{1}{x}\cdot\frac{x}{q}\left|\frac{x\delta}{q} - \tau\right|, \end{aligned} \tag{10}$$

where $\tau = q\beta - [q\beta]$. We want to show that

$$-\frac{1}{\sqrt{5}} \leqslant \frac{x}{q}\left(\frac{x\delta}{q} - \tau\right) < \frac{1}{\sqrt{5}},$$

or

$$\frac{\tau^2}{4\delta} - \frac{1}{\sqrt{5}} \leqslant \frac{x^2\delta}{q^2} - \frac{x\tau}{q} + \frac{\tau^2}{4\delta} < \frac{\tau^2}{4\delta} + \frac{1}{\sqrt{5}}.$$

1) Let us first assume that $\tau^2 \geqslant 4\delta/\sqrt{5}$ so that the left hand side in the above is positive and we have to show that

$$\sqrt{\frac{\tau^2}{4\delta} - \frac{1}{\sqrt{5}}} \leqslant \frac{x\sqrt{\delta}}{q} - \frac{\tau}{2\sqrt{\delta}} < \sqrt{\frac{\tau^2}{4\delta} + \frac{1}{\sqrt{5}}},$$

or

$$\frac{1}{\sqrt{\delta}}\left(\frac{\tau}{2\sqrt{\delta}}+\sqrt{\frac{\tau^2}{4\delta}-\frac{1}{\sqrt{5}}}\right)\leqslant\frac{x}{q}<\frac{1}{\sqrt{\delta}}\left(\frac{\tau}{2\sqrt{\delta}}+\sqrt{\frac{\tau^2}{4\delta}+\frac{1}{\sqrt{5}}}\right).$$

Let

$$\xi_1=\frac{1}{\sqrt{\delta}}\left(\frac{\tau}{2\sqrt{\delta}}+\sqrt{\frac{\tau^2}{4\delta}-\frac{1}{\sqrt{5}}}\right);$$

$$\xi_2=\frac{1}{\sqrt{\delta}}\left(\frac{\tau}{2\sqrt{\delta}}+\sqrt{\frac{\tau^2}{4\delta}-\frac{1}{\sqrt{5}}}\right).$$

We now examine how we can make $\xi_2-\xi_1\geqslant 1$. On simplifying

$$\frac{1}{\sqrt{\delta}}\left(\sqrt{\frac{\tau^2}{4\delta}+\frac{1}{\sqrt{5}}}-\sqrt{\frac{\tau^2}{4\delta}-\frac{1}{\sqrt{5}}}\right)>1$$

(the left hand side is merely $\xi_2-\xi_1$) we obtain $\tau^2<\frac{4}{5}+\delta^2$. Since the numbers involved in the simplification are positive we see that the result is established if $4\delta/\sqrt{5}\leqslant\tau^2<\frac{4}{5}+\delta^2$; that is if

$$2\sqrt{\frac{\delta}{\sqrt{5}}}\leqslant\tau<\sqrt{\frac{4}{5}+\delta^2}\,.$$

We are left with the two cases $\tau^2<4\delta/\sqrt{5}$ and $\sqrt{\frac{4}{5}+\delta^2}\leqslant\tau<1$ to consider.

2) Suppose that $\tau^2<4\delta/\sqrt{5}$. From $\tau>0$ we have

$$\xi=\frac{1}{\sqrt{\delta}}\left(\frac{\tau}{2\sqrt{\delta}}+\sqrt{\frac{\tau^2}{4\delta}+\frac{1}{\sqrt{5}}}\right)>\frac{1}{\sqrt{\delta}}\sqrt{\frac{1}{\sqrt{5}}}>1.$$

Let $\eta>0$, and take $\xi_1=\eta$, $\xi_2=\eta+\xi$ so that $\xi_2-\xi_1=\xi>1$. Therefore the number x in (9) exists, and by our assumption

$$\frac{x}{q}\left(\frac{x\delta}{q}-\tau\right)=\left(\frac{x\sqrt{\delta}}{q}-\frac{\tau}{2\sqrt{\delta}}\right)^2-\frac{\tau^2}{4\delta}>-\frac{1}{\sqrt{5}}.$$

On the other hand we take $y=ax+b$ so that, as x varies in an interval, y^2 attains the maximum value at the end points of the interval, and therefore

$$\begin{aligned}\frac{x}{q}\left(\frac{x\delta}{q}-\tau\right)&=\left(\frac{x\sqrt{\delta}}{q}-\frac{\tau}{2\sqrt{\delta}}\right)^2-\frac{\tau^2}{4\delta}\\&\leqslant\max\left\{\left(\eta\sqrt{\delta}-\frac{\tau}{2\sqrt{\delta}}\right)^2-\frac{\tau^2}{4\delta},\left((\eta+\xi)\sqrt{\delta}-\frac{\tau}{2\sqrt{\delta}}\right)^2-\frac{\tau^2}{4\delta}\right\}\\&=\max\left\{\eta^2\delta-\eta\tau,\left(\sqrt{\frac{\tau^2}{4\delta}+\frac{1}{\sqrt{5}}}+\eta\sqrt{\delta}\right)^2-\frac{\tau^2}{4\delta}\right\}\\&=\frac{1}{\sqrt{5}}+O(\eta).\end{aligned}$$

Since η is arbitrary, the theorem is proved for this case.

3) Suppose that $\sqrt{\frac{4}{5} + \delta^2} \leqslant \tau < 1$. From $\delta < 1/\sqrt{5}$ we have

$$\tau \geqslant \sqrt{\frac{4}{5} + \delta^2} > \sqrt{\left(1 - \frac{1}{\sqrt{5}}\right)^2 + 2\delta\left(1 - \frac{1}{\sqrt{5}}\right) + \delta^2} = 1 - \frac{1}{\sqrt{5}} + \delta$$

or $1 - \tau < 1/\sqrt{5} - \delta$. Let $\eta > 0$. We may specify x and y such that

$$px - qy = [q\beta] + 1, \qquad \eta q \leqslant x < (1 + \eta)q,$$

and similarly to (10) we have

$$|x\vartheta - y - \beta| = \left|\frac{x\delta}{q^2} + \frac{1 - \tau}{q}\right| = \frac{1}{q}\left(\frac{x\delta}{q} + (1 - \tau)\right)$$

$$< \frac{1}{q}\left\{(1 + \eta)\delta + \frac{1}{\sqrt{5}} - \delta\right\} \leqslant \frac{1}{q}(1 + \eta)\frac{1}{\sqrt{5}} < \frac{(1 + \eta)^2}{x\sqrt{5}}.$$

Since η is arbitrary, the theorem is proved. □

Exercise. Let ϑ be an irrational number such that, given any $\varepsilon > 0$, there always exist integers x, y satisfying $|x\vartheta - y| < \varepsilon/x$. Prove that if $\delta > 0$ and β is real, then there exist integers x, y such that $|x\vartheta - y - \beta| < (1 + \delta)/3x$.

10.11 Uniform Distributions and the Uniform Distribution of $n\vartheta$ (mod 1)

Chebyshev's theorem in the last section states that the point set $\{x\vartheta\} = x\vartheta - [x\vartheta], x = 1, 2, 3, \ldots$ is *dense* in the interval $(0, 1)$, in the sense that each point in $(0, 1)$ is a limit point of the set. We may ask about the distribution of this point set in the interval $(0, 1)$. In other words, if (a, b) is a subinterval of $(0, 1)$, then as x takes the values $1, 2, \ldots, n$ does the interval (a, b) receive the "correct proportion" of points? Let us define precisely what we mean by the "correct proportion".

Definition. Let P_i $(i = 1, 2, 3, \ldots)$ be a point set in the interval $(0, 1)$. Let $0 \leqslant a < b \leqslant 1$, and for each positive integer n denote by $N_n(a, b)$ the number of points $P_1, P_2, \ldots, P_n$ that lie in the interval (a, b). If $\lim_{n\to\infty} N_n(a, b)/n = b - a$ always holds, then we say that the point set P_i $(i = 1, 2, 3, \ldots)$ is *uniformly distributed* in $(0, 1)$.

We shall now prove the following

Theorem 11.1. *Let ϑ be irrational. Then the point set $\{x\vartheta\} = x\vartheta - [x\vartheta]$; $x = 1, 2, 3, \ldots$, is uniformly distributed in $(0, 1)$.*

Proof. Let (a, b) be any subinterval of $(0, 1)$. By Theorem 4.1 there are infinitely many pairs of integers $p, q > 0$ such that

$$\vartheta = \frac{p}{q} + \frac{\delta}{q^2}, \qquad |\delta| < 1, \qquad (p, q) = 1.$$

Let u, v be integers satisfying

$$\frac{u-1}{q} < a \leqslant \frac{u}{q} < \frac{v}{q} \leqslant b < \frac{v+1}{q};$$

let $n = rq + s, 0 \leqslant s < q$, and let $0 \leqslant j < r$. Consider now a complete residue system (mod q) namely $jq, jq + 1, \ldots, jq + q - 1$. It is easy to see that

$$\{(jq + k)\vartheta\} = \left\{\frac{kp}{q} + \frac{j\delta}{q} + \frac{k\delta}{q^2}\right\} = \left\{\frac{kp + [j\delta]}{q} + \frac{\delta'}{q}\right\}, \qquad |\delta'| < 2.$$

Since $[j\delta]$ does not depend on k, as k runs over $0, 1, \ldots, q - 1$, the points $pk + [j\delta]$ runs over a complete residue system (mod q). Therefore, in the q numbers $\{(jq + k)\vartheta\}$ those that lie in the interval (a, b) must be more than $v - u - 4$ and less than $v - u + 6$. It follows that as x takes the values $1, 2, \ldots, n$ the number of numbers $\{x\vartheta\}$ that lie in the interval (a, b) exceeds

$$r(v - u - 4) = \frac{n}{q}(v - u - 4) - \frac{s}{q}(v - u - 4)$$

$$\geqslant n(b - a) - \frac{6}{q}n - \frac{v - u - 4}{n}n,$$

but is less than

$$(r + 1)(v - u + 6) \leqslant n\left(\frac{v - u}{q} + \frac{6}{q}\right) + v - u + 6 \leqslant n(b - a)$$

$$+ \frac{6}{q}n + \frac{v - u + 6}{n}n.$$

Let $\varepsilon > 0$. We choose $q > 12/\varepsilon$ and then choose $n > 2(q + 6)/\varepsilon$. It follows that we have $n(b-a) - n\varepsilon \leqslant N_n(a, b) \leqslant n(b-a) + n\varepsilon$. This proves that $\lim_{n\to\infty} N_n(a, b)/n = b - a$. □

10.12 Criteria for Uniform Distributions

Theorem 12.1 (Weyl). *A necessary and sufficient condition for the sequence (x_n), $0 \leqslant x_n \leqslant 1$ to be uniformly distributed in $(0, 1)$ is that the equation*

$$\lim_{n\to\infty} \frac{f(x_1) + \cdots + f(x_n)}{n} = \int_0^1 f(x)\,dx \tag{1}$$

holds for every Riemann integrable function $f(x)$ in $(0, 1)$.

Proof. We first establish the necessity of the condition (1).

1) Let $f(x)$ be defined to be c or 0 according to whether $a \leqslant x \leqslant b$ or not. Then clearly

$$\lim_{n\to\infty} \frac{f(x_1) + \cdots + f(x_n)}{n} = c \lim_{n\to\infty} \frac{N_n(a,b)}{n} = c(b-a),$$

and

$$\int_0^1 f(x)\,dx = c(b-a).$$

Therefore the equation (1) holds for this function $f(x)$.

2) The equation (1) is linear in the sense that if it holds for $f_1, \ldots, f_k$, then it holds for $c_1 f_1 + \cdots + c_k f_k$. From 1) we see that (1) holds for all step functions.

3) It is a simple exercise to show that if f is Riemann integrable, and $\varepsilon > 0$, then there are two step functions $\varphi_\varepsilon(x), \Phi_\varepsilon(x)$ such that $\varphi_\varepsilon(x) \leqslant f(x) \leqslant \Phi_\varepsilon(x)$ and

$$\int_0^1 (\Phi_\varepsilon(t) - \varphi_\varepsilon(t))\,dt < \varepsilon.$$

From 2) we see that (1) holds for $\varphi_\varepsilon(x)$ and $\Phi_\varepsilon(x)$ so that

$$\begin{aligned}\int_0^1 \varphi_\varepsilon(t)\,dt &= \lim_{n\to\infty} \frac{1}{n}(\varphi_\varepsilon(x_1) + \cdots + \varphi_\varepsilon(x_n))\\ &\leqslant \lim_{n\to\infty} \frac{1}{n}(f(x_1) + \cdots + f(x_n))\\ &\leqslant \lim_{n\to\infty} \frac{1}{n}(\Phi_\varepsilon(x_1) + \cdots + \Phi_\varepsilon(x_n)) = \int_0^1 \Phi_\varepsilon(t)\,dt.\end{aligned}$$

Since

$$\int_0^1 \Phi_\varepsilon(t)\,dt \leqslant \int_0^1 f(x)\,dx \leqslant \int_0^1 \Phi_\varepsilon(x)\,dx,$$

it follows that

$$\left| \lim_{n\to\infty} \frac{f(x_1) + \cdots + f(x_n)}{n} - \int_0^1 f(x)\,dx \right| < \varepsilon.$$

The necessity part of the theorem is therefore proved. The sufficiency part is easy:

we let $f(x) = 1$ or 0 according to whether $a \leqslant x \leqslant b$ or otherwise. Then equation (1) becomes $\lim_{n\to\infty} N_n(a,b)/n = b - a$. □

It is very difficult to make a direct application of this theorem. This is because we have to verify (1) for the whole family of Riemann integrable functions. Actually the proof of the theorem shows that we need only restrict ourselves to step functions, and in fact it suffices to have the basis for a linear space of functions which includes the Riemann integrable functions as limits. This is embodied in the next theorem.

Theorem 12.2 (Weyl). *Under the hypothesis of Theorem* 12.1, *another necessary and sufficient condition is that the equation* (1) *holds for the functions* $f(x) = e^{2\pi imx}$ $(m = \pm 1, \pm 2, \ldots)$. *In other words, a necessary and sufficient condition for the sequence* (x_n), $0 \leqslant x \leqslant 1$ *to be uniformly distributed in* $(0, 1)$ *is that*

$$\lim_{n\to\infty} \frac{1}{n}\left|\sum_{v=1}^{n} e^{2\pi imx_v}\right| = 0$$

holds for all $m \neq 0$.

Proof. There is no need to prove the necessity part. For the sufficiency part we define

$$g(x) = \begin{cases} 1, & \text{if } 0 \leqslant x < a, \\ 0, & \text{if } a \leqslant x < 1. \end{cases}$$

Then

$$\lim_{n\to\infty} \frac{g(x_1) + \cdots + g(x_n)}{n} = \lim_{n\to\infty} \frac{N_n(0,a)}{n}.$$

It is clear that we need only prove that

$$\lim_{n\to\infty} \frac{g(x_1) + \cdots + g(x_n)}{n} = a.$$

We now construct a continuous periodic function $g_{\eta,\delta}(x)$ with period 1 to approximate $g(x)$. We define

$$g_{\eta,\delta}(x) = \begin{cases} (x - \eta + \delta)/\delta, & \text{if } \eta - \delta \leqslant x \leqslant \eta, \\ 1, & \text{if } \eta \leqslant x \leqslant a - \eta, \\ -(x - a + \eta - \delta)/\delta, & \text{if } a - \eta \leqslant x \leqslant a - \eta + \delta, \\ 0, & \text{if } a - \eta + \delta \leqslant x \leqslant \eta - \delta + 1. \end{cases}$$

Here $0 < \delta \leqslant \frac{1}{2}\min(a, 1 - a)$, $0 \leqslant \eta \leqslant \delta$. Clearly

$$g_{\delta,\delta}(x) \leqslant g(x) \leqslant g_{0,\delta}(x).$$

Since $g_{\eta,\delta}(x)$ is continuous, it follows that

$$g_{\eta,\delta}(x) = \sum_{n=-\infty}^{\infty} C_n e^{2\pi inx},$$

where

$$C_0 = \int_{\eta-\delta}^{\eta-\delta+1} g_{\eta,\delta}(x)\,dx = a + \delta - 2\eta;$$

and when $n \neq 0$,

$$C_n = \int_{\eta-\delta}^{\eta-\delta+1} e^{-2\pi inx} g_{\eta,\delta}(x)\,dx = \frac{e^{-n\pi ia}}{\delta(n\pi)^2} \sin n\pi(a + \delta - 2\eta) \sin n\pi\delta.$$

It follows that $|C_n| \leqslant 1/\delta(n\pi)^2$ and

$$S_{\eta,\delta}(x) = \frac{g_{\eta,\delta}(x_1) + \cdots + g_{\eta,\delta}(x_k)}{k} = \frac{1}{k} \sum_{j=1}^{k} \sum_{n=-\infty}^{\infty} C_n e^{2\pi inx_j}$$
$$= \frac{1}{k} \sum_{n=-\infty}^{\infty} C_n \sum_{j=1}^{k} e^{2\pi inx_j}.$$

Thus we have

$$S_{\eta,\delta}(x) = C_0 + \sum_{\substack{n=-N \\ n\neq 0}}^{N} C_n \frac{1}{k} \sum_{j=1}^{k} e^{2\pi inx_j} + \sum_{|n|>N} C_n \frac{1}{k} \sum_{j=1}^{k} e^{2\pi inx_j}.$$

We observe that

$$\left| \sum_{|n|>N} C_n \frac{1}{k} \sum_{j=1}^{k} e^{2\pi inx_j} \right| \leqslant \frac{2}{\delta\pi^2} \sum_{n>N} \frac{1}{n^2}.$$

Let $\varepsilon > 0$ and choose N so that the right hand side of this inequality is less than ε. With N fixed we see from

$$\lim_{k\to\infty} \frac{1}{k} \sum_{j=1}^{k} e^{2\pi inx_j} = 0$$

that for all large k,

$$\left| \sum_{\substack{n=-N \\ n\neq 0}}^{N} C_n \frac{1}{k} \sum_{j=1}^{k} e^{2\pi inx_j} \right| < \varepsilon.$$

Thus, given any pair of fixed η, δ we have

$$|S_{\eta,\delta}(x) - (a + \delta - 2\eta)| < 2\varepsilon$$

or

$$\lim_{k\to\infty} S_{\eta,\delta}(x) = a + \delta - 2\eta.$$

Now let

$$S(x) = \frac{g(x_1) + \cdots + g(x_k)}{k}.$$

From $S_{\delta,\delta}(x) \leqslant S(x) \leqslant S_{0,\delta}(x)$ we deduce that

$$a - \delta \leqslant \varliminf_{k\to\infty} S \leqslant \varlimsup_{k\to\infty} S \leqslant a + \delta$$

for any δ. Therefore $\lim_{k\to\infty} S = a$ as required. □

For a clearer description of uniform distribution it is best to use the unit circle to represent the interval. Let $\xi_n = e^{2\pi i x_n}$, $n = 1, 2, \ldots$ so that the sequence (x_n) in $0 \leqslant x_n \leqslant 1$ is now transformed into a sequence on the unit circle. An advantage of using this description is the removal of the special properties of the end points 0, 1 in the interval (0, 1). Take any arc of the unit circle with length $2\pi\alpha$ ($\alpha < 1$). Then any uniformly distributed sequence will have the proportion α of its points on this arc. Moreover, since $e^{2\pi i x_n} = e^{2\pi i (x_n + d)}$, it does not even matter if the sequence (x_n) lie outside the interval (0, 1). In other words we may define uniform distribution of $f(x)$, mod 1 by the uniform distribution of the fractional parts of $f(x)$ in (0, 1). A necessary and sufficient condition for the uniform distribution of $f(x)$, mod 1 is then

$$\lim_{n\to\infty} \frac{1}{n} \sum_{x=1}^{n} e^{2\pi i m f(x)} = 0, \qquad m \neq 0.$$

An interpretation of this condition is that the centre of gravity of the sequence of points $e^{2\pi i m f(x)}$, $x = 1, 2, \ldots, (m \neq 0)$ is the centre of the circle. It is clear that if $f(x)$ is uniformly distributed mod 1, then so is $mf(x)$ for any non-zero integer m.

The most interesting unsolved problem concerning this is whether e^x is uniformly distributed mod 1.

Theorem 12.3. *A necessary and sufficient condition for the uniform distribution of* $f(x)$, mod 1 *is that*

$$\lim_{n\to\infty} \frac{1}{n} \sum_{x=1}^{n} \{f(x) + a\} = \frac{1}{2}, \qquad 0 \leqslant a \leqslant 1.$$

Proof. Necessity. Let $f(x)$ be uniformly distributed, mod 1. Then $f(x) + a$ is also uniformly distributed, mod 1. Therefore we need only establish the case $a = 0$. Let $x_m = \{f(m)\}$. Then, by Theorem 12.1, we have

$$\lim_{n\to\infty} \frac{1}{n} \sum_{x=1}^{n} \{f(x)\} = \int_0^1 x\,dx = \frac{1}{2}.$$

Sufficiency. Let $0 \leqslant b \leqslant 1$. Then

$$\frac{1}{n} \sum_{x=1}^{n} \{f(x) + 1 - b\} = \frac{1}{n} \sum\nolimits_1 (\{f(x)\} + 1 - b) + \frac{1}{n} \sum\nolimits_2 (\{f(x)\} - b),$$

where in $\sum_1$, x runs through those integers $1, 2, \ldots, n$ such that $\{f(x)\} < b$, and in $\sum_2$, x runs through those integers $1, 2, \ldots, n$ such that $\{f(x)\} \geqslant b$. We see therefore that

$$\frac{1}{n} \sum_{x=1}^{n} \{f(x) + 1 - b\} = n^{-1} \sum_{x=1}^{n} \{f(x)\} + n^{-1} N_n(0, b) - b.$$

Letting $n \to \infty$ and observing the hypothesis we see that

$$\lim_{n\to\infty} \frac{1}{n} N_n(0, b) = b$$

as required. □

Chapter 11. Indeterminate Equations

11.1 Introduction

By indeterminate equations we mean equations in which the number of unknowns occurring exceed the number of equations given, and that these unknowns are subject to further constraints such as being integers, or positive integers, or rationals etc. Apart from equations of the first and second degrees, the discussion on indeterminate equations is very scattered. The complicated nature of the subject is illustrated by the fact that Volume II of Dickson's *History of Number Theory* devotes over eight hundred pages on such equations. The study of these equations has a long history. In the third century Diophantus attempted a systematic study and in fact nowadays indeterminate equations are often called *Diophantine equations*. In our country indeterminate equations have an even longer history; for example Soon Go gave the general solution of $x^2 + y^2 = z^2$ in integers x, y, z much earlier than the west.

11.2 Linear Indeterminate Equations

From Theorem 2.6.2 we see that a necessary and sufficient condition for the equation $a_1x_1 + a_2x_2 + \cdots + a_nx_n = N$ to have a solution is that $(a_1, \ldots, a_n) | N$.

Suppose now that $a_1 > 0, \ldots, a_n > 0$, $(a_1, \ldots, a_n) = 1$. We ask for the asymptotic formula for the number of solutions to the equation

$$a_1x_1 + a_2x_2 + \cdots + a_nx_n = N, \qquad x_\nu \geqslant 0 \quad (\nu = 1, 2, \ldots, n). \tag{1}$$

Theorem 2.1. *Let* $(a_1, \ldots, a_n) = 1$, *and denote by* $A(N)$ *the number of solutions to* (1). *Then we have*

$$\lim_{N \to \infty} \frac{A(N)}{N^{n-1}} = \frac{1}{a_1 a_2 \cdots a_n (n-1)!}.$$

Proof. 1) Since $(a_1, \ldots, a_n) = 1$, the number $A(N)$ is the coefficient of x^N in the power series for

$$f(x) = \frac{1}{1 - x^{a_1}} \cdot \frac{1}{1 - x^{a_2}} \cdots \frac{1}{1 - x^{a_n}}.$$

Let $1, \zeta_1, \zeta_2, \ldots, \zeta_t$ be the roots of $(1-x^{a_1})\cdots(1-x^{a_n})=0$, with multiplicities $n, l_1, l_2, \ldots, l_t$ respectively. Since $(a_1, \ldots, a_n)=1$ we have $l_i \leqslant n-1$ $(i=1,2,\ldots,t)$. We have, by partial fractions,

$$f(x)=\frac{A_n}{(1-x)^n}+\cdots+\frac{A_1}{1-x}+\frac{B_{l_1}}{(\zeta_1-x)^{l_1}}+\cdots+\frac{B_1}{\zeta_1-x}$$

$$+\cdots$$

$$+\frac{P_{l_t}}{(\zeta_t-x)^{l_t}}+\cdots+\frac{P_1}{\zeta_t-x}, \tag{2}$$

where $A, B, \ldots, P$ are constants.

2) Denote by $\psi(N)$ the coefficient of x^N in the power series expansion of

$$\frac{A}{(\alpha-x)^l}=A\alpha^{-l}\left(1-\frac{x}{\alpha}\right)^{-l}.$$

Then, by the binomial theorem expansion, we have

$$\psi(N)=A\alpha^{-l}\frac{(-l)(-l-1)\cdots(-l-N+1)}{N!}\left(-\frac{1}{\alpha}\right)^N$$

$$=A\alpha^{-l}\frac{(N+l-1)(N+l-2)\cdots(N+1)}{(l-1)!}\left(\frac{1}{\alpha}\right)^N,$$

so that

$$\lim_{N\to\infty}\frac{\psi(N)\cdot\alpha^{l+N}}{N^{l-1}}=\frac{A}{(l-1)!}. \tag{3}$$

Applying this to the various terms in (2) and observing that $l_i \leqslant n-1$ we see that

$$\lim_{N\to\infty}\frac{A(N)}{N^{n-1}}=\frac{A_n}{(n-1)!}$$

and from (2) we have

$$A_n=\lim_{x\to 1}\frac{(1-x)^n}{(1-x^{a_1})\cdots(1-x^{a_n})}=\frac{1}{a_1\cdots a_n}. \quad \square$$

Theorem 2.2. *Equation* (1) *is always soluble if N is sufficiently large.* $\square$

Exercise. Let $(a, b)=1$, $a>0$, $b>0$. Show that the number of solutions to $ax+by=N$, $x\geqslant 0$, $y\geqslant 0$ is given by

$$\frac{N-(bl+am)}{ab}+1$$

where l and m are the least non-negative solutions to $bl\equiv N \pmod{a}$ and $am\equiv N$ $\pmod{b}$ respectively.

11.3 Quadratic Indeterminate Equations

We shall solve the equation

$$ax^2 + bxy + cy^2 + dx + ey + f = 0. \tag{1}$$

We write $D = b^2 - 4ac$. If $D = 0$, then we multiply (1) by $4a$ giving $(2ax + by)^2 + 4adx + 4aey + 4af = 0$, which is not a difficult equation to solve. Let $2ax + by = t$ so that

$$t^2 + 2(2ae - bd)y + 4af = -2dt,$$

$$(t + d)^2 = 2(bd - 2ae)y + d^2 - 4af.$$

The number t can be obtained from the congruence $(t + d)^2 \equiv d^2 - 4af \pmod{2(bd - 2ae)}$, and so x, y can be solved.

We now assume that $D \neq 0$. Multiplying (1) by D^2 we have

$$aD^2x^2 + bD^2xy + cD^2y^2 + dD^2x + eD^2y + fD^2 = 0. \tag{2}$$

Substituting $Dx = x' + 2cd - be$, $Dy = y' + 2ae - bd$ into (2) we have

$$a(x' + 2cd - be)^2 + b(x' + 2cd - be)(y' + 2ae - bd) + c(y' + 2ae - bd)^2$$
$$+ dD(x' + 2cd - be) + eD(y' + 2ae - bd) + fD^2 = 0,$$

or

$$ax'^2 + bx'y' + cy'^2 = k, \tag{3}$$

where

$$-k = a(2cd - be)^2 + b(2cd - be)(2ae - bd) + c(2ae - bd)^2$$
$$+ dD(2cd - be) + eD(2ae - bd) + fD^2.$$

We see therefore that whether (1) is soluble depends on whether (3) has solutions satisfying

$$x' \equiv be - 2cd, \qquad y' \equiv bd - 2ae \pmod{D}.$$

Our first priority is therefore to solve (3).

11.4 The Solutions to $ax^2 + bxy + cy^2 = k$

We shall solve

$$ax^2 + bxy + cy^2 = k. \tag{1}$$

Let $d = b^2 - 4ac$. We shall assume that d is not a perfect square, and that

$(a, b, c) = 1$. We need only find those solutions satisfying $(x, y) = 1$, and we call these the *proper solutions.*

Theorem 4.1. *Let x, y be a proper solution to (1). Then there are two uniquely determined integers s and r satisfying*

$$xs - yr = 1, \tag{2}$$

and the integer

$$l = (2ax + by)r + (bx + 2cy)s$$

satisfies

$$l^2 \equiv d \pmod{4k}, \qquad 0 \leqslant l < 2k. \tag{3}$$

Proof. Let r_0, s_0 be a solution to (2). Then the general solution to (2) is $r = r_0 + hx$, $s = s_0 + hy$ where h is any integer. Thus

$$l = (2ax + by)r_0 + (bx + 2cy)s_0 + 2h(ax^2 + bxy + cy^2) = l_0 + 2hk,$$

so that we may choose a unique h such that $0 \leqslant l < 2k$. Finally we have

$$\begin{aligned} l^2 &= [(2ax + by)r + (bx + 2cy)s]^2 \\ &= 4(ar^2 + brs + cs^2)(ax^2 + bxy + cy^2) + (b^2 - 4ac)(xs - yr)^2 \\ &\equiv d \pmod{4k}. \quad \square \end{aligned}$$

Theorem 4.2. *Let (x_1, y_1) and (x_2, y_2) be two proper solutions corresponding to the same number l in the previous theorem. Then we have*

$$2ax_1 + (b + \sqrt{d})y_1 = (2ax_2 + (b + \sqrt{d})y_2)\left(\frac{t + u\sqrt{d}}{2}\right), \tag{4}$$

where t and u are integers satisfying

$$t^2 - du^2 = 4. \tag{5}$$

Conversely, if (x_2, y_2) is a proper solution, then the numbers x_1, y_1 defined by (4) also give a proper solution and both solutions correspond to the same number l.

Proof. 1) We first show that

$$\begin{aligned} t &= ((2ax_1 + by_1)(2ax_2 + by_2) - dy_1 y_2)/2ak, \\ u &= -(x_1y_2 - x_2y_1)/k \end{aligned} \tag{6}$$

are the suitable integers; that is we show that t and u are integers satisfying (5). From

$$\frac{t \mp u\sqrt{d}}{2} = \frac{(2ax_1 + by_1)(2ax_2 + by_2) - dy_1y_2 \pm 2a(x_1y_2 - x_2y_1)\sqrt{d}}{4ak}$$
$$= \frac{(2ax_1 + by_1 \mp \sqrt{d}y_1)(2ax_2 + by_2 \pm \sqrt{d}y_2)}{(2ax_1 + by_1 + \sqrt{d}y_1)(2ax_1 + by_1 - \sqrt{d}y_1)}$$
$$= \frac{(2ax_1 + by_1 \mp \sqrt{d}y_1)(2ax_2 + by_2 \pm \sqrt{d}y_2)}{(2ax_2 + by_2 + \sqrt{d}y_2)(2ax_2 + by_2 - \sqrt{d}y_2)},$$

we see that (4) follows. Next from

$$\frac{t^2 - du^2}{4} = \frac{t + \sqrt{d}u}{2} \cdot \frac{t - \sqrt{d}u}{2} = 1,$$

we see that t and u satisfy (5). Also

$$\begin{aligned} 2ax_1 + by_1 &= (2ax_1 + by_1)(s_1x_1 - r_1y_1) \\ &= (2ax_1 + by_1)s_1x_1 - ly_1 + (bx_1 + 2cy_1)s_1y_1 \\ &\equiv -ly_1 \pmod{2k}. \end{aligned} \tag{7}$$

Similarly we have

$$2ax_2 + by_2 \equiv -ly_2 \pmod{2k}.$$

Therefore

$$2a(x_1y_2 - x_2y_1) \equiv 0 \pmod{2k},$$
$$(b + l)(x_1y_2 - x_2y_1) \equiv 0 \pmod{2k}.$$

Similarly we have

$$2c(x_1y_2 - x_2y_1) \equiv 0 \pmod{2k},$$
$$(b - l)(x_1y_2 - x_2y_1) \equiv 0 \pmod{2k}.$$

But

$$(2a, b + l, b - l, 2c) = (2a, 2b, 2c, b + l) \leqslant 2,$$

so that

$$x_1y_2 - x_2y_1 \equiv 0 \pmod{k}.$$

This shows that u is an integer. Therefore t^2 is an integer, and since t is rational, t itself must be an integer.

2) Suppose that

$$2ax_1 + (b + \sqrt{d})y_1 = (2ax_2 + (b + \sqrt{d})y_2)\left(\frac{t + u\sqrt{d}}{2}\right),$$

and $t^2 - du^2 = 4$. Then

$$x_1 = \frac{t - bu}{2}x_2 - cuy_2, \qquad y_1 = aux_2 + \frac{t + bu}{2}y_2.$$

Let r_1, s_1 correspond to the solution x_1, y_1. Then

$$r_2 = \frac{t+bu}{2}r_1 + cus_1, \qquad s_2 = -aur_1 + \frac{t-bu}{2}s_1$$

correspond to the solution x_2, y_2, because

$$1 = x_1s_1 - y_1r_1 = \left(\frac{t-bu}{2}x_2 - cuy_2\right)s_1 - \left(aux_2 + \frac{t+bu}{2}y_2\right)r_1$$

$$= x_2\left(\frac{t-bu}{2}s_1 - aur_1\right) - y_2\left(cus_1 + \frac{t+bu}{2}r_1\right) = x_2s_2 - y_2r_2.$$

Finally, let l_1 and l_2 correspond to (x_1, y_1) and (x_2, y_2) respectively. Then

$$l_1 = 2ax_1r_1 + b(x_1s_1 + y_1r_1) + 2cy_1s_1$$

$$= (2ar_1 + bs_1)\left(\frac{t-bu}{2}x_2 - cuy_2\right) + (br_1 + 2cs_1)\left(aux_2 + \frac{t+bu}{2}y_2\right)$$

$$= \left\{2a\left(r_1\frac{t-bu}{2} + s_1cu\right) + b\left(s_1\frac{t-bu}{2} + r_1au\right)\right\}x_2$$

$$+ \left\{b\left(r_1\frac{t+bu}{2} - s_1cu\right) + 2c\left(s_1\frac{t+bu}{2} - r_1au\right)\right\}y_2$$

$$= 2ax_2r_2 + b(x_2s_2 + y_2r_2) + 2cy_2s_2 = l_2.$$

The theorem is proved. □

We shall now separate our discussion into two cases depending on the sign of d.

Theorem 4.3. *Suppose that $d < 0$. Let*

$$w = \begin{cases} 2 & \text{if } d < -4, \\ 4 & \text{if } d = -4, \\ 6 & \text{if } d = -3. \end{cases}$$

Then there are w proper solutions to (1) *that correspond to the same l.*

Proof. From Theorem 4.2 we see that it suffices to show that the equation $t^2 - du^2 = 4$ has w solutions. If $d < -4$, then clearly $t = \pm 2$, $u = 0$ are the only solutions, so that $w = 2$. If $d = -4$, then $t^2 + 4u^2 = 4$ has the four solutions $t = \pm 2, u = 0$ and $t = 0, u = \pm 1$. Finally if $d = -3$, then $t^2 + 3u^2 = 4$ has the six solutions $t = \pm 1$, $u = \pm 1$; $t = \pm 2$, $u = 0$. □

Theorem 4.4. *Let $d > 0$. Then all the solutions to the equation $x^2 - dy^2 = 4$ can be obtained as follows: Let x_0, y_0 be a solution in which $x_0 + y_0\sqrt{d}$ is least ($x_0 > 0$,*

$y_0 > 0$). *Then all the solutions are given by*

$$\frac{x + y\sqrt{d}}{2} = \pm\left(\frac{x_0 + y_0\sqrt{d}}{2}\right)^n, \qquad n = 0, \pm 1, \pm 2, \ldots .$$

Proof. Since the equation $x^2 - dy^2 = 1$ does possess a solution we see that x_0, y_0 exist. The rest of the proof is the same as that in Theorem 10.9.2. □

Let

$$\varepsilon = \frac{x_0 + y_0\sqrt{d}}{2}, \qquad \bar{\varepsilon} = \frac{x_0 - y_0\sqrt{d}}{2}.$$

Definition. Let $d > 0$. By a *primary solution* to (1) we mean a solution which satisfies

$$2ax + (b - \sqrt{d})y > 0, \qquad 1 \leqslant \left|\frac{2ax + (b + \sqrt{d})y}{2ax + (b - \sqrt{d})y}\right| < \varepsilon^2.$$

If we write $L = 2ax + (b + \sqrt{d})y$, $\bar{L} = 2ax + (b - \sqrt{d})y$, then the condition above becomes

$$\bar{L} > 0, \qquad 1 \leqslant \left|\frac{L}{\bar{L}}\right| < \varepsilon^2.$$

Theorem 4.5. *Let $d > 0$. If the equation* (1) *has proper primary solutions which correspond to the same l, then it has a unique proper primary solution.*

Proof. From Theorem 4.2 we know that if x_0, y_0 is a proper primary solution to (1), then, on denoting by L_0 the associated number L, every proper solution of (1) corresponding to the same l can be represented by $L = \pm L_0\varepsilon^n$. We have

$$\left|\frac{L}{\bar{L}}\right| = \left|\frac{L_0\varepsilon^n}{\bar{L}_0\bar{\varepsilon}^n}\right| = \left|\frac{L_0}{\bar{L}_0}\right|\varepsilon^{2n},$$

so that $1 \leqslant |L/\bar{L}| < \varepsilon^2$ only when $n = 0$, and in this case $\bar{L} = \bar{L}_0 > 0$. □

When $d > 0$ we set $w = 1$.

We can now generalize the definition of a primary solution: When $d > 0$, the definition is as given previously; when $d < 0$ any proper solution is also called a primary solution. Combining Theorems 4.3 and 4.5 we now have

Theorem 4.6. *If, corresponding to the same l, the equation* (1) *has proper primary solutions, then there are w proper primary solutions.* □

Theorem 4.5 suggests that in solving $ax^2 + bxy + cy^2 = k$ there is no need to search for integer points on the whole hyperbola. The primary solution occurs in a finite part of the hyperbola, and having obtained the primary solution we may use the formula $L = \pm L_0\varepsilon^n$ to find all the other solutions. That is, if ε is known, all the solutions can be obtained in a finite number of steps. Specifically, from

$$L_0\bar{L}_0 = 4ak, \qquad \bar{L}_0 > 0, \qquad 1 \leqslant \left|\frac{L_0}{\bar{L}_0}\right| < \varepsilon^2,$$

we see that

$$|\bar{L}_0| \leqslant |L_0| = \sqrt{\left|\frac{L_0\bar{L}_0}{\bar{L}_0}\right|^2} = 2\sqrt{|ak|}\sqrt{\left|\frac{L_0}{\bar{L}_0}\right|} < 2\sqrt{|ak|}\,\varepsilon,$$

or

$$|2\sqrt{d}\,y| = |L_0 - \bar{L}_0| \leqslant |L_0| + |\bar{L}_0| < 4\sqrt{|ak|}\,\varepsilon,$$

giving

$$|y| \leqslant 2\varepsilon\sqrt{|ak|/d}.$$

That is we need only find a solution which satisfies $0 < y \leqslant 2\varepsilon\sqrt{|ak|/d}$ and the rest can be obtained from $L = \pm L_0\varepsilon^n$.

When $a > 0$, $k > 0$ we deduce from $\bar{L} > 0$ and $\bar{L}L > 0$ that $L > 0$, and whence $\bar{L} < L$ so that

$$0 < 2\sqrt{d}\,y = L - \bar{L} \leqslant L = \sqrt{L\bar{L}\frac{L}{\bar{L}}} \leqslant \varepsilon\sqrt{4ak}.$$

Therefore

$$0 < y \leqslant \varepsilon\sqrt{ak/d}.$$

In the actual evaluation of the solutions, this result is better than the previous bound.

Exercise 1. Prove that, under the same hypothesis,

$$0 < y \leqslant \left(\varepsilon - \frac{1}{\varepsilon}\right)\sqrt{\frac{ak}{d}}.$$

Exercise 2. Show that, under the substitution

$$x_1 = \frac{t - bu}{2}x - cuy, \qquad y_1 = aux + \frac{t + bu}{2}y$$

the form $ax^2 + bxy + cy^2$ becomes $ax_1^2 + bx_1y_1 + cy_1^2$.

11.5 Method of Solution

From the above we see that we have to solve the equation $ax^2 + bxy + cy^2 = k$. We now discuss the case when d is positive and not a perfect square. We then have the equation $(2ax + by)^2 - dy^2 = 4ak$. We therefore consider

$$x^2 - dy^2 = \delta k, \qquad k > 0, \qquad \delta = \pm 1. \tag{1}$$

If $k < \sqrt{d}$, then by Theorem 10.8.3, all the solutions to (1) can be obtained from the continued fractions expansion for $\sqrt{d}$, and from periodicity this involves only a finite number of steps.

We now show that if $k > \sqrt{d}$, then we can still reduce it to the case when $k < \sqrt{d}$.

Suppose that x, y is proper solution to (1). Then there are x_1, y_1 such that

$$xy_1 - yx_1 = \delta. \tag{2}$$

Multiplying (1) by $x_1^2 - dy_1^2$ we have

$$(xx_1 - dyy_1)^2 - d(xy_1 - x_1y)^2 = \delta k(x_1^2 - dy_1^2),$$

or

$$(xx_1 - dyy_1)^2 - d = \delta k(x_1^2 - dy_1^2).$$

Let x_0, y_0 be a solution to (2). Then all the solutions to (2) are given by $x_1 = x_0 + tx$, $y_1 = y_0 + ty$ so that $xx_1 - dyy_1 = xx_0 - dyy_0 + (x^2 - dy^2)t = xx_0 - dyy_0 + \delta tk$. We may therefore choose t so that

$$|xx_1 - dyy_1| \leqslant \frac{k}{2}.$$

Let $|xx_1 - dyy_1| = l$. Then

$$x_1^2 - dy_1^2 = \frac{l^2 - d}{\delta k} = \eta h, \qquad \eta = \pm 1, \quad h > 0.$$

Therefore

$$h \leqslant \frac{\max(d, l^2)}{k} < \frac{k^2}{k} = k.$$

From this we see that from a solution to (1) we arrived at a similar equation with a number k which is smaller. If this number is still greater than $\sqrt{d}$ we can repeat the argument. This suggests the following procedure.

We first solve for all those l satisfying $l^2 \equiv d \pmod{k}$, $0 \leqslant l \leqslant k/2$, and we let them be $l_1, l_2, \ldots, l_t$. Set $(l_i^2 - d)/\delta k = \eta_i h_i$, $\eta_i = \pm 1$, $h_i > 0$ and solve the system $x_i^2 - dy_i^2 = \eta_i h_i$ $(1 \leqslant i \leqslant t)$. Suppose that $h_i < \sqrt{d}$. Then we use the method of continued fractions. Let x_i, y_i be a solution. Then

$$x = \frac{-\delta dy_i \pm l_i x_i}{\eta_i h_i}, \qquad y = \frac{-\delta x_i \pm l_i y_i}{\eta_i h_i} \tag{3}$$

is a solution to (1). This is because from

$$\eta_i h_i(x + \sqrt{d}\,y) = (x_i + \sqrt{d}\,y_i)(-\delta\sqrt{d} \pm l_i)$$

we have $x^2 - dy^2 = \delta k$ at once. Further, if x, y in (3) are integers, then they are solutions to (1).

If $h_i > \sqrt{d}$, then we proceed to obtain a specific solution to $x_i^2 - dy_i^2 = \eta_i h_i$. Then all the solutions to (1) can be obtained. We illustrate this with an example.

Example. We wish to solve

$$x^2 - 15y^2 = 61. \tag{4}$$

We first solve $l^2 \equiv 15 \pmod{61}$, $0 \leqslant l \leqslant \frac{61}{2}$. This means solving $l^2 = 15 + 61h$, $l^2 \leqslant 900$, or finding h so that $15 + 61h$ is a square. Letting h run over $0 \leqslant h \leqslant [900/61] = 14$ we see that there is only one suitable h, namely $h = 10$, $l = 25$. We now have to solve

$$x_1^2 - 15y_1^2 = 10. \tag{5}$$

Observing that $10 > \sqrt{15}$ we now consider $l^2 = 15 + 10h$, $l \leqslant \frac{10}{2} = 5$. This gives $h = 1$, $l = 5$ so that we have to solve

$$x_2^2 - 15y_2^2 = 1. \tag{6}$$

From the method of continued fractions, the solutions to (6) are given by $x_2 + \sqrt{15}\,y_2 = \pm(4+\sqrt{15})^n$. Therefore $x_1 + \sqrt{15}\,y_1 = \pm(4+\sqrt{15})^n(5 \pm \sqrt{15})$ and so

$$x + \sqrt{15}\,y = \pm(4 + \sqrt{15})^n(5 \pm \sqrt{15})(25 \pm \sqrt{15})/10.$$

Here the three signs $\pm$ are independent so that either

$$x + \sqrt{15}\,y = \pm(4 + \sqrt{15})^n(14 \pm 3\sqrt{15})$$

or

$$x + \sqrt{15}\,y = \pm(4 + \sqrt{15})^n(11 \pm 2\sqrt{15}).$$

Alternatively we can use the inequality at the end of §4, that is $0 < y < \varepsilon\sqrt{ak/d}$. For this example we have $0 < y \leqslant 7$ and we can construct the following table

y	1	2	3	4	5	6	7
$15(2y-1)$	15	45	75	105	135	165	195
$15y^2$	15	60	135	240	375	540	735
$15y^2+61$	76	121	196	301	436	601	796

Observe that in the second row of this table each term increases by 30, and in the third row the i-th term is the sum of the $(i-1)$-th term and the i-th term of the second row.

Exercise 1. Solve the following indeterminate equations.

(a) $$3x^2 - 8xy + 7y^2 - 4x + 2y = 109,$$

(b) $$3xy + 2y^2 - 4x - 3y = 12,$$

(c) $$9x^2 - 12xy + 4y^2 + 3x + 2y = 12,$$

(d) $$x^2 - 8xy - 17y^2 + 72y - 75 = 0.$$

Exercise 2. Let $k < \sqrt{d}$. Show that the solutions to $ax^2 + bxy + cy^2 = k$ can be obtained from the continued fractions expansions of the roots of the equation $ax^2 + bx + c = 0$. Try and generalize the results in this section.

11.6 Generalization of Soon Go's Theorem

Let us consider the equation $x^2 + y^2 = z^2$. If $(x, y) = d > 1$, then d also divides z. We may therefore assume that $(x, y) = 1$, and we need only consider positive solutions. Next, if x, y are both odd, then $x^2 + y^2 \equiv 2 \pmod 4$, so that z^2 is divisible by 2 but not by 4; since this is impossible we see that x and y must be of opposite parity. We shall assume that x is even.

Theorem 6.1. *The solutions of the equation $x^2 + y^2 = z^2$ satisfying $x > 0$, $y > 0$, $z > 0$, $(x, y) = 1$, $2|x$ are given by*

$$x = 2ab, \qquad y = a^2 - b^2, \qquad z = a^2 + b^2,$$

where a, b are coprime integers of opposite parity satisfying $a > b > 0$. There is a one to one correspondence between (x, y, z) and (a, b). □

On putting $\xi = x/z$, $\eta = y/z$ the equation $x^2 + y^2 = z^2$ becomes $\xi^2 + \eta^2 = 1$ and we deduce from Theorem 6.1 that the unit circle $\xi^2 + \eta^2 = 1$ has infinitely many rational points given by

$$\xi = \frac{2ab}{a^2 + b^2}, \qquad \eta = \frac{a^2 - b^2}{a^2 + b^2}.$$

We generalize the problem and ask if every second degree conic possesses infinitely many rational points. The answer is no; for example the hyperbola $\xi^2 - 3\eta^2 = 2$ has no rational points. For if we put $\xi = x/z$, $\eta = y/z$, $(x, y, z) = 1$ then we have $x^2 - 3y^2 = 2z^2$, so that $x^2 \equiv 2z^2 \pmod 3$, which implies $3|x$ and $3|z$, and whence $3|y$, contradicting $(x, y, z) = 1$. However, we do have the following:

Theorem 6.2. *Let a second degree conic, not a pair of straight lines, have rational coefficients. If the conic has one rational point, then it has infinitely many rational points.*

Proof. We may assume that the conic passes through the origin; otherwise we can translate the origin to the rational point concerned. The conic can be written as $S_2(\xi, \eta) + S_1(\xi, \eta) = 0$, where $S_i(\xi, \eta)$ is homogeneous in ξ and η with degree i. If $S_1(\xi, \eta) \equiv 0$, then the original conic is a pair of straight lines, and if $S_2(\xi, \eta) \equiv 0$, then the original conic is a straight line. Therefore $S_1(\xi, \eta)$ and $S_2(\xi, \eta)$ are not identically zero. Now put $\eta = \zeta\xi$ so that $\xi S_2(1, \zeta) + S_1(1, \zeta) = 0$ giving

$$\xi = -\frac{S_1(1,\zeta)}{S_2(1,\zeta)}, \qquad \eta = -\frac{\zeta S_1(1,\zeta)}{S_2(1,\zeta)}.$$

There are therefore infinitely many rational points. □

Theorem 6.3. *Let A, B, C be rational numbers, not all zero. Suppose that $B^2 - 4AC$ is a square. Then the conic*

$$A\xi^2 + B\xi\eta + C\eta^2 + D\xi + E\eta + F = 0 \tag{1}$$

has infinitely many rational points. In other words, if the asymptotes of a hyperbola has rational points, then the hyperbola has infinitely many rational points; a parabola has infinitely many rational points.

Proof. Write $L^2 = B^2 - 4AC$, so that

$$\begin{aligned} A\xi^2 + B\xi\eta + C\eta^2 &= A\left(\left(\xi + \frac{B}{2A}\eta\right)^2 + \left(\frac{C}{A} - \frac{B^2}{4A^2}\right)\eta^2\right) \\ &= A\left(\xi + \frac{B}{2A}\eta - \frac{L}{2A}\eta\right)\left(\xi + \frac{B}{2A}\eta + \frac{L}{2A}\eta\right). \end{aligned}$$

If $L \neq 0$ we set

$$\xi' = \xi + \frac{B + L}{2A}\eta, \qquad \eta' = \xi - \frac{-B + L}{2A}\eta,$$

and solving for ξ and η and substituting into (1) we have

$$A\xi'\eta' + D'\xi' + E'\eta' + F' = 0,$$

which gives

$$\xi' = -\frac{E'\eta' + F'}{A\eta' + D'}.$$

Therefore (1) has infinitely many rational points.

If $L = 0$ we set $\xi' = \xi + B\eta/2A$, $\eta' = -\eta$ giving $A\xi'^2 + D'\xi' + E'\eta' + F' = 0$. If $E' \neq 0$, then $\eta' = -(A\xi'^2 + D'\xi' + F')/E'$ so that there are infinitely many rational points. If $E' = 0$, then the original curve is not a second degree conic. □

Note: Theorems 6.2 and 6.3 raise the following problem. Let

$$f(x_1, x_2, \ldots, x_n) = 0 \tag{2}$$

be a homogeneous second degree equation in $x_1, x_2, \ldots, x_n$ with integer coefficients, not factorizable into a product of linear terms. We ask if there are infinitely many lattice points satisfying (2). We see from Theorem 6.2 that if $n \geqslant 3$ and if (2) has a non-zero lattice point, then there are infinitely many lattice points. But when does it have a lattice point? For example: $x_1^2 + x_2^2 + \cdots + x_n^2 = 0$ certainly has no non-zero lattice point. We therefore have to assume that $f(\xi_1, \ldots, \xi_n) = 0$ has other real solutions. It can be proved that, under this assumption, and for $n \geqslant 5$, the equation (2) has integer solutions, and indeed infinitely many solutions (this is Mayer's theorem). The result does not hold when $n = 4$. For if $x_1^2 + x_2^2 + x_3^2 - 7x_4^2 = 0$, then we may assume that $(x_1, x_2, x_3, x_4) = 1$. Now from $x_1^2 + x_2^2 + x_3^2 + x_4^2 \equiv 0 \pmod 8$, and $x^2 \equiv 0, 1, 4 \pmod 8$ we can deduce that $2|(x_1, x_2, x_3, x_4)$ which is a contradiction.

11.7 Fermat's Conjecture

Fermat claimed that when $n \geqslant 3$ the equation $x^n + y^n = z^n$ has no positive integer solutions in x, y, z. This has been proved for $2 < n < 125000$, and even this modest amount of result involves some pioneering work by mathematicians.

In order to prove Fermat's claim it suffices to establish the case when $n = 4$ and when n is an odd prime. For if n has an odd prime divisor p, then

$$(x^{n/p})^p + (y^{n/p})^p = (z^{n/p})^p,$$

and if n has no odd prime divisors, then $n = 2^k$ $(k \geqslant 2)$ and

$$(x^{n/4})^4 + (y^{n/4})^4 = (z^{n/4})^4.$$

The case $n = 4$ can be settled using Fermat's method of infinite descent. In fact we have

Theorem 7.1. *The equation $x^4 + y^4 = z^4$ has no positive integer solutions.* □

11.8 Markoff's Equation

We introduced in §10.5 Markoff's equation

$$x^2 + y^2 + z^2 = 3xyz, \tag{1}$$

and we stated the relationship between Markoff numbers and continued fractions. We shall now study this equation.

Theorem 8.1. *Let x_0, y_0, z_0 be a solution to* (1). *Then so is $x_0, y_0, 3x_0y_0 - z_0$.*

Proof.

$$x_0^2 + y_0^2 + (3x_0y_0 - z_0)^2 = x_0^2 + y_0^2 + z_0^2 - 6x_0y_0z_0 + 9x_0^2y_0^2$$
$$= -3x_0y_0z_0 + 9x_0^2y_0^2 = 3x_0y_0(3x_0y_0 - z_0). \quad \square$$

Theorem 8.2. *Every solution of* (1) *can be generated from Theorem* 8.1 *with* $x = y = z = 1$ *as an initial solution.*

Proof. 1) If $x = y = z$, then clearly $x = y = z = 1$.

2) If $x = y \neq z$, then $2x^2 + z^2 = 3x^2z$. Hence $x^2 | z^2$ or $x | z$. Let $z = wx$ so that $2 + w^2 = 3wx$ $(w > 0)$ and hence $w | 2$, giving $w = 1$ or 2. But $x \neq z$ so that $w = 2$ giving $x = 1, y = 1, z = 2$ and this is a solution generated by $(1, 1, 1)$ from Theorem 8.1.

3) We can now assume that $x < y < z$. If we can establish that $3xy - z < z$, then we can reduce the value of $x + y + z$, so that after a finite number of successive steps x, y, z cannot be all different which means that we have reduced the present case to 1) or 2). This is what we shall prove.

From $z^2 - 3xyz + x^2 + y^2 = 0$ we have

$$2z = 3xy \pm \sqrt{9x^2y^2 - 4(x^2 + y^2)}.$$

If

$$2z = 3xy - \sqrt{9x^2y^2 - 4(x^2 + y^2)},$$

then from

$$8x^2y^2 - 4x^2 - 4y^2 = 4x^2(y^2 - 1) + 4y^2(x^2 - 1) > 0$$

we see that

$$2z < 3xy - xy = 2xy,$$

or

$$z < xy.$$

But

$$3xyz = x^2 + y^2 + z^2 < 3z^2,$$

so that $xy < z$ giving a contradiction. Therefore

$$2z = 3xy + \sqrt{9x^2y^2 - 4(x^2 + y^2)} > 3xy$$

as required. $\square$

Example. Starting with $(1, 1, 1)$ we have $(1, 1, 2)$ and then $(1, 2, 5)$; $(1, 5, 13)$; $(2, 5, 29)$. Continuing we have the following table for $x \leqslant y \leqslant z < 1000$.

z	1	2	5	13	29	34	89	169	194	233	433	610	985
y	1	1	2	5	5	13	34	29	13	89	295	233	169
x	1	1	1	1	2	1	1	2	5	1	5^2	1	2

Note: Observe that this is also a method of descent. Fortunately there is no more descent after $x = y = z = 1$. We see therefore that Fermat's method of infinite descent can be used either to prove that there is no solution, or to prove that there are infinitely many solutions.

Exercise 1. Generalize the discussion here to the equation $x_1^2 + x_2^2 + \cdots + x_n^2 = nx_1x_2 \cdots x_n$.

Exercise 2. Solve $x_1^2 + x_2^2 + x_3^2 + x_4^2 = 4x_1x_2x_3x_4$, $x_1 \leqslant x_2 \leqslant x_3 \leqslant x_4 \leqslant 100$.

Exercise 3. Show that the equation $2x^4 - y^4 = z^4$ has infinitely many solutions.

11.9 The Equation $x^3 + y^3 + z^3 + w^3 = 0$

The number 1729 is the smallest positive integer representable as the sum of two cubes in two different ways. That is $1729 = 10^3 + 9^3 = 12^3 + 1^3$. There are other numbers having this property, for example: $2^3 + 34^3 = 15^3 + 33^3$, $9^3 + 15^3 = 2^3 + 16^3$. In fact we even have

$$70^3 + 560^3 = 98^3 + 552^3 = 315^3 + 525^3,$$

$$\begin{aligned} 121170^3 + 969360^3 &= 545275^3 + 908775^3 = 342738^3 + 955512^3 \\ &= 336455^3 + 956305^3, \end{aligned}$$

and $3^4 + 4^3 + 5^3 = 6^3$, $1^3 + 6^3 + 8^3 = 9^3$. The solutions to the equation $x^3 + y^3 + z^3 + w^3 = 0$ present a very interesting problem. Unfortunately we still have not obtained a formula for all the solutions. The *Euler-Binet formula* below provides all the rational solutions.

Theorem 9.1. *The rational solutions to the equation* $W^3 + 3W(X^2 + Y^2 + Z^2) + 6XYZ = 0$ *are given by*

$$W = -6\rho abc, \qquad X = \rho a(a^2 + 3b^2 + 3c^2),$$

$$Y = \rho b(a^2 + 3b^2 + 9c^2), \qquad Z = 3\rho c(a^2 + b^2 + 3c^2).$$

Here $(a, b, c) = 1$, *and* ρ *is a rational number.*

Proof. We rewrite the given equation as

$$\begin{vmatrix} W & 3Z & -3Y \\ -Z & W & 3X \\ Y & -X & W \end{vmatrix} = 0,$$

so that there must be integers a, b, c not all 0 and $(a, b, c) = 1$ such that

$$Wa + 3Zb - 3Yc = 0,$$
$$-Za + Wb + 3Xc = 0,$$
$$Ya - Xb + Wc = 0.$$

Solving these for X, Y, Z, W, the required result follows. □

Let

$$W = \tfrac{1}{2}(\alpha + \beta + \gamma + \delta), \qquad X = \tfrac{1}{2}(\alpha + \beta - \gamma - \delta),$$
$$Y = \tfrac{1}{2}(\alpha - \beta + \gamma - \delta), \qquad Z = \tfrac{1}{2}(\alpha - \beta - \gamma + \delta), \tag{1}$$

so that

$$(\alpha + \beta + \gamma + \delta)^3 + 3(\alpha + \beta + \gamma + \delta)[(\alpha + \beta - \gamma - \delta)^2 + (\alpha - \beta + \gamma - \delta)^2 + (\alpha - \beta - \gamma + \delta)^2] + 6(\alpha + \beta - \gamma - \delta)(\alpha - \beta + \gamma - \delta)(\alpha - \beta - \gamma + \delta) = 0,$$

or

$$\alpha^3 + \beta^3 + \gamma^3 + \delta^3 = 0. \tag{2}$$

Solving (1) we have

$$\alpha = \tfrac{1}{2}(W + X + Y + Z), \qquad \beta = \tfrac{1}{2}(W + X - Y - Z),$$
$$\gamma = \tfrac{1}{2}(W - X + Y - Z), \qquad \delta = \tfrac{1}{2}(W - X - Y + Z),$$

and the solutions to (2) can be obtained from Theorem 9.1.

Theorem 9.2. *Given any positive integer r, there exists a number N which can be represented as a sum of two cubes in r ways.*

Proof. Let ξ_1, η_1 be two fixed rational numbers. Set

$$X = \frac{\xi_1(\xi_1^3 + 2\eta_1^3)}{\xi_1^3 - \eta_1^3}, \qquad Y = \frac{\eta_1(2\xi_1^3 + \eta_1^3)}{\xi_1^3 - \eta_1^3},$$
$$\xi_2 = \frac{X(X^3 - 2Y^3)}{X^3 + Y^3}, \qquad \eta_2 = \frac{Y(2X^3 - Y^3)}{X^3 + Y^3},$$

so that

$$X^3 - Y^3 = \xi_1^3 + \eta_1^3, \qquad \xi_2^3 + \eta_2^3 = X^3 - Y^3. \tag{3}$$

We then have

$$\xi_1^3 + \eta_1^3 = \xi_2^3 + \eta_2^3,$$
$$\frac{X}{Y} = \frac{\xi_1}{2\eta_1}\left(1 + 2\left(\frac{\eta_1}{\xi_1}\right)^3\right)\left(1 + \frac{1}{2}\left(\frac{\eta_1}{\xi_1}\right)^3\right)^{-1},$$
$$\frac{\xi_2}{\eta_2} = \frac{X}{2Y}\left(1 - 2\left(\frac{Y}{X}\right)^3\right)\left(1 - \frac{1}{2}\left(\frac{Y}{X}\right)^3\right)^{-1}.$$

Suppose that $0 < \eta_1/\xi_1 < \varepsilon < \frac{1}{4}$. Then

$$0 < \frac{X}{Y} - \frac{\xi_1}{2\eta_1} = \frac{\frac{3}{4}\left(\frac{\eta_1}{\xi_1}\right)^2}{1 + \frac{1}{2}\left(\frac{\eta_1}{\xi_1}\right)^3} < \frac{3}{4}\left(\frac{\eta_1}{\xi_1}\right)^2 < \frac{3}{4}\varepsilon^2.$$

Therefore $X/Y > \xi_1/2\eta_1 > \frac{1}{2}\varepsilon$, or $Y/X < 2\varepsilon$. Also

$$\left|\frac{\xi_2}{\eta_2} - \frac{X}{2Y}\right| = \frac{\frac{3}{4}\left(\frac{Y}{X}\right)^2}{1 - \frac{1}{2}\left(\frac{Y}{X}\right)^3} < \frac{3}{4}\left(\frac{Y}{X}\right) < \frac{3}{2}\varepsilon,$$

so that

$$\left|\frac{\xi_2}{\eta_2} - \frac{\xi_1}{4\eta_1}\right| \leqslant \left|\frac{\xi_2}{\eta_2} - \frac{X}{2Y}\right| + \frac{1}{2}\left|\frac{X}{Y} - \frac{\xi_1}{2\eta_1}\right| < 2\varepsilon,$$

and hence

$$\frac{\xi_2}{\eta_2} > \frac{\xi_1}{4\eta_1} - 2\varepsilon > \frac{1}{8\varepsilon}, \qquad \frac{\eta_2}{\xi_2} < 8\varepsilon.$$

Continuing this way we have

$$\left|\frac{\xi_3}{\eta_3} - \frac{\xi_2}{4\eta_2}\right| < 2^4\varepsilon, \quad \left|\frac{\xi_4}{\eta_4} - \frac{\xi_3}{4\eta_3}\right| < 2^7\varepsilon, \quad \ldots, \quad \left|\frac{\xi_{s+1}}{\eta_{s+1}} - \frac{\xi_s}{4\eta_s}\right| < 2^{1+3(s-1)}\varepsilon,$$

provided that $2^{3(s-1)}\varepsilon < \frac{1}{4}$.

Therefore, by taking η_1/ξ_1 very small, there are pairs of numbers (ξ_1, η_1), $\ldots, (\xi_r, \eta_r)$ such that

$$\xi_1^3 + \eta_1^3 = \xi_2^3 + \eta_2^3 = \cdots = \xi_r^3 + \eta_r^3,$$

and the ratios

$$\frac{\xi_1}{\eta_1}, 4\frac{\xi_2}{\eta_2}, \ldots, 4^{r-1}\frac{\xi_r}{\eta_r}$$

are nearly equal. Therefore ξ_s/η_s are distinct ratios and the required result follows. □

Exercise 1. Show that the rational solutions to $\alpha^3 + \beta^3 + \gamma^3 + \delta^3 = 0$ can be obtained from

$$\alpha = \sigma(-(\xi - 3\eta)(\xi^2 + 3\eta^2) + 1), \qquad \beta = \sigma((\xi + 3\eta)(\xi^2 + 3\eta^2) - 1),$$
$$\gamma = \sigma((\xi^2 + 3\eta^2)^2 - (\xi + 3\eta)), \qquad \delta = \sigma((\xi^2 + 3\eta^2)^2 - (\xi - 3\eta))$$

where ξ and η are rational numbers. If $\sigma = 1$ and ξ, η are integers, it then follows that $x^3 + y^3 + w^3 + z^3 = 0$ has infinitely many integer solutions. By considering $\alpha = 1, \beta = 12, \gamma = -10, \delta = -9$ show that this method here does not give all the solutions.

Exercise 2. Verify that $y^{12} = (9x^4)^3 + (3xy^3 - 9x^4)^3 + (y^4 - 9x^3y)^3$, and hence show that $5^{12} = 9^3 + 366^3 + 580^3 = 144^3 + 606^3 + 265^3$.

Exercise 3. Use Exercise 2 to prove that there exists n such that the number of non-negative solutions to $n = x^3 + y^3 + z^3$ exceeds $\frac{1}{3}n^{\frac{1}{12}}$.

Exercise 4. Prove that

$$(3a^2 + 5ab - 5b^2)^3 + (4a^2 - 4ab + 6b^2)^3 + (5a^2 - 5ab - 3b^2)^3$$
$$= (6a^2 - 4ab + 4b^2)^3.$$

11.10 Rational Points on a Cubic Surface

The cubic surfaces discussed in this section are non-degenerate.

On dividing equation (2) in §9 by δ^3 and setting $\xi = -\alpha/\delta$, $\eta = -\beta/\delta$, $\zeta = -\gamma/\delta$, we have

$$\xi^3 + \eta^3 + \zeta^3 = 1. \tag{1}$$

In other words, from our results in §9, the cubic surface (1) has infinitely many rational points. We shall generalize this to the most common cubic surfaces.

Before we introduce the difficult method involved we first consider some special examples.

Theorem 10.1. *Let A, B, C be rational numbers with $C \neq 0$. Then the cubic surface*

$$\zeta^2 = \xi^3 + A\xi + B + C\eta^2 \tag{2}$$

has infinitely many rational points.

Proof. We substitute

$$\xi = \eta^2 + T\eta, \qquad \zeta = \eta^3 + \lambda\eta^2 + \mu\eta + \nu \tag{3}$$

into (2) giving

$$(\eta^3 + \lambda\eta^2 + \mu\eta + \nu)^2 = (\eta^2 + T\eta)^3 + A(\eta^2 + T\eta) + B + C\eta^2. \tag{4}$$

On comparing the coefficients of $\eta^6, \eta^5, \eta^4, \eta^3$ we have

$$2\lambda = 3T, \qquad \lambda^2 + 2\mu = 3T^2, \qquad 2(\nu + \lambda\mu) = T^3,$$

giving

$$\lambda = \frac{3}{2}T, \qquad \mu = \frac{3}{8}T^2, \qquad \nu = -\frac{1}{16}T^3.$$

Substituting this into (4) we have the quadratic equation

$$L\eta^2 + M\eta + N = 0, \tag{5}$$

where

$$L = A + C - \mu^2 - 2\lambda\nu = A + C + \frac{3}{64}T^4,$$

$$M = AT - 2\mu\nu = AT + \frac{3}{64}T^5,$$

$$N = B - \nu^2 = B - \frac{1}{256}T^6.$$

The discriminant of (5) is

$$\Delta = M^2 - 4LM = \left[\left(\frac{3}{64}\right)^2 + \frac{3}{64}\cdot\frac{1}{64}\right]T^{10} + \cdots = \frac{3}{1024}T^{10} + \cdots$$

so that Δ cannot be the square of a polynomial with rational coefficients. Therefore the solutions to (5) are given by

$$\eta = \beta_1 \pm \beta_2\sqrt{\Delta}, \qquad \beta_1 = -\frac{M}{2L}, \qquad \beta_2 = \frac{1}{2L}.$$

Substituting this into (3) we have

$$\xi = \alpha_1 \pm \alpha_2\sqrt{\Delta}, \qquad \zeta = \gamma_1 \pm \gamma_2\sqrt{\Delta},$$

where

$$\alpha_2 = (2\beta_1 + T)\beta_2 = \frac{LT - M}{2L^2} = \frac{CT}{2L^2} \neq 0.$$

Let

$$\frac{\xi - \alpha_1}{\alpha_2} = \frac{\eta - \beta_1}{\beta_2} = \frac{\zeta - \gamma_1}{\gamma_2} = \sigma, \tag{6}$$

and substitute this into (2). We then have a cubic equation in σ, with rational functions in T as coefficients, and the leading coefficient α_2^3 is not zero. Since we already know that $\pm\sqrt{\Delta}$ are two of the roots, the remaining root σ_0 must be a rational function of T. Substituting this into (6) we see that ξ, η, ζ can be represented by rational functions of T. However, we still have to prove that the ξ, η, ζ so obtained are not constants since otherwise we may not have infinitely many rational points. If η is a non-zero constant, then $\xi = \eta^2 + T\eta$ is not a constant, and if $\eta = 0$, then, by (3), $\xi = 0$ and $\zeta = v = -T^3/16$ and we see from (2) that this is not possible. Therefore if we substitute σ_0 into (6), then ξ, η, ζ are all rational functions of T and they cannot be all constants. The theorem is proved. □

Theorem 10.2. *Let $f(\xi, \eta)$ be a cubic polynomial with rational coefficients which cannot be transformed into a polynomial with a single variable by a linear transformation. Then the cubic surface*

$$\zeta^2 = f(\xi, \eta) \tag{7}$$

has infinitely many rational points.

Proof. Denote by $f_3(\xi, \eta)$ the homogeneous cubic part of $f(\xi, \eta)$.

1) If $f_3(\xi, 1) = 0$ has a rational root a (similar method for the case of $f_3(1, \eta)$), then $f(\xi + a\eta, \eta) = g(\xi, \eta)$ does not contain a η^3 term. Therefore after the substitution $\xi \to \xi + a\eta$, $\eta \to \eta$, $\zeta \to \zeta$ the equation (7) becomes

$$\zeta^2 = L_1(\xi)\eta^2 + L_2(\xi)\eta + L_3(\xi), \tag{8}$$

where L_i is a polynomial in ξ of degree i. Let $L_1(\xi) = \alpha\xi + \beta$. If $\alpha \neq 0$, then we may take ξ such that $\alpha\xi + \beta = \delta^2 \neq 0$, and the theorem follows from Theorem 6.3.

If $\alpha = 0$ and $\beta = 0$, then (8) is linear in η and the theorem follows by solving η. We now suppose that $\alpha = 0$, $\beta \neq 0$ so that (8) can be written as

$$\zeta^2 = \alpha_1\xi^3 + \alpha_2\xi^2\eta + \beta_1\xi^2 + \beta_2\xi\eta + \beta\eta^2 + \cdots \tag{9}$$

where $\cdots$ represents the linear terms in ξ and η.

Suppose that $\alpha_2 \neq 0$. Let $\alpha_1\xi + \alpha_2\eta = \lambda$ so that

$$\begin{aligned}\zeta^2 &= \lambda\xi^2 + \beta_1\xi^2 + \beta_2\xi\left(\frac{\lambda - \alpha_1\xi}{\alpha_2}\right) + \beta\left(\frac{\lambda - \alpha_1\xi}{\alpha_2}\right)^2 + \cdots \\ &= (\lambda + \beta_1 - \beta_2\alpha_1/\alpha_2 + \beta\alpha_1^2/\alpha_2^2)\xi^2 + \cdots.\end{aligned}$$

We take $\lambda = 1 - \beta_1 + \beta_2\alpha_1/\alpha_2 - \beta\alpha_1^2/\alpha_2^2$, so that

$$\zeta^2 - \xi^2 = (\zeta - \xi)(\zeta + \xi) = A\xi + B.$$

By Theorem 6.3 this surface has infinitely many rational points.

Suppose next that $\alpha_2 = 0$. Then $\alpha_1 \neq 0$, since otherwise $\zeta^2 = f(\xi, \eta)$ is not a cubic surface. Therefore

$$\begin{aligned}\zeta^2 &= \alpha_1\xi^3 + \beta_1\xi^2 + \beta_2\xi\eta + \beta\eta^2 + \cdots \\ &= \beta\eta^2 + (\beta_2\xi + \gamma)\eta + f(\xi) \\ &= \beta\left(\eta + \frac{\beta_2}{2\beta}\xi + \frac{\gamma}{2\beta}\right)^2 + g(\xi).\end{aligned}$$

Here $g(\xi)$ is a cubic polynomial in ξ with leading coefficient α_1. Replacing

$\eta + (\beta_2\xi + \gamma)/2\beta$ by η we have $\zeta^2 = \beta\eta^2 + g(\xi)$. On multiplication by α_1^2 and the simple substitution $\xi' = \alpha_1\xi + A$, $\zeta' = \alpha_1\zeta$ this equation is reduced to (2) and so the theorem follows from Theorem 10.1.

2) Suppose that $f_3(\xi, \eta)$ has no linear rational factor. The equation (7) can be written as

$$\zeta^2 = \alpha\xi^3 + f_1(\eta)\xi^2 + f_2(\eta)\xi + f_3(\eta)$$

where f_i is a polynomial in η with degree i. Replacing ξ by $\xi - f_1(\eta)/3\alpha$ we have the new equation

$$\zeta^2 = \alpha\xi^3 + g_2(\eta)\xi + g_3(\eta).$$

If we multiply both sides by α^2 and replace $\alpha\zeta$, $\alpha\xi$ by ζ, ξ, then we have

$$\zeta^2 = \xi^3 + (A\eta^2 + B\eta + C)\xi + D\eta^3 + E\eta^2 + F\eta + G. \tag{10}$$

As in Theorem 10.1 we substitute

$$\xi = \eta^2 + T\eta, \qquad \zeta = \eta^3 + \lambda\eta^2 + \mu\eta + \nu \tag{11}$$

into (10) giving

$$\begin{aligned}(\eta^3 + \lambda\eta^2 + \mu\eta + \nu)^2 = (\eta^2 + T\eta)^3 &+ (A\eta^2 + B\eta + C)(\eta^2 + T\eta)\\ &+ D\eta^3 + E\eta^2 + F\eta + G.\end{aligned} \tag{12}$$

We choose λ, μ, ν in (12) so that the coefficients for η^5, η^4, η^3 are zero, giving a quadratic equation $L\eta^2 + M\eta + N = 0$. The reader can verify that $L \neq 0$ so that we have $\eta = \beta_1 \pm \beta_2\sqrt{\Delta}$, where β_1, β_2 and Δ are rational functions of T. Substituting this into (11) we have

$$\xi = \alpha_1 \pm \alpha_2\sqrt{\Delta}, \qquad \zeta = \gamma_1 \pm \gamma_2\sqrt{\Delta}.$$

If $\Delta = 0$, then the result follows at once. If $\Delta \neq 0$, we let

$$\frac{\xi - \alpha_1}{\alpha_2} = \frac{\eta - \beta_1}{\beta_2} = \frac{\zeta - \gamma_1}{\gamma_2} = \sigma$$

and obtain from (10) a cubic equation in σ with the non-zero leading coefficient

$$\alpha_2^3 + A\beta_2^2\alpha_2 + D\beta_2^3 \qquad (= f_3(\alpha_2, \beta_2)).$$

Two of the roots of this cubic equation are already known to be $\pm\sqrt{\Delta}$ so that its third root σ_0 is a rational function of T. This means that

$$\xi = \alpha_1 + \alpha_2\sigma_0, \qquad \eta = \beta_1 + \beta_2\sigma_0, \qquad \zeta = \gamma_1 + \gamma_2\sigma_0$$

lie on the cubic surface (10). That ξ, η, ζ are not all constants can be proved as before. □

Theorem 10.3. *Let $S_2(\xi, \eta, \zeta)$ and $T_2(\xi, \eta, \zeta)$ be homogeneous quadratics in ξ, η, ζ. Then the cubic surface*

$$\zeta S_2(\xi, \eta, \zeta) + T_2(\xi, \eta, \zeta) + \zeta = 0 \tag{13}$$

has infinitely many rational points.

Proof. Let us denote the left hand side of (13) by $f(\xi, \eta, \zeta)$. Then we have

$$f(\xi, \eta, \zeta) = (\alpha_1 + \alpha_2\zeta)\xi^2 + (\beta_1 + \beta_2\zeta)\xi\eta + (\gamma_1 + \gamma_2\zeta)\eta^2 + g(\xi, \eta, \zeta), \tag{14}$$

where $g(\xi, \eta, \zeta)$ is linear in ξ and η. In Theorem 6.3 we take $A = \alpha_1 + \alpha_2\zeta$, $B = \beta_1 + \beta_2\zeta$, $C = \gamma_1 + \gamma_2\zeta$ so that

$$B^2 - 4AC = (\beta_1 + \beta_2\zeta)^2 - 4(\alpha_1 + \alpha_2\zeta)(\gamma_1 + \gamma_2\zeta).$$

If $\alpha_2 \neq 0$ (or $\gamma_2 \neq 0$), then take $\zeta = -\alpha_1/\alpha_2$ (or $\zeta = -\gamma_1/\gamma_2$) so that $B^2 - 4AC$ is a square. If $\alpha_2 = \gamma_2 = 0$ and $\beta_2 \neq 0$, then $\delta^2 = (\beta_1 + \beta_2\zeta)^2 - 4\alpha_1\gamma_1$ also has rational roots, by Theorem 6.3. We must also note the following. On the substitution $\zeta = -\alpha_1/\alpha_2$ into (14), all the coefficients for ξ^2, $\xi\eta$, η^2 may be zero. In this case, if the coefficients for ξ and η are not both zero, then ξ (or η) can be represented as a rational function of η and $-\alpha_1/\alpha_2$ (or ξ and $-\alpha_1/\alpha_2$), and the theorem then clearly holds. If the coefficients for ξ and η are both zero, then

$$f(\xi, \eta, \zeta) = (\alpha_1 + \alpha_2\zeta)(\xi^2 + A\xi\eta + B\eta^2 + (C + D\zeta)\xi + (E + F\zeta)\eta + G + H\zeta + J\zeta^2) + K.$$

If we put $\zeta = 0$ in (13), then $f(\xi, \eta, 0)$ is a homogeneous quadratic in ξ and η so that $C = E = 0$ and $\alpha_1 G + K = 0$ in the above, giving

$$f(\xi, \eta, \zeta) = (\alpha_1 + \alpha_2\zeta)(\xi^2 + A\xi\eta + B\eta^2 + D\xi\zeta + F\eta\zeta) + P(\zeta), \quad P(0) = 0. \tag{15}$$

Observe that

$$(\xi + \lambda\zeta)^2 + A(\xi + \lambda\zeta)(\eta + \mu\zeta) + B(\eta + \mu\zeta)^2 + D(\xi + \lambda\zeta)\zeta + F(\eta + \mu\zeta)\zeta$$
$$= \xi^2 + A\xi\eta + B\eta^2 + (2A + A\mu + D)\xi\zeta + (A\lambda + 2B\mu + F)\eta\zeta + \cdots.$$

If $A^2 \neq 4B$, then we may choose λ and μ so that $2\lambda + A\mu + D = 0$, $A\lambda + 2B\mu + F = 0$. We may therefore assume that

$$f(\xi, \eta, \zeta) = (\alpha_1 + \alpha_2\zeta)(\xi^2 + A\xi\eta + B\eta^2) + g(\zeta), \qquad g(0) = 0.$$

Let

$$\zeta = \frac{1}{Z}, \qquad \xi = \frac{X}{Z^2(\alpha_1 + \alpha_2\zeta)}, \qquad \eta = \frac{Y}{Z^2(\alpha_1 + \alpha_2\zeta)}.$$

Then

$$X^2 + AXY + BY^2 + Z^4\left(\alpha_1 + \frac{\alpha_2}{Z}\right)g\left(\frac{1}{Z}\right) = 0.$$

Since $g(0) = 0$, it follows that $Z^4(\alpha_1 + \alpha_2/Z)g(1/Z)$ is a cubic in Z, and the theorem is reduced to Theorem 10.2.

If $A^2 = 4B$, then we write $\xi' = \xi + A\eta/2$, $\eta' = \eta$ so that $f(\xi, \eta, \zeta)$ in (15) is linear in η' and the theorem again follows.

Next, if $\alpha_2 = \beta_2 = \gamma_2 = 0$ and $\beta_1^2 \neq 4\alpha_1\gamma_1$, then, under the transformation $\xi \to \xi + \lambda_1\zeta + \lambda_2\zeta^2$, $\eta \to \eta + \mu_1\zeta + \mu_2\zeta^2$, (14) becomes

$$\alpha_1\xi^2 + \beta_1\xi\eta + \gamma_1\eta^2 + f(\zeta) = 0,$$

where $f(\zeta) = A\zeta^4 + B\zeta^3 + C\zeta^2 + D\zeta$. The further substitution $\xi = X/Z^2$, $\eta = Y/Z^2$, $\zeta = 1/Z$ gives

$$\alpha_1 X^2 + \beta_1 XY + \gamma_1 Y^2 + A + BZ + CZ^2 + DZ^3 = 0.$$

After a further linear substitution this can be reduced to Theorem 10.2.

Finally, if $\alpha_2 = \beta_2 = \gamma_2 = 0$ and $\beta_1^2 = 4\alpha_1\gamma_1$, then the linear substitution $\xi' = \alpha_1\xi + \beta_1\eta/2$, $\eta' = \eta$, $\zeta' = \zeta$ will make the left hand side of (14) linear in η' and so the theorem is proved. □

Theorem 10.4. *If a non-degenerate cubic surface has a rational point, then it has infinitely many rational points.*

Proof. We may assume that the surface passes through the origin so that it can be written as

$$S_3(\xi, \eta, \zeta) + S_2(\xi, \eta, \zeta) + S_1(\xi, \eta, \zeta) = 0, \tag{16}$$

where $S_i(\xi, \eta, \zeta)$ are homogeneous in ξ, η, ζ with degree i.

1) If $S_1(\xi, \eta, \zeta) \equiv 0$, then $S_3(\xi, \eta, \zeta) + S_2(\xi, \eta, \zeta) = 0$, so that

$$\zeta S_3\left(\frac{\xi}{\zeta}, \frac{\eta}{\zeta}, 1\right) + S_2\left(\frac{\xi}{\zeta}, \frac{\eta}{\zeta}, 1\right) = 0,$$

giving $\zeta = -S_2(\alpha, \beta, 1)/S_3(\alpha, \beta, 1)$. Observe that if $S_3(\alpha, \beta, 1) \equiv 0$, then the original surface is not a cubic, and if $S_2(\alpha, \beta, 1) \equiv 0$, then the cubic surface is a degenerated one.

2) If $S_1(\xi, \eta, \zeta) \not\equiv 0$, then under the transformation $S_1(\xi, \eta, \zeta) \to \zeta$ we have

$$S_3(\xi, \eta, \zeta) + S_2(\xi, \eta, \zeta) + \zeta = 0.$$

If $S_3(\xi, \eta, \zeta)$ and $S_2(\xi, \eta, 0)$ are not both identically zero, then we let $\zeta = 0$ giving

$$S_3(\xi, \eta, 0) + S_2(\xi, \eta, 0) = 0, \qquad \eta = -S_2(\xi/\eta, 1, 0)/S_3(\xi/\eta, 1, 0).$$

If $S_2(\xi, \eta, 0) \equiv 0$, then $S_2(\xi, \eta, \zeta) = \zeta L_1(\xi, \eta, \zeta)$. We let $Z = 1/\zeta$, $X = \xi/\zeta$, $Y = \eta/\zeta$ so that

$$S_3(X, Y, 1) + ZL_1(X, Y, 1) + Z^2 = 0$$

which gives

$$(Z + \tfrac{1}{2}L_1(X, Y, 1))^2 = \tfrac{1}{4}L_1^2(X, Y, 1) - S_3(X, Y, 1),$$

and this is included in Theorem 10.2 so that the required result follows.

If $S_3(\xi, \eta, 0) \equiv 0$, we let $S_3(\xi, \eta, \zeta) = \zeta T_2(\xi, \eta, \zeta)$, and this reduces to Theorem 10.3. The theorem is proved. □

Notes

11.1. The problem of the existence of solutions to the famous equation

$$x^2 = y^n + 1, \qquad xy \neq 0$$

has been settled by K. Chao [16]. He proved that, apart from $n = 3$, $x = \pm 3$, $y = 2$, there are no integer solutions.

Chapter 12. Binary Quadratic Forms

12.1 The Partitioning of Binary Quadratic Forms into Classes

Definition. For fixed integers a, b, c the homogeneous quadratic polynomial

$$F = F(x, y) = ax^2 + bxy + cy^2$$

is called a *binary quadratic form*, or simply a *form*, and is denoted by $\{a, b, c\}$. The integer

$$d = b^2 - 4ac$$

is called the *discriminant* of the form.

It is easy to see that

$$d \equiv 0 \quad \text{or } 1 \pmod{4}.$$

Theorem 1.1. *A necessary and sufficient condition for F to be factorized into a product of two linear forms with integer coefficients is that d is a perfect square.*

Proof. 1) Let d be a perfect square, and $a \neq 0$. Then the equation

$$ax^2 + bx + c = a\left\{\left(x + \frac{b}{2a}\right)^2 - \frac{d}{4a^2}\right\} = 0$$

has rational roots, and therefore, by Theorem 1.13.2, the form can be factorized into a product of two linear forms with integer coefficients. If $a = 0$, then clearly $F(x, y) = (bx + cy)y$.

2) If

$$ax^2 + bxy + cy^2 = (rx + sy)(tx + uy),$$

then

$$d = b^2 - 4ac = (st + ru)^2 - 4rt \cdot su = (st - ru)^2.$$

The theorem is proved. □

We shall assume from now on that d is not a perfect square.

If $d < 0$, $a > 0$, then

$$4aF = (2ax + by)^2 + (4ac - b^2)y^2 = (2ax + by)^2 - dy^2,$$

and so $F(x, y) \geqslant 0$ for all x, y, and $F(x, y) = 0$ if and only if $x = y = 0$. We call such a form a *positive definite form*. If $d < 0$, $a < 0$, then $F \leqslant 0$ for all x, y, and we call the form a *negative definite form*. Since a negative definite form becomes a positive definite form on multiplication by -1, we shall only deal with positive definite forms which we shall simply call *definite forms*.

If $d > 0$, then

$$F(1, 0) = a, \qquad F(b, -2a) = ab^2 - b \cdot b \cdot 2a + c \cdot 4a^2 = -da.$$

If $a \neq 0$, then the two values here have different signs. If $c \neq 0$ we can similarly choose two values which have different signs. If $a = c = 0$, then

$$F(1, 1) = b, \qquad F(1, -1) = -b$$

again have different signs. Thus when $d > 0$, the form $F(x, y)$ can take both positive and negative values, and we therefore call such a form an *indefinite form*.

Definition. Let the integer coefficient substitution

$$x = rX + sY, \qquad y = tX + uY, \qquad (ru - st = 1)$$

transform $F(x, y)$ into $G(X, Y)$ — we say that F is transformed into G via $\begin{pmatrix} r & s \\ t & u \end{pmatrix}$. The two forms F and G are then said to be *equivalent*, and we write $F \sim G$ to denote this.

More specifically, let $F = \{a, b, c\}$ and $G = \{a_1, b_1, c_1\}$. Then we have

$$a_1 = ar^2 + brt + ct^2, \tag{1}$$

$$\begin{aligned} b_1 &= 2ars + b(ru + st) + 2ctu \\ &= 2ars + b(1 + 2st) + 2ctu, \end{aligned} \tag{2}$$

$$c_1 = as^2 + bsu + cu^2, \tag{3}$$

and we derive at once

$$\begin{aligned} b_1^2 - 4a_1c_1 &= (2ars + b(ru + st) + 2ctu)^2 \\ &\quad - 4(ar^2 + brt + ct^2)(as^2 + bsu + cu^2) \\ &= (b^2 - 4ac)(ru - st)^2 = b^2 - 4ac = d. \end{aligned}$$

We see therefore that equivalent forms have the same discriminant.

Also, if $d < 0$, $a > 0$, then $a_1 = F(r, t) \geqslant 0$. Since $a_1 = 0$ implies $r = t = 0$ which is impossible we see that $a_1 > 0$. In other words forms which are equivalent to a positive definite form are themselves positive definite.

Theorem 1.2. (i) $F \sim F$ (*reflexive*).

(ii) *If* $F \sim G$, *then* $G \sim F$ (*symmetric*).

(iii) *If* $F \sim G$, $G \sim H$, *then* $F \sim H$ (*transitive*). □

We omit the simple proof for this theorem.

The relation of being equivalent partitions the set of forms with discriminant d into classes, so that all the forms in one class are equivalent among themselves, and two forms from two different classes are not equivalent.

It is clear that forms from the same class represent identical sets of integers. For if $k = G(X, Y)$, then $k = F(rX + sY, tX + uY)$.

12.2 The Finiteness of the Number of Classes

Theorem 2.1. *In every class of forms there is always one which satisfies the condition*

$$|b| \leqslant |a| \leqslant |c|.$$

Proof. Let a be an integer with the least absolute value from the set of nonzero integers representable by forms in the class concerned. Let $\{a_0, b_0, c_0\}$ be any form in the class. Then there exist r, t such that

$$a = a_0r^2 + b_0rt + c_0t^2,$$

and $(r, t) = 1$, since otherwise $a/(r, t)^2$ is also representable by $\{a_0, b_0, c_0\}$, and $|a|/(r, t)^2 < |a|$, which is impossible.

We can fix s and u so that $ru - st = 1$. Then $\{a_0, b_0, c_0\}$ is transformed into $\{a, b', c'\}$ via $\begin{pmatrix} r & s \\ t & u \end{pmatrix}$. Now the transformation $\begin{pmatrix} 1 & h \\ 0 & 1 \end{pmatrix}$ transforms $\{a, b', c'\}$ into $\{a, b, c\}$ where $b = 2ah + b'$. We can choose h so that $|b| \leqslant |a|$.

Since c is representable by $\{a, b, c\}$, and this form also belongs to the class containing $\{a_0, b_0, c_0\}$ it follows that $|c| \geqslant |a|$. (Note that $c \neq 0$, because $c = 0$ implies that d is a perfect square.) □

Theorem 2.2. *The number of classes is finite.*

Proof. 1) $d > 0$ (indefinite). From Theorem 2.1 we have

$$|ac| \geqslant b^2 = d + 4ac > 4ac,$$

so that $ac < 0$. Also

$$4a^2 \leqslant 4|ac| = -4ac = d - b^2 \leqslant d$$

so that

$$|a| \leqslant \frac{\sqrt{d}}{2},$$

and hence, by Theorem 2.1

$$|b| \leqslant \frac{\sqrt{d}}{2}.$$

There are therefore only finitely many possible values for a and b. Since $c = (b^2 - d)/4a$, the required result follows.

2) $d < 0$ (definite). Assuming that $a > 0$ we have, from Theorem 2.1,

$$-d = 4ac - b^2 \geqslant 4a^2 - b^2 \geqslant 3a^2$$

so that

$$0 < a < \sqrt{\frac{|d|}{3}}.$$

As before the required result follows from Theorem 2.1. □

Theorem 2.3. *The number of classes of positive definite forms with discriminant d is equal to the number of sets of integers a, b, c satisfying*

$$b^2 - 4ac = d, \qquad \begin{cases} & -a < b \leqslant a < c, \\ \text{or} & 0 \leqslant b \leqslant a = c. \end{cases} \tag{1}$$

Proof. 1) By Theorem 2.1 there is, in any class, always a form which satisfies

$$-a \leqslant b \leqslant a \leqslant c$$

(because a, c are positive). We have the following extra forms to our concluding result:

$$-a = b, \qquad a < c \qquad \text{and} \qquad -a \leqslant b < 0, \qquad a = c.$$

We now prove that

$$\{a, -a, c\} \sim \{a, a, c\} \qquad \text{and} \qquad \{a, -b, a\} \sim \{a, b, a\}.$$

Since $\{a, -a, c\}$ is transformed into $\{a, a, c\}$ via $\begin{pmatrix} 1 & 1 \\ 0 & 1 \end{pmatrix}$, and $\{a, -b, a\}$ is transformed into $\{a, b, a\}$ via $\begin{pmatrix} 0 & 1 \\ -1 & 0 \end{pmatrix}$, we see that, in any class, there is always a form which satisfies (1).

2) We next prove that any two forms are not equivalent. That is, if $\{a, b, c\} \sim \{a', b', c'\}$ and both satisfy (1), then $a = a'$, $b = b'$, $c = c'$.

We can assume that $a' \leqslant a$. Let $\{a, b, c\}$ be transformed into $\{a', b', c'\}$ via $\begin{pmatrix} r & s \\ t & u \end{pmatrix}$. Then

$$a' = ar^2 + brt + ct^2, \tag{2}$$

$$b' = 2ars + b(ru + st) + 2ctu. \tag{3}$$

From the former we have

$$a \geqslant a' \geqslant ar^2 - a|rt| + at^2 = a(|r| - |t|)^2 + a|rt| \geqslant a|rt|, \tag{4}$$

that is $|rt| \leqslant 1$. If $|rt| = 1$, then $a = a'$. Otherwise $rt = 0$, and then

$$a \geqslant a' \geqslant ar^2 + at^2 = a(r^2 + t^2) \geqslant a$$

so that $a = a'$ also.

Suppose first that $c > a$. Then t must be zero, since otherwise from $ct^2 > at^2$ and (4) we deduce that $a > a$ which is impossible. Therefore $t = 0$, $ru = 1$. Now from (3), we have

$$b' = 2ars + b \equiv b \pmod{2a}.$$

Since $-a < b \leqslant a$ and $-a = -a' < b' \leqslant a' = a$ we arrive at $b = b'$, and hence $c = c'$ at once.

The same conclusion can be obtained if we assume that $c' > a'$ $(= a)$.

It remains to consider the case $a = a' = c = c'$. Here we must have $b = \pm b'$, and from $b \geqslant 0$ and $b' \geqslant 0$ we arrive at $b = b'$. □

Note. The case of the indefinite forms is not this easy.

Definition. We call a form which satisfies (1) a *reduced form*.

Exercise 1. Verify the following table of all the reduced forms for $0 < -d \leqslant 20$.

d	-3	-4	-7	-8	-11	-12		-15		-16		-19	-20	
a	1	1	1	1	1	1	2	1	2	1	2	1	1	2
b	1	0	1	0	1	0	2	1	1	0	0	1	0	2
c	1	1	2	2	3	3	2	4	2	4	2	5	5	3

Exercise 2. Prove that when $d = -48$ there are four reduced forms:

$$\{1, 0, 12\}, \{2, 0, 6\}, \{3, 0, 4\}, \{4, 4, 4\}.$$

12.3 Kronecker's Symbol

Definition. Let $m > 0$, $d \equiv 0$ or $1 \pmod 4$ and d not a perfect square. The *Kronecker's symbol* $\left(\frac{d}{m}\right)$ is defined by

$$\left(\frac{d}{p}\right) = 0, \qquad \text{if } p \mid d;$$

$$\left(\frac{d}{2}\right) = \begin{cases} 1 & \text{if } d \equiv 1 \pmod 8, \\ -1 & \text{if } d \equiv 5 \pmod 8; \end{cases}$$

$$\left(\frac{d}{p}\right) = \text{Legendre's symbol } (p \text{ odd prime}, p \nmid d).$$

If $m = \prod_{r=1}^{v} p_r$ where p_r are primes, then

$$\left(\frac{d}{m}\right) = \prod_{r=1}^{v} \left(\frac{d}{p_r}\right).$$

The following are very easy to prove:

(i) If $(d, m) > 1$, then $\left(\frac{d}{m}\right) = 0$.

(ii) If $(d, m) = 1$, then $\left(\frac{d}{m}\right) = \pm 1$.

(iii) If $m_1 > 0$, $m_2 > 0$, then

$$\left(\frac{d}{m_1 m_2}\right) = \left(\frac{d}{m_1}\right)\left(\frac{d}{m_2}\right).$$

Theorem 3.1. *If $m > 0$, $(m, d) = 1$, then the Kronecker's symbol is given by*

$$\left(\frac{d}{m}\right) = \begin{cases} \left(\frac{m}{|d|}\right), & \text{when } d \text{ is odd} \\ \left(\frac{2}{m}\right)^b (-1)^{\frac{u-1}{2}\frac{m-1}{2}} \left(\frac{m}{|u|}\right), & \text{when } d = 2^b u,\ 2 \nmid u. \end{cases}$$

Here $\left(\frac{m}{|d|}\right), \left(\frac{2}{m}\right), \left(\frac{m}{|u|}\right)$ *are all Jacobi symbols.*

Proof. 1) Let d be odd. From the definition of the Kronecker's symbol and Theorem 3.6.5 we have

$$\left(\frac{d}{m}\right) = \left(\frac{m}{|d|}\right).$$

2) Let $d = 2^b u$, $2 \nmid u$. Then $b \geqslant 2$, and m is odd, so that

$$\left(\frac{d}{m}\right) = \left(\frac{2}{m}\right)^b \left(\frac{u}{m}\right) = \left(\frac{2}{m}\right)^b (-1)^{\frac{u-1}{2}\frac{m-1}{2}} \left(\frac{m}{|u|}\right). \quad \square$$

From this theorem we deduce that

$$\left(\frac{d}{m}\right) = \left(\frac{d}{|d| + m}\right).$$

Therefore we have:

Theorem 3.2. *The Kronecker's symbol* $\left(\frac{d}{m}\right)$ is a real character mod $|d|$. $\square$

Theorem 3.3. *Suppose that $m > 0$, $n > 0$ and $m \equiv -n \pmod{|d|}$. Then*

$$\left(\frac{d}{m}\right) = \begin{cases} \left(\frac{d}{m}\right), & \text{if } d > 0, \\ -\left(\frac{d}{n}\right), & \text{if } d < 0. \end{cases}$$

Proof. Since

$$\left(\frac{d}{m}\right)=\left(\frac{d}{n|d|-n}\right)=\left(\frac{d}{n(|d|-1)}\right)=\left(\frac{d}{n}\right)\left(\frac{d}{|d|-1}\right)$$

it follows from Theorem 3.1 that, when d is odd,

$$\left(\frac{d}{|d|-1}\right)=\left(\frac{|d|-1}{|d|}\right)=\left(\frac{-1}{|d|}\right)=(-1)^{\frac{|d|-1}{2}}$$

$$=\begin{cases} 1, & \text{if } d>0, \\ -1, & \text{if } d<0. \end{cases}$$

When d is even, we let $d=2^b u$, $2\nmid u$, $b\geqslant 2$. Then, from Theorem 3.1, we have

$$\left(\frac{d}{|d|-1}\right)=\left(\frac{2}{|d|-1}\right)^b(-1)^{\frac{u-1}{2}}\left(\frac{|d|-1}{|u|}\right)=(-1)^{\frac{u-1}{2}}\left(\frac{-1}{|u|}\right)$$

$$=(-1)^{\frac{u-1}{2}+\frac{|u|-1}{2}}=\begin{cases} 1, & \text{if } d>0, \\ -1, & \text{if } d<0. \end{cases}$$

The Theorem is proved. □

Theorem 3.4. *Let $k>0$ and $(d,k)=1$. The number of solutions to the congruence*

$$x^2\equiv d \pmod{4k} \tag{1}$$

is equal to

$$2\sum_{f\mid k}\left(\frac{d}{f}\right),$$

where the sum is over all positive square-free divisors f of k.

If x is a solution to (1) then so is $x+2k$. Hence, by the theorem, the number of solutions to

$$x^2\equiv d \pmod{4k},\qquad 0\leqslant x<2k$$

is equal to

$$\sum_{f\mid k}\left(\frac{d}{f}\right).$$

Proof. 1) Let d be odd, so that $d\equiv 1 \pmod 4$ and $(d,4k)=1$. From Theorem 3.5.1 we know that the number of solutions to the congruence $x^2\equiv d \pmod{p^l}$ is

$$\begin{array}{ll} 2, & \text{if } p=2,\ l=2, \\ 2\left(1+\left(\frac{d}{p}\right)\right), & \text{if } p=2,\ l>2, \\ 1+\left(\frac{d}{p}\right), & \text{if } p>2. \end{array}$$

From Theorem 2.8.1 we now see that the number of solutions to (1) is

$$2\prod_{p|k}\left(1+\left(\frac{d}{p}\right)\right)=2\sum_{f|k}\left(\frac{d}{f}\right).$$

2) Let d be even, so that $d \equiv 0 \pmod 4$, and hence k is odd. The congruence $x^2 \equiv d \equiv 0 \pmod 4$ has two solutions, and the congruence $x^2 \equiv d \pmod{p^l}$ has $1 + (\frac{d}{p})$ solutions. Therefore, by Theorem 2.8.1, the number of solutions to (1) is

$$2\prod_{p|k}\left(1+\left(\frac{d}{p}\right)\right)=2\sum_{f|k}\left(\frac{d}{f}\right). \quad \square$$

12.4 The Number of Representations of an Integer by a Form

Definition. If $(a, b, c) = 1$, then we call $\{a, b, c\}$ a *primitive* form. If $(a, b, c) = g > 1$, then we say that $\{a, b, c\}$ is *imprimitive*.

Clearly $\left\{\frac{a}{g}, \frac{b}{g}, \frac{c}{g}\right\}$ is a primitive form with discriminant d/g^2. Also, if $\{a, b, c\} \sim \{a_1, b_1, c_1\}$ then the two forms are either both primitive or both imprimitive.

We denote by $h(d)$ the number of classes of primitive forms with discriminant d. Clearly the number of classes of forms with discriminant d is equal to

$$\sum_{\substack{g^2|d \\ g>0}} h\left(\frac{d}{g^2}\right).$$

From each class of primitive forms we select a representative (for definite forms we consider the primitive positive definite forms) giving a representative system which we denote by

$$F_1, \ldots, F_{h(d)}.$$

Theorem 4.1. *Let $k > 0$, $(k, d) = 1$, and denote by $\psi(k)$ the total number of primary solutions to*

$$k = F_1(x, y), \quad \ldots, \quad k = F_{h(d)}(x, y).$$

Then

$$\psi(k) = w\sum_{n|k}\left(\frac{d}{n}\right).$$

(For the definitions of primary solution and w, see §4 in the previous chapter).

Proof. We begin by considering the solutions to the congruence

$$l^2 \equiv d \pmod{4k}, \qquad 0 \leqslant l < 2k.$$

For a given solution l we can determine an integer m from $l^2 - 4km = d$. This then gives a form $\{k, l, m\}$ which is easily seen to be primitive and with discriminant d. Therefore $\{k, l, m\}$ is equivalent to one and only one F_i. Also, from Theorem 11.4.3, we know that there are w proper primary solutions corresponding to each l. Therefore the total number of proper primary solutions to

$$k = F_1(x, y), \quad \ldots, \quad k = F_{h(d)}(x, y)$$

is

$$w \sum_{f|k} \left(\frac{d}{f}\right).$$

Also the total number of primary solutions is

$$\psi(k) = w \sum_{\substack{g^2|k \\ g>0}} \sum_{f|\frac{k}{g^2}} \left(\frac{d}{f}\right)$$

(since $(k, d) = 1$, so that $((k/g^2), d) = 1$). Since $(g^2, d) = 1$ it follows that

$$\psi(k) = w \sum_{\substack{g^2|k \\ g>0}} \sum_{f|\frac{k}{g^2}} \left(\frac{d}{fg^2}\right) = w \sum_{n|k} \left(\frac{d}{n}\right).$$

(This is because any integer n can be written as fg^2 where f is square-free and $g > 0$. Also $g^2|k$, $f|(k/g^2)$ and $n|k$ are equivalent.) □

Consider now the following application of the theorem. It is easy to prove that $h(-4) = 1$ so that $\psi(k)$ is the number of solutions to $k = x^2 + y^2$. Therefore:

Theorem 4.2. *The number of solutions to $x^2 + y^2 = k$ is equal to four times the difference between the number of divisors of k which are congruent* 1 (mod 4) *and the number which are congruent* 3 (mod 4). □

This agrees completely with Theorem 6.7.5.

Exercise 1. Let m be odd. The number of solutions to $x^2 + y^2 = 2^l m$ is 2σ where σ is the difference between the number of divisors of m which are congruent 1 or 3 (mod 8) and the number which are congruent 5 or 7 (mod 8).

Exercise 2. The number of solutions to $x^2 + xy + y^2 = k$ is $6E(k)$ where $E(k)$ is the number of divisors of k of the form $3h + 1$ subtracting the number of divisors of the form $3h + 2$.

Exercise 3. Let m be odd and consider the number of solutions to the equation $x^2 + 3y^2 = 2^l m$. If l is odd, then this number is zero; if $l = 0$, then this number is $2E(m)$; if l is positive and even, then this number is $6E(m)$. Here $E(m)$ has the same definition as earlier.

Exercise 4. If m is odd, then the equation $x^2 + 3y^2 = 4m$ has $E(m)$ positive odd solutions.

Exercise 5. Let m be odd and consider the number of solutions to the equation $x^2 + 4y^2 = 2^k m$. When $k = 0$, this number is $2E$; when $k = 1$, this number is 0; when $k \geqslant 2$, this number is $2E$. Here E is the number of prime divisors of m congruent 1 (mod 4) subtract the number of divisors of k congruent 3 (mod 4).

Exercise 6. Denote by $e(n)$ the number of divisors of n congruent 1, 2, 4 (mod 7) subtract the number of those congruent 3, 5, 6 (mod 7). The number of solutions to $x^2 + xy + 2y^2 = n > 0$ is then $2e(n)$.

Exercise 7. If m is odd, then $e(2^a m) = (a + 1)e(m)$. Let $3 \nmid t$. If b is odd, then $e(3^b t) = 0$ and if b is even, then $e(3^b t) = e(t)$.

Exercise 8. Let m be positive and odd. The numbers of solutions to $m = x^2 + 7y^2$ and $2m = x^2 + 7y^2$ are $2e(m)$ and 0 respectively. The number of solutions to $4k = x^2 + 7y^2$ is $4e(k)$.

Exercise 9. Let m be positive and odd. Then there are $e(m)$ positive integer solutions to $x^2 + 7y^2 = 8m$.

Exercise 10. The number of solutions to $x^2 + xy + 3y^2 = m > 0$ is twice the difference between the number of divisors of m congruent 1, 3, 4, 5, 9 (mod 11) and the number of those congruent 2, 6, 7, 8, 10 (mod 11).

12.5 The Equivalence of Forms mod q

Let q be a prime number. Suppose that there is an integer valued coefficients substitution

$$x = rX + sY, \qquad y = tX + uY, \qquad (ru - st, q) = 1 \tag{1}$$

such that

$$ax^2 + bxy + cy^2 \equiv a_1 X^2 + b_1 XY + c_1 Y^2 \pmod{q}. \tag{2}$$

Then we say that the two forms $\{a, b, c\}$ and $\{a_1, b_1, c_1\}$ are equivalent mod q. If we denote by d and d_1 the discriminants for $\{a, b, c\}$ and $\{a_1, b_1, c_1\}$, then clearly

$$d_1 \equiv (ru - st)^2 (b^2 - 4ac) \equiv (ru - st)^2 d \pmod{q}. \tag{3}$$

From (3) we see that if $\{a, b, c\}$ and $\{a_1, b_1, c_1\}$ are equivalent mod p, then

$$\left(\frac{d}{p}\right) = \left(\frac{d_1}{p}\right).$$

Let us take q to be a prime $p > 2$. Suppose that the discriminant of $\{a, b, c\}$ is d where $p \nmid d$. Then $\{a, b, c\}$ must be equivalent mod p to a form $\{a_1, 0, c_1\}$. This is because $p \nmid (a, b, c)$, and if $p \nmid a$ then letting

$$X \equiv x + \frac{b}{2a}y, \qquad Y \equiv y \pmod{p}$$

we have

$$ax^2 + bxy + cy^2 \equiv a\left(x + \frac{b}{2a}y\right)^2 - \frac{d}{4a}y^2 \equiv aX^2 - \frac{d}{4a}y^2 \pmod{p},$$

and similarly if $p \nmid c$; if $p|(a, c)$, then taking $x = X + Y$, $y = X - Y$ we have

$$ax^2 + bxy + cy^2 \equiv bxy \equiv bX^2 - bY^2 \pmod{p}.$$

Therefore we can assume from now on that $p|b$ and $p \nmid ac$.

Lemma 1. *If $p \nmid ac$, then there are x, y such that*

$$ax^2 + cy^2 \equiv 1 \pmod{p}.$$

Proof. Let x, y run over $0, 1, \ldots, p - 1$ separately. Then ax^2 and $1 - cy^2$ separately take $(p + 1)/2$ distinct values. Therefore there are x, y such that

$$ax^2 \equiv 1 - cy^2 \pmod{p}$$

as required. □

Let $1 \equiv ar^2 + ct^2 \pmod{p}$ and let s, u be any pair of integers satisfying $p \nmid ru - st$. With s, u fixed, we let

$$b_1 \equiv 2ars + 2ctu, \qquad c_1 \equiv as^2 + cu^2 \pmod{p}$$

so that $\{a, 0, c\} \sim \{1, b_1, c_1\}$ mod p. If d_1 is the discriminant of the second form, then from our discussions we have

$$\{1, b_1, c_1\} \sim \left\{1, 0, -\frac{d_1}{4}\right\} \sim \{1, 0, -d_1\} \pmod{p}.$$

Summarizing we have:

Theorem 5.1. *Let the discriminant of $\{a, b, c\}$ be d, and $p > 2$, $p \nmid d$. Let r be any quadratic non-residue* mod p. *Then*

$$\{a, b, c\} \sim \{1, 0, -1\} \sim \{0, 1, 0\} \pmod{p}$$

if $\left(\frac{d}{p}\right) = 1$, *and*

$$\{a, b, c\} \sim \{1, 0, -r\} \pmod{p}$$

if $\left(\frac{d}{p}\right) = -1$. *Also* $\{1, 0, -1\}$ *and* $\{1, 0, -r\}$ *cannot be equivalent* mod p. □

Corollary. *If p is an odd prime that does not divide d, then any two forms with discriminant d must be equivalent* mod p. □

When $q = 2$ and the forms have odd discriminants we have:

Theorem 5.2. *Any form with an odd discriminant must be equivalent* mod 2 *to exactly one of the following*

$$\{0, 1, 0\}, \{1, 1, 1\}.$$

More specifically, we have

$$\{a, b, c\} \sim \{0, 1, 0\} \pmod{2} \quad \textit{if} \quad 2|ac;$$

$$\{a, b, c\} \sim \{1, 1, 1\} \pmod{2} \quad \textit{if} \quad 2\nmid ac.$$

Proof. Since $2\nmid d$ it follows that $2\nmid b$. Consequently if $2\nmid ac$, then

$$ax^2 + bxy + cy^2 \equiv x^2 + xy + y^2 \pmod{2};$$

if $2|ac$, then either $2|a$ or $2|c$. But if $2|a$ then

$$ax^2 + bxy + cy^2 \equiv xy + cy^2 \equiv y(x + cy) \pmod{2}$$

so that $\{a, b, c\} \sim \{0, 1, 0\} \pmod{2}$, and similarly if $2|c$.

Finally $\{0, 1, 0\}$ and $\{1, 1, 1\}$ cannot be equivalent mod 2 so that the theorem is proved. □

Corollary. *Any two forms with the same odd discriminant must be equivalent* mod 2. □

We next consider the case when p divides the discriminant of the forms.

Lemma 2. *Let n be any given integer. Then there are two integers x, y such that $(x, y) = 1$ and $(F(x, y), n) = 1$.*

Proof. Let q be any prime number. Since $F(x, y)$ is a primitive form, $q\nmid(a, b, c)$. If $q\nmid a$, then $q\nmid F(1, 0)$; if $q\nmid c$, then $q\nmid F(0, 1)$; if $q|(a, c)$ and $q\nmid b$, then $q\nmid F(1, 1)$. Therefore the lemma follows if $n = q$.

Let $q_1, \ldots, q_t$ be all the distinct prime divisors of n. From the above, there are integers x_i, y_i such that $q_i\nmid F(x_i, y_i)$. From the Chinese remainder theorem there are two integers X, Y such that

$$X \equiv x_i \pmod{q_i}, \qquad Y = y_i \pmod{q_i}, \qquad i = 1, 2, \ldots, t.$$

Clearly we have

$$(F(X, Y), n) = 1.$$

Now let $x = X/(X, Y)$, $y = Y/(X, Y)$. Then $(x, y) = 1$ and

$$(F(x, y), n) = 1. \quad \square$$

Consider now $p > 2$, $p \mid d$ where d is the discriminant of the form $\{a, b, c\}$. Since $p \nmid (a, c)$ we may assume that $p \nmid a$. It is easily seen that

$$\{a, b, c\} \sim \{a, 0, 0\} \pmod{p}.$$

Theorem 5.3. *Let $p > 2$ and let the forms $\{a, b, c\}$ and $\{a_1, b_1, c_1\}$ have discriminants d and d_1 respectively where $p|d$, $p|d_1$. A necessary and sufficient condition for $\{a, b, c\}$ and $\{a_1, b_1, c_1\}$ to be equivalent* mod p *is that*

$$\left(\frac{k}{p}\right) = \left(\frac{k_1}{p}\right),$$

where k and k_1 are any integers representable by $\{a, b, c\}$ and $\{a_1, b_1, c_1\}$ respectively and satisfying $(k, d) = 1$, $(k_1, d_1) = 1$.

Proof. That k and k_1 exist follows from Lemma 2. Let $k \equiv ax^2 + bxy + cy^2 \pmod{p}$, $(k, p) = 1$. Then

$$\left(\frac{k}{p}\right) = \left(\frac{ax^2 + bxy + cy^2}{p}\right) = \left(\frac{ax_1^2}{p}\right) = \left(\frac{a}{p}\right).$$

Thus $\left(\frac{k}{p}\right)$ is constant and is equal to $\left(\frac{a}{p}\right)$. Suppose now that $\{a, b, c\}$ and $\{a_1, b_1, c_1\}$ are equivalent mod p. Then, from the definition of equivalence,

$$\left(\frac{k}{p}\right) = \left(\frac{a}{p}\right) = \left(\frac{a_1}{p}\right) = \left(\frac{k_1}{p}\right).$$

Conversely, if $\left(\frac{k}{p}\right) = \left(\frac{k_1}{p}\right)$, then $\left(\frac{a}{p}\right) = \left(\frac{a_1}{p}\right)$ so that there is an integer z such that $a \equiv a_1 z^2 \pmod{p}$ and hence

$$\{a, b, c\} \sim \{a, 0, 0\} \sim \{a_1, 0, 0\} \sim \{a_1, b_1, c_1\} \pmod{p}. \quad \square$$

It remains to consider the situation when $p = 2$ and $2|d$. We first introduce the following symbols:

$$\delta(k) = (-1)^{\frac{k-1}{2}}, \qquad \text{if } \frac{d}{4} \equiv 0 \text{ or } 3 \pmod 4;$$

$$\varepsilon(k) = (-1)^{\frac{k^2-1}{8}}, \qquad \text{if } \frac{d}{4} \equiv 0 \text{ or } 2 \pmod 8;$$

$$\delta(k)\varepsilon(k) = (-1)^{\frac{k-1}{2}+\frac{k^2-1}{8}}, \qquad \text{if } \frac{d}{4} \equiv 0 \text{ or } 6 \pmod 8;$$

where k is an odd integer representable by $\{a, b, c\}$.

Since $2|d$ implies $2|b$ we shall assume that $b = 0$ and consider

$$ax^2 + cy^2, \qquad d = -4ac.$$

Theorem 5.4. *A necessary and sufficient condition for two forms satisfying* $\frac{d}{4} \equiv 3 \pmod 4$ *to be equivalent* mod 4 *is that they should have the same* δ.

Proof. Since $d = -4ac$, it follows that $ac \equiv 1 \pmod 4$, that is $a \equiv c \pmod 4$. If $2 \nmid k$ and k is representable as

$$k \equiv ax^2 + cy^2 \equiv a(x^2 + y^2) \pmod 4,$$

then, since x, y must have the same parity it follows that $k \equiv a \pmod 4$ and hence $\delta(k) = \delta(a)$. The theorem can easily be deduced from this. □

The same method can be used to prove the following theorems:

Theorem 5.5. *A necessary and sufficient condition for two forms satisfying* $\frac{d}{4} \equiv 2 \pmod 8$ *to be equivalent* mod 8 *is that they should have the same* ε. □

Theorem 5.6. *A necessary and sufficient condition for two forms satisfying* $\frac{d}{4} \equiv 6 \pmod 8$ *to be equivalent* mod 8 *is that they should have the same* $\delta\varepsilon$. □

Theorem 5.7. *A necessary and sufficient condition for two forms satisfying* $\frac{d}{4} \equiv 0 \pmod 4$ *to be equivalent* mod 4 *is that they should have the same* δ. □

Theorem 5.8. *A necessary and sufficient condition for two forms satisfying* $\frac{d}{4} \equiv 0 \pmod 8$ *to be equivalent* mod 8 *is that they should have the same* δ *and* ε. □

Exercise 1. Any two forms satisfying $\frac{d}{4} \equiv 2 \pmod 4$ are equivalent mod 4.

Exercise 2. Any two forms satisfying $\frac{d}{4} \equiv 1 \pmod 4$ are equivalent mod 4.

Exercise 3. Any forms satisfying $\frac{d}{4} \equiv 1 \pmod 4$ must be equivalent mod 8 to exactly one of

$$x^2 + 3y^2, \qquad x^2 + 7y^2.$$

Deduce also that any two forms with the same discriminant d which satisfies $\frac{d}{4} \equiv 1 \pmod 4$ must be equivalent mod 8.

Exercise 4. Let q be any positive integer. A necessary and sufficient condition for two quadratic forms to be equivalent mod q is that they have the same character system (see Definition 1 in the next section).

12.6 The Character System for a Quadratic Form and the Genus

It follows at once from the definitions that any two quadratic forms which are equivalent are also equivalent mod q for any q.

Definition 1. Let $p_1, \ldots, p_s$ be the odd prime divisors of d. If $(k, 2d) = 1$ and k is representable by $F(x, y)$ then, from the previous section, we see that

$$\left(\frac{k}{p_i}\right), \delta(k), \varepsilon(k), \delta(k)\varepsilon(k) \tag{1}$$

do not depend on k. We call them the *character system* for $F(x, y)$.

Since two equivalent quadratic forms have the same character system we can speak of the character system of an equivalence class of forms.

Definition 2. If two quadratic forms with the same discriminant d have the same values for each of the characters, then we say that they belong to the same *genus*.

It is easily seen that a genus is formed from various equivalence classes of forms. We shall prove that each genus has the same number of equivalence classes. Since this fact falls more naturally in the study of ideals in a quadratic field we do not give the proof here. The importance of the notion of genus comes from the discussion of the representation of integers by quadratic forms.

Let $F(x, y)$ be a fixed quadratic primitive form. We now discuss the Diophantine equation

$$k = F(x, y). \tag{2}$$

If $h(d) = 1$, then this problem can be solved with Theorem 4.1. But if $h(d) \neq 1$, then we only have certain incomplete results from Theorem 4.1. For example if $\psi(k) = 0$, then (2) has no solutions; but if $\psi(k) \neq 0$, is (2) soluble then? If it is soluble, then how many solutions are there? These questions cannot be answered by Theorem 4.1. The introduction of the notion of genus helps partly to answer these questions.

Example 1. $d = -96$. There are four positive definite reduced primitive forms:

$$\{1, 0, 24\}, \{3, 0, 8\}, \{4, 4, 7\}, \{5, 2, 5\}.$$

From Theorem 4.1 we only know that if k is representable by these four forms, then the total number of solutions is

$$\psi(k) = 2\sum_{n|k}\left(\frac{-96}{n}\right),$$

where n runs over all the positive divisors of k. In order to calculate the character system we first select k coprime with d and representable by the forms. We take

$$k = 1, 11, 7, 5$$

and obtain

Form	$\left(\frac{k}{3}\right)$	$\delta(k)$	$\varepsilon(k)$
$\{1, 0, 24\}$	$+1$	$+1$	$+1$
$\{3, 0, 8\}$	-1	-1	-1
$\{4, 4, 7\}$	$+1$	-1	$+1$
$\{5, 2, 5\}$	-1	$+1$	-1

This table shows that each genus has one equivalence class. Therefore, when $k \equiv 1, 11, 7, 5 \pmod{12}$, $\psi(k)$ represents the number of solutions of the first, the second, the third and the fourth form respectively. More specifically, if $k \equiv 1 \pmod{12}$, then $\psi(k) = 2\sum_{n|k}(-96/n)$ represents the number of solutions to $x^2 + 24y^2 = k$. At the same time we have proved that this equation has no solution if $k \equiv 11, 7, 5 \pmod{12}$.

Example 2. $d = -15$. There are two positive definite reduced primitive forms:

$$\{1, 1, 4\}, \{2, 1, 2\}.$$

Taking $k = 1$ and 17 will give

$$\left(\frac{k}{3}\right) = \left(\frac{k}{5}\right) = 1 \quad \text{and} \quad \left(\frac{k}{3}\right) = \left(\frac{k}{5}\right) = -1.$$

We can then perform the calculations for $k \equiv 1, 4 \pmod{15}$ and $k \equiv 2, 8 \pmod{15}$. We conclude that if $k \equiv 7, 11, 13$ or $14 \pmod{15}$, then k is not representable by either of the two forms. If $k \equiv 1, 4 \pmod{15}$ then there are $2\sum_{n|k}(-15/n)$ ways to represent k by $\{1, 1, 4\}$; if $k \equiv 2, 8 \pmod{15}$, then there are the similar number of ways to represent k by $\{2, 1, 2\}$.

From these two examples we see that if each genus contains only one equivalence class, then the number of solutions to (2) is completely determined when $(k, 2d) = 1$.

We tabulate all the discriminants $d > -400$ in which the genus has only one equivalence class in the following table, where we have also included all the positive definite reduced primitive forms.

Exercise. Study, as in the examples, the cases $d = -20, -24, -32, -35, -51, -75$.

$-d = 3$	1, 1, 1
4	1, 0, 1
7	1, 1, 2
8	1, 0, 2
11	1, 1, 3
12	1, 0, 3
15	1, 1, 4
	2, 1, 2
16	1, 0, 4
19	1, 1, 5
20	1, 0, 5
	2, 2, 3
24	1, 0, 6
	2, 0, 3
27	1, 1, 7
28	1, 0, 7
32	1, 0, 8
	3, 2, 3
35	1, 1, 9
	3, 1, 3
36	1, 0, 9
	2, 2, 5
40	1, 0, 10
	2, 0, 5
43	1, 1, 11
48	1, 0, 12
	3, 0, 4
51	1, 1, 13
	3, 3, 5
52	1, 0, 13
	2, 2, 7
60	1, 0, 15
	3, 0, 5
64	1, 0, 16
	4, 4, 5
67	1, 1, 17
72	1, 0, 18
	2, 0, 9
75	1, 1, 19
	3, 3, 7
84	1, 0, 21
	2, 2, 11
	3, 0, 7
	5, 4, 5
88	1, 0, 22
	2, 0, 11
91	1, 1, 23
	5, 3, 5

$-d = 96$	1, 0, 24
	3, 0, 8
	4, 4, 7
	5, 2, 5
99	1, 1, 25
	5, 1, 5
100	1, 0, 25
	2, 2, 13
112	1, 0, 28
	4, 0, 7
115	1, 1, 29
	5, 5, 7
120	1, 0, 30
	2, 0, 15
	3, 0, 10
	5, 0, 6
123	1, 1, 31
	3, 3, 11
132	1, 0, 33
	2, 2, 17
	3, 0, 11
	6, 6, 7
147	1, 1, 37
	3, 3, 13
148	1, 0, 37
	2, 2, 19
160	1, 0, 40
	4, 4, 11
	5, 0, 8
	7, 6, 7
163	1, 1, 41
168	1, 0, 42
	2, 0, 21
	3, 0, 14
	6, 0, 7
180	1, 0, 45
	2, 2, 23
	5, 0, 9
	7, 4, 7
187	1, 1, 47
	7, 3, 7
192	1, 0, 48
	3, 0, 16
	4, 4, 13
	7, 2, 7

$-d = 195$	1, 1, 49
	3, 3, 17
	5, 5, 11
	7, 1, 7
228	1, 0, 57
	2, 2, 29
	3, 0, 19
	6, 6, 11
232	1, 0, 58
	2, 0, 29
235	1, 1, 59
	5, 5, 13
240	1, 0, 60
	3, 0, 20
	4, 0, 15
	5, 0, 12
267	1, 1, 67
	3, 3, 23
280	1, 0, 70
	2, 0, 35
	5, 0, 14
	7, 0, 10
288	1, 0, 72
	4, 4, 19
	8, 0, 9
	8, 8, 11
312	1, 0, 78
	2, 0, 39
	3, 0, 26
	6, 0, 13
315	1, 1, 79
	5, 5, 17
	7, 7, 13
	9, 9, 11
340	1, 0, 85
	2, 2, 43
	5, 0, 17
	10, 10, 11
352	1, 0, 88
	4, 4, 23
	8, 0, 11
	8, 8, 13
372	1, 0, 93
	2, 2, 47
	3, 0, 31
	6, 6, 17

12.7 The Convergence of the Series $K(d)$

Let

$$K(d) = \sum_{n=1}^{\infty} \left(\frac{d}{n}\right)\frac{1}{n}. \tag{1}$$

This is a very important series. Since $\left(\frac{d}{n}\right)$ is a real character mod $|d|$, it follows from Theorem 7.2.3 that

$$\left|\sum_{a \leqslant n \leqslant b} \left(\frac{d}{n}\right)\right| < |d|.$$

Moreover we see from Theorem 6.8.2 that the series $K(d)$ is convergent.

Theorem 7.1.

$$\lim_{\tau\to\infty} \frac{1}{\tau} \sum_{\substack{1\leqslant k\leqslant\tau \\ (k,d)=1}} \sum_{n|k} \left(\frac{d}{n}\right) = \frac{\varphi(|d|)}{|d|} K(d).$$

Proof. 1) Let $A(\tau; d, n)$ denote the number of positive integers not exceeding τ/n and coprime with d. Then

$$\frac{1}{\tau} \sum_{\substack{1\leqslant k\leqslant\tau \\ (k,d)=1}} \sum_{n|k} \left(\frac{d}{n}\right) = \frac{1}{\tau}\sum_{n=1}^{\infty}\left(\frac{d}{n}\right) \sum_{\substack{1\leqslant k\leqslant\tau \\ (k,d)=1 \\ n|k}} 1 = \frac{1}{\tau}\sum_{n=1}^{\infty}\left(\frac{d}{n}\right) \sum_{\substack{1\leqslant k\leqslant\tau/n \\ (k,d)=1}} 1$$

$$= \sum_{n=1}^{\infty}\left(\frac{d}{n}\right)\frac{A(\tau; d, n)}{\tau}. \tag{2}$$

Since $A(\tau; d, n)$ does not increase as n increases, and

$$\frac{A(\tau; d, n)}{\tau} \leqslant \frac{1}{n},$$

it follows from Theorem 6.8.2 that the series (2) converges uniformly in τ. Also, for fixed n, we have

$$\lim_{\tau\to\infty} \frac{A(\tau; d, n)}{\tau} = \frac{\varphi(|d|)}{|d|}\frac{1}{n}.$$

Therefore

$$\lim_{\tau\to\infty} \frac{1}{\tau} \sum_{\substack{1\leqslant k\leqslant\tau \\ (k,d)=1}} \sum_{n|k} \left(\frac{d}{n}\right) = \sum_{n=1}^{\infty}\left(\frac{d}{n}\right)\lim_{\tau\to\infty}\frac{A(\tau; d, n)}{\tau}$$

$$= \frac{\varphi(|d|)}{|d|}\sum_{n=1}^{\infty}\left(\frac{d}{n}\right)\frac{1}{n}. \quad \square$$

12.8 The Number of Lattice Points Inside a Hyperbola and an Ellipse

Theorem 8.1. *Let $m > 0$ and let there be an ellipse centre at the origin, or a hyperbola centre at the origin (the two curves of the hyperbola together with two lines passing through the origin). Denote by I the (finite) area of the region. Magnify the original figure by $\sqrt{\tau}$ (that is replacing ξ and η by $\xi\sqrt{\tau}$ and $\eta\sqrt{\tau}$), and denote by $U(\tau)$ the number of lattice points in the magnified figure whose coordinates satisfy*

$$\xi \equiv \xi_0 \pmod{m}, \qquad \eta \equiv \eta_0 \pmod{m}.$$

Then

$$\lim_{\tau\to\infty} \frac{U(\tau)}{\tau} = \frac{I}{m^2}.$$

Proof. We form a net in the original figure with the orthogonal lines

$$\xi = \frac{\xi_0 + \gamma m}{\sqrt{\tau}}, \qquad \eta = \frac{\eta_0 + sm}{\sqrt{\tau}}.$$

This gives a net of squares with side length $m/\sqrt{\tau}$.

Denote by $W(\tau)$ the number of squares whose "south-west corners" lie inside the ellipse or the hyperbola. Then clearly

$$U(\tau) = W(\tau).$$

Since the area of each square in the net is m^2/τ it follows at once from the fundamental theorem of calculus that

$$I = \iint d\xi\, d\eta = \lim_{\tau\to\infty} \frac{m^2}{\tau} W(\tau),$$

and hence the required result. □

12.9 The Limiting Average

Denote by $\psi(k, F)$ the number of proper representations of k by F, and let

$$H(\tau, F) = \sum_{\substack{1 \leqslant k \leqslant \tau \\ (k,d)=1}} \psi(k, F), \qquad \tau > 1.$$

The aim of this section is the evaluate

$$\lim_{\tau\to\infty} \frac{1}{\tau} H(\tau, F).$$

Theorem 9.1. *As x, y both run over a complete residue system* mod $|d|$, *there are precisely* $|d|\varphi(|d|)$ *sets of* x, y *such that* $F(x, y)$ *is coprime with* d.

Proof. It suffices to prove that if $p^l | d$, $l > 0$, then there are $p^l\varphi(p^l)$ sets of x, y in a complete residue system mod p^l such that $p \nmid F(x, y)$. For let the standard factorization for $|d|$ be $\prod_i p_i^{l_i}$. Then, since $(d, F(x, y)) = 1$ and $p \nmid F(x, y)$ are equivalent, it follows from the Chinese remainder theorem that, as x, y run over a complete residue system mod $|d|$, there are

$$\prod_{p \mid |d|} p^l\varphi(p^l) = |d|\varphi(|d|)$$

values of $F(x, y)$ which are coprime with d.

Since $(a, b, c) = 1$, we have $p \nmid (a, c)$. We now assume that $p \nmid a$.

1) Suppose that $p > 2$. Since $(p, 4a) = 1$, it follows from

$$4aF = (2ax + by)^2 - dy^2 \not\equiv 0 \pmod p$$

that

$$2ax + by \not\equiv 0 \pmod p,$$

and conversely. For any given value of y (there are p^l values) there are $p - 1$ distinct values for $x \bmod p$, because $p \nmid 2a$. There are thus $p^{l-1}(p - 1) = \varphi(p^l)$ values for $x \bmod p^l$. The required result is proved.

2) Suppose that $p = 2$. Now $2|d$ implies $2|b$. The condition

$$ax^2 + bxy + cy^2 \equiv 1 \pmod 2$$

becomes

$$ax + cy \equiv 1 \pmod 2.$$

Since corresponding to each value of y (there are 2^l values) there are 2^{l-1} values x (mod 2^l) which satisfy the above equation, the theorem is proved. □

Theorem 9.2. *We have*

$$\lim_{\tau\to\infty} \frac{H(\tau, F)}{\tau} = \begin{cases} \dfrac{2\pi}{\sqrt{|d|}} \dfrac{\varphi(|d|)}{|d|}, & \text{if } d < 0, \\[2ex] \dfrac{\log \varepsilon}{\sqrt{d}} \dfrac{\varphi(d)}{d}, & \text{if } d > 0. \end{cases}$$

Proof. If $d < 0$, we let $U(\tau) = U(\tau, F, x_0, y_0)$ denote the number of solutions to

$$0 \leqslant F(x, y) \leqslant \tau,$$

$$x \equiv x_0 \pmod{|d|}, \qquad y \equiv y_0 \pmod{|d|}.$$

If $d > 0$, then we let $U(\tau) = U(\tau, F, x_0, y_0)$ denote the number of solutions to

$$0 \leqslant F(x, y) \leqslant \tau, \qquad \bar{L} > 0, \qquad 1 \leqslant \left|\frac{L}{\bar{L}}\right| < \varepsilon^2,$$

$$x \equiv x_0 \pmod{|d|}, \qquad y \equiv y_0 \pmod{|d|}.$$

Here the definitions for L, $\bar{L}$, ε are the same as §11.4.

Let x_0, y_0 both run over the complete residue system mod $|d|$ such that $(F(x_0, y_0), d) = 1$. Then

$$\sum_{\substack{(x_0, y_0) \\ (F(x_0, y_0), d) = 1}} U(\tau) = \sum_{\substack{1 \leqslant k \leqslant \tau \\ (k, d) = 1}} \psi(k, F) = H(\tau, F),$$

and hence

$$\lim_{\tau \to \infty} \frac{H(\tau, F)}{\tau} = \lim_{\tau \to \infty} \frac{1}{\tau} \sum_{\substack{(x_0, y_0) \\ (F(x_0, y_0), d) = 1}} U(\tau).$$

By Theorem 9.1 we see that our theorem follows if we can prove that, for each set of x_0, y_0, we have

$$\lim_{\tau \to \infty} \frac{U(\tau)}{\tau} = \begin{cases} \dfrac{2\pi}{\sqrt{|d|}} \dfrac{1}{d^2}, & \text{if } d < 0, \\[2ex] \dfrac{\log \varepsilon}{\sqrt{d}} \dfrac{1}{d^2}, & \text{if } d > 0. \end{cases}$$

Also, by Theorem 8.1, we need now only evaluate the area for the ellipse $F(x, y) \leqslant 1$, $(d < 0)$, and the area for the hyperbola $0 \leqslant F(x, y) \leqslant 1, \tau > 0, 1 \leqslant \left|\frac{L}{\bar{L}}\right| < \varepsilon^2$ $(d > 0)$.

1) Suppose that $d < 0$. It is well known that the area of the ellipse $ax^2 + bxy + cy^2 \leqslant 1$ is $2\pi/\sqrt{|d|}$. The theorem is therefore proved.

2) Suppose that $d > 0$, and we may assume that $a > 0$. Since

$$L = 2ax + (b + \sqrt{d})y, \qquad \bar{L} = 2ax + (b - \sqrt{d})y,$$

so that

$$L\bar{L} = 4a(ax^2 + bxy + cy^2),$$

and hence $L > 0$.

The required area for the hyperbola is

$$I = \iint dx\, dy$$

where the integration is over $L\bar{L} \leqslant 4a$, $\bar{L} > 0$, $1 \leqslant L/\bar{L} < \varepsilon^2$. We make the substitution

$$\frac{L}{2\sqrt{a}} = \rho, \qquad \frac{\bar{L}}{2\sqrt{a}} = \sigma$$

whose Jacobian has the value

$$\begin{vmatrix} \dfrac{\partial \rho}{\partial x} & \dfrac{\partial \rho}{\partial y} \\ \dfrac{\partial \sigma}{\partial x} & \dfrac{\partial \sigma}{\partial y} \end{vmatrix} = \frac{1}{2\sqrt{a}} \cdot \frac{1}{2\sqrt{a}} \begin{vmatrix} 2a & b + \sqrt{d} \\ 2a & b - \sqrt{d} \end{vmatrix} = -\sqrt{d}.$$

Therefore

$$I = \frac{1}{\sqrt{d}} \iint d\rho\, d\sigma,$$

where the integration is over $\rho\sigma \leqslant 1$, $\sigma > 0$, $\sigma \leqslant \rho < \varepsilon^2\sigma$. This is the region formed by the two straight lines from the points (1, 1) and $(\varepsilon, 1/\varepsilon)$ to (0, 0) together with the rectangular hyperbola joining the points (1, 1) and $(\varepsilon, 1/\varepsilon)$. Therefore

$$\begin{aligned} \sqrt{d}\, I &= \int_0^1 d\rho \int_{\rho/\varepsilon^2}^{\rho} d\sigma + \int_1^{\varepsilon} d\rho \int_{\rho/\varepsilon^2}^{1/\rho} d\sigma \\ &= \int_0^1 \left(\rho - \frac{\rho}{\varepsilon^2}\right) d\rho + \int_1^{\varepsilon} \left(\frac{1}{\rho} - \frac{\rho}{\varepsilon^2}\right) d\rho \\ &= \int_0^1 \rho\, d\rho + \int_1^{\varepsilon} \frac{d\rho}{\rho} - \int_0^{\varepsilon} \frac{\rho}{\varepsilon^2}\, d\rho = \log \varepsilon. \end{aligned}$$

This gives

$$I = \frac{\log \varepsilon}{\sqrt{d}},$$

and the theorem is proved. □

12.10 The Class Number: An Analytic Expression

Theorem 10.1.

$$h(d) = \begin{cases} \dfrac{w\sqrt{|d|}}{2\pi} K(d), & \textit{if } d < 0, \\ \dfrac{\sqrt{d}}{\log \varepsilon} K(d), & \textit{if } d > 0. \end{cases}$$

Proof. Let

$$F_1, \ldots, F_{h(d)}$$

be a representative system. From Theorem 4.1 we have

$$\sum_F H(\tau, F) = \sum_{\substack{1 \leqslant k \leqslant \tau \\ (k,d)=1}} \sum_F \psi(k, F)$$

$$= \sum_{\substack{1 \leqslant k \leqslant \tau \\ (k,d)=1}} \psi(k) = w \sum_{\substack{1 \leqslant k \leqslant \tau \\ (k,d)=1}} \sum_{n|k} \left(\frac{d}{n}\right).$$

From Theorem 7.1 and Theorem 9.2 we have

$$h(d) \begin{Bmatrix} 2\pi \\ \log \varepsilon \end{Bmatrix} \frac{\varphi(|d|)}{|d|^{\frac{3}{2}}} = w \frac{\varphi(|d|)}{|d|} K(d) \begin{cases} \text{if} & d < 0, \\ \text{if} & d > 0, \end{cases}$$

as required. □

Therefore our problem becomes that of the determination of the sum of the series

$$K(d) = \sum_{n=1}^{\infty} \frac{1}{n} \left(\frac{d}{n}\right).$$

12.11 The Fundamental Discriminants

Definition. By a *fundamental discriminant* we mean a discriminant d which has no odd prime square divisor, and d is odd or $d \equiv 8$ or $12 \pmod{16}$.

For example: 5, 8, 12, 13, 17, 21, 24, 28, 29, ... are fundamental discriminants.

Theorem 11.1. *Each discriminant d is uniquely expressible as fm^2 where f is a fundamental discriminant.*

Proof. 1) If d is odd, then we let m^2 be the largest square that divides d. Write $d = fm^2$ for the required result.

2) If d is even, then we first write $d = qr^2$ where r^2 is the largest square that divides d. Clearly $2|r$. If $q \equiv 1 \pmod 4$, then q is a fundamental discriminant. If $q \equiv 2$ or $3 \pmod 4$, then we take $f = 4q$ so that from $4q \equiv 8$ or $12 \pmod{16}$ we see that f is a fundamental discriminant.

3) *Uniqueness.* Let $d = fm^2$, $m > 0$ and f be a fundamental discriminant. If f is odd, then f has no square divisor so that m^2 is the largest square divisor of d. If f is even, then $f \equiv 8$ or $12 \pmod{16}$, hence $4 \nmid f/4$ and therefore $(2m)^2$ is the largest square divisor of d. From this we see that the uniqueness property follows. □

Theorem 11.2. *Let $d = fm^2$ be the representation in Theorem 11.1. Then*

$$K(d) = \prod_{p|m} \left(1 - \left(\frac{f}{p}\right)\frac{1}{p}\right) K(f).$$

Proof. We have

$$K(d) = \sum_{n=1}^{\infty} \left(\frac{d}{n}\right)\frac{1}{n} = \sum_{n=1}^{\infty} \left(\frac{m^2 f}{n}\right)\frac{1}{n}$$
$$= \sum_{\substack{n=1 \\ (m,n)=1}}^{\infty} \left(\frac{f}{n}\right)\frac{1}{n}.$$

Let the standard factorization of m be $p_1^{l_1} \cdots p_s^{l_s}$. Then from Theorem 1.7.1 we have

$$K(d) = K(f) - \sum_{p_i | m} \left(\frac{f}{p_i}\right)\frac{1}{p_i} K(f)$$
$$+ \sum_{\substack{p_i | m, p_j | m \\ p_i \neq p_j}} \left(\frac{f}{p_i p_j}\right)\frac{1}{p_i p_j} K(f) - + \cdots$$
$$= \prod_{p|m} \left(1 - \left(\frac{f}{p}\right)\frac{1}{p}\right) K(f). \quad \square$$

We see from this theorem that we need only determine the values for $K(f)$.

Exercise. Show that if d is a fundamental discriminant then $\left(\frac{d}{n}\right)$ is a real primitive character mod $|d|$.

12.12 The Class Number Formula

We now assume that d is a fundamental discriminant. Let

$$\sqrt{\xi} = \begin{cases} +\sqrt{\xi}, & \text{if } \xi \text{ is positive} \\ i\sqrt{|\xi|}, & \text{if } \xi \text{ is negative.} \end{cases}$$

Theorem 12.1. *If* $0 < \varphi < 2\pi$, *then*

$$\sum_{n=1}^{\infty} \frac{\sin n\varphi}{n} = \frac{\pi}{2} - \frac{\varphi}{2},$$

and

$$\sum_{n=1}^{\infty} \frac{\cos n\varphi}{n} = -\log\left(2\sin\frac{\varphi}{2}\right).$$

Proof. From $0 < \varphi < 2\pi$ we have*

$$\sum_{n=1}^{\infty} \frac{e^{in\varphi}}{n} = -\log(1 - e^{i\varphi})$$

---

* The rigorous proof of this requires Abel's theorem.

$$= -\log\left(2\sin\frac{\varphi}{2}\right) + i\arctan\left(\cot\frac{\varphi}{2}\right)$$

$$= -\log\left(2\sin\frac{\varphi}{2}\right) + i\left(\frac{\pi}{2} - \frac{\varphi}{2}\right).$$

The required results follow from taking the real and imaginary parts of the equation. □

Theorem 12.2. *If d is a fundamental discriminant, then*

$$K(d) = \begin{cases} -\dfrac{1}{\sqrt{d}} \displaystyle\sum_{r=1}^{d-1} \left(\frac{d}{r}\right) \log\sin\frac{\pi r}{d}, & \text{if } d > 0, \\ -\dfrac{\pi}{|d|^{\frac{3}{2}}} \displaystyle\sum_{r=1}^{|d|-1} \left(\frac{d}{r}\right) r, & \text{if } d < 0. \end{cases}$$

Proof. From character sums we have

$$\sum_{r=1}^{|d|-1} \left(\frac{d}{r}\right) e^{2\pi i n r/|d|} = \left(\frac{d}{n}\right)\sqrt{d}.$$

(If d is a fundamental discriminant, then $\left(\frac{d}{r}\right)$ is a primitive character.) Therefore

$$\sqrt{d}\,K(d) = \sum_{n=1}^{\infty} \left(\frac{d}{n}\right) \frac{\sqrt{d}}{n} = \sum_{n=1}^{\infty} \frac{1}{n} \sum_{r=1}^{|d|-1} \left(\frac{d}{r}\right) e^{\frac{2\pi i}{|d|} nr}$$

$$= \sum_{r=1}^{|d|-1} \left(\frac{d}{r}\right) \sum_{n=1}^{\infty} \frac{1}{n} e^{\frac{2\pi i}{|d|} nr}.$$

1) If $d > 0$, then on taking the real parts of the above equation we have

$$\sqrt{d}\,K(d) = \sum_{r=1}^{d-1} \left(\frac{d}{r}\right) \sum_{n=1}^{\infty} \frac{1}{n} \cos\frac{2\pi nr}{d}$$

$$= -\sum_{r=1}^{d-1} \left(\frac{d}{r}\right) \log\left(2\sin\frac{\pi r}{d}\right)$$

$$= -\sum_{r=1}^{d-1} \left(\frac{d}{r}\right) \log\sin\frac{\pi r}{d}$$

(since $\log 2 \sum_{r=1}^{d-1} \left(\frac{d}{r}\right) = 0$).

2) If $d < 0$, then on taking the imaginary parts we have

$$\sqrt{|d|}\,K(d) = \sum_{r=1}^{|d|-1} \left(\frac{d}{r}\right) \sum_{n=1}^{\infty} \frac{1}{n} \sin\frac{2\pi rn}{|d|}$$

$$= \sum_{r=1}^{|d|-1} \left(\frac{d}{r}\right) \left(\frac{\pi}{2} - \frac{\pi r}{|d|}\right) = -\frac{\pi}{|d|} \sum_{r=1}^{|d|-1} \left(\frac{d}{r}\right) r. \quad \square$$

From Theorem 12.2 and Theorem 10.1 we deduce at once:

Theorem 12.3. *Let d be a fundamental discriminant. Then for $d > 0$, we have*

$$\varepsilon^{h(d)} = \prod_t \sin\frac{\pi t}{d} \Big/ \prod_s \sin\frac{\pi s}{d};$$

and for $d < 0$, we have

$$h(d) = \frac{w}{2|d|}\left(\sum_t t - \sum_s s\right),$$

where s runs over those r $(0 < r < |d|)$ satisfying $\left(\frac{d}{r}\right) = 1$, and t those r satisfying $\left(\frac{d}{r}\right) = -1$. □

Theorem 12.4. *Let d be a negative fundamental discriminant. Then*

$$h(d) = \frac{w}{2\left(2 - \left(\frac{d}{2}\right)\right)} \sum_{r=1}^{\left[\frac{|d|}{2}\right]} \left(\frac{d}{r}\right).$$

Proof. By Theorem 12.1 we have, for $2\pi < \varphi < 4\pi$,

$$\sum_{n=1}^{\infty} \frac{\sin n\varphi}{n} = \sum_{n=1}^{\infty} \frac{\sin n(\varphi - 2\pi)}{n} = \frac{\pi}{2} - \left(\frac{\varphi - 2\pi}{2}\right) = \frac{\pi}{2} - \frac{\varphi}{2} + \pi.$$

Following the proof of Theorem 12.2 we have

$$\begin{aligned}\sqrt{d}\,K(d)\left(\frac{d}{2}\right) &= \sum_{n=1}^{\infty} \frac{1}{n}\left(\frac{d}{2n}\right)\sqrt{d}\\ &= \sum_{n=1}^{\infty} \frac{1}{n} \sum_{r=1}^{|d|-1} \left(\frac{d}{r}\right) e^{\frac{2\pi i}{|d|}\cdot 3nr}.\end{aligned}$$

On comparing the imaginary parts we have

$$\begin{aligned}\sqrt{|d|}\,K(d)\left(\frac{d}{2}\right) &= \sum_{n=1}^{\infty} \frac{1}{n} \sum_{r=1}^{|d|-1} \left(\frac{d}{r}\right) \sin\frac{4\pi nr}{|d|}\\ &= \sum_{r=1}^{|d|-1} \left(\frac{d}{r}\right) \sum_{n=1}^{\infty} \frac{1}{n} \sin\frac{4\pi nr}{|d|}\\ &= \sum_{1 \le r \le \frac{1}{2}|d|} \left(\frac{d}{r}\right)\left(\frac{\pi}{2} - \frac{2\pi r}{|d|}\right) + \sum_{\frac{1}{2}|d| < r < |d|} \left(\frac{d}{r}\right)\left(\frac{\pi}{2} - \frac{2\pi r}{|d|} + \pi\right).\end{aligned}$$

(Note: When $r = |d|/2$, the sum of the infinite series is 0 and not $-\pi/2$. But then $\left(\frac{d}{r}\right) = 0$ so that no harm is done.) Therefore

$$\sqrt{|d|}\,K(d)\left(\frac{d}{2}\right) = -\frac{2\pi}{|d|}\sum_{r=1}^{|d|-1}\left(\frac{d}{r}\right)r + \pi \sum_{\frac{1}{2}|d|<r<|d|}\left(\frac{d}{r}\right)$$

$$= 2\sqrt{|d|}\,K(d) + \pi \sum_{\frac{1}{2}|d|<r<|d|}\left(\frac{d}{r}\right)$$

$$= 2\sqrt{|d|}\,K(d) - \pi \sum_{1\leqslant r\leqslant\frac{1}{2}|d|}\left(\frac{d}{r}\right),$$

so that

$$\sqrt{|d|}\left(2-\left(\frac{d}{2}\right)\right)K(d) = \pi \sum_{1\leqslant r\leqslant\frac{1}{2}|d|}\left(\frac{d}{r}\right).$$

The required result follows. □

Exercise 1. Use the method here to prove directly that

$$|d|\sum_{r=1}^{[\frac{1}{2}|d|]}\left(\frac{d}{r}\right) = \left(2-\left(\frac{d}{2}\right)\right)\sum_{r=1}^{|d|}\left(\frac{d}{r}\right)r.$$

Exercise 2. Let $p \equiv 3 \pmod 4$. Then, in the interval between 0 and $p/2$, there are more quadratic residues than non-residues. If $p \equiv 1 \pmod 4$, then they are the same in number.

12.13 The Least Solution to Pell's Equation

We now consider an application of our result. Let $d > 1$ and $d \equiv 0$ or $1 \pmod 4$. Also let x_0, y_0 be the solution to

$$x^2 - dy^2 = 4$$

in which $x_0 + \sqrt{d}\,y_0$ is least ($x_0 > 0$, $y_0 > 0$) and let

$$\varepsilon = \frac{x_0 + \sqrt{d}\,y_0}{2}.$$

The aim of this section is to prove that

$$\varepsilon < d^{\sqrt{d}}.$$

Let $d = m^2 f$, where f is a fundamental discriminant.

Theorem 13.1. *Let $f > 0$, and A^* be the least non-negative integer $\equiv A \pmod f$. Then*

$$\frac{1}{A^*+1}\left|\sum_{a=1}^{A}\sum_{n=1}^{a}\left(\frac{f}{n}\right)\right| \leqslant \frac{1}{2}\left(\sqrt{f} - \frac{A^*+1}{\sqrt{f}}\right).$$

Proof. We can prove, from Theorem 3.3, that

$$\sum_{a=1}^{f}\sum_{n=1}^{a}\left(\frac{f}{n}\right)=0,$$

so that

$$\sum_{a=1}^{A}\sum_{n=1}^{a}\left(\frac{f}{n}\right)=\sum_{a=1}^{A^*}\sum_{n=1}^{a}\left(\frac{f}{n}\right).$$

Also we can use the same method as in Theorem 7.9.2 to prove that

$$\frac{1}{A^*+1}\left|\sum_{a=1}^{A^*}\sum_{n=1}^{f}\left(\frac{f}{n}\right)\right|\leqslant\frac{1}{2}\left(\sqrt{f}-\frac{A^*+1}{\sqrt{f}}\right),$$

and so the required result follows. □

Theorem 13.2. *Let $d>1$. Then*

$$\left|\sum_{a=1}^{A}\sum_{n=1}^{a}\left(\frac{d}{n}\right)\right|\leqslant A\sqrt{d}.$$

Proof. A direct computation gives

$$\left(\frac{d}{n}\right)=\left(\frac{m}{n}\right)^2\left(\frac{f}{n}\right)=\begin{cases}\left(\frac{f}{n}\right), & \text{if } (m,n)=1\\ 0, & \text{if } (m,n)>1.\end{cases}$$

Therefore

$$\begin{aligned}\left|\sum_{a=1}^{A}\sum_{n=1}^{a}\left(\frac{d}{n}\right)\right| &= \left|\sum_{a=1}^{A}\sum_{\substack{n=1\\(n,m)=1}}^{a}\left(\frac{f}{n}\right)\right|\\ &=\left|\sum_{a=1}^{A}\left(\sum_{n=1}^{a}\left(\frac{f}{n}\right)-\prod_{p|m}\left(\frac{f}{p}\right)\sum_{n=1}^{\left[\frac{a}{p}\right]}\left(\frac{f}{n}\right)\right.\right.\\ &\quad\left.\left.+\sum_{\substack{p_1p_2|m\\p_1\neq p_2}}\left(\frac{f}{p_1p_2}\right)\sum_{n=1}^{\left[\frac{a}{p_1p_2}\right]}\left(\frac{f}{n}\right)-\cdots\right)\right|\\ &\leqslant\sum_{k|m}\left|\sum_{a=1}^{A}\sum_{n=1}^{\left[\frac{a}{k}\right]}\left(\frac{f}{n}\right)\right|\\ &\leqslant\sum_{k|m}\left\{k\left|\sum_{b=1}^{\left[\frac{A}{k}\right]-1}\sum_{n=1}^{b}\left(\frac{f}{n}\right)\right|+k\left|\sum_{n=1}^{\left[\frac{A}{k}\right]}\left(\frac{f}{n}\right)\right|\right\}\\ &\leqslant\sum_{k|m}k\left\{\frac{1}{2}\left[\frac{A}{k}\right]\sqrt{f}+\left[\frac{A}{k}\right]\right\}\leqslant A\sum_{k|m}\left(\frac{1}{2}\sqrt{f}+1\right)\\ &\leqslant A\sqrt{f}m=A\sqrt{d}.\end{aligned}$$

(Since $f \geqslant 5$ implies $1 < \frac{1}{2}\sqrt{f}$; also $\sum_{k|m} 1 \leqslant m$.) □

Theorem 13.3. *Let $d \geqslant 5$. Then*

$$K(d) < \tfrac{1}{2}\log d + 1.$$

Proof. We define

$$S(n) = \sum_{a=1}^{n} \sum_{k=1}^{a} \left(\frac{d}{k}\right),$$

for $n \geqslant 1$, and we let $S(-1) = S(0) = 0$. We then have

$$S(n) - S(n-1) = \sum_{k=1}^{n} \left(\frac{d}{k}\right),$$

$$S(n) - 2S(n-1) + S(n-2) = \left(\frac{d}{n}\right), \qquad n \geqslant 1.$$

Therefore

$$\begin{aligned} K(d) &= \sum_{n=1}^{\infty} \left(\frac{d}{n}\right)\frac{1}{n} = \sum_{n=1}^{\infty} \frac{1}{n}\{S(n) - 2S(n-1) + S(n-2)\} \\ &= \sum_{n=1}^{\infty} S(n)\left\{\frac{1}{n} - \frac{2}{n+1} + \frac{1}{n+2}\right\} \\ &= \sum_{n=1}^{\infty} \frac{2S(n)}{n(n+1)(n+2)}. \end{aligned}$$

Let

$$S_1 = \sum_{n=1}^{A-1} \frac{2S(n)}{n(n+1)(n+2)}, \qquad S_2 = \sum_{n=A}^{\infty} \frac{2S(n)}{n(n+1)(n+2)}.$$

Since $|S(n)| \leqslant n(n+1)/2$, it follows that

$$\begin{aligned} |S_1| &\leqslant \sum_{n=1}^{A-1} \frac{1}{n+2} = \sum_{n=1}^{A-1} \frac{1}{n} - \frac{3}{2} + \frac{1}{A} + \frac{1}{A+1} \\ &\leqslant \log(A-1) + \gamma - \frac{3}{2} + \frac{1}{A} + \frac{1}{A+1}^{*}. \end{aligned}$$

Also, by Theorem 13.2, we have

$$|S_2| \leqslant \sum_{n=A}^{\infty} \frac{2\sqrt{d}}{(n+1)(n+2)} = \frac{2\sqrt{d}}{A+1}.$$

We choose $A = [2\sqrt{d}] + 1$, so that

---

* Here $\gamma = 0.5772\ldots$ is Euler's constant. That $\sum_{n=1}^{x} \frac{1}{n} \leqslant \log x + \gamma$ can be proved from the fact that $(\sum_{n=1}^{x} \frac{1}{n} - \log x)$ increases to the limit γ.

$$|K(d)| \leqslant |S_1| + |S_2| \leqslant \log(A-1) + \gamma - \frac{3}{2} + \frac{1}{A} + \frac{2\sqrt{d}+1}{A+1} < \frac{1}{2}\log d + 1.$$

(because $d \geqslant 5$). □

Theorem 13.4. *We always have*

$$\log \varepsilon < \sqrt{d}(\tfrac{1}{2}\log d + 1).$$

Proof. From Theorem 10.1 we have

$$1 \leqslant h(d) = \frac{\sqrt{d}}{\log \varepsilon} K(d).$$

The required result follows from Theorem 13.3. □

Theorem 13.5 (Schur). *We always have*

$$\log \varepsilon < \sqrt{d}\log d.$$

Proof. The result follows from the previous theorem if $d > e^2$. But if $d < e^2$, then $d = 5$ and now $\varepsilon = (3 + \sqrt{5})/2$ giving $\log \varepsilon < \sqrt{5}\log 5$. □

Note: It was conjectured by Gauss that $h(d) \to \infty$ as $|d| \to \infty$. This is a famous difficult problem. In 1934 Heilbronn proved that $h(d) \to \infty$ as $d \to -\infty$. The following year Siegel proved that

$$\lim_{d \to -\infty} \frac{\log h(d)}{\log |d|} = \frac{1}{2},$$

so that we know the order of $h(d)$ as $d \to -\infty$.

Whether $h(d) \to \infty$ as $d \to \infty$ or not is still a difficult unsolved problem. Corresponding to this, Siegel's result is that

$$\lim_{d \to \infty} \frac{\log(h(d)\log \varepsilon)}{\log d} \quad \frac{1}{2}.$$

But unfortunately we do not know enough about $\log \varepsilon$, so that we cannot settle whether $h(d)$ tends to infinity or not.

We shall give the two results of Siegel in §15.

12.14 Several Lemmas

We shall prove the Heilbronn-Siegel Theorem in the next section. In that proof we shall require some knowledge of complex function theory and certain simple properties of the ζ-function established in Chapter 9. We shall, for convenience, give all the necessary requirements for the next section here.

1) The theorem in complex function theory that we shall need is

Theorem 14.1 (Cauchy's inequality). *Let*

$$f(s) = \sum_{n=0}^{\infty} a_n(s-a)^n$$

be analytic in $|s-a| \leqslant r$, *and let* $|f(s)| \leqslant M$. *Then*

$$|a_n| \leqslant Mr^{-n} \qquad (n = 0, 1, 2, \ldots).$$

(The proof is given in any complex function theory textbooks). □

2) Concerning the ζ-function we shall require:

Theorem 14.2. $\zeta(s)$ $(s = \sigma + it)$ *is analytic in the half plane* $\sigma > 0$ *apart from the simple pole at* $s = 1$ *where it has residue* 1. *Also*

$$\left|\zeta(s) - \frac{1}{s-1}\right| \leqslant \frac{|s|}{\sigma} \qquad (\sigma > 0). \tag{1}$$

(See Theorem 9.2.1.) □

3) We now introduce another function

$$L_d(s) = \sum_{n=1}^{\infty} \left(\frac{d}{n}\right)\frac{1}{n^s} \qquad (\sigma > 0),$$

where d is a discriminant. Clearly

$$L_d(1) = K(d).$$

Theorem 14.3. $L_d(s)$ *is analytic in the right hand half plane* $\sigma > 0$ *and satisfies*

$$|L_d(s)| < \frac{|d| \cdot |s|}{\sigma} \qquad (\sigma > 0), \tag{2}$$

and

$$0 < L_d(1) < 2 + \log|d|. \tag{3}$$

Proof. Let n_1, n_2 be any two positive integers and $n_2 > n_1$. Since

$$\left|\sum_{n_1 \leqslant n \leqslant m} \left(\frac{d}{n}\right)\right| < |d|$$

for any $m > n_1$, it follows from Theorem 6.8.1 that

$$\left|\sum_{n_1 \leqslant n \leqslant n_2} \left(\frac{d}{n}\right)\frac{1}{n^s}\right| \leqslant |d| \left(\sum_{n_1 \leqslant n \leqslant n_2 - 1} \left|\frac{1}{n^s} - \frac{1}{(n+1)^s}\right| + \left|\frac{1}{n_2^s}\right|\right).$$

Also

$$\left|\frac{1}{n^s} - \frac{1}{(n+1)^s}\right| = \left|s \int_n^{n+1} x^{-s-1}\,dx\right| \leqslant |s| \int_n^{n+1} x^{-\sigma-1}\,dx,$$

so that

$$\left|\sum_{n_1 \leqslant n \leqslant n_2} \left(\frac{d}{n}\right)\frac{1}{n^s}\right| \leqslant |d| \cdot |s| \left(\int_{n_1}^{n_2} x^{-\sigma-1}\,dx + |s|^{-1} n_2^{-\sigma}\right)$$

$$\leqslant \frac{|d| \cdot |s|}{\sigma} n_1^{-\sigma} \qquad (\sigma > 0). \tag{4}$$

From (4) we see that, given any $\sigma_0 > 0$, the series for $L_d(s)$ is uniformly convergent in any finite region in the half plane $\sigma \geqslant \sigma_0$. Since σ_0 can be any small positive number it follows that $L_d(s)$ is analytic in the half plane $\sigma > 0$.

We now let $n_1 = 1$ and $n_2 \to \infty$ so that (2) follows.

From Theorem 10.1 and $h(d) \geqslant 1$, $\log \varepsilon > 0$, we see that

$$L_d(1) = K(d) > 0.$$

Separating the sum for $L_d(1)$ into two parts:

$$L_d(1) = \sum_{n=1}^{\infty} \left(\frac{d}{n}\right)\frac{1}{n} = \sum_{n=1}^{|d|} \left(\frac{d}{n}\right)\frac{1}{n} + \sum_{n=|d|+1}^{\infty} \left(\frac{d}{n}\right)\frac{1}{n},$$

we see that the first part satisfies

$$\left|\sum_{n=1}^{|d|} \left(\frac{d}{n}\right)\frac{1}{n}\right| \leqslant \sum_{n=1}^{|d|} \frac{1}{n} < 1 + \log|d|,$$

while, by (4), the second part satisfies

$$\left|\sum_{n=|d|+1}^{\infty} \left(\frac{d}{n}\right)\frac{1}{n}\right| \leqslant \frac{|d|}{|d|+1} < 1,$$

so that the theorem is proved. □

12.15 Siegel's Theorem

Theorem 15.1. *Let d and d_1 be two discriminants, and*

$$f(s) = \zeta(s) L_d(s) L_{d_1}(s) L_{dd_1}(s).$$

Then, for $\sigma > 1$, we have

$$f(s) = \sum_{n=1}^{\infty} a_n n^{-s}, \qquad a_1 = 1, \qquad a_n \geqslant 0 \qquad (n = 2, 3, \ldots).$$

Proof. When $\sigma > 1$ the two series

$$\sum_{n=1}^{\infty} \frac{1}{n^s} \qquad \text{and} \qquad \sum_{n=1}^{\infty} \left(\frac{d}{n}\right)\frac{1}{n^s}$$

are absolutely convergent. Also $\frac{1}{n^s}$ and $\left(\frac{d}{n}\right)\frac{1}{n^s}$ are both completely multiplicative functions. Therefore, by Theorem 5.4.4,

$$\zeta(s) = \prod_p \left(1 - \frac{1}{p^s}\right)^{-1}, \qquad L_d(s) = \prod_p \left(1 - \left(\frac{d}{p}\right)\frac{1}{p^s}\right)^{-1} \quad (\sigma > 1).$$

If we let

$$g(s,p) = \left\{(1 - p^{-s})\left(1 - \left(\frac{d}{p}\right)p^{-s}\right)\left(1 - \left(\frac{d_1}{p}\right)p^{-s}\right)\left(1 - \left(\frac{dd_1}{p}\right)p^{-s}\right)\right\}^{-1},$$

then we have

$$f(s) = \prod_p g(s,p) \qquad (\sigma > 1). \tag{1}$$

Now $\left(\frac{d}{p}\right), \left(\frac{d_1}{p}\right), \left(\frac{dd_1}{p}\right)$ can only take the values $0, 1,$ and -1. When $\left(\frac{d}{p}\right) = \left(\frac{d_1}{p}\right) = 1,$

$$g(s,p) = (1 - p^{-s})^{-4} = \frac{1}{6}\sum_{m=0}^{\infty} (m+1)(m+2)(m+3)p^{-ms};$$

when

$$\left(\frac{d}{p}\right) = -1, \ \left(\frac{d_1}{p}\right) = \pm 1, \qquad \text{or} \qquad \left(\frac{d_1}{p}\right) = -1, \ \left(\frac{d}{p}\right) = \pm 1,$$

we have

$$g(s,p) = (1 - p^{-2s})^{-2} = \sum_{m=0}^{\infty} (m+1)p^{-2ms};$$

when one of $\left(\frac{d}{p}\right), \left(\frac{d_1}{p}\right)$ is 0 and the other is 0, 1, or -1, we have

$$g(s,p) = (1 - p^{-s})^{-1} = \sum_{m=0}^{\infty} p^{-ms};$$

$$g(s,p) = (1 - p^{-s})^{-2} = \sum_{m=0}^{\infty} (m+1)p^{-ms};$$

$$g(s,p) = (1 - p^{-2s})^{-1} = \sum_{m=0}^{\infty} p^{-2ms}.$$

Therefore, in all cases and for any prime p, we have $a_1 = 1$, $a_{p^m} \geqslant 0$ $(m = 1, 2, \ldots)$

such that

$$g(s,p) = \sum_{m=0}^{\infty} a_{p^m} p^{-ms} \qquad (\sigma > 1). \tag{2}$$

From (1) and (2) we have

$$f(s) = \prod_p \left(\sum_{m=0}^{\infty} a_{p^m} p^{-ms} \right) \qquad (\sigma > 1). \tag{3}$$

Suppose now that the standard factorization of n is $p_1^{q_1} \cdots p_l^{q_l}$. We define

$$a_n = a_{p_1^{q_1}} \cdots a_{p_l^{q_l}},$$

so that a_n is defined for all natural numbers n, and is a multiplicative function satisfying $a_1 = 1$, $a_n \geqslant 0$. Again, by Theorem 5.4.4 and (3) we see that

$$f(s) = \sum_{n=1}^{\infty} a_n n^{-s} \qquad (\sigma > 1),$$

where a_n has the requirement stated in the theorem. □

Theorem 15.2. *Let d and d_1 be two fundamental discriminants, $|d| > |d_1| > 1$. Then dd_1 is a discriminant. Let $f(s)$ be defined in Theorem 15.1 and let*

$$\rho = L_d(1) L_{d_1}(1) L_{dd_1}(1).$$

Then, for $0 < \delta < a < 1$ (δ is any fixed positive number less than 1), we have

$$f(a) > \frac{1}{2} - \frac{C_1 \rho}{1-a} |dd_1|^{C_2(1-a)},$$

where C_1, C_2 are positive constants depending on δ.

Proof. The function $f(s) - \rho/(s-1)$ is analytic in $|s-2| < 1$ and has the Taylor expansion

$$f(s) - \frac{\rho}{s-1} = \sum_{m=0}^{\infty} (b_m - \rho)(2-s)^m, \tag{4}$$

where

$$b_0 = f(2), \qquad b_m = (-1)^m \frac{f^{(m)}(2)}{m!} \qquad (m = 1, 2, \ldots).$$

By Theorem 15.1, we have $f(2) \geqslant 1$, and

$$(-1)^m f^{(m)}(2) = \sum_{n=1}^{\infty} a_n n^{-2} \log^m n \geqslant 0 \qquad (m = 1, 2, \ldots),$$

that is

$$b_0 \geqslant 1, \qquad b_m \geqslant 0 \qquad (m = 1, 2, \ldots). \tag{5}$$

From Theorems 14.2 and 14.3 we know that $f(s) - \rho/(s-1)$ is analytic in the right hand half plane $\sigma > 0$, so that the expansion (4) actually holds for $|s-2| < 2$. We now apply Theorem 14.1 to give an upper bound for $|b_m - \rho|$. For this purpose, we first seek an upper bound for $|f(s) - \rho/(s-1)|$ on the circle $|s-2| = (2-\delta)/\xi$ where ξ is a number satisfying $0 < \xi < 1$ and $1 < (2-\delta)/\xi < 2$. From Theorem 14.2 and Theorem 14.3 we have

$$|f(s)| \leqslant \left(\frac{1}{|s-1|} + \frac{|s|}{\sigma}\right)\left(\frac{|s|}{\sigma}\right)^3 |dd_1|^2 \qquad (s \neq 1, \sigma > 0). \tag{6}$$

Since $|s-2| = (2-\delta)/\xi$ we have

$$\frac{|s|}{\sigma} \leqslant \left(2 + \frac{2-\delta}{\xi}\right)\Big/\left(2 - \frac{2-\delta}{\xi}\right), \qquad \frac{1}{|s-1|} \leqslant \left(\frac{2-\delta}{\xi} - 1\right)^{-1},$$

and hence

$$|f(s)| \leqslant C_3 |dd_1|^2 \qquad \left(|s-2| = \frac{2-\delta}{\xi}\right),$$

where C_3 is a positive constant depending only on δ and ξ. Also, from Theorem 14.3, we have $|\rho| \leqslant |dd_1|^2$, so that

$$\left| f(s) - \frac{\rho}{s-1} \right| \leqslant C_4 |dd_1|^2, \qquad |s-2| = \frac{2-\delta}{\xi}, \tag{7}$$

and, from the maximum modulus theorem, we see that (7) also holds for $|s-2| \leqslant (2-\delta)/\xi$. Therefore, from Theorem 14.1 we have that

$$|b_m - \rho| \leqslant C_4 |dd_1|^2 \left(\frac{\xi}{2-\delta}\right)^m, \qquad m = 0, 1, 2, \ldots . \tag{8}$$

We can now obtain a lower bound for $f(a)$ from the expansion (4). We have

$$f(a) = \frac{\rho}{a-1} + \sum_{m=0}^{m_0-1} (b_m - \rho)(2-a)^m + \sum_{m=m_0}^{\infty} (b_m - \rho)(2-a)^m,$$

and, by (5), we have

$$\sum_{m=0}^{m_0-1} (b_m - \rho)(2-a)^m \geqslant 1 - \sum_{m=0}^{m_0-1} \rho(2-a)^m = 1 - \rho\frac{(2-a)^{m_0} - 1}{1-a},$$

while, by (8), we have

$$\begin{aligned}\sum_{m=m_0}^{\infty} (b_m - \rho)(2-a)^m &\geqslant -C_4 |dd_1|^2 \sum_{m=m_0}^{\infty} \left(\frac{\xi}{2-\delta}\right)^m (2-\delta)^m \\ &= -C_4 |dd_1|^2 \xi^{m_0}(1-\xi)^{-1} = -C_5 |dd_1|^2 \xi^{m_0},\end{aligned}$$

giving

$$f(a) \geqslant 1 - \rho\frac{(2-a)^{m_0}}{1-a} - C_5 |dd_1|^2 \xi^{m_0}. \tag{9}$$

We now choose

$$m_0 = \left[\frac{\log(2C_5|dd_1|^2)}{-\log\xi}\right] + 1,$$

so that

$$m_0 < \frac{2\log|dd_1|}{-\log\xi} + C_6 \qquad (C_6 > 1),$$

and

$$C_5|dd_1|^2\xi^{m_0} < \frac{1}{2},$$

$$\begin{aligned}(2-a)^{m_0} &< 2^{C_7}|dd_1|^{\frac{2}{-\log\xi}\log(2-a)} \leqslant 2^{C_7}|dd_1|^{\frac{2}{-\log\xi}(1-a)}\\ &= C_1|dd_1|^{C_2(1-a)},\end{aligned}$$

and the theorem follows by substituting these into (9). □

Theorem 15.3 (Siegel). *Let d be a fundamental discriminant and $\varepsilon > 0$. Then*

$$\frac{1}{L_d(1)} = O(|d|^{\varepsilon}).$$

Proof. We can assume that $0 < \varepsilon < \frac{1}{2}$. Let

$$\begin{aligned} f(s) &= \zeta(s)L_d(s)L_{d_1}(s)L_{dd_1}(s),\\ \rho &= L_d(1)L_{d_1}(1)L_{dd_1}(1), \end{aligned} \tag{10}$$

where d_1 is chosen as follows: If there is a fundamental discriminant d_1 such that $L_{d_1}(\sigma)$ has a zero in $1-\varepsilon < \sigma < 1$, then we take this d_1 to be the d_1 in (10) and we denote by a any zero of $L_{d_1}(\sigma)$ in this interval, so that $f(a) = 0$. If there is no fundamental discriminant d_1 such that $L_{d_1}(\sigma)$ has a zero in $1-\varepsilon < \sigma < 1$, then we take any fundamental discriminant d_1. In this case, if $f(\sigma)$ has zeros in $1-\varepsilon < \sigma < 1$, then we take a to be any one of its zeros so that $f(a) = 0$; if $f(\sigma)$ has no zero either in the interval, then we take a to be any point in the interval $1-\varepsilon < \sigma < 1$. For this last case, since $f(\sigma)$ has no zero, it has a fixed sign. Moreover, $\rho > 0$ by Theorem 14.3 and since $f(s) - \rho/(s-1)$ is analytic in the right hand half plane we see that $f(\sigma) \to -\infty$ as σ tends to 1 from the left, and we deduce that $f(\sigma)$ is negative in $1-\varepsilon < \sigma < 1$. Therefore, no matter how we choose d_1 or a, we always have

$$f(a) \leqslant 0. \tag{11}$$

Let $|d| > |d_1|$. From Theorem 15.2 (taking $\delta = \frac{1}{2}$ so that $0 < \delta < 1 - \varepsilon < a < 1$), we have

$$\frac{C_1}{1-a}L_d(1)L_{d_1}(1)L_{dd_1}(1)|dd_1|^{C_2(1-a)} > \frac{1}{2},$$

where C_1 and C_2 are absolute positive constants. Therefore

$$\frac{1}{L_d(1)} < \frac{2C_1}{1-a} L_{d_1}(1) L_{dd_1}(1) |dd_1|^{C_2(1-a)} = C L_{dd_1}(1) |d|^{C_2(1-a)},$$

where

$$C = \frac{2C_1}{1-a} L_{d_1}(1) |d_1|^{C_2(1-a)}$$

is a constant which does not depend on d. When $|d| > |d_1| > 1$ we have

$$L_{dd_1}(1) \leqslant 2 + \log|dd_1| < 2(1 + \log|d|)$$

and since $1 - a < \varepsilon$ we have therefore that

$$\frac{1}{L_d(1)} < 2C(1 + \log|d|)|d|^{C_2\varepsilon} = O(|d|^{(C_2+1)\varepsilon}),$$

and since ε is arbitrary the theorem is proved. □

Theorem 15.4. *If d is a discriminant, then*

$$\lim_{d \to -\infty} \frac{\log h(d)}{\log|d|} = \frac{1}{2},$$

$$\lim_{d \to \infty} \frac{\log(h(d) \log \varepsilon)}{\log d} = \frac{1}{2}.$$

Proof. 1) If d is a fundamental discriminant, then given $\eta > 0$ we have, by Theorem 15.3 and Theorem 14.3,

$$C_8|d|^{-\eta} < K(d) \leqslant 2 + \log|d| < C_9|d|^{\eta}, \tag{12}$$

and so by Theorem 10.1

$$C_{10}|d|^{\frac{1}{2}-\eta} \leqslant h(d) \begin{Bmatrix} 1 \\ \log \varepsilon \end{Bmatrix} \leqslant C_{11}|d|^{\frac{1}{2}+\eta}$$

which is the required result.

2) If d is not a fundamental discriminant and $d = fm^2$, where f is a fundamental discriminant, then from

$$K(d) = \prod_{p|m} \left(1 - \left(\frac{f}{p}\right)\frac{1}{p}\right) K(f),$$

$$\prod_{p|m} \left(1 - \left(\frac{f}{p}\right)\frac{1}{p}\right) \leqslant C_{12} m^{\eta},$$

$$\prod_{p|m} \left(1 - \left(\frac{f}{p}\right)\frac{1}{p}\right) \geqslant \prod_{p|m} \left(1 - \frac{1}{p}\right) = \frac{\varphi(m)}{m} \geqslant C_{13} m^{-\eta},$$

we have

$$C_{13}m^{-\eta}K(f) \leqslant K(d) \leqslant C_{12}m^{\eta}K(f).$$

From (12) we arrive at

$$C_{14}|d|^{\frac{1}{2}-\eta} \leqslant |d|^{\frac{1}{2}}K(d) \leqslant C_{15}|d|^{\frac{1}{2}+\eta}$$

and the theorem follows from Theorem 10.1. $\square$

Notes

12.1. The method of D. A. Burgess (see Note 7.2) can be used to give an improved estimate on the least solution $\varepsilon = (x_0 + \sqrt{d}\,y_0)/2$ to Pell's equation $x^2 - dy^2 = 4$, $d > 0$, $d \equiv 0$ or 1 (mod 4). The result is: corresponding to every $\delta > 0$ there exists a constant $c(\delta)$ such that

$$\log \varepsilon < (\tfrac{1}{4} + \delta)\sqrt{d}\log d$$

whenever $d > c(\delta)$ (see Y. Wang [63]).

Chapter 13. Unimodular Transformations

13.1 The Complex Plane

Let $z = x + yi$ be a complex number which is represented by a point P on a plane with coordinates (x, y). From the origin O we construct a directed line to P and we call this line the vector $\overrightarrow{OP}$. There is a bijection between z and P so that every complex number now corresponds to a vector from the origin.

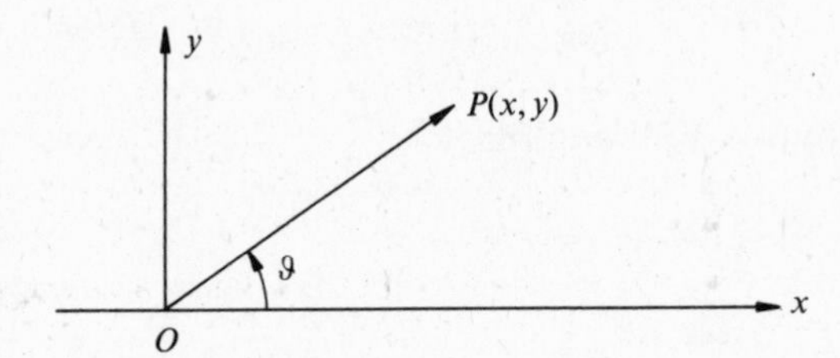

The distance from O to P, also known as the *length* of the vector $\overrightarrow{OP}$, is given by $\rho = \sqrt{x^2 + y^2}$, and is the same as the absolute value of z. The angle ϑ measured from the positive x-axis to the vector $\overrightarrow{OP}$, is called the *argument* of z. We have

$$x = \rho \cos \vartheta, \qquad y = \rho \sin \vartheta$$

and (ρ, ϑ) are referred to as the *polar coordinates* of the point (x, y). Clearly we have

$$z = x + yi = \rho(\cos \vartheta + i \sin \vartheta) = \rho e^{i\vartheta}.$$

We usually write

$$|z| = \sqrt{x^2 + y^2}, \qquad \arg z = \vartheta.$$

The circle centre c with radius r ($\geqslant 0$) can be represented by the equation

$$|z - c| = r,$$

and the particular circle $|z| = 1$ is called the *unit circle*.

We next investigate the bilinear transformation

$$z' = \frac{az + b}{cz + d}, \tag{1}$$

where a, b, c, d are (in general complex) constants, and $ad - bc \neq 0$. This transformation maps a point z ($\neq -d/c$) in the plane into another point z'. Corresponding to the point $z = -d/c$ we introduce an ideal point, called the *point at infinity*, for its image and we write $z' = \infty$. Our discussion is concerned with the plane together with this ideal point. This is often called the *extended complex plane*, but in this chapter we shall simply call it the complex plane. Corresponding to the point $z = \infty$ we have the image $z' = a/c$.

If we solve (1) for z, we have

$$z = \frac{-dz' + b}{cz' - a},$$

which is also a bilinear transformation known as the *inverse transformation* of (1). We see therefore that the transformation (1) is a bijection from the complex plane onto itself.

Let us place a sphere on the complex plane with point of contact at the origin. We may refer to this point of contact as the "south-pole", and the point on the sphere which is diametrically opposite to this as the "north-pole". Consider a line joining a point z on the plane to the "north-pole". This line crosses the sphere at a point, and if we map the point z onto this point and the point at infinity onto the "north-pole" we see at once that this sets up a bijection between the complex plane and the surface of the sphere. This replacement of the abstract notion of the complex plane with an ideal point by the concrete notion of the surface of a sphere is due to Riemann, and we often call the sphere here the *Riemann sphere*.

13.2 Properties of the Bilinear Transformation

Corresponding to a bilinear transformation A:

$$z' = \frac{az + b}{cz + d}, \tag{1}$$

there is a matrix

$$\begin{pmatrix} a & b \\ c & d \end{pmatrix} \tag{2}$$

whose determinant is $ad - bc$ ($\neq 0$), which we call the determinant of the transformation. Note that different matrices may correspond to the same transformation, since

$$\begin{pmatrix} a\rho & b\rho \\ c\rho & d\rho \end{pmatrix}, \qquad \rho \neq 0$$

all represent the same transformation (1). However it is not difficult to prove that, apart from this situation, there is no other matrix which corresponds to the transformation (1). We can choose ρ so that $\rho^2(ad - bc) = 1$ so that there is always a unit determinant matrix to represent the bilinear transformation A. It is easy to show that there are only two unit determinant matrices which correspond to a given bilinear transformation, namely the matrices

$$\begin{pmatrix} \pm a, & \pm b \\ \pm c, & \pm d \end{pmatrix}.$$

Let there be another bilinear transformation B:

$$z'' = \frac{a'z' + b'}{c'z' + d'}, \tag{3}$$

so that we have the bilinear transformation C:

$$\begin{aligned} z'' &= \frac{a'(az + b) + b'(cz + d)}{c'(az + b) + d'(cz + d)} \\ &= \frac{(a'a + b'c)z + a'b + b'd}{(c'a + d'c)z + c'b + d'd}. \end{aligned} \tag{4}$$

with corresponding matrix

$$\begin{pmatrix} a'a + b'c & a'b + b'd \\ c'a + d'c & c'b + d'd \end{pmatrix}$$

known as the *product* of the two matrices $\begin{pmatrix} a' & b' \\ c' & d' \end{pmatrix}$ and $\begin{pmatrix} a & b \\ c & d \end{pmatrix}$, and we write

$$\begin{pmatrix} a'a + b'c & a'b + b'd \\ c'a + d'c & c'b + d'd \end{pmatrix} = \begin{pmatrix} a' & b' \\ c' & d' \end{pmatrix} \begin{pmatrix} a & b \\ c & d \end{pmatrix}.$$

The transformation (4) is also referred to as the product of the transformation (3) and (1) and we write $C = BA$. Note however that BA is not necessarily the same as AB. We denote by A^{-1} the inverse transformation to A.

The transformation

$$z' = z$$

is called the identity transformation and is denoted by E. We have $AA^{-1} = A^{-1}A = E$.

Definition 1.* Let a set of bilinear transformations have the following three properties: (i) it contains the identity transformation, (ii) the product of any two transformations in the set is also in the set, (iii) the inverse of any transformation is also in the set. Then we say that the set of transformations form a *group*.

Example 1. The set of all bilinear transformations form a group.

Example 2. The set of all bilinear transformations with real coefficients form a group.

* The three properties here are interrelated, but they suffice for our purpose of keeping matters simple and easy.

Example 3. The set of all bilinear transformations with real coefficients and positive determinants form a group.

Example 4. The set of all bilinear transformations with integer coefficients a, b, c, d satisfying $ad - bc = \pm 1$ form a group.

Example 5. The set of bilinear transformations with complex integer (that is $a = a' + a''i, a', a''$ integers) coefficients form a group.

Definition 2. If the image of z_0 under the transformation A is z_0 itself, then we call z_0 a *fixed point* of A.

In general a bilinear transformation has two distinct fixed points (from $z' = z$). They are the two roots of the quadratic equation

$$cz^2 + (d - a)z - b = 0. \tag{5}$$

If z_1, z_2 are the two roots of this equation, then we can rewrite the transformation in the standard form

$$\frac{z' - z_1}{z' - z_2} = \lambda \frac{z - z_1}{z - z_2}. \tag{6}$$

Taking $z = \infty$ so that $z' = a/c$ we can specify λ as

$$\lambda = \frac{a - cz_1}{a - cz_2}.$$

It is easy to show that λ satisfies the quadratic equation

$$\lambda + \frac{1}{\lambda} = \frac{a^2 + d^2 + 2bc}{ad - bc} = \frac{(a + d)^2}{ad - bc} - 2. \tag{7}$$

If $|\lambda| = 1$, $\lambda \neq 1$, then we say that the transformation is *elliptic*.

If λ is real and not equal to ± 1, then we say that the transformation is *hyperbolic*.

If λ is complex and $|\lambda| \neq 1$, then we say that the transformation is *loxodromic*.

If $c = 0$ and $d - a \neq 0$, then one of the fixed points is the point at infinity. Taking $z_2 = \infty$ equation (6) then becomes

$$z' - z_1 = \lambda(z - z_1). \tag{8}$$

If the two fixed points coincide, that is $z_1 = z_2$, then

$$(a - d)^2 + 4bc = 0$$

or

$$(a + d)^2 + 4(bc - ad) = 0. \tag{9}$$

A transformation satisfying this condition is said to be *parabolic*. Substituting (9) into (7) gives $\lambda = 1$ and the standard equation (6) becomes

$$\frac{1}{z' - z_1} = \frac{1}{z - z_1} + k$$

where $z_1 = (a - d)/2c$, $k = 2c/(a + d)$. In particular when $c = 0$, $a - d = 0$, this fixed point becomes the point at infinity and the transformation then becomes

$$z' = z + k, \qquad k = b/a.$$

If on the repeated applications of a transformation the product becomes the identical transformation then we call the transformation a transformation of *finite order*. In this case, the *period* of the transformation is defined to be the least number of applications required to result in the identical transformation. Repeated applications of (10) and (6) give

$$\frac{1}{z' - z_1} = \frac{1}{z - z_1} + nk,$$

$$\frac{z' - z_1}{z' - z_2} = \lambda^n \frac{z - z_1}{z - z_2}$$

so that the parabolic, hyperbolic and loxodromic transformations are not of finite order. Only for elliptic transformations do we have $\lambda^n = 1$ and the period is the least positive integer n such that $\lambda^n = 1$.

When $n = 2$ so that $\lambda = -1$ we call the transformation an *involution*.

13.3 Geometric Properties of the Bilinear Transformation

Definition. The *cross ratio* of the four points z_1, z_2, z_3, z_4 is defined by

$$(z_1 z_2 z_3 z_4) = \frac{z_3 - z_1}{z_2 - z_3} \Big/ \frac{z_4 - z_1}{z_2 - z_4}.$$

Theorem 3.1. *The cross ratio is invariant under a bilinear transformation.*

Proof. Let

$$z'_i = \frac{az_i + b}{cz_i + d},$$

so that

$$z'_i - z'_j = \frac{(ad - bc)(z_i - z_j)}{(cz_i + d)(cz_j + d)},$$

and hence

$$(z'_1 z'_2 z'_3 z'_4) = (z_1 z_2 z_3 z_4). \quad \square$$

Given any three points z_1, z_2, z_3 there exists a bilinear transformation which maps them onto any three specified points z_1', z_2', z_3'. This transformation can be written down explicitly by

$$\frac{z' - z_1'}{z' - z_2'} = \frac{z_3' - z_1'}{z_3' - z_2'} \frac{z_3 - z_2}{z_3 - z_1} \frac{z - z_1}{z - z_2}, \tag{1}$$

or

$$\frac{z_3' - z_2'}{z_3' - z_1'} \frac{z' - z_1'}{z' - z_2'} = \frac{z_3 - z_2}{z_3 - z_1} \frac{z - z_1}{z - z_2},$$

or

$$(z_1'z_2'z_3'z') = (z_1z_2z_3z). \tag{2}$$

If there is a bilinear transformation with the above property, then by Theorem 3.1 after z having been specified, z' must satisfy (2). That is, z' is uniquely determined. Therefore a bilinear transformation with the above property is unique. In other words, (2) is a general form for a bilinear transformation.

Let A_1, A_2, A_3, P be the points representing z_1, z_2, z_3, z respectively. Then we have

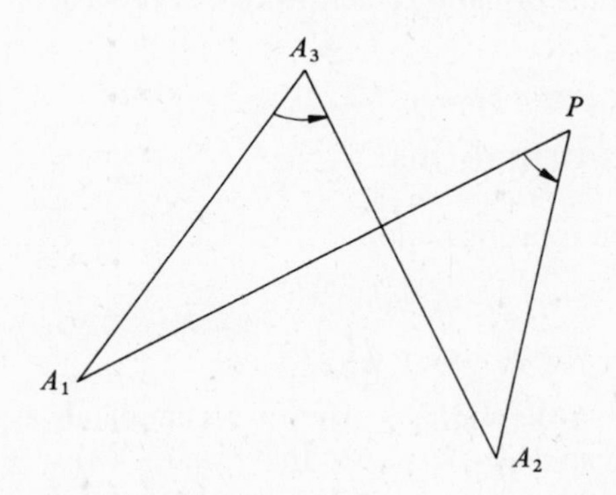

$$\arg(z_1z_2z_3z) = \angle A_1PA_2 - \angle A_1A_3A_2,$$

where the direction of the signed angle is as shown in the diagram. From this we see that if the cross ratio is a real number, then

$$\angle A_1PA_2 - \angle A_1A_3A_2$$

must be a multiple of π, and hence P lies on the circle through the three points A_1, A_2, A_3.

If $(z_1z_2z_3z)$ is real, then by (2), $(z_1'z_2'z_3'z')$ is also real, so that as z describes the circle through z_1, z_2, z_3, the point z' will describe the circle through z_1', z_2', z_3', and conversely. We have therefore proved that a bilinear transformation maps circles into circles. Note however that, in the present context, a straight line is interpreted as a circle with infinite radius.

Theorem 3.2. *A bilinear transformation preserves the angle of intersection between two circles. That is, if two circles intersect with angle ϑ, then the two image circles of a bilinear transformation also intersect with angle ϑ.*

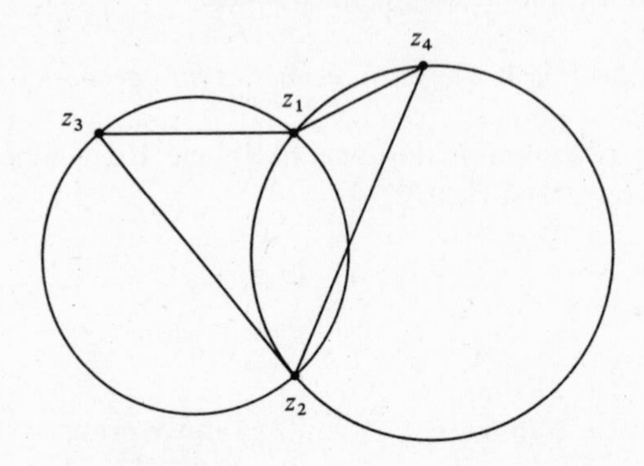

Proof. Let the two circles intersect at z_1 and z_2, and take two points z_3, z_4 in the neighbourhood of z_1 on the two circles. The argument of the cross ratio $\arg(z_3z_4z_1z_2)$ is $\angle z_3z_2z_4 - \angle z_3z_1z_4$. As z_3 and z_4 both tend to z_1 this gives the value of the angle of intersection for the two circles. Since the cross ratio is invariant under the bilinear transformation, the theorem is proved. □

13.4 Real Transformations

We now consider the transformation

$$z' = \frac{az + b}{cz + d}, \qquad ad - bc \neq 0,$$

where a, b, c, d are real numbers. Here we cannot always choose ρ so that $\rho^2(ad - bc) = 1$; we can only choose ρ so that $\rho^2(ad - bc) = \pm 1$. From now on we shall assume that

$$ad - bc = \pm 1.$$

The set of all real bilinear transformations with determinant 1 form a group which we denote by $\mathfrak{R}$. Clearly members of this group map the real axis onto itself. Moreover, given any three real numbers, there is a member which maps them onto any three specified real numbers.

Theorem 4.1. *Members of $\mathfrak{R}$ map the upper half plane* (*that is $y > 0$*) *onto itself.*

Proof. Let $z' = x' + iy'$, $z = x + iy$, $\bar{z} = x - iy$. Then

$$2iy' = \frac{az + b}{cz + d} - \frac{a\bar{z} + b}{c\bar{z} + d} = \frac{2(ad - bc)iy}{|cz + d|^2}, \tag{1}$$

and the required result follows. □

Definition 1. A semicircle centred on the x-axis lying in the upper half plane is called a *geodesic*.

From Theorem 4.1 and Theorem 3.2 we have:

Theorem 4.2. *Members of $\mathfrak{R}$ transform geodesics into geodesics.* □

Let z_1, z_2 be any two points in the upper half plane. If a member of $\mathfrak{R}$ maps z_1, z_2 into z'_1, z'_2 respectively, then clearly

$$(z_1\bar{z}_1z_2\bar{z}_2) = (z'_1\bar{z}'_1z'_2\bar{z}'_2)$$

or

$$\left|\frac{z_2 - z_1}{\bar{z}_1 - z_2}\right|^2 = \left|\frac{z'_2 - z'_1}{\bar{z}'_1 - z'_2}\right|^2.$$

Take $z_2 = z + \Delta z$, $z_1 = z$ and letting $\Delta z \to 0$ we have

$$\left|\frac{dz}{2y}\right|^2 = \left|\frac{dz'}{2y'}\right|^2,$$

or

$$\frac{dx^2 + dy^2}{y^2} = \frac{dx'^2 + dy'^2}{y'^2}.$$

From this we see that the metric

$$\frac{\sqrt{dx^2 + dy^2}}{y} \tag{2}$$

is invariant under transformations in $\mathfrak{R}$. The area

$$\frac{dx\,dy}{y^2} \tag{3}$$

which corresponds to this metric is also invariant under transformations in $\mathfrak{R}$. Readers who are not familiar with differential geometry can prove the invariance of (2) and (3) under members of $\mathfrak{R}$ by a direct method.

Theorem 4.3. *Let z_1, z_2 be two points on the upper half plane and let C be a smooth curve lying in the upper half plane joining z_1 and z_2. Then the value of the integral*

$$\int_C \frac{\sqrt{dx^2 + dy^2}}{y}$$

is minimum when C is (part of) a geodesic.

Proof. Construct a circle centre on the x-axis passing through z_1 and z_2. Denote its centre by $(t, 0)$ so that the circle is described by

$$x = t + \rho \cos \vartheta, \qquad y = \rho \sin \vartheta.$$

Let $\vartheta = \vartheta_1$ and ϑ_2 when $z = z_1$ and z_2 respectively. Now the curve C can be described by

$$\left.\begin{aligned} x &= t + \rho(\vartheta)\cos\vartheta \\ y &= \rho(\vartheta)\sin\vartheta \end{aligned}\right\} \qquad \rho(\vartheta_1) = \rho(\vartheta_2), \qquad 0 < \vartheta_1 < \vartheta_2 < \pi,$$

and hence

$$\int_C \frac{\sqrt{dx^2 + dy^2}}{y} = \int_{\vartheta_1}^{\vartheta_2} \frac{\sqrt{(\rho'(\vartheta)\cos\vartheta - \rho(\vartheta)\sin\vartheta)^2 + (\rho'(\vartheta)\sin\vartheta + \rho(\vartheta)\cos\vartheta)^2}}{\rho(\vartheta)\sin\vartheta}\, d\vartheta$$

$$= \int_{\vartheta_1}^{\vartheta_2} \sqrt{1 + \left(\frac{\rho'(\vartheta)}{\rho(\vartheta)}\right)^2}\, \frac{d\vartheta}{\sin\vartheta}$$

$$\geqslant \int_{\vartheta_1}^{\vartheta_2} \frac{d\vartheta}{\sin\vartheta} = \log \frac{\tan\frac{1}{2}\vartheta_2}{\tan\frac{1}{2}\vartheta_1}.$$

This shows that the values of the integral is minimum when and only when $\rho'(\vartheta) = 0$, that is when $\rho(\vartheta) = \rho$ is constant. □

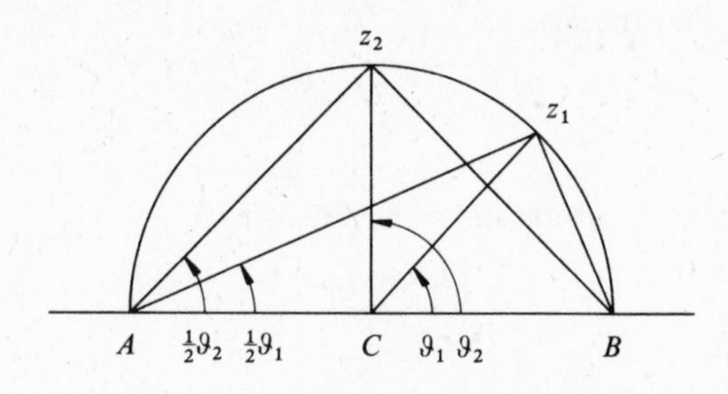

Figure 1

The above proof actually gives the minimum value of the integral along the geodesic. We can interpret the value geometrically as follows: Let the geodesic through z_1, z_2 intersect the x-axis at the points A, B with its centre at C (see Fig. 1). Then we have

$$\tan\frac{1}{2}\vartheta_1 = \frac{Bz_1}{z_1A}, \qquad \tan\frac{1}{2}\vartheta_2 = \frac{Bz_2}{z_2A}.$$

Therefore

$$\log\left(\frac{\tan\frac{1}{2}\vartheta_2}{\tan\frac{1}{2}\vartheta_1}\right) = \log|(BAz_2z_1)|.$$

Definition 2. The minimum value of the integral in Theorem 4.3 is called the *non-Euclidean distance* between the two points z_1 and z_2.

Definition 3. In this chapter the curvilinear triangular region between three geodesics will be called a *triangle*.

Theorem 4.4. *The non-Euclidean area* $\iint \frac{dx\,dy}{y^2}$ *of a triangle ABC is given by* $\pi - \angle A - \angle B - \angle C$.

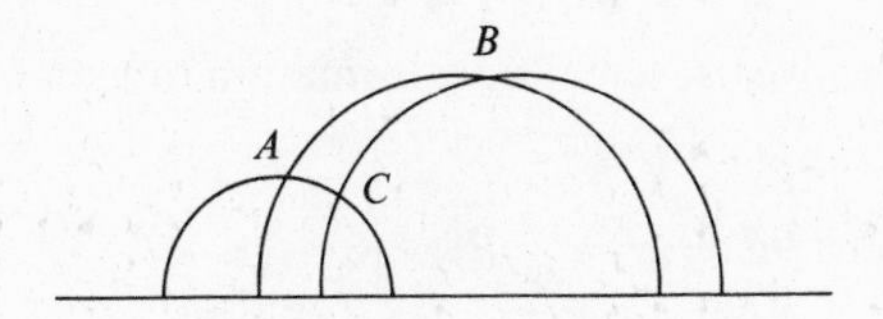

Figure 2

Proof. 1) We first consider the case $\angle B = \angle C = 0$ (see Fig. 3). It is not difficult to prove that there

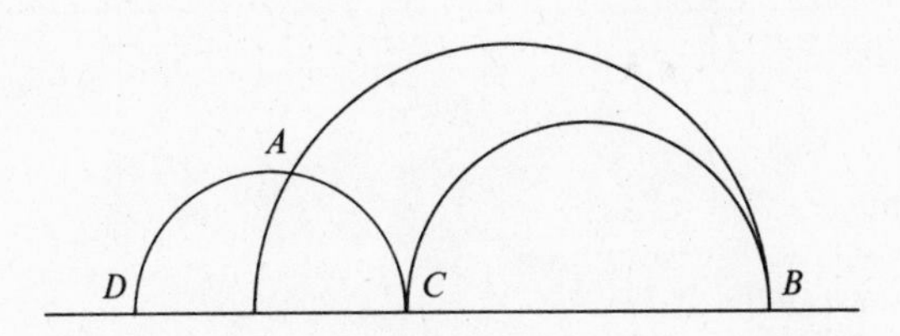

Figure 3

is a real bilinear transformation which maps B to the point at infinity, C to the point 1, D to the point -1 (or C to -1 and D to 1), and that the determinant is positive*. Thus Fig. 3 is transformed into Fig. 4. Let the coordinate for A be (x_0, y_0). Then

$$\int_{x_0}^{1} \int_{\sqrt{1-x^2}}^{\infty} \frac{dx\,dy}{y^2} = \int_{x_0}^{1} \frac{dx}{\sqrt{1-x^2}} = \sin^{-1} x \Big|_{x_0}^{1} = \frac{\pi}{2} - \sin^{-1} x_0 = \pi - \angle A.$$

---

* The real transformation which maps B, C, D into ∞, ± 1, ∓ 1 is given by

$$z' = \pm \frac{(D - 2B + C)z + (BC - 2DC + BD)}{(C - D)z + (D - C)B},$$

and the value of the determinant is $\pm 2(D - C)(C - B)(B - D)$.

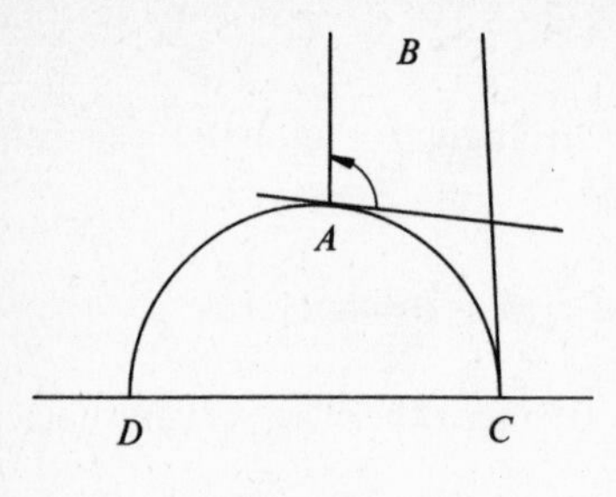

Figure 4

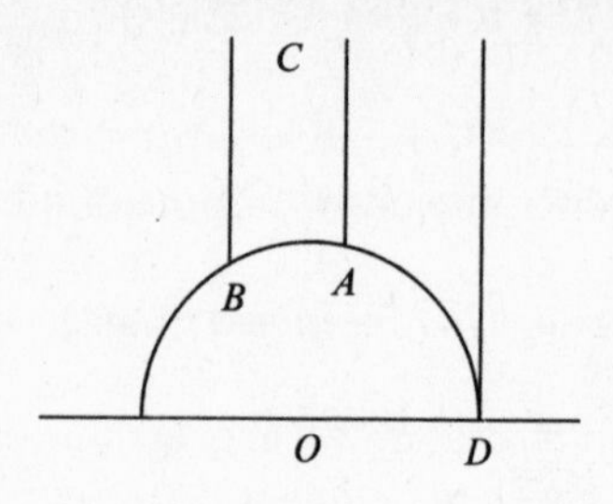

Figure 5

2) If $\angle C = 0$, then we use a real transformation to map C to ∞, giving Fig. 5. From 1) we have

$$\Delta ABC = \Delta BDC - \Delta ADC = (\pi - \angle B) - (\pi - (\pi - \angle A)) = \pi - \angle A - \angle B.$$

3) If none of $\angle A$, $\angle B$, $\angle C$ is zero as in Fig. 6,

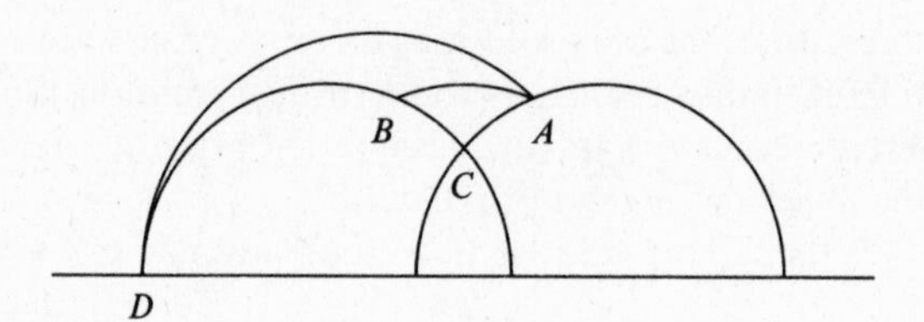

Figure 6

then, by 2), we have

$$\begin{aligned} \Delta ABC &= \Delta ADC - \Delta ABD \\ &= (\pi - \angle C - \angle A - \angle BAD) - [\pi - (\pi - \angle B) - \angle BAD] \\ &= \pi - \angle A - \angle B - \angle C. \quad \square \end{aligned}$$

From this theorem we see that the sum of the interior angles of a triangle is at most two right angles, and its value can be any number between 0 and π. What we have described here is a model of the famous Lobachevskian geometry which is an important tool in the study of modular functions.

13.5 Unimodular Transformations

Definition. Let a, b, c, d be integers satisfying $ad - bc = 1$. Then the transformation

$$z' = \frac{az + b}{cz + d} \tag{1}$$

is called a *unimodular transformation*.

It is easy to see that unimodular transformations form a group.

From (7) in §2 we have

$$\lambda + \lambda^{-1} = (a + d)^2 - 2.$$

The discriminant of this quadratic equation is

$$[(a + d)^2 - 2]^2 - 4 = (a + d)^2[(a + d)^2 - 4].$$

In our discussion we may assume that $a + d \geqslant 0$, since otherwise we can replace a, b, c, d by $-a$, $-b$, $-c$, $-d$.

1) If $a + d > 2$, then the transformation is hyperbolic and there are two real fixed points. These two fixed points are the roots of the quadratic equation

$$cz^2 + (d - a)z - b = 0.$$

The condition for this quadratic equation to have rational roots is that

$$(d - a)^2 + 4bc = (a + d)^2 - 4 = u^2,$$

where u is an integer. Since the only solutions for $x^2 - y^2 = 4$ are $x = \pm 2, y = 0$ it follows that the fixed points of a hyperbolic transformation must be irrational numbers which are the roots of a quadratic equation with rational coefficients. We call such numbers *quadratic algebraic numbers*.

2) If $a + d = 2$, then $\lambda = 1$ and we have the parabolic transformation

$$\frac{1}{z' - (a - 1)/c} = \frac{1}{z - (a - 1)/c} + c.$$

If $c = 0$, then $a = d = 1$ and we have

$$z' = z + b.$$

The former has the rational number $(a - 1)/c$ as the fixed point while the latter has ∞ as the fixed point.

3) If $a + d = 1$, then $\lambda^2 + \lambda + 1 = 0$ and so λ is $\rho = e^{2\pi i/3} = (-1 + \sqrt{-3})/2$ or ρ^2. The fixed points are then given by

$$z_1 = \frac{a + \rho^2}{c}, \qquad z_2 = \frac{a + \rho}{c},$$

and the standard form for the transformation is

$$\frac{z' - (a + \rho^2)/c}{z' - (a + \rho)/c} = \rho \frac{z - (a + \rho^2)/c}{z - (a + \rho)/c}.$$

This is an elliptic transformation whose period is 3. Replacing ρ by ρ^2 will give another elliptic transformation with period 3.

4) If $a + d = 0$, then the equation for λ is $(\lambda + 1)^2 = 0$ so that $\lambda = -1$, and the fixed points are the roots of

$$cz^2 - 2az - b = 0,$$

that is

$$z = \frac{a \pm i}{c}.$$

The standard form for the transformation is

$$\frac{z' - (a+i)/c}{z' - (a-i)/c} = -\frac{z - (a+i)/c}{z - (a-i)/c}.$$

This is an elliptic transformation with period 2.

Summarizing we have:

Theorem 5.1. *If $a + d = 0$, then the unimodular transformation* (1) *is an involution; if $a + d = \pm 1$, then it is a transformation with period* 3; *if $a + d = \pm 2$ then it is parabolic and its fixed point is either a rational number or the point at infinity; if $|a + d| > 2$, then it is hyperbolic and its fixed points are real quadratic algebraic numbers.* □

13.6 The Fundamental Region

Definition 1. Let z, z' be two points on the upper half plane. Suppose that there is a unimodular transformation which maps z into z'. Then we say that z and z' are *equivalent*, and we write $z \sim z'$.

Clearly we have (i) $z \sim z$; (ii) if $z \sim z'$, then $z' \sim z$; (iii) if $z \sim z'$, $z' \sim z''$, then $z \sim z''$.

We shall consider the following region in the upper half plane:

$$D: \begin{cases} -\frac{1}{2} \leqslant x < \frac{1}{2}, \\ x^2 + y^2 > 1 & \text{when} \quad x > 0, \\ x^2 + y^2 \geqslant 1 & \text{when} \quad x \leqslant 0. \end{cases}$$

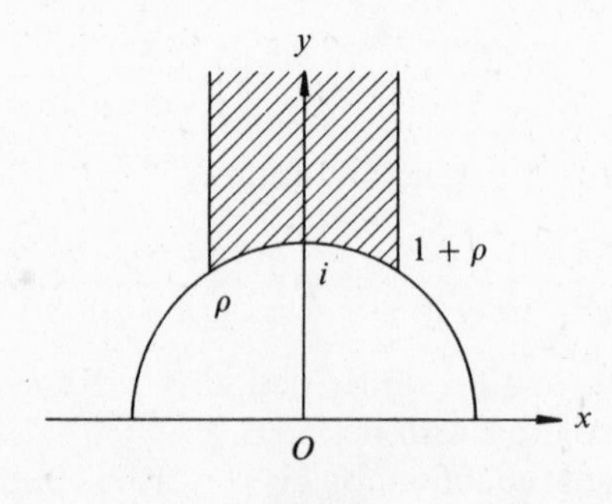

Figure 7

Definition 2. We call the points in D *reduced points*, and the region D the *fundamental region*. This region D is a triangle with interior angles $0, \pi/3, \pi/3$.

Theorem 6.1. *No two reduced points are equivalent.*

Proof. Let z, z' be two distinct reduced points and suppose that

$$z' = \frac{az + b}{cz + d}.$$

Then, by (1) in §4, we have

$$y' = \frac{y}{|cz + d|^2}.$$

We have

$$\begin{aligned}|cz + d|^2 &= c^2 z\bar{z} + cd(z + \bar{z}) + d^2 \\ &= c^2(x^2 + y^2) + 2cdx + d^2 \\ &\geqslant c^2 - |cd| + d^2 > 1\end{aligned}$$

where we must exclude the exceptional cases: $c = \pm 1, d = 0$, or $c = 0, d = \pm 1$, or $c = d = 1$. Therefore, apart from these exceptional cases, we always have

$$y' < y.$$

When $c = d = 1$, $|cz + d|^2 = 1$ only when $z = \rho$. From $a - b = 1$ and $\rho^2 + \rho + 1 = 0$, we have

$$z' = \frac{a\rho + b}{\rho + 1} = -\frac{a\rho + b}{\rho^2} = -a\rho^2 - b\rho = -\rho^2 + b.$$

Therefore $I(z') = \sqrt{3}/2$. If $z' \in D$, then $z' = \rho$ which contradicts with z, z' being distinct points.

We also have

$$z = \frac{dz' - b}{-cz' + a}$$

so that

$$y < y';$$

again we must exclude the cases: $c = \pm 1$, $a = 0$, or $c = 0, a = \pm 1$. Since we cannot have both $y > y'$ and $y < y'$, it follows that we need only examine the following two cases: (i) $c = 0, a = d = 1$; (ii) $c = 1$, $a = d = 0$.

In the first case we have

$$z' = z + b, \qquad b \neq 0$$

so that $x' = x + b$, and hence $|x' - x| \geqslant 1$ and therefore z, z' cannot be both in D.

In the second case we have $b = -1$ so that

$$z' = -\frac{1}{z}.$$

Thus if $|z| > 1$, then $|z'| < 1$ and so z' cannot be a reduced point, and similarly if $|z'| > 1$, then z cannot be a reduced point. If $|z| = 1$, then z must lie on the arc of the circle from ρ to i, and z' $(= -1/z)$ lies on the arc of the circle from $\rho + 1$ to i. If $z \neq i$, then z' cannot be a reduced point, and if $z = i$, then $z' = i = z$, contradicting our assumption. □

Theorem 6.2. *The number of points in the rectangle $-\frac{1}{2} \leqslant x < \frac{1}{2}, y \geqslant \gamma$ $(\gamma > 0)$ which are equivalent to a fixed point is finite. That is, if we partition the rectangle into sets of mutually equivalent points, then each set has only a finite number of points.*

Proof. Let $z = x + iy$ and

$$z' = \frac{az + b}{cz + d}.$$

Then we have

$$y' = \frac{y}{|cz + d|^2} = \frac{y}{c^2(x^2 + y^2) + 2cdx + d^2}.$$

If $y' \geqslant \gamma$, then

$$c^2(x^2 + y^2) + 2cdx + d^2 \leqslant \frac{y}{\gamma},$$

or

$$(cx + d)^2 + c^2y^2 \leqslant \frac{y}{\gamma},$$

and clearly there can only be a finite number of integers c, d satisfying this.

Let (c', d') be any such pair of integers, and $(c', d') = 1$. Then all the solutions (a, b) of the equation $ad' - bc' = 1$ can be represented by

$$a = a' + mc', \qquad b = b' + md'$$

where a', b' is a fixed solutions (that is $a'd' - b'c' = 1$), and m is any integer. Thus

$$z' = \frac{az + b}{cz + d} = \frac{a'z + b'}{c'z + d'} + m.$$

There can only be one m such that $-\frac{1}{2} \leqslant x' < \frac{1}{2}$. Therefore corresponding to each pair (c', d') with $(c', d') = 1$ there is only one set a, b such that $-\frac{1}{2} \leqslant x' < \frac{1}{2}$. Therefore the number of points in the rectangle which are equivalent to z is finite. □

Theorem 6.3. *Every point in the upper half plane is equivalent to a unique reduced point.*

Proof. Let $z = x_0 + iy_0$, $y_0 > 0$. We take the unique integer m satisfying

$$-\tfrac{1}{2} \leqslant x_0 + m < \tfrac{1}{2},$$

and let

$$z' = z + m.$$

If $|z'| > 1$, then z' is a reduced point and there is nothing more to prove. If $|z'| = 1$ and z' lies on the arc from ρ to i, then it is a reduced point, and if it lies on the arc from $1 + \rho$ to i, then the transformation $-1/z$ will give the former situation. If $|z'| < 1$, then we let

$$z'' = -\frac{1}{z'} \quad \text{and} \quad y'' = \frac{y_0}{|z'|^2} > y_0.$$

Choose m' such that

$$z''' = z'' + m', \qquad -\tfrac{1}{2} \leqslant x''' < \tfrac{1}{2}.$$

If z''' is still not a reduced point, then we use the same method and construct $z^{\text{iv}} = -1/z'''$.

In this way we obtain $z', z''', \ldots$ all lying in the rectangle

$$-\tfrac{1}{2} \leqslant x < \tfrac{1}{2}, \qquad y > y_0.$$

From Theorem 6.2 we see that there can only be a finite number of such points.

Therefore every point must be equivalent to a reduced point. Also, from Theorem 6.1, there cannot be two equivalent reduced points. The theorem is proved. □

In order to appreciate the significance of this theorem the reader should try to give direct proofs of the following two exercises which are immediate consequences of the theorem.

Exercise 1. All the points

$$z = \frac{a + i}{c}, \qquad a^2 + bc + 1 = 0$$

are equivalent to i.

Exercise 2. All the points

$$z = \frac{a + \rho}{c}, \qquad a(1 - a) - bc = 1$$

are equivalent to ρ.

13.7 The Net of the Fundamental Region

Theorem 7.1. *Suppose that z is not a fixed point of any unimodular transformation. Let U, V be two distinct unimodular transformations. Then*

$$Uz \neq Vz,$$

where Uz represents the image of z under U.

Proof. If $Uz = Vz$, then $z = U^{-1}Vz$, so that z is a fixed point of a unimodular transformation. □

Theorem 7.2. *The set of all triangular images of the fundamental region forms a covering of the upper half plane with no overlaps.*

Proof. The first part of the theorem follows from Theorem 6.3. If U and V are two distinct unimodular transformations whose triangular images of D overlap, then the mapping $U^{-1}V$ must map D into a triangular region which overlaps with D. Let z be a point in this overlap. Then there must be a point in D which is equivalent to z, and this is impossible if z is in D. □

We can explain this theorem in terms of tile covering. In ordinary space we can cover regions without overlaps using equal size square tiles, and by this we mean that each tile can be "translated" from one place to another which is occupied by another tile. Here the fundamental region is the shape of our new tile, and "translation" is now a unimodular transformation. The above theorem then tells us that such a tile can be used to cover the upper half plane with no overlaps. This is the interpretation of non-Euclidean geometry, and with this language the notion of a fundamental region becomes clearer, and generalization becomes easier.

We can alter the definition of a fundamental region as follows: Any region in the upper half plane is called a fundamental region if

(i) any point must be equivalent to a point in the region;
(ii) no two points in the region are equivalent.

Take any point z in the upper half plane which is not a fixed point of any unimodular transformation. Construct the points $z_1, z_2, \ldots$ which are equivalent to z, and then construct the perpendicular bisectors of (z, z_i), that is those points which have the two equal non-Euclidean distances from z and z_i. Discard the part on the side of z_i. Then the remaining part is a fundamental region. (The reader should supply the proof for this, and also determine the fundamental region corresponding to $z = 2i$.)

We remark that besides being important theoretically Lobachevskian geometry also has useful applications in number theory and in function theory.

We note the following: The fixed points of an elliptic transformation with period 2 lie on the lines joining vertices with angles $\pi/3$. The fixed points of an elliptic transformation with period 3 are the common vertices of six triangles. The fixed points of parabolic transformations are those points with infinitely many lines through them. The fixed points of a hyperbolic transformation cannot be vertices of any triangle (and it is even clearer that they cannot lie on the sides).

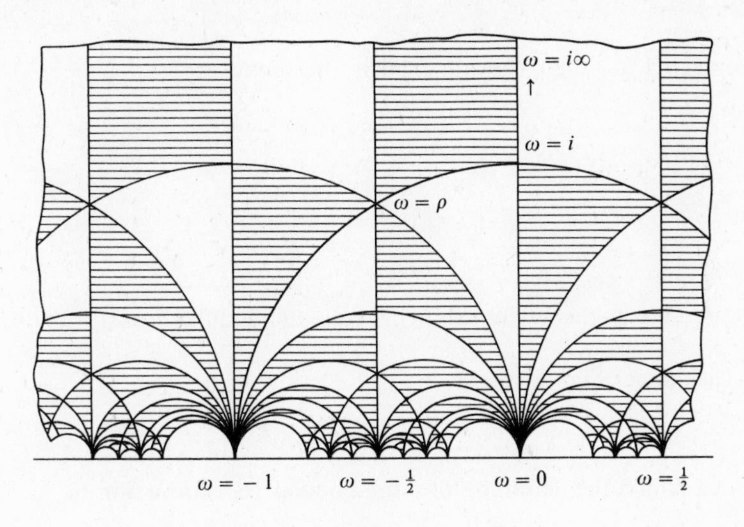

Figure 8

13.8 The Structure of the Modular Group

Let us denote by S and T the transformations $z' = z + 1$ and $z' = -1/z$ respectively. Then S^{-1} denotes the transformation $z' = z - 1$. The three transformations transform a fundamental region into the three neighbouring regions, and conversely the transformation which maps a neighbouring region into a fundamental region must be one of S, T or S^{-1}.

Let M be any unimodular transformation, and z be any point in the fundamental region D. We join z to M_z by a curve not passing through any vertices. Suppose that the various regions that this curve crosses are

$$D, D_1, D_2, \ldots, D_n(= MD).$$

Also, denote by M_i the unimodular transformation which maps D into D_i. Now $M_1 = S, T$ or S^{-1}. Suppose that M_k can be represented as a product of S and T. Since M_k^{-1} maps D_k into D, and D_{k+1} is a neighbouring region to D_k, it follows that M_k^{-1} maps D_{k+1} into M'_{k+1}, a neighbouring region of D. But D'_{k+1} can be mapped into D via a transformation M'^{-1} $(= S, T$ or $S^{-1})$. That is

$$M'^{-1}M_k^{-1}D_{k+1} = M'^{-1}D'_{k+1} = D$$

or

$$M_kM'D = D_{k+1}.$$

Therefore $M_{k+1} = M_k M'$ can be represented as a product of S and T, and hence so can M itself. We have therefore proved:

Theorem 8.1. *Any unimodular transformation is representable as a product of S and T.* □

Theorem 8.1 has the following explicit interpretation: If

$$M = S^{m_1} T S^{m_2} T S^{m_s} \cdots T S^{m_v},$$

then

$$z' = m_1 - \frac{1}{m_2} - \frac{1}{m_3} - \frac{1}{m_4} - \cdots - \frac{1}{m_v + z}.$$

This shows clearly the relationship between unimodular transformations and continued fractions.

It is easy to see that

$$T^2 = E, \qquad (ST)^3 = E.$$

Note. If we extend the definition of a unimodular transformation to:

$$z' = \frac{az + b}{cz + d}, \qquad ad - bc = \pm 1,$$

then we can have the result

$$z' = m_1 + \frac{1}{m_2} + \frac{1}{m_3} + \cdots + \frac{1}{m_v + z}.$$

13.9 Positive Definite Quadratic Forms

Let ω be any complex number in the upper half plane, ρ be a positive real number, and consider the quadratic form

$$F(x, y) = \rho(x - \omega y)(x - \bar{\omega} y) = \rho x^2 - \rho(\omega + \bar{\omega})xy + \rho\omega\bar{\omega}y^2.$$

If we apply the unimodular transformation

$$\omega = \frac{a\omega' + b}{c\omega' + d},$$

then the above becomes

$$\begin{aligned} &\rho((c\omega' + d)x - (a\omega' + b)y)((c\bar{\omega}' + d)x - (a\bar{\omega}' + b)y)/|c\omega' + d|^2 \\ &= \rho(dx - by - \omega'(-cx + ay))(dx - by - \bar{\omega}'(-cx + ay))/|c\omega' + d|^2, \end{aligned}$$

or

$$\rho(X - \omega' Y)(X - \bar{\omega}' Y)/|c\omega' + d|^2,$$

where

$$X = dx - by, \qquad Y = -cx + ay.$$

Therefore

$$\{\rho, -\rho(\omega + \bar{\omega}), \rho\omega\bar{\omega}\} \sim \left\{\frac{\rho}{|c\omega' + d|^2}, -\frac{\rho(\omega' + \bar{\omega}')}{|c\omega' + d|^2}, \frac{\rho\omega'\bar{\omega}'}{|c\omega' + d|^2}\right\}. \tag{1}$$

We also note that

$$\omega - \bar{\omega} = \frac{\omega' - \bar{\omega}'}{|c\omega' + d|^2}.$$

Starting from any positive definite form

$$\{\alpha, \beta, \gamma\}$$

where α, β, γ are real ($\alpha > 0$) and $\beta^2 - 4\alpha\gamma < 0$, we have, by comparing the left hand side of (1), that

$$\rho = \alpha, \qquad \omega = \frac{-\beta + \sqrt{\beta^2 - 4\alpha\gamma}}{2\alpha}.$$

Assuming that ω' is in the fundamental region, then from (1) we have that

$$-1 \leqslant \omega' + \bar{\omega}' < 1, \qquad \begin{cases} \omega'\bar{\omega}' > 1, & \text{if } \omega' + \bar{\omega}' > 0, \\ \omega'\bar{\omega}' \geqslant 1, & \text{if } \omega' + \bar{\omega}' \leqslant 0. \end{cases}$$

Substituting $\{\alpha', \beta', \gamma'\}$ into the right hand side of (1) we then have that

$$-1 < \frac{\beta'}{\alpha'} \leqslant 1, \qquad \begin{cases} \dfrac{\gamma'}{\alpha'} > 1, & \text{if } \beta' < 0, \\ \dfrac{\gamma'}{\alpha'} \geqslant 1, & \text{if } \beta' \geqslant 0. \end{cases}$$

Hence

$$-\alpha' < \beta' \leqslant \alpha' < \gamma'$$

or

$$0 \leqslant \beta' \leqslant \alpha' \leqslant \gamma'.$$

This generalizes Theorem 12.2.3 to real positive definite quadratic forms.

Exercise 1. Find the standard representations of quadratic forms which are invariant under a non-identical unimodular transformation. (Answer: $x^2 + y^2$, $x^2 + xy + y^2$).

13.10 Indefinite Quadratic Forms

We now consider the indefinite quadratic form

$$F = \{a, b, c\} = ax^2 + bxy + cy^2 = a(x - \omega_1 y)(x - \omega_2 y), \qquad d = b^2 - 4ac > 0,$$

with real coefficients. We suppose that $a > 0$, and that ω_1, ω_2 are irrational. We construct the circle with the line segment joining ω_1, ω_2 as diameter. This circle has the equation

$$a(x^2 + y^2) + bx + c = 0,$$

and is known as the fundamental circle for the form. Observe that this circle intersects with infinitely many triangles. For, by Theorem 6.2, corresponding to each rational point $-d'/c'$ there are infinitely many unimodular transformations which map this point to the point at infinity. That is there are infinitely many triangles having this rational point as a vertex. Our observation now follows from the fact that there are infinitely many rational points in the neighbourhood of a real number. If, among the triangles that the fundamental circle intersects, there is one which is a fundamental region, then we call the quadratic form a reduced quadratic form. It can be deduced from Theorem 6.3 that every indefinite quadratic form must be equivalent to a reduced quadratic form.

A fundamental circle of a reduced quadratic form must contain ρ or $1 + \rho$, that is at least one of the points $(\frac{1}{2}, \sqrt{3}/2), (-\frac{1}{2}, \sqrt{3}/2)$ belongs to the circle. This means that

$$a\left(a \pm \frac{b}{2} + c\right) < 0, \tag{1}$$

and with the substitution $c = (b^2 - d)/4a$, we have

$$4a^2 \pm 2ab + b^2 < d$$

or

$$3a^2 + (a \pm b)^2 < d. \tag{2}$$

Spreading out towards the left and the right along the arcs of a fundamental region D_0 we can list the triangles intersected by the fundamental circle as

$$\ldots, D_{-2}, D_{-1}, D_0, D_1, D_2, \ldots .$$

Let M_i be the unimodular transformation which maps D_0 into D_i. Then the quadratic form F_i, obtained from F via M_i, is equivalent to F, and we now have a chain of indefinite quadratic forms

$$\ldots, F_{-2}, F_{-1}, F, F_1, F_2, \ldots . \tag{3}$$

Since M_i^{-1} is a real transformation, the two real roots of F are mapped into the real roots of F_i. Therefore M_i^{-1} maps the fundamental circle for F into the fundamental circle for F_i. But the fundamental circle for F passes through D_i, so that the fundamental circle for F_i passes through D_0. This means that the chain of quadratic forms in (3) are reduced quadratic forms.

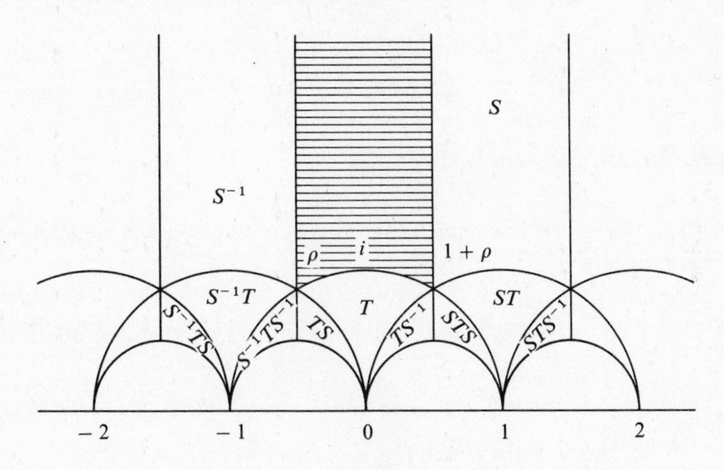

Figure 9

The region D_1 is a neighbouring region of D_0 unless the fundamental circle passes through the vertex $1 + \rho$, in which case D_1 may be the region described as STS or $S^{-1}TS^{-1}$ in Fig. 9. We see therefore that the neighbouring members of the chain of forms can be obtained from one of the transformations

$$S, S^{-1}, T, STS, S^{-1}TS^{-1}.$$

That is, if $\{a, b, c\}$ is a form in the chain, then the form immediately before, or after, it must be one of the following five forms:

$$\{a, \pm 2a + b, a \pm b + c\}, \{c, -b, a\}, \{a \pm b + c, b \pm 2c, c\}.$$

We now take one step further and consider quadratic forms with integer coefficients. We see from (2) that there are only finitely many reduced quadratic forms (with a given discriminant), and therefore there are only finitely many distinct forms in the chain.

Theorem 10.1. *The hyperbolic transformation with the two roots ω_1, ω_2 as fixed points are given by*

$$\begin{pmatrix} \frac{1}{2}(t - bu) & -cu \\ au & \frac{1}{2}(t + bu) \end{pmatrix} \tag{4}$$

where

$$t^2 - du^2 = 4.$$

Moreover, there are no other unimodular transformations with these properties.

Proof. The fixed points of the hyperbolic transformation (4) are the roots of the equation

$$aux^2 + \left(\frac{t+bu}{2} - \frac{t-bu}{2}\right)x + cu = 0$$

or

$$ax^2 + bx + c = 0.$$

The last part of the theorem is easy. □

Theorem 10.2. *The hyperbolic transformation* (4) *leaves the form* $\{a, b, c\}$ *unaltered, and there are no others.*

Proof. It is easy to verify that the coefficients for x^2, xy, y^2 in the quadratic form

$$a(\tfrac{1}{2}(t-bu)x - cuy)^2 + b(\tfrac{1}{2}(t-bu)x - cuy)(aux + \tfrac{1}{2}(t+bu)y)$$
$$+ c(aux + \tfrac{1}{2}(t+bu)y)^2$$

are precisely a, b, c. □

Theorem 10.3. *There is periodicity in the chain* (3).

Proof. We already noted that there are only finitely many distinct forms in (3). Denote by m the least positive integer such that $F_m = F$. Let M be the unimodular transformation which maps F into F_m. Then M^{-1} leaves the fundamental circle invariant so that M^{-1} maps D_{m+1} into D_1 giving $F_{m+1} = F_1, \ldots$. □

Example. $d = 37 \times 4$. The chain beginning with $(1, 0, -37)$ is

$$(1,0,-37),(1,2,-36),(1,4,-33),(1,6,-28),(1,8,-21),$$
$$(1,10,-12),(1,12,-1),(-1,-12,1),(-1,-10,12),\ldots,$$
$$(-1,12,1),(1,-12,-1),(1,-10,-12),\ldots,(1,-2,-36).$$

The chain beginning with $(3, 2, -12)$ is

$$(3,2,-12),(3,8,-7),(4,-6,-7),(4,2,-9),$$
$$(4,10,-3),(-3,-10,4),(-3,-4,11),(-3,2,12),$$
$$(-3,8,7),(-4,-6,7),(-4,2,9),(-4,10,3),$$
$$(3,-10,-4),(3,-4,-11).$$

13.11 The Least Value of an Indefinite Quadratic Form

We now return to quadratic forms with real coefficients. Admitting the wider definition, that is allowing also $ad - bc = -1$, we can be even more explicit than our previously described result.

Theorem 11.1. *Every indefinite quadratic form must be equivalent to a form whose fundamental circle has a diameter with end points satisfying* $-1 < \omega_1' < 0$ *and* $\omega_2' > 1$.

Proof. From the previous section every indefinite quadratic form must be equivalent to a reduced quadratic form. Consider the fundamental circle for this reduced quadratic form. If it intersects with the arc from ρ to i, then there are three cases:

1) $-1 < \omega_1' < 0$, $\omega_2' > 1$;
2) $\omega_1' < -1$, $0 < \omega_2' < 1$;
3) $-1 < \omega_2' < 0$, and so $\omega_1' < -2$.

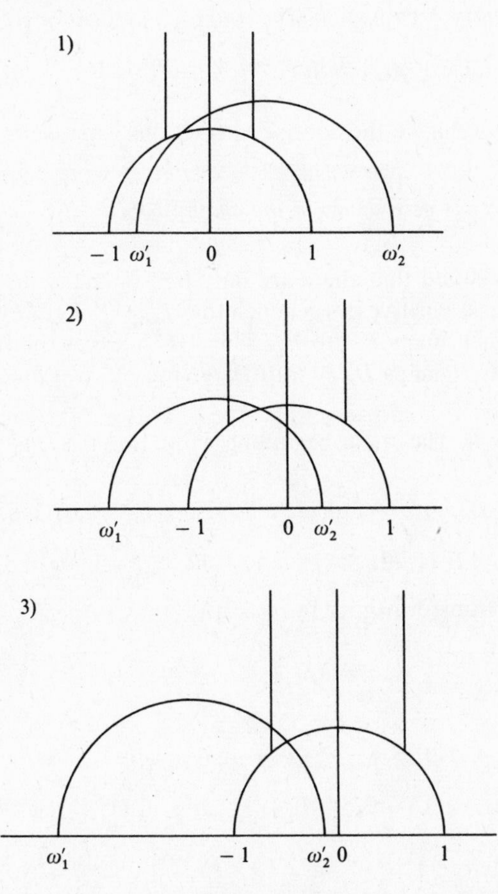

There is nothing to prove for 1), and 2) follows at once with the transformation

$z' = z + 1$. For case 3) we let

$$z'' = -z' - 1$$

so that

$$-1 < \omega_2'' < 0 \quad \text{and} \quad \omega_1'' = -\omega_1' - 1 > 1,$$

giving the required result.

If the fundamental circle intersects with the arc from i to $1 + \rho$, then the transformation $z' = -z$ reduces this to the previous situation. If it does not intersect with this arc, then there must be a unimodular transformation $z' = z + m$ which reduces this to either of the two previous situations. The theorem is proved. □

Suppose now that

$$F_0 = \alpha_0 x_0^2 + \beta_0 x_0 y_0 + \gamma_0 y_0^2$$

whose roots $\omega_1^{(0)}, \omega_2^{(0)}$ satisfy

$$\omega_1^{(0)} > 1, \qquad -1 < \omega_2^{(0)} < 0.$$

Let $\omega_1^{(0)}$ and $-1/\omega_1^{(0)}$ have the continued fractions expansions

$$\omega_1^{(0)} = d_1 + \frac{1}{d_2 +}\frac{1}{d_3 + \cdots}, \qquad -\frac{1}{\omega_2^{(0)}} = d_0 + \frac{1}{d_{-1} +}\frac{1}{d_{-2} + \cdots}.$$

The transformation

$$x_0 = d_1 x_1 + y_1, \qquad y_0 = x_1$$

transforms F_0 into

$$F_1 = \alpha_1 x_1^2 + \beta_1 x_1 y_1 + \gamma_1 y_1^2,$$

with roots

$$\omega_1^{(1)} = d_2 + \frac{1}{d_3 + \cdots}, \qquad -\frac{1}{\omega_2^{(1)}} = d_1 + \frac{1}{d_0 + \cdots}.$$

In general we transform F_{i-1} by

$$x_{i-1} = d_i x_i + y_i, \qquad y_{i-1} = x_i$$

into

$$F_i = \alpha_i x_i^2 + \beta_i x_i y_i + \gamma_i y_i^2$$

with roots

$$\omega_1^{(i)} = d_{i+1} + \frac{1}{d_{i+2} + \cdots}, \qquad -\frac{1}{\omega_2^{(i)}} = d_i + \frac{1}{d_{i-1} +}\frac{1}{d_{i-2} + \cdots}.$$

The difference between the two roots is

$$\frac{\sqrt{d}}{\alpha_i} = \omega_1^{(i)} - \omega_2^{(i)} = \left(d_{i+1} + \frac{1}{d_{i+2}+}\cdots\right) + \left(\frac{1}{d_i+}\frac{1}{d_{i-1}+}\cdots\right).$$

Denote by $L(F)$ the least value of

$$|\alpha_0 x_0^2 + \beta_0 x_0 y_0 + \gamma_0 y_0^2|$$

for all integers (x_0, y_0). Then clearly we have

$$L(F) \leqslant |\alpha_i| = \frac{\sqrt{d}}{\left(d_{i+1} + \frac{1}{d_{i+2}+}\cdots\right) + \left(\frac{1}{d_i+}\frac{1}{d_{i-1}+}\cdots\right)}.$$

Let

$$\min_i \left(\left(d_{i+1} + \frac{1}{d_{i+2}+}\cdots\right) + \left(\frac{1}{d_i+}\frac{1}{d_{i-1}+}\cdots\right)\right) = U,$$

so that

$$L(F) \leqslant \frac{\sqrt{d}}{U}$$

has infinitely many solutions.

If all the d_i are 1, then from

$$1 + \frac{1}{1+}\frac{1}{1+}\cdots = \frac{1}{2}(1+\sqrt{5}), \qquad \frac{1}{1+}\frac{1}{1+}\cdots = \frac{1}{2}(\sqrt{5}-1),$$

we have $U = \sqrt{5}$. Therefore

$$|ax^2 + bxy + cy^2| \leqslant \frac{\sqrt{d}}{\sqrt{5}}$$

has infinitely many solutions.

But if $\omega_1 = \frac{1}{2}(1+\sqrt{5})$, $\omega_2 = -\frac{1}{2}(\sqrt{5}-1)$, then

$$F = (x^2 - xy - y^2)\sqrt{\frac{d}{5}},$$

and, for all integers x, y,

$$|F(x, y)| \geqslant \sqrt{\frac{d}{5}}.$$

Also if there is one $d_i \geqslant 3$, then

$$[d_i, d_{i+1}, \ldots] + [0, d_{i-1}, \ldots] \geqslant 3 > \sqrt{5},$$

so that

$$L(F) \leqslant \frac{\sqrt{d}}{3}.$$

Finally, if there is one $d_i = 2$, then $(1 \leqslant d_{i-1} \leqslant 2)$

$$[2, d_{i+1}, d_{i+2}, \ldots] \geqslant 2$$

and

$$[0, d_{i-1}, \ldots] \geqslant \frac{1}{d_{i-1}} \frac{1}{+ d_{i-2}} \geqslant \frac{1}{2+} \frac{1}{1} = \frac{1}{3},$$

giving

$$[d_i, d_{i+1}, \ldots] + [0, d_{i-1}, \ldots] \geqslant 2 + \tfrac{1}{3} > \sqrt{5}.$$

We have therefore:

Theorem 11.2. *We always have*

$$L(F) \leqslant \sqrt{\frac{d}{5}}.$$

Moreover if

$$L(F) = \sqrt{\frac{d}{5}},$$

then F is equivalent to

$$\sqrt{\frac{d}{5}}(x^2 \pm xy - y^2). \quad \square$$

Chapter 14. Integer Matrices and Their Applications

14.1 Introduction

We first discuss 2×2 matrices in order to describe the content of this chapter. Part of this discussion already appeared in Chapter 13 but for the sake of completeness and ease of understanding we shall repeat slightly.

Throughout this section by a matrix we mean the square matrix

$$M = \begin{pmatrix} a & b \\ c & d \end{pmatrix}, \tag{1}$$

where a, b, c, d are integers, and we call them the elements of the matrix M. If all the elements are zero, then we call the matrix a *null* matrix and we denote it by 0. The quantity

$$ad - bc$$

is called the *determinant* of the matrix M. If the determinant is ± 1, then we call M a *modular matrix*, and if the determinant is 1, then we call M a *positive modular matrix*. If the determinant is zero, then we say that M is *singular*, otherwise we say that M is *non-singular*.

The *product* of two matrices

$$A = \begin{pmatrix} a & b \\ c & d \end{pmatrix}, \qquad B = \begin{pmatrix} a_1 & b_1 \\ c_1 & d_1 \end{pmatrix}$$

is defined to be the matrix

$$\begin{pmatrix} aa_1 + bc_1 & ab_1 + bd_1 \\ ca_1 + dc_1 & cb_1 + dd_1 \end{pmatrix}, \tag{2}$$

which is denoted by AB. It is clear that the determinant of AB is the product of the determinants of A and B, and hence the product of two (positive) modular matrices is a (positive) modular matrix.

Let k be an integer. We define

$$k \cdot \begin{pmatrix} a & b \\ c & d \end{pmatrix} = \begin{pmatrix} ka & kb \\ kc & kd \end{pmatrix}.$$

The matrix

$$I = \begin{pmatrix} 1 & 0 \\ 0 & 1 \end{pmatrix}$$

is called the *unit matrix*. For any matrix M we always have $MI = IM = M$. If $AB = I$, then we call B the inverse of A and we denote it by A^{-1}. It is easy to see that the inverse of a modular matrix $A = \begin{pmatrix} a & b \\ c & d \end{pmatrix}$ always exists, and that

$$A^{-1} = \pm \begin{pmatrix} d & -b \\ -c & a \end{pmatrix},$$

where we take the positive sign if and only if A is a positive modular matrix. It is also clear that $AA^{-1} = A^{-1}A = I$. Again, from taking the determinants of both sides of the equation $AB = I$, we see that if A has an inverse, then A must be a modular matrix. Therefore a necessary and sufficient condition for the existence of A^{-1} is that A be a modular matrix.

There are two very important positive modular matrices, namely

$$S = \begin{pmatrix} 1 & 1 \\ 0 & 1 \end{pmatrix}, \tag{3}$$

and

$$T = \begin{pmatrix} 0 & 1 \\ -1 & 0 \end{pmatrix}. \tag{4}$$

It is easy to verify that, for any integer m,

$$S^m = \begin{pmatrix} 1 & m \\ 0 & 1 \end{pmatrix} \tag{5}$$

and

$$T^2 = -I. \tag{6}$$

Theorem 1.1. *Any positive modular matrix can be expressed as a product of the matrices S and T. In other words the group of positive modular matrices can be generated by S and T.*

Proof. Let

$$M = \begin{pmatrix} a & b \\ c & d \end{pmatrix} \tag{7}$$

be a positive modular matrix. If $a = 0$, then $b \neq 0$, and so from

$$\begin{pmatrix} 0 & b \\ c & d \end{pmatrix} T = \begin{pmatrix} -b & 0 \\ -d & c \end{pmatrix}$$

we see that we may assume that $a \neq 0$. Again from

$$MT^2 = -M$$

we may further assume that $a > 0$. We can also suppose that

$$0 \leqslant b < a, \tag{8}$$

since we can choose an integer q such that $0 \leqslant aq + b < a$, and the matrix

$$\begin{pmatrix} a & b \\ c & d \end{pmatrix}\begin{pmatrix} 1 & q \\ 0 & 1 \end{pmatrix} = \begin{pmatrix} a & aq+b \\ c & cq+d \end{pmatrix} \tag{9}$$

satisfies the condition (8).

We now proceed by induction on a. If $a = 1$, then, by (8), $b = 0$ and hence $d = 1$, giving

$$\begin{pmatrix} 1 & 0 \\ c & 1 \end{pmatrix} = \begin{pmatrix} 0 & 1 \\ -1 & 0 \end{pmatrix}\begin{pmatrix} 1 & -c \\ 0 & 1 \end{pmatrix}\begin{pmatrix} 0 & -1 \\ 1 & 0 \end{pmatrix} = TS^{-c}T^{-1}$$

so that the matrix (7) is a product of S and T.

Suppose now that, for $0 < a < k$, all matrices (7) satisfying the condition (8) are products of S and T. Then the positive modular matrix

$$\begin{pmatrix} k & l \\ s & t \end{pmatrix}, \qquad 0 \leqslant l < k$$

(since $k > 1$, we see that l must be positive) satisfies

$$\begin{pmatrix} k & l \\ s & t \end{pmatrix}\begin{pmatrix} 0 & -1 \\ 1 & 0 \end{pmatrix} = \begin{pmatrix} l & -k \\ t & -s \end{pmatrix}$$

and we see, from the method of (9), that the right hand side of this equation is a product of S and T. The inductive argument is complete. □

Note: Positive modular matrices can also be expressed as a product of

$$\begin{pmatrix} 1 & 1 \\ 0 & 1 \end{pmatrix} \quad \text{and} \quad \begin{pmatrix} 1 & 0 \\ 1 & 1 \end{pmatrix}. \tag{10}$$

This is because

$$\begin{pmatrix} 0 & 1 \\ -1 & 0 \end{pmatrix} = \begin{pmatrix} 1 & 1 \\ 0 & 1 \end{pmatrix}\begin{pmatrix} 1 & 0 \\ 1 & 1 \end{pmatrix}^{-1}\begin{pmatrix} 1 & 1 \\ 0 & 1 \end{pmatrix}.$$

Theorem 1.2. *Any modular matrix can be expressed as a product of the matrices*

$$\begin{pmatrix} 0 & 1 \\ 1 & 0 \end{pmatrix} \quad \textit{and} \quad \begin{pmatrix} 1 & 1 \\ 0 & 1 \end{pmatrix}. \tag{11}$$

That is the group of modular matrices can be generated by these two matrices.

Proof. If a modular matrix M is not positive, then

$$M\begin{pmatrix} 0 & 1 \\ 1 & 0 \end{pmatrix}$$

is positive. It follows from the note above that any modular matrix is expressible as a product of the three matrices

$$\begin{pmatrix} 1 & 1 \\ 0 & 1 \end{pmatrix}, \begin{pmatrix} 1 & 0 \\ 1 & 1 \end{pmatrix}, \begin{pmatrix} 0 & 1 \\ 1 & 0 \end{pmatrix}.$$

But

$$\begin{pmatrix} 1 & 0 \\ 1 & 1 \end{pmatrix} = \begin{pmatrix} 0 & 1 \\ 1 & 0 \end{pmatrix}\begin{pmatrix} 1 & 1 \\ 0 & 1 \end{pmatrix}\begin{pmatrix} 0 & 1 \\ 1 & 0 \end{pmatrix}$$

so that the theorem follows. □

Definition 1. Let M and N be two matrices. Suppose that there is a modular matrix U such that

$$M = UN.$$

Then we say that N is *left associated* to M, and we denote this by $M \overset{L}{=} N$.

Clearly left association has the following three properties: (i) $M \overset{L}{=} M$ (*reflexive*); (ii) if $M \overset{L}{=} N$, then $N \overset{L}{=} M$ (*symmetric*); (iii) if $M \overset{L}{=} N$, $N \overset{L}{=} P$, then $M \overset{L}{=} P$ (*transitive*).

A similar definition can be given for *right association*.

Theorem 1.3. *Any matrix is left associated to a matrix of the form*

$$\begin{pmatrix} a & 0 \\ c & d \end{pmatrix}, \qquad a \geqslant 0, \qquad d \geqslant 0; \tag{12}$$

if $a > 0$, *then* $0 \leqslant c < a$.

Proof. Corresponding to the matrix

$$M = \begin{pmatrix} a & b \\ c & d \end{pmatrix}$$

there are integers r, s such that

$$rb + sd = 0, \qquad (r, s) = 1.$$

Now there are integers u, v such that $rv - su = 1$ so that

$$U = \begin{pmatrix} r & s \\ u & v \end{pmatrix}$$

is a positive modular matrix, and

$$UM = \begin{pmatrix} a_1 & 0 \\ c_1 & d_1 \end{pmatrix}.$$

If $a_1 \leqslant 0$, then we multiply this matrix by $\begin{pmatrix} -1 & 0 \\ 0 & 1 \end{pmatrix}$ which will give a matrix with

$a_1 \geqslant 0$, and similarly we can make $d_1 \geqslant 0$. Therefore every matrix is left associated to a matrix of the form

$$\begin{pmatrix} a & 0 \\ c & d \end{pmatrix}, \qquad a \geqslant 0, \qquad d \geqslant 0.$$

If $a > 0$, then we can choose q so that $0 \leqslant qa + c < a$, and from

$$\begin{pmatrix} 1 & 0 \\ q & 1 \end{pmatrix}\begin{pmatrix} a & 0 \\ c & d \end{pmatrix} = \begin{pmatrix} a & 0 \\ qa + c & d \end{pmatrix}$$

we see that the theorem is proved. □

Definition 2. We call the matrix in (12) *the normal form of Hermite.*

Theorem 1.4. *The normal form of Hermite for a non-singular matrix is unique.*

Proof. We first note that the normal form of Hermite $\begin{pmatrix} a & 0 \\ c & d \end{pmatrix}$ for a non-singular matrix cannot have a or d equal to zero. Now if

$$\begin{pmatrix} s & t \\ u & v \end{pmatrix}\begin{pmatrix} a & 0 \\ c & d \end{pmatrix} = \begin{pmatrix} a_1 & 0 \\ c_1 & d_1 \end{pmatrix}, \qquad sv - tu = \pm 1,$$

then, from $td = 0$, we have $t = 0$. Also, from $sa = a_1 > 0$, $vd = d_1 > 0$ and $sv = \pm 1$ we see that $s = v = 1$. Finally, from $ua + c = c_1$, $0 \leqslant c < a$, $0 \leqslant c_1 < a_1 = a$ we see that $u = 0$. The theorem is proved. □

Exercise. Investigate the situation for a singular matrix.

Definition 3. Let there be two modular matrices U and V such that

$$UMV = N.$$

Then we say that M and N are *equivalent*, and we write $M \sim N$. Clearly, being equivalent has the three properties of being reflexive, symmetric, and transitive.

Theorem 1.5. *Any matrix is equivalent to a matrix of the form*

$$\begin{pmatrix} a_1 & 0 \\ 0 & a_1 a_2 \end{pmatrix}, \qquad a_1 \geqslant 0, \qquad a_2 \geqslant 0. \tag{13}$$

Proof. Consider the matrix

$$M = \begin{pmatrix} a & b \\ c & d \end{pmatrix}.$$

Since the theorem becomes trivial if M is the null matrix we can assume that $a \neq 0$, and indeed we can even assume that $a > 0$. We first prove that M must be equivalent to a matrix of the form

$$\begin{pmatrix} a_1 & b_1 \\ c_1 & d_1 \end{pmatrix}, \qquad a_1|(b_1, c_1, d_1).$$

We use induction on a, the case $a = 1$ being trivial. When $a > 1$ and $a \nmid b$, we can choose q so that $0 < aq + b < a$ and we consider

$$\begin{pmatrix} a & b \\ c & d \end{pmatrix}\begin{pmatrix} q & 1 \\ 1 & 0 \end{pmatrix} = \begin{pmatrix} aq + b & * \\ * & * \end{pmatrix},$$

where the leading element is a positive integer less than a. If $a|b$ and $a \nmid c$, then we choose q' such that $0 < aq' + c < a$, and we consider

$$\begin{pmatrix} q' & 1 \\ 1 & 0 \end{pmatrix}\begin{pmatrix} a & b \\ c & d \end{pmatrix} = \begin{pmatrix} aq' + c & * \\ * & * \end{pmatrix}$$

where the leading element is once again a positive integer less than a. Finally, if $a|(b, c)$, but $a \nmid d$, we let $c = c'a$ so that

$$\begin{pmatrix} 1 & 1 \\ 0 & 1 \end{pmatrix}\begin{pmatrix} 1 & 0 \\ -c' & 1 \end{pmatrix}\begin{pmatrix} a & b \\ c & d \end{pmatrix} = \begin{pmatrix} a & (1 - c')b + d \\ * & * \end{pmatrix},$$

and $a \nmid \{(1 - c')b + d\}$ which reduces back to the case when $a \nmid b$. The inductive argument is now complete.

Now $a_1|(b_1, c_1, d_1)$. We let $b_1 = a_1 b_2, c_1 = a_1 c_2$, and $d_1 = a_1 d_2$, and we consider

$$\begin{pmatrix} 1 & 0 \\ -c_2 & 1 \end{pmatrix}\begin{pmatrix} a_1 & a_1 b_2 \\ a_1 c_2 & a_1 d_2 \end{pmatrix}\begin{pmatrix} 1 & -b_2 \\ 0 & 1 \end{pmatrix} = \begin{pmatrix} a_1 & 0 \\ 0 & a_1(d_2 - b_2 c_2) \end{pmatrix}$$

where we can assume that $a_1 > 0$, since otherwise we can multiply by $\begin{pmatrix} -1 & 0 \\ 0 & 1 \end{pmatrix}$. Similarly we can assume that $a_2 = d_2 - b_2 c_2 \geqslant 0$. The theorem is proved. □

Definition 4. We call the matrix in (13) the *normal form of Smith*.

We summarize our result as follows: By Theorem 1.2 any modular matrix is a product of the matrices

$$\begin{pmatrix} 0 & 1 \\ 1 & 0 \end{pmatrix}, \quad \begin{pmatrix} 1 & 1 \\ 0 & 1 \end{pmatrix}.$$

From

$$\begin{pmatrix} 0 & 1 \\ 1 & 0 \end{pmatrix}\begin{pmatrix} a & b \\ c & d \end{pmatrix} = \begin{pmatrix} c & b \\ a & d \end{pmatrix}$$

and

$$\begin{pmatrix} a & b \\ c & d \end{pmatrix}\begin{pmatrix} 0 & 1 \\ 1 & 0 \end{pmatrix} = \begin{pmatrix} b & a \\ d & c \end{pmatrix}$$

we see that the effect of multiplying by $\begin{pmatrix} 0 & 1 \\ 1 & 0 \end{pmatrix}$ or its inverse is merely the interchanging of the two rows or the two columns of the matrix. Again, from

$$\begin{pmatrix} 1 & \pm 1 \\ 0 & 1 \end{pmatrix}\begin{pmatrix} a & b \\ c & d \end{pmatrix} = \begin{pmatrix} a \pm c & b \pm d \\ c & d \end{pmatrix}$$

and

$$\begin{pmatrix} a & b \\ c & d \end{pmatrix}\begin{pmatrix} 1 & \pm 1 \\ 0 & 1 \end{pmatrix} = \begin{pmatrix} a & b \pm a \\ c & d \pm c \end{pmatrix},$$

we see that the effect of multiplying by $\begin{pmatrix} 1 & 1 \\ 0 & 1 \end{pmatrix}$ or by its inverse $\begin{pmatrix} 1 & -1 \\ 0 & 1 \end{pmatrix}$ is the addition or subtraction of the second row to the first row, or the first column to the second column of the matrix. We call these operations here the elementary transformations of the matrix. We can therefore restate Theorem 1.5 as follows: We can use elementary transformations to reduce a given matrix to the normal form of Smith.

Now the greatest common factor of the elements of a matrix is invariant under an elementary transformation, and so from Theorem 1.5 we have $(a, b, c, d) = a_1$. Also

$$\begin{vmatrix} a & b \\ c & d \end{vmatrix} = ad - bc = \pm a_1^2 a_2.$$

Therefore we have

Theorem 1.6. *The normal form of Smith for a given matrix is unique.* □

14.2 The Product of Matrices

Let $a_{11}, a_{12}, \ldots, a_{mn}$ be integers. We call the array

$$A = \begin{pmatrix} a_{11} & \cdots & a_{1n} \\ a_{21} & \cdots & a_{2n} \\ \cdots & \cdots & \cdots \\ a_{m1} & \cdots & a_{mn} \end{pmatrix}$$

an m by n matrix and we sometimes denote it by $A^{(m,n)}$. If $m = n$, then we denote it by $A^{(n)}$ and we call it a square matrix of order n. Let B be an n by l matrix

$$B = \begin{pmatrix} b_{11} & \cdots & b_{1l} \\ b_{21} & \cdots & b_{2l} \\ \cdots & \cdots & \cdots \\ b_{n1} & \cdots & b_{nl} \end{pmatrix}.$$

We define the *product matrix* of A and B by

$$AB = C = \begin{pmatrix} c_{11} & \cdots & c_{1l} \\ c_{21} & \cdots & c_{2l} \\ \cdots & \cdots & \cdots \\ c_{m1} & \cdots & c_{ml} \end{pmatrix}, \qquad c_{rs} = \sum_{t=1}^{n} a_{rt} b_{ts}$$

$$(r = 1, \ldots, m;\ s = 1, \ldots, l). \tag{1}$$

We see from the definition that the product matrix of A and B exists only when the number of columns in A is the same as the number of rows in B. Note also that, when AB and BA both exist, they may be different. If $AB = BA$, then we say that A and B *commute*. However we always have $(AB)D = A(BD)$ whenever either side of this equation exists.

If A and B are square matrices, then the determinant of AB is the product of the determinants of A and B. A square matrix whose determinant is zero is called a *singular* matrix, otherwise we call it a *non-singular* matrix. *Modular matrices* are those square matrices whose determinants equal ± 1 and *positive modular matrices* are those whose determinants equal 1. Clearly the product of two (positive) modular matrices is a (positive) modular matrix.

The square matrix

$$A = \begin{pmatrix} \lambda_1 & 0 & \cdots & 0 \\ 0 & \lambda_2 & \cdots & 0 \\ \cdots & \cdots & \cdots & \cdots \\ 0 & 0 & \cdots & \lambda_n \end{pmatrix}$$

where each element not on the main diagonal is zero is called a *diagonal matrix*, and we denote it simply by $\Lambda = [\lambda_1, \lambda_2, \ldots, \lambda_n]$. If $\lambda_1 = \lambda_2 = \cdots = \lambda_n = 1$ then,

$$\Lambda = I = \begin{pmatrix} 1 & 0 & \cdots & 0 \\ 0 & 1 & \cdots & 0 \\ \cdots & \cdots & \cdots & \cdots \\ 0 & 0 & \cdots & 1 \end{pmatrix}$$

and we call I the *unit matrix*. Clearly we have $AI = IA = A$ for any square matrix A of order n.

If the square matrices A and B satisfy $AB = I$, then we call B the *inverse* of A and we denote it by A^{-1}.

Consider a square matrix A $(= A^{(n)})$. By the *cofactor* of the element a_{ij} we mean the determinant of the square matrix of order $(n - 1)$ obtained by removing the i-th row and the j-th column of A. If we attach the sign $(-1)^{i+j}$ to the cofactor of a_{ij} then we call it the *algebraic cofactor* of a_{ij} and we denote it by A_{ij}. Let

$$A_0 = \begin{pmatrix} A_{11} & A_{21} & \cdots & A_{n1} \\ A_{12} & A_{22} & \cdots & A_{n2} \\ \cdots & \cdots & \cdots & \cdots \\ A_{1n} & A_{2n} & \cdots & A_{nn} \end{pmatrix},$$

that is the matrix obtained from A by replacing each element a_{rs} with the algebraic cofactor A_{rs} of a_{rs}, is called the *adjoint matrix* of A. It is not difficult to prove that

$$AA_0 = A_0 A = aI,$$

where a is the determinant of A. It follows that if A is a modular matrix, then its inverse exists, and that $A^{-1} = \pm A_0$. Conversely, if A has an inverse, then it must be a modular matrix.

If $AB = I$, then from $B = (\pm A_0 A)B = \pm A_0(AB) = \pm A_0$ we see that the inverse is unique and that $AA^{-1} = A^{-1}A = I$. Also, if A and B both have inverses, then $(AB)^{-1} = B^{-1}A^{-1}$.

A 1 by n matrix $(x_1, \ldots, x_n)$, where we no longer restrict the elements to be integers, is called a *vector*, and we write $x = (x_1, \ldots, x_n)$. We should take care that this notation here is not to be confused with the greatest common factor symbol $(x_1, \ldots, x_n) = d$. We shall use the convention that $(x_1, \ldots, x_n)$ by itself always represents a vector, while $(x_1, \ldots, x_n) = d$ means the greatest common factor of $x_1, \ldots, x_n$. Also we shall always use the letters x and y to denote a vector with n terms.

The equation

$$y = xB \qquad (B = B^{(n,l)}) \tag{2}$$

represents the system of linear equations

$$y_i = \sum_{j=1}^{n} x_j b_{ji}, \qquad 1 \leqslant i \leqslant l.$$

If $n = l$ and B is non-singular, then (2) is called a *transformation*. Corresponding to integers $x_1, \ldots, x_n$ the transformation gives integers $y_1, \ldots, y_n$, but not conversely. However, if B is a modular matrix, then when $y_1, \ldots, y_n$ are integers, the numbers $x_1, \ldots, x_n$ must also be integers. In this case we call (2) a *modular transformation*.

Example 1. Let $r \neq 1$, and $y_1 = -x_r$, $y_r = x_1$, $y_i = x_i$ $(i \neq 1,\ i \neq r)$. This is a modular transformation whose corresponding matrix is obtained from I by multiplying the first row by -1 and then interchanging it with the r-th row (or multiplying the r-th column by -1 and then interchanging it with the first column). We denote this matrix by E_r so that

$$E_r = \left(\begin{array}{cccccc} 0 & 0 & \cdots & 1 & \cdots & 0 \\ 0 & 1 & \cdots & 0 & \cdots & 0 \\ \cdots & \cdots & \cdots & \cdots & \cdots & \cdots \\ -1 & 0 & \cdots & 0 & \cdots & 0 \\ \cdots & \cdots & \cdots & \cdots & \cdots & \cdots \\ 0 & 0 & \cdots & 0 & \cdots & 1 \end{array}\right) r. \tag{3}$$

$$\qquad\qquad\qquad r$$

Example 2. Let $r \neq 1$, and $y_i = x_i$ $(i \neq r)$, $y_r = x_r + x_1$. This too is a modular transformation and its corresponding matrix is

$$V_r = \begin{pmatrix} 1 & 0 & \cdots & 1 & \cdots & 0 \\ 0 & 1 & \cdots & 0 & \cdots & 0 \\ \cdots & \cdots & \cdots & \cdots & \cdots & \cdots \\ 0 & 0 & \cdots & 0 & \cdots & 1 \end{pmatrix}, \qquad (4)$$
$$\hphantom{V_r = 1\ 0\ \cdots\ } r$$

that is the matrix obtained from I by adding the r-th row to the first row (or adding the first column to the r-th column).

It is easy to prove that V_r is representable as a product of V_2 and E_i. In fact, if $r > 2$, then

$$V_r = E_2 E_r E_2 V_2 E_2 E_r E_2. \qquad (5)$$

The proof is as follows: Let

$$t = \begin{pmatrix} t_1 \\ t_2 \\ \vdots \\ t_n \end{pmatrix},$$

so that

$$E_2 t = \begin{bmatrix} t_2 \\ -t_1 \\ t_3 \\ \vdots \\ t_n \end{bmatrix}, \qquad E_r E_2 t = \begin{bmatrix} t_r \\ -t_1 \\ t_3 \\ \vdots \\ -t_2 \\ \vdots \\ t_n \end{bmatrix} r, \ldots,$$

$$E_2 E_r E_2 V_2 E_2 E_r E_2 t = \begin{pmatrix} t_1 + t_r \\ t_2 \\ \vdots \\ t_n \end{pmatrix}.$$

But

$$V_r t = \begin{pmatrix} t_1 + t_r \\ t_2 \\ \vdots \\ t_n \end{pmatrix},$$

so that (5) follows.

Example 3. For fixed distinct r and s we let $y_i = x_i\ (i \neq s)$ and $y_s = x_s + x_r$. Then this is also a modular transformation whose matrix is obtained from I by adding the s-th row to the r-th row (or adding the r-th column to the s-th column). We denote this matrix by V_{rs} so that

$$V_{rs} = \begin{pmatrix} 1 & 0 & \cdots & 0 & \cdots & 0 & \cdots & 0 \\ \cdots & \cdots & \cdots & \cdots & \cdots & \cdots & \cdots & \cdots \\ 0 & 0 & \cdots & 1 & \cdots & 1 & \cdots & 0 \\ \cdots & \cdots & \cdots & \cdots & \cdots & \cdots & \cdots & \cdots \\ 0 & 0 & \cdots & 0 & \cdots & 1 & \cdots & 0 \\ \cdots & \cdots & \cdots & \cdots & \cdots & \cdots & \cdots & \cdots \\ 0 & 0 & \cdots & 0 & \cdots & 0 & \cdots & 1 \end{pmatrix} \begin{matrix} \\ \\ r \\ \\ s \\ \\ \\ \end{matrix}. \tag{6}$$

(columns r and s)

When $s > 1$, $V_{rs} = E_r^{-1}V_sE_r$, and $V_{r1} = E_r^{-1}V_r^{-1}E_r$. Therefore V_{rs} can also be represented as a product of V_2 and $E_2, \ldots, E_n$.

The matrices V_{rs} ($1 \leqslant r \leqslant n$, $1 \leqslant s \leqslant n$, $r \neq s$) together with all the products formed by them forms a group which we denote by $\mathfrak{M}_n$. We saw, from the note following Theorem 1.1, that the group $\mathfrak{M}_2$, generated by the matrices $V_{21} = \begin{pmatrix} 1 & 0 \\ 1 & 1 \end{pmatrix}$ and $V_{12} = \begin{pmatrix} 1 & 1 \\ 0 & 1 \end{pmatrix}$, is identical with the group of all 2 by 2 positive modular matrices. We now prove the corresponding result for n by n positive modular matrices.

Theorem 2.1. *The group $\mathfrak{M}_n$ is the group of all n by n positive modular matrices.*

It is clear that each matrix in $\mathfrak{M}_n$ is a positive modular matrix so that we only have to prove that every positive modular matrix is in $\mathfrak{M}_n$, that is every positive modular matrix can be expressed as a product of the matrices V_{rs}. For this purpose we shall first establish the following two theorems.

Theorem 2.2. *If $(x_1, \ldots, x_n) = d$, then there exists $U \in \mathfrak{M}_n$ such that*

$$(x_1, \ldots, x_n)U = (d, 0, \ldots, 0).$$

Proof. Consider first the case $n = 2$. If $(x_1, x_2) = d$, then there are integers r and s such that

$$rx_1 + sx_2 = d, \qquad (r, s) = 1.$$

We take $u = -x_2/d$, $v = x_1/d$ so that

$$vx_2 + ux_1 = 0, \qquad vr - us = 1.$$

Thus

$$(x_1, x_2)\begin{pmatrix} r & u \\ s & v \end{pmatrix} = (d, 0)$$

and $P = \begin{pmatrix} r & u \\ s & v \end{pmatrix}$ is a positive modular matrix. Since we already know that $P \in \mathfrak{M}_2$ by the note following Theorem 1.1, the case $n = 2$ is proved.

We now proceed by induction on n. Let $(x_{n-1}, x_n) = d_1$, so that there exists $P \in \mathfrak{M}_2$ such that

$$(x_{n-1}, x_n)P = (d_1, 0).$$

Let

$$V^{(n)} = \begin{bmatrix} 1 & 0 & \cdots & 0 & 0 \\ 0 & 1 & \cdots & 0 & 0 \\ \cdots & \cdots & \cdots & \cdots & \cdots \\ 0 & 0 & \cdots & r & u \\ 0 & 0 & \cdots & s & v \end{bmatrix} = \begin{pmatrix} I^{(n-2)} & 0 \\ 0 & P \end{pmatrix}.$$

It is easy to see that $V^{(n)} \in \mathfrak{M}_n$ and that

$$(x_1, \ldots, x_n)V^{(n)} = (x_1, \ldots, x_{n-2}, d_1, 0).$$

From the induction hypothesis we have $V^{(n-1)} \in \mathfrak{M}_{n-1}$ and that

$$(x_1, \ldots, x_{n-2}, d_1)V^{(n-1)} = (d, 0, \ldots, 0).$$

We now let

$$V_1^{(n)} = \begin{pmatrix} V^{(n-1)} & 0 \\ 0 & 1 \end{pmatrix}$$

so that

$$(x_1, \ldots, x_n)V^{(n)}V_1^{(n)} = (d, 0, \ldots, 0).$$

It is easy to see that

$$U = V^{(n)}V_1^{(n)} \in \mathfrak{M}_n$$

so that the theorem is proved. □

Theorem 2.3. *Let* $(a_{11}, a_{12}, \ldots, a_{1n}) = d$. *Then there is a matrix in* $\mathfrak{M}_n$ *whose first row is*

$$\left(\frac{a_{11}}{d}, \frac{a_{12}}{d}, \ldots, \frac{a_{1n}}{d}\right).$$

Proof. By Theorem 2.2 there is a matrix U in $\mathfrak{M}_n$ such that

$$(a_{11}, a_{12}, \ldots, a_{1n})U = (d, 0, \ldots, 0)$$

and so the matrix U^{-1} is a suitable candidate. □

Proof of Theorem 2.1. The case $n = 2$ is already established. We now use induction on n. Let

$$A = \begin{pmatrix} a_{11} & a_{12} & \cdots & a_{1n} \\ a_{21} & a_{22} & \cdots & a_{2n} \\ \cdots & \cdots & \cdots & \cdots \\ a_{n1} & a_{n2} & \cdots & a_{nn} \end{pmatrix}$$

be any positive modular matrix. Clearly $(a_{11}, a_{12}, \ldots, a_{1n}) = 1$. On multiplying A by the matrix U in Theorem 2.3 we have

$$AU = \begin{pmatrix} 1 & 0 & \cdots & 0 \\ a'_{21} & a'_{22} & \cdots & a'_{2n} \\ \cdots & \cdots & \cdots & \cdots \\ a'_{n1} & a'_{n2} & \cdots & a'_{nn} \end{pmatrix}.$$

The matrix

$$V = \begin{pmatrix} 1 & 0 & 0 & \cdots & 0 \\ -a'_{21} & 1 & 0 & \cdots & 0 \\ -a'_{31} & 0 & 1 & \cdots & 0 \\ \cdots & \cdots & \cdots & \cdots & \cdots \\ -a'_{n1} & 0 & 0 & \cdots & 1 \end{pmatrix}$$

is in $\mathfrak{M}_n$, and

$$VAU = \begin{pmatrix} 1 & 0 & 0 & \cdots & 0 \\ 0 & a'_{22} & a'_{23} & \cdots & a'_{2n} \\ \cdots & \cdots & \cdots & \cdots & \cdots \\ 0 & a'_{n2} & a'_{n3} & \cdots & a'_{nn} \end{pmatrix}.$$

From the induction hypothesis, the matrix

$$\begin{pmatrix} a'_{22} & a'_{23} & \cdots & a'_{2n} \\ \cdots & \cdots & \cdots & \cdots \\ a'_{n2} & a'_{n3} & \cdots & a'_{nn} \end{pmatrix}$$

is in $\mathfrak{M}_{n-1}$, and so the matrix

$$\begin{pmatrix} 1 & 0 & 0 & \cdots & 0 \\ 0 & a'_{22} & a'_{23} & \cdots & a'_{2n} \\ \cdots & \cdots & \cdots & \cdots & \cdots \\ 0 & a'_{n2} & a'_{n3} & \cdots & a'_{nn} \end{pmatrix}$$

is in $\mathfrak{M}_n$. From (7) we see that the theorem follows. □

14.3 The Number of Generators for Modular Matrices

We proved in §1 that any 2 by 2 positive modular matrix can be expressed as a product of the matrices $V_{21} = \begin{pmatrix} 1 & 0 \\ 1 & 1 \end{pmatrix}$ and $V_{12} = \begin{pmatrix} 1 & 1 \\ 0 & 1 \end{pmatrix}$. We now discuss the general case, and ask for the matrices whose products give all possible n by n positive modular matrices – that is we want to know the generators of the group $\mathfrak{M}_n$.

From the definition for $\mathfrak{M}_n$ any matrix in it is a product of V_{rs}, and from the previous section we know that each V_{rs} is expressible as a product of the following n matrices:

$$E_2 = \begin{bmatrix} 0 & 1 & 0 & \cdots & 0 \\ -1 & 0 & 0 & \cdots & 0 \\ 0 & 0 & 1 & \cdots & 0 \\ \cdots & \cdots & \cdots & \cdots & \cdots \\ 0 & 0 & 0 & \cdots & 1 \end{bmatrix}, \ldots, \quad E_n = \begin{bmatrix} 0 & 0 & 0 & \cdots & 1 \\ 0 & 1 & 0 & \cdots & 0 \\ 0 & 0 & 1 & \cdots & 0 \\ \cdots & \cdots & \cdots & \cdots & \cdots \\ -1 & 0 & 0 & \cdots & 0 \end{bmatrix},$$

$$V_2 = \begin{bmatrix} 1 & 1 & 0 & \cdots & 0 \\ 0 & 1 & 0 & \cdots & 0 \\ 0 & 0 & 1 & \cdots & 0 \\ \cdots & \cdots & \cdots & \cdots & \cdots \\ 0 & 0 & 0 & \cdots & 1 \end{bmatrix}.$$

Thus $\mathfrak{M}_n$ can be generated by the n matrices $E_2, E_3, \ldots, E_n, V_2$.

Let

$$U_1 = \begin{pmatrix} 0 & 0 & \cdots & 0 & (-1)^{n-1} \\ 1 & 0 & \cdots & 0 & 0 \\ \cdots & \cdots & \cdots & \cdots & \cdots \\ 0 & 0 & \cdots & 1 & 0 \end{pmatrix}.$$

It is not difficult to prove that each of $E_2, E_3, \ldots, E_n$ is expressible as a product of U_1 and E_2. In fact, we have

$$\begin{aligned} E_r &= (E_2U_1)^{r-2}E_2(E_2U_1)^{n-r+1}, && \text{if } n \text{ is even,} \\ E_r &= (E_2^{-1}U_1)^{r-2}E_2(E_2^{-1}U_1)^{n-r+1}, && \text{if } n \text{ is odd, } r \text{ is even,} \\ E_r &= (E_2^{-1}U_1)^{r-2}E_2^{-1}(E_2^{-1}U_1)^{n-r+1}, && \text{if } n \text{ and } r \text{ are odd.} \end{aligned} \tag{1}$$

The proof of (1) is similar to that of (2.5).

Thus $\mathfrak{M}_n$ can be generated by the three matrices U_1, V_2, E_2. If we write

$$U^* = \begin{bmatrix} 1 & 0 & 0 & \cdots & 0 \\ 1 & 1 & 0 & \cdots & 0 \\ 0 & 0 & 1 & \cdots & 0 \\ \cdots & \cdots & \cdots & \cdots & \cdots \\ 0 & 0 & 0 & \cdots & 1 \end{bmatrix},$$

then it is easy to verify that $E_2 = U^{*-1}V_2U^{*-1}$, so that $\mathfrak{M}_n$ can also be generated by the three matrices

$$U_1 = \begin{bmatrix} 0 & 0 & \cdots & 0 & (-1)^{n-1} \\ 1 & 0 & \cdots & 0 & 0 \\ 0 & 1 & \cdots & 0 & 0 \\ \cdots & \cdots & \cdots & \cdots & \cdots \\ 0 & 0 & \cdots & 1 & 0 \end{bmatrix},$$

$$U_2 = V_2 = \begin{bmatrix} 1 & 1 & 0 & \cdots & 0 \\ 0 & 1 & 0 & \cdots & 0 \\ 0 & 0 & 1 & \cdots & 0 \\ \cdots & \cdots & \cdots & \cdots & \cdots \\ 0 & 0 & 0 & \cdots & 1 \end{bmatrix}, \qquad U^* = \begin{bmatrix} 1 & 0 & 0 & \cdots & 0 \\ 1 & 1 & 0 & \cdots & 0 \\ 0 & 0 & 1 & \cdots & 0 \\ \cdots & \cdots & \cdots & \cdots & \cdots \\ 0 & 0 & 0 & \cdots & 1 \end{bmatrix} \tag{2}$$

When $n = 2$ we saw that $\mathfrak{M}_2$ can actually be generated by the *two* matrices $U_1 = \begin{pmatrix} 0 & -1 \\ 1 & 0 \end{pmatrix}$ and $U_2 = \begin{pmatrix} 1 & 1 \\ 0 & 1 \end{pmatrix}$. We now ask whether $\mathfrak{M}_n$ $(n \geqslant 3)$ can also be generated by U_1 and U_2; that is whether U^* is expressible as a product of U_1 and U_2. We first examine the cases $n = 3$ and 4.

(1) For $n = 3$, we have

$$U_1 = \begin{pmatrix} 0 & 0 & 1 \\ 1 & 0 & 0 \\ 0 & 1 & 0 \end{pmatrix}, \qquad U_2 = \begin{pmatrix} 1 & 1 & 0 \\ 0 & 1 & 0 \\ 0 & 0 & 1 \end{pmatrix}, \qquad U^* = \begin{pmatrix} 1 & 0 & 0 \\ 1 & 1 & 0 \\ 0 & 0 & 1 \end{pmatrix}.$$

In the following we call the position for the i-th row and the j-th column the "position (i, j)". Consider the operation of multiplying U_2 by U_1 on the left and U_1^{-1} on the right. We see from

$$S = U_1 U_2 U_1^{-1} = \begin{pmatrix} 1 & 0 & 0 \\ 0 & 1 & 1 \\ 0 & 0 & 1 \end{pmatrix},$$

$$T = U_1^2 U_2 (U_1^{-1})^2 = U_1^{-1} U_2 U_1 = \begin{pmatrix} 1 & 0 & 0 \\ 0 & 1 & 0 \\ 1 & 0 & 1 \end{pmatrix},$$

$$U_1^3 U_2 (U_1^{-1})^3 = U_2 = \begin{pmatrix} 1 & 1 & 0 \\ 0 & 1 & 0 \\ 0 & 0 & 1 \end{pmatrix},$$

that successive applications of the above operation will leave the elements in the main diagonal invariant, whereas the element 1 not on the main diagonal will take up the successive positions $(1, 2)$, $(2, 3)$, $(3, 1)$. Similarly the elements in the three positions $(3, 2)$, $(1, 3)$, $(2, 1)$ will be permuted along a rail as shown in the diagram.

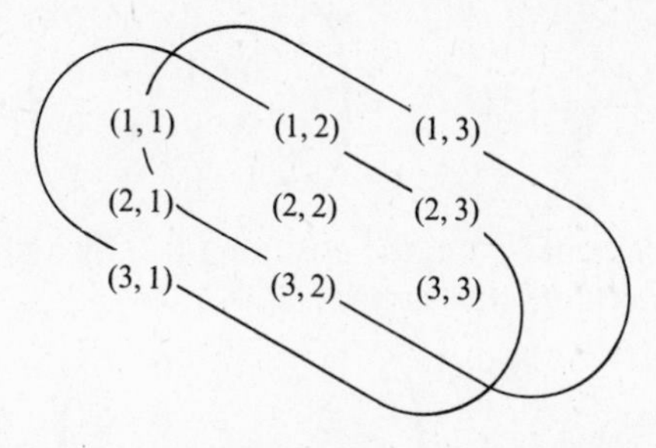

Consequently in order to obtain the element 1 in the position $(2, 1)$ we have first to produce this element in one of the positions $(1, 3)$ or $(3, 2)$. Now if we multiply

T by U_2^{-1} on the left and U_2 on the right, it will give rise to the element 1 in the position (3, 2); that is

$$U_2^{-1}TU_2 = \begin{pmatrix} 1 & 0 & 0 \\ 0 & 1 & 0 \\ 1 & 1 & 1 \end{pmatrix}.$$

The operation of multiplying by U_1^{-1} on the left and U_1 on the right will make the element 1 in the position (3, 2) in the matrix $U_2^{-1}TU_2$ move to the position (2, 1), that is

$$W = U_1^{-1}U_2^{-1}TU_2U_1 = \begin{pmatrix} 1 & 0 & 0 \\ 1 & 1 & 1 \\ 0 & 0 & 1 \end{pmatrix}.$$

Therefore we need only to annihilate the element 1 in the position (2, 3) to give the required matrix U^*, and this can be accomplished by multiplying by S^{-1} on the left; that is

$$S^{-1}W = \begin{pmatrix} 1 & 0 & 0 \\ 1 & 1 & 0 \\ 0 & 0 & 1 \end{pmatrix} = U^*.$$

Therefore, for $n = 3$, we have

$$U^* = U_1U_2^{-1}U_1U_2^{-1}U_1^{-1}U_2U_1U_2U_1. \tag{3}$$

(2) For $n = 4$ we have

$$U_1 = \begin{pmatrix} 0 & 0 & 0 & -1 \\ 1 & 0 & 0 & 0 \\ 0 & 1 & 0 & 0 \\ 0 & 0 & 1 & 0 \end{pmatrix}, \quad U_2 = \begin{pmatrix} 1 & 1 & 0 & 0 \\ 0 & 1 & 0 & 0 \\ 0 & 0 & 1 & 0 \\ 0 & 0 & 0 & 1 \end{pmatrix},$$

$$U^* = \begin{pmatrix} 1 & 0 & 0 & 0 \\ 1 & 1 & 0 & 0 \\ 0 & 0 & 1 & 0 \\ 0 & 0 & 0 & 1 \end{pmatrix}.$$

Similarly to the case $n = 3$, we start with

$$T = U_1^{-1}U_2U_1 = \begin{pmatrix} 1 & 0 & 0 & 0 \\ 0 & 1 & 0 & 0 \\ 0 & 0 & 1 & 0 \\ -1 & 0 & 0 & 1 \end{pmatrix}.$$

We can produce the element -1 in the position (4, 2) by multiplying T by U_2^{-1} on the left and U_2 on the right; that is

$$U_2^{-1}TU_2 = \begin{pmatrix} 1 & 0 & 0 & 0 \\ 0 & 1 & 0 & 0 \\ 0 & 0 & 1 & 0 \\ -1 & -1 & 0 & 1 \end{pmatrix}.$$

Again, the operation of multiplying by U_1^{-1} on the left and U_1 on the right will move the element -1 from the position $(4,2)$ to the position $(3,1)$; that is

$$U_1^{-1}(U_2^{-1}TU_2)U_1 = \begin{pmatrix} 1 & 0 & 0 & 0 \\ 0 & 1 & 0 & 0 \\ -1 & 0 & 1 & 1 \\ 0 & 0 & 0 & 1 \end{pmatrix}. \tag{4}$$

Performing the first operation of multiplying by U_2^{-1} on the left and U_2 on the right will now produce the element -1 in the position $(3,2)$; that is

$$U_2^{-1}(U_1^{-1}U_2^{-1}TU_2U_1)U_2 = \begin{pmatrix} 1 & 0 & 0 & 0 \\ 0 & 1 & 0 & 0 \\ -1 & -1 & 1 & 1 \\ 0 & 0 & 0 & 1 \end{pmatrix}.$$

Performing the second operation of multiplying by U_1^{-1} on the left and U_1 on the right will now move the element -1 in the position $(3,2)$ to the position $(2,1)$; that is

$$W = U_1^{-1}(U_2^{-1}U_1^{-1}U_2^{-1}TU_2U_1U_2)U_1 = \begin{pmatrix} 1 & 0 & 0 & 0 \\ -1 & 1 & 1 & 1 \\ 0 & 0 & 1 & 0 \\ 0 & 0 & 0 & 1 \end{pmatrix}.$$

At this point we observe that the elements of the matrix below the main diagonal matches those of U^{*-1}, and the problem now is the anihilation of the elements 1 above the main diagonal.

From (4) we have

$$S = U_1^{-1}(U_1^{-1}U_2^{-1}TU_2U_1)U_1 = \begin{pmatrix} 1 & 0 & 0 & 0 \\ 0 & 1 & 1 & 1 \\ 0 & 0 & 1 & 0 \\ 0 & 0 & 0 & 1 \end{pmatrix}$$

and hence

$$S^{-1}W = \begin{pmatrix} 1 & 0 & 0 & 0 \\ -1 & 1 & 0 & 0 \\ 0 & 0 & 1 & 0 \\ 0 & 0 & 0 & 1 \end{pmatrix} = U^{*-1}.$$

Therefore, for $n = 4$, we have

$$U^{*-1} = U_1^{-1}U_1^{-1}U_2^{-1}U_1^{-1}U_2^{-1}U_1U_2U_1U_1U_1^{-1}U_2^{-1}U_1^{-1}U_2^{-1}U_1^{-1} \times U_2U_1U_2U_1U_2U_1. \tag{5}$$

If we write $U = U_2U_1$, then (3) and (5) become

$$U^* = U_1^{-1}U^{-1}U_1U_1U^{-1}U_1^{-1}U^2 \qquad (n = 3),$$
$$U^{*-1} = U_1^{-1}(U^{-1})^2U_1UU_1(U^{-1})^2U_1^{-1}U^3 \qquad (n = 4), \tag{6}$$

and in general we have

$$U^{*(-1)^{n-1}} = U_1^{-1}(U^{-1})^{n-2}U_1U^{n-3}U_1(U^{-1})^{n-2}U_1^{-1}U^{n-1}. \tag{7}$$

The reader can follow the proof of (2.5) to prove (7). Therefore we have

Theorem 3.1. *The group $\mathfrak{M}_n$ of positive modular matrices can be generated by the two matrices*

$$U_1 = \begin{pmatrix} 0 & 0 & \cdots & 0 & (-1)^{n-1} \\ 1 & 0 & \cdots & 0 & 0 \\ \cdots & \cdots & \cdots & \cdots & \cdots \\ 0 & 0 & \cdots & 1 & 0 \end{pmatrix}, \qquad U_2 = \begin{pmatrix} 1 & 1 & 0 & \cdots & 0 \\ 0 & 1 & 0 & \cdots & 0 \\ 0 & 0 & 1 & \cdots & 0 \\ \cdots & \cdots & \cdots & \cdots & \cdots \\ 0 & 0 & 0 & \cdots & 1 \end{pmatrix}.$$

In other words, any positive modular matrix is expressible as a product of U_1 and U_2. □

Any modular matrix which is not positive will become so on multiplying by

$$U_3 = \begin{pmatrix} -1 & 0 & \cdots & 0 \\ 0 & 1 & \cdots & 0 \\ \cdots & \cdots & \cdots & \cdots \\ 0 & 0 & \cdots & 1 \end{pmatrix}.$$

Therefore we have

Theorem 3.2. *The group of all modular matrices can be generated by the three matrices U_1, U_2 and U_3. In other words any modular matrix is expressible as a product of the matrices U_1, U_2 and U_3.* □

14.4 Left Association

Definition 1. Let A and B be two square matrices. Suppose that there is a modular matrix U such that

$$A = UB.$$

Then we say that B is *left associated* to A and we write $A \overset{L}{=} B$.

Clearly left association is reflexive, symmetric and transitive.

Theorem 4.1. *Any square matrix is left associated to a matrix of the form*

$$\begin{pmatrix} b_{11} & 0 & 0 & \cdots & 0 & 0 \\ b_{21} & b_{22} & 0 & \cdots & 0 & 0 \\ \cdots & \cdots & \cdots & \cdots & \cdots & \cdots \\ b_{n-1,1} & b_{n-1,2} & b_{n-1,3} & \cdots & b_{n-1,n-1} & 0 \\ b_{n1} & b_{n2} & b_{n3} & \cdots & b_{n,n-1} & b_{nn} \end{pmatrix}, \tag{1}$$

where $b_{\nu\nu} \geqslant 0$. Also if $b_{\nu\nu} > 0$, then $0 \leqslant b_{i\nu} < b_{\nu\nu}$ $(i > \nu)$.

Proof. The case $n = 2$ has already been proved (Theorem 1.3). We now proceed by induction on n. Let

$$A = \begin{pmatrix} a_{11} & a_{12} & \cdots & a_{1n} \\ a_{21} & a_{22} & \cdots & a_{2n} \\ \cdots & \cdots & \cdots & \cdots \\ a_{n1} & a_{n2} & \cdots & a_{nn} \end{pmatrix}$$

be any square matrix. If there is a non-zero element in the last column of the matrix A, then we let $(a_{1n}, a_{2n}, \ldots, a_{nn}) = b_{nn}$. There are integers $b_1, b_2, \ldots, b_n$ such that

$$b_1 a_{1n} + b_2 a_{2n} + \cdots + b_n a_{nn} = b_{nn}, \qquad (b_1, b_2, \ldots, b_n) = 1.$$

By Theorem 2.3 there is a modular matrix V whose first row is $(b_1, b_2, \ldots, b_n)$. We interchange the first row of V with its n-th row to give a modular matrix U whose n-th row is $(b_1, b_2, \ldots, b_n)$. We then have

$$A \overset{L}{=} UA = \begin{pmatrix} a'_{11} & a'_{12} & \cdots & a'_{1n} \\ a'_{21} & a'_{22} & \cdots & a'_{2n} \\ \cdots & \cdots & \cdots & \cdots \\ a'_{n1} & a'_{n2} & \cdots & b_{nn} \end{pmatrix}.$$

It is easy to see that $a'_{1n}, \ldots, a'_{n-1,n}$ are linear combinations of $a_{1n}, a_{2n}, \ldots, a_{nn}$ and are therefore divisible by b_{nn}. Therefore

$$A \overset{L}{=} \begin{bmatrix} 1 & 0 & \cdots & 0 & -\dfrac{a'_{1n}}{b_{nn}} \\ 0 & 1 & \cdots & 0 & -\dfrac{a'_{2n}}{b_{nn}} \\ \cdots & \cdots & \cdots & \cdots & \cdots \\ 0 & 0 & \cdots & 0 & 1 \end{bmatrix} \begin{pmatrix} a'_{11} & a'_{12} & \cdots & a'_{1n} \\ a'_{21} & a'_{22} & \cdots & a'_{2n} \\ \cdots & \cdots & \cdots & \cdots \\ a'_{n1} & a'_{n2} & \cdots & b_{nn} \end{pmatrix}$$

$$= \begin{bmatrix} a''_{11} & \cdots & a''_{1,n-1} & 0 \\ a''_{21} & \cdots & a''_{2,n-1} & 0 \\ \cdots & \cdots & \cdots & \cdots \\ a''_{n-1,1} & \cdots & a''_{n-1,n-1} & 0 \\ a''_{n1} & \cdots & a''_{n,n-1} & b_{nn} \end{bmatrix}. \tag{2}$$

The above still holds even when all the elements in the last column of A are zero, except that we have $b_{nn} = 0$. It follows from the induction hypothesis that

$$A \overset{L}{=} \begin{bmatrix} b_{11} & 0 & \cdots & 0 & 0 \\ b_{21} & b_{22} & \cdots & 0 & 0 \\ \cdots & \cdots & \cdots & \cdots & \cdots \\ b_{n-1,1} & b_{n-1,2} & \cdots & b_{n-1,n-1} & 0 \\ b'_{n1} & b'_{n2} & \cdots & b'_{n,n-1} & b_{nn} \end{bmatrix},$$

where $b_{vv} \geqslant 0$, $b_{iv} = 0\ (i < v)$, and if $b_{vv} > 0$, then $0 \leqslant b_{iv} < b_{vv}$ $(1 \leqslant v < i \leqslant n-1)$.

If $b_{n-1,n-1} > 0$, then there exists an integer q_{n-1} such that

$$0 \leqslant q_{n-1}b_{n-1,n-1} + b'_{n,n-1} < b_{n-1,n-1}.$$

Therefore

$$A \overset{L}{=} \begin{pmatrix} b_{11} & 0 & \cdots & 0 & 0 \\ b_{21} & b_{22} & \cdots & 0 & 0 \\ \cdots & \cdots & \cdots & \cdots & \cdots \\ b_{n-1,1} & b_{n-1,2} & \cdots & b_{n-1,n-1} & 0 \\ b''_{n1} & b''_{n2} & \cdots & b''_{n,n-1} & b_{nn} \end{pmatrix},$$

where $b''_{ni} = q_{n-1}b_{n-1,i} + b'_{ni}$ $(1 \leqslant i \leqslant n-1)$, $0 \leqslant b''_{n,n-1} < b_{n-1,n-1}$.

The theorem follows from repeated applications of this. □

Definition 2. We call a square matrix of the form (1) the *normal form of Hermite*.

Exercise. Prove that the normal form of Hermite for a non-singular square matrix is unique.

14.5 Invariant Factors and Elementary Divisors

Definition 1. Let A $(= A^{(m,n)})$ and B $(= B^{(m,n)})$ be two matrices. Suppose that there are two modular matrices U $(= U^{(m)})$, V $(= V^{(n)})$ such that

$$A = UBV.$$

Then we say that A and B are *equivalent* and we write $A \sim B$.

Clearly equivalence has the three properties of being reflexive, symmetric and transitive.

Theorem 5.1. *Any matrix A $(= A^{(m,n)})$ must be equivalent to a matrix of the form*

$$\begin{pmatrix} d_1 & 0 & 0 & \cdots & 0 & 0 & \cdots & 0 \\ 0 & d_1d_2 & 0 & \cdots & 0 & 0 & \cdots & 0 \\ 0 & 0 & d_1d_2d_3 & \cdots & 0 & 0 & \cdots & 0 \\ \cdots & \cdots & \cdots & \cdots & \cdots & \cdots & \cdots & \cdots \\ 0 & 0 & 0 & \cdots & d_1d_2\cdots d_m & 0 & \cdots & 0 \end{pmatrix} \quad (m \leqslant n) \qquad (1)$$

or

$$\begin{pmatrix} d_1 & 0 & \cdots & 0 \\ 0 & d_1d_2 & \cdots & 0 \\ \cdots & \cdots & \cdots & \cdots \\ 0 & 0 & \cdots & d_1d_2\cdots d_n \\ 0 & 0 & \cdots & 0 \\ \cdots & \cdots & \cdots & \cdots \\ 0 & 0 & \cdots & 0 \end{pmatrix} \quad (m \geqslant n) \qquad (2)$$

where $d_i \geqslant 0$.

Proof. Let $A = (a_{11}, a_{12}, \ldots, a_{1k})$ be a 1 by k matrix where k is any positive integer $(k > 1)$. By Theorem 2.2 there is a positive modular matrix U such that

$$AU = (d, 0, 0, \ldots, 0)$$

and so the required result is proved. Also, from

$$U' \begin{pmatrix} a_{11} \\ a_{12} \\ \vdots \\ a_{1k} \end{pmatrix} = \begin{pmatrix} d \\ 0 \\ \vdots \\ 0 \end{pmatrix},$$

where U' is the transposed matrix of U, we see that the theorem also holds for k by 1 matrices.

We now proceed by induction on the number of rows of the matrix A. Let A be any given matrix. If $A = 0$, then the result is trivial. If $A \neq 0$, then we may assume that $a_{11} \neq 0$ and indeed we can even assume that $a_{11} > 0$. We first prove that A must be equivalent to a matrix of the form:

$$A \sim A_1 = \begin{pmatrix} a'_{11} & a'_{12} & \cdots & a'_{1n} \\ a'_{21} & a'_{22} & \cdots & a'_{2n} \\ \cdots & \cdots & \cdots & \cdots \\ a'_{m1} & a'_{m2} & \cdots & a'_{mn} \end{pmatrix}, \qquad a'_{11} | a'_{ii} \qquad (1 \leqslant i \leqslant m,\ 1 \leqslant j \leqslant n).$$

This is clearly so if $a_{11} = 1$. When $a_{11} > 1$, if $a_{11} \nmid a_{i_0 j_0}$ then we can move $a_{i_0 j_0}$ to one of the positions occupied by a_{12}, a_{21}, a_{22}, by means of row or column interchanging. Therefore, using the method of proof for Theorem 1.5, we can change the leading element to a positive integer which is less than a_{11}, and an inductive argument completes the first part of proof.

Now from

$$\begin{bmatrix} 1 & 0 & \cdots & 0 \\ -\dfrac{a'_{21}}{a'_{11}} & 1 & \cdots & 0 \\ \cdots & \cdots & \cdots & \cdots \\ -\dfrac{a'_{m1}}{a'_{11}} & 0 & \cdots & 1 \end{bmatrix} \begin{pmatrix} a'_{11} & a'_{12} & \cdots & a'_{1n} \\ a'_{21} & a'_{22} & \cdots & a'_{2n} \\ \cdots & \cdots & \cdots & \cdots \\ a'_{m1} & a'_{m2} & \cdots & a'_{mn} \end{pmatrix}$$

$$\times \begin{bmatrix} 1 & -\dfrac{a'_{12}}{a'_{11}} & \cdots & -\dfrac{a'_{1n}}{a'_{11}} \\ 0 & 1 & \cdots & 0 \\ \cdots & \cdots & \cdots & \cdots \\ 0 & 0 & \cdots & 1 \end{bmatrix} = \begin{pmatrix} a'_{11} & 0 & \cdots & 0 \\ 0 & a''_{22} & \cdots & a''_{2n} \\ \cdots & \cdots & \cdots & \cdots \\ 0 & a''_{m2} & \cdots & a''_{mn} \end{pmatrix},$$

we have

$$A \sim \begin{pmatrix} a'_{11} & 0 & \cdots & 0 \\ 0 & a''_{22} & \cdots & a''_{2n} \\ \cdots & \cdots & \cdots & \cdots \\ 0 & a''_{m2} & \cdots & a''_{nn} \end{pmatrix}.$$

Therefore, from the induction hypothesis, we have

$$A \sim \begin{pmatrix} a'_{11} & 0 & 0 & \cdots & 0 & 0 & \cdots & 0 \\ 0 & d'_2 & 0 & \cdots & 0 & 0 & \cdots & 0 \\ \multicolumn{8}{c}{\dotfill} \\ 0 & 0 & 0 & \cdots & d'_2 \cdots d'_m & 0 & \cdots & 0 \end{pmatrix} \qquad (m \leqslant n) \tag{4}$$

or

$$A \sim \begin{pmatrix} a'_{11} & 0 & \cdots & 0 \\ 0 & d'_2 & \cdots & 0 \\ \multicolumn{4}{c}{\dotfill} \\ 0 & 0 & \cdots & d'_2 \cdots d'_n \\ 0 & 0 & \cdots & 0 \\ \multicolumn{4}{c}{\dotfill} \\ 0 & 0 & \cdots & 0 \end{pmatrix} \qquad (m \geqslant n). \tag{5}$$

Since $a'_{11} | a'_{ii}$, and d'_2 is a linear combination of the elements of A_1, it follows that $a'_{11} | d'_2$. If we let $a'_{11} = d_1$, $d'_2 = d_1 d_2$, $d'_3 = d_3$, $d'_4 = d_4, \ldots$, then the theorem follows from (4) and (5). □

Definition 2. We call matrices of the form (1) or (2) the *normal forms of Smith.*

In the proof of Theorem 5.1 the operations that we use are: the interchange of rows (or columns), the addition of an integer multiple of a row (or column) to another row (or column); the multiplication by -1 to a row (or column). We call these operations the *elementary operations* of matrices. We can therefore restate Theorem 5.1 as follows: any matrix can be reduced to the normal form of Smith by elementary operations.

After the interchange of two rows (or columns) or the multiplication by -1 to a row (or column), the i by i sub-determinants of the resulting matrix are either the same as the i by i sub-determinants of the original matrix, or differ by their signs only. Again if we add an integer multiple of a row (or column) to another row (or column) the i by i sub-determinants of the resulting matrix are either the same as the i by i sub-determinants of the original matrix, or the i by i sub-determinants with the addition of an integer multiple of i by i sub-determinants. It follows that the greatest common factor of all the i by i sub-determinants of a matrix is invariant under any elementary transformation. Therefore we have

Theorem 5.2. *Let $A \sim B$. Then the greatest common factors of the i by i sub-determinants of the two matrices A and B are the same.* □

Meanwhile we see from (1) and (2) that

$$h_i = d_1 \cdot d_1 d_2 \cdots d_1 \cdots d_i,$$

are the greatest common factors of the i by i sub-determinants of A. Therefore we have

Theorem 5.3. *The normal form of Smith for a matrix is unique.* □

Definition 3. Let the non-zero elements of the normal form of Smith in (1) or (2) for a matrix A be

$$d_1, d_1 d_2, \ldots, d_1, \ldots, d_k \qquad (k \leqslant \min(m, n)).$$

We call these numbers the *invariant factors* of A of orders $1, 2, \ldots, k$ respectively. The number k is called the *rank* of the matrix A. Let

$$d_1 \cdots d_i = p_1^{e_{i1}} \cdots p_{l_i}^{e_{il_i}} \qquad (e_{ij} > 0,\ 1 \leqslant i \leqslant k,\ l_{i-1} \leqslant l_i)$$

be the standard prime factorization of an invariant factor. We call the prime power $p_j^{e_{ij}}$ an *elementary divisor* of the matrix A.

It is easy to see that the indices of the elementary divisors satisfy:

$$e_{kj} \geqslant e_{k-1,j} \geqslant e_{k-2,j} \geqslant \cdots \qquad (1 \leqslant j \leqslant l).$$

It also follows from the definition that if two matrices have the same invariant factors, then they have the same rank and the same elementary divisors. Conversely if the ranks are the same and the elementary divisors are the same, then the invariant factors are the same. Therefore we have

Theorem 5.4. *A necessary and sufficient condition for two m by n matrices to be equivalent is that they should have the same rank and the same elementary divisors.* □

14.6 Applications

Let us consider the solutions to the system of linear equations

$$y_i = \sum_{j=1}^{n} x_j a_{ji} \qquad (1 \leqslant i \leqslant m, \qquad n \geqslant m), \tag{1}$$

with integer coefficients, and given integers y_i—that is we consider the integer solutions to

$$y = xA, \qquad y = (y_1, \ldots, y_m), \qquad x = (x_1, \ldots, x_n),$$

$$A = \begin{pmatrix} a_{11} & a_{12} & \cdots & a_{1m} \\ a_{21} & a_{22} & \cdots & a_{2m} \\ \cdots & \cdots & \cdots & \cdots \\ a_{n1} & a_{n2} & \cdots & a_{nm} \end{pmatrix}. \tag{2}$$

We saw in the previous section that there are two modular matrices $U\ (= U^{(n)})$ and $V\ (= V^{(m)})$ such that

$$UAV = \begin{pmatrix} d_1 & 0 & \cdots & 0 \\ 0 & d_1 d_2 & \cdots & 0 \\ \cdots & \cdots & \cdots & \cdots \\ 0 & 0 & \cdots & d_1 \cdots d_m \\ 0 & 0 & \cdots & 0 \\ \cdots & \cdots & \cdots & \cdots \\ 0 & 0 & \cdots & 0 \end{pmatrix} = D. \tag{3}$$

We now let

$$yV = y^* = (y'_1, \ldots, y'_m), \qquad xU^{-1} = x^* = (x'_1, \ldots, x'_n),$$

so that, from (2),

$$y^* = x^* D, \tag{4}$$

or

$$y'_i = d_1 d_2 \cdots d_i x'_i \qquad (1 \leqslant i \leqslant m). \tag{5}$$

A necessary and sufficient condition for (1) to have a solution is that (5) has a solution. If $d_1 \cdots d_k \neq 0, d_{k+1} = 0$, then a necessary and sufficient condition for (5) to have a solution is that

$$d_1 \cdots d_i | y'_i \quad (1 \leqslant i \leqslant k), \qquad y'_{k+1} = \cdots = y'_m = 0. \tag{6}$$

From (3) we have

$$\begin{pmatrix} U & 0 \\ 0 & 1 \end{pmatrix} \begin{pmatrix} A \\ y \end{pmatrix} V = \begin{pmatrix} D \\ y^* \end{pmatrix}. \tag{7}$$

Now, if (6) holds, then we have, by (7), that

$$\begin{pmatrix} A \\ y \end{pmatrix} \sim \begin{pmatrix} D \\ 0 \end{pmatrix}; \tag{8}$$

conversely, if (8) holds, then

$$\begin{pmatrix} D \\ y^* \end{pmatrix} \sim \begin{pmatrix} D \\ 0 \end{pmatrix},$$

and from Theorem 5.2 we have

$$d_1 | y'_1, d_1 d_2 | y'_2, \ldots, d_1 \cdots d_k | y'_k, \qquad y'_{k+1} = \cdots = y'_m = 0,$$

which is formula (6). Therefore a necessary and sufficient condition for (1) to have a solution is that (8) holds; that is, we have

Theorem 6.1. *A necessary and sufficient condition for the system* (1) *to have a solution is that there are two matrices A and $\begin{pmatrix} A \\ y \end{pmatrix}$ with the same invariant factors.* □

If (5) holds, then we have

$$x_1' = \frac{y_1'}{d_1}, \quad x_2' = \frac{y_2'}{d_1 d_2}, \quad \ldots, \quad x_k' = \frac{y_k'}{d_1 \cdots d_k}. \tag{9}$$

This means that $x_1', x_2', \ldots, x_k'$ are uniquely determined, and $x_{k+1}', \ldots, x_n'$ can be any integers. Thus, if $t_1, \ldots, t_{n-k}$ are $n - k$ arbitrary integers, then

$$\begin{aligned} x_i &= \sum_{j=1}^{k} x_j' u_{ji} + \sum_{l=1}^{n-k} t_l u_{k+l,i} \\ &= x_i^{(0)} + \sum_{l=1}^{n-k} t_l u_{k+l,i} \qquad (1 \leqslant i \leqslant n), \end{aligned} \tag{10}$$

where $x_1^{(0)}, \ldots, x_n^{(0)}$ is set of solution to (1) when $t_1 = t_2 = \cdots = t_{n-k} = 0$.

14.7 Matrix Factorizations and Standard Prime Matrices

Definition 7.1. Let A and B be two square matrices, and suppose that there is a matrix C such that $A = CB$. Then we call B a *right divisor* of A, or we say that B *right-divides* A, and we write $B|A$.

Clearly we have (i) $A|A$; (ii) if $A|B$ and $B|C$, then $A|C$. We can define a *left divisor* and *left-divide* similarly.

Definition 7.2. Let A be a non-singular square matrix which is not a modular matrix. Suppose that for any factorization $A = BC$, we always have either B or C a modular matrix. Then we call A an *irreducible matrix* or a *prime matrix*. Otherwise we call A a *composite matrix*.

Let A be a non-singular square matrix. By Theorem 5.1 there are two modular matrices U and V such that

$$A = U[d_1, d_1 d_2, \ldots, d_1 \cdots d_n]V. \tag{1}$$

It is easy to decompose $[d_1, d_1 d_2, \ldots, d_1 \cdots d_n]$ into prime matrices. More specifically its factors are of the form $P = [1, \ldots, 1, p, 1, \ldots, 1]$ where p is a prime number, and the number of such factors is the number of prime factors in $d_1 \cdot d_1 d_2 \cdot \cdots \cdot d_1 d_2 \cdots d_n$. Therefore we have

$$A = UP_1 P_2 \cdots P_s V, \qquad P = [1, \ldots, 1, p, 1, \ldots, 1], \tag{2}$$

where any two P can be interchanged. Consequently we have the following two theorems.

Theorem 7.1. *A necessary and sufficient condition for a square matrix to be a prime matrix is that its determinant is a prime number.* □

Theorem 7.2. *Any composite square matrix can be factorized into a product of a finite number of prime matrices, and the number of factors is equal to the number of prime divisors of the determinant of the matrix.* □

This type of factorization does not possess the uniqueness property. This is because we can always insert WW^{-1} (where W is a modular matrix) in between two factors P_i, P_{i+1} so that P_iW and $W^{-1}P_{i+1}$ are now different factors from P_i and P_{i+1}. However, if we impose certain restrictions on the form of the factors, then we may have a sort of uniqueness theorem.

Definition 7.3. If a prime matrix is expressible as $U^{-1}[1,\ldots,1,p]U$, where U is a modular matrix, then we call it a *standard prime matrix*.

It is clear that every prime matrix must be left associated to a standard prime matrix.

We now rewrite (2) as the following:

$$A = UV(V^{-1}P_1V)(V^{-1}P_2V)\cdots(V^{-1}P_sV), \tag{3}$$

where any two $V^{-1}PV$ can be interchanged, and they are all standard prime matrices. Therefore we have:

Theorem 7.3. *Any composite square matrix must be left associated with a product of a finite number of interchangeable standard prime matrices.* □

Definition 7.4. By the *standard factorization* of A we mean

$$A = W(V^{-1}P_1V)(V^{-1}P_2V)\cdots(V^{-1}P_sV), \tag{4}$$

where W and V are modular matrices, and $P_1,\ldots,P_s$ are of the form in (2). It is clear that $P_1,\ldots,P_s$ are uniquely determined by A, apart from ordering.

Before we prove our uniqueness theorem we need the following definition:

Definition 7.5. Let A be a non-singular square matrix. A modular matrix U satisfying

$$AUA_0 \equiv 0 \pmod{|A|}$$

is called an *adjoint modular matrix* of A. Here A_0 is the adjoint matrix of A, and $|A|$ is the absolute value of the determinant of A.

Since the elements of the matrix AUA_0 are all multiples of $|A|$, it follows that the elements of the matrix $(1/|A|)AUA_0$ are integers. Moreover, on taking the determinant, we see that it is actually a modular matrix.

Theorem 7.4. *The set of all adjoint modular matrices of A forms a group.*

Proof. Let U and V be adjoint modular matrices of A. From

$$AUA_0AVA_0 = \pm |A| \cdot AUVA_0 \equiv 0 \pmod{|A|^2}$$

we see that UV is an adjoint modular matrix of A. Also, from

$$|A|AIA_0 = \pm AUA_0AU^{-1}A_0 \equiv 0 \pmod{|A|^2},$$

we have

$$\frac{1}{|A|}AUA_0 \cdot AU^{-1}A_0 \equiv 0 \pmod{|A|},$$

and $(1/|A|)AUA_0$ is a modular matrix, so that U^{-1} is also an adjoint modular matrix of A. The theorem therefore follows. □

Definition 7.6. The group of adjoint modular matrices of A is called the *adjoint group* of A.

Theorem 7.5. *Let*

$$A = W_1(V_1^{-1}P_1V_1)(V_1^{-1}P_2V_1)\cdots(V_1^{-1}P_sV_1) \tag{5}$$

be any standard factorization of A. Then there exists an adjoint modular matrix U of A such that $V_1 = VU$, $W_1 = (\pm 1/|A|)AU^{-1}A_0WU$ where W and V are the matrices in (4).

Proof. From (4) and (5) we have

$$A = WV^{-1}P_1P_2\cdots P_sV = W_1V_1^{-1}P_1P_2\cdots P_sV_1,$$
$$AV^{-1}V_1 = WV^{-1}V_1W_1^{-1}A.$$

It follows easily that $U = V^{-1}V_1$ is an adjoint modular matrix of A, and that

$$\frac{\pm 1}{|A|}AUA_0 = WUW_1^{-1}. \quad \square$$

This theorem therefore gives the relationship between any two standard factorizations of A. Concerning the interchangeability of the standard prime matrices we have the following two theorems:

Theorem 7.6. *Let $P = [1, \ldots, 1, p]$ and $Q = U^{-1}[1, \ldots, 1, q]U$ be two interchangeable standard prime matrices. Then Q must be of the form*

$$Q = \begin{pmatrix} Q_1 & 0 \\ 0 & r \end{pmatrix}, \tag{6}$$

where $r = q$ or 1. Also, if $r = q$, then $Q_1 = I$, and if $r = 1$, then Q_1 is a standard prime matrix.

Proof. Let

$$Q = \begin{pmatrix} Q_1 & x \\ y & r \end{pmatrix}, \qquad x = \begin{pmatrix} a_1 \\ \vdots \\ a_{n-1} \end{pmatrix}, \qquad y = (b_1, \ldots, b_{n-1}).$$

From $PQ = QP$ we have

$$\begin{pmatrix} Q_1 & x \\ py & pr \end{pmatrix} = \begin{pmatrix} Q_1 & xp \\ y & rp \end{pmatrix}. \tag{7}$$

It follows at once that

$$x = \begin{pmatrix} 0 \\ \vdots \\ 0 \end{pmatrix}, \qquad y = (0, \ldots, 0).$$

Next, let

$$U = \begin{pmatrix} U_1 & x_1 \\ y_1 & u \end{pmatrix}$$

so that, from $UQ = [1, \ldots, 1, q]U$, we have

$$\begin{pmatrix} U_1 Q_1 & x_1 r \\ y_1 Q_1 & ur \end{pmatrix} = \begin{pmatrix} U_1 & x_1 \\ q y_1 & qu \end{pmatrix}. \tag{8}$$

If $u \neq 0$, then we have $r = q$. In this case, from $x_1 r = x_1$, we deduce that

$$x_1 = \begin{pmatrix} 0 \\ \vdots \\ 0 \end{pmatrix}$$

and hence $u = \pm 1$, and that U_1 is a modular matrix. From $U_1 Q_1 = U_1$ it follows that $Q_1 = I$.

If $u = 0$, then

$$x \neq \begin{pmatrix} 0 \\ \vdots \\ 0 \end{pmatrix},$$

so that $r = 1$. From $U_1 Q_1 = U_1$ and Q_1 cannot be I, it follows that U_1 is singular. By Theorem 5.1 there are two modular matrices V_1 and W_1 such that $V_1 U_1 W_1 = [\lambda_1, \ldots, \lambda_{n-2}, 0]$, $\lambda_i \geqslant 0$. Therefore, if we let

$$V = \begin{pmatrix} V_1 & 0 \\ 0 & 1 \end{pmatrix}, \qquad W = \begin{pmatrix} W_1 & 0 \\ 0 & 1 \end{pmatrix},$$

then

$$X = VUW = \begin{pmatrix} V_1 U_1 W_1 & V_1 x_1 \\ y_1 W_1 & 0 \end{pmatrix}$$

$$= \begin{pmatrix} \lambda_1 & 0 & \cdots & 0 & 0 & c_1 \\ 0 & \lambda_2 & \cdots & 0 & 0 & c_2 \\ \cdots & \cdots & \cdots & \cdots & \cdots & \cdots \\ 0 & 0 & \cdots & \lambda_{n-2} & 0 & c_{n-2} \\ 0 & 0 & \cdots & 0 & 0 & c_{n-1} \\ d_1 & d_2 & \cdots & d_{n-2} & d_{n-1} & 0 \end{pmatrix}.$$

From $|c_{n-1}d_{n-1}\lambda_1 \cdots \lambda_{n-2}| = |X| = 1$ we see that $\lambda_1 = \cdots = \lambda_{n-2} = 1$, $c_{n-1} = \pm 1$, $d_{n-1} = \pm 1$; here $|X|$ denotes the absolute value of the determinant of X.

Next we let

$$Y = \begin{pmatrix} 1 & 0 & \cdots & 0 & \mp c_1 & 0 \\ 0 & 1 & \cdots & 0 & \mp c_2 & 0 \\ \cdots & \cdots & \cdots & \cdots & \cdots & \cdots \\ 0 & 0 & \cdots & 1 & \mp c_{n-2} & 0 \\ 0 & 0 & \cdots & 0 & 1 & 0 \\ 0 & 0 & \cdots & 0 & 0 & 1 \end{pmatrix},$$

$$Z = \begin{pmatrix} 1 & 0 & \cdots & 0 & 0 & 0 \\ \cdots & \cdots & \cdots & \cdots & \cdots & \cdots \\ 0 & 0 & \cdots & 1 & 0 & 0 \\ \mp d_1 & \mp d_2 & \cdots & \mp d_{n-2} & 1 & 0 \\ 0 & 0 & \cdots & 0 & 0 & 1 \end{pmatrix} = \begin{pmatrix} Z_1 & 0 \\ 0 & 1 \end{pmatrix},$$

where in the matrices Y and Z the ambiguous signs are determined by the opposite signs of c_{n-1} and d_{n-1} respectively. We then have

$$XYZ = \begin{pmatrix} 1 & 0 & \cdots & 0 & 0 & 0 \\ 0 & 1 & \cdots & 0 & 0 & 0 \\ \cdots & \cdots & \cdots & \cdots & \cdots & \cdots \\ 0 & 0 & \cdots & 1 & 0 & 0 \\ 0 & 0 & \cdots & 0 & 0 & c_{n-1} \\ 0 & 0 & \cdots & 0 & d_{n-1} & 0 \end{pmatrix}.$$

It follows from

$$XW^{-1}QW = VUQW = V[1, \ldots, 1, q]UW = [1, \ldots, 1, q]X,$$

that

$$YXZZ^{-1}W^{-1}QWZ = Y[1, \ldots, 1, q]XZ = [1, \ldots, 1, q]YXZ,$$

or

$$(WZ)^{-1}Q(WZ) = (YXZ)^{-1}[1, \ldots, 1, q]YXZ = [1, \ldots, 1, q, 1].$$

Therefore we have

$$(W_1Z_1)^{-1}Q_1(W_1Z_1) = [1, \ldots, 1, q].$$

This proves that Q_1 is a standard prime matrix. $\square$

Theorem 7.7. *Corresponding to any set of interchangeable standard prime matrices $P_1, \ldots, P_s$, there is a modular matrix U such that $U^{-1}P_iU$ are all diagonal matrices.*

Proof. The theorem is trivial if $s = 1$. We now proceed by induction and assume that the theorem holds when the number of matrices is less than s.

Corresponding to P_s, there is a modular matrix U_s such that $U_s^{-1}P_sU_s = [1, \ldots, 1, p_s]$. Let

$$U_s^{-1}P_iU_s = Q_i, \qquad i = 1, 2, \ldots, s.$$

It is clear that these Q are interchangeable standard prime matrices. By Theorem 7.6 we have

$$Q_i = \begin{pmatrix} Q_i^* & 0 \\ 0 & r_i \end{pmatrix}, \qquad 1 \leqslant i \leqslant s,$$

where $r_i = p_i$ or 1. Also if $r_i = p_i$, then $Q_i^* = I$, and Q_i is of diagonal form; if $r_i = 1$, then Q_i^* is a standard prime matrix. Since the matrices Q are interchangeable we may assume that $r_1 = r_2 = \cdots = r_t = 1$, $r_{t+1} = p_{t+1}, \ldots, r_s = p_s$ $(0 \leqslant t \leqslant s - 1)$. If $t = 0$, then the theorem follows at once. Otherwise from the induction hypothesis, corresponding to the interchangeable standard prime matrices $Q_1^*, \ldots, Q_t^*$ there is a modular matrix U^* such that $U^{*-1}Q_i^*U^*$ $(1 \leqslant i \leqslant t)$ are of diagonal form. Now take

$$U_1 = \begin{pmatrix} U^* & 0 \\ 0 & 1 \end{pmatrix},$$

so that $U_1^{-1}Q_iU_1$ $(1 \leqslant i \leqslant s)$ are all of diagonal form. The theorem follows on taking $U = U_sU_1$. □

Exercise. Examine the properties of the adjoint group of the matrix $A = [d_1, d_1d_2, \ldots, d_1 \cdots d_n]$.

14.8 The Greatest Common Factor and the Least Common Multiple

Definition 8.1. Let A and B be two square matrices, not both equal to 0. Let D be a common right divisor of A and B such that any common right divisor is also a right divisor of D. Then we call D a *right greatest common divisor* of A and B.

Suppose that A and B are both right divisors of the square matrix $M(\neq 0)$, and that M is a right divisor of any square matrix having both A and B as right divisors. Then we call M a *left least common multiple* of A and B.

Similar definitions for left greatest common divisors and right least common multiples can be given. We shall only discuss right greatest common divisors and left least common multiples and, for the sake of simplicity, we shall call them greatest common divisors and least common multiples.

We define the sum of the two matrices $A = (a_{ij})$ and $B = (b_{ij})$ by

$$A + B = (a_{ij} + b_{ij}).$$

Theorem 8.1. *Let A and B be two square matrices which are not both* 0. *Then their greatest common divisor D exists. Moreover there are square matrices P and Q such that*

$$PA + QB = D.$$

Proof. Consider the $2n$ by n matrix

$$C = \begin{pmatrix} A \\ B \end{pmatrix}.$$

By Theorem 5.1 there are two modular matrices $U\ (= U^{(2n)})$, $V\ (= V^{(n)})$ such that

$$UCV = \begin{pmatrix} D_1 \\ 0 \end{pmatrix}, \qquad D_1 = [d_1, d_1 d_2, \ldots, d_1 d_2 \cdots d_n].$$

We denote by

$$U = \begin{pmatrix} U_{11} & U_{12} \\ U_{21} & U_{22} \end{pmatrix},$$

where U_{ij} are n by n matrices. Then, from the above, we have

$$\begin{pmatrix} U_{11} & U_{12} \\ U_{21} & U_{22} \end{pmatrix} \begin{pmatrix} A \\ B \end{pmatrix} = \begin{pmatrix} D_1 \\ 0 \end{pmatrix} V^{-1} = \begin{pmatrix} D \\ 0 \end{pmatrix}, \tag{1}$$

and so

$$U_{11} A + U_{12} B = D \tag{2}$$

and hence any right divisor of A and B must be a right divisor of D.

Also, if we let

$$\begin{pmatrix} U_{11} & U_{12} \\ U_{21} & U_{22} \end{pmatrix}^{-1} = \begin{pmatrix} X_{11} & X_{12} \\ X_{21} & X_{22} \end{pmatrix}$$

where X_{ij} are n by n matrices, then from (1) we have

$$\begin{pmatrix} A \\ B \end{pmatrix} = \begin{pmatrix} X_{11} & X_{12} \\ X_{21} & X_{22} \end{pmatrix} \begin{pmatrix} D \\ 0 \end{pmatrix},$$

and so

$$A = X_{11} D, \qquad B = X_{21} D,$$

and therefore D is a greatest common divisor of A and B. On taking $U_{11} = P$, $U_{12} = Q$ the theorem is proved. □

Theorem 8.2. *Let the square matrices A and B have a non-singular greatest common divisor D. Then any greatest common divisor of A and B must be of the form UD, where U is a modular matrix.*

Proof. Let D_1 be any greatest common divisor. Then, from the definition, we have $D = RD_1$ and $D_1 = SD$, and hence $D = RSD$. On taking the determinants we see that R and S are modular matrices. □

We now consider least common multiples. If the two matrices are both singular, then a least common multiple need not exist. For example

$$\begin{pmatrix} 1 & 0 \\ 0 & 0 \end{pmatrix} \quad \text{and} \quad \begin{pmatrix} 1 & 1 \\ 1 & 1 \end{pmatrix}$$

has no least common multiple. This is because every right divisor of $\begin{pmatrix} 1 & 0 \\ 0 & 0 \end{pmatrix}$ must take the form $\begin{pmatrix} a & 0 \\ c & 0 \end{pmatrix}$, and every right divisor of $\begin{pmatrix} 1 & 1 \\ 1 & 1 \end{pmatrix}$ must take the form $\begin{pmatrix} a & a \\ c & c \end{pmatrix}$, and these two forms are equal only when $a = c = 0$. However, we have the following:

Theorem 8.3. *Let A and B be two non-singular square matrices. Then their least common multiple M exists. Moreover, M is non-singular, and any least common multiple is of the form UM where U is a modular matrix.*

Proof. From (1) we have

$$U_{21}A + U_{22}B = 0.$$

We let

$$M = U_{21}A = -U_{22}B$$

so that M is a common multiple of A and B. We now prove that M is a least common multiple. Let M_1 be any common multiple of A and B. Then a greatest common divisor M_2 of M and M_1 is also a common multiple of A and B. Let

$$M = HM_2, \qquad M_2 = KA = LB$$

so that

$$U_{21}A = HKA, \qquad -U_{22}B = HLB. \tag{4}$$

Denote by A_0 and B_0 the adjoint matrices of A and B, so that $AA_0 = aI$, and $BB_0 = bI$ where a and b are the determinants of A and B. Since A, B are non-singular, we have $a \neq 0$, $b \neq 0$ and, from (4),

$$U_{21} = HK, \qquad -U_{22} = HL.$$

Therefore we have, from (3), that

$$I = U_{21}X_{12} + U_{22}X_{22} = H(KX_{12} - LX_{22}),$$

so that H is a modular matrix and H^{-1} exists. We see from

$$M_1 = GM_2 = GH^{-1}M$$

that M is a least common multiple.

We next prove that M is non-singular. From the definition of a least common multiple it suffices to prove that A and B have a non-singular common multiple. From Theorem 5.1 there are two modular matrices U_1, V_1 such that

$$U_1AV_1 = [d'_1, d'_1d'_2, \dots, d'_1 \cdots d'_n].$$

Let

$$M^* = d'_1 \cdots d'_nU_1B.$$

It is easily seen that M^* is a non-singular square matrix, and that

$$M^* = U_1BV_1[d'_2 \cdots d'_n, d'_3 \cdots d'_n, \dots, 1]U_1A.$$

This matrix M^* thus serves our purpose.

If M_3 is any least common multiple, then from the definition, we have

$$M = EM_3, \qquad M_3 = FM$$

and so

$$M = EFM, \qquad I = EF;$$

thus E, F are modular matrices, and the theorem is proved. □

Theorem 8.4. *Let A be a square matrix. Then, corresponding to each non-zero integer m, there are two square matrices R and Q such that*

(1) $A = mQ$ *or*

(2) $A = mQ + R$, *and* $0 < |R| < |m|^n$,

where $|R|$ denotes the absolute value of the determinant of R.

Proof. By Theorem 5.1 there are two matrices U and V such that

$$A = U[d_1, d_1d_2, \dots, d_1 \cdots d_n]V \qquad (d_i \geqslant 0,\ 1 \leqslant i \leqslant n).$$

There are integers q_i and r_i (> 0) such that

$$d_1 \cdots d_i = mq_i + r_i, \qquad 0 < r_i \leqslant |m| \qquad (1 \leqslant i \leqslant n).$$

Let

$$Q_1 = [q_1, q_2, \dots, q_n], \qquad R_1 = [r_1, r_2, \dots, r_n],$$

so that

$$A = U(mQ_1 + R_1)V. \tag{5}$$

If $r_i = |m|$ $(1 \leqslant i \leqslant n)$, then $R_1 = |m|I = \pm mI$ so that, from (5), we have

$$A = mU(Q_1 \pm I)V = mQ$$

which proves (1).

If there exists j such that $0 < r_j < |m|$, then $0 < |R_1| = r_1 r_2 \cdots r_n < |m|^n$, and so, from (5), we have

$$A = mUQ_1V + UR_1V = mQ + R$$

and

$$|R| = |UR_1V| = |R_1|.$$

This proves (2). □

Theorem 8.5. *Let B be a non-singular square matrix. Then, corresponding to any square matrix A, there exist two square matrices Q and C such that*
(1) *$A = QB$, or*
(2) *$A = QB + C$, and $0 < |C| < |B|$.*

Proof. Let B_0 be the adjoint matrix of B so that $BB_0 = B_0B = bI$, where b is the determinant of B. From the previous theorem there are two square matrices Q and R such that

$$AB_0 = bQ \tag{6}$$

or

$$AB_0 = bQ + R, \qquad 0 < |R| < |b|^n. \tag{7}$$

If we multiply equation (6) by B, then we have, from $b \neq 0$, that

$$A = QB$$

which is (1). Also, from (7),

$$R = AB_0 - bQ = AB_0 - QBB_0 = (A - QB)B_0 = CB_0.$$

Therefore, from

$$A = QB + (A - QB) = QB + C$$

and

$$0 < |C| = \frac{|R|}{|B_0|} < \frac{|b|^n}{|b|^{n-1}} = |b| = |B|$$

we arrive at (2). □

14.9 Linear Modules

Let $x_1, \ldots, x_n$ represent n indeterminants (variables). We denote by $\mathfrak{D} = \{x_1, \ldots, x_n\}$ the set of all linear forms

$$y = a_1 x_1 + \cdots + a_n x_n$$

with integer coefficients $a_1, \ldots, a_n$. If

$$y' = a_1' x_1 + \cdots + a_n' x_n$$

is another linear form in $\mathfrak{D}$, then we define

$$y \pm y' = (a_1 \pm a_1')x_1 + \cdots + (a_n \pm a_n')x_n.$$

Definition 9.1. Let $\mathfrak{M}$ be a subset of $\mathfrak{D}$ with the property that if y_1, y_2 are in $\mathfrak{M}$, then so are $y_1 \pm y_2$. Then we call $\mathfrak{M}$ a module.

Clearly $\mathfrak{D}$ itself is a module. The subset of linear forms $0, \pm x_1, \pm 2x_1, \ldots$ also form a module. The module formed by the subset whose only member is $0 = 0x_1 + \cdots + 0x_n$ will be excluded from our discussions.

Definition 9.2. Suppose that the module $\mathfrak{M}$ contains the forms $y_1, \ldots, y_l$ such that any form $\mathfrak{M}$ can be expressed uniquely as

$$b_1 y_1 + \cdots + b_l y_l,$$

where $b_1, \ldots, b_l$ are integers. Then we say that $\mathfrak{M}$ has *dimension* l, and we call $y_1, \ldots, y_l$ a *basis* for $\mathfrak{M}$.

It follows at once from the definition that $y_1, \ldots, y_l$ are linearly independent – that is $a_1 y_1 + \cdots + a_l y_l = 0$ implies $a_1 = \cdots = a_l = 0$.

Theorem 9.1. *Every module has a basis and has dimension at most n.*

Proof. Let $l (\leqslant n)$ be the integer such that, for each member of $\mathfrak{M}$, the coefficients of $x_{l+1}, \ldots, x_n$ are all zero, but there is a member of $\mathfrak{M}$ whose coefficient of x_l is not zero. It follows that the set of coefficients of x_l forms a non-zero integral modulus. We denote by b_l the least positive integer in this integral modulus, and we let the corresponding linear form in $\mathfrak{M}$ be

$$y_l = b_1 x_1 + \cdots + b_l x_l.$$

Now the coefficient of x_l for any member y of $\mathfrak{M}$ must be a multiple of b_l so that

$$y = y' + g y_l$$

where g is an integer, and y' is a linear form of the indeterminants $x_1, \ldots, x_{l-1}$.

Consider now the set of all such forms y'. We can determine an integer l' ($\leqslant l-1$) such that, for each y', the coefficients of $x_{l'+1}, \ldots, x_{l-1}$ are all zero, but there is a y' whose coefficient of $x_{l'}$ is not zero. As before we can determine a linear form

$$y_{l'} = b'_1 x_1 + \cdots + b'_{l'} x_{l'},$$

where $b'_{l'}$ is the least positive coefficient of $x_{l'}$ among all forms y'. Let $y' = y'' + g' y_{l'}$, where g' is an integer and y'' is a linear form in $x_1, \ldots, x_{l'-1}$. Proceeding inductively we see that $\mathfrak{M}$ has a basis $y_l, y'_l, \ldots$ with at most n members. The theorem is proved. □

Theorem 9.2. *The dimension of a module is independent of the choice of bases.*

Proof. Let $y_1, \ldots, y_l$ and $z_1, \ldots, z_{l'}$ be any two bases for a module $\mathfrak{M}$ and suppose, if possible, that $l \neq l'$. We may assume that $l > l'$.

From the definition of a basis there are integers a_{ij} and b_{ij} such that

$$\begin{pmatrix} y_1 \\ \vdots \\ y_l \end{pmatrix} = \begin{pmatrix} a_{11} & \cdots & a_{1l'} & 0 & \cdots & 0 \\ a_{21} & \cdots & a_{2l'} & 0 & \cdots & 0 \\ \cdots & \cdots & \cdots & \cdots & \cdots & \cdots \\ a_{l1} & \cdots & a_{ll'} & 0 & \cdots & 0 \end{pmatrix} \begin{bmatrix} z_1 \\ \vdots \\ z_{l'} \\ 0 \\ \vdots \\ 0 \end{bmatrix}$$

$$\begin{bmatrix} z_1 \\ \vdots \\ z_{l'} \\ 0 \\ \vdots \\ 0 \end{bmatrix} = \begin{bmatrix} b_{11} & \cdots & b_{1l} \\ \cdots & \cdots & \cdots \\ b_{l'1} & \cdots & b_{l'l} \\ 0 & \cdots & 0 \\ \cdots & \cdots & \cdots \\ 0 & \cdots & 0 \end{bmatrix} \begin{pmatrix} y_1 \\ \vdots \\ y_l \end{pmatrix},$$

where (a_{ij}) and (b_{ij}) are $l \times l$ square matrices. Therefore

$$\begin{pmatrix} y_1 \\ \vdots \\ y_l \end{pmatrix} = \begin{pmatrix} a_{11} & \cdots & a_{1l'} & 0 & \cdots & 0 \\ a_{21} & \cdots & a_{2l'} & 0 & \cdots & 0 \\ \cdots & \cdots & \cdots & \cdots & \cdots & \cdots \\ a_{l1} & \cdots & a_{ll'} & 0 & \cdots & 0 \end{pmatrix} \begin{bmatrix} b_{11} & \cdots & b_{1l} \\ \cdots & \cdots & \cdots \\ b_{l'1} & \cdots & b_{l'l} \\ 0 & \cdots & 0 \\ \cdots & \cdots & \cdots \\ 0 & \cdots & 0 \end{bmatrix} \begin{pmatrix} y_1 \\ \vdots \\ y_l \end{pmatrix}.$$

But $y_1, \ldots, y_l$ are linearly independent, so that $(a_{ij})(b_{ij}) = I$. Since the determinant of the left hand side is zero, we have a contradiction and therefore the theorem is proved. □

From now on we shall only consider modules with dimension n. Let $y_1, \ldots, y_n$ be a basis for a module $\mathfrak{M}$. Then

$$\begin{pmatrix} y_1 \\ y_2 \\ \vdots \\ y_n \end{pmatrix} = \begin{pmatrix} a_{11} & a_{12} & \cdots & a_{1n} \\ a_{21} & a_{22} & \cdots & a_{2n} \\ \cdots & \cdots & \cdots & \cdots \\ a_{n1} & a_{n2} & \cdots & a_{nn} \end{pmatrix} \begin{pmatrix} x_1 \\ x_2 \\ \vdots \\ x_n \end{pmatrix}. \tag{1}$$

Therefore corresponding to each n dimensional module and its basis $y_1, \ldots, y_n$, there is a square matrix

$$A = (a_{ij}) = \begin{pmatrix} a_{11} & a_{12} & \cdots & a_{1n} \\ a_{21} & a_{22} & \cdots & a_{2n} \\ \cdots & \cdots & \cdots & \cdots \\ a_{n1} & a_{n2} & \cdots & a_{nn} \end{pmatrix}, \tag{2}$$

which is a non-singular because $y_1, \ldots, y_n$ are linearly independent. Conversely, corresponding to each non-singular square matrix A, we can determine a set of linearly independent linear forms $y_1, \ldots, y_n$ which can then be used as a basis to determine an n dimensional module $\mathfrak{M}'$. This then sets up a relationship between n dimensional modules and non-singular square matrices of order n. We now ask: What is the relationship between the two matrices corresponding to the two different bases of a module?

Let $z_1, \ldots, z_n$ be another basis for $\mathfrak{M}$ with the corresponding matrix $B = (b_{ij})$ so that

$$\begin{pmatrix} z_1 \\ z_2 \\ \vdots \\ z_n \end{pmatrix} = \begin{pmatrix} b_{11} & b_{12} & \cdots & b_{1n} \\ b_{21} & b_{22} & \cdots & b_{2n} \\ \cdots & \cdots & \cdots & \cdots \\ b_{n1} & b_{n2} & \cdots & b_{nn} \end{pmatrix} \begin{pmatrix} x_1 \\ x_2 \\ \vdots \\ x_n \end{pmatrix}.$$

Since both $y_1, \ldots, y_n$ and $z_1, \ldots, z_n$ are bases, there are two square matrices $U = (u_{ij})$, $V = (v_{ij})$ such that

$$\begin{pmatrix} y_1 \\ \vdots \\ y_n \end{pmatrix} = U \begin{pmatrix} z_1 \\ \vdots \\ z_n \end{pmatrix}, \quad \begin{pmatrix} z_1 \\ \vdots \\ z_n \end{pmatrix} = V \begin{pmatrix} y_1 \\ \vdots \\ y_n \end{pmatrix},$$

and so

$$\begin{pmatrix} y_1 \\ \vdots \\ y_n \end{pmatrix} = UV \begin{pmatrix} y_1 \\ \vdots \\ y_n \end{pmatrix}.$$

Since $y_1, \ldots, y_n$ are linearly independent, we deduce that $UV = I$, that is U and V are modular matrices. Now

$$\begin{pmatrix} z_1 \\ \vdots \\ z_n \end{pmatrix} = V \begin{pmatrix} y_1 \\ \vdots \\ y_n \end{pmatrix} = VA \begin{pmatrix} x_1 \\ \vdots \\ x_n \end{pmatrix},$$

so that

$$B = VA. \tag{3}$$

Therefore matrices corresponding to the same module are left associated. Conversely, two non-singular square matrices which are left associated correspond to the same module. If we partition the family of all non-singular matrices of order n into classes by left association, then each class represents a module, and modules represented by distinct classes are different. We may therefore speak of "the matrix A associated with $\mathfrak{M}$" to mean that A is a member of the class of matrices which represent $\mathfrak{M}$.

From Theorem 4.1 we see that, for an n dimensional module $\mathfrak{M}$, we can select a basis $y_1, \ldots, y_n$ such that

$$\begin{aligned} y_1 &= a_{11}x_1, \\ y_2 &= a_{21}x_1 + a_{22}x_2, \\ &\cdots\cdots\cdots\cdots\cdots\cdots \\ y_n &= a_{n1}x_1 + a_{n2}x_2 + \cdots + a_{nn}x_n, \end{aligned} \tag{4}$$

where $a_{\nu\nu} > 0$ $(1 \leqslant \nu \leqslant n)$, and $0 \leqslant a_{\mu\nu} < a_{\nu\nu}$ $(\mu > \nu)$. This is a standard form for a basis, or a *standard basis.*

Theorem 9.3. *Let $\mathfrak{M}$ and $\mathfrak{N}$ be two modules. A necessary and sufficient condition for $\mathfrak{N}$ to be included in $\mathfrak{M}$ is that the matrix associated with $\mathfrak{M}$ is a right divisor of the matrix associated with $\mathfrak{N}$.*

Proof. Let the bases for $\mathfrak{M}$ and $\mathfrak{N}$ be $y_1, \ldots, y_n$ and $z_1, \ldots, z_n$ and let the associated matrices be $A = (a_{ij})$ and $B = (b_{ij})$ respectively. If $\mathfrak{N}$ is included in $\mathfrak{M}$, then

$$\begin{pmatrix} z_1 \\ z_2 \\ \vdots \\ z_n \end{pmatrix} = (c_{ij}) \begin{pmatrix} y_1 \\ y_2 \\ \vdots \\ y_n \end{pmatrix} = (c_{ij})(a_{ij}) \begin{pmatrix} x_1 \\ x_2 \\ \vdots \\ x_n \end{pmatrix} = (b_{ij}) \begin{pmatrix} x_1 \\ x_2 \\ \vdots \\ x_n \end{pmatrix}$$

so that $B = CA$. Conversely, if $B = CA$, then

$$\begin{pmatrix} z_1 \\ z_2 \\ \vdots \\ z_n \end{pmatrix} = CA \begin{pmatrix} x_1 \\ x_2 \\ \vdots \\ x_n \end{pmatrix} = C \begin{pmatrix} y_1 \\ y_2 \\ \vdots \\ y_n \end{pmatrix}$$

so that $\mathfrak{N}$ is included in $\mathfrak{M}$. □

Definition 9.3. Suppose that the difference between two linear forms z_1 and z_2 is a member of the module $\mathfrak{M}$. Then we say that z_1 and z_2 are *congruent* mod $\mathfrak{M}$, and we write $z_1 \equiv z_2 \pmod{\mathfrak{M}}$.

The relation of being congruent mod $\mathfrak{M}$ is reflexive, symmetric and transitive, so that the family of all linear forms is partitioned into equivalence classes mod $\mathfrak{M}$. The number of such classes is called the *norm* of $\mathfrak{M}$, and is denoted by $N(\mathfrak{M})$, the existence of which has yet to be proved. Clearly $\mathfrak{M}$ itself is an equivalence class.

Theorem 9.4. *Let A correspond to the module $\mathfrak{M}$. Then*

$$N(\mathfrak{M}) = |A|.$$

Proof. Since the matrices associated with $\mathfrak{M}$ have the same absolute value for their determinants, we may assume that the basis chosen is the standard basis in (4). Any linear form

$$y = a_1x_1 + \cdots + a_{n-1}x_{n-1} + a_nx_n$$

gives another one with $0 \leqslant a_n < a_{nn}$, by subtracting a suitable multiple of $y_n = a_{n1}x_1 + \cdots + a_{nn}x_n$. We may further subtract multiples of $y_{n-1} = a_{n-1,1}x_1 + \cdots + a_{n-1,n-1}x_{n-1}$ so that $0 \leqslant a_{n-1} < a_{n-1,n-1}$, and so on. Thus every linear form is congruent to a linear form

$$a_1x_1 + \cdots + a_nx_n, \qquad 0 \leqslant a_\nu < a_{\nu\nu} \qquad (1 \leqslant \nu \leqslant n).$$

The total number of such linear forms is $a_{11}a_{22}\cdots a_{nn} = |A|$, and moreover no two such linear forms are congruent. The theorem is proved. □

From Theorem 9.3 and Theorem 9.4 we have

Theorem 9.5. Let $\mathfrak{N} \subset \mathfrak{M}$ *and let A and B be matrices associated with $\mathfrak{M}$ and $\mathfrak{N}$ respectively. Then, in the partitioning of $\mathfrak{M}$ into congruent classes* mod $\mathfrak{N}$, *the number of classes is equal to*

$$\frac{N(\mathfrak{N})}{N(\mathfrak{M})} = \frac{|B|}{|A|}. \quad \square$$

The set $\mathfrak{D} = \{x_1, \ldots, x_n\}$ with indeterminants $x_1, \ldots, x_n$ can also be represented by other indeterminants. If we let

$$\begin{pmatrix} x_1 \\ \vdots \\ x_n \end{pmatrix} = \begin{pmatrix} u_{11} & \cdots & u_{1n} \\ \cdots & \cdots & \cdots \\ u_{n1} & \cdots & u_{nn} \end{pmatrix} \begin{pmatrix} x'_1 \\ \vdots \\ x'_n \end{pmatrix},$$

where $U = (u_{ij})$ is a modular matrix, then $\mathfrak{D}$ can also be represented by $x'_1, \ldots, x'_n$; that is $\mathfrak{D} = \{x_1, \ldots, x_n\} = \{x'_1, \ldots, x'_n\}$.

Let a module $\mathfrak{M}$, together with its basis $y_1, \ldots, y_n$ corresponding to the indeterminants $x_1, \ldots, x_n$ have the associated matrix A. We now consider the associated matrix corresponding to a change of indeterminants to $x'_1, \ldots, x'_n$. From

$$\begin{pmatrix} y_1 \\ \vdots \\ y_n \end{pmatrix} = A \begin{pmatrix} x_1 \\ \vdots \\ x_n \end{pmatrix} = AU \begin{pmatrix} x'_1 \\ \vdots \\ x'_n \end{pmatrix},$$

we see that the required matrix corresponding to the indeterminants $x'_1, \ldots, x'_n$ is AU. This means that the relation of right association corresponds to the change of indeterminants, or the change of basis for the representation of $\mathfrak{D}$. Also, from (3) we see that the relation of left association corresponds to the change of basis for the module. It therefore follows from Theorem 5.1 that each fixed n dimensional module $\mathfrak{M}$, after suitable changes of bases for the module and for $\mathfrak{D}$, has an associated matrix which is a diagonal matrix

$$[d_1, d_1 d_2, \ldots, d_1 \cdots d_n] \qquad (d_1 > 0, \ldots, d_n > 0).$$

From Theorem 7.2 and Theorem 9.5 we have

Theorem 9.6. *Let $\mathfrak{M}$ be an n dimensional module. Then there is a chain*

$$\mathfrak{M} = \mathfrak{M}_0 \subset \mathfrak{M}_1 \subset \cdots \subset \mathfrak{M}_l = \mathfrak{D} \tag{5}$$

such that

$$\frac{N(\mathfrak{M}_{i-1})}{N(\mathfrak{M}_i)} \qquad (1 \leqslant i \leqslant l)$$

are prime numbers. □

The set of forms belonging to both the modules $\mathfrak{M}_1$ and $\mathfrak{M}_2$ is also a module which is called the *intersection* of $\mathfrak{M}_1$ and $\mathfrak{M}_2$, and we denote it by $\mathfrak{M}_m$. Also the set of forms obtained from addition and subtraction of members belonging to $\mathfrak{M}_1$ and $\mathfrak{M}_2$ forms a module which is called the *sum* of $\mathfrak{M}_1$ and $\mathfrak{M}_2$, and we denote it by $\mathfrak{M}_d$. We then have:

Theorem 9.7. *Let the matrices M_1, M_2, M_m, M_d be associated with the modules $\mathfrak{M}_1$, $\mathfrak{M}_2$, $\mathfrak{M}_m$, $\mathfrak{M}_d$ respectively. Then M_d is a least common multiple of M_1 and M_2, and M_m is a greatest common divisor of M_1 and M_2.*

Proof. Since $\mathfrak{M}_m \subseteq \mathfrak{M}_1$ and $\mathfrak{M}_m \subseteq \mathfrak{M}_2$, we have

$$M_m = A_1 M_1, \qquad M_m = A_2 M_2.$$

If $M_3 = B_1 M_1 = B_2 M_2$ is a common multiple of M_1 and M_2, and $\mathfrak{M}_3$ is the module with which the matrix M_3 is associated, then

$$\mathfrak{M}_3 \subseteq \mathfrak{M}_1, \qquad \mathfrak{M}_3 \subseteq \mathfrak{M}_2$$

and hence

$$\mathfrak{M}_3 \subseteq \mathfrak{M}_m, \qquad M_3 = C M_m.$$

Thus M_m is a least common multiple of M_1 and M_2. The proof that M_d is a greatest common divisor of M_1 and M_2 is similar. □

Chapter 15. p-adic Numbers

15.1 Introduction

The purpose of this chapter is to introduce the theory of p-adic numbers due to Hensel. This theory has extensive applications in number theory, algebraic geometry and algebraic functions, and is an important theory in the study of modern algebra. Before we give the rigorous definitions we give a simple introduction as to how we obtain the p-adic numbers. We recall the method of solution to the congruence

$$f(x) \equiv 0 \pmod{p^l} \tag{1}$$

which we discussed in chapter 2; here $f(x)$ is a polynomial with integer coefficients and p is a prime number. Our method was first to solve the congruence

$$f(x) \equiv 0 \pmod{p}. \tag{2}$$

If (2) has a solution a_0 $(0 \leqslant a_0 < p)$ and $f'(a_0) \not\equiv 0 \pmod{p}$, then we let $x = a_0 + py$, and we consider the congruence

$$f(a_0 + py) \equiv 0 \pmod{p^2}, \qquad 0 \leqslant y < p,$$

or

$$\frac{f(a_0)}{p} + f'(a_0)y \equiv 0 \pmod{p}, \qquad 0 \leqslant y < p.$$

This congruence uniquely determines y which is then denoted by a_1 so that

$$x = a_0 + a_1 p, \qquad 0 \leqslant a_0 < p, \qquad 0 \leqslant a_1 < p,$$

is a solution to

$$f(x) \equiv 0 \pmod{p^2}.$$

In general, if

$$x = x_0 = a_0 + a_1 p + a_2 p^2 + \cdots + a_{l-2} p^{l-2}, \qquad 0 \leqslant a_v < p$$

is a solution to $f(x) \equiv 0 \pmod{p^{l-1}}$ and $f'(x_0) \not\equiv 0 \pmod{p}$, then let $x = x_0 + p^{l-1}y$ and consider the congruence

$$f(x_0 + p^{l-1}y) \equiv 0 \pmod{p^l}, \qquad 0 \leqslant y < p,$$

or

$$\frac{f(x_0)}{p^{l-1}} + f'(x_0)y \equiv 0 \pmod{p}, \qquad 0 \leqslant y < p.$$

This then uniquely determines y which is then denoted by a_{l-1} and

$$x = a_0 + a_1 p + \cdots + a_{l-1}p^{l-1}, \qquad 0 \leqslant a_v < p$$

is a solution to (1).

This method can be continued indefinitely, and we can now form a power series in p:

$$a_0 + a_1 p + \cdots + a_l p^l + \cdots, \qquad 0 \leqslant a_v < p. \tag{3}$$

We call this power series a p-adic solution to the equation

$$f(x) = 0.$$

As we know in applying the method of successive approximations to obtain the real solution to $f(x) = 0$, the accuracy of the approximations increases with the number of iterations. Here a similar situation occurs in that if we solve the successive congruences

$$f(x) \equiv 0 \pmod{p},$$
$$f(x) \equiv 0 \pmod{p^2},$$
$$\cdots$$
$$f(x) \equiv 0 \pmod{p^l}$$
$$\cdots$$

for the p-adic solutions of $f(x) = 0$, then the larger the number l, the closer the solution

$$x = a_0 + a_1 p + \cdots + a_{l-1}p^{l-1}, \qquad 0 \leqslant a_v < p$$

is to the p-adic solution.

Abstractly we call the power series (3) a p-adic number. However these are not all the p-adic numbers because we should also allow a finite number of negative powers of p in the expansion – that is a p-adic number may be represented by

$$a_{-n}p^{-n} + \cdots + a_0 + a_1 p + \cdots + a_l p^l + \cdots, \qquad 0 \leqslant a_v < p. \tag{4}$$

This is analogous to the well known fact that every positive real number has the decimal expansion

$$a_{-n}10^n + \cdots + a_0 + a_1 10^{-1} + \cdots + a_l 10^{-l} + \cdots, \qquad 0 \leqslant a_v < 10.$$

Let two p-adic numbers be represented by

$$a_{-n}p^{-n} + \cdots + a_0 + a_1 p + \cdots + a_l p^l + \cdots, \qquad 0 \leqslant a_v < p,$$

$$b_{-m}p^{-m} + \cdots + b_0 + b_1 p + \cdots + b_l p^l + \cdots, \qquad 0 \leqslant b_v < p$$

where $n \geqslant m$. We can define the sum and the difference by

$$a_{-n}p^{-n} + \cdots + a_{-m-1}p^{-m-1} + (a_{-m} \pm b_{-m})p^{-m} + \cdots + (a_0 \pm b_0)$$
$$+ (a_1 \pm b_1)p + \cdots + (a_l \pm b_l)p^l + \cdots,$$

that is by the corresponding addition and subtraction of the coefficients for the power series. If on addition the coefficient is now greater than or equal to p, then we increase the next coefficient by 1. For example, if $a_v + b_v \geqslant p$, then we let $(a_v + b_v)p^v = (a_v + b_v - p)p^v + p^{v+1}$ and we add p^{v+1} to the next term. Similarly, if on subtraction a coefficient becomes negative, then we can "borrow" a unit from the previous term: for example, if $a_v - b_v < 0$, then we rewrite $(a_v - b_v)p^v + (a_{v+1} + b_{v+1})p^{v+1} + \cdots$ as $(a_v - b_v + p)p^v + (a_{v+1} - b_{v+1} - 1)p^{v+1} + \cdots$. In this way we can always end up with a set of non-negative coefficients which are less than p.

Two p-adic numbers can also be multiplied together via the formal multiplication of power series, again making allowance for shifting the terms so that all the coefficients are non-negative and less than p.

Example 1. A 5-adic solution to the equation $3x = 2$ is

$$4 + 1 \cdot 5 + 3 \cdot 5^2 + 1 \cdot 5^3 + 3 \cdot 5^4 + \cdots,$$

where, apart from the first term, all the coefficients are periodically 1 and 3.

The reader can, of course, verify this by the method of solution to the various congruences. However, the following also shows that the power series is indeed a 5-adic solution to the equation:

$$\begin{aligned}
3 \cdot (4 &+ 1 \cdot 5 + 3 \cdot 5^2 + 1 \cdot 5^3 + 3 \cdot 5^4 + \cdots) \\
&= 12 + 3 \cdot 5 + 9 \cdot 5^2 + 3 \cdot 5^3 + 9 \cdot 5^4 + \cdots \\
&= 2 + (2 + 3) \cdot 5 + 9 \cdot 5^2 + 3 \cdot 5^3 + 9 \cdot 5^4 + \cdots \\
&= 2 + 10 \cdot 5^2 + 3 \cdot 5^3 + 9 \cdot 5^4 + \cdots \\
&= 2 + 5 \cdot 5^3 + 9 \cdot 5^4 + \cdots \\
&= 2 + 10 \cdot 5^4 + \cdots \\
&= \cdots \\
&\cdots \\
&= 2.
\end{aligned}$$

Example 2. A 3-adic solution to the equation $x^2 = 7$ is

$$1 + 1 \cdot 3 + 1 \cdot 3^2 + 0 \cdot 3^3 + 2 \cdot 3^4 + \cdots.$$

Let a be a rational number. We call the p-adic solution to the equation $x = a$ the p-adic representation of a.

The p-adic solution to the simplest equation $x = d$ (d a positive integer) can be obtained as follows: Divide d by p and denote by q_0 and d_0 the quotient and the remainder respectively, so that

$$d = d_0 + q_0 p, \qquad 0 \leqslant d_0 < p.$$

Next divide q_0 by p to obtain q_1 and d_1 as before so that

$$d = d_0 + d_1 p + q_1 p^2, \qquad 0 \leqslant d_0, \qquad d_1 < p.$$

Continuing in this way we have eventually the unique p-adic representation for d, namely

$$d_0 + d_1 p + d_2 p^2 + \cdots + d_l p^l, \qquad 0 \leqslant d_v < p.$$

This is, of course, the method of obtaining the representation of d in the number base p, and if p is not restricted to be a prime, say $p = 10$, then we have the usual decimal representation. Therefore the p-adic representation of an integer is the same as the representation of the integer in the number base p.

Exercise 1. Find another 3-adic solution to the equation $x^2 = 7$.

Exercise 2. Find a 7-adic solution to $x^2 + x + 1 = 0$.

Exercise 3. Find a 3-adic solution to $9x^2 = 7$. (Suggestion: Set $y = 3x$ and then find a 3-adic solution, say y_0, to the equation $y^2 = 7$. Now $x_0 = 3^{-1} y_0$ is a 3-adic solution to the original equation.)

15.2 The Definition of a Valuation

In the previous section we did not discuss the convergence problem associated with the power series

$$a_{-n} p^{-n} + \cdots + a_0 + a_1 p + \cdots + a_l p^l + \cdots, \qquad 0 \leqslant a_v < p.$$

In fact it is obvious that this power series is never convergent in the ordinary sense. We now introduce a new notion whereby the power series above can be given a rigorous definition. We call this notion a *valuation*, and it is an abstraction from the ordinary notion of the absolute value of a real number.

Definition. Let φ be a function from the rationals to the rationals satisfying the following conditions:

1) $\varphi(a) \geqslant 0$ with equality if and only if $a = 0$;
2) $\varphi(ab) = \varphi(a)\varphi(b)$;
3) $\varphi(a + b) \leqslant \varphi(a) + \varphi(b)$.

Then we call φ a *valuation.*

We see easily that φ possesses the following properties: $\varphi(1) = \varphi(-1) = 1$, $\varphi(-a) = \varphi(a)$, and for any positive integer n, $\varphi(n) \leqslant n$. Also, from 3) we have

$$|\varphi(a) - \varphi(b)| \leqslant \varphi(a + b).$$

Example 1. Let $\varphi(0) = 0$, and $\varphi(a) = 1$ if $a \neq 0$. Then clearly φ is a valuation. We call this the *identical valuation* and we denote it by φ_0.

Example 2. Let $\varphi(a) = |a|$, the usual absolute value. Then clearly φ is a valuation.

Example 3. Let p be a fixed prime number. Any non-zero rational number a can be uniquely written as

$$a = \frac{r}{s}p^n, \qquad s > 0,$$

where r, s are coprime integers, $p \nmid rs$ and n is an integer which may be positive, negative or zero. We now define

$$\varphi(a) = p^{-n}, \qquad a \neq 0; \qquad \varphi(0) = 0.$$

We shall prove that φ is a valuation, called the p-adic valuation and we write $\varphi(a) = |a|_p$.

The condition 1) obviously holds. Next, if

$$a = \frac{r_1}{s_1}p^m, \qquad b = \frac{r_2}{s_2}p^n \qquad (s_1 > 0, s_2 > 0),$$

where $(r_1, s_1) = (r_2, s_2) = 1, p \nmid r_1 s_1 r_2 s_2$, then

$$ab = \frac{r_1 r_2}{s_1 s_2}p^{m+n}$$

giving

$$|ab|_p = p^{-(m+n)} = |a|_p \cdot |b|_p,$$

which proves 2). Finally, assume that $m \leqslant n$, we have

$$a + b = \frac{r_1 s_2 + r_2 s_1 p^{n-m}}{s_1 s_2}p^m, \tag{1}$$

and since $p \nmid s_1 s_2$, it follows that $|a + b|_p \leqslant p^{-m} = |a|_p$ and hence, by 1), $|a + b|_p \leqslant |a|_p + |b|_p$, which proves 3).

In this example we actually proved that

$$|a + b|_p \leqslant \max(|a|_p, |b|_p).$$

In fact we can show further that if $|a|_p \neq |b|_p$, then

$$|a + b|_p = \max(|a|_p, |b|_p).$$

For, we may assume that $m < n$ in (1) so that, from $p \nmid (r_1 s_2 + r_2 s_1 p^{n-m})$ (because $p \nmid r_1 s_1 r_2 s_2$), we arrive at

$$|a + b|_p = p^{-m} = |a|_p = \max(|a|_p, |b|_p).$$

15.3 The Partitioning of Valuations into Classes

Definition 3.1. Let φ and φ' be two valuations. Suppose $\varphi(a) < \varphi(b)$ if and only if $\varphi'(a) < \varphi'(b)$. Then we say that φ and φ' are *equivalent*.

Let $s > 0$ and φ be a valuation. Then φ^s satisfies 1) and 2), but not always 3). However, if $0 < s \leqslant 1$, then 3) is also satisfied. (This follows from the inequality

$$(x + y)^s - y^s = s \int_0^x (t + y)^{s-1}\, dt \leqslant sxy^{s-1} \leqslant x^s$$

valid for $0 \leqslant x \leqslant y, 0 < s \leqslant 1$.) Let us denote φ^s $(0 < s \leqslant 1)$ by φ', so that φ' is now a valuation and it is easy to see that φ and φ' are equivalent.

Theorem 3.1. *Let φ be a non-identical valuation and let φ' be an equivalent valuation. Then there exists $s > 0$ such that $\varphi' = \varphi^s$.*

Proof. Since $\varphi \neq \varphi_0$ there exists a rational a_0 such that $0 < \varphi(a_0) < 1$ (if $\varphi(a_0) > 1$, then, by 2), $\varphi(a_0^{-1}) < 1$). Let a be any non-zero rational number and consider the set of pairs of positive integers (m, n) satisfying $\varphi(a_0^m) < \varphi(a^n)$, that is $(\varphi(a_0))^m < (\varphi(a))^n$ or

$$\frac{m}{n} > \frac{\log \varphi(a)}{\log \varphi(a_0)}. \tag{1}$$

We may view $(\log \varphi(a))/\log \varphi(a_0)$ as a lower bound for the set of rational numbers satisfying (1). If φ' and φ are equivalent, then $(\log \varphi'(a))/\log \varphi'(a_0)$ also acts as a lower bound for this set of rational numbers. It follows that, for any rational $a \neq 0$,

$$\frac{\log \varphi(a)}{\log \varphi(a_0)} = \frac{\log \varphi'(a)}{\log \varphi'(a_0)}.$$

This means that there exists a positive constant s, depending only on φ and φ', such that

$$\frac{\log \varphi'(a)}{\log \varphi(a)} = \frac{\log \varphi'(a_0)}{\log \varphi(a_0)} = s > 0$$

holds for all rational $a \neq 0$. Therefore $\varphi'(a) = \varphi^s(a)$. □

Definition 3.2. Let φ be a valuation and suppose that there exists a positive integer $n_0 > 1$ such that $\varphi(n_0) > 1$. Then we call φ an *Archimedian valuation*. A *non-Archimedian valuation* φ is one such that $\varphi(n) \leqslant 1$ for all positive integers n.

The valuation $\varphi(a) = |a|$ is Archimedian, the identical valuation φ_0 and the p-adic valuation $\varphi(a) = |a|_p$ are non-Archimedian.

15.4 Archimedian Valuations

Theorem 4.1. *Any Archimedian valuation is equivalent to the absolute value valuation.*

Proof. Let φ be an Archimedian valuation and let n, n' be two integers greater than 1. We represent n' as

$$n' = a_0 + a_1 n + a_2 n^2 + \cdots + a_v n^v, \qquad 0 \leqslant a_i < n, \qquad a_v \neq 0.$$

Then

$$\varphi(n') \leqslant \varphi(a_0) + \varphi(a_1)\varphi(n) + \varphi(a_2)\varphi(n^2) + \cdots + \varphi(a_v)\varphi(n^v).$$

From $\varphi(a_i) \leqslant a_i < n$ $(i = 0, 1, \ldots, v)$, we see that

$$\begin{aligned} \varphi(n') &\leqslant n(1 + \varphi(n) + \varphi(n)^2 + \cdots + \varphi(n)^v) \\ &\leqslant n(1 + v)\max(1, \varphi(n)^v). \end{aligned}$$

From the representation of n' we know that $n^v \leqslant n'$ so that $v \leqslant \log n'/\log n$, and hence

$$\varphi(n') \leqslant n\left(1 + \frac{\log n'}{\log n}\right)\max(1, \varphi(n)^{\log n'/\log n}).$$

Substituting n'^h for n' we have

$$\varphi(n')^h \leqslant n\left(1 + h\frac{\log n'}{\log n}\right)\max(1, \varphi(n)^{h \log n'/\log n}),$$

or

$$\varphi(n') \leqslant \left(n\left(1 + h\frac{\log n'}{\log n}\right)\right)^{1/h} \max(1, \varphi(n)^{\log n'/\log n}).$$

Letting $h \to \infty$ we have

$$\varphi(n') \leqslant \max(1, \varphi(n)^{\log n'/\log n}). \tag{1}$$

This holds for all $n, n' > 1$.

By the Archimedian property there exists $n_0 > 1$ such that $\varphi(n_0) > 1$. Therefore $1 < \max(1, \varphi(n)^{\log n_0/\log n})$ and whence $1 < \varphi(n)^{\log n_0/\log n}$. Therefore $\varphi(n) > 1$ whenever $n > 1$ and (1) may be rewritten as

$$\varphi(n') \leqslant \varphi(n)^{\log n'/\log n},$$

or

$$\frac{\log \varphi(n')}{\log n'} \leqslant \frac{\log \varphi(n)}{\log n}.$$

By the symmetry of n and n' we deduce that

$$\frac{\log \varphi(n')}{\log n'} = \frac{\log \varphi(n)}{\log n},$$

and this implies the existence of a positive constant s, depending only on φ, such that

$$\frac{\log \varphi(n)}{\log n} = s > 0, \qquad n > 1.$$

Therefore $\varphi(n) = n^s$. Also, from $\varphi(n) \leqslant n$ we know that $s \leqslant 1$.

Next, from $\varphi(-n) = \varphi(n)$ we see that $\varphi(n) = |n|^s$ for all integers n such that $|n| > 1$. Finally, from 2) we see that, for all rational numbers a,

$$\varphi(a) = |a|^s, \qquad 0 < s \leqslant 1. \quad \square$$

15.5 Non-Archimedian Valuations

We saw in §2 that for the p-adic valuation $\varphi(a) = |a|_p$ we have $|a + b|_p \leqslant \max(|a|_p, |b|_p)$ with equality when $|a|_p \neq |b|_p$. We now prove that all non-Archimedian valuations share this property.

Theorem 5.1. *Let φ be a non-Archimedian valuation. Then*

3') $$\varphi(a + b) \leqslant \max(\varphi(a), \varphi(b)).$$

Also, if $\varphi(a) \neq \varphi(b)$, then,

3'') $$\varphi(a + b) = \max(\varphi(a), \varphi(b)).$$

Conversely, if a valuation φ satisfies 3') then φ is non-Archimedian.

Proof. From the Binomial theorem

$$(a+b)^n = a^n + \binom{n}{1} a^{n-1}b + \cdots + \binom{n}{n-1} ab^{n-1} + b^n,$$

and the inequality $\varphi(n) \leqslant 1$, which holds for a non-Archimedian valuation φ and positive integers n, we see that

$$\begin{aligned}\varphi((a+b)^n) &\leqslant \varphi(a)^n + \varphi(a)^{n-1}\varphi(b) + \cdots + \varphi(a)\varphi(b)^{n-1} + \varphi(b)^n \\ &\leqslant (n+1)\max(\varphi(a)^n, \varphi(b)^n),\end{aligned}$$

or

$$\varphi(a+b) \leqslant (n+1)^{1/n} \max(\varphi(a), \varphi(b)),$$

and 3′) follows from this by letting $n \to \infty$.

If $\varphi(a) \neq \varphi(b)$, we may assume that $\varphi(b) < \varphi(a)$. Then, from 3′), we have $\varphi(a+b) \leqslant \varphi(a)$. Now, if $\varphi(a+b) < \varphi(a)$, then by 3′) we have

$$\varphi(a) = \varphi((a+b) - b) \leqslant \max(\varphi(a+b), \varphi(b)) < \varphi(a)$$

which is impossible. Therefore

$$\varphi(a+b) = \varphi(a) = \max(\varphi(a), \varphi(b)).$$

Conversely, suppose that a valuation φ satisfies 3′). Then, for any positive integer n, we have

$$\varphi(n) = \varphi(1 + 1 + \cdots + 1) \leqslant \varphi(1) = 1,$$

so that φ is non-Archimedian. □

From this theorem we see that in order to prove that φ is a non-Archimedian valuation, it suffices to establish 1), 2) and 3′). Moreover, if φ is a non-Archimedian valuation, then φ^s $(s > 0)$ is always a valuation regardless of whether $s \leqslant 1$. This is because, from 3′),

$$\varphi^s(a+b) \leqslant \max(\varphi^s(a), \varphi^s(b)) \leqslant \varphi^s(a) + \varphi^s(b)$$

which gives 3).

Given any non-Archimedian valuation φ we put

$$w(a) = -\log \varphi(a),$$

where the base of the logarithm is any real number greater than 1. The choice of the base has little relevance because φ^s $(s > 0)$ is also a non-Archimedian valuation. From the properties of φ we see that w has the following properties.

i) If $a \neq 0$, then $w(a)$ is a real number, and $w(0) = \infty$;

ii) $w(ab) = w(a) + w(b)$;

iii) $w(a+b) \geqslant \min(w(a), w(b))$;

iv) $w(a+b) = \min(w(a), w(b))$ if $w(a) \neq w(b)$.

If φ is not the identical valuation, then there must be a rational a_0 such that $0 < w(a_0) < \infty$. We also note that $w(1) = 0$, $w(-a) = w(a)$ and $w(n) \geqslant 0$ for integers n.

Theorem 5.2. *The following is a necessary and sufficient condition for the equivalence of two non-identical non-Archimedean valuations φ and φ'. There exists $s > 0$ such that, for every rational $a \neq 0$,*

$$w'(a) = sw(a),$$

where $w'(a) = -\log \varphi'(a)$ and $w(a) = -\log \varphi(a)$. □

Theorem 5.3. *Every non-identical non-Archimedean valuation φ is equivalent to some p-adic valuation $|a|_p$.*

Proof. First $w(n) \geqslant 0$ for integers n, and from $\varphi \neq \varphi_0$ there exists an integer $m \neq 1$ such that $w(m) > 0$. We next show that the set of integers satisfying this inequality forms a modulus. This is easy since if $w(n) > 0$ and $w(n') > 0$, then $w(n \pm n') \geqslant \min(w(n), w(n') > 0$ by iii). From Theorem 1.4.3 we know that there exists a least positive integer g in the modulus such that g divides every member of the modulus. Obviously $g > 1$, and we now prove that g is a prime number. Suppose, if possible, that $g = g'g''$ $(g' > 1, g'' > 1)$. Then

$$w(g) = w(g'g'') = w(g') + w(g'').$$

Since $w(g)$ is positive and $w(g')$, $w(g'')$ are non-negative, it follows that at least one of $w(g')$ and $w(g'')$ is positive. But g' and g'' are less than g and this contradicts with g being the least positive integer in the modulus. Therefore g is a prime number which we shall now denote by p. We have now proved that $w(n) = 0$ if $p \nmid n$, and $w(n) > 0$ if $p | n$.

Corresponding to any rational number $a \neq 0$ we have the unique representation $a = (r/s)p^l$ $(s > 0)$, where r, s are coprime integers, $p \nmid rs$ and l is an integer. Therefore

$$w(a) = w\left(\frac{r}{s}\right) + lw(p) = w(r) - w(s) + lw(p) = lw(p).$$

Now

$$w'(a) = -\log |a|_p = l \log p,$$

so that

$$w(a) = \frac{w(p)}{\log p} w'(a).$$

Let $s = w(p)/\log p$ and the result follows from Theorem 5.2. □

15.6 The φ-Extension of the Rationals

Readers who are familiar with Cantor's method for the construction of real numbers in mathematical analysis should have no difficulty with this and the next section.

Let φ be a valuation, and we shall write $\{a_n\}$ for a sequence $a_1, a_2, \ldots, a_n, \ldots$ of rational numbers.

Definition 6.1. By a *fundamental sequence*, or a *φ-convergent sequence*, we mean a sequence $\{a_n\}$ which satisfies the following condition: Given any rational number $\varepsilon > 0$, there exists a positive integer $N\,(= N(\varepsilon))$ such that $\varphi(a_m - a_n) < \varepsilon$ whenever $m, n > N$.

For example, the constant sequence, where $a_1 = a_2 = \cdots = a_n = \cdots = a$, is a fundamental sequence which we shall denote by $\{a\}$.

If $\{a_n\}$ is a fundamental sequence, then there exists A such that $\varphi(a_n) \leqslant A$ for all n.

We define the sum, the difference and the product of two sequences by

$$\{a_n\} \pm \{b_n\} = \{a_n \pm b_n\}, \qquad \{a_n\} \cdot \{b_n\} = \{a_n b_n\}.$$

From

$$\varphi((a_m \pm b_m) - (a_n \pm b_n)) \leqslant \varphi(a_m - a_n) + \varphi(b_m - b_n)$$

and

$$\begin{aligned}\varphi(a_m b_m - a_n b_n) &= \varphi(a_m(b_m - b_n) + b_n(a_m - a_n)) \\ &\leqslant \varphi(a_m)\varphi(b_m - b_n) + \varphi(b_n)\varphi(a_m - a_n),\end{aligned}$$

we see at once that the sum, the difference and the product of two fundamental sequences are fundamental.

Definition 6.2. Let $\{a_n\}$ be a sequence such that there exists a rational number a with the following property: Given any rational number $\varepsilon > 0$, there exists a positive integer $N\,(= N(\varepsilon))$ such that $\varphi(a_n - a) < \varepsilon$ whenever $n > N$. Then we say that $\{a_n\}$ has the *φ-limit* a, and we write $\varphi\text{-}\lim_{n\to\infty} a_n = a$.

Obviously the φ-limit of $\{a\}$ is a. From $\varphi(a_m - a_n) \leqslant \varphi(a_m - a) + \varphi(a_n - a)$ we see that the existence of a φ-limit implies the sequence being fundamental. Note, however, that the converse does not follow—that is, not every fundamental sequence possesses a φ-limit.

Let $\{a_n\}$ and $\{b_n\}$ have the φ-limits a and b. Then the sum, the difference and the product also have φ-limits, namely $a + b$, $a - b$ and ab respectively. Also, if φ-$\lim_{n\to\infty} a_n = a$, then $\lim_{n\to\infty} \varphi(a_n) = \varphi(a)$.

Definition 6.3. By a *null sequence* we mean a sequence having the φ-limit 0. We denote by $\overline{\{0\}}$ the class of all null sequences.

Example 1. If $\varphi(a) = |a|$, then $\{a_n = 1/n\}$ is a null sequence.

Example 2. If $\varphi(a) = |a|_p$, then $\{a_n = p^n\}$ is a null sequence.

It is easy to prove that the sum of two null sequences is a null sequence; so is the product of a null sequence and a fundamental sequence.

We now define the quotient of two sequences. Let $\{b_n\}$ be a non-null sequence. Then the quotient $\{a_n\}/\{b_n\}$ is defined to be the sequence $\{a_n b_n^{-1}\}$. Observe that since $\{b_n\}$ is not a null sequence we may discard those terms which are zero without affecting the discussion.

If $\{a_n\}$ is a fundamental sequence but not a null sequence, then there exists a positive rational number c and a positive natural number N such that $\varphi(a_n) > c > 0$ whenever $n > N$. It is not difficult to deduce from this that the quotient $\{a_n\}/\{b_n\}$ ($\{b_n\}$ not null) of the fundamental sequences is a fundamental sequence.

Definition 6.4. Let $\{a_n\}$ and $\{b_n\}$ be two fundamental sequences whose difference $\{a_n - b_n\}$ is a null sequence. Then we say that $\{a_n\}$ and $\{b_n\}$ are *congruent* and we write $\{a_n\} \equiv \{b_n\} (\bmod \overline{\{0\}})$.

Being congruent is an equivalence relation and the set of fundamental sequences is now partitioned into equivalence classes. From each class we may select a fundamental sequence $\{a_n\}$ to represent the class $\overline{\{a_n\}}$.

We can now define the sum, the difference, the product and the quotient of two classes $\overline{\{a_n\}}$ and $\overline{\{b_n\}}$. We let $\{a_n\}$ and $\{b_n\}$ be the representatives respectively and we define $\overline{\{a_n\}} \pm \overline{\{b_n\}} = \overline{\{a_n \pm b_n\}}$, $\overline{\{a_n\}} \cdot \overline{\{b_n\}} = \overline{\{a_n b_n\}}$, and when $\overline{\{b_n\}} \neq \overline{\{0\}}$ we define $\overline{\{a_n\}} \cdot \overline{\{b_n\}}^{-1} = \overline{\{a_n b_n^{-1}\}}$. It is easy to verify that the definitions are independent of the choices of the representatives.

The aggregate of these classes is called the *φ-extension* of the rationals, and each class is called a number in the φ-extension. When $\varphi(a) = |a|$, the φ-extension coincides with the set of real numbers. When $\varphi(a) = |a|_p$ we call the φ-extension the set of p-adic numbers. This then gives a rigorous definition of a p-adic number. Our next task is to give a concrete representation of a p-adic number.

The aggregate of classes contains the class $\overline{\{a\}}$ (a rational), and each fundamental sequence in the class is φ-convergent to the same rational number a, that is, a is their φ-limit. We shall write $\overline{\{a\}} = a$, since now there is a one-to-one correspondence between these classes and the set of rational numbers. Since there are fundamental sequences which are not φ-convergent to any rational number we see that the φ-extension of the rationals is an aggregate which is larger than the set of all rational numbers.

In general we let $\overline{\{a_n\}}$ be the number to which the fundamental sequence in it φ-converges. That is, we define

$$\varphi\text{-}\lim_{n\to\infty} a_n = \overline{\{a_n\}}.$$

We should add that, when $\{a_n\}$ and $\{a'_n\}$ belong to the same class, $\varphi\text{-}\lim_{n\to\infty} a_n = \varphi\text{-}\lim_{n\to\infty} a'_n$.

In the above discussion the valuation is defined only in the field of the rationals. We shall now extend its definition to the φ-extension of the rationals.

Definition 6.5. $\varphi(\overline{\{a_n\}}) = \lim_{n\to\infty} \varphi(a_n)$.

We should point out that in this definition, $\varphi(\overline{\{a_n\}})$ is independent of the choice of $\{a_n\}$. That is, if $\{a_n\} \equiv \{a_n'\} \pmod{\overline{\{0\}}}$, then $\lim_{n\to\infty} \varphi(a_n) = \lim_{n\to\infty} \varphi(a_n')$. This can easily be proved from $\varphi(a_n) - \varphi(a_n') \leqslant \varphi(a_n - a_n')$.

It is convenient to use Greek letters $\alpha, \beta, \gamma, \ldots$ to denote the classes. We have the following three properties for $\varphi(\alpha)$: 1) $\varphi(\alpha) \geqslant 0$ with equality if and only if $\alpha = \overline{\{0\}}$; 2) $\varphi(\alpha\beta) = \varphi(\alpha)\varphi(\beta)$; 3) $\varphi(\alpha + \beta) \leqslant \varphi(\alpha) + \varphi(\beta)$.

Exercise 1. Show that equivalent valuations give the same extension of the rationals.

Exercise 2. Let φ be a non-Archimedian valuation. Prove that $\{a_n\}$ is convergent if and only if $\lim_{n\to\infty} \varphi(a_{n+1} - a_n) = 0$.

15.7 The Completeness of the Extension

In the previous section we constructed the φ-extension of the rationals from the fundamental sequences of rational numbers and we saw that the φ-extension is larger than the set of rationals. We then extended the domain of definition of φ from the rationals to that of the φ-extension, giving a definition of $\varphi(\alpha)$ where α is a class of fundamental sequences. We now ask the following: If we repeat the process to obtain another φ-extension from the φ-extension already obtained, do we have a still larger aggregate than the first φ-extension? If the answer is no, then we say that the extension is *complete*. In order to discuss this we have to consider sequences $\{\alpha_l\}$ of classes, and define the terms fundamental sequences, classes, φ-limit, null sequences etc. all over again. It turns out that the φ-extension is complete, but we shall omit the proof.

Theorem 7.1. *The φ-extension of the rationals is complete in the sense that every fundamental* (*or φ-convergent*) *sequence $\{\alpha_l\}$ has a φ-limit.* □

15.8 The Representation of p-adic Numbers

In this section we let $\varphi(a) = |a|_p$, and we examine the representation of p-adic numbers.

1) We first consider the p-adic representation of a rational number

$$\frac{a}{b}, \qquad (a, b) = 1, \qquad p \nmid b.$$

For this we examine the solution of the congruence

$$bx \equiv a \pmod{p^l}, \qquad 0 \leqslant x < p^l.$$

Denoting the solution by x_l we have

$$\left|\frac{a}{b} - x_l\right|_p \leqslant p^{-l}.$$

Therefore

$$\varphi\text{-}\lim_{l \to \infty} \left(\frac{a}{b} - x_l\right) = 0,$$

or

$$\frac{a}{b} = \varphi\text{-}\lim_{l \to \infty} x_l.$$

We already defined, in §1,

$$x_l = a_0 + a_1 p + \cdots + a_{l-1} p^{l-1}, \qquad 0 \leqslant a_v < p.$$

From

$$\begin{aligned}\varphi(x_l - x_{l'}) &= \varphi(a_l p^l + \cdots + a_{l'-1} p^{l'-1}) \\ &\leqslant p^{-l}\varphi(a_l) + \cdots + p^{-(l'-1)}\varphi(a_{l'-1}) \\ &\leqslant p^{-l} + \cdots + p^{-(l'-1)} = \frac{\dfrac{1}{p^l} - \dfrac{1}{p^{l'}}}{1 - \dfrac{1}{p}} < \varepsilon \quad (l' > l > L(\varepsilon)),\end{aligned}$$

we know that $\{x_l\}$ is φ-convergent. This means that the limit

$$a_0 + a_1 p + \cdots + a_{l-1} p^{l-1} + \cdots, \qquad 0 \leqslant a_v < p$$

in the φ-extension is the p-adic representation of the rational number a/b $(p \nmid b)$.

2) We next deal with the p-adic representation for the rational number

$$\frac{a}{b}, \qquad (a, b) = 1, \qquad p^m \| b, \qquad m \geqslant 0.$$

The general p-adic representation of a rational number is the power series

$$p^{-m}(a_0 + a_1 p + \cdots + a_l p^l + \cdots), \qquad 0 \leqslant a_v < p, \qquad m \geqslant 0. \tag{1}$$

If, for this power series (1), we have

$$a_{l+v} = a_{l+v+t} = a_{l+v+2t} = \cdots = a_{l+v+nt} = \cdots \quad (v = 1, 2, \ldots, t),$$

where l and t are fixed integers, $t \geqslant 1$, then we say that (1) is periodic, and in this case we may rewrite it as

$$p^{-m}((a_0 + a_1 p + \cdots + a_l p^l) + p^{l+1}(a_{l+1} + a_{l+2}p + \cdots + a_{l+t}p^{t-1}) + p^{l+t+1}(a_{l+1} + a_{l+2}p + \cdots + a_{l+t}p^{t-1}) + \cdots),$$

or simply

$$p^{-m}(A + p^{l+1}B + p^{l+t+1}B + p^{l+2t+1}B + \cdots),$$

where

$$A = a_0 + a_1 p + \cdots + a_l p^l, \qquad B = a_{l+1} + a_{l+2}p + \cdots + a_{l+t}p^{t-1}.$$

Theorem 8.1. *The p-adic representation of a rational number is a periodic power series in p; conversely a periodic power series in p is a rational number.*

Proof. 1) If

$$\alpha = p^{-m}\{A + p^{l+1}B + p^{l+t+1}B + p^{l+2t+1}B + \cdots),$$

where

$$A = a_0 + a_1 p + \cdots + a_l p^l, \qquad B = a_{l+1} + a_{l+2}p + \cdots + a_{l+t}p^{t-1},$$

then

$$\begin{aligned} \alpha p^m - A &= p^{l+1}B + p^{l+t+1}B + p^{l+2t+1}B + \cdots \\ &= p^{l+1}B(1 + p^t + p^{2t} + \cdots). \end{aligned}$$

Now

$$1 + p^t + p^{2t} + \cdots + p^{kt} = \frac{1 - p^{(k+1)t}}{1 - p^t},$$

$$\left| \frac{1}{1 - p^t} - \frac{1 - p^{(k+1)t}}{1 - p^t} \right|_p = p^{-(k+1)t} < \varepsilon \qquad (k \geqslant k_0),$$

so that

$$1 + p^t + p^{2t} + \cdots + p^{kt} + \cdots = \frac{1}{1 - p^t}.$$

Therefore

$$\alpha p^m - A = p^{l+1}B\frac{1}{1 - p^t},$$

or

$$\alpha = p^{-m}A + p^{l+1-m}B\frac{1}{1 - p^t},$$

so that α is a rational number.

2) We first consider the rational number

$$\alpha = \frac{r}{s}, \qquad |\alpha| < 1, \qquad (r, s) = 1, \qquad s > 0, \qquad r < 0, \qquad p \nmid s. \tag{2}$$

Let the index of $p \pmod s$ be t, that is t is the least positive integer satisfying $p^t \equiv 1 \pmod s$. Let $1 - p^t = ms$, $m < 0$, so that

$$\alpha = \frac{r}{s} = \frac{mr}{1 - p^t}.$$

Since $|\alpha| < 1$, the number mr has the representation

$$mr = b_0 + b_1 p + \cdots + b_{t-1}p^{t-1}, \qquad 0 \leqslant b_i < p.$$

Therefore

$$\begin{aligned} \alpha &= (b_0 + b_1 p + \cdots + b_{t-1}p^{t-1})(1 + p^t + p^{2t} + \cdots) \\ &= (b_0 + b_1 p + \cdots + b_{t-1}p^{t-1}) + p^t(b_0 + b_1 p + \cdots + b_{t-1}p^{t-1}) + \cdots, \end{aligned}$$

which shows that α has a periodic power series in p as representation.

Next, let α be any positive rational, say $\alpha = a/b$, $(a, b) = 1$, $p^m \| b$. Then α has the representation

$$p^m \alpha = a_0 + a_1 p + \cdots + a_v p^v + \frac{r}{s}, \qquad 0 \leqslant a_i < p,$$

where r/s is either 0 or a number satisfying the conditions in (2). Therefore α again has a periodic power series in p as its representation.

Finally if $-\alpha$ is a negative rational number, then we first obtain the representation of the positive number α, and then use the representation

$$0 = p + (p - 1)p + (p - 1)p^2 + \cdots$$

to obtain a representation for $-\alpha$ by subtraction. Therefore $-\alpha$ also has a periodic power series in p as representation. □

Having obtained the representations for the rationals we now consider the general situation. We first show that the power series

$$\alpha = p^{-m}(a_0 + a_1 p + a_2 p^2 + \cdots), \qquad 0 \leqslant a_v < p, \quad m \geqslant 0, \tag{3}$$

does represent a p-adic number. We let

$$x_l = p^{-m}(a_0 + a_1 p + a_2 p^2 + \cdots + a_{l-1}p^{l-1}).$$

Since $\{x_l\}$ is φ-convergent, it follows that its φ-limit

$$\alpha = p^{-m}(a_0 + a_1 p + a_2 p^2 + \cdots), \qquad 0 \leqslant a_v < p, \quad m \geqslant 0$$

does represent a p-adic number.

Therefore a power series in p of the form (3) represents a p-adic number. We now ask: How is a given p-adic number represented? From the previous section we know that any p-adic number is a limit of a φ-convergent sequence $\{a_l\}$ in the φ-extension of the rationals. But any rational number a_l can be represented as

$$a_l = p^{-m_l}(a_0^{(l)} + a_1^{(l)}p + \cdots), \qquad 0 \leqslant a_v^{(l)} < p.$$

Our problem is thus solved if we can show that the limit of $\{a_l\}$, in the φ-extension of the rationals, also has this representation.

Corresponding to each positive integer t, there exists a positive integer L $(= L(t))$ such that

$$|a_l - a_{l'}|_p < \frac{1}{p^t}$$

whenever $l, l' > L$. This shows that, when $l > L$, the first $t + k$ $(k \geqslant 0)$ terms of the power series in p representing $a_l, a_{l+1}, a_{l+2}, \ldots$ must be equal. Since t can be arbitrarily large the required result follows.

We have proved that all the power series in p of the form (3), finite or infinite, together give the whole set of p-adic numbers.

15.9 Application

Although the notion of a p-adic number is introduced as such only in this chapter, it has appeared several times already in this book. An example of this was pointed out in the beginning of this chapter. The generalization of this example is known as Hensel's Lemma.

Theorem 9.1 (Hensel). *Let $f(x)$ be a polynomial with integer coefficients, and $f(x) \equiv g_0(x)h_0(x) \pmod{p}$, where $g_0(x)$ and $h_0(x)$ are coprime polynomials. Then, among the p-adic numbers, there are two polynomials $g(x)$, $h(x)$ such that $g(x) \equiv g_0(x)$, $h(x) \equiv h_0(x) \pmod{p}$, and $f(x) = g(x)h(x)$.*

Proof. Let $g_l(x)$, $h_l(x)$ be two polynomials satisfying

$$g_l(x) \equiv g_0(x), \qquad h_l(x) \equiv h_0(x) \pmod{p^l}$$

and

$$f(x) \equiv g_l(x)h_l(x) \pmod{p^l}.$$

Clearly g_l and h_l are coprime $(\bmod\, p)$. Let

$$g_{l+1}(x) = g_l(x) + p^l\varphi(x)$$

and

$$h_{l+1}(x) = h_l(x) + p^l\psi(x),$$

so that we have

$$g_{l+1}(x)h_{l+1}(x) \equiv g_l(x)h_l(x) + p^l(\varphi(x)h_l(x) + \psi(x)g_l(x)) \pmod{p^{l+1}}.$$

Let

$$\frac{f(x) - g_l(x)h_l(x)}{p^l} \equiv t(x) \pmod{p}.$$

Since $g_l(x)$ and $h_l(x)$ are coprime (mod p), there are two polynomials $\varphi(x)$ and $\psi(x)$ such that $t(x) \equiv \varphi(x)h_l(x) + \psi(x)g_l(x) \pmod{p}$. Therefore

$$\begin{aligned} f(x) - g_{l+1}(x)h_{l+1}(x) &\equiv f(x) - g_l(x)h_l(x) - p^l(\varphi(x)h_l(x) + \psi(x)g_l(x)) \\ &\equiv p^l(t(x) - \varphi(x)h_l(x) - \psi(x)g_l(x)) \equiv 0 \pmod{p^{l+1}}. \end{aligned}$$

Since the degree of $t(x)$ does not exceed the degree of $g_l(x)h_l(x)$ we may assume that the degrees of $\varphi(x)$ and $\psi(x)$ do not exceed the degrees of $g_l(x)$ and $h_l(x)$ respectively. The coefficients of $g_l(x)$ and $h_l(x)$ are φ-convergent, and so they converge to $g(x)$ and $h(x)$ respectively. The theorem is proved. □

Note: Lemma 7.10.1 can be given an interpretation in p-adic numbers.

Chapter 16. Introduction to Algebraic Number Theory

16.1 Algebraic Numbers

Definition 1.1. By an *algebraic number* we mean a number ϑ which is a root of the algebraic equation

$$f(x) = a_n x^n + a_{n-1} x^{n-1} + \cdots + a_0 = 0, \tag{1}$$

where the coefficients a_r are rational numbers.

Examples of algebraic numbers are $\sqrt{2}$, $i = \sqrt{-1}$ and the rational numbers themselves. By clearing the denominators of all the fractions a_r in equation (1) we obtain an algebraic equation with integer coefficients. From now on we shall call an ordinary integer a *rational integer* to distinguish it from an *algebraic integer*, which we shall define later. We see therefore that algebraic numbers may also be defined as the roots of algebraic equations with rational integer coefficients.

If the equation (1) is irreducible and $a_n \neq 0$, then we call $n = \partial^0 f$ the degree of the algebraic number ϑ. For example, rational numbers have degree 1, and the number i has degree 2.

Let the equation (1) be irreducible and let $\vartheta^{(1)}, \vartheta^{(2)}, \ldots, \vartheta^{(n)}$ be all its roots. From Theorem 4.2.2 we know that $\vartheta^{(j)}$ are distinct, and if $\vartheta^{(j)}$ satisfies a rational coefficient equation $g(x) = 0$, then so do the remaining $n - 1$ numbers. We see therefore that the degree of an algebraic number is uniquely determined.

Theorem 1.1. *The sum, the difference, the product and the quotient (not dividing by zero) of two algebraic numbers are algebraic.* □

With the aid of the symmetric polynomial theorem, the proof of this theorem is only a simple exercise, and we shall omit many of the proofs in this chapter.

Definition 1.2. If the irreducible algebraic equation defining ϑ has rational integer coefficients and leading coefficient 1, then we call ϑ an *algebraic integer*.

Examples of algebraic integers are $\sqrt{2}$, i, $(1 + \sqrt{5})/2$, and the rational integers themselves.

Theorem 1.2. *Any algebraic integer which is rational must be a rational integer.* □

Theorem 1.3. *The sum, the difference and the product of two algebraic integers are algebraic integers.* □

Theorem 1.4. *Let ϑ be an algebraic number. Then there exists a natural number q such that $q\vartheta$ is an algebraic integer.* □

Definition 1.3. If ϑ and ϑ^{-1} are both algebraic integers, then we call ϑ a *unit*.

Examples of units are i and $3 - 2\sqrt{2}$.

Theorem 1.5. *A necessary and sufficient condition for ϑ to be a unit is that ϑ satisfies an algebraic equation with rational integer coefficients, and with leading coefficient* 1 *and last coefficient* ± 1. □

16.2 Algebraic Number Fields

Definition 2.1. Let F be a set of complex numbers with at least two distinct members. Suppose that, given any two members in F, their sum, difference, product and quotient (not dividing by 0) are also members of F. Then we call F a *number field*, or simply a *field*.

An example of a field is the set of rational numbers which we shall, from now on, denote by R. It is clear that every number field must contain the rational field R.

Theorem 2.1. *Let ϑ be an algebraic number of degree n. Then the set of numbers of the form*

$$a_0 + a_1\vartheta + a_2\vartheta^2 + \cdots + a_{n-1}\vartheta^{n-1}, \tag{1}$$

where a_k are rational numbers, forms a field. Moreover numbers represented by (1) *are all distinct.* □

Definition 2.2. The field in Theorem 2.1 is called the *single extension* of R by ϑ, and we shall denote it by $R(\vartheta)$.

For example, $R(i)$ is the field of numbers of the form $a + ib$ where a and b are rational numbers.

Theorem 2.2. *If $\vartheta \neq 0$, then $R(\vartheta)$ is largest set of numbers obtained from ϑ by means of addition, subtraction, multiplication and division (except by* 0*).* □

Definition 2.3. Let $\vartheta_1, \ldots, \vartheta_l$ be algebraic numbers. The field obtained from addition, subtraction, multiplication and division (except by 0) of these numbers is called a *finite extension* of R and is denoted by $R(\vartheta_1, \ldots, \vartheta_l)$.

Theorem 2.3. *Every finite extension of R is a single extension. That is, given any finite extension* $R(\vartheta_1, \ldots, \vartheta_l)$, *there exists an algebraic number* ϑ *such that* $R(\vartheta) = R(\vartheta_1, \ldots, \vartheta_l)$. □

From this theorem we need only consider single extensions $R(\vartheta)$, which we now call *algebraic number fields*. We also call the degree of ϑ the *degree* of the field $R(\vartheta)$. For example, $R(i)$ is a quadratic field, and R is the only field with degree 1.

Theorem 2.4. *Let D run over all the rational integers not equal to 1 with no square divisors. Then* $R(\sqrt{D})$ *runs over all the quadratic fields.* □

16.3 Basis

In this section $R(\vartheta)$ denotes an algebraic number field of degree n. We set $\vartheta = \vartheta^{(1)}$, and let $\vartheta^{(2)}, \ldots, \vartheta^{(n)}$ denote the remaining $n - 1$ roots of the irreducible equation defining ϑ.

From the previous section we see that each number $\alpha \in R(\vartheta)$ is representable as

$$\alpha = a(\vartheta) = a_0 + a_1\vartheta + \cdots + a_{n-1}\vartheta^{n-1}$$

where a_j are rational numbers.

Definition 3.1. Let $\alpha^{(1)} = \alpha$. We put $\alpha^{(k)} = a(\vartheta^{(k)})$, $k = 2, 3, \ldots, n$ and we call them the *conjugates* of α. We also call the numbers

$$S(\alpha) = \alpha^{(1)} + \cdots + \alpha^{(n)} = a(\vartheta^{(1)}) + \cdots + a(\vartheta^{(n)}),$$
$$N(\alpha) = \alpha^{(1)} \cdots \alpha^{(n)} = a(\vartheta^{(1)}) \cdots a(\vartheta^{(n)}),$$

the *trace* and the *norm* of α respectively.

It is easy to see that $S(\alpha + \beta) = S(\alpha) + S(\beta)$ and $N(\alpha\beta) = N(\alpha)N(\beta)$. Also, from the symmetric polynomial theorem, we see that $S(\alpha)$ and $N(\alpha)$ are rational numbers, and if α is rational then $S(\alpha) = n\alpha$ and $N(\alpha) = \alpha^n$. Next, if α is an algebraic integer, then so are $\alpha^{(i)}$, and hence so are $S(\alpha)$ and $N(\alpha)$; but $S(\alpha)$ and $N(\alpha)$ are known to be rational so that they must be rational integers.

If α is a unit, then from $N(\alpha)N(\alpha^{-1}) = N(\alpha\alpha^{-1}) = N(1) = 1$ and the fact that $N(\alpha)$, $N(\alpha^{-1})$ are rational integers we deduce that $N(\alpha) = \pm 1$. Conversely, if α is an algebraic integer and $N(\alpha) = \pm 1$, then $\alpha^{-1} = \pm \alpha^{(2)} \cdots \alpha^{(n)}$ is also an algebraic integer and so α must be a unit. Therefore a necessary and sufficient condition for an algebraic integer α to be a unit is that $N(\alpha) = \pm 1$.

Theorem 3.1. *Let* $\alpha \in R(\vartheta)$ *and let the irreducible equation satisfied by* α *be* $h(x) = 0$, $\partial^0 h = l$. *Also, let* $g(x) = \prod_{\nu=1}^{n} (x - \alpha^{(\nu)})$. *Then* $g(x)$ *is a polynomial with rational coefficients, and* $g(x) = c(h(x))^{n/l}$, *where* $l \mid n$ *and* c *is a rational number.*

Proof. That $g(x)$ is a polynomial with rational coefficients follows at once from the symmetric polynomial theorem. Let $\alpha = a(\vartheta)$. Then from $h(\alpha) = 0$ we have $h(\alpha^{(\nu)}) = h(a(\vartheta^{(\nu)}) = 0$, so that every root of $g(x) = 0$ is also a root of $h(x) = 0$. Since $h(x)$ is an irreducible polynomial we must have $h(x)|g(x)$. Let $g(x) = h(x)g_1(x)$. If $g_1(x)$ is a constant, then the required result is proved; otherwise $g_1(x)$ has zeros and these must also be zeros of $h(x)$, so that $h(x)|g_1(x)$. Let $g_1(x) = h(x)g_2(x)$. We can repeat the argument, and since the degree of $g(x)$ is finite we finally obtain $g(x) = c(h(x))^{n/l}$. □

From this theorem we see that if α is an algebraic number of degree l, then there are l distinct numbers among $\alpha^{(1)}, \ldots, \alpha^{(n)}$ and each of them occurs n/l times.

Definition 3.2. Suppose that there exists a set of numbers $\alpha_1, \ldots, \alpha_m$ in $R(\vartheta)$ such that any number in $R(\vartheta)$ is uniquely representable as $a_1\alpha_1 + \cdots + a_m\alpha_m$ where a_j $(1 \leqslant j \leqslant m)$ are rational numbers. Then we call $\alpha_1, \ldots, \alpha_m$ a *basis* for $R(\vartheta)$.

It is easy to see that no one of $\alpha_1, \ldots, \alpha_m$ is expressible as a linear combination of the other $m - 1$ numbers with rational coefficients. From Theorem 2.1 we know that $1, \vartheta, \ldots, \vartheta^{n-1}$ forms a basis for $R(\vartheta)$, so that basis certainly exists. Following the proof of Theorem 14.9.2 the reader can easily prove

Theorem 3.2. *Every basis for $R(\vartheta)$ has precisely n elements.* □

Let $\alpha_1, \ldots, \alpha_n$ and $\beta_1, \ldots, \beta_n$ be two bases for $R(\vartheta)$. Then, from the definition of a basis, there are rational numbers a_{jk} $(1 \leqslant j, k \leqslant n)$ such that

$$\alpha_j = \sum_{k=1}^{n} a_{jk}\beta_k \qquad (1 \leqslant j \leqslant n), \qquad |a_{jk}| = \begin{vmatrix} a_{11} & \cdots & a_{1n} \\ \cdots & \cdots & \cdots \\ a_{n1} & \cdots & a_{nn} \end{vmatrix} \neq 0.$$

Definition 3.3. Let $\alpha_1, \ldots, \alpha_n \in R(\vartheta)$. By the *discriminant* of $\alpha_1, \ldots, \alpha_n$ we mean the number

$$\Delta(\alpha_1, \ldots, \alpha_n) = \begin{vmatrix} \alpha_1^{(1)} & \cdots & \alpha_n^{(1)} \\ \cdots & \cdots & \cdots \\ \alpha_1^{(n)} & \cdots & \alpha_n^{(n)} \end{vmatrix}^2.$$

Theorem 3.3. *The discriminant $\Delta(\alpha_1, \ldots, \alpha_n)$ possesses the following properties:*

1) *$\Delta(\alpha_1, \ldots, \alpha_n)$ is a rational number; and if $\alpha_1, \ldots, \alpha_n$ are algebraic integers, then $\Delta(\alpha_1, \ldots, \alpha_n)$ is a rational integer.*

2) *Let $\alpha_1, \ldots, \alpha_n$ and $\beta_1, \ldots, \beta_n$ be two bases for $R(\vartheta)$, and $a_j = \sum_{k=1}^{n} a_{jk}\beta_k$ $(1 \leqslant j \leqslant n)$. Then*

$$\Delta(\alpha_1, \ldots, \alpha_n) = |a_{jk}|^2 \Delta(\beta_1, \ldots, \beta_n).$$

3) *A necessary and sufficient condition for $\alpha_1, \ldots, \alpha_n$ to be a basis for $R(\vartheta)$ is that $\Delta(\alpha_1, \ldots, \alpha_n) \neq 0$.* □

Theorem 3.4. *Suppose that, among the numbers $\vartheta^{(1)}, \ldots, \vartheta^{(n)}$, r_1 of them are real, and r_2 pairs of them are complex conjugates $(r_1 + 2r_2 = n)$. Then, for any basis $\alpha_1, \ldots, \alpha_n$*

of $R(\vartheta)$, we have

$$(-1)^{r_2}\Delta(\alpha_1, \ldots, \alpha_n) > 0.$$

Proof. From Theorem 3.3 we need only examine the case $\alpha_1 = 1$, $\alpha_2 = \vartheta, \ldots,$ $\alpha_n = \vartheta^{n-1}$. Now

$$\Delta(1, \vartheta, \ldots, \vartheta^{n-1}) = \left(\prod_{1 \leqslant j < k \leqslant n} (\vartheta^{(j)} - \vartheta^{(k)})\right)^2.$$

Let us denote by $\bar{\vartheta}$ the complex conjugate of ϑ. When $\vartheta^{(k)} \neq \bar{\vartheta}^{(j)}$ we have $((\vartheta^{(j)} - \vartheta^{(k)})(\bar{\vartheta}^{(j)} - \bar{\vartheta}^{(k)}))^2 > 0$, and $(\vartheta^{(j)} - \bar{\vartheta}^{(j)})^2 < 0$. Therefore

$$(-1)^{r_2}\Delta(1, \vartheta, \ldots, \vartheta^{n-1}) > 0.$$

16.4 Integral Basis

In the remaining part of this chapter we shall use the word integer to mean an algebraic integer.

Definition 4.1. Let $\omega_1, \ldots, \omega_m$ be m integers in $R(\vartheta)$. If every integer in $R(\vartheta)$ can be expressed uniquely as $a_1\omega_1 + \cdots + a_m\omega_m$, where $a_1, \ldots, a_m$ are rational integers, then we call $\omega_1, \ldots, \omega_m$ an *integral basis* for $R(\vartheta)$.

Theorem 4.1. *Integral basis exists. More specifically let $\omega_1, \ldots, \omega_n$ be a basis where ω_j $(1 \leqslant j \leqslant n)$ are integers such that $|\Delta(\omega_1, \ldots, \omega_n)|$ is least. Then $\omega_1, \ldots, \omega_n$ is an integral basis.*

Proof. We can choose a natural number q so that $q\vartheta$ is an integer, and now $1, q\vartheta,$ $(q\vartheta)^2, \ldots, (q\vartheta)^{n-1}$ are integers which form a basis for $R(\vartheta)$. Therefore a basis $\alpha_1, \ldots, \alpha_n$ consisting of integers certainly exists.

We shall now prove the set $\omega_1, \ldots, \omega_n$ of integers forming a basis which makes $|\Delta(\alpha_1, \ldots, \alpha_n)|$ least is an integral basis. Suppose the contrary. Then there exists an integer $\omega = a_1\omega_1 + \cdots + a_n\omega_n$, where some a_i is not a rational integer. We may assume without loss that a_1 is not a rational integer, say $a_1 = g + t$ where g is a rational integer and $0 < t < 1$. Then $\omega_1' = \omega - g\omega_1 = t\omega_1 + a_2\omega_2 + \cdots + a_n\omega_n$ is also an integer, and $\omega_1', \omega_2, \ldots, \omega_n$ still forms a basis for $R(\vartheta)$. But

$$|\Delta(\omega_1', \omega_2, \ldots, \omega_n)| = t^2|\Delta(\omega_1, \ldots, \omega_n)| < |\Delta(\omega_1, \ldots, \omega_n)|,$$

contradicting the minimal property of $|\Delta(\omega_1, \ldots, \omega_n)|$. The theorem is proved. □

From this theorem we see that an integral basis is a basis, so that each integral basis consists of n elements.

Theorem 4.2. *All integral basis have the same discriminant. That is, if $\omega_1, \ldots, \omega_n$ and $\omega_1', \ldots, \omega_n'$ are two integral bases, then $\Delta(\omega_1, \ldots, \omega_n) = \Delta(\omega_1', \ldots, \omega_n')$.* □

Definition 4.2. By the *discriminant of the field* $R(\vartheta)$ we mean the discriminant of its integral basis. We shall denote the discriminant of $R(\vartheta)$ by $\Delta(R(\vartheta))$ or simply Δ.

Theorem 4.3 (Stickelberger). *The discriminant of a field satisfies* $\Delta \equiv 0$ *or* $1 \pmod 4$.

Proof. Let $i_1, \ldots, i_n$ be a permutation of $1, 2, \ldots, n$, and let $\delta_{i_1, \ldots, i_n}$ be 1 or -1 depending on whether $i_1, \ldots, i_n$ is an even or odd permutation. Then, from the expansion of a determinant, we have

$$\begin{vmatrix} \omega_1^{(1)} & \cdots & \omega_1^{(n)} \\ \cdots & \cdots & \cdots \\ \omega_n^{(1)} & \cdots & \omega_n^{(n)} \end{vmatrix} = \sum_{(i_1, \ldots, i_n)} \delta_{i_1, \ldots, i_n} \omega_1^{(i_1)} \cdots \omega_n^{(i_n)}$$

$$= \sum_{(i_1, \ldots, i_n)} \omega_1^{(i_1)} \cdots \omega_n^{(i_n)} + 2\eta = a + 2\eta,$$

where η is an algebraic integer, and $a = \sum_{(i_1, \ldots, i_n)} \omega_1^{(i_1)} \cdots \omega_n^{(i_n)}$ is a symmetric function of $\vartheta^{(1)}, \ldots, \vartheta^{(n)}$, so that a is rational and hence a rational integer. Therefore

$$\Delta = (a + 2\eta)^2 = a^2 + 4\eta(\eta + a).$$

Since the integer $\eta(\eta + a) = (\Delta - a^2)/4$ is rational, it is a rational integer. Therefore $\Delta \equiv a^2 \equiv 0$ or $1 \pmod 4$. $\square$

We shall now examine the quadratic field $R(\sqrt{D})$ where D is a square-free rational integer. Each number in $R(\sqrt{D})$ is representable as $\alpha = (a + b\sqrt{D})/2$ where a, b are rational numbers. The trace and the norm of α are given by

$$S(\alpha) = a, \qquad N(\alpha) = \frac{a^2 - b^2 D}{4}.$$

Theorem 4.4. *In the quadratic field* $R(\sqrt{D})$, *a necessary and sufficient condition for* α *to be an integer is that* a, b *are both rational integers satisfying*

$$a \equiv b \pmod 2, \qquad \text{when} \quad D \equiv 1 \pmod 4;$$

$$a \equiv b \equiv 0 \pmod 2, \qquad \text{when} \quad D \equiv 2, 3 \pmod 4. \tag{1}$$

Proof. Since, in a quadratic field, α is an integer if and only if $S(\alpha)$, $N(\alpha)$ are rational integers, the sufficiency of the condition (1) follows at once.

Conversely, if α is an integer, then a and $(a^2 - b^2 D)/4$ are rational integers, so that

$$b^2 D = a^2 - 4\left(\frac{a^2 - b^2 D}{4}\right)$$

is also a rational integer. Since D is square-free, the number b must be rational. The necessity of the condition (1) now follows from $a^2 - b^2 D \equiv 0$. $\square$

When $D \equiv 1 \pmod 4$, $(1+\sqrt{D})/2$ is an integer in $R(\sqrt{D})$. From

$$\frac{a+b\sqrt{D}}{2} = \frac{a-b}{2} + b\frac{1+\sqrt{D}}{2}$$

and

$$\begin{vmatrix} 1 & 1 \\ \sqrt{D} & -\sqrt{D} \end{vmatrix}^2 = 4D, \qquad \begin{vmatrix} 1 & 1 \\ \dfrac{1+\sqrt{D}}{2} & \dfrac{1-\sqrt{D}}{2} \end{vmatrix}^2 = D,$$

we have the following:

Theorem 4.5. *Let D be a square-free rational integer, and let*

$$\Delta = \begin{cases} D \\ 4D \end{cases}, \qquad \omega = \begin{cases} \dfrac{1+\sqrt{D}}{2}, & \text{when} \quad D \equiv 1 \pmod 4, \\ \sqrt{D}, & \text{when} \quad D \equiv 2, 3 \pmod 4. \end{cases}$$

Then Δ is the discriminant of $R(\sqrt{D})$, and $1, \omega$ is an integral basis. The numbers 1, $(\Delta + \sqrt{\Delta})/2$ also form an integral basis. □

From this theorem we see that, in a quadratic field, we may choose an integer ω such that $1, \omega$ form an integral basis. This is not true in general; that is, if $R(\vartheta)$ is a field of degree $n \geqslant 3$, we may not always find an integer ω such that $1, \omega, \ldots, \omega^{n-1}$ is an integral basis for $R(\vartheta)$.

Example. Let α be a zero of $f(x) = x^3 - x^2 - 2x - 8$. We shall prove that no integer ω, with the property that $1, \omega, \omega^2$ is an integral basis for $R(\alpha)$, exists.

Since $\pm 1, \pm 2, \pm 4, \pm 8$ are not zeros of $f(x)$, we know that $f(x)$ is irreducible so that $R(\alpha)$ is definitely a cubic field. It is easy to show that $\Delta(1, \alpha, \alpha^2) = -4 \times 503$.

Since $\beta = 4/\alpha$ is a zero of $g(y) = y^3 + y^2 + 2y - 8$, it follows that β is an integer in $R(\alpha)$. Let us denote by α' and α'' the two remaining zeros of $f(x)$. Then

$$\Delta(1, \alpha, \beta) = \begin{vmatrix} 1 & \alpha & 4/\alpha \\ 1 & \alpha' & 4/\alpha' \\ 1 & \alpha'' & 4/\alpha'' \end{vmatrix}^2 = \frac{4^2}{(N(\alpha))^2} \begin{vmatrix} 1 & \alpha & \alpha^2 \\ 1 & \alpha' & \alpha'^2 \\ 1 & \alpha'' & \alpha''^2 \end{vmatrix}^2$$

$$= \frac{4^2}{(N(\alpha))^2} \Delta(1, \alpha, \alpha^2) = -503.$$

Since $\Delta(1, \alpha, \beta) \neq 0$, the numbers $1, \alpha, \beta$ form a basis. Indeed $1, \alpha, \beta$ must be an integral basis for $R(\alpha)$, since otherwise the discriminant Δ of the field must satisfy $|\Delta| < 503$, and from Theorem 3.3 there exists a natural number $a \neq 1$ such that $-503 = a^2 \Delta$, which is impossible because 503 is a prime number.

Now let ω be any integer in $R(\alpha)$. Then there are rational integers a, b, c such that $\omega = a + b\alpha + c\beta$. Now

$$\alpha^2 = \alpha + 2 + \frac{8}{\alpha} = 2 + \alpha + 2\beta,$$

$$\beta^2 = -\beta - 2 + \frac{8}{\beta} = -2 + 2\alpha - \beta,$$

so that

$$\begin{aligned}\omega^2 &= a^2 + b^2(2 + \alpha + 2\beta) + c^2(-2 + 2\alpha - \beta) + 2ab\alpha + 8bc + 2ac\beta \\ &= (a^2 + 2b^2 - 2c^2 + 8bc) + (b^2 + 2c^2 + 2ab)\alpha + (2b^2 - c^2 + 2ac)\beta,\end{aligned}$$

and hence

$$\Delta(1, \omega, \omega^2) = \begin{vmatrix} 1 & a & a^2 + 2b^2 - 2c^2 + 8bc \\ 0 & b & b^2 + 2c^2 + 2ab \\ 0 & c & 2b^2 - c^2 + 2ac \end{vmatrix}^2 \cdot \Delta(1, \alpha, \beta) \equiv 0 \pmod{4 \cdot 503}.$$

Therefore 1, ω, ω^2 cannot be an integral basis for $R(\alpha)$.

16.5 Divisibility

Definition 5.1. Let α and β be two integers. Suppose that there exists an integer γ such that $\alpha = \beta\gamma$. Then we say that β *divides* α and we write $\beta \mid \alpha$. We also say β is a divisor of α, or that α is a *multiple of* β.

Theorem 5.1. *Let*

$$g(x) = \alpha_l x^l + \cdots + \alpha_0, \qquad \alpha_l \neq 0,$$
$$h(x) = \beta_m x^m + \cdots + \beta_0, \quad \beta_m \neq 0,$$

where the numbers α, β *are integers, and let*

$$g(x)h(x) = \gamma_{l+m}x^{l+m} + \cdots + \gamma_0.$$

If there exists an integer δ *satisfying* $\delta \mid \gamma_u$ $(0 \leqslant u \leqslant l + m)$, *then* $\delta \mid \alpha_v \beta_w$ $(0 \leqslant v \leqslant l,$ $0 \leqslant w \leqslant m)$. □

The consideration of divisibility leads naturally to the problem of factorization of algebraic integers and the uniqueness of factorization.

However, the factorization of integers in the field of all algebraic numbers has little meaning since an integer may be a product of infinitely many integers. For example $2 = 2^{\frac{1}{2}} \times 2^{\frac{1}{4}} \times 2^{\frac{1}{8}} \cdots$. From this we see that we must somehow restrict the domain of the divisors, and therefore we only discuss the factorization problem within a certain algebraic field $R(\vartheta)$.

Next, there may be infinitely many units in an algebraic field. If ε is a unit, then every integer may be written as $\alpha = \varepsilon \cdot \varepsilon^{-1}\alpha$, and therefore α has infinitely many factorizations whenever $R(\vartheta)$ has infinitely many units. For example, the numbers $(1+\sqrt{2})^n$ ($n=\pm 1, \pm 2, \ldots$) are all units in $R(\sqrt{2})$ so that integers in $R(\sqrt{2})$ have infinitely many factorizations. In order to avoid this difficulty we introduce the notion of *association*.

Definition 5.2. Two integers α, β which differ only from a unit divisor are called *associates* of each other.

Being associates is an equivalence relation.

Definition 5.3. Let α be an integer in $R(\vartheta)$. If there exist non-unit integers β, γ such that $\alpha=\beta\gamma$, then we say that α is *non-prime*; otherwise we call α a *prime* in $R(\vartheta)$.

Theorem 5.2. *Every algebraic integer in* $R(\vartheta)$ *can be factorized into a product of primes in* $R(\vartheta)$.

Proof. If α is a prime, then there is nothing to prove. If $\alpha=\beta\gamma$ wher β, γ are not units, then $|N(\alpha)|=|N(\beta)|\cdot|N(\gamma)|$. Since β, γ are not units the natural numbers $|N(\beta)|$, $|N(\gamma)|$ are proper divisors of $|N(\alpha)|$, so that $|N(\alpha)|>|N(\beta)|>1$ and $|N(\alpha)|>|N(\gamma)|>1$. The proof can now be completed by induction on $|N(\alpha)|$. □

It remains to consider the uniqueness of the factorization, and this is an important problem in algebraic number theory. We shall now examine the quadratic field $R(\sqrt{-5})$ and show that there is no unique factorization.

Since $-5\equiv 3 \pmod 4$, every integer in the field takes the form $\alpha=a+b\sqrt{-5}$ where a, b are rational integers. We shall show that $2, 3, 1\pm\sqrt{-5}$ are primes in the field, and that 2, 3 are not associates of $1\pm\sqrt{-5}$, so that from $6=2\cdot 3=(1+\sqrt{-5})(1-\sqrt{-5})$ we see that there is no unique factorization in $R(\sqrt{-5})$.

First 2, 3 cannot be associates of $1\pm\sqrt{-5}$ because $|N(2)|=4$, $|N(3)|=9$ and $|N(1\pm\sqrt{-5})|=6$. Next, if 2 is non-prime in $R(\sqrt{-5})$, we let

$$2=\alpha\beta, \qquad |N(\alpha)|>1, \qquad |N(\beta)|>1.$$

Write $\alpha=a+b\sqrt{-5}$. Then, from $|N(2)|=4$, we have $|N(\alpha)|=a^2+5b^2=2$ and this is impossible. Therefore 2 is a prime in $R(\sqrt{-5})$. Similarly $3, 1\pm\sqrt{-5}$ are also primes in $R(\sqrt{-5})$.

In order to overcome this problem Kummer invented the notion of *ideals*.

16.6 Ideals

We shall now consider a fixed algebraic number field $R(\vartheta)$ of degree n.

Definition 6.1. Let $\alpha_1, \ldots, \alpha_q$ be any q integers in $R(\vartheta)$. The set of integers of the form

$$\eta_1\alpha_1+\cdots+\eta_q\alpha_q$$

where $\eta_1, \ldots, \eta_q$ are integers in $R(\vartheta)$ is called an *ideal* generated by $\alpha_1, \ldots, \alpha_q$, and is denoted by $[\alpha_1, \ldots, \alpha_q]$.

We shall use the capital Gothic letters $\mathfrak{A}$, $\mathfrak{B}$, $\mathfrak{C}$, $\mathfrak{D}, \ldots$ to denote ideals.

Definition 6.2. An ideal $[\alpha]$ generated by a single integer α is called a *principle ideal.*

The set $[0]$ containing only the integer 0 is an ideal, but we shall assume that our ideals are distinct from $[0]$. The ideal $[1]$ contains all the integers in $R(\vartheta)$, and is called the *unit ideal* which we shall denote by $\mathfrak{O}$.

Theorem 6.1. *Ideals possess the following properties:*
1) *If α, β are in the ideal, then so are $\alpha \pm \beta$;*
2) *If α is in the ideal and η is an integer in $R(\vartheta)$, then $\eta\alpha$ is in the ideal.* □

We see from this theorem that if $1 \in \mathfrak{A}$, then $\mathfrak{A} = [1]$.

Definition 6.3. Let $\mathfrak{A} = [\alpha_1, \ldots, \alpha_q]$ and $\mathfrak{B} = [\beta_1, \ldots, \beta_r]$ be two ideals. If $\mathfrak{A}$ and $\mathfrak{B}$ contain exactly the same integers in $R(\vartheta)$, then we say that they are equal and we write $\mathfrak{A} = \mathfrak{B}$.

Theorem 6.2. *A necessary and sufficient condition for two ideals $[\alpha_1, \ldots, \alpha_q]$ and $[\beta_1, \ldots, \beta_r]$ to be equal is that there are integers ξ_{ij}, η_{ji} $(1 \leqslant i \leqslant q, 1 \leqslant j \leqslant r)$ such that*

$$\alpha_i = \sum_{j=1}^{r} \xi_{ij}\beta_j, \qquad \beta_j = \sum_{i=1}^{q} \eta_{ji}\alpha_i.$$

In particular, if $[\alpha] = [\beta]$, then α and β are associates. □

Let $\alpha_1, \ldots, \alpha_q$ be any q rational integers with greatest common factor d. Then there are rational integers $x_1, \ldots, x_q$ such that $d = x_1\alpha_1 + \cdots + x_q\alpha_q$, and hence, in the rational number field, $[\alpha_1, \ldots, \alpha_q] = [d]$. In other words there are only principal ideals of the rational number field. On the other hand we know from our discussion in the last section that, in $R(\sqrt{-5})$, the ideal $[2, 1 + \sqrt{-5}]$ cannot be reduced to a principal ideal, so that non-principal ideals exist.

Definition 6.4. Let $\mathfrak{A} = [\alpha_1, \ldots, \alpha_q]$ and $\mathfrak{B} = [\beta_1, \ldots, \beta_r]$ be two ideals. We call the ideal $[\alpha_1\beta_1, \ldots, \alpha_1\beta_r, \alpha_2\beta_1, \ldots, \alpha_2\beta_r, \ldots, \alpha_q\beta_r]$ the *product of* $\mathfrak{A}$ and $\mathfrak{B}$; we shall denote it by $\mathfrak{A} \cdot \mathfrak{B}$.

Theorem 6.3. *The product of $\mathfrak{A}$ and $\mathfrak{B}$ is independent of the choices α_i, β_i. That is, if*

$$\mathfrak{A} = [\alpha_1, \ldots, \alpha_q] = [\alpha_1', \ldots, \alpha_s'], \qquad \mathfrak{B} = [\beta_1, \ldots, \beta_r] = [\beta_1', \ldots, \beta_t'],$$

then

$$\begin{aligned} \mathfrak{A} \cdot \mathfrak{B} &= [\alpha_1\beta_1, \ldots, \alpha_1\beta_r, \alpha_2\beta_1, \ldots, \alpha_2\beta_r, \ldots, \alpha_q\beta_r] \\ &= [\alpha_1'\beta_1', \ldots, \alpha_1'\beta_t', \alpha_2'\beta_1', \ldots, \alpha_2'\beta_t', \ldots, \alpha_s'\beta_t']. \end{aligned}$$ □

This can easily be proved from the definition of equality for ideals. Also we have $\mathfrak{O} \cdot \mathfrak{A} = \mathfrak{A}$ for any ideal $\mathfrak{A}$, and that multiplication of ideals is commutative and associative. We can then use induction to define $\mathfrak{A}_1 \cdots \mathfrak{A}_m$ and $\mathfrak{A}^m$, where m is a natural number, and show that the usual rules of indices hold.

Definition 6.5. Let $\mathfrak{A}$, $\mathfrak{B}$ be two ideals. Suppose that there exists an ideal $\mathfrak{C}$ such that $\mathfrak{A} = \mathfrak{B}\mathfrak{C}$. Then we say that $\mathfrak{B}$ *divides* $\mathfrak{A}$ and we write $\mathfrak{B}|\mathfrak{A}$. We call $\mathfrak{B}$, $\mathfrak{C}$ the *divisors* of $\mathfrak{A}$.

Clearly we have: 1) if $\mathfrak{C}|\mathfrak{B}$, $\mathfrak{B}|\mathfrak{A}$, then $\mathfrak{C}|\mathfrak{A}$; 2) if $\mathfrak{B}|\mathfrak{A}$ and $\mathfrak{D}$ is any ideal, then $\mathfrak{B}\mathfrak{D}|\mathfrak{A}\mathfrak{D}$; 3) for any ideal $\mathfrak{A}$ we always have $\mathfrak{O}|\mathfrak{A}$, $\mathfrak{A}|\mathfrak{A}$.

Theorem 6.4. *If* $\mathfrak{B}|\mathfrak{A}$ *then* $\mathfrak{A}$ *is a subset of* $\mathfrak{B}$.

Proof. Let $\mathfrak{A} = \mathfrak{B}\mathfrak{C}$ where $\mathfrak{B} = [\beta_1, \ldots, \beta_r]$, $\mathfrak{C} = [\gamma_1, \ldots, \gamma_r]$. Then each α in $\mathfrak{A}$ is of the form

$$\alpha = \sum_{j=1}^{r} \sum_{k=1}^{s} \eta_{jk}\beta_j\gamma_k = \sum_{j=1}^{r} \left(\sum_{k=1}^{s} \eta_{jk}\gamma_k \right) \beta_j$$

where η_{jk} are integers in the field, and hence α lies in $\mathfrak{B}$. □

We shall see in the next section that the converse of this theorem also holds; that is, if every integer in $\mathfrak{A}$ lies in $\mathfrak{B}$, then $\mathfrak{B}$ must divide $\mathfrak{A}$. From Theorem 6.4 we see that if $\mathfrak{A}|\mathfrak{O}$ then $\mathfrak{A} = \mathfrak{O}$.

16.7 Unique Factorization Theorem for Ideals

Theorem 7.1. *Given any ideal* $\mathfrak{A}$ *there exists an ideal* $\mathfrak{B}$ *such that the product* $\mathfrak{A} \cdot \mathfrak{B}$ *is a principal ideal* $[a]$ *generated by a natural number* a.

Proof. Suppose first that $\mathfrak{A}$ is a principal ideal, say $[\alpha]$. Then we take $\mathfrak{B} = [\alpha^{(2)}, \ldots, \alpha^{(n)}]$ where $\alpha^{(2)}, \ldots, \alpha^{(n)}$ are the conjugates of α, and with $a = |N(\alpha)|$ we see at once that $\mathfrak{A} \cdot \mathfrak{B} = [\alpha\alpha^{(2)} \cdots \alpha^{(n)}] = [a]$.

Suppose now that $\mathfrak{A} = [\alpha_l, \ldots, \alpha_0]$ is not a principal ideal. Let $f(x) = \alpha_l x^l + \cdots + \alpha_0$, and set

$$g(x) = \beta_m x^m + \cdots + \beta_0 \qquad (m = (n-1)l)$$

so that

$$f(x)g(x) = \prod_{j=1}^{n} (\alpha_l^{(j)} x^l + \cdots + \alpha_0^{(j)}) = c_{l+m}x^{l+m} + \cdots + c_0,$$

where c_i are rational integers, so that β_j are integers in $R(\vartheta)$. Now put $\mathfrak{B} = [\beta_m, \ldots, \beta_0]$ and $a = (c_{l+m}, \ldots, c_0)$, and we shall prove that $\mathfrak{A} \cdot \mathfrak{B} = [a]$.

Since $a|c_k$ $(0 \leqslant k \leqslant l+m)$, it follows from Theorem 5.1 that $a|\alpha_\mu\beta_\nu$ $(0 \leqslant \mu \leqslant l,$ $0 \leqslant \nu \leqslant m)$, and hence $\alpha_\mu\beta_\nu$ are numbers in $[a]$. Conversely, from the greatest common factor definition for a, there are rational integers $d_{l+m}, \ldots, d_0$ such that $a = c_{l+m}d_{l+m} + \cdots + c_0 d_0$. Also, from

$$c_k = \sum_{\substack{\mu+\nu=k \\ 0 \leqslant \mu \leqslant l \\ 0 \leqslant \nu \leqslant m}} \alpha_\mu\beta_\nu \qquad (0 \leqslant k \leqslant l+m),$$

we have

$$a = \sum_{\mu=0}^{l} \sum_{\nu=0}^{m} \eta_{\mu\nu}\alpha_\mu\beta_\nu,$$

where $\eta_{\mu\nu}$ are integers in $R(\vartheta)$, and so a lies in $\mathfrak{A} \cdot \mathfrak{B}$. Therefore $\mathfrak{A} \cdot \mathfrak{B} = [a]$. □

Theorem 7.2. *If* $\mathfrak{A} \cdot \mathfrak{C} = \mathfrak{A} \cdot \mathfrak{D}$, *then* $\mathfrak{C} = \mathfrak{D}$.

Proof. We choose $\mathfrak{B}$ and a natural number a so that $\mathfrak{A} \cdot \mathfrak{B} = [a]$. This then gives $[a] \cdot \mathfrak{C} = [a] \cdot \mathfrak{D}$, and this equation means that the set of integers in $\mathfrak{C}$, when multiplied by a, coincides with the set of integers in $\mathfrak{D}$ when multiplied by a. Therefore $\mathfrak{C} = \mathfrak{D}$. □

Theorem 7.3. *If* $\mathfrak{C}$ *is a subset of* $\mathfrak{A}$, *then* $\mathfrak{A}|\mathfrak{C}$.

Proof. We take $\mathfrak{B}$ and a so that $\mathfrak{A} \cdot \mathfrak{B} = [a]$. Then the ideal $\mathfrak{B} \cdot \mathfrak{C}$ is a subset of $\mathfrak{A} \cdot \mathfrak{B} = [a]$ so that we may write $\mathfrak{B} \cdot \mathfrak{C} = [a\gamma_1, \ldots, a\gamma_q] = [a] \cdot [\gamma_1, \ldots, \gamma_q] = \mathfrak{B} \cdot \mathfrak{A} \cdot [\gamma_1, \ldots, \gamma_q]$, which gives $\mathfrak{C} = \mathfrak{A} \cdot [\gamma_1, \ldots, \gamma_q]$. □

From this theorem and Theorem 6.4 we see that a necessary and sufficient condition for $\mathfrak{B}|\mathfrak{A}$ is that $\mathfrak{A}$ is a subset of $\mathfrak{B}$.

We shall now consider the factorization of ideals and the uniqueness problem associated with it.

Definition 7.1. By a *prime ideal* we mean an ideal with only two divisors, namely $\mathfrak{O}$ and the ideal itself. We shall denote a prime ideal by $\mathfrak{P}$.

It is easy to see that in the rational field the prime ideals are $[p]$, where p is an ordinary rational prime.

Theorem 7.4. *Given any two ideals* $\mathfrak{A}$ *and* $\mathfrak{B}$, *there exists a unique ideal* $\mathfrak{D}$ *with the properties*: 1) $\mathfrak{D}|\mathfrak{A}$, $\mathfrak{D}|\mathfrak{B}$; 2) *if* $\mathfrak{D}_1|\mathfrak{A}$, $\mathfrak{D}_1|\mathfrak{B}$, *then* $\mathfrak{D}_1|\mathfrak{D}$. *Furthermore each integer in* $\mathfrak{D}$ *is expressible as* $\alpha + \beta$ *where* $\alpha \in \mathfrak{A}$, $\beta \in \mathfrak{B}$. □

Definition 7.2. We call the ideal $\mathfrak{D}$ the *greatest common divisor* of $\mathfrak{A}$ and $\mathfrak{B}$, and we denote it by $(\mathfrak{A}, \mathfrak{B})$. We also define $(\mathfrak{A}_1, \ldots, \mathfrak{A}_{m-1}, \mathfrak{A}_m) = ((\mathfrak{A}_1, \ldots, \mathfrak{A}_{m-1}), \mathfrak{A}_m)$. If $(\mathfrak{A}, \mathfrak{B}) = \mathfrak{O}$ then we say that $\mathfrak{A}$, $\mathfrak{B}$ are *coprime*.

It is easy to see that if $(\mathfrak{A}, \mathfrak{B}) = \mathfrak{D}$, then $(\mathfrak{A}\mathfrak{C}, \mathfrak{B}\mathfrak{C}) = \mathfrak{D}\mathfrak{C}$ for any ideal $\mathfrak{C}$.

Theorem 7.5. *Let $\mathfrak{P}$ be a prime ideal. Suppose that $\mathfrak{P}|\mathfrak{A}\mathfrak{B}$ and $\mathfrak{P}\nmid\mathfrak{A}$. Then $\mathfrak{P}|\mathfrak{B}$.*

Proof. Since $\mathfrak{P}\nmid\mathfrak{A}$ we have $(\mathfrak{P}, \mathfrak{A}) = \mathfrak{O}$ and so $(\mathfrak{P}\mathfrak{B}, \mathfrak{A}\mathfrak{B}) = \mathfrak{B}$. Since $\mathfrak{P}|\mathfrak{A}\mathfrak{B}$ we now have $\mathfrak{P}|\mathfrak{B}$. □

Theorem 7.6. *Every ideal has finitely many distinct divisors.*

Proof. Given the ideal $\mathfrak{A}$ we choose $\mathfrak{B}$ and a natural number a such that $\mathfrak{A} \cdot \mathfrak{B} = [a]$. Therefore $\mathfrak{A}$ contains a, and any divisor of $\mathfrak{A}$ also contains a. Thus it suffices to show that there is at most a finite number of ideals containing a fixed natural number.

Let $\mathfrak{M} = [\alpha_1, \ldots, \alpha_m]$ be an ideal which contains a, and let $\omega_1, \ldots, \omega_n$ be an integral basis for $R(\vartheta)$ so that each α_j can be written as $\alpha_j = g_{j1}\omega_1 + \cdots + g_{jn}\omega_n$ $(1 \leqslant j \leqslant m)$, where g_{jk} are rational integers. Now set

$$g_{jk} = aq_{jk} + r_{jk} \qquad (0 \leqslant r_{jk} < a),$$

$$\beta_j = \sum_{k=1}^{n} q_{jk}\omega_k, \qquad \gamma_j = \sum_{k=1}^{n} r_{jk}\omega_k,$$

so that $\alpha_j = a\beta_j + \gamma_j$. Since a lies in $\mathfrak{M}$, we have

$$\mathfrak{M} = [a\beta_1 + \gamma_1, \ldots, a\beta_m + \gamma_m, a] = [\gamma_1, \ldots, \gamma_m, a].$$

Since there is at most a finite number of sets $\gamma_1, \ldots, \gamma_m$ the required result follows. □

Theorem 7.7 (Fundamental theorem for ideals). *Any ideal $\mathfrak{A}$ distinct from $\mathfrak{O}$ can be factorized into a product of prime ideals. Furthermore, apart from the ordering of the factors, this factorization is unique.*

Proof. Since each ideal has at most a finite number of divisors we can use induction on the number of divisors of $\mathfrak{A}$.

We first establish the existence of a factorization. If $\mathfrak{A}$ is a prime ideal, then there is nothing more to prove; otherwise we let $\mathfrak{A} = \mathfrak{B}\mathfrak{C}$ $(\mathfrak{B} \neq \mathfrak{O}, \mathfrak{C} \neq \mathfrak{O})$. Since the numbers of divisors of $\mathfrak{B}$ and of $\mathfrak{C}$ are less than that of $\mathfrak{A}$, the required result follows by induction.

We now prove the uniqueness of the factorization. Suppose that

$$\mathfrak{A} = \mathfrak{P}_1\mathfrak{P}_2 \cdots \mathfrak{P}_l = \mathfrak{P}'_1\mathfrak{P}'_2 \cdots \mathfrak{P}'_m, \qquad m \geqslant 1, \quad l \geqslant 1.$$

If $\mathfrak{A}$ is a prime ideal, then $l = m = 1$ and there is nothing to prove. If $\mathfrak{A}$ is not a prime ideal, then $l > 1$, $m > 1$. Since $\mathfrak{P}_1|\mathfrak{P}'_1 \cdots \mathfrak{P}'_m$, there must be a $\mathfrak{P}'_j$ $(1 \leqslant j \leqslant m)$ such that $\mathfrak{P}_1 = \mathfrak{P}'_j$. We may assume without loss that $j = 1$ so that $\mathfrak{P}_2 \cdots \mathfrak{P}_l = \mathfrak{P}'_2 \cdots \mathfrak{P}'_m$, and the required result follows from the induction hypothesis. □

We have proved that every ideal distinct from $\mathfrak{O}$ can be written as $\mathfrak{P}_1^{a_1}\mathfrak{P}_2^{a_2} \cdots \mathfrak{P}_r^{a_r}$ where $\mathfrak{P}_j$ are distinct prime ideals, and a_j are natural numbers. The representation is unique apart from the ordering of $\mathfrak{P}_j$.

16.8 Basis for Ideals

Let $\omega_1, \ldots, \omega_n$ be an integral basis for $R(\vartheta)$, and let $\mathfrak{A}$ be any ideal of $R(\vartheta)$. Since each member of $\mathfrak{A}$ is representable as a linear combination of $\omega_1, \ldots, \omega_n$ with rational integer coefficients we see, from Theorem 6.1, that $\mathfrak{A}$ can be viewed as a linear module. Also, corresponding to the ideal $\mathfrak{A}$, there is an ideal $\mathfrak{B}$ and a natural number a such that $\mathfrak{A}\mathfrak{B} = [a]$, so that $a\omega_1, \ldots, a\omega_n$ all lie in $\mathfrak{A}$; and since these n numbers are linearly independent we see that $\mathfrak{A}$ is actually a linear module of dimension n. From our discussion in Chapter 14, section 9, this module $\mathfrak{A}$ must have a basis, and every basis must have exactly n integers. In particular, we have:

Theorem 8.1. *Let $\mathfrak{A}$ be an ideal of $R(\vartheta)$. Then we can find n integers $\alpha_1, \ldots, \alpha_n$ in $\mathfrak{A}$ such that*

$$\alpha_1 = a_{11}\omega_1,$$

$$\alpha_2 = a_{21}\omega_1 + a_{22}\omega_2,$$

$$\cdots$$

$$\alpha_n = a_{n1}\omega_1 + a_{n2}\omega_2 + \cdots + a_{nn}\omega_n,$$

where a_{ij} are rational integers, $a_{ij} > 0$ $(1 \leqslant i \leqslant n)$, $0 \leqslant a_{ji} < a_{ii}$ $(1 \leqslant i < j \leqslant n)$, and $\alpha_1, \ldots, \alpha_n$ form a standard basis for $\mathfrak{A}$. □

Let $\alpha_1, \ldots, \alpha_n$ and $\beta_1, \ldots, \beta_n$ be two basis for $\mathfrak{A}$ and let

$$\alpha_i = \sum_{j=1}^{n} u_{ij}\beta_j \qquad (i = 1, \ldots, n).$$

Then the coefficient matrix (u_{ij}) must be a modular matrix so that $\Delta(\alpha_1, \ldots, \alpha_n) = \Delta(\beta_1, \ldots, \beta_n)$. Thus the discriminant of a basis of an ideal is independent of the choice of the basis so that we may write this as $\Delta(\mathfrak{A})$.

We shall now examine the standard basis for ideals of the quadratic fields $R(\sqrt{D})$. Let $1, \omega$ be an integral basis for $R(\sqrt{D})$; the definition of ω is given in Theorem 4.5. From Theorem 8.1 we can find two integers a, $b + c\omega$ to form a standard basis. Here a, b, c are rational integers and we may suppose that $a > 0$, $c > 0$, $0 \leqslant b < a$. However we should note that not all pairs of integers of the above form always form a basis for the ideal; there are other conditions on a, b, c.

It is easy to see that a, $b + c\omega$ form a standard basis for a certain ideal only when $a\omega$, $\omega(b + c\omega)$ are representable as $xa + y(b + c\omega)$, where x, y are rational integers. From $a\omega = xa + y(b + c\omega)$ we have $a = yc$, $ax + by = 0$, so that $c | a$, $c | b$. Let $a = cm$, $b = cn$. Then from

$$\begin{aligned} c(n + \omega)\omega &= c(n + \omega)(n + \omega + \omega') - c(n + \omega)(n + \omega') \\ &= -cN(n + \omega) + c(n + \omega)(n + S(\omega)), \end{aligned}$$

where $S(\omega)$ and $N(n + \omega)$ represent the trace and the norm of ω and $n + \omega$ respectively, we see that a necessary and sufficient condition for cm, $c(n + \omega)$ to be a standard basis for a certain ideal is that

$$N(n + \omega) \equiv 0 \pmod{m}. \tag{1}$$

From Theorem 4.5 we see that (1) is equivalent to

$$\Delta \equiv \begin{cases} (2n+1)^2 \pmod{4m}, & \text{if} \quad D \equiv 1 \pmod{4}; \\ (2n)^2 \pmod{4m}, & \text{if} \quad D \equiv 2, 3 \pmod{4}. \end{cases} \tag{2}$$

Therefore we have:

Theorem 8.2. *A necessary and sufficient condition for a pair of integers cm, $c(n + \omega)$ ($c > 0, m > 0, 0 \leqslant n < m$) to be a standard basis for a certain ideal of $R(\sqrt{D})$ is that either* (1) *or* (2) *holds.* □

16.9 Congruent Relations

Definition 9.1. If $\mathfrak{A} | [\alpha]$, then we say that $\mathfrak{A}$ *divides* α, and we write $\mathfrak{A} | \alpha$. It is easy to see that $\mathfrak{A} | \alpha$ means that α is in $\mathfrak{A}$.

We can follow the discussion in Chapter 14, section 9, and define a congruent relation on the integers of the field $R(\vartheta)$ with respect to an ideal.

Definition 9.2. If $\mathfrak{A} | \alpha - \beta$, where α, β are integers in $R(\vartheta)$, then we say that α and β are *congruent modulo* $\mathfrak{A}$, and we write $\alpha \equiv \beta \pmod{\mathfrak{A}}$.

The integers of the field $R(\vartheta)$ are now partitioned into equivalence classes, called the residue classes modulo $\mathfrak{A}$. We shall denote by $N(\mathfrak{A})$ the number of these residue classes, and we call $N(\mathfrak{A})$ the norm of $\mathfrak{A}$. From Theorem 14.9.3 we have:

Theorem 9.1. *Let $\omega_1, \ldots, \omega_n$ be an integral basis for $R(\vartheta)$, and let $\alpha_1, \ldots, \alpha_n$ be any basis for the ideal $\mathfrak{A}$. If $\alpha_i = \sum_{j=1}^{n} a_{ij}\omega_j$, then $N(\mathfrak{A})$ is equal to the absolute value of the determinant of the coefficients, that is $N(\mathfrak{A}) = \|a_{ij}\|$.* □

From this theorem we deduce at once:

Theorem 9.2. *Let Δ be the discriminant of $R(\vartheta)$, and $\Delta(\mathfrak{A})$ be the discriminant of the basis for $\mathfrak{A}$. Then we have $\Delta(\mathfrak{A}) = (N(\mathfrak{A}))^2 \Delta$.* □

Theorem 9.3. *The norm of a principal ideal $[\alpha]$ satisfies $N([\alpha]) = |N(\alpha)|$.* □

Theorem 9.4. $N(\mathfrak{A}\mathfrak{B}) = N(\mathfrak{A})N(\mathfrak{B})$.

Proof. Since $\mathfrak{A}$ contains $\mathfrak{A}\mathfrak{B}$, from Theorem 14.9.4, the members of $\mathfrak{A}$ are partitioned into residue classes modulo $\mathfrak{A}\mathfrak{B}$, and the number of classes is equal to $N(\mathfrak{A}\mathfrak{B})/N(\mathfrak{A})$. It remains to prove that the number of classes is also equal to $N(\mathfrak{B})$.

Let $\beta_1, \ldots, \beta_{N(\mathfrak{B})}$ denote the residue classes mod $\mathfrak{B}$. There exists an integer $\alpha \in \mathfrak{A}$ such that $([\alpha], \mathfrak{A}\mathfrak{B}) = \mathfrak{A}$. Now $\alpha\beta_1, \ldots, \alpha\beta_{N(\mathfrak{B})}$ all lie in $\mathfrak{A}$, and if $j \neq k$ $(1 \leqslant j, k \leqslant n)$, then $\alpha\beta_j \not\equiv \alpha\beta_k \pmod{\mathfrak{A}\mathfrak{B}}$.

From $([\alpha], \mathfrak{A}\mathfrak{B}) = \mathfrak{A}$, we know that corresponding to any γ in $\mathfrak{A}$, there are integers η, δ such that $\gamma = \eta\alpha + \delta$, $\delta \in \mathfrak{A}\mathfrak{B}$. Also, corresponding to the integer η, there is an integer β and a natural number j $(1 \leqslant j \leqslant N(\mathfrak{B}))$ such that $\eta = \beta_j + \beta$ so that $\gamma = \alpha\beta_j + \alpha\beta + \delta \equiv \alpha\beta_j \pmod{\mathfrak{A}\mathfrak{B}}$. This shows that every member of $\mathfrak{A}$ must be congruent to exactly one of $\alpha\beta_1, \ldots, \alpha\beta_{N(\mathfrak{B})}$ modulo $\mathfrak{A}\mathfrak{B}$, and therefore the number of classes concerned must be equal to $N(\mathfrak{B})$ as required. □

Theorem 9.5. *Let $\mathfrak{P}$ be a prime ideal, and let α be any integer not divisible by $\mathfrak{P}$. Then* $\alpha^{N(\mathfrak{P})-1} \equiv 1 \pmod{\mathfrak{P}}$.

Proof. Let $0, \pi_1, \pi_2, \ldots, \pi_{N(\mathfrak{P})-1}$ denote the residue classes mod $\mathfrak{P}$. Since $\mathfrak{P} \nmid \alpha$, the numbers $0, \alpha\pi_1, \alpha\pi_2, \ldots, \alpha\pi_{N(\mathfrak{P})-1}$ also represent the residue classes mod $\mathfrak{P}$. Therefore

$$\alpha^{N(\mathfrak{P})-1}\pi_1\pi_2 \cdots \pi_{N(\mathfrak{P})-1} \equiv \pi_1\pi_2 \cdots \pi_{N(\mathfrak{P})-1} \pmod{\mathfrak{P}},$$

and the theorem follows. □

16.10 Prime Ideals

Theorem 10.1. *Every prime ideal $\mathfrak{P}$ must divide a rational prime p. Moreover, p is the least positive rational integer in $\mathfrak{P}$ so that it is unique.*

Proof. From Theorem 7.1 there must exist a rational integer a such that $\mathfrak{P}|[a]$. Let $a = \Pi p$ be its factorization, so that there must be a prime p such that $\mathfrak{P}|[p]$, or $\mathfrak{P}|p$.

Suppose, if possible, there exists a positive rational integer b such that $b < p$ and $\mathfrak{P}|b$. Then $b \in \mathfrak{P}$ so that $(p, b) = 1$ also lie in $\mathfrak{P}$ giving $\mathfrak{P} = [1]$, which is impossible. Therefore p is the least positive rational integer in $\mathfrak{P}$. □

Let the prime ideal factorization for $[p]$ be $\mathfrak{P}_1\mathfrak{P}_2 \cdots \mathfrak{P}_t$. Then, on taking the norm, we have $p^n = N([p]) = N(\mathfrak{P}_1)N(\mathfrak{P}_2) \cdots N(\mathfrak{P}_t)$. It follows that the norm of a prime ideal must be a prime power. If $N(\mathfrak{P}) = p^f$, then we call f the *degree* of $\mathfrak{P}$.

Concerning the factorization of $[p]$ there is the following important theorem which we shall not prove.

Theorem 10.2 (Dedekind's discriminant theorem). *A necessary and sufficient condition for $\mathfrak{P}^2|p$ is that $p|\Delta$.* □

Let us examine the factorization of $[p]$ in the quadratic field $R(\sqrt{D})$. Clearly there can only be the following three possibilities. 1) $[p] = \mathfrak{P}$; 2) $[p] = \mathfrak{P}\mathfrak{Q}$, $\mathfrak{P} \neq \mathfrak{Q}$, $N(\mathfrak{P}) = N(\mathfrak{Q}) = p$; 3) $[p] = \mathfrak{P}^2$, $N(\mathfrak{P}) = p$. Concerning the factorization of $[p]$ in a quadratic field, we have:

Theorem 10.3. *Let Δ be the discriminant of $R(\sqrt{D})$. Then* 1), 2) *or* 3) *in the above holds according to* $(\frac{\Delta}{p}) = -1, +1$, *or* 0. *Here* $(\frac{\Delta}{p})$ *is the Kronecker's symbol.*

Proof. If $\mathfrak{P}$ is a prime divisor of $[p]$, and $N(\mathfrak{P}) = p$, then either $[p] = \mathfrak{P}\mathfrak{Q}$ or $[p] = \mathfrak{P}^2$. Let cm, $c(n+\omega)$ be a standard basis for the ideal. Then $N(\mathfrak{P}) = c^2 m = p$, so that $c = 1, m = p$. From (2) in section 8, we now see that $(\frac{\Delta}{p})$ is either $+1$ or 0.

Let us suppose, conversely, that $(\frac{\Delta}{p}) = +1$ or 0. We first consider the case $p \neq 2$.

1) If $(\frac{\Delta}{p}) = 1$, then there exists a such that $p \nmid a$ and $\Delta \equiv a^2 \pmod{p}$. Since $p \neq 2$, we have $(p, 2a) = 1$ so that

$$[p, a + \sqrt{\Delta}][p, a - \sqrt{\Delta}] = [p]\left[p, a + \sqrt{\Delta}, a - \sqrt{\Delta}, \frac{a^2 - \Delta}{p}\right]$$

$$= [p]\left[p, a + \sqrt{\Delta}, 2a, \frac{a^2 - \Delta}{p}, 1\right] = [p].$$

Also $[p, a + \sqrt{\Delta}] \neq [p, a - \sqrt{\Delta}]$, since otherwise we have $[p, a + \sqrt{\Delta}] = [p, a - \sqrt{\Delta}] = [p, a + \sqrt{\Delta}, 2a] = [1]$ and this is impossible; $[p, a + \sqrt{\Delta}]$ and $[p, a - \sqrt{\Delta}]$ are not $\mathfrak{O}$. Therefore, when $p \neq 2$ and $(\frac{\Delta}{p}) = 1$, $[p]$ is the product of two distinct prime ideals.

2) If $(\frac{\Delta}{p}) = 0$, then $p|\Delta$, so that

$$[p, \sqrt{\Delta}]^2 = [p, \sqrt{\Delta}][p, \sqrt{\Delta}] = [p]\left[p, \sqrt{\Delta}, \frac{\Delta}{p}\right].$$

But $\Delta = D$ or $4D$, $p \neq 2$ and D is square-free, so that $(p, \frac{\Delta}{p}) = 1$ and hence $[p] = [p, \sqrt{\Delta}]^2$. That is, if $p \neq 2$ and $(\frac{\Delta}{p}) = 0$, then $[p]$ is the square of a prime ideal.

Let us now consider the case $p = 2$. Since $(\frac{\Delta}{2}) \neq -1$ we must have $D \equiv 2, 3 \pmod{4}$ or $D \equiv 1 \pmod{8}$. As before we can prove:

3) When $D \equiv 2 \pmod{4}$, we have $(\frac{\Delta}{2}) = 0$ and $[2] = [2, \sqrt{D}]^2$;

4) When $D \equiv 3 \pmod{4}$, we have $(\frac{\Delta}{2}) = 0$ and $[2] = [2, 1 + \sqrt{D}]^2$;

5) When $D \equiv 1 \pmod{8}$, we have $(\frac{\Delta}{2}) = 1$ and

$$[2] = \left[2, \frac{1 + \sqrt{D}}{2}\right] \cdot \left[2, \frac{1 - \sqrt{D}}{2}\right].$$

Since the two factors here are distinct, [2] is now the product of two distinct prime ideals. □

Theorem 10.3 establishes Dedekind's discriminant theorem for quadratic fields. We shall now examine a specific example for a cubic field.

Let α be a zero of $f(x) = x^3 - x^2 - 2x - 8$. We saw in §4 that $R(\alpha)$ is a cubic field with discriminant 503, that $1, \alpha, \beta = 4/\alpha$ form an integral basis, and that β is a zero of $g(y) = y^3 + y^2 + 2y - 8$.

We now consider the factorization of [503] in $R(\alpha)$. Let $\mathfrak{P}, \mathfrak{Q}, \mathfrak{R}$ denote prime ideals of $R(\alpha)$. Then the factorization of [503] must take one of the following five situations:

1) $[503] = \mathfrak{P}\mathfrak{Q}\mathfrak{R}$; $\mathfrak{P}$, $\mathfrak{Q}$, $\mathfrak{R}$ distinct and $N(\mathfrak{P}) = N(\mathfrak{Q}) = N(\mathfrak{R}) = 503$;
2) $[503] = \mathfrak{P}^2\mathfrak{Q}$; $\mathfrak{P} \neq \mathfrak{Q}$ and $N(\mathfrak{P}) = N(\mathfrak{Q}) = 503$;
3) $[503] = \mathfrak{P}^3$; $N(\mathfrak{P}) = 503$;
4) $[503] = \mathfrak{P}\mathfrak{Q}$; $N(\mathfrak{P}) = 503$, $N(\mathfrak{Q}) = 503^2$;
5) $[503] = \mathfrak{P}$; $N(\mathfrak{P}) = 503^3$.

In each of the first four situations, [503] has a prime divisor $\mathfrak{P}$ with norm 503. Let us first examine these four situations. Let a_0, $b_0 + b_1\alpha$, $c_0 + c_1\alpha + c_2\beta$ be a standard basis for $\mathfrak{P}$ so that $b_0 < a_0$, $c_0 < a_0$, $c_1 < b_1$. Also, since $a_0\alpha$, $a_0\beta$ lie in $\mathfrak{P}$ we have, in addition, that $b_1 \leqslant a_0$, $c_2 \leqslant a_0$, and from $N(\mathfrak{P}) = a_0b_1c_2 = 503$, we obtain $a_0 = 503$, $b_1 = 1$, $c_2 = 1$, $c_1 = 0$. Therefore $\mathfrak{P}$ must take the form $[503, a + \alpha, b + \beta]$, and $503, a + \alpha, b + \beta$ form a standard basis for $\mathfrak{P}$.

Since $a + \alpha, b + \beta \in \mathfrak{P}$ and $N(\mathfrak{P}) = 503$, we have $N(a + \alpha) \equiv N(b + \beta) \equiv 0 \pmod{503}$. But $a + \alpha$ and $b + \beta$ are the roots of $f(x - a) = 0$ and $g(y - b) = 0$ respectively so that $N(a + \alpha) = |f(-a)|$ and $N(b + \beta) = |g(-b)|$. Therefore a and b satisfy the cubic congruences $a^3 + a^2 - 2a + 8 \equiv 0 \pmod{503}$ and $b^3 - b^2 + 2b + 8 \equiv 0 \pmod{503}$, which give the solutions $a \equiv 149, 149, 204$ and $b \equiv 395, 395, 217 \pmod{503}$. Therefore $\mathfrak{P}$ must be one of the following four ideals:

$$[503, 149 + \alpha, 395 + \beta], \qquad [503, 204 + \alpha, 217 + \beta],$$

$$[503, 149 + \alpha, 217 + \beta], \qquad [503, 204 + \alpha, 395 + \beta].$$

The third ideal is not $\mathfrak{P}$, since otherwise

$$\alpha(217 + \beta) - 217(149 + \alpha) + 65(503) = 4 - 217 \cdot 149 + 65 \cdot 503 = 366$$

would be in $\mathfrak{P}$, and from $(366, 503) = 1$ we would have $\mathfrak{P} = \mathfrak{O}$. Similarly the fourth ideal is not $\mathfrak{P}$. Next, from

$$(149 + \alpha)\alpha = -46(503) + 150(149 + \alpha) + 2(395 + \beta),$$

$$(149 + \alpha)\beta = -117(503) + 149(395 + \beta),$$

$$(395 + \beta)\alpha = -117(503) + 395(149 + \alpha),$$

$$(395 + \beta)\beta = -310(503) + 2(149 + \alpha) + 394(395 + \beta),$$

we see that $503, 149 + \alpha, 395 + \beta$ do form a standard basis for the prime ideal $[503, 149 + \alpha, 395 + \beta]$. Similarly $503, 204 + \alpha, 217 + \beta$ do form a standard basis for the prime ideal $[503, 204 + \alpha, 217 + \beta]$. Finally the two ideals $[503, 149 + \alpha, 395 + \beta]$ and $[503, 204 + \alpha, 217 + \beta]$ are distinct divisors of the ideal [503] and we therefore conclude that our situation 2) is the only possibility, and computation shows that actually

$$[503] = [503, 149 + \alpha, 395 + \beta]^2 \cdot [503, 204 + \alpha, 217 + \beta].$$

16.11 Units

We have the following result on units: Among all the units in $R(\vartheta)$ we can choose $r = r_1 + r_2 - 1$ of them, say $\varepsilon_1, \ldots, \varepsilon_r$, such that every unit is representable as $\rho\varepsilon_1^{l1} \cdots \varepsilon_r^{lr}$ $(l = 0, \pm 1, \pm 2, \ldots)$; here ρ is a certain root of unity in $R(\vartheta)$.

Here we shall only concern ourselves with quadratic fields $R(\sqrt{D})$. Let a unit be $x + y\omega$ so that $N(x + y\omega) = \pm 1$. We need therefore to solve these equations in rational integers for the units in $R(\sqrt{D})$. Now

$$N(x + y\omega) = (x + y\omega)(x + y\omega')$$

$$= \begin{cases} \left(x + \dfrac{y}{2}\right)^2 - \dfrac{y^2}{4}D, & \text{if } D \equiv 1 \pmod 4, \\ x^2 - y^2 D, & \text{if } D \equiv 2, 3 \pmod 4). \end{cases}$$

When $D < 0$, the equations $(2x + y)^2 - y^2 D = 4$ and $x^2 - y^2 D = 1$ have only finitely many solutions, so that $R(\sqrt{D})$ can have only finitely many units. In fact if we denote by w the number of units in $R(\sqrt{D})$, it is not difficult to show that $w = 6$, 4 or 2 according to whether $\Delta = -3, -4$ or $\Delta \leqslant -7$.

Consider next $D > 0$. Now the equations $(2x + y)^2 - y^2 D = \pm 4$ and $x^2 - y^2 D = \pm 1$ are the Pell equations we considered in Chapter 10. Therefore there exists a unit η in $R(\sqrt{D})$ such that any unit in $R(\sqrt{D})$ is representable as $\pm \eta^n$, $n = 0, \pm 1, \pm 2, \ldots$. This number η is called the *fundamental unit* of $R(\sqrt{D})$.

16.12 Ideal Classes

Definition 12.1. Let $\mathfrak{A}$ and $\mathfrak{B}$ be two ideals. Suppose that there exist two principal ideals $[\alpha]$ and $[\beta]$ such that $[\alpha]\mathfrak{A} = [\beta]\mathfrak{B}$. Then we say that the two ideals $\mathfrak{A}$ and $\mathfrak{B}$ belong to the same *ideal class*, and we write $\mathfrak{A} \sim \mathfrak{B}$.

It is easy to see that being in the same ideal class is an equivalence relation, and moreover we have 1) $\mathfrak{A} \sim \mathfrak{O}$ if and only if $\mathfrak{A}$ is a principal ideal; 2) if $\mathfrak{A} \sim \mathfrak{B}$ and $\mathfrak{C} \sim \mathfrak{D}$, then $\mathfrak{A}\mathfrak{C} \sim \mathfrak{B}\mathfrak{D}$; 3) if $\mathfrak{A}\mathfrak{C} \sim \mathfrak{B}\mathfrak{C}$ then $\mathfrak{A} \sim \mathfrak{B}$. The ideals of $R(\vartheta)$ are now partitioned into classes called *ideal classes*.

Theorem 12.1. *The number of ideal classes of $R(\vartheta)$ is finite.*

Proof. It suffices to show that there exists a positive number M, depending only on $R(\vartheta)$, such that every class contains an ideal $\mathfrak{B}$ satisfying $N(\mathfrak{B}) \leqslant M$. This is because there can only be finitely many ideals having a given norm.

Let $\mathfrak{C}$ be any ideal of $R(\vartheta)$. We already know that there exists an ideal $\mathfrak{A}$ such that $\mathfrak{A}\mathfrak{C} \sim \mathfrak{O}$, and if we can choose an ideal $\mathfrak{B}$ such that $\mathfrak{A}\mathfrak{B} \sim \mathfrak{O}$ and $N(\mathfrak{B}) \leqslant M$, then our theorem is proved. This is because $\mathfrak{A}\mathfrak{B} \sim \mathfrak{A}\mathfrak{C}$ so that $\mathfrak{B} \sim \mathfrak{C}$.

Let $\omega_1, \ldots, \omega_n$ be an integral basis for $R(\vartheta)$ and let

$$M = \prod_{s=1}^{n} (|\omega_1^{(s)}| + \cdots + |\omega_n^{(s)}|).$$

We define the natural number k by $k^n \leqslant N(\mathfrak{A}) < (k+1)^n$. Among the $(k+1)^n$ integers $x_1\omega_1 + \cdots + x_n\omega_n$ $(x_m = 0, 1, \ldots, k)$ there are at least two which are congruent modulo $\mathfrak{A}$, say

$$y_1\omega_1 + \cdots + y_n\omega_n \equiv z_1\omega_1 + \cdots + z_n\omega_n \pmod{\mathfrak{A}};$$

here $0 \leqslant y_m \leqslant k$, $0 \leqslant z_m \leqslant k$, and we now have the non-zero integer

$$\alpha = (y_1 - z_1)\omega_1 + \cdots + (y_n - z_n)\omega_n$$

in $\mathfrak{A}$. Since $|y_m - z_m| \leqslant k$ it follows that

$$|N(\alpha)| = \left| \prod_{s=1}^{n} \sum_{m=1}^{n} (y_m - z_m)\omega_m^{(s)} \right| \leqslant \prod_{s=1}^{n} \sum_{m=1}^{n} k|\omega_m^{(s)}| = k^n M \leqslant M \cdot N(\mathfrak{A}).$$

Since α is in $\mathfrak{A}$ we see that $\mathfrak{A}|[\alpha]$, and we may write $[\alpha] = \mathfrak{A}\mathfrak{B}$ which gives $N(\mathfrak{A})N(\mathfrak{B}) = |N(\alpha)| \leqslant M \cdot N(\mathfrak{A})$ or $N(\mathfrak{B}) \leqslant M$ as required. □

Theorem 12.2. *Let h be the number of ideal classes of $R(\vartheta)$. Then, for any ideal $\mathfrak{A}$, we have $\mathfrak{A}^h \sim \mathfrak{O}$.*

Proof. Let $\mathfrak{A}_1, \ldots, \mathfrak{A}_h$ be ideals that belong to different classes. Then so are $\mathfrak{A}\mathfrak{A}_1, \ldots, \mathfrak{A}\mathfrak{A}_h$ and hence $\mathfrak{A}_1 \cdots \mathfrak{A}_h \sim (\mathfrak{A}\mathfrak{A}_1) \cdots (\mathfrak{A}\mathfrak{A}_h)$, or $\mathfrak{A}^h \sim \mathfrak{O}$. □

16.13 Quadratic Fields and Quadratic Forms

Let Δ be the discriminant of the quadratic field $R(\sqrt{D})$. We shall now establish the relationship between the ideal classes of $R(\sqrt{D})$ and the classes of quadratic forms having discriminant Δ.

Let $\mathfrak{A}$ be an ideal of $R(\sqrt{D})$ and let α_1, α_2 be a basis for $\mathfrak{A}$ satisfying

$$\alpha_1\alpha_2' - \alpha_1'\alpha_2 = N(\mathfrak{A})\sqrt{\Delta}, \tag{1}$$

where α_1', α_2' are the conjugates of α_1, α_2.

Corresponding to $\mathfrak{A}$ we construct the quadratic form

$$F(x, y) = \frac{N(\alpha_1 x + \alpha_2 y)}{N(\mathfrak{A})} = \frac{(\alpha_1 x + \alpha_2 y)(\alpha_1' x + \alpha_2' y)}{N(\mathfrak{A})} = ax^2 + bxy + cy^2.$$

Since $a = N(\alpha_1)/N(\mathfrak{A})$, $b = (N(\alpha_1 + \alpha_2) - N(\alpha_1) - N(\alpha_2))/N(\mathfrak{A})$, $c = N(\alpha_2)/N(\mathfrak{A})$, and α_1, α_2, $\alpha_1 + \alpha_2$ are in $\mathfrak{A}$ we see that a, b, c are rational integers. Also, the discriminant of $F(x, y)$ is $b^2 - 4ac = (\alpha_1\alpha_2' - \alpha_1'\alpha_2)^2/N(\mathfrak{A})^2 = \Delta$. We say that $F(x,y)$ is a quadratic form belonging to $\mathfrak{A}$.

When $\Delta < 0$ the quadratic field $R(\sqrt{D})$ is imaginary so that $a > 0$ and $F(x, y)$ is positive definite. Also, it is not difficult to see that as α_1, α_2 run through the basis for $\mathfrak{A}$ satisfying (1) we obtain all the quadratic forms equivalent to F.

Theorem 13.1. *Every indefinite or positive definite quadratic form $F(x,y) = ax^2 + bxy + cy^2$ with rational integer coefficients and discriminant Δ belongs to an ideal $\mathfrak{A}$ with basis α_1, α_2.*

Proof. We first show that a, $(b - \sqrt{\Delta})/2$ form a basis for the ideal $\mathfrak{M} = [a, (b - \sqrt{\Delta})/2]$. Observe that $(b - \sqrt{\Delta})/2$ satisfies the equation $x(b - x) = ac$ so that it is an integer. Also we have $\omega = (s(\omega) + \sqrt{\Delta})/2$, where $s(\omega) = 0$ or 1, and

$$a\omega = \frac{s(\omega) + b - (b - \sqrt{\Delta})}{2}a = \frac{s(\omega) + b}{2}a - a\frac{b - \sqrt{\Delta}}{2},$$

$$\frac{b - \sqrt{\Delta}}{2}\omega = \frac{b - \sqrt{\Delta}}{2} \cdot \frac{s(\omega) - b + b + \sqrt{\Delta}}{2} = \frac{b^2 - \Delta}{4a}a + \frac{s(\omega) - b}{2} \cdot \frac{b - \sqrt{\Delta}}{2},$$

where $(s(\omega) \pm b)/2$ and $(b^2 - \Delta)/4a$ are rational integers, so that a, $(b - \sqrt{\Delta})/2$ do indeed form a basis for $\mathfrak{M}$.

If $a > 0$ we take $\mathfrak{A} = \mathfrak{M}$, $\alpha_1 = a$, $\alpha_2 = (b - \sqrt{\Delta})/2$, and from $N(\mathfrak{M}) = a$ we have the quadratic form

$$\frac{(ax + \frac{1}{2}(b - \sqrt{\Delta})y)(ax + \frac{1}{2}(b + \sqrt{\Delta})y)}{a} = ax^2 + bxy + cy^2,$$

so that $\mathfrak{M}$ is the required ideal.

If $a < 0$, then, since the quadratic form is not negative definite, $\Delta > 0$ and we now take $\mathfrak{A} = \sqrt{\Delta}\mathfrak{M}$, $\alpha_1 = a\sqrt{\Delta}$ and $a_2 = (b - \sqrt{\Delta})\sqrt{\Delta}/2$. It is easy to see that α_1, α_2 form a basis for $\mathfrak{A}$ satisfying (1). Also $N(\mathfrak{A}) = -a\Delta$ and we can now construct the quadratic form

$$\frac{-\Delta(ax + \frac{1}{2}(b - \sqrt{\Delta})y)(ax + \frac{1}{2}(b + \sqrt{\Delta})y)}{-a\Delta} = ax^2 + bxy + cy^2.$$

The theorem is proved. □

From the above we see that if F belongs to $\mathfrak{A}$, then every quadratic form equivalent to F also belongs to $\mathfrak{A}$. However, given a quadratic form F, there may be two different ideals $\mathfrak{A}$ and $\mathfrak{B}$ to which F belongs. This then establishes a relationship between $\mathfrak{A}$ and $\mathfrak{B}$.

Definition 13.1. Let $\mathfrak{A}$ and $\mathfrak{B}$ be two ideals. Suppose that there are integers α and β such that $[\alpha]\mathfrak{A} = [\beta]\mathfrak{B}$ and $N(\alpha\beta) > 0$. Then we say that $\mathfrak{A}$ and $\mathfrak{B}$ are *equivalent in the narrower sense*, and we write $\mathfrak{A} \simeq \mathfrak{B}$.

It is clear that being equivalent in the narrower sense is a special case of being equivalent.

Theorem 13.2. *Equivalent quadratic forms belong to ideals which are equivalent in the narrower sense. Conversely, quadratic forms belonging to ideals which are equivalent in the narrower sense are equivalent forms.* □

Let h_0 denote the number of ideal classes (not in the narrower sense), and let h denote the number of classes under the narrower sense of equivalence. Assume that the discriminant of the field concerned is Δ. Then h is the class number of quadratic forms with discriminant Δ.

If $\mathfrak{A} \sim \mathfrak{B}$ then either $\mathfrak{A} \simeq \mathfrak{B}$ or $\mathfrak{A} \simeq [\sqrt{\Delta}]\mathfrak{B}$, and we deduce that $h \leqslant 2h_0$. In fact, if $\mathfrak{A} \sim \mathfrak{B}$, then there are integers α, β such that $[\alpha]\mathfrak{A} = [\beta]\mathfrak{B}$.

(i) If $\Delta < 0$, then $N(\alpha\beta) > 0$ so that $\mathfrak{A} \simeq \mathfrak{B}$, and whence $h_0 = h$.

(ii) If $\Delta > 0$ and the fundamental unit η satisfies $N(\eta) = -1$, then $[\alpha]\mathfrak{A} = [\beta]\mathfrak{B} = [\eta\beta]\mathfrak{B}$ and one of $N(\alpha\beta)$, $N(\alpha\beta\eta)$ must be positive, so that we still have $\mathfrak{A} \simeq \mathfrak{B}$ and $h_0 = h$.

(iii) If $\Delta > 0$ and the fundamental unit η satisfies $N(\eta) = 1$ then $\mathfrak{A}$ cannot be equivalent in the narrower sense to both $\mathfrak{B}$ and $\mathfrak{B}[\sqrt{\Delta}]$, so that $h_0 = h/2$.

Therefore we have

$$h_0 = \begin{cases} h, & \text{if } \Delta < 0 \text{ or } \Delta > 0, \ N(\eta) = -1; \\ \dfrac{h}{2}, & \text{if } \Delta > 0, \ N(\eta) = +1. \end{cases}$$

Also if we replace d by D in Theorem 11.4.4 and define ε accordingly, then

$$\varepsilon = \begin{cases} \eta^2, & \text{if } \Delta > 0, \ N(\eta) = -1; \\ \eta, & \text{if } \Delta > 0, \ N(\eta) = +1. \end{cases}$$

Again, from our results on the class number in Chapter 12 we have:

Theorem 13.3. *Let h_0 denote the number of ideal classes. Then*

$$h_0 = \frac{W}{2\left(2 - \left(\frac{\Delta}{2}\right)\right)} \sum_{s=1}^{[\frac{1}{2}|\Delta|]} \left(\frac{\Delta}{s}\right), \qquad \textit{if } \Delta < 0,$$

$$\eta^{h_0} = \prod_{s=1}^{[\frac{1}{2}(\Delta-1)]} \left(\sin\frac{s\pi}{\Delta}\right)^{-\left(\frac{\Delta}{s}\right)}, \qquad \textit{if } \Delta > 0. \ \square$$

Example 1. In $R(i)$ we have $\Delta = -4$, $W = 4$ so that

$$h_0 = \frac{4}{2(2-0)} \sum_{s=1}^{2} \left(\frac{-4}{s}\right) = 1.$$

Example 2. In $R(\sqrt{-3})$ we have $\Delta = -3$, $W = 6$ so that

$$h_0 = \frac{6}{2(2-(-1))} \sum_{s=1}^{1} \left(\frac{-3}{s}\right) = 1.$$

Example 3. In $R(\sqrt{-5})$ we have $\Delta = -20$, $W = 2$ so that

$$h_0 = \frac{2}{2(2-0)} \sum_{s=1}^{10} \left(\frac{-20}{s}\right) = 2.$$

Example 4. In $R(\sqrt{-19})$ we have $\Delta = -19$, $W = 2$ so that

$$h_0 = \frac{2}{2(2-(-1))} \sum_{s=1}^{9} \left(\frac{-19}{s}\right) = 1.$$

Example 5. In $R(\sqrt{2})$ we have $\Delta = 8$, $\varepsilon = 3 + 2\sqrt{2}$. Since $\eta = 1 + \sqrt{2}$ has norm -1 and $\eta^2 = \varepsilon$, η is a fundamental unit. Also

$$(1+\sqrt{2})^{h_0} = \prod_{s=1}^{3} \left(\sin\frac{\pi s}{8}\right)^{-\left(\frac{8}{s}\right)} = \sin\frac{3\pi}{8} \Big/ \sin\frac{\pi}{8} = (1+\sqrt{2}),$$

so that $h_0 = 1$.

16.14 Genus

Let $R(\sqrt{D})$ be a fixed quadratic field with discriminant Δ, and we shall assume in this section that the ideal classes are derived from the equivalence relation on ideals being equivalent in the narrower sense.

Definition 14.1. If a quadratic form $F(x, y)$ belongs to an ideal $\mathfrak{A}$ then we call the character system for $F(x, y)$ (see Definition 12.6.1) the *character system* for $\mathfrak{A}$. That is, if $p_1, \ldots, p_s$ are the odd prime divisors of Δ, we take an integer α in $\mathfrak{A}$ so that $(N(\alpha)/N(\mathfrak{A}), 2\Delta) = 1$ and we call

$$\left(\frac{N(\alpha)/N(\mathfrak{A})}{p_i}\right) \qquad (i = 1, \ldots, s)$$

and

$$\delta(\alpha) = (-1)^{\frac{1}{2}\left[\frac{N(\alpha)}{N(\mathfrak{A})} - 1\right]}, \qquad \text{if} \quad D = \frac{\Delta}{4} \equiv 3 \pmod 4;$$

$$\varepsilon(\alpha) = (-1)^{\frac{1}{8}\left[\left(\frac{N(\alpha)}{N(\mathfrak{A})}\right)^2 - 1\right]}, \qquad \text{if} \quad \frac{\Delta}{4} \equiv 2 \pmod 8;$$

$$\delta(\alpha)\,\varepsilon(\alpha), \qquad \text{if} \quad \frac{\Delta}{4} \equiv 6 \pmod 8$$

the character system for $\mathfrak{A}$.

Since ideals belonging to the same class have the same character system we may speak of the character system for an ideal class.

Definition 14.2. Two ideal classes with the same character system are said to belong to the same *genus*. There is now a one-to-one correspondence between ideal classes in the quadratic field $R(\sqrt{D})$ and classes of primitive forms having discriminant Δ.

Theorem 14.1. *The values of the character system for $\mathfrak{A}\mathfrak{B}$ correspond to the products of the values of the character systems for $\mathfrak{A}$, $\mathfrak{B}$.*

Proof. If α, β belong to $\mathfrak{A}$, $\mathfrak{B}$ respectively, then $\alpha\beta$ belongs to $\mathfrak{A}\mathfrak{B}$. Also

$$\frac{N(\alpha)}{N(\mathfrak{A})}\cdot\frac{N(\beta)}{N(\mathfrak{B})}=\frac{N(\alpha\beta)}{N(\mathfrak{A}\mathfrak{B})},$$

and

$$\frac{N(\alpha\beta)}{N(\mathfrak{A}\mathfrak{B})}-1\equiv\frac{N(\alpha)}{N(\mathfrak{A})}-1+\frac{N(\beta)}{N(\mathfrak{B})}-1 \pmod 4,$$

$$\left(\frac{N(\alpha\beta)}{N(\mathfrak{A}\mathfrak{B})}\right)^2-1\equiv\left(\frac{N(\alpha)}{N(\mathfrak{A})}\right)^2-1+\left(\frac{N(\beta)}{N(\mathfrak{B})}\right)^2-1 \pmod{16},$$

and if

$$\left(\frac{N(\alpha)}{N(\mathfrak{A})},2\Delta\right)=1,\quad \left(\frac{N(\beta)}{N(\mathfrak{B})},2\Delta\right)=1,\quad \text{then}\quad \left(\frac{N(\alpha\beta)}{N(\mathfrak{A}\mathfrak{B})},2\Delta\right)=1.$$

The theorem is proved. □

From this theorem we deduce at once:

1) The character system for the product of two classes is the product of the two character systems.

2) If $\{\mathfrak{A}\}$ and $\{\mathfrak{B}\}$ belong to a genus, and $\{\mathfrak{A}_1\}\{\mathfrak{B}_1\}$ belong to a genus, then $\{\mathfrak{A}\mathfrak{A}_1\}$ and $\{\mathfrak{B}\mathfrak{B}_1\}$ also belong to a genus.

Definition 14.3. We call the class to which the unit ideal $\mathfrak{O}$ belongs the *principal class*, and the genus to which the principal class belongs the *principal genus*. Also, if $\mathfrak{A}\mathfrak{B}=[a]$ where a is a natural number, then we call $\{\mathfrak{B}\}$ the inverse of the class $\{\mathfrak{A}\}$.

From Theorem 7.1 we see that the inverse of any ideal class always exists. Also $\{\mathfrak{O}\}\{\mathfrak{A}\}=\{\mathfrak{A}\}$. Since the values of the character system for the principal class, as well as for all the classes in the principal genus, are all 1, it follows that the product of any two classes in the principal genus, and the inverse of any class in the principal genus, are classes in the principal genus. (The family of all ideal classes forms a group with respect to class multiplication, and the sub-family of ideal classes in the principal genus forms a sub-group.)

Theorem 14.2. *Every genus has the same number of classes.*

Proof. We let $\mathfrak{J}$ be the principal genus, and we let $\mathfrak{J}\{\mathfrak{A}\}$ denote the family of classes obtained from the product of classes in $\mathfrak{J}$ with $\{\mathfrak{A}\}$. We put all the ideal classes into various families

$$\mathfrak{J},\ \mathfrak{J}\{\mathfrak{A}_2\},\ \mathfrak{J}\{\mathfrak{A}_3\},\ldots,\ \mathfrak{J}\{\mathfrak{A}_g\}, \tag{1}$$

where $\{\mathfrak{A}_i\}$ is any class not belonging to $\mathfrak{J}$, $\mathfrak{J}\{\mathfrak{A}_2\},\ldots,\mathfrak{J}\{\mathfrak{A}_{i-1}\}$. It is easy to see that there is no ideal class which belongs to two of the families in (1).

From Theorem 14.1 we know that in each family in (1) all the classes belong to the same genus, and distinct families belong to different genera, so that each family in (1) forms a genus. Since any two classes in $\mathfrak{I}\{\mathfrak{A}_i\}$ are distinct the theorem is proved. □

16.15 Euclidean Fields and Simple Fields

Definition 15.1. If $h_0 = 1$, then we call the field a *simple field.*

It is clear that, in a simple field, every ideal is a principal ideal. Therefore we have:

Theorem 15.1. *The unique factorization theorem holds for integers in a simple field.* □

There is a type of simple fields, called Euclidean fields, having properties which are very similar to those of the rational field.

Definition 15.2. If, corresponding to any two integers ξ, η $(\eta \neq 0)$ in $R(\sqrt{D})$, there exist two integers κ, λ such that

$$\xi = \kappa\eta + \lambda \qquad |N(\lambda)| < |N(\eta)|, \tag{1}$$

then we call $R(\sqrt{D})$ an *Euclidean field.*

An alternative definition is:

Definition 15.3. If, corresponding to any δ in $R(\sqrt{D})$, there exists an integer κ such that

$$|N(\delta - \kappa)| < 1, \tag{2}$$

then we call $R(\sqrt{D})$ an *Euclidean field.*

Theorem 15.2. *Every Euclidean field is a simple field.*

Proof. Let $R(\sqrt{D})$ be Euclidean. In order to prove that $R(\sqrt{D})$ is simple it suffices to show that every ideal is a principal ideal. Let $\mathfrak{A}$ be any ideal in $R(\sqrt{D})$ and let α_1, α_2 be a basis for $\mathfrak{A}$, and we may assume without loss that $0 < |N(\alpha_1)| \leqslant |N(\alpha_2)|$. Since $R(\sqrt{D})$ is Euclidean there are integers α_2' and β_2 such that $\alpha_2 = \alpha_2'\alpha_1 + \beta_2$, $|N(\beta_2)| < |N(\alpha_1)|$. If $\beta_2 \neq 0$, then there are α_1' and β_1 such that $\alpha_1 = \alpha_1'\beta_2 + \beta_1$, $|N(\beta_1)| < |N(\beta_2)|$. Continuing with the argument, which must terminate after a finite number of steps because $|N(\alpha_1)|$ is a natural number, we arrive at an integer α such that $\mathfrak{A} = [\alpha_1, \alpha_2] = [\alpha]$. The theorem is proved. □

Theorem 15.3. *There are only five quadratic imaginary Euclidean fields, namely* $R(\sqrt{-1})$, $R(\sqrt{-2})$, $R(\sqrt{-3})$, $R(\sqrt{-7})$ *and* $R(\sqrt{-11})$.

Proof. 1) Let $D \equiv 2, 3 \pmod 4$. Put $\delta = r + s\sqrt{D}, \kappa = x + y\sqrt{D}$. Then the condition (2) becomes: corresponding to any pair of rational numbers r, s there are rational integers x, y such that

$$|(r-x)^2 - D(s-y)^2| < 1. \tag{3}$$

Setting $r = s = \frac{1}{2}$ the condition (3) gives $\frac{1}{4} + |D|\frac{1}{4} < 1$, or $|D| < 3$. Therefore $R(\sqrt{D})$ cannot be Euclidean if $D \leqslant -3$.

On the other hand, if r, s are given rational numbers we can always find rational integers x, y such that $|r-x| \leqslant \frac{1}{2}$, $|s-y| \leqslant \frac{1}{2}$ so that corresponding to $D = -1, -2$, the inequalities $|(r-x)^2 - D(s-y)^2| \leqslant \frac{1}{4} + |D|\frac{1}{4} < 1$ hold so that $R(\sqrt{-1})$ and $R(\sqrt{-2})$ are Euclidean.

2) Let $D \equiv 1 \pmod 4$. Put $\delta = r + s\sqrt{D}$, $\kappa = x + y(1+\sqrt{D})/2$ so that

$$\left|\left(r - x - \frac{y}{2}\right)^2 - D\left(s - \frac{y}{2}\right)^2\right| < 1.$$

Setting $r = s = \frac{1}{4}$ we have $\frac{1}{16} + \frac{1}{16}|D| < 1$ or $|D| < 15$. Therefore there can only be the three Euclidean fields $R(\sqrt{-3})$, $R(\sqrt{-7})$ and $R(\sqrt{-11})$, and these fields are indeed Euclidean because, given rational numbers r, s we may choose rational integers x, y such that $|2s - y| \leqslant \frac{1}{2}$, $|r - x - (y/2)| \leqslant \frac{1}{2}$, and therefore when $D = -3, -7, -11$,

$$\left|\left(r - x - \frac{y}{2}\right)^2 - D\left(s - \frac{y}{2}\right)^2\right| \leqslant \frac{1}{4} + \frac{|D|}{16} \leqslant \frac{15}{16} < 1. \qquad \square$$

In §13 we calculated the class number for $R(\sqrt{-19})$ to be 1. We see therefore that there are simple fields which are not Euclidean.

From Theorem 12.15.4 we know that there are only finitely many imaginary fields which are simple. The question then is exactly how many? It is not difficult to prove that $R(\sqrt{D})$ is simple when

$$D = -1, -2, -3, -7, -11, -19, -43, -67, -163.$$

It has also been proved that there is at most one more value of D, and that if it exists, then $D < -5 \cdot 10^9$. (In fact no extra D exists; see Notes.)

Concerning real Euclidean fields we have:

Theorem 15.4. *The field* $R(\sqrt{D})$ *is a real Euclidean field only when*

$$D = 2, 3, 5, 6, 7, 11, 13, 17, 19, 21, 29, 33, 37, 41, 57, 73. \qquad \square$$

Various Chinese mathematicians, including the author, made contributions to this problem, which in principle was eventually settled by Davenport. The proof of the theorem is beyond the scope of this book.

16.16 Lucas's Criterion for the Determination of Mersenne Primes

We first sharpen Theorem 9.5 for the quadratic field $R(\sqrt{D})$, $D > 0$. From Theorem 10.3 we know that all the prime ideals can be separated into three classes according to whether $(\frac{\Delta}{p}) = 0, +1$ or -1. We shall write q for a prime number satisfying $(\frac{\Delta}{q}) = 1$ so that $q = \mathfrak{Q}\bar{\mathfrak{Q}}$; we write r for a prime number satisfying $(\frac{\Delta}{r}) = -1$ so that r itself is a prime ideal in $R(\sqrt{D})$. From Theorem 9.5 we have, if $Q \nmid \alpha$, then

$$\alpha^{q-1} \equiv 1 \pmod{\mathfrak{Q}}, \tag{1}$$

and if $r \nmid \alpha$ then

$$\alpha^{r^2-1} \equiv 1 \pmod{r}. \tag{2}$$

Theorem 16.1. *Suppose that q, r are not 2. If $q \nmid \alpha$, then*

$$\alpha^{q-1} \equiv 1 \pmod{q}, \tag{3}$$

and if $r \nmid \alpha$, then

$$\alpha^{r+1} \equiv N(\alpha) \pmod{r}. \tag{4}$$

Observe that (1) and (3) are equivalent, and that (2) follows (4).

Proof. Let $\alpha = a + b(\Delta + \sqrt{\Delta})/2$ where a, b are rational integers. Let p be an odd prime so that, from Fermat's theorem,

$$\alpha^p \equiv a^p + b^p \frac{\Delta^p + (\sqrt{\Delta})^p}{2^p} \equiv a + \frac{b}{2}(\Delta + \Delta^{\frac{p-1}{2}}\sqrt{\Delta})$$

$$\equiv a + \frac{b}{2}\left(\Delta + \left(\frac{\Delta}{p}\right)\sqrt{\Delta}\right) \pmod{p}.$$

Therefore if $p = q$, then $\alpha^q \equiv \alpha \pmod{q}$ which gives (3), and if $p = r$, then $\alpha^r \equiv \bar{\alpha} \pmod{r}$ which gives (4). □

Now let p be an odd prime and we shall examine the nature of the Mersenne number $M = M_p = 2^p - 1$. If there exists $\Delta > 0$ such that

$$\left(\frac{\Delta}{M}\right) = -1$$

and there exists a unit ε in $R(\sqrt{\Delta})$ satisfying $N(\varepsilon) = -1$, then we let

$$r_m = \varepsilon^{2^m} + \varepsilon'^{2^m},$$

where ε' is the conjugate of ε.

Theorem 16.2. *A necessary and sufficient condition for M to be prime is that*

$$r_{p-1} \equiv 0 \pmod{M}. \tag{6}$$

Proof. 1) Assume that M is a prime. From (5) we know that M is of the type r, and so from Theorem 16.1 we have $\varepsilon^{M+1} \equiv -1 \pmod{M}$ and therefore

$$\varepsilon^{2^{p-1}} + \varepsilon'^{2^{p-1}} = \varepsilon'^{2^{p-1}}(\varepsilon^{2^p} + 1) \equiv \varepsilon'^{2^{p-1}}(\varepsilon^{M+1} + 1) \equiv 0 \pmod{M}.$$

2) Assume that M is composite, say $M = q_1 \cdots q_s r_1 \cdots r_t$. From (5) we know that at least one of the prime divisors of M is of type r. If (6) holds, then $M | r_{p-1}$ or

$$\varepsilon^{2^{p-1}} + \varepsilon'^{2^{p-1}} \equiv 0 \pmod{M},$$

and hence

$$\varepsilon^{2^p} \equiv -1 \pmod{M}, \tag{7}$$

and on squaring

$$\varepsilon^{2^{p+1}} \equiv 1 \pmod{M}. \tag{8}$$

Let $\mathfrak{P}$ be a prime ideal divisor of M and let l be the least positive integer satisfying $\varepsilon^l \equiv 1 \pmod{\mathfrak{P}}$. Then, by (8), $l | 2^{p+1}$, and so by (7), $l = 2^{p+1}$.

If $\mathfrak{P}$ is a divisor of a certain q, then $\varepsilon^{q-1} \equiv 1 \pmod{\mathfrak{P}}$ by Theorem 16.1, and hence $2^{p+1} | q - 1$, which is impossible because q cannot exceed M.

If $\mathfrak{P}$ is a certain r, then $\varepsilon^{r+1} \equiv -1 \pmod{r}$ by Theorem 16.1. This then gives $2^{p+1} | 2(r + 1)$ and so $r = 2^p m - 1$. But $r \leqslant M$ so that $m = 1, r = M$. That is M must be prime after all. □

Example. Take $\Delta = 5$, $\varepsilon = (1 + \sqrt{5})/2$ so that

$$r_{p-1} = (\tfrac{1}{2}(1 + \sqrt{5}))^{2^{p-1}} + (\tfrac{1}{2}(1 - \sqrt{5}))^{2^{p-1}}.$$

If we take $p = 7$, $M_p = 127$, then the residues mod 127 for r_m $(m = 1, 2, 3, 4, 5, 6)$ are $3, 7, 47, 48, 16, 0$. Therefore 127 is a prime. Of course the full power of the theorem is not revealed in this specific example. However, with the aid of electronic computers, the same method can be used to show, for example, that the 687 digit number $M_{2281} = 2^{2281} - 1$ is prime. Indeed all the large known Mersenne primes are found by essentially the same type of method.

16.17 Indeterminate Equations

The invention of the theory of ideals to tackle Fermat's problem is an important development in algebraic number theory. From the standpoint of mathematics this theory is far more important than that of settling a difficult problem. Let p be an odd prime and $\rho = e^{2\pi i/p}$. If we can prove that

$$\xi^p + \eta^p + \zeta^p = 0, \qquad \xi\eta\zeta \neq 0$$

has no integer solutions in the field $R(\rho)$, then obviously Fermat's Last Theorem is established. The expression $\xi^p + \eta^p$ can be factorized into linear terms in $R(\rho)$ so that the problem is easier to start with. Indeed this is Kummer's starting point in his research on Fermat's problem, but the principal difficulty lies with the absence of a unique factorization theorem. It is for this reason that Kummer invented his theory of ideals which has now become an indispensable part of mathematics.

It is not easy to understand Kummer's method. That is, even if we assume that there is unique factorization in $R(\rho)$, we still need a deep theorem of Kummer's before we can settle Fermat's problem. The theorem concerned is as follows: A necessary and sufficient condition for a unit ε in $R(\rho)$ to be a p-th power of another unit is that ε is congruent to a rational number mod $(1 - \rho)^p$. We can only consider two simple examples in this book.

Theorem 17.1. *The equation*

$$\xi^4 + \eta^4 = \tau^2, \qquad \xi\eta\tau \neq 0 \tag{1}$$

has no solution in integers in $R(\sqrt{-1})$.

Proof. The unique factorization theorem holds in the field $R(\sqrt{-1})$, that is every ideal is a principal ideal. We may therefore assume without loss that $(\xi, \eta) = 1$.

1) Let $\lambda = 1 - i$. Then λ is irreducible, and $\lambda^2 = -2i$ and $2 = i(1-i)^2$ are associates. Also $N(2) = 4$ so that every integer in $R(\sqrt{-1})$ must be congruent to one of the four numbers $0, 1, i, 1 - i \pmod 2$.

Since $0, 1 - i$ are divisible by λ, any integer α not divisible by λ must satisfy $\alpha \equiv 1$ or $i \pmod{\lambda^2}$ so that $\alpha = 1 + \beta\lambda^2$ or $\alpha = i + \beta\lambda^2$, and hence

$$\alpha^4 \equiv 1 \pmod{\lambda^6}. \tag{2}$$

Now let ξ, η, τ satisfy (1). Suppose, if possible, that ξ, η are not divisible by λ. From (2) and (1) we have $2 \equiv \tau^2 \pmod{\lambda^6}$. Since $2 = \lambda^2 i$ we see that $\lambda | \tau$. Write $\tau = \lambda\gamma$ so that $\lambda \nmid \gamma$, and $i\lambda^2 \equiv \lambda^2\gamma^2 \pmod{\lambda^6}$ or $\gamma^2 \equiv i \pmod{\lambda^4}$. On squaring this we deduce from (2) that $1 \equiv \gamma^4 \equiv -1 \pmod{\lambda^4}$, which is impossible. Therefore one of ξ, η is divisible by λ. By symmetry we may assume that $\lambda | \xi$, and we now write $\xi = \lambda^n\delta$, $n \geqslant 1$, $\lambda \nmid \delta$, so that we have

$$\lambda^{4n}\delta^4 = \tau^2 - \eta^4, \qquad n \geqslant 1, \quad \lambda \nmid \delta\eta, \quad (\delta, \eta) = 1.$$

2) We now prove a more general result, namely that there are no integers δ, τ, η in $R(\sqrt{-1})$ such that

$$\varepsilon\lambda^{4n}\delta^4 = \tau^2 - \eta^4, \qquad \varepsilon \text{ unit}, \quad \lambda \nmid \delta\eta, \quad (\delta, \eta) = 1, \quad n \geqslant 1. \tag{3}$$

The proof is divided into two steps. In the first step we show that if (3) is soluble then n must be at least 2; in the second step we show that if (3) is soluble for a certain n, then it is soluble for $n - 1$ also. The theorem therefore follows from this contradiction.

If (3) holds for integers δ, τ, η then $\lambda \nmid \tau$. Since $N(\lambda) = 2$ we see that $\tau \equiv 1 \pmod{\lambda}$. Let $\tau = 1 + \mu\lambda$ so that on squaring we have

$$\tau^2 = 1 + 2\mu\lambda + \mu^2\lambda^2 \equiv 1 + \mu^2\lambda^2 \pmod{\lambda^3}.$$

Also, by (2),

$$\eta^4 \equiv 1 \pmod{\lambda^6} \tag{4}$$

so that, by (3),

$$0 \equiv \varepsilon\lambda^{4n}\delta^4 \equiv \tau^2 - \eta^4 \equiv \mu^2\lambda^2 \pmod{\lambda^3}.$$

Thus $\lambda | \mu$ and we may write

$$\tau = 1 + v\lambda^2,$$
$$\tau^2 = 1 + 2v\lambda^2 + v^2\lambda^4 = 1 + \lambda^4 v(i + v). \tag{5}$$

Since v, $i + v$ form a complete residue system mod λ we have $v(i + v) \equiv 0 \pmod{\lambda}$ giving $\tau^2 \equiv 1 \pmod{\lambda^5}$. From (3) and (4) we deduce that $\varepsilon\lambda^{4n}\delta^4 \equiv \tau^2 - \eta^4 \equiv 0 \pmod{\lambda^5}$, and we conclude that $n \geqslant 2$.

Now assume that δ, τ, η satisfy (3) with $n \geqslant 2$. Then $\varepsilon\lambda^{4n}\delta^4 = (\tau - \eta^2)(\tau + \eta^2)$. From (5) we have $\tau \equiv 1 \pmod{\lambda^2}$, and on the other hand, since $\lambda \nmid \eta$ we have

$$\eta^2 = (1 + \kappa\lambda)^2 \equiv 1 \pmod{\lambda^2}. \tag{6}$$

Thus $\tau - \eta^2 \equiv 0 \pmod{\lambda^2}$ and $\tau + \eta^2 \equiv 2 \equiv 0 \pmod{\lambda^2}$. Since

$$\left(\frac{\tau - \eta^2}{\lambda^2}, \frac{\tau + \eta^2}{\lambda^2}\right) = \left(\frac{\tau - \eta^2}{\lambda^2}, \tau, \eta^2\right) = 1,$$

it follows from

$$\varepsilon\lambda^{4(n-1)}\delta^4 = \frac{\tau - \eta^2}{\lambda^2}\frac{\tau + \eta^2}{\lambda^2} \tag{7}$$

that $\lambda^{4(n-1)}$ must divide one of these two divisors. We may assume that $\lambda^{4(n-1)}$ actually divides the latter divisor, since otherwise we may replace η by $i\eta$. From (7) we have

$$\frac{\tau - \eta^2}{\lambda^2} = \varepsilon_1\sigma^4, \qquad \frac{\tau + \eta^2}{\lambda^2} = \varepsilon_2\lambda^{4(n-1)}\varphi^4 \qquad (\lambda \nmid \varphi\sigma, (\sigma, \varphi) = 1),$$

where ε_1, ε_2 are two units. Thus

$$i\eta^2 = \frac{2\eta^2}{\lambda^2} = \varepsilon_2\lambda^{4(n-1)}\varphi^4 - \varepsilon_1\sigma^4,$$

or

$$\eta^2 - \varepsilon_3\sigma^4 = \varepsilon_4\lambda^{4(n-1)}\varphi^4,$$

where $\varepsilon_3 = -\varepsilon_1/i$, $\varepsilon_4 = \varepsilon_2/i$ are also units.

Since $n \geqslant 2$, $\lambda \nmid \sigma$ we see from (2) that $\eta^2 \equiv \varepsilon_3 \pmod{\lambda^4}$ and hence, by (6), $1 \equiv \varepsilon_3 \pmod{\lambda^2}$. Therefore ε_3 is either $+1$ or -1 and not $\pm i$, that is

$$\varepsilon_4\lambda^{4(n-1)}\varphi^4 = \eta^2 \mp \sigma^4, \qquad \lambda \nmid \varphi\sigma, \quad (\varphi, \sigma) = 1.$$

If we take the negative sign here then our second step follows at once, and if we take the positive sign then the same result is obtained by replacing η by $i\eta$. □

Theorem 17.2. *The equation*

$$\xi^3 + \eta^3 + \zeta^3 = 0, \qquad \xi\eta\zeta \neq 0 \tag{8}$$

has no solution in integers in $R(\rho)$, $\rho = (-1 + \sqrt{-3})/2$.

Proof. Since $R(\rho)$ is a simple field we may assume that $(\xi, \eta) = 1$.

1) Let $\lambda = 1 - \rho$, so that $1 - \rho^2 = -\rho^2(1 - \rho) = -\rho^2\lambda$ and $N(\lambda) = -\rho^2\lambda^2 = 3$. Therefore λ is irreducible and all the integers are partitioned into three classes represented by $0, 1, -1$. Therefore, if $\lambda \nmid \xi$, then $\xi \equiv \pm 1 \pmod{\lambda}$. We shall now show that

$$\xi^3 \equiv \pm 1 \pmod{\lambda^4}. \tag{9}$$

We need only consider the $+$ sign case, since otherwise we may replace ξ by $-\xi$. Let $\xi = 1 + \beta\lambda$ so that

$$\begin{aligned}\xi^3 - 1 &= (\xi - 1)(\xi - \rho)(\xi - \rho^2) = \beta\lambda(\beta\lambda + 1 - \rho)(\beta\lambda + 1 - \rho^2)\\ &= \beta\lambda(\beta\lambda + \lambda)(\beta\lambda - \rho^2\lambda) = \lambda^3\beta(\beta + 1)(\beta - \rho^2).\end{aligned}$$

Since β, $\beta + 1$, $\beta - \rho^2$ are incongruent mod λ, and $N(\lambda) = 3$ there must be one of them which is divisible by λ. We deduce that if $\lambda \nmid \eta$, then

$$\eta^3 \equiv \pm 1 \pmod{\lambda^4}. \tag{10}$$

Now if $\lambda \nmid \xi\eta\zeta$, then $0 \equiv \xi^3 + \eta^3 + \zeta^3 \equiv \pm 1 \pm 1 \pm 1 \pmod{\lambda^3}$. The possible choices are ± 1, ± 3 and none of them is divisible by λ^3, so that one of ξ, η, ζ must be divisible by λ. Let it be $\zeta = \lambda^n\gamma$, $n \geqslant 1$, $\lambda \nmid \gamma$ so that

$$\xi^3 + \eta^3 + \lambda^{3n}\gamma^3 = 0, \qquad (\xi, \eta) = 1, \quad \lambda \nmid \gamma, \quad n \geqslant 1.$$

2) We shall now prove a more general result, namely that

$$\xi^3 + \eta^3 + \varepsilon\lambda^{3n}\gamma^3 = 0, \qquad (\xi, \eta) = 1, \quad \lambda \nmid \gamma, \quad n \geqslant 1, \tag{11}$$

where ε is a unit, has no integer solutions in $R(\rho)$. As in the proof of Theorem 17.1 we separate into two steps where, in the first step, we show that if (11) has a solution, then $n \geqslant 2$, and in the second step we show that if (11) has a solution, then n may be replaced by $n - 1$ and there is still a solution. The theorem then follows by this contradiction.

If (11) has a solution, then by (10)

$$-\varepsilon\lambda^{3n}\gamma^3 \equiv \xi^3 + \eta^3 \equiv \pm 1 \pm 1 \pmod{\lambda^4}.$$

Since $+1+1$ and $-1-1$ are not divisible by λ we see that $-\varepsilon\lambda^{3n}\gamma^3 \equiv 0 \pmod{\lambda^4}$, and so $n \geqslant 2$.

Suppose that ξ, η, γ are solutions to (11). From $1 \equiv \rho \equiv \rho^2 \pmod{\lambda}$ we deduce that $\xi + \eta \equiv \xi + \rho\eta \equiv \xi + \rho^2\eta \pmod{\lambda}$ and hence $-\varepsilon\lambda^{3n}\gamma^3 = \xi^3 + \eta^3 = (\xi + \eta)(\xi + \rho\eta)(\xi + \rho^2\eta)$ where the three divisors are all multiples of λ.

It is not difficult to show that $(\xi + \eta)/\lambda$, $(\xi + \rho\eta)/\lambda$, $(\xi + \rho^2\eta)/\lambda$ are pairwise coprime. In fact, for example, from $(\xi + \eta) - (\xi + \rho\eta) = \lambda\eta$ and $\rho(\xi + \eta) - (\xi + \rho\eta) = -\lambda\xi$ we see that $(\xi + \eta)/\lambda$ and $(\xi + \rho\eta)/\lambda$ are coprime. Thus one of the three divisors in the factorization

$$-\varepsilon\lambda^{3(n-1)}\gamma^3 = \frac{\xi + \eta}{\lambda}\,\frac{\xi + \rho\eta}{\lambda}\,\frac{\xi + \rho^2\eta}{\lambda}$$

must be a multiple of $\lambda^{3(n-1)}$, and we may assume that it is $(\xi + \eta)/\lambda$ since otherwise we can replace η by $\rho\eta$ or $\rho^2\eta$. Hence

$$\xi + \eta = \varepsilon_1\lambda^{3n-2}\mu^3, \qquad \xi + \rho\eta = \varepsilon_2\lambda\nu^3, \qquad \xi + \rho^2\eta = \varepsilon_3\lambda\sigma^3, \tag{12}$$

where $\varepsilon_1, \varepsilon_2, \varepsilon_3$ are units and μ, ν, σ are pairwise coprime integers not divisible by λ.

From (12) we have

$$0 = \xi + \eta + \rho(\xi + \rho\eta) + \rho^2(\xi + \rho^2\eta) = \varepsilon_1\lambda^{3n-2}\mu^3 + \rho\varepsilon_2\lambda\nu^3 + \rho^2\varepsilon_3\lambda\sigma^3,$$

giving

$$\nu^3 + \varepsilon_4\sigma^3 + \varepsilon_5\lambda^{3(n-1)}\mu^3 = 0, \qquad (\nu, \sigma) = 1, \quad \lambda \nmid \mu \tag{13}$$

where $\varepsilon_4, \varepsilon_5$ are also units. From (13) we have $\nu^3 + \varepsilon_4\sigma^3 \equiv 0 \pmod{\lambda^2}$ and here, by (10), $\pm 1 \pm \varepsilon_4 \equiv 0 \pmod{\lambda^2}$. Among the units $\pm 1, \pm\rho, \pm\rho^2$ only $\varepsilon_4 = \pm 1$ can satisfy this congruence. Hence $\varepsilon_4 = \pm 1$ and we see that (13) is the same as (11) with n replaced by $n - 1$. The theorem is proved. □

16.18 Tables

We conclude this chapter with two tables displaying all the quadratic fields $R(\sqrt{D})$ with $-100 < D \leqslant 100$. We list their integral basis, discriminants, ideal classes and the quadratic forms associated with the ideal classes together with their character systems. We also display the continued fraction representations for ω and the fundamental units in the second table. More precisely:

In Table I, the first column is the value for D. The second column is ω (see the definition in Theorem 4.5). The third column is the discriminant Δ. The fourth column displays the ideal classes of $R(\sqrt{D})$. The fifth column indicates the relationship between the ideal classes. The sixth column displays the quadratic forms representing the classes of forms corresponding to the ideal classes. The seventh column is the character systems associated with these classes of forms.

In Table II, the first two columns are as before. The third column displays the continued fractions expansion representing ω when D is square-free and representing $\sqrt{D}$ when D is not square-free. The fourth column is the discriminant Δ. The fifth column displays $x + y\sqrt{D}$ when D is square-free and it is the fundamental unit η of $R(\sqrt{D})$; when D is not square-free it displays the least positive integer solutions to $x^2 - y^2D = \pm 1$ (if $x^2 - y^2D = -1$ is soluble, then $x + y\sqrt{D}$ satisfies $x^2 - y^2D = -1$, otherwise x, y satisfy $x^2 - y^2D = +1$). The sixth column is $N(x + y\sqrt{D})$. The last four columns are the same as the last four columns in Table I.

Table I

D	ω	Δ	Ideal classes	Relations	Quadratic forms	Character systems
-1	$\sqrt{-1}$	-2^2	(1)	1	x^2+y^2	$+1$
-2	$\sqrt{-2}$	-2^3	(1)	1	$2x^2+y^2$	$+1$
-3	$\frac{1+\sqrt{-3}}{2}$	-3	(1)	1	x^2+xy+y^2	$+1$
-5	$\sqrt{-5}$	$-2^2\cdot 5$	(1)	A^2	$5x^2+y^2$	$+1, +1$
			$(2, 1+\sqrt{-5})$	A	$3x^2+2xy+2y^2$	$-1, -1$
-6	$\sqrt{-6}$	$-2^3\cdot 3$	(1)	A^2	$6x^2+y^2$	$+1, +1$
			$(2, \sqrt{-6})$	A	$3x^2+2y^2$	$-1, -1$
-7	$\frac{1+\sqrt{-7}}{2}$	-7	(1)	1	$2x^2+xy+y^2$	$+1$
-10	$\sqrt{-10}$	$-5\cdot 2^3$	(1)	A^2	$10x^2+y^2$	$+1, +1$
			$(2, \sqrt{-10})$	A	$5x^2+2y^2$	$-1, -1$
-11	$\frac{1+\sqrt{-11}}{2}$	-11	(1)	1	$3x^2+xy+y^2$	$+1$
-13	$\sqrt{-13}$	$-2^2\cdot 13$	(1)	A^2	$13x^2+y^2$	$+1, +1$
			$(2, 1+\sqrt{-13})$	A	$7x^2+2xy+2y^2$	$-1, -1$
-14	$\sqrt{-14}$	$-7\cdot 2^3$	(1)	l^4	$14x^2+y^2$	$+1, +1$
			$(3, 2+\sqrt{-14})$	l^3	$6x^2+4xy+3y^2$	$-1, -1$
			$(2, \sqrt{-14})$	l^2	$7x^2+2y^2$	$+1, +1$
			$(3, 1+\sqrt{-14})$	l	$5x^2+2xy+3y^2$	$-1, -1$
-15	$\frac{1+\sqrt{-15}}{2}$	$-3\cdot 5$	(1)	A^2	$4x^2+xy+y^2$	$+1, +1$
			$(2, 1+\omega)$	A	$3x^2+3xy+2y^2$	$-1, -1$
-17	$\sqrt{-17}$	$-2^2\cdot 17$	(1)	l^4	$17x^2+y^2$	$+1, +1$
			$(3, 2+\sqrt{-17})$	l^3	$7x^2+4xy+3y^2$	$-1, -1$
			$(2, 1+\sqrt{-17})$	l^2	$9x^2+2xy+2y^2$	$+1, +1$
			$(3, 1+\sqrt{-17})$	l	$6x^2+2xy+3y^2$	$-1, -1$
-19	$\frac{1+\sqrt{-19}}{2}$	-19	(1)	1	$5x^2+xy+y^2$	$+1$
-21	$\sqrt{-21}$	$-3\cdot 2^2\cdot 7$	(1)	$A^2A_1^2$	$+21x^2+y^2$	$+1, +1, +1$
			$(5, 3+\sqrt{-21})$	AA_1	$6x^2+6xy+5y^2$	$-1, -1, +1$
			$(3, \sqrt{-21})$	A_1	$7x^2+3y^2$	$+1, -1, -1$
			$(2, 1+\sqrt{-21})$	A	$11x^2+2xy+2y^2$	$-1, +1, -1$
-22	$\sqrt{-22}$	$-2^3\cdot 11$	(1)	A^2	$22x^2+y^2$	$+1, +1$
			$(2, \sqrt{-22})$	A	$11x^2+2y^2$	$-1, -1$

Table I (continued)

D	ω	Δ	Ideal classes	Relations	Quadratic forms	Character systems
-23	$\frac{1+\sqrt{-23}}{2}$	-23	(1)	l^3	$6x^2+xy+y^2$	$+1$
			$\left(2, 1+\frac{1+\sqrt{-23}}{2}\right)$	l^2	$4x^2+3xy+2y^2$	$+1$
			$\left(2, \frac{1+\sqrt{-23}}{2}\right)$	l	$3x^2+2xy+2y^2$	$+1$
-26	$\sqrt{-26}$	$-2^3\cdot 13$	(1)	l^6	$26x^2+y^2$	$+1, +1$
			$(5, 3+\sqrt{-26})$	l^5	$7x^2+6xy+5y^2$	$-1, -1$
			$(3, 1+\sqrt{-26})$	l^4	$9x^2+2xy+3y^2$	$+1, +1$
			$(2, \sqrt{-26})$	l^3	$13x^2+2y^2$	$-1, -1$
			$(3, 2+\sqrt{-26})$	l^2	$10x^2+4xy+3y^2$	$+1, +1$
			$(5, 2+\sqrt{-26})$	l	$6x^2+4xy+5y^2$	$-1, -1$
-29	$\sqrt{-29}$	$-2^2\cdot 29$	(1)	l^6	$29x^2+y^2$	$+1, +1$
			$(3, 2+\sqrt{-29})$	l^5	$11x^2+4xy+3y^2$	$-1, -1$
			$(5, 4+\sqrt{-29})$	l^4	$9x^2+8xy+5y^2$	$+1, +1$
			$(2, 1+\sqrt{-29})$	l^3	$15x^2+2xy+2y^2$	$-1, -1$
			$(5, 1+\sqrt{-29})$	l^2	$6x^2+2xy+5y^2$	$+1, +1$
			$(3, 1+\sqrt{-29})$	l	$10x^2+2xy+3y^2$	$-1, -1$
-30	$\sqrt{-30}$	$-2^3\cdot 3\cdot 5$	(1)	$A^2A_1^2$	$30x^2+y^2$	$+1, +1, +1$
			$(2, \sqrt{-30})$	AA_1	$15x^2+2y^2$	$-1, -1, +1$
			$(3, \sqrt{-30})$	A_1	$10x^2+3y^2$	$+1, -1, -1$
			$(5, \sqrt{-30})$	A	$6x^2+5y^2$	$-1, +1, -1$
-31	$\frac{1}{2}(1+\sqrt{-31})$	-31	(1)	l^3	$8x^2+xy+y^2$	$+1$
			$(2, \omega)$	l^2	$4x^2+xy+2y^2$	$+1$
			$(2, 1+\omega)$	l	$5x^2+3xy+2y^2$	$+1$
-33	$\sqrt{-33}$	$-2^2\cdot 3\cdot 11$	(1)	$A^2A_1^2$	$33x^2+y^2$	$+1, +1, +1$
			$(2, 1+\sqrt{-33})$	AA_1	$17x^2+2xy+2y^2$	$-1, -1, +1$
			$(3, \sqrt{-33})$	A_1	$11x^2+3y^2$	$-1, +1, -1$
			$(6, 3+\sqrt{-33})$	A	$7x^2+6xy+6y^2$	$+1, -1, -1$
-34	$\sqrt{-34}$	$-2^3\cdot 17$	(1)	l^4	$34x^2+y^2$	$+1, +1$
			$(5, 4+\sqrt{-34})$	l^3	$10x^2+8xy+5y^2$	$-1, -1$
			$(2, \sqrt{-34})$	l^2	$17x^2+2y^2$	$+1, +1$
			$(5, 1+\sqrt{-34})$	l	$7x^2+2xy+5y^2$	$-1, -1$
-35	$\frac{1}{2}(1+\sqrt{-35})$	$-5\cdot 7$	(1)	A^2	$9x^2+xy+y^2$	$+1, +1$
			$\left(5, \frac{5+\sqrt{-35}}{2}\right)$	A	$3x^2+5xy+5y^2$	$-1, -1$

Table I (continued)

D	ω	Δ	Ideal classes	Relations	Quadratic forms	Character systems
-37	$\sqrt{-37}$	$-2^2\cdot 37$	(1)	A^2	$37x^2+y^2$	$+1, +1$
			$(2, 1+\sqrt{-37})$	A	$19x^2+2xy+2y^2$	$-1, -1$
-38	$\sqrt{-38}$	$-2^3\cdot 19$	(1)	l^6	$38x^2+y^2$	$+1, +1$
			$(3, 2+\sqrt{-38})$	l^5	$14x^2+4xy+3y^2$	$-1, -1$
			$(7, 2+\sqrt{-38})$	l^4	$6x^2+4xy+7y^2$	$+1, +1$
			$(2, \sqrt{-38})$	l^3	$19x^2+2y^2$	$-1, -1$
			$(7, 5+\sqrt{-38})$	l^2	$9x^2+10xy+7y^2$	$+1, +1$
			$(3, 1+\sqrt{-38})$	l	$13x^2+2xy+3y^2$	$-1, -1$
-39	$\frac{1}{2}(1+\sqrt{-39})$	$-3\cdot 13$	(1)	l^4	$10x^2+xy+y^2$	$+1, +1$
			$(2, 1+\omega)$	l^3	$6x^2+3xy+2y^2$	$-1, -1$
			$(3, 1+\omega)$	l^2	$4x^2+3xy+3y^2$	$+1, +1$
			$(2, \omega)$	l	$5x^2+xy+2y^2$	$-1, -1$
-41	$\sqrt{-41}$	$-2^2\cdot 41$	(1)	l^8	$41x^2+y^2$	$+1, +1$
			$(3, 2+\sqrt{-41})$	l^7	$15x^2+4xy+3y^2$	$-1, -1$
			$(5, 3+\sqrt{-41})$	l^6	$10x^2+6xy+5y^2$	$+1, +1$
			$(7, 6+\sqrt{-41})$	l^5	$11x^2+12xy+7y^2$	$-1, -1$
			$(2, 1+\sqrt{-41})$	l^4	$21x^2+2xy+2y^2$	$+1, +1$
			$(7, 1+\sqrt{-41})$	l^3	$6x^2+2xy+7y^2$	$-1, -1$
			$(5, 2+\sqrt{-41})$	l^2	$9x^2+4xy+5y^2$	$+1, +1$
			$(3, 1+\sqrt{-41})$	l	$14x^2+2xy+3y^2$	$-1, -1$
-42	$\sqrt{-42}$	$-3\cdot 2^3\cdot 7$	(1)	$A^2A_1^2$	$42x^2+y^2$	$+1, +1, +1$
			$(7, \sqrt{-42})$	AA_1	$6x^2+7y^2$	$+1, -1, -1$
			$(3, \sqrt{-42})$	A_1	$14x^2+3y^2$	$-1, -1, +1$
			$(2, \sqrt{-42})$	A	$21x^2+2y^2$	$-1, +1, -1$
-43	$\frac{1}{2}(1+\sqrt{-43})$	-43	(1)	1	$11x^2+xy+y^2$	$+1$
-46	$\sqrt{-46}$	$-2^3\cdot 23$	(1)	l^4	$46x^2+y^2$	$+1, +1$
			$(5, 3+\sqrt{-46})$	l^3	$11x^2+6xy+5y^2$	$-1, -1$
			$(2, \sqrt{-46})$	l^2	$23x^2+2y^2$	$+1, +1$
			$(5, 2+\sqrt{-46})$	l	$10x^2+4xy+5y^2$	$-1, -1$
-47	$\frac{1}{2}(1+\sqrt{-47})$	-47	(1)	l^5	$12x^2+xy+y^2$	$+1$
			$(2, \omega)$	l^4	$6x^2+xy+2y^2$	$+1$
			$(3, 2+\omega)$	l^3	$6x^2+5xy+3y^2$	$+1$
			$(3, \omega)$	l^2	$4x^2+xy+3y^2$	$+1$
			$(2, 1+\omega)$	l	$7x^2+3xy+2y^2$	$+1$
-51	$\frac{1}{2}(1+\sqrt{-51})$	$-3\cdot 17$	(1)	A^2	$13x^2+xy+y^2$	$+1, +1$
			$(3, 1+\omega)$	A	$5x^2+3xy+3y^2$	$-1, -1$

Table I (continued)

D	ω	Δ	Ideal classes	Relations	Quadratic forms	Character systems
-53	$\sqrt{-53}$	$-2^2\cdot 53$	(1)	l^6	$53x^2+y^2$	$+1, +1$
			$(3, 2+\sqrt{-53})$	l^5	$19x^2+4xy+3y^2$	$-1, -1$
			$(9, 8+\sqrt{-53})$	l^4	$13x^2+16xy+9y^2$	$+1, +1$
			$(2, 1+\sqrt{-53})$	l^3	$27x^2+2xy+2y^2$	$-1, -1$
			$(9, 1+\sqrt{-53})$	l^2	$6x^2+2xy+9y^2$	$+1, +1$
			$(3, 1+\sqrt{-53})$	l	$18x^2+2xy+3y^2$	$-1, -1$
-55	$\frac{1}{2}(1+\sqrt{-55})$	$-5\cdot 11$	(1)	l^4	$14x^2+xy+y^2$	$+1, +1$
			$(2, 1+\omega)$	l^3	$8x^2+3xy+2y^2$	$-1, -1$
			$(5, 2+\omega)$	l^2	$4x^2+5xy+5y^2$	$+1, +1$
			$(2, \omega)$	l	$7x^2+xy+2y^2$	$-1, -1$
-57	$\sqrt{-57}$	$-3\cdot 2^2\cdot 19$	(1)	$A^2A_1^2$	$57x^2+y^2$	$+1, +1, +1$
			$(2, 1+\sqrt{-57})$	AA_1	$29x^2+2xy+2y^2$	$-1, -1, +1$
			$(3, \sqrt{-57})$	A_1	$19x^2+3y^2$	$+1, -1, -1$
			$(6, 3+\sqrt{-57})$	A	$11x^2+6xy+6y^2$	$-1, +1, -1$
-58	$\sqrt{-58}$	$-2^3\cdot 29$	(1)	A^2	$58x^2+y^2$	$+1, +1$
			$(2, \sqrt{-58})$	A	$29x^2+2y^2$	$-1, -1$
-59	$\frac{1}{2}(1+\sqrt{-59})$	-59	(1)	l^3	$15x^2+xy+y^2$	$+1$
			$\left(3, \frac{5+\sqrt{-59}}{2}\right)$	l^2	$7x^2+5xy+3y^2$	$+1$
			$\left(3, \frac{1+\sqrt{-59}}{2}\right)$	l	$5x^2+xy+3y^2$	$+1$
-61	$\sqrt{-61}$	$-2^2\cdot 61$	(1)	l^3	$61x^2+y^2$	$+1, +1$
			$(5, 3+\sqrt{-61})$	l^2	$14x^2+6xy+5y^2$	$+1, +1$
			$(5, 2+\sqrt{-61})$	l	$13x^2+4xy+5y^2$	$+1, +1$
			$(7, 4+\sqrt{-61})$	Al^2	$11x^2+8xy+7y^2$	$-1, -1$
			$(7, 3+\sqrt{-61})$	Al	$10x^2+6xy+7y^2$	$-1, -1$
			$(2, 1+\sqrt{-61})$	A	$31x^2+2xy+2y^2$	$-1, -1$
-62	$\sqrt{-62}$	$-2^3\cdot 31$	(1)	l^8	$62x^2+y^2$	$+1, +1$
			$(3, 2+\sqrt{-62})$	l^7	$22x^2+4xy+3y^2$	$-1, -1$
			$(7, 1+\sqrt{-62})$	l^6	$9x^2+2xy+7y^2$	$+1, +1$
			$(11, 2+\sqrt{-62})$	l^5	$6x^2+4xy+11y^2$	$-1, -1$
			$(2, \sqrt{-62})$	l^4	$31x^2+2y^2$	$+1, +1$
			$(11, 9+\sqrt{-62})$	l^3	$13x^2+18xy+11y^2$	$-1, -1$
			$(7, 6+\sqrt{-62})$	l^2	$14x^2+12xy+7y^2$	$+1, +1$
			$(3, 1+\sqrt{-62})$	l	$21x^2+2xy+3y^2$	$-1, -1$

Table I (continued)

D	ω	Δ	Ideal classes	Relations	Quadratic forms	Character systems
-65	$\sqrt{-65}$	$-2^2\cdot 5\cdot 13$	(1)	l^4	$65x^2+y^2$	$+1, +1, +1$
			$(3, 2+\sqrt{-65})$	l^3	$23x^2+4xy+3y^2$	$-1, +1, -1$
			$(9, 4+\sqrt{-65})$	l^2	$9x^2+8xy+9y^2$	$+1, +1, +1$
			$(3, 1+\sqrt{-65})$	l	$22x^2+2xy+3y^2$	$-1, +1, -1$
			$(11, 10+\sqrt{-65})$	Al^3	$15x^2+20xy+11y^2$	$+1, -1, -1$
			$(2, 1+\sqrt{-65})$	Al^2	$33x^2+2xy+2y^2$	$-1, -1, +1$
			$(11, 1+\sqrt{-65})$	Al	$6x^2+2xy+11y^2$	$+1, -1, -1$
			$(5, \sqrt{-65})$	A	$13x^2+5y^2$	$-1, -1, +1$
-66	$\sqrt{-66}$	$-2^3\cdot 3\cdot 11$	(1)	l^4	$66x^2+y^2$	$+1, +1, +1$
			$(5, 3+\sqrt{-66})$	l^3	$15x^2+6xy+5y^2$	$-1, +1, -1$
			$(3, \sqrt{-66})$	l^2	$22x^2+3y^2$	$+1, +1, +1$
			$(5, 2+\sqrt{-66})$	l	$14x^2+4xy+5y^2$	$-1, +1, -1$
			$(7, 2+\sqrt{-66})$	Al^2	$10x^2+4xy+7y^2$	$+1, -1, -1$
			$(11, \sqrt{-66})$	Al^2	$6x^2+11y^2$	$-1, -1, +1$
			$(7, 5+\sqrt{-66})$	Al	$13x^2+10xy+7y^2$	$+1, -1, -1$
			$(2, \sqrt{-66})$	A	$33x^2+2y^2$	$-1, -1, +1$
-67	$\frac{1}{2}(1+\sqrt{-67})$	-67	(1)	1	$17x^2+xy+y^2$	$+1$
-69	$\sqrt{-69}$	$-2^2\cdot 3\cdot 23$	(1)	l^4	$69x^2+y^2$	$+1, +1, +1$
			$(7, 6+\sqrt{-69})$	l^3	$15x^2+12xy+7y^2$	$+1, -1, -1$
			$(6, 3+\sqrt{-69})$	l^2	$13x^2+6xy+6y^2$	$+1, +1, +1$
			$(7, 1+\sqrt{-69})$	l	$10x^2+2xy+7y^2$	$+1, -1, -1$
			$(5, 1+\sqrt{-69})$	Al^3	$14x^2+2xy+5y^2$	$-1, -1, +1$
			$(3, \sqrt{-69})$	Al^2	$23x^2+3y^2$	$-1, +1, -1$
			$(5, 4+\sqrt{-69})$	Al	$17x^2+8xy+5y^2$	$-1, -1, +1$
			$(2, 1+\sqrt{-69})$	A	$35x^2+2xy+2y^2$	$-1, +1, -1$
-70	$\sqrt{-70}$	$-2^3\cdot 5\cdot 7$	(1)	$A^2A_1^2$	$70x^2+y^2$	$+1, +1, +1$
			$(7, \sqrt{-70})$	AA_1	$10x^2+7y^2$	$-1, -1, +1$
			$(5, \sqrt{-70})$	A_1	$14x^2+5y^2$	$+1, -1, -1$
			$(2, \sqrt{-70})$	A	$35x^2+2y^2$	$-1, +1, -1$
-71	$\frac{1}{2}(1+\sqrt{-71})$	-71	(1)	l^7	$71x^2+y^2$	$+1$
			$\left(2, \frac{3+\sqrt{-71}}{2}\right)$	l^6	$10x^2+3xy+2y^2$	$+1$
			$\left(5, \frac{7+\sqrt{-71}}{2}\right)$	l^5	$6x^2+7xy+5y^2$	$+1$

Table I (continued)

D	ω	Δ	Ideal classes	Relations	Quadratic forms	Character systems
-71			$\left(3, \frac{5+\sqrt{-71}}{2}\right)$	l^4	$8x^2+5xy+3y^2$	$+1$
			$\left(3, \frac{1+\sqrt{-71}}{2}\right)$	l^3	$6x^2+xy+3y^2$	$+1$
			$\left(5, \frac{3+\sqrt{-71}}{2}\right)$	l^2	$4x^2+3xy+5y^2$	$+1$
			$\left(2, \frac{1+\sqrt{-71}}{2}\right)$	l	$9x^2+xy+2y^2$	$+1$
-73	$\sqrt{-73}$	$-2^2\cdot 73$	(1)	l^4	$73x^2+y^2$	$+1, +1$
			$(7, 5+\sqrt{-73})$	l^3	$14x^2+10xy+7y^2$	$-1, -1$
			$(2, 1+\sqrt{-73})$	l^2	$37x^2+2xy+2y^2$	$+1, +1$
			$(7, 2+\sqrt{-73})$	l	$11x^2+4xy+7y^2$	$-1, -1$
-74	$\sqrt{-74}$	$-2^3\cdot 37$	(1)	l^5	$74x^2+y^2$	$+1, +1$
			$(11, 6+\sqrt{-74})$	l^4	$10x^2+12xy+11y^2$	$+1, +1$
			$(3, 1+\sqrt{-74})$	l^3	$25x^2+2xy+3y^2$	$+1, +1$
			$(3, 2+\sqrt{-74})$	l^2	$26x^2+4xy+3y^2$	$+1, +1$
			$(11, 5+\sqrt{-74})$	l	$9x^2+10xy+11y^2$	$+1, +1$
			$(5, 4+\sqrt{-74})$	Al^4	$18x^2+8xy+5y^2$	$-1, -1$
			$(6, 4+\sqrt{-74})$	Al^3	$15x^2+8xy+6y^2$	$-1, -1$
			$(6, 2+\sqrt{-74})$	Al^2	$13x^2+4xy+6y^2$	$-1, -1$
			$(5, 1+\sqrt{-74})$	Al	$15x^2+2xy+5y^2$	$-1, -1$
			$(2, \sqrt{-74})$	A	$37x^2+2y^2$	$-1, -1$
-77	$\sqrt{-77}$	$-2^2\cdot 7\cdot 11$	(1)	l^4	$77x^2+y^2$	$+1, +1, +1$
			$(3, 2+\sqrt{-77})$	l^3	$27x^2+4xy+3y^2$	$-1, +1, -1$
			$(14, 7+\sqrt{-77})$	l^2	$9x^2+14xy+14y^2$	$+1, +1, +1$
			$(3, 1+\sqrt{-77})$	l	$26x^2+2xy+3y^2$	$-1, +1, -1$
			$(6, 5+\sqrt{-77})$	Al^3	$17x^2+10xy+6y^2$	$-1, -1, +1$
			$(7, \sqrt{-77})$	Al^2	$11x^2+7y^2$	$+1, -1, -1$
			$(6, 1+\sqrt{-77})$	Al	$13x^2+2xy+6y^2$	$-1, -1, +1$
			$(2, 1+\sqrt{-77})$	A	$39x^2+2xy+2y^2$	$+1, -1, -1$
-78	$\sqrt{-78}$	$-2^3\cdot 3\cdot 13$	(1)	$A^2A_1^2$	$78x^2+y^2$	$+1, +1, +1$
			$(2, \sqrt{-78})$	AA_1	$39x^2+2y^2$	$-1, -1, +1$
			$(13, \sqrt{-78})$	A_1	$6x^2+13y^2$	$+1, -1, -1$
			$(3, \sqrt{-78})$	A	$26x^2+3y^2$	$-1, +1, -1$

Table I (continued)

D	ω	Δ	Ideal classes	Relations	Quadratic forms	Character systems
-79	$\frac{1}{2}(1+\sqrt{-79})$	-79	(1)	l^5	$20x^2+xy+y^2$	$+1$
			$\left(2,\frac{1+\sqrt{-79}}{2}\right)$	l^4	$10x^2+xy+2y^2$	$+1$
			$\left(5,\frac{9+\sqrt{-79}}{2}\right)$	l^3	$8x^2+9xy+5y^2$	$+1$
			$\left(5,\frac{1+\sqrt{-79}}{2}\right)$	l^2	$4x^2+xy+5y^2$	$+1$
			$\left(2,\frac{3+\sqrt{-79}}{2}\right)$	l	$11x^2+3xy+2y^2$	$+1$
-82	$\sqrt{-82}$	$-2^3\cdot 41$	(1)	l^4	$82x^2+y^2$	$+1,+1$
			$(7,4+\sqrt{-82})$	l^2	$14x^2+8xy+7y^2$	$-1,-1$
			$(2,\sqrt{-82})$	l^2	$41x^2+2y^2$	$+1,+1$
			$(7,3+\sqrt{-82})$	l	$13x^2+6xy+7y^2$	$-1,-1$
-83	$\frac{1}{2}(1+\sqrt{-83})$	-83	(1)	l^3	$21x^2+xy+y^2$	$+1$
			$\left(3,\frac{5+\sqrt{-83}}{2}\right)$	l^2	$9x^2+5xy+3y^2$	$+1$
			$\left(3,\frac{1+\sqrt{-83}}{2}\right)$	l	$7x^2+xy+3y^2$	$+1$
-85	$\sqrt{-85}$	$-2^2\cdot 5\cdot 17$	(1)	$A^2A_1^2$	$85x^2+y^2$	$+1,+1,+1$
			$(5,\sqrt{-85})$	AA_1	$17x^2+5y^2$	$-1,-1,+1$
			$(10,5+\sqrt{-85})$	A_1	$11x^2+10xy+10y^2$	$+1,-1,-1$
			$(2,1+\sqrt{-85})$	A	$43x^2+2xy+2y^2$	$-1,+1,-1$
-86	$\sqrt{-86}$	$-2^3\cdot 43$	(1)	l^{10}	$86x^2+y^2$	$+1,+1$
			$(3,2+\sqrt{-86})$	l^9	$30x^2+4xy+3y^2$	$-1,-1$
			$(9,2+\sqrt{-86})$	l^8	$10x^2+4xy+9y^2$	$+1,+1$
			$(5,2+\sqrt{-86})$	l^7	$18x^2+4xy+5y^2$	$-1,-1$
			$(17,13+\sqrt{-86})$	l^6	$15x^2+26xy+17y^2$	$+1,+1$
			$(2,\sqrt{-86})$	l^5	$43x^2+2$	$-1,-1$
			$(17,4+\sqrt{-86})$	l^4	$6x^2+8xy+17y^2$	$+1,+1$
			$(5,3+\sqrt{-86})$	l^3	$19x^2+6xy+5y^2$	$-1,-1$
			$(9,7+\sqrt{-86})$	l^2	$15x^2+14xy+9y^2$	$+1,+1$
			$(3,1+\sqrt{-86})$	l	$29x^2+2xy+3y^2$	$-1,-1$
-87	$\frac{1}{2}(1+\sqrt{-87})$	$-3\cdot 29$	(1)	l^6	$22x^2+xy+y^2$	$+1,+1$
			$\left(2,\frac{3+\sqrt{-87}}{2}\right)$	l^5	$12x^2+3xy+2y^2$	$-1,-1$

Table I (continued)

D	ω	Δ	Ideal classes	Relations	Quadratic forms	Character systems
-87			$\left(7, \frac{5+\sqrt{-87}}{2}\right)$	l^4	$4x^2+5xy+7y^2$	$+1, +1$
			$\left(3, \frac{3+\sqrt{-87}}{2}\right)$	l^3	$8x^2+3xy+3y^2$	$-1, -1$
			$\left(7, \frac{9+\sqrt{-87}}{2}\right)$	l^2	$6x^2+9xy+7y^2$	$+1, +1$
			$\left(2, \frac{1+\sqrt{-87}}{2}\right)$	l	$11x^2+xy+2y^2$	$-1, -1$
-89	$\sqrt{-89}$	$-2^2 \cdot 89$	(1)	l^{12}	$89x^2+y^2$	$+1, +1$
			$(3, 2+\sqrt{-89})$	l^{11}	$31x^2+4xy+3y^2$	$-1, -1$
			$(17, 9+\sqrt{-89})$	l^{10}	$10x^2+18xy+17y^2$	$+1, +1$
			$(7, 3+\sqrt{-89})$	l^9	$14x^2+6xy+7y^2$	$-1, -1$
			$(5, 4+\sqrt{-89})$	l^8	$21x^2+8xy+5y^2$	$+1, +1$
			$(6, 1+\sqrt{-89})$	l^7	$15x^2+2xy+6y^2$	$-1, -1$
			$(2, 1+\sqrt{-89})$	l^6	$45x^2+2xy+2y^2$	$+1, +1$
			$(6, 5+\sqrt{-89})$	l^5	$19x^2+10xy+6y^2$	$-1, -1$
			$(5, 1+\sqrt{-89})$	l^4	$18x^2+2xy+5y^2$	$+1, +1$
			$(7, 4+\sqrt{-89})$	l^3	$15x^2+8xy+7y^2$	$-1, -1$
			$(17, 8+\sqrt{-89})$	l^2	$9x^2+16xy+17y^2$	$+1, +1$
			$(3, 1+\sqrt{-89})$	l	$30x^2+2xy+3y^2$	$-1, -1$
-91	$\frac{1+\sqrt{-91}}{2}$	$-7 \cdot 13$	(1)	A^2	$23x^2+xy+y^2$	$+1, +1$
			$\left(7, \frac{7+\sqrt{-91}}{2}\right)$	A	$5x^2+7xy+7y^2$	$-1, -1$
-93	$\sqrt{-93}$	$-2^2 \cdot 3 \cdot 31$	(1)	$A^2A_1^2$	$93x^2+y^2$	$+1, +1, +1$
			$(6, 3+\sqrt{-93})$	AA_1	$17x^2+6xy+6y^2$	$-1, -1, +1$
			$(3, \sqrt{-93})$	A_1	$31x^2+3y^2$	$+1, -1, -1$
			$(2, 1+\sqrt{-93})$	A	$47x^2+2xy+2y^2$	$-1, +1, -1$
-94	$\sqrt{-94}$	$-2^2 \cdot 47$	(1)	l^8	$94x^2+y^2$	$+1, +1$
			$(5, 4+\sqrt{-94})$	l^7	$22x^2+8xy+5y^2$	$-1, -1$
			$(7, 5+\sqrt{-94})$	l^6	$17x^2+10xy+7y^2$	$+1, +1$
			$(11, 4+\sqrt{-94})$	l^5	$10x^2+8xy+11y^2$	$-1, -1$
			$(2, \sqrt{-94})$	l^4	$47x^2+2y^2$	$+1, +1$
			$(11, 7+\sqrt{-94})$	l^3	$13x^2+14xy+11y^2$	$-1, -1$

Table I (continued)

D	ω	Δ	Ideal classes	Relations	Quadratic forms	Character systems
-94			$(7, 2+\sqrt{-94})$	l^2	$14x^2+4xy+7y^2$	$+1, +1$
			$(5, 1+\sqrt{-94})$	l	$19x^2+2xy+5y^2$	$-1, -1$
-95	$\frac{1}{2}(1+\sqrt{-95})$	$-5\cdot 19$	(1)	l^8	$24x^2+xy+y^2$	$+1, +1$
			$\left(2, \frac{1+\sqrt{-95}}{2}\right)$	l^7	$12x^2+xy+2y^2$	$-1, -1$
			$\left(4, \frac{1+\sqrt{-95}}{2}\right)$	l^6	$6x^2+xy+4y^2$	$+1, +1$
			$\left(3, \frac{5+\sqrt{-95}}{2}\right)$	l^5	$10x^2+5xy+3y^2$	$-1, -1$
			$\left(5, \frac{5+\sqrt{-95}}{2}\right)$	l^4	$6x^2+5xy+5y^2$	$+1, +1$
			$\left(3, \frac{1+\sqrt{-95}}{2}\right)$	l^3	$8x^2+xy+3y^2$	$-1, -1$
			$\left(4, \frac{7+\sqrt{-95}}{2}\right)$	l^2	$9x^2+7xy+4y^2$	$+1, +1$
			$\left(2, \frac{3+\sqrt{-95}}{2}\right)$	l	$13x^2+3xy+2y^2$	$-1, -1$
-97	$\sqrt{-97}$	$-2^2\cdot 97$	(1)	l^4	$97x^2+y^2$	$+1, +1$
			$(7, 6+\sqrt{-97})$	l^3	$19x^2+12xy+7y^2$	$-1, -1$
			$(2, 1+\sqrt{-97})$	l^2	$49x^2+2xy+2y^2$	$+1, +1$
			$(7, 1+\sqrt{-97})$	l	$14x^2+2xy+7y^2$	$-1, -1$

Table II

D	ω	Continued fractions	Δ	$x+y\sqrt{D}$	$N(x+y\sqrt{D})$	Ideal classes	Relations	Quadratic forms	Character systems
2	$\sqrt{2}$	$[1,\dot{2}]$	2^3	$1+\sqrt{2}$	-1	(1)	1	$-2x^2+y^2$	$+1$
3	$\sqrt{3}$	$[1,\dot{1},\dot{2}]$	$3\cdot 2^2$	$2+\sqrt{3}$	$+1$	(1)	1	$-3x^2+y^2$	$+1, +1$
								$-x^2+3y^2$	$-1, -1$
5	$\frac{1}{2}(1+\sqrt{5})$	$[1,1]$	5	ω	-1	(1)	1	$-x^2+xy+y^2$	$+1$
6	$\sqrt{6}$	$[2,\dot{2},\dot{4}]$	$3\cdot 2^3$	$5+2\sqrt{6}$	$+1$	(1)	1	$-6x^2+y^2$	$+1, +1$
								$-x^2+6y^2$	$-1, -1$
7	$\sqrt{7}$	$[2,\dot{1},1,1,\dot{4}]$	$2^2\cdot 7$	$8+3\sqrt{7}$	$+1$	(1)	1	$-7x^2+y^2$	$+1, +1$
								$-x^2+7y^2$	$-1, -1$
8		$[2,\dot{1},\dot{4}]$		$3+\sqrt{8}$	$+1$				
10	$\sqrt{10}$	$[3,\dot{6}]$	$5\cdot 2^3$	$3+\sqrt{10}$	-1	(1)	A^2	$-10x^2+y^2$	$+1, +1$
						$(2,\sqrt{10})$	A	$-5x^2+2y^2$	$-1, -1$
11	$\sqrt{11}$	$[3,\dot{3},\dot{6}]$	$2^2\cdot 11$	$10+3\sqrt{11}$	$+1$	(1)	1	$-11x^2+y^2$	$+1, +1$
								$-x^2+11y^2$	$-1, -1$
12		$[3,\dot{2},\dot{6}]$		$7+2\sqrt{12}$	$+1$				
13	$\frac{1}{2}(1+\sqrt{13})$	$[2,\dot{3}]$	13	$1+\omega$	-1	(1)	1	$-3x^2+xy+y^2$	$+1$
14	$\sqrt{14}$	$[3,\dot{1},2,1,\dot{6}]$	$7\cdot 2^3$	$15+4\sqrt{14}$	$+1$	(1)	1	$-14x^2+y^2$	$+1, +1$
								$-x^2+14y^2$	$-1, -1$
15	$\sqrt{15}$	$[3,\dot{1},\dot{6}]$	$3\cdot 2^2\cdot 5$	$4+\sqrt{15}$	$+1$	(1)	A^2	$-15x^2+y^2$	$+1, +1, +1$
								$-x^2+15y^2$	$-1, +1, -1$
						$(2, 1+\sqrt{15})$	A	$-7x^2+2xy+2y^2$	$-1, -1, +1$
								$-2x^2-2xy+7y^2$	$+1, -1, -1$
17	$\frac{1}{2}(1+\sqrt{17})$	$[2,\dot{1},1,\dot{3}]$	17	$3+2\omega$	-1	(1)	1	$-4x^2+xy+y^2$	$+1$

Table II (continued)

D	ω	Continued fractions	Δ	$x+y\sqrt{D}$	$N(x+y\sqrt{D})$	Ideal classes	Relations	Quadratic forms	Character systems
18		$[4,\dot{4},\dot{8}]$		$17+4\sqrt{18}$	$+1$				
19	$\sqrt{19}$	$[4,\dot{2},1,3,1,2,\dot{8}]$	$2^2\cdot 19$	$170+39\sqrt{19}$	$+1$	(1)	1	$-19x^2+y^2$	$+1,+1$
								$-x^2+19y^2$	$-1,-1$
20		$[4,\dot{2},\dot{8}]$		$9+2\sqrt{20}$	$+1$				
21	$\frac{1}{2}(1+\sqrt{21})$	$[2,\dot{1},\dot{3}]$	$3\cdot 7$	$2+\omega$	$+1$	(1)	1	$-5x^2+xy+y^2$	$+1,+1$
								$-x^2-xy+5y^2$	$-1,-1$
22	$\sqrt{22}$	$[4,\dot{1},2,4,2,1,\dot{8}]$	$2^3\cdot 11$	$197+42\sqrt{22}$	$+1$	(1)	1	$-22x^2+y^2$	$+1,+1$
								$-x^2+22y^2$	$-1,-1$
23	$\sqrt{23}$	$[4,\dot{1},3,1,\dot{8}]$	$2^2\cdot 23$	$24+5\sqrt{27}$	$+1$	(1)	1	$-23x^2+y^2$	$+1,+1$
								$-x^2+23y^2$	$-1,-1$
24		$[4,\dot{1},\dot{8}]$		$5+\sqrt{24}$	$+1$				
26	$\sqrt{26}$	$[5,\dot{1}\dot{0}]$	$2^3\cdot 13$	$5+\sqrt{26}$	-1	(1)	A^2	$-26x^2+y^2$	$+1,+1$
						$(2,\sqrt{26})$	A	$-13x^2+2y^2$	$-1,-1$
27		$[5,\dot{5},\dot{1}\dot{0}]$		$26+5\sqrt{27}$	$+1$				
28		$[5,\dot{3},2,3,\dot{1}\dot{0}]$		$127+24\sqrt{28}$	$+1$				
29	$\frac{1}{2}(1+\sqrt{29})$	$[3,\dot{5}]$	29	$2+\omega$	-1	(1)	1	$-7x^2+xy+y^2$	$+1$
30	$\sqrt{30}$	$[5,\dot{2},\dot{1}\dot{0}]$	$3\cdot 5\cdot 2^3$	$11+2\sqrt{30}$	$+1$	(1)	A^2	$-30x^2+y^2$	$+1,+1,+1$
								$-x^2+30y^2$	$-1,+1,-1$
						$(2,\sqrt{30})$	A	$-15x^2+2y^2$	$-1,-1,+1$
								$-2x^2+15y^2$	$+1,-1,-1$
31	$\sqrt{31}$	$[5,\dot{1},1,3,5,3,1,1,\dot{1}\dot{0}]$	$2^2\cdot 31$	$1520+273\sqrt{31}$	$+1$	(1)	1	$-31x^2+y^2$	$+1,+1$
								$-x^2+31y^2$	$-1,-1$
32		$[5,\dot{1},1,1,\dot{1}\dot{0}]$		$17+3\sqrt{32}$	$+1$				

Table II (continued)

D	ω	Continued fractions	Δ	$x+y\sqrt{D}$	$N(x+y\sqrt{D})$	Ideal classes	Relations	Quadratic forms	Character systems
33	$\frac{1}{2}(1+\sqrt{33})$	$[3,\dot{2},1,2,\dot{5}]$	$3\cdot 11$	$19+8\omega$	$+1$	(1)	1	$-8x^2+xy+y^2$	$+1, +1$
								$-x^2-xy+8y^2$	$-1, -1$
34	$\sqrt{34}$	$[5,\dot{1},4,1,\dot{1}0]$	$2^3\cdot 17$	$35+6\sqrt{34}$	$+1$	(1)	A^2	$-34x^2+y^2$	$+1, +1$
								$-x^2+34y^2$	$+1, +1$
						$(3, 1+\sqrt{34})$	A	$-11x^2+2xy+3y^2$	$-1, -1$
								$-3x^2-2xy+11y^2$	$-1, -1$
35	$\sqrt{35}$	$[5,\dot{1},\dot{1}0]$	$2^2\cdot 5\cdot 7$	$6+\sqrt{35}$	$+1$	(1)	A^2	$-35x^2+y^2$	$+1, +1, +1$
								$-x^2+35y^2$	$+1, -1, -1$
						$(2, 1+\sqrt{35})$	A	$-17x^2+2xy+2y^2$	$-1, +1, -1$
								$-2x^2-2xy+17y^2$	$-1, -1, +1$
37	$\frac{1}{2}(1+\sqrt{37})$	$[3,\dot{1},1,\dot{5}]$	37	$5+2\omega$	-1	(1)	1	$-9x^2+xy+y^2$	$+1$
38	$\sqrt{38}$	$[6,\dot{6},\dot{1}2]$	$2^3\cdot 19$	$37+6\sqrt{38}$	$+1$	(1)	1	$-38x^2+y^2$	$+1, +1$
								$-x^2+38y^2$	$-1, -1$
39	$\sqrt{39}$	$[6,\dot{4},\dot{1}2]$	$3\cdot 2^2\cdot 13$	$25+4\sqrt{39}$	$+1$	(1)	A^2	$-39x^2+y^2$	$+1, +1, +1$
								$-x^2+39y^2$	$-1, +1, -1$
						$(2, 1+\sqrt{39})$	A	$-19x^2+2xy+2y^2$	$-1, -1, +1$
								$-2x^2-2xy+19y^2$	$+1, -1, -1$
40		$[6,\dot{3},\dot{1}2]$		$19+3\sqrt{40}$	$+1$				
41	$\frac{1}{2}(1+\sqrt{41})$	$[3,\dot{1},2,2,1,\dot{5}]$	41	$27+10\omega$	-1	(1)	1	$-10x^2+xy+y^2$	$+1$
42	$\sqrt{42}$	$[6,\dot{2},\dot{1}2]$	$3\cdot 2^3\cdot 7$	$13+2\sqrt{42}$	$+1$	(1)	A^2	$-42x^2+y^2$	$+1, +1, +1$
								$-x^2+42y^2$	$-1, -1, +1$
						$(2, \sqrt{42})$	A	$-21x^2+2y^2$	$-1, +1, -1$

Table II (continued)

D	ω	Continued fractions	Δ	$x+y\sqrt{D}$	$N(x+y\sqrt{D})$	Ideal classes	Relations	Quadratic forms	Character systems
42								$-2x^2+21y^2$	$+1, -1, -1$
43	$\sqrt{43}$	$[6, \dot{1}, 1, 3, 1, 5, 1, 3, 1, 1, \dot{12}]$	$2^2 \cdot 43$	$3482+531\sqrt{43}$	$+1$	(1)	1	$-43x^2+y^2$	$+1, +1$
								$-x^2+43y^2$	$-1, -1$
44		$[6, \dot{1}, 1, 1, 2, 1, 1, 1, \dot{12}]$		$199+30\sqrt{44}$	$+1$				
45		$[6, \dot{1}, 2, 2, 2, 1, \dot{12}]$		$161+24\sqrt{45}$	$+1$				
46	$\sqrt{46}$	$[6, \dot{1}, 3, 1, 1, 2, 6, 2, 1, 1, 3, 1, \dot{12}]$	$2^3 \cdot 23$	$24335+3588\sqrt{46}$	$+1$	(1)	1	$-46x^2+y^2$	$+1, +1$
								$-x^2+46y^2$	$-1, -1$
47	$\sqrt{47}$	$[6, \dot{1}, 5, 1, \dot{12}]$	$2^2 \cdot 47$	$48+7\sqrt{47}$	$+1$	(1)	1	$-47x^2+y^2$	$+1, +1$
								$-x^2+47y^2$	$-1, -1$
48		$[6, \dot{1}, \dot{12}]$		$7+\sqrt{49}$	$+1$				
50		$[7, \dot{14}]$		$7+\sqrt{50}$	-1				
51	$\sqrt{51}$	$[7, \dot{7}, \dot{14}]$	$3 \cdot 2^2 \cdot 17$	$50+7\sqrt{51}$	$+1$	(1)	A^2	$-51x^2+y^2$	$+1, +1, +1$
								$-x^2+51y^2$	$-1, +1, -1$
						$(3, \sqrt{51})$	A	$-17x^2+3y^2$	$+1, -1, -1$
								$-3x^2+17y^2$	$-1, -1, +1$
52		$[7, \dot{4}, 1, 2, 1, 4, \dot{14}]$		$649+90\sqrt{52}$	$+1$				
53	$\frac{1}{2}(1+\sqrt{53})$	$[4, \dot{7}]$	53	$3+\omega$	-1	(1)	1	$-13x^2+xy+y^2$	$+1$
54		$[7, \dot{2}, 1, 6, 1, 2, \dot{14}]$		$485+66\sqrt{54}$	$+1$				
55	$\sqrt{55}$	$[7, \dot{2}, 2, 2, \dot{14}]$	$2^2 \cdot 5 \cdot 11$	$89+12\sqrt{55}$	$+1$	(1)	A^2	$-55x^2+y^2$	$+1, +1, +1$
								$-x^2+55y^2$	$+1, -1, -1$
						$(2, 1+\sqrt{55})$	A	$-27x^2+2xy+2y^2$	$-1, -1, +1$
								$-2x^2-2xy+27y^2$	$-1, +1, -1$

Table II (continued)

D	ω	Continued fractions	Δ	$x+y\sqrt{D}$	$N(x+y\sqrt{D})$	Ideal classes	Relations	Quadratic forms	Character systems
56		$[7, \dot{2}, \dot{14}]$		$15+2\sqrt{56}$	$+1$				
57	$\frac{1}{2}(1+\sqrt{57})$	$[4, \dot{3}, 1, 1, 1, 3, \dot{7}]$	$3\cdot 19$	$131+40\omega$	$+1$	(1)	1	$-14x^2+xy+y^2$	$+1, +1$
								$-x^2-xy+14y^2$	$-1, -1$
58	$\sqrt{58}$	$[7, \dot{1}, 1, 1, 1, 1, 1, \dot{14}]$	$2^3\cdot 29$	$99+13\sqrt{58}$	-1	(1)	A^2	$-58x^2+y^2$	$+1, +1$
						$(2, \sqrt{58})$	A	$-29x^2+2y^2$	$-1, -1$
59	$\sqrt{59}$	$[7, \dot{1}, 2, 7, 2, 1, \dot{14}]$	$2^2\cdot 59$	$530+69\sqrt{59}$	$+1$	(1)	1	$-59x^2+y^2$	$+1, +1$
								$-x^2+59y^2$	$-1, -1$
60		$[7, \dot{1}, 2, 1, \dot{14}]$		$31+4\sqrt{60}$	$+1$				
61	$\frac{1}{2}(1+\sqrt{61})$	$[4, \dot{2}, 2, \dot{7}]$	61	$17+5\omega$	-1	(1)	1	$-15x^2+xy+y^2$	$+1$
62	$\sqrt{62}$	$[7, \dot{1}, 6, 1, \dot{14}]$	$2^3\cdot 31$	$63+8\sqrt{62}$	$+1$	(1)	1	$-62x^2+y^2$	$+1, +1$
								$-x^2+62y^2$	$-1, -1$
63		$[7, \dot{1}, \dot{14}]$		$8+\sqrt{64}$	$+1$				
65	$\frac{1}{2}(1+\sqrt{65})$	$[4, \dot{1}, 1, \dot{7}]$	$5\cdot 13$	$7+2\omega$	-1	(1)	A^2	$-16x^2+xy+y^2$	$+1, +1$
						$\left(5, 2+\frac{1+\sqrt{65}}{2}\right)$	A	$-2x^2+5xy+5y^2$	$-1, -1$
66	$\sqrt{66}$	$[8, \dot{8}, \dot{16}]$	$3\cdot 2^3\cdot 11$	$65+8\sqrt{66}$	$+1$	(1)	A^2	$-66x^2+y^2$	$+1, +1, +1$
								$-x^2+66y^2$	$-1, -1, +1$
						$(3, \sqrt{66})$	A	$-22x^2+3y^2$	$-1, +1, -1$
								$-3x^2+22y^2$	$+1, -1, -1$
67	$\sqrt{67}$	$[8, \dot{5}, 2, 1, 1, 7, 1, 1, 2, 5, \dot{16}]$	$2^2\cdot 67$	$48842+5967\sqrt{67}$	$+1$	(1)	1	$-67x^2+y^2$	$+1, +1$
								$-x^2+67y^2$	$-1, -1$
68		$[8, \dot{4}, \dot{16}]$		$33+4\sqrt{68}$	$+1$				

Table II (continued)

D	ω	Continued fractions	Δ	$x+y\sqrt{D}$	$N(x+y\sqrt{D})$	Ideal classes	Relations	Quadratic forms	Character systems
69	$\frac{1}{2}(1+\sqrt{69})$	$[4, \dot{1}, 1, 1, \dot{7}]$	$3\cdot 23$	$11+3\omega$	$+1$	(1)	1	$-17x^2+xy+y^2$	$+1, +1$
								$-x^2-xy+17y^2$	$-1, -1$
70	$\sqrt{70}$	$[8, \dot{2}, 1, 2, 1, 2, \dot{1}\dot{6}]$	$5\cdot 7\cdot 2^3$	$251+30\sqrt{70}$	$+1$	(1)	A^2	$-70x^2+y^2$	$+1, +1, +1$
								$-x^2+70y^2$	$+1, -1, -1$
						$(2, \sqrt{70})$	A	$-35x^2+2y^2$	$-1, +1, -1$
								$-2x^2+35y^2$	$-1, -1, +1$
71	$\sqrt{71}$	$[8, \dot{2}, 2, 1, 7, 1, 2, 2, \dot{1}\dot{6}]$	$2^2\cdot 71$	$3480+413\sqrt{71}$	$+1$	(1)	1	$-71x^2+y^2$	$+1, +1$
								$-x^2+71y^2$	$-1, -1$
72		$[8, \dot{2}, \dot{1}\dot{6}]$		$17+2\sqrt{72}$	$+1$				
73	$\frac{1}{2}(1+\sqrt{73})$	$[4, \dot{1}, 3, 2, 1, 1, 2, 3, 1, \dot{7}]$	73	$943+250\omega$	-1	(1)	1	$-18x^2+xy+y^2$	$+1$
74	$\sqrt{74}$	$[8, \dot{1}, 1, 1, 1, \dot{1}\dot{6}]$	$2^3\cdot 37$	$43+5\sqrt{74}$	-1	(1)	A^2	$-74x^2+y^2$	$+1, +1$
						$(2, \sqrt{74})$	A	$-37x^2+2y^2$	$-1, -1$
75		$[8, \dot{1}, 1, 1, \dot{1}\dot{6}]$		$26+3\sqrt{75}$	$+1$				
76		$[8, \dot{1}, 2, 1, 1, 5, 4, 5, 1, 1, 2, 1, \dot{1}\dot{6}]$		$57799+6630\sqrt{76}$	$+1$				
77	$\frac{1}{2}(1+\sqrt{77})$	$[4, \dot{1}, \dot{7}]$	$7\cdot 11$	$4+\omega$	$+1$	(1)	1	$-19x^2+xy+y^2$	$+1, +1$
								$-x^2-xy+19y^2$	$-1, -1$
78	$\sqrt{78}$	$[8, \dot{1}, 4, 1, \dot{1}\dot{6}]$	$3\cdot 2^3\cdot 13$	$53+6\sqrt{78}$	$+1$	(1)	A^2	$-78x^2+y^2$	$+1, +1, +1$
								$-x^2+78y^2$	$-1, +1, -1$
						$(2, \sqrt{78})$	A	$-39x^2+2y^2$	$-1, -1, +1$
								$-2x^2+39y^2$	$+1, -1, -1$
79	$\sqrt{79}$	$[8, \dot{1}, 7, 1, \dot{1}\dot{6}]$	$2^2\cdot 79$	$80+9\sqrt{79}$	$+1$	(1)	I^2	$-79x^2+y^2$	$+1, +1$
								$-x^2+79y^2$	$-1, -1$

Table II (continued)

D	ω	Continued fractions	Δ	$x+y\sqrt{D}$	$N(x+y\sqrt{D})$	Ideal classes	Relations	Quadratic forms	Character systems
79						$(3, 2+\sqrt{79})$	l^2	$-25x^2+4xy+3y^2$	$-1, -1$
								$-3x^2-4xy+25y^2$	$+1, +1$
						$(3, 1+\sqrt{79})$	l	$-26x^2+2xy+3y^2$	$-1, -1$
								$-3x^2-2xy+26y^2$	$+1, +1$
80		$[8, \dot{1}, \dot{1}6]$		$9+\sqrt{80}$	$+1$				
82	$\sqrt{82}$	$[9, \dot{1}8]$	$2^3\cdot 41$	$9+\sqrt{82}$	-1	(1)	l^4	$-82x^2+y^2$	$+1, +1$
						$(3, 1+\sqrt{82})$	l^2	$-27x^2+2xy+3y^2$	$-1, -1$
						$(2, \sqrt{82})$	l^2	$-41x^2+2y^2$	$+1, +1$
						$(3, 2+\sqrt{82})$	l	$-26x^2+4xy+3y^2$	$-1, -1$
83	$\sqrt{83}$	$[9, \dot{9}, \dot{1}8]$	$2^2\cdot 83$	$82+9\sqrt{83}$	$+1$	(1)	1	$-83x^2+y^2$	$+1, +1$
								$-x^2+83y^2$	$-1, -1$
84		$[9, \dot{6}, \dot{1}8]$		$55+6\sqrt{84}$	$+1$				
85	$\frac{1}{2}(1+\sqrt{85})$	$[5, \dot{9}]$	$5\cdot 17$	$4+\omega$	-1	(1)	A^2	$-21x^2+xy+y^2$	$+1, +1$
						$\left(5, 2+\frac{1+\sqrt{85}}{2}\right)$	A	$-3x^2+5xy+5y^2$	$-1, -1$
86	$\sqrt{86}$	$[9, \dot{3}, 1, 1, 1, 8, 1, 1, 1, 3, \dot{1}8]$	$2^3\cdot 43$	$10405+1122\sqrt{86}$	$+1$	(1)	1	$-86x^2+y^2$	$+1, +1$
								$-x^2+86y^2$	$-1, -1$
87	$\sqrt{87}$	$[9, \dot{3}, \dot{1}8]$	$3\cdot 2^2\cdot 29$	$28+3\sqrt{87}$	$+1$	(1)	A^2	$-87x^2+y^2$	$+1, +1, +1$
								$-x^2+87y^2$	$-1, +1, -1$
						$(2, 1+\sqrt{87})$	A	$-43x^2+2xy+2y^2$	$-1, -1, +1$
								$-2x^2-2xy+43y^2$	$+1, -1, -1$

Table II (continued)

D	ω	Continued fractions	Δ	$x+y\sqrt{D}$	$N(x+y\sqrt{D})$	Ideal classes	Relations	Quadratic forms	Character systems
88		$[9,\dot{2},1,1,1,2,\dot{1}8]$		$197+21\sqrt{88}$	$+1$				
89	$\frac{1}{2}(1+\sqrt{89})$	$[5,\dot{4},1,1,1,1,4,\dot{9}]$	89	$447+106\omega$	-1	(1)	1	$-22x^2+xy+y^2$	$+1$
90		$[9,\dot{2},\dot{1}8]$		$19+2\sqrt{90}$	$+1$				
91	$\sqrt{91}$	$[9,\dot{1},1,5,1,5,1,1,\dot{1}8]$	$2^2\cdot 7\cdot 13$	$1574+165\sqrt{91}$	$+1$	(1)	A^2	$-91x^2+y^2$	$+1,+1,+1$
								$-x^2+91y^2$	$-1,+1,-1$
						$(2,1+\sqrt{91})$	A	$-45x^2+2xy+2y^2$	$+1,-1,-1$
								$-2x^2-2xy+45y^2$	$-1,-1,+1$
92		$[9,\dot{1},1,2,4,2,1,1,\dot{1}8]$		$1151+120\sqrt{92}$	$+1$				
93	$\frac{1}{2}(1+\sqrt{93})$	$[5,\dot{3},\dot{9}]$	$3\cdot 31$	$13+3\omega$	$+1$	(1)	1	$-23x^2+xy+y^2$	$+1,+1$
								$-x^2-xy+23y^2$	$-1,-1$
94	$\sqrt{94}$	$[9,\dot{1},2,3,1,1,5,1,8,1,5,$ $1,1,3,2,1,\dot{1}8]$	$2^3\cdot 47$	$2143295+221064\sqrt{94}$	$+1$	(1)	1	$-94x^2+y^2$	$+1,+1$
								$-x^2+94y^2$	$-1,-1$
95	$\sqrt{95}$	$[9,\dot{1},2,1,\dot{1}8]$	$2^2\cdot 5\cdot 19$	$39+4\sqrt{95}$	$+1$	(1)	A^2	$-95x^2+y^2$	$+1,+1,+1$
								$-x^2+95y^2$	$+1,-1,-1$
						$(2,1+\sqrt{95})$	A	$-47x^2+2xy+2y^2$	$-1,-1,+1$
								$-2x^2-2xy+47y^2$	$-1,+1,-1$
96		$[9,\dot{1},3,1,\dot{1}8]$		$49+5\sqrt{96}$	$+1$				
97	$\frac{1}{2}(1+\sqrt{97})$	$[5,\dot{2},2,1,4,4,1,2,2,\dot{9}]$	97	$5035+1138\omega$	-1	(1)	1	$-24x^2+xy+y^2$	$+1$
98		$[9,\dot{1},8,1,\dot{1}8]$		$99+10\sqrt{98}$	$+1$				
99		$[9,\dot{1},\dot{1}8]$		$10+\sqrt{99}$	$+1$				

Notes

16.1. The problem concerning the number of imaginary quadratic fields that are simple fields was solved by H. M. Stark [55] and A. Baker [3] independently; see also [4].

Chapter 17. Algebraic Numbers and Transcendental Numbers

17.1 The Existence of Transcendental Numbers

A real number can be represented as a point on a straight line, so that a collection of real numbers is sometimes called a point set. For example, $\{1/n : n = 1, 2, \ldots\}$ is a point set, the set of rational numbers in the interval (a, b) is a point set.

Definition 1.1. Let A, B be two point sets. Suppose that there exists a one-to-one correspondence between A and B (that is, there exists a bijection from A to B). Then we say that A and B are *equipotent*, or A and B have the same *cardinal number*.

Being equipotent is an equivalence relation.

Definition 1.2. Any set which is equipotent to the set of natural numbers is called *enumerable*. A *countable* set is one which is either finite or enumerable.

The set of natural numbers itself is, of course, enumerable; so is the set $\{1/n : n = 1, 2, \ldots\}$. Any sequence of numbers is a countable set.

Theorem 1.1. *The countable union of countable sets is countable.*

Proof. Let the countable sets be $M_1, M_2, \ldots$ where $M_i = (\alpha_{i1}, \ldots, \alpha_{ij}, \ldots)$. On displaying the union as the rectangular array

$$\begin{array}{ccccc} \alpha_{11} & \alpha_{12} & \alpha_{13} & \alpha_{14} & \cdots \\ & \swarrow & \swarrow & & \\ \alpha_{21} & \alpha_{22} & \alpha_{23} & \cdots & \\ & \swarrow & & & \\ \alpha_{31} & \alpha_{32} & \cdots & & \\ \cdots & & & & \end{array}$$

we can form the sequence $(\alpha_{11}, \alpha_{12}, \alpha_{21}, \alpha_{13}, \alpha_{22}, \alpha_{31}, \alpha_{14}, \ldots)$ following the arrows. The theorem is proved. $\square$

Theorem 1.2. *The set of rational numbers is countable.*

Proof. From Theorem 1.1 it suffices to show that the rational numbers in the interval $[0, 1]$ is countable. We first arrange the reduced fractions in $[0, 1]$

according to the size of their denominators, and when two fractions have the same denominator, we then arrange them according to the size of their numerators. This then gives the sequence

$$\frac{0}{1}, \frac{1}{1}, \frac{1}{2}, \frac{1}{3}, \frac{2}{3}, \frac{1}{4}, \frac{3}{4}, \frac{1}{5}, \frac{2}{5}, \frac{3}{5}, \frac{4}{5}, \ldots$$

and so the theorem is proved. □

Theorem 1.3. *The set of real numbers in the interval* (0, 1) *is not countable.*

Proof. Suppose the contrary and let $\alpha_1, \alpha_2, \alpha_3, \ldots$ be an enumeration of the real numbers in (0, 1). Each α_i has a decimal expansion

$$\alpha_i = 0 \cdot a_{i1}a_{i2} \cdots a_{in} \cdots, \qquad 0 \leqslant a_{in} \leqslant 9.$$

We define a real number $\beta = 0 \cdot b_1 b_2 \cdots b_n \cdots$ by setting

$$b_i = \begin{cases} a_{ii} + 1, & \text{if } \ 0 \leqslant a_{ii} \leqslant 5, \\ a_{ii} - 1, & \text{if } \ 6 \leqslant a_{ii} \leqslant 9. \end{cases}$$

We note that β is a real number in the interval (0, 1) which is different from α_i, for every i, because they differ at the i-th decimal place. This gives the required contradiction. (Observe that a terminating decimal may have two decimal representations, for example $0.12 = 0.11999\ldots$. However the decimal representation of β does not contain any 0 or 9.) □

Exercise 1. Determine the position of the reduced fraction a/b in the proof of Theorem 1.2.

Exercise 2. Show that a subset of a countable set is countable.

In the previous chapter we defined an algebraic number ξ to be a root of the equation

$$a_n\xi^n + a_{n-1}\xi^{n-1} + \cdots + a_0 = 0$$

where $a_n, a_{n-1}, \ldots, a_0$ are rational integers. If this equation is irreducible and $a_n \neq 0$, then ξ is called an algebraic number of degree n, and if $a_n = 1$ then ξ is called an algebraic integer of degree n.

Theorem 1.4. *The set of all algebraic numbers is countable.*

Proof. Let $N = n + |a_n| + |a_{n-1}| + \cdots + |a_0|$ so that $N \geqslant 2$. Corresponding to each fixed N there can only be a finite number of polynomial equations, and each equation has only a finite number of roots so that the number of algebraic numbers corresponding to N is also finite. We denote by E_N this set of algebraic numbers and consider the sequence $E_2, E_3, \ldots, E_N, \ldots$. Let E'_N be the subset of E_N whose members are not already members of $E_2, \ldots, E_{N-1}$. We then form the sequence of finite sets $E_2, E'_3, E'_4, \ldots$. From Theorem 1.1 the union of these sets is countable and the required result is proved. □

Definition 1.3. A number which is not algebraic is called a *transcendental number*.

Theorem 1.5. *Transcendental numbers exist.*

Proof. From Theorem 1.3 and Exercise 2 we know that the set of all real numbers is uncountable. Since the set of real algebraic numbers is countable the required result is proved. □

17.2 Liouville's Theorem and Examples of Transcendental Numbers

Theorem 2.1 (Liouville). *Any algebraic number of degree n is not approximable by rationals to an order greater than n. That is, if ξ is an algebraic number of degree n, then to every $\delta > 0$ and $A > 0$ the inequality*

$$\left|\xi - \frac{p}{q}\right| < \frac{A}{q^{n+\delta}} \tag{1}$$

has only finitely many rational integer solutions in p and q.

Proof. Suppose that ξ satisfies the equation

$$f(\xi) = a_n\xi^n + a_{n-1}\xi^{n-1} + \cdots + a_0 = 0.$$

There exists $M = M(\xi)$ such that $|f'(y)| < M$ for all y in the interval $\xi - 1 < y < \xi + 1$. If p/q $(q > 0)$ is a rational number near ξ we may assume that $\xi - 1 < p/q < \xi + 1$ and $f(p/q) \neq 0$, so that

$$\left|f\left(\frac{p}{q}\right)\right| = \frac{|a_np^n + a_{n-1}p^{n-1}q + \cdots + a_0q^n|}{q^n} \geqslant \frac{1}{q^n};$$

also

$$f\left(\frac{p}{q}\right) = f\left(\frac{p}{q}\right) - f(\xi) = \left(\frac{p}{q} - \xi\right)f'(\eta),$$

where η lies between p/q and ξ. Therefore we have

$$\left|\xi - \frac{p}{q}\right| = \frac{\left|f\left(\frac{p}{q}\right)\right|}{|f'(\eta)|} > \frac{1}{Mq^n}.$$

It follows that given $\delta > 0$ and $A > 0$, the inequality (1) can have at most a finite number of solutions in p and q. □

We can now construct two transcendental numbers using this theorem.

Theorem 2.2. *The two numbers*

$$\xi = \frac{1}{10} + \frac{1}{10^{2!}} + \frac{1}{10^{3!}} + \cdots$$

and

$$\xi = \frac{1}{10 +} \frac{1}{10^{2!} +} \frac{1}{10^{3!} +} \cdots$$

are transcendental.

Proof. 1) Let

$$\alpha_n = \frac{1}{10} + \frac{1}{10^{2!}} + \cdots + \frac{1}{10^{n!}} = \frac{p}{q}, \qquad q = 10^{n!}$$

so that

$$0 < \xi - \frac{p}{q} = \frac{1}{10^{(n+1)!}} + \cdots < \frac{2}{10^{(n+1)!}} = \frac{2}{q^{n+1}},$$

where n can be arbitrarily large. Therefore, by Theorem 2.1, ξ cannot be algebraic.

2) Let

$$\xi = \frac{1}{10 +} \frac{1}{10^{2!} +} \frac{1}{10^{3!} +} \cdots = [0, a_1, a_2, a_3, \ldots],$$

and let p_n/q_n be its n-th convergent. Then

$$\left|\xi - \frac{p_n}{q_n}\right| < \frac{1}{q_n q_{n+1}} < \frac{1}{a_{n+1} q_n^2} < \frac{1}{a_{n+1}}.$$

Now $a_{n+1} = 10^{(n+1)!}$, and

$$q_1 < a_1 + 1, \qquad \frac{q_{n+1}}{q_n} = a_{n+1} + \frac{q_{n-1}}{q_n} < a_{n+1} + 1 \qquad (n \geqslant 1),$$

so that

$$\begin{aligned} q_n &< (a_1 + 1)(a_2 + 1) \cdots (a_n + 1) \\ &< \left(1 + \frac{1}{10}\right)\left(1 + \frac{1}{10^2}\right) \cdots \left(1 + \frac{1}{10^n}\right) a_1 a_2 \cdots a_n \\ &< 2 a_1 a_2 \cdots a_n = 2 \cdot 10^{1! + 2! + \cdots + n!} < 10^{2 \cdot n!} = a_n^2. \end{aligned}$$

Therefore

$$\left|\xi - \frac{p_n}{q_n}\right| < \frac{1}{a_{n+1}} = \frac{1}{a_n^{n+1}} < \frac{1}{a_n^n} < \frac{1}{q_n^{\frac{1}{2}n}},$$

so that, as before, ξ must be transcendental. □

Exercise. Construct an uncountable set of transcendental numbers. (Suggestion: Show that if (a_n) is an increasing sequence of natural numbers, then

$$\frac{1}{10^{a_1}} + \frac{1}{10^{a_2 \cdot 2!}} + \frac{1}{10^{a_3 \cdot 3!}} + \cdots$$

is a transcendental number.)

17.3 Roth's Theorem on Rational Approximations to Algebraic Numbers

Liouville's theorem can be made much sharper. Let us denote by κ the least positive number with the following property: given any real algebraic number ξ of degree n ($\geqslant 2$) and given $v > \kappa$, the inequality

$$\left| \xi - \frac{p}{q} \right| < \frac{1}{q^v}$$

has only finitely many rational integer solutions in p and q ($q > 0$). From Liouville's theorem we see at once that $\kappa \leqslant n$. Liouville's theorem was successively improved by Thue, Siegel and Dyson who proved that $\kappa \leqslant \frac{1}{2}n + 1$, $\kappa \leqslant \min_{1 \leqslant s \leqslant n-1} \cdot (s + n/(s + 1))$, and $\kappa \leqslant \sqrt{2n}$ respectively. Finally in 1955, Roth proved that $\kappa \leqslant 2$, and this is the best possible result since given any irrational number ξ there are always infinitely many integers p, q ($q > 0$) such that $|\xi - p/q| < 1/q^2$.

Theorem 3.1 (Roth). *Let ξ be any irrational algebraic number and let $\delta > 0$. Then the inequality*

$$\left| \xi - \frac{p}{q} \right| < \frac{1}{q^{2+\delta}}$$

has only finitely many rational integer solutions in p and q ($q > 0$). □

The proof of this celebrated theorem can be found in *An Introduction to Diophantine Approximation* by J. W. S. Cassels, Cambridge University Press, 1957.

17.4 Application of Roth's Theorem

Theorem 4.1. *Let $n \geqslant 3$ and let*

$$f(x, y) = b_0 x^n + b_1 x^{n-1} y + \cdots + b_n y^n$$

be an irreducible homogeneous polynomial with rational integer coefficients. Suppose that

$$g(x, y) = \sum_{r+s \leqslant n-3} g_{rs} x^r y^s$$

is a polynomial with degree at most $n-3$ and with rational coefficients. Then the equation

$$f(x,y) = g(x,y) \tag{1}$$

has only finitely many solutions in integers (x,y).

Proof. We need only consider the case $|x| \leqslant |y|$. If $y = 0$, then there is at most one solution. Suppose now that $y > 0$. Let $\alpha_1, \ldots, \alpha_n$ be the roots of the equation $f(x,1) = 0$, and let $G = \max(|g_{rs}|)$. From (1) we have

$$\begin{aligned} |b_0(x-\alpha_1 y)\cdots(x-\alpha_n y)| &\leqslant G(1+2y+\cdots+(n-2)y^{n-3}) \\ &\leqslant n^2 G y^{n-3}. \end{aligned} \tag{2}$$

Therefore there exists ν such that

$$|x - \alpha_\nu y| < c_1 y^{1-\frac{3}{n}};$$

here c_1, and in what follows, c_2, c_3, c_4, c_5 are positive constants.

When $\mu \neq \nu$, and y is greater than a sufficiently large c_2, we have

$$|x - \alpha_\mu y| = |(\alpha_\nu - \alpha_\mu)y + (x - \alpha_\nu y)| > c_3 y - c_1 y^{1-\frac{3}{n}} > c_4 y, \tag{3}$$

so that by (2) and (3), $|x - \alpha_\nu y| < c_5/y^2$ or

$$\left|\alpha_\nu - \frac{x}{y}\right| < \frac{c_5}{y^3}.$$

By Roth's theorem this inequality has only finitely many integer solutions, and the situation when $y \leqslant c_2$ is also clear. The theorem is proved. □

Theorem 4.2 (Thue). *Let $n \geqslant 3$ and let*

$$g(x,y) = b_0 x^n + b_1 x^{n-1} y + \cdots + b_n y^n$$

be an irreducible homogeneous polynomial with rational integer coefficients. If a is a rational integer, then the equation $g(x,y) = a$ has only finitely many integer solutions in x, y.

Proof. This is a special case of Theorem 4.1. □

Theorem 4.3 (Thue). *In the previous theorem if $a \neq 0$, then the hypothesis that $g(x,y)$ be irreducible can be relaxed, provided that $g(x,y)$ is not the n-th power of a linear form or the $n/2$-th power of a quadratic form.*

Proof. If $g(z) = g(z,1)$ is irreducible then there is nothing more to discuss. If $g(z)$ is a power of an irreducible polynomial $h(z)$ with degree $m \geqslant 3$, so that $g(z) = (h(z))^{n/m}$, then the problem becomes that of the solutions to the equation $y^m h(x/y) = a^{m/n}$. If

$a^{m/n}$ is a rational integer, then the problem is reduced to that in Theorem 4.2, and if $a^{m/n}$ is not a rational integer the equation concerned has no solutions. We now suppose that $g(z) = g_1(z)g_2(z)$ where $g_1(z)$ and $g_2(z)$ are integer coefficient polynomials with degrees r and s respectively, and with no common factors. The problem is now transformed into that of the solutions to

$$y^r g_1\left(\frac{x}{y}\right) = a_1, \qquad y^s g_2\left(\frac{x}{y}\right) = a_2, \qquad a = a_1 a_2.$$

Given a, the number of pairs a_1, a_2 is finite. If there is a solution $y \neq 0, \pm 1$, then $y^r g_1(z) = a_1, y^s g_2(z) = a_2$ have common roots so that

$$a_2^r (g_1(z))^s = a_1^s (g_2(z))^r.$$

But $g_1(z)$ and $g_2(z)$ have no common factor and $g_2(z) \neq \pm a_2, g_1(z) \neq \pm a_1$ so that this equation cannot hold. The case $y = 0, \pm 1$ is trivial. □

Note: The condition $a \neq 0$ is necessary, since the equation $x^3 - y^3 = 0$ has infinitely many solutions. Also $(sx + ty)^n = a^n, (x^2 - 2y^2)^l = 1$ both have infinitely many solutions so that the remaining conditions are also necessary.

17.5 Application of Thue's Theorem

Theorem 5.1 (Landau-Ostrowski-Thue). *Let* $n \geqslant 3, b^2 - 4ac \neq 0, a \neq 0, d \neq 0$. *Then the indeterminate equation*

$$ay^2 + by + c = dx^n \tag{1}$$

has only finitely many solutions.

Proof. We rewrite (1) as $(2ay + b)^2 - (b^2 - 4ac) = 4adx^n$, and we see that if the equation $y_1^2 - (b^2 - 4ac) = 4adx^n$ has only finitely many solutions, then the same holds for the original equation. We can therefore assume that $a = 1, b = 0$, that is, it suffices to prove that if $n \geqslant 3$, $k \neq 0$, $l \neq 0$, then

$$y^2 - k = lx^n \tag{2}$$

has only finitely many solutions.

1) Suppose that $k = m^2$ so that (2) becomes

$$(y - m)(y + m) = lx^n.$$

If $x = 0$, then $y = \pm m$. Suppose now that $x \neq 0$ so that $y \neq \pm m$. If $p \nmid 2ml$, then p cannot be a prime divisor of both $y + m$ and $y - m$. Therefore $y + m$ and $y - m$ have the representations

$$y + m = \pm p_1^{r_1} \cdots p_j^{r_j} z^n = qz^n, \qquad 0 \leqslant r_i \leqslant n-1,$$
$$y - m = \pm p_1^{s_1} \cdots p_j^{s_j} w^n = tw^n, \qquad 0 \leqslant s_i \leqslant n-1,$$

where $p_1, \ldots, p_j$ are the prime divisors of $2ml$, and q, t may take only finitely many non-zero values.

Corresponding to a set $q \neq 0, t \neq 0$, the function $f(z) = qz^n - t$ has no repeated zeros, so that the hypothesis to Theorem 4.3 is satisfied. It follows that the indeterminate equation $qz^n - tw^n = 2m$ has only finitely many sets of non-zero solutions, and so the theorem is proved.

2) Suppose that k is not a perfect square and let

$$\vartheta = \begin{cases} \sqrt{k}, & k > 0, \\ i\sqrt{|k|}, & k < 0. \end{cases}$$

We need only consider the solutions to (2) with $x > 0$, that is the solutions to

$$y^2 - k = lx^n, \qquad x > 0. \tag{3}$$

Let x, y be any set of solutions to (3). By Theorem 6.10.5 there are integers r and q such that

$$\left| \frac{y}{x} - \frac{r}{q} \right| < \frac{1}{q\sqrt{x}}, \qquad 0 < q \leqslant \sqrt{x}. \tag{4}$$

Let $qr - rx = s$, so that

$$|s| < \sqrt{x} \tag{5}$$

and

$$s \equiv qy \pmod{x}. \tag{6}$$

Also, let

$$t = \left(\frac{s^2 - q^2 k}{x} \right)^n;$$

since k is not a perfect square, $t \neq 0$. From (6) and (3) we have

$$s^2 - q^2 k \equiv q^2(y^2 - k) \equiv q^2 l x^n \equiv 0 \pmod{x}$$

so that t is an integer. Also, from

$$|t| \leqslant \left(\frac{s^2 + q^2|k|}{x} \right)^n < \left(\frac{x + x|k|}{x} \right)^n = (1 + |k|)^n,$$

we see that, corresponding to any fixed n and k, the integer t may take only finitely many values.

We next define $\beta = (s - q\vartheta)^n (y + \vartheta)/x^n$, $\xi = s + q\vartheta$ so that

$$t(y + \vartheta) = \beta \xi^n. \tag{7}$$

From

$$(s - q\vartheta)^n = (q(y - \vartheta) - rx)^n = (A_1 + A_2\vartheta)(y - \vartheta) + (-1)^n r^n x^n,$$

we have

$$x^n\beta = (s - q\vartheta)^n(y + \vartheta) = (A_1 + A_2\vartheta)(y^2 - k) + (-1)^n r^n x^n(y + \vartheta)$$
$$= (A_1 + A_2\vartheta)lx^n + (A_3 + A_4\vartheta)x^n = (A_5 + A_6\vartheta)x^n,$$

or

$$\beta = A_5 + A_6\vartheta, \tag{8}$$

where $A_1, A_2, \ldots, A_6$ are integers depending on x, y.

From

$$|y| + |\vartheta| \leqslant \sqrt{|k| + |l|x^n} + \sqrt{|k|} \leqslant x^{n/2}(\sqrt{|k| + |l|} + \sqrt{|k|})$$

and

$$|s| + q|\vartheta| \leqslant \sqrt{x} + \sqrt{x}\sqrt{|k|} = \sqrt{x}(1 + \sqrt{|k|}),$$

we have

$$|A_5 \pm A_6\vartheta| = \left|\frac{(s \mp q\vartheta)^n(y \pm \vartheta)}{x^n}\right|$$
$$\leqslant (1 + \sqrt{|k|})^n(\sqrt{|k| + |l|} + \sqrt{|k|}).$$

Therefore from

$$A_5 = \tfrac{1}{2}((A_5 + A_6\vartheta) + (A_5 - A_6\vartheta))$$

and

$$A_6 = \frac{1}{2\vartheta}((A_5 + A_6\vartheta) - (A_5 - A_6\vartheta)),$$

we see that, given fixed integers n, k, l the numbers A_5, A_6 may only take a finite number of distinct values, that is the number of values that β may take is finite. We already know that t may take only finitely many values so that, given fixed n, k, l there are only finitely many equations (7).

From (7) we have

$$t(y + \vartheta) = (A_5 + A_6\vartheta)(s + q\vartheta)^n,$$
$$t(y - \vartheta) = (A_5 - A_6\vartheta)(s - q\vartheta)^n.$$

Therefore

$$2t = \frac{1}{\vartheta}[(A_5 + A_6\vartheta)(s + q\vartheta)^n - (A_5 - A_6\vartheta)(s - q\vartheta)^n]. \tag{9}$$

Given n, t ($\neq 0$), A_5, A_6, if we can show that the right hand side of the above equation is a polynomial $g(s, q)$ satisfying the hypothesis of Theorem 4.3, then (9) has only finitely many sets of solutions (s, q) so that from (7) we see that there are only finitely many different values for y, and hence the equation (2) also has only finitely many solutions.

It remains therefore to show that $g(s, q)$ satisfies the hypothesis of Theorem 4.3, and for this it suffices to show that

$$f(z) = \frac{1}{\vartheta}[(A_5 + A_6\vartheta)(z + \vartheta)^n - (A_5 - A_6\vartheta)(z - \vartheta)^n]$$

has no repeated zeros. But this is a simple matter. For if $f(z) = 0, f'(z) = 0$ have a common solution $z = z_0$, then z_0 must satisfy

$$\frac{A_5 + A_6\vartheta}{A_5 - A_6\vartheta} = \left(\frac{z_0 - \vartheta}{z_0 + \vartheta}\right)^n = \left(\frac{z_0 - \vartheta}{z_0 + \vartheta}\right)^{n-1}.$$

Since $(z - \vartheta)/(z + \vartheta) \neq 1$, this is impossible. The theorem is proved. □

Exercise 1. Let n be an odd integer greater than 1. Arrange the integers which are either a square or an n-th power into an increasing sequence (z_r). Prove that $z_{r+1} - z_r \to \infty$ as $r \to \infty$.

Exercise 2. Let $\langle\xi\rangle = \min(\xi - [\xi], [\xi] + 1 - \xi)$. Prove that

$$\lim_{\substack{x \to \infty \\ x \neq k^2}} x^{n/2}\langle x^{n/2}\rangle = \infty.$$

17.6 The Transcendence of e

The existence of transcendental numbers has been established, and indeed our proof shows that almost all real numbers are transcendental since the algebraic numbers form a countable set. We now turn to the problem of the transcendence of specific given numbers such as e, π or sin 1. This problem is much more difficult. In this and the next section we prove the transcendence of e and π, but up to the present no one has established the transcendence, or otherwise, of $e + \pi$. Similarly we do not know the transcendence of Euler's constant

$$\gamma = \lim_{n \to \infty}\left(1 + \frac{1}{2} + \cdots + \frac{1}{n} - \log n\right),$$

and in fact we do not even know whether it is rational or irrational. This is a part of the famous seventh problem of Hilbert and the other part is the main theme of sections 8 and 9.

Theorem 6.1. *The number e is irrational.*

Proof. We shall prove that e^{-1} is irrational. Let $e^{-1} = \sigma_n + \rho_n$ where

$$\sigma_n = \sum_{k=0}^{n} \frac{(-1)^k}{k!}, \qquad \rho_n = \sum_{k=n+1}^{\infty} \frac{(-1)^k}{k!}.$$

It is easy to see that

$$0 < (-1)^{n+1}\rho_n = \frac{1}{(n+1)!} - \frac{1}{(n+2)!} + \cdots < \frac{1}{(n+1)!}.$$

Therefore

$$0 < n!\rho_n(-1)^{n+1} < \frac{1}{n+1} < 1,$$

so that

$$n!e^{-1} = n!\sigma_n + n!\rho_n(-1)^{n+1}$$

cannot be an integer. □

Theorem 6.2. *Let*

$$f(x) = \sum_{m=0}^{n} a_m x^m,$$

$$F(x) = \sum_{k=0}^{m} f^{(k)}(x), \qquad F(0)e^x - F(x) = Q(x).$$

Then

$$|Q(x)| \leqslant e^{|x|} \sum_{m=0}^{n} |a_m||x|^m.$$

Proof. We have the following identity:

$$\begin{aligned} F(x) &= \sum_{k=0}^{n} \sum_{m=k}^{n} a_m \frac{m!}{(m-k)!} x^{m-k} \\ &= \sum_{m=0}^{n} a_m \sum_{k=0}^{m} \frac{m!}{(m-k)!} x^{m-k} = \sum_{m=0}^{n} a_m \sum_{k=0}^{m} \frac{m!}{k!} x^k. \end{aligned}$$

In particular we have

$$F(0) = \sum_{m=0}^{n} a_m m!.$$

Therefore

$$\begin{aligned} |Q(x)| &= \left| \sum_{m=0}^{n} a_m \sum_{k=0}^{\infty} \frac{m!}{k!} x^k - \sum_{m=0}^{n} a_m \sum_{k=0}^{m} \frac{m!}{k!} x^k \right| \\ &= \left| \sum_{m=0}^{n} a_m \sum_{k=m+1}^{\infty} \frac{m!}{k!} x^k \right| \\ &\leqslant \sum_{m=0}^{n} |a_m| \sum_{k=m+1}^{\infty} |x|^k/(k-m)! \\ &= \sum_{m=0}^{n} |a_m|\,|x|^m \sum_{l=1}^{\infty} \frac{|x|^l}{l!} \leqslant e^{|x|} \sum_{m=0}^{n} |a_m|\,|x|^m. \end{aligned}$$

□

Theorem 6.3 (Hermite). *The number e is transcendental.*

Proof. Suppose that e is a zero of $P(x)$ where

$$P(x) = \sum_{h=0}^{m} g_h x^h, \qquad g_0 \neq 0, \qquad m > 0,$$

and g_h are rational integers. Let p be a prime number greater than $\max(m, |g_0|)$, and let

$$f(x) = \frac{x^{p-1} \prod_{h=1}^{m} (h - x)^p}{(p-1)!} = \sum_{k=0}^{n} a_k x^k \qquad (a_k = a_k(p)).$$

Since h is a zero of $f(x)$ of order p we can write

$$f(x) = \frac{(m!)^p x^{p-1} + A_p x^p + \cdots}{(p-1)!} = \frac{B_{p,h}(x-h)^p + B_{p+1,h}(x-h)^{p+1} + \cdots}{(p-1)!},$$

where A, B are rational integers. We can now construct $F(x)$ and $Q(x)$ according to Theorem 6.2, and we have

$$0 = F(0)P(e) = F(0) \sum_{h=0}^{m} g_h e^h = \sum_{h=0}^{m} g_h F(h) + \sum_{h=0}^{m} g_h Q(h). \tag{1}$$

We already know that

$$\sum_{h=0}^{m} g_h F(h) = g_0 \sum_{k=0}^{n} f^{(k)}(0) + \sum_{h=1}^{m} g_h \sum_{k=0}^{n} f^{(k)}(h)$$

$$= g_0((m!)^p + pA_p + \cdots) + \sum_{h=1}^{m} g_h(pB_{p,h} + p(p+1)B_{p+1,h} + \cdots)$$

is a rational integer. Observe that, on the right hand side of this equation, $p \nmid g_0(m!)^p$ and the remaining terms are all multiples of p, so that

$$\left| \sum_{h=0}^{m} g_h F(h) \right| \geqslant 1.$$

If we can prove that there exists a prime $p > \max(m, |g_0|)$ such that

$$\left| \sum_{h=0}^{m} g_h Q(h) \right| < 1,$$

then the required contradiction will be obtained from equation (1). By Theorem 6.2 it suffices to prove that, for any fixed x, we have

$$\sum_{k=0}^{n} |a_k|\,|x|^k \to 0 \qquad \text{as} \qquad p \to \infty,$$

but this is easy because, as $p \to \infty$,

$$\sum_{k=0}^{n} |a_k|\,|x|^k \leqslant \frac{|x|^{p-1} \prod_{h=1}^{m} (h + |x|)^p}{(p-1)!} \to 0. \qquad \square$$

17.7 The Transcendence of π

Theorem 7.1. *The number π is irrational.*

Proof (Niven). Suppose the contrary, so that $\pi = a/b$ where a, b are positive integers. Let

$$f(x) = \frac{x^n(a - bx)^n}{n!}$$

and

$$F(x) = f(x) - f^{(2)}(x) + f^{(4)}(x) - \cdots + (-1)^n f^{(2n)}(x).$$

It is easy to see that $f(x)$ together with its derivatives are integers when $x = 0$ and π, so that $F(0)$ and $F(\pi)$ are integers. Now

$$\frac{d}{dx}(F'(x)\sin x - F(x)\cos x) = (F''(x) + F(x))\sin x = f(x)\sin x,$$

so that

$$\int_0^\pi f(x)\sin x\, dx = F(\pi) - F(0) \tag{1}$$

is an integer. But, for $0 < x < \pi$ and n sufficiently large, we have

$$0 < f(x)\sin x < \frac{\pi^n a^n}{n!} < \frac{1}{\pi},$$

so that

$$0 < \int_0^\pi f(x)\sin x\, dx < 1,$$

giving a contradiction to (1) being an equation in integers. □

Theorem 7.2 (Lindemann). *The number π is transcendental.*

Proof. We shall prove that $i\pi$ is not algebraic. Suppose the contrary and let $i\pi$ satisfy the equation

$$f(x) = ax^m + a_1 x^{m-1} + \cdots = 0, \qquad a > 0$$

so that $ai\pi$ satisfies the equation

$$a^{m-1} f\left(\frac{x}{a}\right) = x^m + a_1 x^{m-1} + \cdots = 0.$$

Our supposition that $i\pi$ is algebraic implies that $ai\pi$ is also algebraic. The required contradiction will therefore follow if we can prove that $ai\pi$ does not satisfy any equation of the form

$$P(y) = y^m + k_{m-1}y^{m-1} + \cdots + k_0 = 0, \qquad m > 0.$$

Let

$$P(y) = \prod_{h=1}^{m} (y - a\alpha_h).$$

Since $1 + e^{i\pi} = 0$, it suffices to show that

$$R = \prod_{h=1}^{m} (e^0 - e^{\alpha_h}) \neq 0.$$

Now R can be written as

$$R = c + \sum e^{\alpha} + \sum e^{\alpha+\alpha'} + \cdots = c + e^{\beta_1} + e^{\beta_2} + \cdots + e^{\beta_r},$$

where c is the number of the 2^m terms in which the power of the exponential is zero, and $\beta_1, \beta_2, \ldots, \beta_r$ are non-zero numbers.

Let p be a prime greater than $\max(c, a, \prod_{h=1}^{r} a|\beta_h|)$ and define $f(x)$ by

$$f(x) = \frac{(ax)^{p-1} \prod_{h=1}^{r} (ax - a\beta_h)^p}{(p-1)!} = \sum_{k=0}^{n} a_k x^k.$$

Similarly to the proof of Theorem 6.3 we have

$$f(x) = \frac{A_{p-1}x^{p-1} + A_p x^p + \cdots}{(p-1)!} = \frac{\gamma_{p,h}(x - \beta_h)^p + \gamma_{p+1,h}(x - \beta_h)^{p+1} + \cdots}{(p-1)!},$$

where A_l, being symmetric functions of $a\beta_1, \ldots, a\beta_r$ and hence symmetric functions of $a\alpha_1, \ldots, a\alpha_h$, are rational integers and $A_{p-1} \not\equiv 0 \pmod{p}$.

On the construction of the corresponding $F(x)$ and $Q(x)$ we have

$$F(0)R = F(0)\left(c + \sum_{h=1}^{r} e^{\beta_h}\right) = cF(0) + \sum_{h=1}^{r} F(\beta_h) + \sum_{h=1}^{r} Q(\beta_h),$$

so that

$$cF(0) = c(A_{p-1} + pA_p + \cdots)$$

is a rational integer which is not a multiple of p. Also

$$\begin{aligned}\sum_{h=1}^{r} F(\beta_h) &= \sum_{h=1}^{r} (p\gamma_{p,h} + p(p+1)\gamma_{p+1,h} + \cdots) \\ &= p\sum_{h=1}^{r} \gamma_{p,h} + p(p+1)\sum_{h=1}^{r} \gamma_{p+1,h} + \cdots \\ &= pc_p + p(p+1)c_{p+1} + \cdots,\end{aligned}$$

where $c_p, c_{p+1}, \ldots$, being symmetric functions of $a\beta_1, \ldots, a\beta_r$, are integers. It follows that $\sum_{h=1}^{r} F(\beta_h)$ is a multiple of p and whence

$$\left| cF(0) + \sum_{h=1}^{r} F(\beta_h) \right| \geqslant 1.$$

It only remains to show that, for sufficiently large p,

$$\left| \sum_{h=1}^{r} Q(\beta_h) \right| < 1.$$

But, as $p \to \infty$,

$$\sum_{k=1}^{n} |a_k| \, |x|^k \leqslant \frac{(a|x|)^{p-1} \sum_{h=1}^{r} (a|x| + a|\beta_h|)^p}{(p-1)!} \to 0$$

so that the result follows from Theorem 6.2. □

Remark. This theorem settles the problem of "squaring the circle" – it is impossible to construct a square equal in area to a given circle, using only straight edge and compass.

Exercise 1. Prove that $\sinh \xi$ is transcendental whenever ξ is rational.

Exercise 2. Prove that $\sin 1$ is transcendental by proving that e^i is transcendental.

17.8 Hilbert's Seventh Problem

In the year 1900 Hilbert gave a list of 23 unsolved problems which he believed to be worthy of the attention of mathematicians in the twentieth century. We already mentioned the first part of his seventh problem, and the remaining part is the following: Let α and β be algebraic numbers with $\alpha \neq 0, 1$ and β irrational. Does it follow that α^β is transcendental? As specific examples he asked for the proofs of the transcendence of $2^{\sqrt{2}}$ and $e^\pi = (-1)^{-i}$.

In 1929 the Russian mathematician A. O. Gelfond made an important contribution to the solution of this problem. He proved the transcendence of e^π and pointed out that his method can be used to settle Hilbert's problem when β lies in an imaginary quadratic field. In 1930 Kusmin used Gelfond's method to settle the case when β lies in a real quadratic field and proved in particular that $2^{\sqrt{2}}$ is transcendental. Then in 1934 the complete solutions to Hilbert's problem were given independently by Gelfond and Schneider.

It may be of some interest to recall that, when discussing this problem, Hilbert was of the opinion that the solution would not be available before the solutions to the Riemann's hypothesis and Fermat's last theorem. It seems therefore that it is very difficult to judge the difficulty of an unsolved problem before a solution is available.

Let K be an algebraic number field of degree h, and let $\beta_1, \ldots, \beta_h$ be an integer basis, so that every integer in K has the unique representation $a_1\beta_1 + \cdots + a_h\beta_h$ where $a_1, \ldots, a_h$ are rational integers. We shall denote by $\overline{|\alpha|}$ the maximum of the modulus of the conjugates $\alpha^{(i)}$ $(1 \leqslant i \leqslant h)$ of α, that is

$$\overline{|\alpha|} = \max_{1 \leqslant i \leqslant h} |\alpha^{(i)}|.$$

In the following we let c, c_1, c_2 be natural numbers depending on K and its basis $\beta_1, \ldots, \beta_h$. It is easy to show that if α is an algebraic integer with $\alpha = a_1\beta_1 + \cdots + a_h\beta_h$, then $|a_i| \leqslant c\overline{|\alpha|}$.

Lemma 8.1. *Let $0 < M < N$, and a_{jk} be rational integers satisfying $|a_{jk}| \leqslant A$ $(A \geqslant 1$, $1 \leqslant j \leqslant M, 1 \leqslant k \leqslant N)$. Then there exists a set of rational integers $x_1, \ldots, x_N$, not all zero, satisfying*

$$a_{j1}x_1 + \cdots + a_{jN}x_N = 0, \qquad 1 \leqslant j \leqslant M, \tag{1}$$

and

$$|x_k| \leqslant [(NA)^{\frac{M}{N-M}}], \qquad 1 \leqslant k \leqslant N. \tag{2}$$

Proof. Let

$$y_j = a_{j1}x_1 + \cdots + a_{jN}x_N, \qquad 1 \leqslant j \leqslant M,$$

so that this defines a mapping from rational integers $(x_1, \ldots, x_N)$ to rational integers $(y_1, \ldots, y_M)$. We write

$$H = [(NA)^{\frac{M}{N-M}}] \qquad \text{so that} \qquad NA < (H+1)^{\frac{N-M}{M}},$$

and hence

$$NAH + 1 \leqslant NA(H+1) < (H+1)^{\frac{N}{M}}. \tag{3}$$

For any set of integers $(x_1, \ldots, x_N)$ satisfying

$$0 \leqslant x_k \leqslant H, \qquad 1 \leqslant k \leqslant N \tag{4}$$

we have

$$-B_jH \leqslant y_j \leqslant C_jH, \qquad B_j + C_j \leqslant NA,$$

where $-B_j$ and C_j represent respectively the sum of the negative and positive coefficients of y_j, so that the number of values assumed by y_j cannot exceed $NAH + 1$. The number of sets $(x_1, \ldots, x_N)$ satisfying (4) is $(H+1)^N$ and the corresponding number of sets $(y_1, \ldots, y_M)$ is at most $(NAH+1)^M$. It follows from (3) that there must be two sets $(x'_1, \ldots, x'_N)$ and $(x''_1, \ldots, x''_N)$ which correspond to the same set $(y_1, \ldots, y_M)$. Let $x_k = x'_k - x''_k$ $(1 \leqslant k \leqslant N)$ so that $(x_1, \ldots, x_N)$ is now the required set satisfying (1) and (2). □

Lemma 8.2. *Let $0 < p < q$, and let α_{kl} $(1 \leqslant k \leqslant p, 1 \leqslant l \leqslant q)$ be integers in K satisfying $\overline{|\alpha_{kl}|} \leqslant A$. Then there exists a set of algebraic integers $\xi_1, \ldots, \xi_q$ in K, not all zero, satisfying*

$$\alpha_{k1}\xi_1 + \cdots + \alpha_{kq}\xi_q = 0, \qquad 1 \leqslant k \leqslant p \tag{5}$$

and

$$\overline{|\xi_l|} < c_1(1 + (c_1 qA)^{p/(q-p)}), \qquad 1 \leqslant l \leqslant q. \tag{6}$$

Proof. Let $\xi_l = x_{l1}\beta_1 + \cdots + x_{lh}\beta_h$ $(1 \leqslant l \leqslant q)$ where $x_{l1}, \ldots, x_{lh}$ are rational integers. Let

$$\alpha_{kl}\beta_r = a_{klr1}\beta_1 + \cdots + a_{klrh}\beta_h \tag{7}$$

where $a_{klr1}, \ldots, a_{klrh}$ are also rational integers. For $1 \leqslant k \leqslant p$ we have, from (5), that

$$\begin{aligned} 0 &= \sum_{l=1}^{q} \alpha_{kl}\xi_l = \sum_{l=1}^{q} \alpha_{kl} \sum_{r=1}^{h} x_{lr}\beta_r \\ &= \sum_{r=1}^{h} \sum_{l=1}^{q} x_{lr} \sum_{u=1}^{h} a_{klru}\beta_u = \sum_{u=1}^{h} \left(\sum_{r=1}^{h} \sum_{l=1}^{q} a_{klru}x_{lr} \right) \beta_u. \end{aligned}$$

Since $\beta_1, \ldots, \beta_h$ are linearly independent we have the hp number of equations

$$\sum_{r=1}^{h} \sum_{l=1}^{q} a_{klru}x_{lr} = 0, \qquad 1 \leqslant u \leqslant h, \qquad 1 \leqslant k \leqslant p, \tag{8}$$

with hq number of unknowns.

From (7) and our remark preceeding Lemma 8.1 we see that $|a_{klru}| \leqslant c \max_{1 \leqslant i \leqslant h} \overline{|\beta_i|}\, A \leqslant c_2 A$. It now follows from Lemma 8.1 that the system (8) has a non-trivial set of solutions in rational integers satisfying

$$|x_{lr}| \leqslant 1 + (hqc_2 A)^{p/(q-p)}, \qquad 1 \leqslant l \leqslant q, \qquad 1 \leqslant r \leqslant h.$$

Therefore

$$\begin{aligned} \overline{|\xi_l|} &\leqslant |x_{l1}|\,\overline{|\beta_1|} + \cdots + |x_{lh}|\,\overline{|\beta_h|} \\ &\leqslant c_2 h(1 + (hqc_2 A)^{p/(q-p)}). \end{aligned}$$

Taking $c_1 = c_2 h$ the lemma is proved. □

17.9 Gelfond's Proof

Let α and β be algebraic numbers with $\alpha \neq 0, 1$ and β irrational, and we have to prove that α^β is transcendental. Suppose the contrary, so that $\gamma = \alpha^\beta = e^{\beta \log \alpha}$ (where $\log \alpha$ may be any fixed value of the logarithm of α) is also algebraic. We shall derive a contradiction.

Suppose that α, β, γ lie in an algebraic field with degree h. Let

$$m = 2h + 2, \qquad n = \frac{q^2}{2m}$$

where $q^2 = t$ is a square of a natural number and is a multiple of $2m$. Also, let $\rho_1, \rho_2, \ldots, \rho_t$ represent the t numbers

$$(a + b\beta)\log\alpha, \qquad 1 \leqslant a \leqslant q, \qquad 1 \leqslant b \leqslant q.$$

We introduce the integral function

$$R(x) = \eta_1 e^{\rho_1 x} + \cdots + \eta_t e^{\rho_t x}, \tag{1}$$

where the coefficients $\eta_1, \ldots, \eta_t$ are determined by the following conditions. We solve the system of mn homogeneous linear equations

$$(\log\alpha)^{-k} R^{(k)}(l) = 0, \qquad 0 \leqslant k \leqslant n-1, \quad 1 \leqslant l \leqslant m, \tag{2}$$

in the $t = 2mn$ unknowns $\eta_1, \ldots, \eta_t$. The coefficients of this system are numbers in K and

$$(\log\alpha)^{-k}((a + b\beta)\log\alpha)^k e^{l(a+b\beta)\log\alpha} = (a + b\beta)^k \alpha^{al} \gamma^{bl},$$

$$1 \leqslant l \leqslant m, \qquad 1 \leqslant a, \qquad b \leqslant q, \qquad 0 \leqslant k \leqslant n-1.$$

Let $c_1, c_2, \ldots$ denote natural numbers which are independent of n. There exists c_1 such that $c_1\alpha$, $c_1\beta$ and $c_1\gamma$ are all integers in K, so that on multiplying each of the coefficients of the system by $c_1^{n-1} c_1^{mq} c_1^{mq} = c_1^{n-1+2mq}$ $(\leqslant c_2^n)$, the resulting coefficients become integers in K. Moreover the absolute value of the conjugates of the various coefficients is at most

$$c_2^n (q + q\overline{|\beta|})^{n-1} \overline{|\alpha|}^{mq} \overline{|\gamma|}^{mq} \leqslant c_3^n n^{\frac{1}{2}(n-1)}.$$

It follows from Lemma 8.2 that there is a non-trivial set of integers solutions $\eta_1, \ldots, \eta_t$ in K such that

$$\overline{|\eta_k|} \leqslant c_4^n n^{\frac{1}{2}(n+1)}, \qquad 1 \leqslant k \leqslant t.$$

Since the numbers $\rho_1, \ldots, \rho_t$ are distinct, the function $R(x)$ is not identically zero. For suppose otherwise; then on expanding the right hand side of (1) we have

$$\eta_1 \rho_1^k + \eta_2 \rho_2^k + \cdots + \eta_t \rho_t^k = 0, \qquad k = 0, 1, 2, \ldots$$

which implies $\eta_1 = \eta_2 = \cdots = \eta_t = 0$, a contradiction. Thus we see from (2) that

$$R(x) = a_{n,l}(x - l)^n + a_{n+1,l}(x - l)^{n+1} + \cdots, \qquad 1 \leqslant l \leqslant m, \tag{3}$$

where $a_{n,l}, a_{n+1,l}, \ldots$ are not all zero. Hence there must be a natural number r such that $R^{(k)}(l) = 0, 0 \leqslant k \leqslant r-1, 1 \leqslant l \leqslant m$. But for $1 \leqslant l_0 \leqslant m$ we have $R^{(r)}(l_0) \neq 0$ so that we see from (3) that $r \geqslant n$.

Let us now examine the number

$$(\log \alpha)^{-r} R^{(r)}(l_0) = \rho \neq 0. \tag{4}$$

This number lies in K, and $c_1^{r+2mq}\rho$ is an integer in K so that

$$|N(\rho)| > c_1^{-h(r+2mq)} > c_5^{-r}. \tag{5}$$

On the other hand

$$\overline{|\rho|} \leqslant t c_4^n n^{\frac{1}{2}(n+1)} (c_6 q)^r c_7^q \leqslant c_8^r r^{r+\frac{3}{2}}. \tag{6}$$

We now determine a suitable upper bound for $|\rho|$. We apply Cauchy's integral formula to the function

$$S(z) = r! \frac{R(z)}{(z-l_0)^r} \prod_{\substack{k=1\\k\neq l_0}}^{m} \left(\frac{l_0-k}{z-k}\right)^r.$$

We then have

$$\rho = (\log \alpha)^{-r} S(l_0) = (\log \alpha)^{-r} \frac{1}{2\pi i} \int_C \frac{S(z)}{z-l_0}\, dz, \tag{7}$$

where C is the circle $|z| = m(1 + r/q)$, so that l_0 ($\leqslant m$) lies inside C. As z varies on the circle we have

$$\begin{aligned}
|R(z)| &\leqslant t \max_{1\leqslant k\leqslant t} |\eta_k| e^{(q+q|\beta|)\log|\alpha|\cdot m\left(1+\frac{r}{q}\right)}\\
&\leqslant t c_4^n n^{\frac{1}{2}(n+1)} c_9^{r+q} \leqslant c_{10}^r r^{\frac{1}{2}(r+3)},
\end{aligned}$$

$$|z-l_0| \geqslant |z| - |l_0| \geqslant m\left(1+\frac{r}{q}\right) - m = \frac{mr}{q},$$

$$|z-k| \geqslant \frac{mr}{q}, \qquad 1 \leqslant k \leqslant m,$$

$$\left|(z-l_0)^{-r} \prod_{\substack{k=1\\k\neq l_0}}^{m} \left(\frac{l_0-k}{z-k}\right)^r\right| \leqslant c_{11}^r \left(\frac{q}{r}\right)^{mr},$$

$$|S(x)| \leqslant r! c_{10}^r r^{\frac{1}{2}(r+3)} c_{11}^r \left(\frac{q}{r}\right)^{mr} \leqslant c_{12}^r r^{\frac{1}{2}r(3-m)+\frac{3}{2}}.$$

From (7) we now have

$$\begin{aligned}
|\rho| &\leqslant \frac{1}{2\pi} |(\log \alpha)^{-r}| \int_C \left|\frac{S(z)}{z-l_0}\right| |dz|\\
&\leqslant |(\log \alpha)^{-r}|\, m\left(1+\frac{r}{q}\right) c_{12}^r r^{\frac{1}{2}r(3-m)+\frac{3}{2}} \frac{q}{mr} \leqslant c_{13}^r r^{\frac{1}{2}r(3-m)+\frac{3}{2}}.
\end{aligned} \tag{8}$$

From (6) and (8) we have

$$|N(\rho)| \leqslant c_{14}^{r} r^{(h-1)(r+\frac{3}{2})+\frac{1}{2}(3-m)r+\frac{3}{2}}.$$

Replacing m by $2h + 2$ we now have

$$|N(\rho)| \leqslant c_{14}^{r} r^{-\frac{1}{2}r+\frac{3}{2}h}$$

and from (5) we deduce that

$$r^{\frac{1}{2}r-\frac{3}{2}h} < c_{14}^{r} c_{5}^{r} = c_{15}^{r}.$$

Since $r \geqslant n$, this cannot hold for sufficiently large n, and the required contradiction is obtained. □

Notes

17.1. The proof of Roth's theorem has been omitted in this English edition (see J. W. S. Cassels [15]). W. M. Schmidt [51], [52] has given the following important generalization of this famous theorem:

Let $\alpha_1, \ldots, \alpha_n$ be real algebraic numbers such that $1, \alpha_1, \ldots, \alpha_n$ are linearly independent over that rational field R. Then, given any $\varepsilon > 0$, the inequality

$$\langle q_1\alpha_1 + \cdots + q_n\alpha_n \rangle \{(1 + |q_1|) \cdots (1 + |q_n|)\}^{1+\varepsilon} < 1$$

has at most a finite number of sets of integer solutions $q_1, \ldots, q_n$.

17.2. A. Baker [2] has made the following important improvement on Thue's theorem: Let $g(x, y)$ be a homogeneous irreducible polynomial of degree n ($\geqslant 3$) with rational integer coefficients, and let m be a positive integer. Then all the integer solutions to the equation $g(x, y) = m$ can be effectively determined. More specifically, if H exceeds the absolute values of all the coefficients of $g(x, y)$, then all the integer solutions to $g(x, y) = m$ must satisfy

$$\max(|x|, |y|) < \exp\{(nH)^{(10n)^5} + (\log m)^{2n+2}\}.$$

17.3. A. Baker [1] has made the following important generalizations of the Gelfond-Schneider theorem:

(i) If $\alpha_1, \ldots, \alpha_n, \beta_0, \beta_1, \ldots, \beta_n$ are non-zero algebraic numbers, then $e^{\beta_0}\alpha_1^{\beta_1} \cdots \alpha_n^{\beta_n}$ is transcendental.

ii) Let $\alpha_1, \ldots, \alpha_n$ be algebraic numbers not equal to 0 or 1, and let $\beta_1, \ldots, \beta_n$ be algebraic numbers such that $1, \beta_1, \ldots, \beta_n$ are linearly independent over the rational field R. Then $\alpha_1^{\beta_1} \cdots \alpha_n^{\beta_n}$ is transcendental.

Chapter 18. Waring's Problem and the Problem of Prouhet and Tarry

18.1 Introduction

In the year 1770 Waring wrote the following in his Meditationes Algebraicae:

Every positive integer is the sum of four squares, nine cubes, nineteen biquadrates, and so on.

We may interpret "and so on" to mean that there exists an integer $s(k)$ such that every positive integer is the sum of $s(k)$ k-th powers. Well over a hundred years later Hilbert gave the first proof of the existence of $s(k)$ for every positive integer k.

We can restate the problem more precisely as follows: Let k be any fixed positive integer. We ask if there exists an integer $s = s(k)$ such that, for any $n > 0$, the equation

$$n = x_1^k + \cdots + x_s^k, \qquad x_v \geqslant 0 \tag{1}$$

is always soluble in integers x_v. We now denote by $g(k)$ the least of all integers s with this property. Then Waring's statement becomes:

$$\text{“}g(2) = 4, \qquad g(3) = 9, \qquad g(4) = 19, \qquad \text{and so on.”}$$

We also denote by $G(k)$ the least number s with the property that (1) is soluble for all sufficiently large n. Then clearly we have

$$G(k) \leqslant g(k),$$

but in actual fact there is a great difference between the two numbers.

In this chapter we only prove some very special results. The proof of the Waring-Hilbert theorem (that is $g(k) < \infty$) is given in the next chapter. The proof, which Khintchin described as one of the three pearls in number theory, is due to Linnik and is much simpler than the original proof by Hilbert.

18.2 Lower Bounds for $g(k)$ and $G(k)$

Theorem 2.1. $g(k) \geqslant 2^k + [(\frac{3}{2})^k] - 2.$

Proof. Let $q = [(\frac{3}{2})^k]$ and consider

$$n = 2^k q - 1 < 3^k.$$

This number n can only be the sum of the powers 1^k and 2^k, and in fact the least s for the decomposition is given by

$$n = (q-1)2^k + (2^k - 1)1^k,$$

that is, n requires $(q-1)$ lots of 2^k and $2^k - 1$ lots of 1^k, giving

$$g(k) \geqslant 2^k + q - 2. \quad \square$$

From this theorem we see at once that

$$g(2) \geqslant 4, \qquad g(3) \geqslant 9, \qquad g(4) \geqslant 19, \qquad g(5) \geqslant 37, \ldots.$$

Theorem 2.2. *If $k \geqslant 2$, then $G(k) \geqslant k + 1$.*

Proof. Denote by $A(N)$ the number of positive integers not exceeding N which are expressible in the form

$$x_1^k + \cdots + x_k^k, \qquad x_v \geqslant 0.$$

We may suppose that $x_1, \ldots, x_k$ are arranged so that

$$0 \leqslant x_1 \leqslant x_2 \leqslant \cdots \leqslant x_k \leqslant [N^{1/k}].$$

Hence $A(N)$ cannot exceed the number of solutions to this set of inequalities, that is

$$A(N) \leqslant \sum_{x_k=0}^{[N^{1/k}]} \sum_{x_{k-1}=0}^{x_k} \sum_{x_{k-2}=0}^{x_{k-1}} \cdots \sum_{x_1=0}^{x_2} 1.$$

We claim that the sum on the right hand side is

$$B(N) = \frac{1}{k!}([N^{1/k}] + 1)([N^{1/k}] + 2) \cdots ([N^{1/k}] + k).$$

We can use induction to prove this. The claim clearly holds when $k = 1$, and so it remains to prove that

$$\sum_{x=0}^{y} \binom{x+k-1}{k-1} = \binom{y+k}{k},$$

and this is easy to establish.

When $N \to \infty$,

$$B(N) \sim \frac{N}{k!} < \frac{2}{3}N,$$

and so, for sufficiently large N, we have

$$A(N) < \tfrac{2}{3}N.$$

This shows that the $A(N)$ numbers concerned cannot include all the positive integers less than N, and therefore

$$G(k) \geqslant k + 1. \quad \square$$

For certain values of k it is possible to raise the lower bound slightly by congruence considerations. For example: since

$$x^4 \equiv 0 \quad \text{or } 1 \pmod{16}$$

it follows that numbers of the form $16m + 15$ require at least 15 biquadrates as their sum, and therefore $G(4) \geqslant 15$. But if

$$16n = x_1^4 + \cdots + x_{15}^4$$

then $2|(x_1, \ldots, x_{15})$ so that

$$n = x_1'^4 + \cdots + x_{15}'^4.$$

Moreover, 31 cannot be the sum of fewer than 16 biquadrates so that $16 \cdot 31$ cannot be the sum of 15 biquadrates. We see therefore that

$$G(4) \geqslant 16.$$

In general we can use this argument to prove:

Theorem 2.3. *If $k = 2^\vartheta \geqslant 4$, then $G(k) \geqslant 4k$.*

Proof. We have already proved the case $\vartheta = 2$. Suppose now that $\vartheta > 2$. It is not difficult to see that

$$x^k \equiv 0 \quad \text{or } 1 \pmod{4k}.$$

Let n be any odd number. Suppose that $2^{\vartheta+2}n$ is the sum of $2^{\vartheta+2} - 1$ or fewer k-th powers. Then each of these k-th powers must be even, and hence is a multiple of 2^k. But $2^k > 2^{\vartheta+2}$ and $2 \nmid n$ so that our supposition is impossible. Therefore the theorem is proved. $\square$

18.3 Cauchy's Theorem

Let $q > 1$. In this section we discuss the condition for the solubility of the congruence

$$x_1^k + \cdots + x_s^k \equiv n \pmod{q}.$$

From the Chinese remainder theorem we see that we can restrict our discussion to the congruence

$$x_1^k + \cdots + x_s^k \equiv 0 \pmod{p^l}, \qquad (1)$$

where p is a prime number. Since $n = n - 1 + 1^k$ we may also assume in what follows that $p \nmid n$.

We first prove the following:

Theorem 3.1 (Cauchy). *Let $x_1, \ldots, x_m$ be m incongruent numbers* (mod q) *and $y_1, \ldots, y_n$ be n incongruent numbers (mod q). Suppose that there exists y_i such that $(y_i - y_j, q) = 1$ whenever $j \neq i$. Then the number of incongruent numbers (mod q) represented by $x_u + y_v$ $(1 \leqslant u \leqslant m,\ 1 \leqslant v \leqslant n)$ is at least*

$$\min(m + n - 1, q).$$

Proof. The theorem is trivial if $n = 1$. Suppose therefore that $n \geqslant 2$ and we may also assume that $i = 1$. We use an inductive argument.

Let $z_1, \ldots, z_t$ be incongruent numbers (mod q) of the form $x_i + y_j$. If $t = q$ the required result is established. We suppose therefore that $t < q$ and we denote by X, Y, Z the sets $x_1, \ldots, x_m$; $y_1, \ldots, y_n$; $z_1, \ldots, z_t$ respectively.

Consider $x_1 + y_1 + \lambda(y_n - y_1)$. When $\lambda = 0, 1$ all such numbers belong to Z. Since $(q, y_n - y_1) = 1$ there must exist a λ_0 such that $x_1 + y_1 + (\lambda_0 - 1)(y_n - y_1) \in Z$ and $x_1 + y_1 + \lambda_0(y_n - y_1) \notin Z$. Let $\delta = x_1 + y_1 + \lambda_0(y_n - y_1) + y_1$. Then $\delta - y_1 \notin Z$ and $\delta - y_n \in Z$. We can arrange $y_1, \ldots, y_n$ so that

$$\begin{cases} \delta - y_s \notin Z & (1 \leqslant s \leqslant r), \\ \delta - y_{s'} \in Z & (r < s' \leqslant n). \end{cases}$$

Clearly $r \leqslant n - 1$. Write

$$Z' = \{z : z = x_u + y_s;\ u = 1, 2, \ldots, m;\ s = 1, 2, \ldots, r\}.$$

Then $\delta - y_{s'} \notin Z'$; otherwise from $\delta - y_{s'} = x_u + y_s$ we have $\delta - y_s = x_u + y_{s'} \in Z$. If we denote by t' the number of incongruent numbers (mod q) in Z', then $t' \leqslant t - (n - r)$. On the other hand, from the induction hypothesis we have $t' \geqslant m + r - 1$. Therefore $t \geqslant m + n - 1$. □

Definition. Suppose that $p^\tau \| k$. Then we define

$$\gamma = \begin{cases} \tau + 1, & \text{when } p > 1; \\ \tau + 2, & \text{when } p = 2. \end{cases}$$

Theorem 3.2. *If the congruence*

$$x^k \equiv a \pmod{p^\gamma}, \quad p \nmid a \tag{2}$$

is soluble, then, the congruence

$$x^k \equiv a \pmod{p^l}$$

is also soluble whenever $l > \gamma$.

Proof. Let y be a solution to (2), and g be a primitive root of p^l (if $p = 2$, we take $g = 5$). We fix $b \geqslant 0$ so that

$$a \equiv y^k g^b \pmod{p^l} \tag{3}$$

and hence $g^b \equiv 1 \pmod{p^\gamma}$. Therefore $p^\tau(p-1)|b$. Let $b = p^\tau(p-1)b_1$. We can clearly replace the exponent b in (3) by

$$b + hp^{l-1}(p-1) = p^\tau(p-1)(b_1 + hp^{l-\tau-1})$$

where h is any integer. Let $k = p^\tau k_1$, $(k_1, p) = 1$. We can then take h so that

$$b_1 + hp^{l-r-1} \equiv 0 \pmod{k_1}.$$

We then have

$$a \equiv y^k g^b \equiv y^k g^{b + hp^{l-1}(p-1)} \equiv y^k g^{h_1 k} \pmod{p^l}.$$

The theorem is proved. □

Theorem 3.3. *Suppose that the congruence* (1) *has a solution when* $l = \gamma$. *Then it has a solution when* $l > \gamma$.

Proof. By hypothesis there are $y_1, \dots, y_s$ such that

$$y_1^k + \cdots + y_s^k \equiv n \pmod{p^\gamma}.$$

Since $p \nmid n$ there must be a y, which we may take as y_1, such that $p \nmid y_1$, and so from

$$y_1^k \equiv n - y_2^k - \cdots - y_s^k \pmod{p^\gamma}$$

and Theorem 3.2, there exists x_1 such that

$$x_1^k + y_2^k + \cdots + y_s^k \equiv n \pmod{p^l}. \quad \square$$

Theorem 3.4. *If* $k = 2^\tau$, *then* (1) *is always soluble with* $s \geqslant 4k$; *if* $k \neq 2^\tau$, *then* (1) *is always soluble with* $s \geqslant 3k + 1$.

Proof. Clearly it suffices to consider the case $l \geqslant \gamma$, and by Theorem 3.3 we need only consider the case $l = \gamma$.

1) If $k = 2^\tau$, then $p^\gamma = 2^{\tau+2} = 4k$. The congruence

$$x_1^k + \cdots + x_s^k \equiv n \pmod{2^\gamma}$$

is clearly soluble when $s \geqslant 4k$.

2_1) $p = 2$, $k = 2^\tau k_0$, $k_0 > 1$, $2 \nmid k_0$. Here $k \geqslant \frac{3}{4} \cdot 2^\gamma$ so that, when $s \geqslant 3k > 2^\gamma$, (1) is soluble.

2_2) $p > 2$, $p - 1 | k$. Here $k \geqslant p^\tau(p-1) > p^\gamma/3$ so that, when $s \geqslant 3k > p^\gamma$, (1) is soluble.

2_3) $p > 2$, $(p-1) \nmid k$, $p \nmid k$. Here $\gamma = 1$. From $p - 1 \nmid k$, Theorems 3.7.2 and 3.7.3, we see that as x runs over a set of reduced residues mod p, x^k gives

$$d = \frac{p-1}{(k,p-1)} > 1$$

incongruent numbers $(\bmod p)$. From Theorem 3.1, $x_1^k + \cdots + x_s^k (p \nmid x_1, \ldots, x_s)$ gives

$$\min(d + (d-1)(s-1), p)$$

incongruent numbers $(\bmod p)$. When

$$s \geqslant 2k > \frac{p-1}{\frac{1}{2}d} \geqslant \frac{p-1}{d-1},$$

we have

$$\min(d + (d-1)(s-1), p) = p$$

so that the theorem follows.

2_4) $p > 2$, $(p-1) \nmid k$, $k = p^\tau k_0$, $p \nmid k_0$. From

$$x^{p^\tau k_0} \equiv x^{k_0} \pmod p$$

and $(p-1) \nmid k_0$, we see that x^k runs over at least $(p-1)/(p-1, k_0)$ (>1) incongruent numbers $(\bmod p)$. Therefore

$$x_1^k + \cdots + x_s^k, \qquad p \nmid x_1, \ldots, x_s$$

gives

$$\min\left(\frac{p-1}{(p-1,k_0)} + \left(\frac{p-1}{(p-1,k_0)} - 1\right)(s-1), p^\gamma\right)$$

incongruent numbers $\bmod p^\gamma$. From

$$s - 1 \geqslant 3k \geqslant \frac{2pk}{p-1} \geqslant \frac{p^\gamma}{\dfrac{1}{2}\dfrac{p-1}{(k_0,p-1)}} \geqslant \frac{p^\gamma - 1}{\dfrac{p-1}{(k_0,p-1)} - 1},$$

we see that $x_1^k + \cdots + x_s^k (p \nmid x_1, \ldots, x_s)$ gives p^γ incongruent numbers. The proof of the theorem is complete. □

18.4 Elementary Methods

In the study of Waring's problem an elementary method usually gives rather poor results. We now introduce several examples which prove the existence of upper bounds for $G(k)$ and $g(k)$ for some special k. Sometimes we can even determine explicitly such an upper bound, but such a result will not be sharp. From Theorem 8.7.8 we already have that $g(2) = 4$.

Theorem 4.1. $g(4) \leqslant 50$.

Proof. We start with the identity

$$\begin{aligned}6(a^2 + b^2 + c^2 + d^2)^2 = &(a + b)^4 + (a - b)^4 + (c + d)^4 + (c - d)^4 \\ &+ (a + c)^4 + (a - c)^4 + (b + d)^4 + (b - d)^4 \\ &+ (a + d)^4 + (a - d)^4 + (b + c)^4 + (b - c)^4.\end{aligned}$$

Since $a^2 + b^2 + c^2 + d^2$ can represent any positive integer, it follows that the left hand side of the identity represents $6x^2$ where x is any integer. Now any integer n can be written as

$$n = 6N + r, \qquad r = 0, 1, 2, 3, 4, 5$$

so that

$$n = 6(x_1^2 + x_2^2 + x_3^2 + x_4^2) + r.$$

By the identity $6x_1^2$ is representable as a sum of 12 biquadrates. Therefore n is the sum of at most $4 \times 12 + 5 = 53$ biquadrates.

We take one further step. Any $n \geqslant 81$ is expressible as

$$n = 6N + t$$

where $N \geqslant 0$, and $t = 0, 1, 2, 81, 16$ and 17 corresponding to $n \equiv 0, 1, 2, 3, 4, 5 \pmod 6$. But

$$1 = 1^4, \qquad 2 = 1^4 + 1^4, \qquad 81 = 3^4, \qquad 16 = 2^4, \qquad 17 = 2^4 + 1.$$

Therefore, following the method above, if $n \geqslant 81$, then it is the sum of $4 \times 12 + 2 = 50$ biquadrates.

We can deal with $n \leqslant 80$ easily: If $n \leqslant 50$, then trivially $n = n \cdot 1^4$. If $50 < n \leqslant 80$, then $n = 3 \cdot 2^4 + (n - 48) \cdot 1^4$ and this is the sum of $3 + n - 48 < 50$ biquadrates. □

The same method together with the identity

$$\begin{aligned}5040(a^2 + b^2 + c^2 + d^2)^4 = &6\sum (2a)^8 + 60\sum (a \pm b)^8 + \sum (2a \pm b \pm c)^8 \\ &+ 6\sum (a \pm b \pm c \pm d)^8,\end{aligned} \tag{2}$$

can be used to prove that $g(8) < \infty$. In this identity there are 840 8th powers on the right hand side, and since every $n \leqslant 5039$ is expressible as a sum of at most 273 numbers 1^8 and 2^8, we see that

$$g(8) \leqslant 840g(4) + 273 \leqslant 42273.$$

Theorem 4.2. $G(3) \leqslant 13$.

Proof. We start with the identity

$$\sum_{i=1}^{4} ((z^3 + x_i)^3 + (z^3 - x_i)^3) = 8z^9 + 6z^3(x_1^2 + x_2^2 + x_3^2 + x_4^2). \tag{1}$$

If a number is expressible as

$$8z^9 + 6mz^3, \qquad 0 \leqslant m \leqslant z^6, \tag{2}$$

then from (1) this number must be a sum of 8 cubes; this is because m is expressible as $x_1^2 + x_2^2 + x_3^2 + x_4^2$, and $x_i \leqslant z^3$.

Let z be a positive integer congruent to 1 (mod 6). We denote by I_z the interval

$$\varphi(z) = 11z^9 + (z^3 + 1)^3 + 125z^3 \leqslant n \leqslant 14z^9 = \psi(z). \tag{3}$$

Clearly, for sufficiently large z, we have

$$\varphi(z + 6) < \psi(z), \tag{4}$$

that is the intervals I_z overlap. Hence, for sufficiently large n, there must exist z such that (3) holds. We define r, s and N as follows:

$$n \equiv 6r \pmod{z^3}, \qquad 1 \leqslant r \leqslant z^3,$$
$$n \equiv s + 4 \pmod 6, \qquad 0 \leqslant s \leqslant 5,$$
$$N = (r + 1)^3 + (r - 1)^3 + 2(z^3 - r)^3 + (sz)^3.$$

Then

$$0 < N < (z^3 + 1)^3 + 3z^9 + 125z^3 = \varphi(z) - 8z^9 \leqslant n - 8z^9,$$

so that

$$8z^9 < n - N < 14z^9. \tag{5}$$

We now prove that $n - N$ is expressible as (2). Now

$$n - N \equiv 6r - (r + 1)^3 - (r - 1)^3 + 2r^3 \equiv 0 \equiv 8z^9 \pmod{z^3},$$

and

$$n - N \equiv s + 4 - (r + 1) - (r - 1) - 2(z^3 - r) - sz$$
$$\equiv s + 4 - z(s + 2) \equiv 2 \equiv 8 \equiv 8z^9 \pmod 6,$$

so that $n - N - 8z^9$ is a multiple of $6z^3$, that is

$$n = N + 8z^9 + 6mz^3.$$

The theorem then follows from $0 \leqslant m \leqslant z^6$, which is a consequence of (5). □

Theorem 4.3. $g(3) \leqslant 13$.

Proof. 1) First we prove that $\varphi(z+6) \leqslant \psi(z)$ for $z \geqslant 373$, or when $t \geqslant 379$ we have that

$$11t^9 + (t^3+1)^3 + 125t^3 \leqslant 14(t-6)^9,$$

or

$$14\left(1 - \frac{6}{t}\right)^9 \geqslant 12 + \frac{3}{t^3} + \frac{128}{t^6} + \frac{1}{t^9}. \tag{6}$$

Now $(1-\delta)^m \geqslant 1 - m\delta$ whenever $0 < \delta < 1$, so that

$$\left(1 - \frac{6}{t}\right)^9 \geqslant 1 - \frac{54}{t}.$$

We see therefore that (6) will follow if we have

$$14\left(1 - \frac{54}{t}\right) \geqslant 12 + \frac{3}{t^3} + \frac{128}{t^6} + \frac{1}{t^9},$$

and hence if we have

$$2(t - 7 \times 54) \geqslant \frac{3}{t^2} + \frac{128}{t^5} + \frac{1}{t^8}.$$

But $t \geqslant 379 = 7 \times 54 + 1$ so that (6) is proved.

So when $z \geqslant 373$ the various intervals I_z overlap. This means that when

$$n \geqslant 14(373)^9,$$

n must lie in one of the intervals I_z. Also $10^{25} > 14\,(373)^9$ so that any integer exceeding 10^{25} is expressible as a sum of 13 cubes.

2) We next prove that numbers not exceeding 10^{25} are also sums of 13 cubes. From tables it is known that all numbers up to 40000 are sums of 8 cubes with the two exceptions of 23 and 239 which require 9 cubes. Thus, if $240 \leqslant n \leqslant 40000$ then n is the sum of 8 cubes. Now if $N \geqslant 1$ and $m = [N^{1/3}]$, then

$$N - m^3 = (N^{1/3})^3 - m^3 \leqslant 3N^{2/3}(N^{1/3} - m) < 3N^{2/3}.$$

Suppose that

$$240 \leqslant n \leqslant 10^{25},$$

and let

$$n = 240 + N, \qquad 0 \leqslant N < 10^{25}.$$

Then

$$\begin{array}{lll} N = m^3 + N_1, & m = [N^{1/3}], & 0 < N_1 < 3N^{2/3}, \\ N_1 = m_1^3 + N_2, & m_1 = [N_1^{1/3}], & 0 < N_2 < 3N_1^{2/3}, \\ \cdots\cdots & \cdots\cdots & \cdots\cdots \\ N_4 = m_4^3 + N_5, & m_4 = [N_4^{1/3}], & 0 < N_5 < 3N_4^{2/3}. \end{array}$$

Therefore

$$n = 240 + N = 240 + N_5 + m^3 + m_1^3 + m_2^3 + m_3^3 + m_4^3.$$

From

$$\begin{aligned} 0 < N_5 &\leqslant 3N_4^{2/3} \leqslant 3(3N_3^{2/3})^{2/3} \leqslant \cdots \\ &\leqslant 3 \cdot 3^{2/3} \cdot 3^{(2/3)^2} 3^{(2/3)^3} 3^{(2/3)^4} N^{(2/3)^5} \\ &= 27\left(\frac{N}{27}\right)^{(2/3)^5} < 27\left(\frac{10^{25}}{27}\right)^{(2/3)^5} < 35000, \end{aligned}$$

we have

$$240 \leqslant 240 + N_5 < 35240 < 40000,$$

and so $240 + N_5$ is the sum of 8 cubes. The theorem therefore follows. □

From the identity

$$60(a^2 + b^2 + c^2 + d^2)^3 = \sum (a \pm b \pm c)^6 + 2\sum (a \pm b)^6 + 36\sum a^6,$$

we can prove that $g(6) \leqslant 184g(3) + 59 \leqslant 2451$.

18.5 The Easier Problem of Positive and Negative Signs

Denote by $v(k)$ the least natural number s such that, for any integer n, the equation

$$n = \pm x_1^k \pm x_2^k \pm \cdots \pm x_s^k$$

is soluble. Here x_i are integers and we can attach any $\pm$ sign for them. Clearly

$$v(k) \leqslant g(k).$$

Indeed the problem of the existence of $v(k)$ is quite easy.

Theorem 5.1. $v(k) \leqslant 2^{k-1} + \frac{1}{2}k!$.

We shall require the following:

Theorem 5.2. *Let*

$$\Delta f(x) = f(x + 1) - f(x), \qquad \Delta^{m+1} f(x) = \Delta(\Delta^m f(x)).$$

Then

$$\Delta^{k-1} x^k = k!x + d$$

where d is an integer.

If $f(x)$ is a polynomial of degree k with leading coefficient a, then $\Delta f(x)$ is a polynomial of degree $k - 1$ with leading coefficient ka. The theorem follows from repeated applications of this fact. □

Proof of Theorem 5.1. We can consider $\Delta^{k-1}x^k$ as a sum of 2^{k-1} terms of $\pm x^k$. Now any integer n is expressible as

$$n - d = k!x + l, \qquad |l| \leqslant \tfrac{1}{2}k!$$

that is

$$n = \Delta k^{-1}x^k + l.$$

From

$$2^{k-1} + l \leqslant 2^{k-1} + \tfrac{1}{2}k!$$

the theorem follows. □

Theorem 5.3. $v(k) \leqslant G(k) + 1$.

Proof. We take y sufficiently large so that $n + y^k$ exceeds a certain sufficiently large number. From the definition of $G(k)$ we have

$$n + y^k = x_1^k + \cdots + x_{G(k)}^k,$$

and hence the result. □

Theorem 5.4. $v(2) = 3$.

Proof. We have, by Theorem 5.1, $v(2) \leqslant 3$. But 6 cannot be a sum of two squares and moreover the difference between two squares is either odd or a multiple of 4. Therefore $v(2) > 2$. □

Theorem 5.5. $v(3)$ *is either* 4 *or* 5.

Proof. Since $n^3 - n$ is divisible by 6 we can let $n^3 - n = 6x$. Now

$$n = n^3 - (x + 1)^3 - (x - 1)^3 + 2x^3$$

so that $v(3) \leqslant 5$.

Next

$$y^3 \equiv 0, 1 \quad \text{or} \quad -1 \pmod 9$$

so that, if $n = 9m \pm 4$, then n cannot be the sum of three cubes. Therefore $v(3) \geqslant 4$. □

It is an open problem whether $v(3)$ is 4 or 5, and it is conjectured that $v(3) = 4$. It has been verified by Chao Ko that every integer with absolute value $\leqslant 100$ is the sum of four cubes.

Theorem 5.6. $v(4)$ *is either* 9 *or* 10.

Proof. Consider the identities:

$$48x + 4 = 2(2x + 3)^4 + (2x + 6)^4 + 2(2x^2 + 8x + 11)^4 - (2x^2 + 8x + 10)^4 - (2x^2 + 8x + 12)^4;$$

$$48x - 14 = 2(2x + 5)^4 + (2x + 8)^4 + (x^2 + 6x + 9)^4 + (x^2 + 6x + 12)^4 - (x^2 + 6x + 8)^4 - (x^2 + 6x + 13)^4;$$

$$24x = (4y + 11)^4 + (2y - 87)^4 + (y - 9)^4 + (y - 41)^4 + (y - 83)^4 + (y + 125)^4 + (y^2 + 603)^4 + (y^2 + 625)^4 - (y^2 + 602)^4 - (y^2 + 626)^4,$$

where $y = x - 10319691$;

$$24x - 8 = (4y + 11)^4 + (2y - 87)^4 + (y + 883)^4 + (y - 933)^4 + (y - 975)^4 + (y + 1017)^4 + (y^2 + 39851)^4 + (y^2 + 39873)^4 - (y^2 + 39850)^4 - (y^2 + 39874)^4,$$

where $y = x - 120858614086$.

We can replace x by $-x$ and obtain similar identities for $48x - 4$, $48x + 14$, $24x + 8$. Now it is easy to see that if n is a multiple of 8, then n is the sum of 10 biquadrates. If n is not a multiple of 8, then we can write n as

$$n = 48z + \gamma, \qquad -24 < \gamma < 24,$$

and we can easily prove that there are integers x_1, x_2, x_3 such that

$$\gamma \pm x_1^4 \pm x_2^4 \pm x_3^4 \equiv \pm 4, \pm 14 \pmod{48}.$$

This proves that $v(4) \leqslant 10$.

From $\pm y^4 \equiv 0, \pm 1 \pmod{16}$ we see that numbers of the form $16x + 8$ require at least 8 biquadrates to represent them, and that they must all take the same sign. But numbers such as 24, 104 cannot be expressed this way. Therefore $v(4) \geqslant 9$, and the proof of the theorem is complete. □

18.6 Equal Power Sums Problem

Let $N(k)$ be the least integer s such that there exist $x_1, \ldots, x_s$; $y_1, \ldots, y_s$, but $y_1, \ldots, y_s$ not merely a permutation of $x_1, \ldots, x_s$, with the property that

$$\begin{aligned} x_1 + \cdots + x_s &= y_1 + \cdots + y_s, \\ &\cdots\cdots\cdots \\ x_1^k + \cdots + x_s^k &= y_1^k + \cdots + y_s^k. \end{aligned} \qquad (1)$$

We also let $M(k)$ denote the least s so that (1) holds, and furthermore,

$$x_1^{k+1} + \cdots + x_s^{k+1} \neq y_1^{k+1} + \cdots + y_s^{k+1}. \tag{2}$$

Theorem 6.1. $M(k) \geqslant N(k) \geqslant k + 1$.

Proof. From

$$\begin{gathered} x_1 + \cdots + x_k = y_1 + \cdots + y_k, \\ \cdots\cdots\cdots\cdots\cdots\cdots \\ x_1^k + \cdots + x_k^k = y_1^k + \cdots + y_k^k, \end{gathered}$$

we have

$$(x - x_1) \cdots (x - x_k) = (x - y_1) \cdots (x - y_k)$$

so that $y_1, \ldots, y_k$ is only a permutation of $x_1, \ldots, x_k$. □

Theorem 6.2. $N(k) \leqslant M(k) \leqslant 2^k$.

Proof. Let $x_1, \ldots, x_s; y_1, \ldots, y_s$ be solutions to (1) and (2). Then

$$\sum_{i=1}^{s} ((x_i + d)^h + y_i^h) = \sum_{i=1}^{s} (x_i^h + (y_i + d)^h), \qquad 1 \leqslant h \leqslant k + 1, \tag{3}$$

$$\sum_{i=1}^{s} ((x_i + d)^{k+2} + y_i^{k+2}) \neq \sum_{i=1}^{s} (x_i^{k+2} + (y_i + d)^{k+2}). \tag{4}$$

The proofs of these two formulae follow from the expansions of (3), (4) and applying (1), (2).

Thus, if $M(k)$ exists, then taking $s = M(k)$ we have $M(k + 1) \leqslant 2M(k)$. But $M(1) = N(1) = 2$, so that the theorem follows by mathematical induction. □

Theorem 6.3. $N(k) \leqslant \frac{1}{2}k(k + 1) + 1$.

Proof. Suppose that $n > s!s^k$. Let a_i $(i = 1, 2, \ldots, s)$ run over $1, 2, \ldots, n$. Then there are n^s sets $a_1, a_2, \ldots, a_s$. Each fixed set $a_1, a_2, \ldots, a_s$ has $s!$ permutations. It follows that among the n^s sets $a_1, a_2, \ldots, a_s$ there are at least $n^s/s!$ sets in which every set is a permutation of a certain other set.

Let

$$s_h(a) = a_1^h + a_2^h + \cdots + a_s^h, \qquad h = 1, 2, \ldots, k.$$

Then

$$s \leqslant s_h(a) \leqslant sn^h.$$

Therefore there are at most

$$\prod_{h=1}^{k} (sn^h - s + 1) < s^k n^{\frac{1}{2}k(k+1)}$$

sets of different

$$s_1(a), s_2(a), \ldots, s_k(a). \tag{5}$$

Take $s = \frac{1}{2}k(k+1)+1$. Then, from $n > s!s^k$, we have

$$s^k n^{\frac{1}{2}k(k+1)} = s^k n^{s-1} < \frac{n^s}{s!}.$$

Therefore there are at least two different sets $a_1, a_2, \ldots, a_s$ such that (5) takes the same values. Since these two sets are not permutations of each other, it follows that $N(k) \leqslant s$, and the theorem is proved. □

We now write

$$[a_1, \ldots, a_s]_k = [b_1, \ldots, b_s]_k$$

to represent (1) and (2).

From Theorem 6.1 and the following examples, we have:

Theorem 6.4. *If* $k \leqslant 9$, *then* $M(k) = N(k) = k + 1$.

$$[0,3]_1 = [1,2]_1,$$

$$[1,2,6]_2 = [0,4,5]_2,$$

$$[0,4,7,11]_3 = [1,2,9,10]_3,$$

$$[1,2,10,14,18]_4 = [0,4,8,16,17]_4,$$

$$[0,4,9,17,22,26]_5 = [1,2,12,14,24,25]_5,$$

$$[0,18,27,58,64,89,101]_6 = [1,13,38,44,75,84,102]_6,$$

$$[0,4,9,23,27,41,46,50]_7 = [1,2,11,20,30,39,48,49]_7,$$

$$[0,24,30,83,86,133,157,181,197]_8 = [1,17,41,65,112,115,168,174,198]_8,$$

$$[0,3083,3301,11893,23314,24186,35607,44199,44417,47500]_9$$
$$= [12,2865,3519,11869,23738,23762,35631,43981,44635,47488]_9. \quad \square$$

18.7 The Problem of Prouhet and Tarry

In this and the next sections we shall prove that

$$M(k) \leqslant (k+1)\left(\left[\frac{\log\frac{1}{2}(k+2)}{\log\left(1+\frac{1}{k}\right)}\right]+1\right) \sim k^2 \log k.$$

Actually our eventual result gives even more than this. We first prove several lemmas. In this and the next sections the constants $c_1, c_2, \ldots$ as well as the constant implied by the O-symbol depend only on k. Moreover, $c_1, c_2, \ldots$ are positive.

Theorem 7.1 (Bunyakovsky-Schwarz). *Let a_i, b_i $(i = 1, 2, \ldots, n)$ be real numbers. Then*

$$\left(\sum_{i=1}^{n} a_i b_i\right)^2 \leqslant \left(\sum_{i=1}^{n} a_i^2\right)\left(\sum_{i=1}^{n} b_i^2\right),$$

with equality sign only when

$$\frac{a_1}{b_1} = \frac{a_2}{b_2} = \cdots = \frac{a_n}{b_n}.$$

Proof. The required result follows at once from

$$\sum a_i^2 \sum b_i^2 - \left(\sum a_i b_i\right)^2 = \sum_{i<j} (a_i b_j - b_i a_j)^2 \geqslant 0. \quad \square$$

Theorem 7.2. *Let H be any given number. Then there exists a set of positive integers $a_1, a_2, \ldots, a_k$ depending only on k and H such that the product of the terms of the main diagonal of the determinant*

$$D_k = \begin{vmatrix} 1 & \cdots & 1 \\ a_1 & \cdots & a_k \\ \\ a_1^{k-1} & \cdots & a_k^{k-1} \end{vmatrix}$$

is greater than H times the sum of the absolute values of the remaining terms of the expansion of D_k.

Proof. We use mathematical induction to prove the theorem. Suppose that $j \leqslant k$ and let $\varphi_j(a_1, \ldots, a_j)$ denote the product of the terms of the main diagonal of D_j minus the sum of the absolute values of the remaining terms of the expansion of D_j. Then clearly we have

$$\varphi_j(a_1, \ldots, a_j) = a_j^{j-1} \varphi_{j-1}(a_1, \ldots, a_{j-1}) - H\psi(a_1, \ldots, a_j),$$

where ψ is a polynomial in a_j of degree $j - 2$. From the induction hypothesis we can take $a_1, \ldots, a_{j-1}$ so that $\varphi_{j-1}(a_1, \ldots, a_{j-1}) > 0$. For this set $a_1, \ldots, a_{j-1}$ we can clearly set a_j so that $\varphi_j > 0$. But $\varphi_1(a_1) = 1$, so that the theorem is proved. $\square$

Theorem 7.3. *Let $a_1, \ldots, a_k$ be a set of positive integers satisfying Theorem 7.2. Let $Q \geqslant 1$ and $X_1, \ldots, X_k$ be positive integers belonging to the intervals*

$$a_i Q \leqslant X_i \leqslant 2a_i Q \qquad (i = 1, 2, \ldots, k).$$

Denote by N the number of sets $(X_1, \ldots, X_k)$ such that

$$X_1^k + \cdots + X_k^k,\ X_1^{k-1} + \cdots + X_k^{k-1}, \ldots, X_1 + \cdots + X_k$$

lie in intervals with lengths

$$O(Q^{k-1}), O(Q^{k-2}), \ldots, O(Q), O(1)$$

respectively. Then

$$N = O(1).$$

Proof. Let $(X_1, \ldots, X_k)$ and $(X'_1, \ldots, X'_k)$ be two sets which satisfy the conditions of the theorem. Then

$$X_1^k - X_1'^k + \cdots + X_k^k - X_k'^k = O(Q^{k-1}),$$
$$\cdots\cdots\cdots\cdots\cdots\cdots\cdots\cdots$$
$$X_1 - X'_1 + \cdots + X_k - X'_k = O(1).$$

Let $Y_i = X_i - X'_i$. Then

$$A_{11}Y_1 + \cdots + A_{1k}Y_k = O(Q^{k-1}),$$
$$\cdots\cdots\cdots\cdots\cdots\cdots\cdots\cdots$$
$$A_{k1}Y_1 + \cdots + A_{kk}Y_k = O(1),$$

so that

$$A_{ij} = X_j^{k-i} + X_j^{k-i-1}X'_j + \cdots + X_j'^{k-i} \qquad (1 \leqslant i, j \leqslant k).$$

Thus

$$(k - i + 1)(a_j Q)^{k-i} \leqslant A_{ij} \leqslant (k - i + 1)(2a_j Q)^{k-i}.$$

The ratio of the product of the terms of the main diagonal of the determinant $|A_{k-i+1,j}|$ to that of D_k in the previous theorem is clearly greater than

$$k!Q^{k-1+k-2+\cdots+2+1} = k!Q^{\frac{1}{2}k(k-1)}.$$

Also the ratio of the absolute value of each remaining term in the expansion of $|A_{k-i+1,j}|$ to the corresponding absolute value term for D_k is smaller than

$$2^{\frac{1}{2}k(k-1)}k!Q^{\frac{1}{2}k(k-1)}.$$

We now take $H = 2^{\frac{1}{2}k(k-1)}$ in Theorem 7.2, so that we have

$$\|A_{zj}\| \geqslant c_1 Q^{\frac{1}{2}k(k-1)}.$$

It is then easy to see that

$$\begin{vmatrix} O(Q^{k-1}) & A_{12} \cdots A_{1k} \\ \cdots\cdots & \cdots\cdots \\ O(1) & A_{k2} \cdots A_{kk} \end{vmatrix} = O(Q^{\frac{1}{2}k(k-1)}).$$

Therefore

$$Y_1 = O(1).$$

Similarly we have

$$Y_2 = O(1), \quad \dots, \quad Y_k = O(1).$$

The theorem is proved. □

Theorem 7.4. *Suppose that the conditions in Theorem 7.3 are satisfied. Let* $\lambda_1 \geqslant 0$, $\lambda_2 \geqslant 0, \dots, \lambda_k \geqslant 0$. *Then the number of sets* $(X_1, \dots, X_k)$ *such that*

$$X_1^k + \cdots + X_k^k, X_1^{k-1} + \cdots + X_k^{k-1}, \dots, X_1 + \cdots + X_k$$

lie in intervals with lengths

$$O(Q^{k+\lambda_k-1}), O(Q^{k+\lambda_{k-1}-2}), \dots, O(Q^{\lambda_1})$$

respectively is

$$O(Q^{\lambda_1+\cdots+\lambda_k}).$$

Proof. Since an interval with length $O(Q^{k-i+\lambda_{k-i+1}})$ can be divided into $O(Q^{\lambda_{k-i+1}})$ intervals with lengths $O(Q^{k-i})$, the required result follows at once from Theorem 7.3. □

Now let $\beta = k/(k+1)$ and $a_1, \dots, a_{k+1}$ be a set of positive integers satisfying the conditions of Theorem 7.2 (where we have replaced k by $k+1$). We suppose that

$$a_u Q^{\beta^{v-1}} \leqslant y_{uv} \leqslant 2a_u Q^{\beta^{v-1}} \qquad (1 \leqslant u \leqslant k+1,\ 1 \leqslant v \leqslant l).$$

Denote by $r(n_1, \dots, n_k)$ the number of solutions to the system

$$\sum_{u=1}^{k+1} \sum_{v=1}^{l} y_{uv}^h = n_h \qquad (1 \leqslant h \leqslant k).$$

We now prove the following theorems:

Theorem 7.5. *There exists a set of integers* $N_1, \dots, N_k$ *such that*

$$r(N_1, \dots, N_k) \geqslant c_1 Q^{(k+1)^2(1-\beta^l) - \frac{1}{2}k(k+1)}.$$

Proof. The numbers of different sets (y_{uv}) must be

$$\geqslant \frac{1}{2} \prod_{u=1}^{k+1} \prod_{v=1}^{l} a_u Q^{\beta^{v-1}} \geqslant c_2 Q^{(k+1)(1+\beta+\cdots+\beta^{l-1})}$$
$$= c_2 Q^{(k+1)^2(1-\beta^l)}.$$

Since $|n_h| \leqslant c_3 Q^h$, the number of different sets (n_h) is

$$\leqslant c_4 Q^{1+2+\cdots+k} = c_4 Q^{\frac{1}{2}k(k+1)}.$$

Therefore there must be a set of integers $N_1, \ldots, N_k$ such that

$$r(N_1, \ldots, N_k) \geqslant \frac{c_2}{c_4} Q^{(k+1)^2(1-\beta^l) - \frac{1}{2}k(k+1)}. \quad \square$$

Theorem 7.6. *The number of solutions to the system*

$$\sum_{u=1}^{k+1} \sum_{v=1}^{l} y_{uv}^h = N_h \qquad (1 \leqslant h \leqslant k+1)$$

is at most

$$c_5 Q^{\frac{1}{2}k(k+1)(1-\beta^l)}.$$

Proof. From

$$\sum_{u=1}^{k+1} y_{u1}^h = N_h - \sum_{u=1}^{k+1} \sum_{v=2}^{l} y_{uv}^h \qquad (1 \leqslant h \leqslant k+1)$$

and

$$a_u Q^{\beta^{v-1}} \leqslant y_{uv} \leqslant 2a_u Q^{\beta^{v-1}} \qquad (1 \leqslant u \leqslant k+1,\ 1 \leqslant v \leqslant l),$$

we see that

$$y_{11}^{k+1} + \cdots + y_{k+1,1}^{k+1}, y_{11}^k + \cdots + y_{k+1,1}^k, \ldots, y_{11} + \cdots + y_{k+1,1}$$

lie in intervals with lengths

$$O(Q^{(k+1)\beta}), \quad O(Q^{k\beta}), \quad \ldots, \quad O(Q^{\beta})$$

respectively. We take $\lambda_u = u\beta - (u-1) \geqslant 0$ in Theorem 7.4. Then, from

$$\sum_{u=1}^{k+1} \{u\beta - (u-1)\} = \tfrac{1}{2}\beta(k+1)(k+2) - \tfrac{1}{2}k(k+1) = \tfrac{1}{2}k,$$

we see that the number of sets $(y_{11}, \ldots, y_{k+1,1})$ is $O(Q^{k/2})$.

Corresponding to each fixed set $(y_{11}, \ldots, y_{k+1,1})$ the sums

$$y_{12}^{k+1} + \cdots + y_{k+1,2}^{k+1}, \ldots, y_{12} + \cdots + y_{k+1,2}$$

clearly lie in intervals with lengths

$$O(Q^{(k+1)\beta^2}), \quad O(Q^{k\beta^2}), \quad \ldots, \quad O(Q^{\beta^2})$$

respectively. Replacing Q by Q^β in Theorem 7.4, we see that the number of different sets $y_{12}, \ldots, y_{k+1,2}$ is $O(Q^{k\beta/2})$. Continuing this way the theorem is proved. $\square$

18.8 Continuation

Theorem 8.1. *Denote by $W(k,j)$ the least integer s such that the system*

$$\sum_{i=1}^{s} x_{i1}^h = \sum_{i=1}^{s} x_{i2}^h = \cdots = \sum_{i=1}^{s} x_{ij}^h \qquad (1 \leqslant h \leqslant k),$$

$$\sum_{i=1}^{s} x_{ip}^{k+1} \neq \sum_{i=1}^{s} x_{iq}^{k+1}, \qquad (p \neq q,\ 1 \leqslant p,\ q \leqslant j)$$

is soluble in integers. Then

$$W(k,j) \leqslant (k+1)\left(\left[\frac{\log \frac{1}{2}(k+2)}{\log\left(1+\frac{1}{k}\right)}\right]+1\right).$$

Proof. This theorem is an immediate consequence of the following theorem.

Theorem 8.2. *Let*

$$s \geqslant (k+1)\left(\left[\frac{\log \frac{1}{2}(k+2)}{\log\left(1+\frac{1}{k}\right)}\right]+1\right).$$

Then, given any j, there are integers

$$N_1, \ldots, N_k;\ M_1, \ldots, M_j \qquad (M_{t_1} \neq M_{t_2}, \text{ if } t_1 \neq t_2)$$

such that the system

$$R_t(1 \leqslant t \leqslant j): \begin{cases} \sum_{i=1}^{s} x_{it}^h = N_h & (1 \leqslant h \leqslant k), \\ \sum_{i=1}^{s} x_{it}^{k+1} = M_t & (x_{it} \geqslant 0) \end{cases}$$

is soluble.

Proof. Let $r(N_1, \ldots, N_h)$ be as defined in the previous section. By Theorem 7.5 there are $N_1, \ldots, N_h$ such that

$$r(N_1, \ldots, N_h) \geqslant c_1 Q^{(k+1)^2(1-\beta^l)-\frac{1}{2}k(k+1)}.$$

Corresponding to a set of solutions (y_{uv}) to the system

$$\sum_{u=1}^{k+1} \sum_{v=1}^{l} y_{uv}^h = N_h \qquad (1 \leqslant h \leqslant k)$$

there is clearly a number M such that

$$\sum_{u=1}^{k+1} \sum_{v=1}^{l} y_{uv}^{k+1} = M.$$

If such an M has only e ($\leqslant j-1$) different values, say $M_1, M_2, \ldots, M_e$, then, by Theorem 7.6, the number of solutions to the e-system

$$\prod_i (1 \leqslant i \leqslant e): \begin{cases} \sum_{u=1}^{k+1} \sum_{v=1}^{l} y_{uv}^h = N_h \qquad (1 \leqslant h \leqslant k), \\ \sum_{u=1}^{k+1} \sum_{v=1}^{l} y_{uv}^{k+1} = M_i \end{cases}$$

is at most $c_5 e Q^{\frac{1}{2}k(k+1)(1-\beta^l)}$. From the definition of M_i the number of solutions to this e-system is at least $r(N_1, \ldots, N_k)$. On the other hand, if we take

$$l > \left\{\log \frac{1}{2}(k+2) \Big/ \log\left(1 + \frac{1}{k}\right)\right\},$$

then, for large Q, we have

$$c_5 e Q^{\frac{1}{2}k(k+1)(1-\beta^l)} < c_1 Q^{(k+1)^2(1-\beta^l) - \frac{1}{2}k(k+1)} \leqslant r(N_1, \ldots, N_k)$$

giving a contradiction. Our theorem is proved. □

Notes

18.1. Concerning the value of $g(k)$ in Waring's problem there is the following result: When $k > 6$ and

$$(\tfrac{3}{2})^k - [(\tfrac{3}{2})^k] \leqslant 1 - (\tfrac{1}{2})^k \{[(\tfrac{3}{2})^k] + 3\},$$

we have

$$g(k) = 2^k + [(\tfrac{3}{2})^k] - 2.$$

(See Hua [30].) Moreover K. Mahler [41] proved that there exists a constant k_0 such that the above inequality holds whenever $k > k_0$. Unfortunately the method which is based on Roth's theorem is ineffective in the sense that it does not allow us to make a computation for the value of k_0.

J. R. Chen [18] proved that $g(5) = 37$. R. Balasubramanian proved that $19 \leqslant g(4) \leqslant 21$ (see [5]).

18.2. I. M. Vinogradov [61] has improved on his own result on $G(k)$ in Waring's problem: For sufficiently large k we have

$$G(k) < k(2 \log k + 4 \log\log k + 2 \log\log\log k + 13).$$

Chapter 19. Schnirelmann Density

19.1 The Definition of Density and its History

The purpose of this chapter is to prove the following two important results:

"*There exists a positive integer c such that every positive integer is the sum of at most c primes.*"

"*Let k be any positive integer. Then there exists a positive integer* c_k *(depending only on k) such that every positive integer is the sum of at most* c_k *k-th powers.*"

These two results are obviously related to the Goldbach problem and the Waring problem. Indeed we can even say that these two results are the most fundamental first steps towards these two famous problems. We shall call them the Goldbach-Schnirelmann theorem and the Waring-Hilbert theorem respectively.

In this chapter we introduce the notion of density created by Schnirelmann. This notion is extremely elementary, and yet it allows us to establish the two historic results. Our proof of the Goldbach-Schnirelmann theorem differs slightly from Schnirelmann's original proof in that we replace the application of Brun's sieve method by Selberg's sieve method.

Again our proof of the Waring-Hilbert theorem is not the original proof due to Hilbert, nor that due to Schnirelmann. We shall give instead the proof by Linnik, given in 1943, with some simplifications and modifications.

In both these proofs the notion of Schnirelmann density occupies an important place. The definition of density is as follows:

Definition 1. Let $\mathfrak{A}$ denote a set of (distinct) non-negative integers a. Denote by $A(n)$ the number of positive integers in $\mathfrak{A}$ which do not exceed n; that is

$$A(n) = \sum_{1 \leqslant a \leqslant n} 1.$$

Suppose that there exists a positive number α such that $A(n) \geqslant \alpha n$ for every positive integer n. Then we say that the set $\mathfrak{A}$ has positive density, or that $\mathfrak{A}$ is a *positive density set*. The greatest α with this property is then called the *density* of $\mathfrak{A}$.

Obviously we have the following simple properties:

(i) Since $A(n) \leqslant n$, it follows that $\alpha \leqslant 1$.

(ii) If $\alpha = 1$, then $A(n) = n$ for all n and so $\mathfrak{A}$ must include all the positive integers.

Exercise. Let $\tau \geqslant 1$. Determine the density of the set $1 + [\tau(n-1)]$, $n = 1, 2, \ldots$.

19.2 The Sum of Sets and its Density

We now introduce the symbols $\mathfrak{B}$, b, $B(n)$, β and $\mathfrak{C}$, c, $C(n)$, γ. The definitions for them are analogous to those for $\mathfrak{A}$, a, $A(n)$, α: that is $b \in \mathfrak{B}$, $B(n) = \sum_{1 \leqslant b \leqslant n} 1$, and β is the density of the positive density set $\mathfrak{B}$.

Definition. The set of integers of the form $a + b$ ($a \in \mathfrak{A}, b \in \mathfrak{B}$) is called the *sum* of the sets $\mathfrak{A}$, $\mathfrak{B}$, and is denoted by $\mathfrak{C}$. We also write $\mathfrak{A} + \mathfrak{B} = \mathfrak{C}$.

Theorem 2.1. *Let* $0 \in \mathfrak{A}$ *and* $\mathfrak{C} = \mathfrak{A} + \mathfrak{B}$. *Then* $\gamma \geqslant \alpha + \beta - \alpha\beta$.

Proof. Since $\beta > 0$ we see that $1 \in \mathfrak{B}$. The following three types of numbers are positive integers in $\mathfrak{C}$; they are all different and are at most n.

(i) In $\mathfrak{B}$ we write $b_1 = 1, b_2, \ldots, b_{B(n)}$, the numbers being arranged in increasing order. Since $0 \in \mathfrak{A}$ we see that $b_1, b_2, \ldots, b_{B(n)}$ are members of $\mathfrak{C}$, and that there are $B(n)$ such members.

(ii) Corresponding to any v where $1 \leqslant v \leqslant B(n) - 1$, the various numbers $a + b_v$, with $a \in \mathfrak{A}$ and $1 \leqslant a \leqslant b_{v+1} - b_v - 1$, are distinct positive integers not exceeding n in the set $\mathfrak{C}$. This is because

$$a + b_v \leqslant (b_{v+1} - b_v - 1) + b_v = b_{v+1} - 1 \leqslant b_{B(n)} - 1 \leqslant n - 1$$

and

$$a + b_v \geqslant 1 + b_v,$$

so that

$$1 + b_v \leqslant a + b_v \leqslant b_{v+1} - 1.$$

It is clear that the two types of numbers in (i) and (ii) are mutually distinct. For each fixed v ($1 \leqslant v \leqslant B(n) - 1$), there are $A(b_{v+1} - b_v - 1)$ such numbers $a + b_v$ in $\mathfrak{C}$.

(iii) For $a \in \mathfrak{A}$, $1 \leqslant a \leqslant n - b_{B(n)}$, the numbers $a + b_{B(n)}$ are distinct positive integers not exceeding n in the set $\mathfrak{C}$. Since $a + b_{B(n)} \geqslant 1 + b_{B(n)}$ we see that these numbers of type (iii) are different from those in types (i) and (ii), and there are $A(n - b_{B(n)})$ such numbers $a + b_{B(n)}$.

From the results of (i), (ii) and (iii) we have

$$\begin{aligned} C(n) &\geqslant B(n) + \sum_{v=1}^{B(n)-1} A(b_{v+1} - b_v - 1) + A(n - b_{B(n)}) \\ &\geqslant B(n) + \sum_{v=1}^{B(n)-1} \alpha(b_{v+1} - b_v - 1) + \alpha(n - b_{B(n)}) \\ &= B(n) + \alpha\{b_{B(n)} - b_1 - (B(n) - 1) + n - b_{B(n)}\} \\ &= B(n) + \alpha\{n - B(n)\} \geqslant (1 - \alpha)\beta n + \alpha n \\ &= n(\alpha + \beta - \alpha\beta), \end{aligned}$$

and hence

$$\frac{C(n)}{n} \geqslant \alpha + \beta - \alpha\beta, \qquad \gamma \geqslant \alpha + \beta - \alpha\beta. \quad \square$$

Note: This theorem is not the best concerning the density of the sum of sets. The sharpest result should be $\gamma \geqslant \min(1, \alpha + \beta)$, a theorem proved by Mann in 1942. The proof of Mann's theorem is more complicated, and since there is no fundamental improvement concerned with the applications to the principal results in this chapter, we do not include it in this book. Let us now take $\mathfrak{A}$ and $\mathfrak{B}$ both to be sets of positive integers congruent to 1 mod q, and assume also that $0 \in \mathfrak{A}$. Then $\mathfrak{A} + \mathfrak{B}$ include all the positive integers congruent to 1, 2 mod q. Obviously the densities of $\mathfrak{A}$ and $\mathfrak{B}$ are $1/q$ while the density of $\mathfrak{A} + \mathfrak{B}$ is $2/q$. Therefore the result of Mann cannot be improved.

Theorem 2.2. *Let* $0 \in \mathfrak{A}$ *and* $\alpha + \beta \geqslant 1$. *Then* $\gamma = 1$; *that is the set* $\mathfrak{C} = \mathfrak{A} + \mathfrak{B}$ *contains all the positive integers.*

Proof. Suppose that $\gamma < 1$, so that there is a least positive integer $n \notin \mathfrak{C}$. From $\beta > 0$ we deduce that $1 \in \mathfrak{B}$, and from $0 \in \mathfrak{A}$ we further deduce that $1 \in \mathfrak{C}$, so that $n \geqslant 2$. Again, from $0 \in \mathfrak{A}$ we have $n \notin \mathfrak{B}$ also.

Consider the following natural numbers a and $n - b$ not exceeding $n - 1$:

$$a, \qquad 1 \leqslant a \leqslant n - 1, \qquad a \in \mathfrak{A},$$

$$n - b, \qquad 1 \leqslant b \leqslant n - 1, \qquad b \in \mathfrak{B}.$$

Each a is different from $n - b$, because $a = n - b$ would give $n = a + b \in \mathfrak{C}$, a contradiction. Also since both a and $n - b$ do not exceed $n - 1$, the number of such numbers is at most $n - 1$.

On the other hand the number of such numbers a and $n - b$ is $A(n - 1) + B(n - 1)$. From

$$A(n - 1) \geqslant \alpha(n - 1)$$

and

$$B(n - 1) = B(n) \geqslant \beta n > \beta(n - 1),$$

we arrive at

$$A(n - 1) + B(n - 1) > \alpha(n - 1) + \beta(n - 1) = (\alpha + \beta)(n - 1) \geqslant n - 1,$$

contradicting our earlier claim that the number of numbers a and $n - b$ does not exceed $n - 1$. The theorem is proved. $\square$

Theorem 2.3. *Let* $\mathfrak{A}$ *have density* $\alpha < 1$ *and let*

$$s_0 = 2\left[\frac{\log 2}{-\log(1 - \alpha)}\right] + 2.$$

If $0 \in \mathfrak{A}$, then every positive integer is the sum of s_0 numbers in $\mathfrak{A}$. If $0 \notin \mathfrak{A}$, then every positive integer is the sum of at most s_0 numbers in $\mathfrak{A}$.

Proof. The second part of the theorem follows trivially from the first part by artificially inserting 0 into the set $\mathfrak{A}$. We now prove the first part of the theorem – that is we shall assume that $0 \in \mathfrak{A}$.

Let $\mathfrak{A}_h = \mathfrak{A} + \cdots + \mathfrak{A}$, the h-fold sum of $\mathfrak{A}$. We denote by α_h the density of $\mathfrak{A}_h$ and we proceed to prove by induction that $\alpha_h \geqslant 1 - (1 - \alpha)^h$. When $h = 1$ we have $\alpha_1 = \alpha$. From $\mathfrak{A}_h = \mathfrak{A} + \mathfrak{A}_{h-1}$, Theorem 2.1 and the induction hypothesis, namely

$$\alpha_{h-1} \geqslant 1 - (1 - \alpha)^{h-1},$$

we have

$$\begin{aligned} \alpha_h &\geqslant \alpha + \alpha_{h-1} - \alpha\alpha_{h-1} = \alpha + (1 - \alpha)\alpha_{h-1} \\ &\geqslant \alpha + (1 - \alpha)\{1 - (1 - \alpha)^{h-1}\} \\ &= 1 - (1 - \alpha)^h. \end{aligned}$$

Therefore $\alpha_h \geqslant 1 - (1 - \alpha)^h$ holds for $h = 1, 2, 3, \ldots$. Now

$$\frac{s_0}{2} = \left[\frac{\log 2}{-\log(1 - \alpha)}\right] + 1 > \frac{\log 2}{-\log(1 - \alpha)},$$

so that

$$(1 - \alpha)^{s_0/2} \leqslant (1 - \alpha)^{\frac{\log 2}{-\log(1-\alpha)}} = e^{-\frac{\log 2}{\log(1-\alpha)} \cdot \log(1-\alpha)} = \tfrac{1}{2},$$

and hence

$$\alpha_{s_0/2} \geqslant 1 - (1 - \alpha)^{s_0/2} \geqslant 1 - \tfrac{1}{2} = \tfrac{1}{2}.$$

Since $0 \in \mathfrak{A}_{s_0/2}$ the set $\mathfrak{A}_{s_0} = \mathfrak{A}_{s_0/2} + \mathfrak{A}_{s_0/2}$ must, by Theorem 2.2, include all the positive integers and therefore every positive integer is expressible as the sum of s_0 members of $\mathfrak{A}$. □

Theorem 2.4. *Let $\mathfrak{A}^*$ be a collection of non-negative integers, with multiplicity of membership being allowed. Let $\mathfrak{A}$ be the largest set from $\mathfrak{A}^*$ without multiplicity of membership. Let $r(a)$ denote the multiplicity of a in $\mathfrak{A}$. Suppose that*

$$\frac{1}{n} \frac{\left(\sum\limits_{1 \leqslant a \leqslant n} r(a)\right)^2}{\sum\limits_{1 \leqslant a \leqslant n} r^2(a)} \geqslant \alpha' \qquad (> 0),$$

holds for all $n \geqslant 1$. Then $\mathfrak{A}$ has a positive density $\alpha \geqslant \alpha'$.

Proof. From the Bunyakovsky-Schwarz inequality (Theorem 18.7.1) we have

$$\left(\sum_{1 \leqslant a \leqslant n} r(a)\right)^2 \leqslant \sum_{1 \leqslant a \leqslant n} r^2(a) \sum_{1 \leqslant a \leqslant n} 1^2 = A(n) \sum_{1 \leqslant a \leqslant n} r^2(a),$$

so that

$$\frac{A(n)}{n} \geqslant \frac{1}{n}\left(\sum_{1 \leqslant a \leqslant n} r(a)\right)^2 \bigg/ \sum_{1 \leqslant a \leqslant n} r^2(a) \geqslant \alpha'.$$

The theorem is proved. □

19.3 The Goldbach-Schnirelmann Theorem

In §§3–5, the letters $c_1, c_2, \ldots$ denote absolute positive constants. The purpose of §§3–5 is to prove

Theorem 3.1. *There exists a positive integer c such that every integer greater than* 1 *is the sum of at most c prime numbers.*

We define $\mathfrak{A}^*$ to be the collection of numbers 1 together with $p_1 + p_2$ where p_1, p_2 run through all the prime numbers. Note that members of $\mathfrak{A}^*$ may have multiplicity. We also define $\mathfrak{A}$ to be the largest set from $\mathfrak{A}^*$ without multiplicity of membership. In order to prove Theorem 3.1 it suffices to prove

Theorem 3.2. $\mathfrak{A}$ *has positive density* c_1.

By Theorem 2.3 any positive integer m is expressible as a sum of at most s_0 members of $\mathfrak{A}$ (that is, a sum of terms involving 1 and numbers of the form $p_1 + p_2$). This implies that m is the sum of at most $2s_0$ numbers which are primes or 1. Therefore, for any $n > 2$, we have $n = 2 + (n - 2) = 2 + b \cdot 1 + \sum p$, where the number of primes p being summed is at most $2s_0 - b$. Since $2 + b$ is expressible as a sum of at most $b + 1$ primes, it follows that n is expressible as a sum of at most $2s_0 + 1$ primes. Therefore Theorem 3.1 follows from Theorem 3.2.

We now let $r(1) = 1$ and $r(a)$ be the multiplicity of a in the collection $\mathfrak{A}^*$, that is

$$r(a) = \begin{cases} 1, & \text{if } a = 1, \\ \displaystyle\sum_{p_1 + p_2 = a} 1, & \text{if } a \geqslant 2. \end{cases}$$

Following Theorem 2.4 our aims are to find a lower bound for $\sum_{1 \leqslant a \leqslant n} r(a)$ and an upper bound for $\sum_{1 \leqslant a \leqslant n} r^2(a)$. The former task is not at all difficult while the latter task is the main concern of the next section.

Theorem 3.3. *If* $n \geqslant 2$, *then*

$$\sum_{1 \leqslant a \leqslant n} r(a) \geqslant c_2 \frac{n^2}{\log^2 n}. \tag{1}$$

Proof. Let $n \geqslant 4$. From Theorem 5.6.2 we have

$$\sum_{1 \leqslant a \leqslant n} r(a) = 1 + \sum_{4 \leqslant a \leqslant n} \sum_{p_1 + p_2 = a} 1$$

$$\geqslant \sum_{p_1, p_2 \leqslant n/2} 1 = \pi^2(\tfrac{1}{2}n)$$

$$\geqslant \left(c_3 \frac{n}{2} \Big/ \log \frac{n}{2}\right)^2 \geqslant \frac{c_3^2}{4} \frac{n^2}{\log^2 n}.$$

If $n = 2$ or 3, then $\sum r(a) = 1$, so that the theorem follows by taking

$$c_2 = \min\left(\frac{c_3^2}{4}, \frac{\log^2 2}{4}, \frac{\log^2 3}{9}\right). \quad \square$$

From Theorem 2.4 and $r(1) = 1$ we see that the crux of the matter is now to prove

Theorem 3.4. *If $n \geqslant 2$, then*

$$\sum_{1 \leqslant a \leqslant n} r^2(a) \leqslant c_4 \frac{n^3}{\log^4 n}. \tag{2}$$

In other words if Theorem 3.4 is proved, then from

$$\frac{1}{n} \frac{\left(\sum_{1 \leqslant a \leqslant n} r(a)\right)^2}{\sum_{1 \leqslant a \leqslant n} r^2(a)} \geqslant \frac{1}{n} \frac{(c_2 n^2/\log^2 n)^2}{c_4 n^3/\log^4 n} = \frac{c_2^2}{c_4}$$

and Theorem 2.4, we see that Theorem 3.2 follows.

It remains therefore to prove Theorem 3.4.

19.4 Selberg's Inequality

Although we can do without the following theorem in this section, nevertheless the reader should know the result

Theorem 4.1. *Let $a_i > 0$ $(i = 1, 2, \ldots, n)$ and b_i $(i = 1, 2, \ldots, n)$ be fixed real numbers. The minimum value of $\sum_{i=1}^n a_i x_i^2$, subject to the constraint $\sum_{i=1}^n b_i x_i = 1$, is given by $(\sum_{i=1}^n b_i^2/a_i)^{-1}$. Moreover, the minimum value is attained when*

$$x_i = \frac{\dfrac{b_i}{a_i}}{\sum_{i=1}^n \dfrac{b_i^2}{a_i}}, \qquad i = 1, 2, \ldots, n.$$

Proof. From the Bunyakovsky-Schwarz inequality (Theorem 18.7.1) we have

$$\left(\sum_{i=1}^n a_i x_i^2\right)\left(\sum_{i=1}^n b_i^2 a_i^{-1}\right) \geqslant \left(\sum_{i=1}^n x_i b_i\right)^2 = 1.$$

Therefore

$$\sum_{i=1}^{n} a_i x_i^2 \geqslant \frac{1}{\sum_{i=1}^{n} b_i^2 a_i^{-1}}. \tag{1}$$

Again, by Theorem 18.7.1, we know that the condition for equality in (1) is that there exists a real number t_0 such that

$$\sqrt{a_i}\, x_i = t_0 b_i \frac{1}{\sqrt{a_i}} \qquad (i = 1, 2, \ldots, n),$$

that is

$$x_i = b_i a_i^{-1} t_0 \qquad (i = 1, 2, \ldots, n).$$

Hence

$$1 = \sum_{i=1}^{n} b_i x_i = \sum_{i=1}^{n} b_i^2 a_i^{-1} t_0,$$

or

$$t_0 = \frac{1}{\sum_{i=1}^{n} b_i^2 a_i^{-1}}.$$

Therefore

$$x_i = \frac{b_i a_i^{-1}}{\sum_{i=1}^{n} b_i^2 a_i^{-1}} \qquad (i = 1, 2, \ldots, n). \tag{2}$$

The theorem is proved. □

Theorem 4.2 (A. Selberg). *Let $\{b\}$ be a set of M integers such that the number of integers b in the set which are divisible by a positive integer k is*

$$\sum_{\substack{b \\ k|b}} 1 = g(k)M + R(k), \tag{1}$$

where $R(k)$ is a certain remainder term, and $g(k)$ is a positive valued multiplicative function satisfying $g(p) < 1$.

Let N_ξ denote the number of integers b in $\{b\}$ which are not divisible by any prime $\leqslant \xi$. Then

$$N_\xi \leqslant \frac{M}{\sum_{1 \leqslant k \leqslant \xi} \frac{\mu^2(k)}{f(k)}} + \sum_{1 \leqslant k_1, k_2 \leqslant \xi} \lambda_{k_1} \lambda_{k_2} R\left\{\frac{k_1 k_2}{(k_1, k_2)}\right\},$$

where

$$f(k) = \sum_{d|k} \mu(d) \Big/ g\left(\frac{k}{d}\right)^*, \tag{4}$$

$$\lambda_k = \frac{\mu(k)}{f(k)g(k)} \sum_{\substack{1 \leqslant m \leqslant \xi/k \\ (m,k)=1}} \frac{\mu^2(m)}{f(m)} \Big/ \sum_{1 \leqslant m \leqslant \xi} \frac{\mu^2(m)}{f(m)}. \tag{5}$$

Proof. Let $1 = \lambda_1, \lambda_2, \ldots, \lambda_{[\xi]}$ be real numbers. Since the least common multiple of k_1, k_2 is $k_1k_2/(k_1, k_2)$, we have, from (3), that

$$N_\xi = \sum_{p|b \Rightarrow p > \xi} {}_b \, 1 = \sum_{p|b \Rightarrow p > \xi} {}_b \, 1 \Bigg(\sum_{\substack{k|b \\ 1 \leqslant k \leqslant \xi}} \lambda_k \Bigg)^2 \leqslant \sum_b \Bigg(\sum_{\substack{k|b \\ 1 \leqslant k \leqslant \xi}} \lambda_k \Bigg)^2$$

$$= \sum_{1 \leqslant k_1, k_2 \leqslant \xi} \lambda_{k_1}\lambda_{k_2} \sum_{\substack{k_1|b \\ k_2|b}} {}_b \, 1 = \sum_{1 \leqslant k_1, k_2 \leqslant \xi} \lambda_{k_1}\lambda_{k_2} \sum_{\frac{k_1k_2}{(k_1,k_2)} \Big| b} {}_b \, 1$$

$$= \sum_{1 \leqslant k_1, k_2 \leqslant \xi} \lambda_{k_1}\lambda_{k_2} \left(g\left\{\frac{k_1k_2}{(k_1, k_2)}\right\} M + R\left\{\frac{k_1k_2}{(k_1, k_2)}\right\}\right),$$

where $p|b \Rightarrow p > \xi$ means that the prime divisors of b are greater than ξ. From Theorem 6.2.4 we have

$$N_\xi \leqslant MQ + \sum_{1 \leqslant k_1, k_2 \leqslant \xi} \lambda_{k_1}\lambda_{k_2} R\left\{\frac{k_1k_2}{k_1, k_2}\right\}, \tag{6}$$

$$Q = \sum_{1 \leqslant k_1, k_2 \leqslant \xi} \lambda_{k_1}\lambda_{k_2} \frac{g(k_1)g(k_2)}{g\{(k_1, k_2)\}}.$$

From (4) and Theorem 6.4.1, we have

$$Q = \sum_{1 \leqslant k_1, k_2 \leqslant \xi} \lambda_{k_1}\lambda_{k_2} g(k_1)g(k_2) \sum_{d|(k_1,k_2)} f(d)$$

$$= \sum_{1 \leqslant d \leqslant \xi} f(d) \sum_{\substack{1 \leqslant k_1 \leqslant \xi \\ d|k_1}} \lambda_{k_1} g(k_1) \sum_{\substack{1 \leqslant k_2 \leqslant \xi \\ d|k_2}} \lambda_{k_2} g(k_2)$$

$$= \sum_{1 \leqslant d \leqslant \xi} f(d) \Bigg\{ \sum_{\substack{1 \leqslant k \leqslant \xi \\ d|k}} \lambda_k g(k) \Bigg\}^2. \tag{7}$$

From (5) and Theorem 6.2.1 we see that $\lambda_1 = 1$ (the reader can use Theorem 4.1 to prove that Q is actually the minimum for this choice of $\lambda_1, \ldots, \lambda_{[\xi]}$).

Let

$$s = \sum_{1 \leqslant m \leqslant \xi} \frac{\mu^2(m)}{f(m)}. \tag{8}$$

From Theorem 6.2.2 we see that $f(n)$ is also multiplicative so that, by (5) we have

---

* When k is square-free, $f(k) = (1/g(k)) \prod_{p|k} (1 - g(p)) > 0$.

$$\lambda_k g(k) = \frac{\mu(k)}{sf(k)} \sum_{\substack{1 \leqslant m \leqslant \xi/k \\ (m,k)=1}} \frac{\mu^2(m)}{f(m)} = \sum_{\substack{1 \leqslant m \leqslant \xi/k \\ (m,k)=1}} \mu(m) \frac{\mu(mk)}{sf(mk)}$$

$$= \sum_{1 \leqslant m \leqslant \xi/k} \mu(m) \frac{\mu(mk)}{sf(mk)},$$

and hence, by Theorem 6.3.2,

$$\frac{\mu(m)}{sf(m)} = \sum_{1 \leqslant k \leqslant \xi/m} \lambda_{km} g(km) = \sum_{\substack{1 \leqslant r \leqslant \xi \\ m|r}} \lambda_r g(r).$$

Therefore, by (7) and (8) we have

$$Q = \sum_{1 \leqslant d \leqslant \xi} f(d) \left\{\frac{\mu(d)}{sf(d)}\right\}^2 = \frac{1}{s^2} \sum_{1 \leqslant d \leqslant \xi} \frac{\mu^2(d)}{f(d)} = \frac{s}{s^2} = \frac{1}{s}.$$

The required result follows from (6) and (8). □

Theorem 4.3. *Let the conditions in Theorem 4.2 hold. If $g_1(n)$ is a completely multiplicative function, and $g_1(p) = g(p)$, then*

$$N_\xi \leqslant \frac{M}{\sum_{1 \leqslant k \leqslant \xi} g_1(k)} + \sum_{1 \leqslant k_1, k_2 \leqslant \xi} \left| R\left\{\frac{k_1 k_2}{(k_1, k_2)}\right\} \right| \prod_{p|k_1} \{1 - g_1(p)\}^{-1} \prod_{p|k_2} \{1 - g_1(p)\}^{-1}.$$

We first establish the following:

Theorem 4.4. *Let $f(n)$ be a completely multiplicative function satisfying $0 \leqslant f(p) < 1$. If $\beta_n \geqslant 0$, then*

$$\sum_{1 \leqslant n \leqslant \xi} \beta_n f(n) \prod_{p|k_n} \{1 - f(p)\}^{-1} \geqslant \sum_{1 \leqslant n \leqslant \xi} f(n) \sum_{\substack{m|n \\ p|\frac{n}{m} \Rightarrow p|k_m}} \beta_m,$$

where $p|\frac{n}{m} \Rightarrow p|k_m$ means that n/m has only the prime divisors of k_m.

Proof.

$$\sum_{1 \leqslant n \leqslant \xi} \beta_n f(n) \prod_{p|k_n} \{1 - f(p)\}^{-1} = \sum_{1 \leqslant n \leqslant \xi} \beta_n f(n) \prod_{p|k_n} \sum_{m=0}^{\infty} \{f(p)\}^m$$

$$= \sum_{1 \leqslant n \leqslant \xi} \beta_n f(n) \prod_{p|k_n} \sum_{m=0}^{\infty} f(p^m) = \sum_{1 \leqslant n \leqslant \xi} \beta_n f(n) \sum_{\substack{r=1 \\ p|r \Rightarrow p|k_n}}^{\infty} f(r)$$

$$= \sum_{1 \leqslant n \leqslant \xi} \beta_n \sum_{\substack{r=1 \\ p|r \Rightarrow p|k_n}}^{\infty} f(nr) = \sum_{1 \leqslant n \leqslant \xi} \beta_n \sum_{\substack{s=1 \\ n|s \\ p|\frac{s}{n} \Rightarrow p|k_n}}^{\infty} f(s)$$

$$= \sum_{s=1}^{\infty} f(s) \sum_{\substack{1 \leqslant n \leqslant \xi \\ n|s \\ p|\frac{s}{n} \Rightarrow p|k_n}} \beta_n \geqslant \sum_{1 \leqslant s \leqslant \xi} f(s) \sum_{\substack{1 \leqslant n \leqslant \xi \\ n|s \\ p|\frac{s}{n} \Rightarrow p|k_n}} \beta_n$$

$$= \sum_{1 \leqslant s \leqslant \xi} f(s) \sum_{\substack{n|s \\ p|\frac{s}{n} \Rightarrow p|k_n}} \beta_n. \quad \square$$

Proof of Theorem 4.3. We have, by (4),

$$f(p) = \frac{\mu(1)}{g(p)} + \frac{\mu(p)}{g(1)} = \frac{1}{g(p)} - 1 = \frac{1 - g(p)}{g(p)}.$$

If k is square-free, then, by Theorem 6.2.2,

$$\frac{\mu^2(k)}{f(k)} = \mu^2(k) \prod_{p|k} \frac{g_1(p)}{1 - g_1(p)} = \mu^2(k) \frac{\prod_{p|k} g_1(p)}{\prod_{p|k} \{1 - g_1(p)\}}$$

$$= \mu^2(k) g_1(k) \prod_{p|k} \{1 - g_1(p)\}^{-1}. \tag{9}$$

The above still holds when $k = 1$ and when k has square divisors. Therefore, by Theorem 4.4,

$$\sum_{1 \leqslant k \leqslant \xi} \frac{\mu^2(k)}{f(k)} = \sum_{1 \leqslant k \leqslant \xi} \mu^2(k) g_1(k) \prod_{p|k} \{1 - g_1(p)\}^{-1}$$

$$\geqslant \sum_{1 \leqslant k \leqslant \xi} g_1(k) \sum_{\substack{m|k \\ p|\frac{k}{m} \Rightarrow p|m}} \mu^2(m).$$

Let d_k be the greatest square-free divisor of k, so that $d_k|k$. If $p \left| \frac{k}{d_k} \right.$, then $p|k$ and so $p|d_k$. Therefore d_k is a number satisfying the condition on m, so that

$$\sum_{1 \leqslant k \leqslant \xi} \frac{\mu^2(k)}{f(k)} \geqslant \sum_{1 \leqslant k \leqslant \xi} g_1(k). \tag{10}$$

From (9) we see that $\mu^2(k)/f(k) \geqslant 0$ and so, by (5) and (9) we have that

$$|\lambda_k| \leqslant \frac{\mu^2(k)}{f(k)g(k)} = \frac{\mu^2(k)}{f(k)g_1(k)} \leqslant \prod_{p|k} \{1 - g_1(p)\}^{-1}.$$

When $k = 1$ or k is square-free, $g(k) = g_1(k)$, and if k is not square-free, $\mu(k) = 0$; therefore the above holds for all k. The theorem now follows from (10) and Theorem 4.2. $\square$

Theorem 4.5. *Let $A \geqslant 0$, $M \geqslant 3$ and denote by $\pi(A; M)$ the number of primes between A and $A + M$. Then*

$$\pi(A; M) \leqslant \frac{2M}{\log M}\left\{1 + O\left(\frac{\log\log M}{\log M}\right)\right\}.$$

The implied constant here is independent of A and M.

Proof. Let

$$S(A; M) = \sum_{A + \sqrt{M} < p \leqslant A + M} 1,$$

so that

$$\pi(A; M) = \sum_{A < p \leqslant A + \sqrt{M}} 1 + S(A; M) \leqslant \sqrt{M} + S(A; M). \tag{11}$$

We now take $\{b\}$ to be the set of all integers n satisfying $A < n \leqslant A + M$. With the notation of Theorem 4.3 we have

$$S(A; M) \leqslant N_\xi, \qquad 1 < \xi \leqslant \sqrt{M} \tag{12}$$

uniformly in A. We now estimate N_ξ. From

$$\sum_{\substack{k|b \\ A < b \leqslant A + M}} 1 = \left[\frac{A + M}{k}\right] - \left[\frac{A}{k}\right] = \frac{M}{k} + R(k), \qquad |R(k)| \leqslant 1,$$

we have $g_1(k) = 1/k$, giving

$$\sum_{1 \leqslant k \leqslant \xi} g_1(k) = \log \xi + O(1).$$

From Theorem 5.9.3 we have

$$\prod_{p|k} (1 - g_1(p))^{-1} = \prod_{p|k}\left(1 - \frac{1}{p}\right)^{-1} \leqslant \prod_{p \leqslant k}\left(1 - \frac{1}{p}\right)^{-1} = O(\log k).$$

Hence

$$\sum_{1 \leqslant k_1, k_2 \leqslant \xi} \left| R\left(\frac{k_1 k_2}{(k_1, k_2)}\right)\right| \prod_{p|k_1} (1 - g_1(p))^{-1} \prod_{p|k_2} (1 - g_1(p))^{-1}$$

$$= O\left(\sum_{1 \leqslant k_1, k_2 \leqslant \xi} \log k_1 \log k_2\right) = O(\xi^2 \log^2 \xi).$$

Therefore

$$N_\xi \leqslant \frac{M}{\log \xi + O(1)} + O(\xi^2 \log^2 \xi).$$

Taking

$$\xi = M^{\frac{1}{4}}/\log^2 M,$$

we have

$$N_{M^{\frac{1}{4}}/\log^2 M} \leqslant \frac{2M}{\log M}\left(1 + O\left(\frac{\log\log M}{\log M}\right)\right).$$

Substituting this into (11) and (12), the required result follows. □

19.5 The Proof of the Goldbach-Schnirelmann Theorem

Theorem 5.1. *If* $a \geqslant 2$, *then*

$$r(a) \leqslant c_5 \frac{a}{\log^2 a} \sum_{k|a} \frac{\mu^2(k)}{k}.$$

Proof. When $a = 2$ or 3 we have $r(a) = 0$, and if a is odd and $a = p_1 + p_2$ then either p_1 or p_2 is 2 so that $r(a) \leqslant 2$. Since the theorem is trivial for these values of a we shall assume that $a \geqslant 4$ and a is even in the following. We have

$$r(a) = \sum_{p_1+p_2=a} 1 \leqslant \sum_{\substack{p_1+p_2=a \\ p_1,p_2>\sqrt{a}}} 1 + \sum_{\substack{p_1+p_2=a \\ p_1\leqslant\sqrt{a}}} 1 + \sum_{\substack{p_1+p_2=a \\ p_2\leqslant\sqrt{a}}} 1 \leqslant S(a) + 2\sqrt{a}, \tag{1}$$

where

$$S(a) = \sum_{\substack{p_1+p_2=a \\ p_1,p_2>\sqrt{a}}} 1.$$

We now define a set of integers $b_c = c(a-c)$ $(c = 1, 2, \ldots, a)$. If $p_1 + p_2 = a$ and $p_1, p_2 > \sqrt{a}$, then $p_1(a-p_1) = p_2(a-p_2) = p_1p_2$ is not divisible by any prime $\leqslant \sqrt{a}$. Using the notation of §4, we have

$$S(a) \leqslant N_\xi, \qquad 1 < \xi \leqslant \sqrt{a}. \tag{2}$$

Denote by $M(k)$ the number of solutions to the congruence

$$x(a-x) \equiv 0 \pmod{k}, \qquad 0 \leqslant x < k.$$

Then

$$\sum_{k|b} 1 = \sum_{\substack{c=1 \\ c(a-c)\equiv 0 (\bmod k)}}^{a} 1 = \left[\frac{a}{k}\right] M(k) + T(k),$$

where $0 \leqslant T(k) \leqslant M(k)$. Therefore

$$\sum_{k|b} 1 \leqslant \frac{M(k)}{k} a + M(k)$$

and

$$\sum_{k|b} 1 \geqslant \left[\frac{a}{k}\right] M(k) > \left(\frac{a}{k} - 1\right) M(k) = \frac{M(k)}{k} a - M(k).$$

Let

$$g(k) = \frac{M(k)}{k}, \tag{3}$$

so that

$$\sum_{\substack{b \\ k|b}} 1 = g(k)a + R(k), \tag{4}$$

where

$$|R(k)| \leqslant M(k) \leqslant k. \tag{5}$$

From Theorem 2.8.1 we see that $M(k)$ is a multiplicative function of k and hence so is $g(k)$. Also

$$M(p) = \begin{cases} 1, & p|a \\ 2, & p\nmid a. \end{cases} \tag{6}$$

Therefore, by (3),

$$g_1(p) = g(p) = \begin{cases} \dfrac{1}{p}, & p|a, \\[2ex] \dfrac{2}{p}, & p\nmid a. \end{cases} \tag{7}$$

Since $2|a$ we have $g(2) = \frac{1}{2}$ and therefore $0 < g(p) < 1$ so that Theorem 4.3 is applicable. If $k = p_1^{a_1} \cdots p_r^{a_r}$, then from (3) and (6) we have

$$\begin{aligned} g_1(k) = \prod_{s=1}^{r} \{g_1(p_s)\}^{a_s} = \prod_{s=1}^{r} \frac{\{M(p_s)\}^{a_s}}{p_s^{a_s}} &= \frac{1}{k} \prod_{\substack{s=1 \\ p_s\nmid a}}^{r} 2^{a_s} \\ &\geqslant \frac{1}{k} \prod_{\substack{s=1 \\ p_s\nmid a}}^{r} (1 + a_s) = \frac{h(k)}{k}, \end{aligned}$$

where

$$h(p_1^{a_1} \cdots p_r^{a_r}) = \prod_{\substack{s=1 \\ p_s\nmid a}}^{r} (1 + a_s), \qquad p_1, \ldots, p_r \text{ distinct primes.} \tag{8}$$

From Theorem 4.4 we have

$$\begin{aligned} \prod_{p|a} \left(1 - \frac{1}{p}\right)^{-1} \sum_{1 \leqslant k \leqslant \xi} g_1(k) &\geqslant \sum_{1 \leqslant k \leqslant \xi} h(k) \frac{1}{k} \prod_{p|a} \left(1 - \frac{1}{p}\right)^{-1} \\ &\geqslant \sum_{1 \leqslant k \leqslant \xi} \frac{1}{k} \sum_{\substack{m|k \\ p \left| \frac{k}{m} \Rightarrow p|a \right.}} h(m). \end{aligned}$$

If we write $k = p_1^{a_1} \cdots p_t^{a_t} q_1^{b_1} \cdots q_u^{b_u}$, where p_T and q_U are all distinct primes and $p_T|a$, $q_U\nmid a$, then m can take all the numbers of the form

$$m = \frac{k}{p_1^{c_1} \cdots p_t^{c_t}} = p_1^{a_1 - c_1} \cdots p_t^{a_t - c_t} q_1^{b_1} \cdots q_u^{b_u},$$

where $0 \leqslant c_1 \leqslant a_1, \ldots, 0 \leqslant c_t \leqslant a_t$. For such m we have, by (8), that

$$h(m) = (1 + b_1) \cdots (1 + b_u).$$

Therefore, by Exercise 6.5.1, we have

$$\begin{aligned}
\prod_{p|a}\left(1 - \frac{1}{p}\right)^{-1} \sum_{1 \leqslant k \leqslant \xi} g_1(k) &\geqslant \sum_{1 \leqslant k \leqslant \xi} \frac{1}{k} \sum_{c_1=0}^{a_1} \cdots \sum_{c_t=0}^{a_t} (1 + b_1) \cdots (1 + b_u) \\
&= \sum_{1 \leqslant k \leqslant \xi} \frac{1}{k}(1 + a_1) \cdots (1 + a_t)(1 + b_1) \cdots (1 + b_u) \\
&= \sum_{1 \leqslant k \leqslant \xi} \frac{d(k)}{k} \geqslant c_6 \log^2 \xi.
\end{aligned}$$

Hence

$$\begin{aligned}
\sum_{1 \leqslant k \leqslant \xi} g_1(k) &\geqslant c_6 \log^2 \xi \prod_{p|a}\left(1 - \frac{1}{p}\right) = c_6 \log^2 \xi \prod_{p|a}\left(1 - \frac{1}{p^2}\right)\prod_{p|a}\left(1 + \frac{1}{p}\right)^{-1} \\
&\geqslant c_6 \log^2 \xi \prod_{p}\left(1 - \frac{1}{p^2}\right)\prod_{p|a}\left(1 + \frac{1}{p}\right)^{-1} \\
&\geqslant c_7 \log^2 \xi \left\{\sum_{k|a} \frac{\mu^2(k)}{k}\right\}^{-1}. \qquad (9)
\end{aligned}$$

Secondly, if $k = \prod_{p|k} p^c$, then from

$$\begin{aligned}
\prod_{p|k} \{1 - g_1(p)\}^{-1} &\leqslant \{1 - g_1(2)\}^{-1}\{1 - g_1(3)\}^{-1} \prod_{5 \leqslant p|k} \{1 - g_1(p)\}^{-1} \\
&\leqslant 2 \cdot 3 \prod_{5 \leqslant p|k} \left(1 - \frac{2}{5}\right)^{-1} < 6 \prod_{p|k} (1 + c) = 6d(k) \leqslant 6k,
\end{aligned}$$

together with Theorem 4.3, (5) and (9), we have that

$$\begin{aligned}
S(a) \leqslant N_\xi &\leqslant \frac{1}{c_7} \cdot \frac{a}{\log^2 \xi} \sum_{k|a} \frac{\mu^2(k)}{k} + \sum_{1 \leqslant k_1, k_2 \leqslant \xi} \frac{k_1 k_2}{(k_1, k_2)} 6k_1 \cdot 6k_2 \\
&\leqslant \frac{1}{c_7} \cdot \frac{a}{\log^2 \xi} \sum_{k|a} \frac{\mu^2(k)}{k} + 36\xi^6.
\end{aligned}$$

We take $\xi = a^{\frac{1}{10}}$, and the theorem follows from (1). □

Proof of Theorem 3.4. When $n \geqslant 2$, we have

$$\begin{aligned}
\sum_{1 \leqslant a \leqslant n} r^2(a) &\leqslant 1 + \sum_{4 \leqslant a \leqslant n} c_5^2 \frac{a^2}{\log^4 a} \sum_{k_1|a} \frac{\mu^2(k_1)}{k_1} \sum_{k_2|a} \frac{\mu^2(k_2)}{k_2} \\
&\leqslant 1 + c_5^2 \frac{n^2}{\log^4 n} \sum_{4 \leqslant a \leqslant n} \sum_{\substack{k_1|a \\ k_2|a}} \frac{1}{k_1 k_2}
\end{aligned}$$

$$\leqslant 1 + c_5^2 \frac{n^2}{\log^4 n} \sum_{1 \leqslant k_1, k_2 \leqslant n} \frac{1}{k_1 k_2} \sum_{\substack{1 \leqslant a \leqslant n \\ \frac{k_1 k_2}{(k_1, k_2)} \mid a}} 1$$

$$\leqslant 1 + c_5^2 \frac{n^2}{\log^4 n} \sum_{1 \leqslant k_1, k_2 \leqslant n} \frac{1}{k_1 k_2} \cdot \frac{n}{\frac{k_1 k_2}{(k_1, k_2)}}.$$

Since $(k_1, k_2) \leqslant \min\{k_1, k_2\} \leqslant \sqrt{k_1 k_2}$, it follows that

$$\sum_{1 \leqslant a \leqslant n} r^2(a) \leqslant 1 + c_5^2 \frac{n^2}{\log^4 n} \sum_{1 \leqslant k_1, k_2 \leqslant n} \frac{n}{(k_1 k_2)^{3/2}}$$

$$\leqslant 1 + c_5^2 \frac{n^3}{\log^4 n} \left(\sum_{k=1}^{\infty} \frac{1}{k^{3/2}} \right)^2$$

$$\leqslant c_4 \frac{n^3}{\log^4 n}.$$

The theorem is proved. □

Exercise 1. Let x, k, l be positive integers, and $(l, k) = 1$. Denote by $\pi(x; k, l)$ the number of primes in the arithmetic progression $kn + l$ $(n = 1, 2, \ldots)$ not exceeding x, and let $0 < \delta < 1$. Prove that, for $k < x^{\delta}$,

$$\pi(x; k, l) \leqslant \frac{2x}{\varphi(k) \log \frac{x}{k}} \left(1 + O\left(\frac{(\log\log x)^2}{\log x} \right) \right),$$

where the implied constant depends at most on δ.

Exercise 2. When p, $p + 2$ are both primes, we call them a pair of "prime twins". Denote by $Z_2(N)$ the number of pairs of "prime twins" not exceeding N. Prove that

$$Z_2(N) \leqslant c_8 \frac{N}{\log^2 N},$$

and that the series

$$\sum_{p^*} \frac{1}{p^*},$$

where the summation is over all "prime twins" p^*, is convergent.

19.6 The Waring-Hilbert Theorem

In §§6 – 7 the letters $c, c_1, c_2, \ldots$ denote positive constants depending only on k. The constants implied by the O-symbol also depend at most on k. The purpose of §§6 – 7 is to prove

Theorem 6.1 (Hilbert). *Corresponding to each positive integer k there exists a positive integer c such that every positive integer is the sum of at most c k-th powers.*

We now define $\mathfrak{A}_t^*$ to be the collection of integers $x_1^k + \cdots + x_t^k$ where each x_m runs over all the non-negative integers. We define $\mathfrak{A}_t$ to be the largest set of distinct elements from $\mathfrak{A}_t^*$. Let

$$c_1 = c_1(k) = \tfrac{1}{2}8^{k-1}.$$

The proof of Theorem 6.1 is divided into sections of a chain:

Theorem 6.2. *If $k \geqslant 2$, then $\mathfrak{A}_{c_1}$ has positive density.*

We see that Theorem 6.1 can be deduced at once from Theorem 2.3 and Theorem 6.2.

We define $r(a)$ to be the number of solutions to

$$x_1^k + \cdots + x_{c_1}^k = a, \qquad x_m \geqslant 0.$$

We first prove:

Theorem 6.3. *If $n \geqslant 1$, then*

$$\sum_{1 \leqslant a \leqslant n} r(a) \geqslant c_2(k) n^{c_1/k}.$$

Proof. Clearly we can assume that $n > c_1$. Then

$$\begin{aligned}\sum_{1 \leqslant a \leqslant n} r(a) &= -1 + \sum_{0 \leqslant a \leqslant n} \sum_{\substack{x_1^k + \cdots + x_{c_1}^k = a \\ x_m \geqslant 0}} 1 \\ &\geqslant -1 + \sum_{0 \leqslant x_1 \leqslant (n/c_1)^{1/k}} \cdots \sum_{0 \leqslant x_{c_1} \leqslant (n/c_1)^{1/k}} 1 \\ &\geqslant \left(\frac{n}{c_1}\right)^{c_1/k} - 1 \geqslant c_3(k) n^{c_1/k}. \quad \square\end{aligned}$$

From Theorem 6.3 and Theorem 2.4 we see that the vital link in the chain is to prove:

Theorem 6.4. *If $k \geqslant 2$ and $n \geqslant 1$, then*

$$\sum_{1 \leqslant a \leqslant n} r^2(a) \leqslant c_4(k) n^{2c_1/k - 1}.$$

If this theorem is proved, then Theorem 6.2 follows at once from Theorem 2.4 and Theorem 6.3.

We now transform Theorem 6.4 to the following:

Theorem 6.5. *If $k \geqslant 2$ and $P \geqslant 1$, then*

$$\int_0^1 \left| \sum_{x=0}^{P} e^{2\pi i x^k \alpha} \right|^{2c_1} d\alpha \leqslant c_5(k) P^{2c_1 - k}.$$

We take $P = [n^{1/k}]$ so that, when n is large, $c_1 P^k > n$. We also note that, for any integer q,

$$\int_0^1 e^{2\pi i q \alpha}\, d\alpha = \begin{cases} 1, & \text{if } q = 0, \\ 0, & \text{if } q \neq 0. \end{cases}$$

From Theorem 6.5 we have

$$\begin{aligned} \sum_{1 \leqslant a \leqslant n} r^2(a) &\leqslant \sum_{0 \leqslant a \leqslant c_1 P^k} \Bigg(\sum_{\substack{x_1^k + \cdots + x_{c_1}^k = a \\ 0 \leqslant x_i \leqslant P \\ 1 \leqslant i \leqslant c_1}} 1 \Bigg)^2 \\ &= \int_0^1 \Bigg| \sum_{0 \leqslant a \leqslant c_1 P^k} e^{2\pi i a \alpha} \sum_{\substack{x_1^k + \cdots + x_{c_1}^k = a \\ 0 \leqslant x_i \leqslant P \\ 1 \leqslant i \leqslant c_1}} 1 \Bigg|^2 d\alpha \\ &= \int_0^1 \left| \sum_{x_1 = 0}^{P} \cdots \sum_{x_{c_1} = 0}^{P} e^{2\pi i (x_1^k + \cdots + x_{c_1}^k)\alpha} \right|^2 d\alpha \\ &= \int_0^1 \left| \sum_{x=0}^{P} e^{2\pi i x^k \alpha} \right|^{2c_1} d\alpha \leqslant c_5(k) P^{2c_1 - k} \\ &\leqslant c_4(k) n^{2c_1/k - 1} \end{aligned}$$

giving Theorem 6.4. Our aim therefore is to prove Theorem 6.5.

Exercise. Deduce Theorem 6.5 from Theorem 6.4.

19.7 The Proof of the Waring-Hilbert Theorem

Theorem 7.1. *Let X, $Y \geqslant 1$, n be an integer, and $q(n)$ denote the number of integer solutions to*

$$x_1 y_1 + x_2 y_2 = n \qquad (|x_m| \leqslant X,\ |y_m| \leqslant Y,\ m = 1, 2). \tag{1}$$

Then

$$q(n) \leqslant \begin{cases} 27 X^{3/2} Y^{3/2}, & \text{if } n = 0; \\ 60 XY \sum_{d \mid n} \dfrac{1}{d}, & \text{if } n \neq 0. \end{cases} \tag{2}$$

Proof. 1) $n = 0$. Here the values taken by x_1, x_2 and y_1 cannot exceed $2X + 1$, $2X + 1$ and $2Y + 1$. When x_1, x_2, y_1 are specified, y_2 can only take one value. Therefore

$$q(0) \leqslant (2X + 1)^2(2Y + 1) \leqslant (3X)^2(3Y) = 27X^2Y,$$

and similarly $q(0) \leqslant 27XY^2$, and hence

$$q(0) \leqslant \min(27X^2Y, 27XY^2) \leqslant \sqrt{27X^2Y \cdot 27XY^2} = 27X^{3/2}Y^{3/2}.$$

2) $n \neq 0$. We can assume without loss that $X \leqslant Y$. Let $q_1(n)$ be the number of integer solutions to

$$x_1y_1 + x_2y_2 = n \qquad ((x_1, x_2) = 1, |x_2| \leqslant |x_1| \leqslant X, |y_m| \leqslant Y, m = 1, 2). \quad (3)$$

Clearly $x_1 \neq 0$, since otherwise $x_2 = 0$ giving $n = 0$, contradicting our present hypothesis. Next, for a fixed set x_1, x_2 with $(x_1, x_2) = 1$, $|x_2| \leqslant |x_1| \leqslant X$ we denote by $q_2(n; x_1, x_2)$ the number of integer solutions in y_1, y_2 for (3). From Theorem 1.8.2 we see that (3) is soluble, and if y_1', y_2' is a set of solutions, then all the solutions are given by

$$y_1 = y_1' + tx_2, \qquad y_2 = y_2' - tx_1, \qquad t \text{ integer}.$$

Therefore

$$|t| = \left|\frac{y_2' - y_2}{x_1}\right| \leqslant \frac{Y + Y}{|x_1|} = \frac{2Y}{|x_1|},$$

and hence the number of values taken by t does not exceed

$$2 \cdot \frac{2Y}{|x_1|} + 1 \leqslant \frac{4Y + X}{|x_1|} \leqslant \frac{5Y}{|x_1|},$$

that is

$$q_2(n; x_1, x_2) \leqslant \frac{5Y}{|x_1|}.$$

Therefore

$$q_1(n) \leqslant \sum_{1 \leqslant |x_1| \leqslant X} \sum_{|x_2| \leqslant |x_1|} \frac{5Y}{|x_1|} \leqslant 5Y \sum_{1 \leqslant |x_1| \leqslant X} \frac{2|x_1| + 1}{|x_1|}$$
$$\leqslant 5Y \cdot 3 \cdot 2X = 30XY.$$

It follows that, with the condition $(x_1, x_2) = 1$, the number of solutions to (1) does not exceed $2 \cdot 30XY = 60XY$.

Next, if $(x_1, x_2) = d \neq 1$, $d | n$, then we let $x_1' = x_1/d$, $x_2' = x_2/d$, so that we now seek the number of integer solutions to

$$x_1'y_1 + x_2'y_1 = \frac{n}{d} \qquad \left(|x_m'| \leqslant \frac{X}{d}, |y_m| \leqslant Y, m = 1, 2, (x_1', x_2') = 1\right)$$

and we see from the above that this number does not exceed $60\frac{X}{d} \cdot Y$.

Therefore, when $n \neq 0$,

$$q(n) \leqslant 60XY \sum_{d|n} \frac{1}{d}.$$

The proof of the theorem is complete. □

Theorem 6.5 is obviously a consequence of the following

Theorem 7.2. *Let $k \geqslant 2$, and $f(x)$ be a polynomial with degree k having integer valued coefficients*:

$$f(x) = a_k x^k + a_{k-1}x^{k-1} + \cdots + a_1 x + a_0,$$

$$a_k = O(1), \quad a_{k-1} = O(P), \quad \ldots, \quad a_1 = O(P^{k-1}), \quad a_0 = O(P^k).$$

Then

$$\int_0^1 \left| \sum_{x=0}^{P} e^{2\pi i f(x)\alpha} \right|^{8^{k-1}} d\alpha = O(P^{8^{k-1}-k}). \tag{4}$$

Proof. When $k = 2$, the left hand side of (4) is the number of integer solutions to

$$f(x_1) + f(x_2) - f(y_1) - f(y_2) = f(x_3) + f(x_4) - f(y_3) - f(y_4)$$

$$(f(x) = a_2 x^2 + a_1 x + a_0, \qquad a_2 = O(1), \qquad a_1 = O(P), \qquad a_0 = O(P^2)),$$

$$0 \leqslant x_m, y_m \leqslant P, \qquad 1 \leqslant m \leqslant 4. \tag{5}$$

Let $x_i - y_i = z_i$, $a_2(x_i + y_i) + a_1 = w_i$ $(1 \leqslant i \leqslant 4)$. We see that the number of solutions to (5) does not exceed the number of integer solutions to

$$z_1 w_1 + z_2 w_2 = z_3 w_3 + z_4 w_4 \qquad (z_i = O(P),\ w_i = O(P),\ 1 \leqslant i \leqslant 4). \tag{6}$$

If we denote by $q(n)$ the number of integer solutions to

$$z_1 w_1 + z_2 w_2 = n$$

($z_i = O(P)$, $w_i = O(P)$, $m = 1, 2$ where the constants implied by the O-symbol are the same as those of (6)), then the number of solutions to (6) is $\sum_{|n| \leqslant c_6 P^2} q^2(n)$. From Theorem 7.1, we have

$$\sum_{|n| \leqslant c_6 P^2} q(n)^2 = O(P^6) + O\left(\sum_{1 \leqslant n \leqslant c_6 P^2} \left(P^2 \sum_{d|n} \frac{1}{d} \right)^2 \right)$$

$$= O(P^6) + O\left(P^4 \sum_{1 \leqslant d_1, d_2 \leqslant c_6 P^2} \frac{1}{d_1 d_2} \sum_{\substack{\frac{d_1 d_2}{(d_1, d_2)} \mid n \\ 1 \leqslant n \leqslant c_6 P^2}} 1 \right)$$

$$= O(P^6) + O\left(P^4 \sum_{d_1=1}^{\infty} \sum_{d_2=1}^{\infty} \frac{P^2}{(d_1 d_2)^{3/2}}\right)$$

$$= O(P^6),$$

and the required result (4) follows.

Suppose now that $k \geqslant 3$. We proceed by mathematical induction, and assume as induction hypothesis that the theorem holds when k is replaced by $k - 1$. From

$$\left|\sum_{x=0}^{P} e^{2\pi i f(x)\alpha}\right|^2 = \sum_{x=0}^{P} e^{-2\pi i f(x)\alpha} \sum_{-x \leqslant h \leqslant P-x} e^{2\pi i f(x+h)\alpha}$$

$$= \sum_{0<|h|\leqslant P}{}' \sum_{x=0}^{P}{}' e^{2\pi i h\varphi(x,h)\alpha} + P, \tag{7}$$

where $\sum'$ means that the summation is over those integers in the relevant part of the set in the interval, and $\varphi(x, h) = \frac{1}{h}(f(x+h) - f(x))$ $(h \neq 0)$. Viewing $\varphi(x, h)$ as a polynomial in the variable x, we see that $\varphi(x, h)$ is a polynomial of degree $k - 1$ satisfying the conditions of the theorem. Writing

$$a_h = \sum_{x=0}^{P}{}' e^{2\pi i h\varphi(x,h)\alpha},$$

we have

$$\left|\sum_{x=0}^{P} e^{2\pi i f(x)\alpha}\right|^{2\cdot 8^{k-2}} \leqslant 2^{8^{k-2}} \max\left(\left|\sum_{0<|h|\leqslant P}{}' a_h\right|^{8^{k-2}}, P^{8^{k-2}}\right).$$

If

$$\left|\sum_{0<|h|\leqslant P}{}' a_h\right| \leqslant P,$$

then clearly the required result follows. Otherwise we repeatedly use the Bunyakovsky-Schwarz inequality to give

$$2^{-8^{k-2}}\left|\sum_{x=0}^{P} e^{2\pi i f(x)\alpha}\right|^{2\cdot 8^{k-2}} \leqslant \left|\sum_{0<|h|\leqslant P}{}' a_h\right|^{8^{k-2}} \leqslant \left\{\sum_{0<|h|\leqslant P}{}' 1 \cdot \sum_{0<|h|\leqslant P}{}' |a_h|^2\right\}^{2^{3(k-2)-1}}$$

$$\leqslant \left\{\left(\sum_{0<|h|\leqslant P}{}' 1\right)^{2^2-1} \sum_{0<|h|\leqslant P}{}' |a_h|^{2^2}\right\}^{2^{3(k-2)-2}} \leqslant \cdots$$

$$\leqslant \left\{\left(\sum_{0<|h|\leqslant P}{}' 1\right)^{2^{3(k-2)-1}-1} \sum_{0<|h|\leqslant P}{}' |a_h|^{2^{3(k-2)-1}}\right\}^2$$

$$\leqslant (3P)^{8^{k-2}-1} \sum_{0<|h|\leqslant P}{}' |a_h|^{8^{k-2}}$$

$$= O\left(P^{8^{k-2}-1} \sum_{0<|h|\leqslant P}{}' \left|\sum_{x=0}^{P}{}' e^{2\pi i h\varphi(x,h)\alpha}\right|^{8^{k-2}}\right). \tag{8}$$

Let

$$\left|\sum_{x=0}^{P}{}' e^{2\pi i h\varphi(x,h)\alpha}\right|^{8^{k-2}} = \sum_n A(n)e^{2\pi i h n\alpha}. \tag{9}$$

From $0 \leqslant x \leqslant P$, we deduce that

$$n = 0\left(\max_{0 \leqslant x \leqslant P} |\varphi(x, h)|\right) = O(P^{k-1}).$$

From (9) and the induction hypothesis, we have

$$|A(n)| = \left|\int_0^1 \left|\sum_{x=0}^{P}{}' e^{2\pi i\varphi(x,h)\beta}\right|^{8^{k-2}} e^{-2\pi i n\beta}\, d\beta\right|$$

$$\leqslant \int_0^1 \left|\sum_{x=0}^{P} e^{2\pi i\varphi(x,h)\beta}\right|^{8^{k-2}} d\beta = O(P^{8^{k-2}-(k-1)}).$$

We raise (8) to the 4th power, and then integrate with respect to α over $0 \leqslant \alpha \leqslant 1$ giving

$$\int_0^1 \left|\sum_{x=0}^{P} e^{2\pi i f(x)\alpha}\right|^{8^{k-1}} d\alpha$$

$$= O\left(P^{4\cdot 8^{k-2}-4} \int_0^1 \left(\sum_{0<|h|\leqslant P}{}' \left|\sum_{x=0}^{P}{}' e^{2\pi i h\varphi(x,h)\alpha}\right|^{8^{k-2}}\right)^4 d\alpha\right)$$

$$= O\left(P^{4\cdot 8^{k-2}-4} \sum_{\substack{n_1h_1+n_2h_2=n_3h_3+n_4h_4\\ 0<|h_i|\leqslant P\\ n_i=O(P^{k-1})\\ i=1,2,3,4}} A(n_1)A(n_2)A(n_3)A(n_4)\right)$$

$$= O(P^{4\cdot 8^{k-2}-4}P^{3k}P^{4\cdot 8^{k-2}-4(k-1)}) = O(P^{8^{k-1}-k}).$$

The proof of the theorem is complete. □

Notes

19.1. Bombieri's theorem on $\pi(x; k, l)$ (see Note 5.4) can also be used to prove the Goldbach-Schnirelmann theorem. There is a long list of references on this problem in a paper by R. C. Vaughan [58] where he used a mean value theorem on $\pi(x; k, l)$ by Davenport and Halberstam to prove that every sufficiently large odd number is a sum of 5 odd primes and thus every sufficiently large even number is a sum of at most 6 primes. Vaughan also proved that *every* even number is a sum of at most 27 primes.

Chapter 20. The Geometry of Numbers

20.1 The Two Dimensional Situation

In this section we restrict ourselves to the two dimensional space to give a brief description of the fundamental results in the chapter.

Definition 1. Let c be a simple closed curve in the plane and denote by R the region within c. Suppose that, given any two points of R, the midpoint* also lies in R. Then we call R a *convex region.*

For example: Circles, ellipses, parallelograms and regular n-sided polygons are all convex regions.

The area of a convex region always exists. (We can define it as the supremum of the sums of the areas of networks of small squares whose vertices lie in the region.) Throughout the chapter we make use of various properties concerning convexity, and strictly speaking we need some knowledge of measure theory and topology to give the proper justifications. The reader, however, can understand without difficulty the fundamental content of the chapter without this knowledge, and in any case when it comes to our specific applications and examples the results can be obtained without this knowledge.

Theorem 1.1 (Minkowski's Fundamental Theorem). *Let R be a convex region symmetrical about the origin with area greater than* 4. *Then R must contain a lattice point different from the origin.* (A lattice point is a point with integral coordinates.)

Proof (Hajös). Denote by $S_{2r,2s}$ the square with side length 2 centre at the even coordinate point $(2r, 2s)$. If $S_{2r,2s}$ intersects with R, then we translate this intersection to the square $S_{0,0}$ with the transformation $x - 2r = x'$, $y - 2s = y'$. This process moves the whole of R into $S_{0,0}$ and since the region R has an area greater than 4 there must be at least two points of R which have gone into the same point in $S_{0,0}$. Let these two points come from the two different squares $S_{2r,2s}$ and $S_{2r',2s'}$ so that the original coordinates of the two points are

$$(x_0 + 2r, y_0 + 2s) \qquad \text{and} \qquad (x_0 + 2r', y_0 + 2s').$$

Since R is symmetrical about the origin, the point $(-(x_0 + 2r'), (y_0 + 2s'))$ also

---

* It is not difficult to prove that, in this case, the whole line joining the two points lies in R.

lies in R. Since R is also convex, the midpoint of the two points

$$(x_0 + 2r, y_0 + 2s) \qquad \text{and} \qquad (-x_0 - 2r', -y_0 - 2s'),$$

that is, the point

$$\left(\frac{x_0 + 2r - (x_0 + 2r')}{2}, \frac{y_0 + 2s - (y_0 + 2s')}{2}\right) = (r - r', s - s')$$

must lie in R. The theorem is proved. □

It is not difficult to deduce the following:

Theorem 1.2. *If we change the hypothesis in Theorem* 1.1 *by replacing "greater than* 4" *with "at least* 4", *then the conclusion becomes "Then there must be a lattice point different from the origin which lies in R or on its boundary".*

Application 1. Take R to be the parallelogram

$$|\xi| \leqslant b, \qquad |\eta| \leqslant c, \tag{1}$$

where

$$\xi = \alpha x + \beta y, \qquad \eta = \gamma x + \delta y, \qquad \alpha\delta - \beta\gamma = \Delta(\neq 0),$$

and α, β, γ, δ are real numbers. The inequalities (1) define a parallelogram symmetrical about the origin, and is therefore a convex region. The area of this region is equal to

$$A = \iint\limits_{\substack{|\xi| \leqslant b \\ |\eta| \leqslant c}} dx\,dy = \iint\limits_{\substack{|\xi| \leqslant b \\ |\eta| \leqslant c}} \left|\frac{\partial(x, y)}{\partial(\xi, \eta)}\right| d\xi\,d\eta = \frac{1}{|\Delta|} \iint\limits_{\substack{|\xi| \leqslant b \\ |\eta| \leqslant c}} d\xi\,d\eta = \frac{4bc}{|\Delta|}.$$

Therefore, if $(4bc/|\Delta|) \geqslant 4$, then there is a lattice point different from $(0, 0)$ which satisfies (1). That is:

Theorem 1.3. *If* $b > 0, c > 0, bc \geqslant |\Delta|$, *then there must be a lattice point* (x, y) $(\neq (0, 0))$ *satisfying* (1). □

Taking in particular $\alpha = \delta = 1$, $\gamma = 0$. Then there is a lattice point (x, y) $(\neq (0, 0))$ such that

$$|x + \beta y| \leqslant b, \qquad |y| \leqslant \frac{1}{b},$$

that is

$$\left|\beta + \frac{x}{y}\right| \leqslant \frac{b}{|y|} \leqslant \frac{1}{y^2}.$$

This is Theorem 6.10.6.

Application 2. Take R to be the ellipse

$$\xi^2 + \eta^2 \leqslant r^2. \tag{2}$$

This clearly satisfies the hypothesis of Theorem 1.1. The area of the region in (2) is given by

$$\iint\limits_{\xi^2+\eta^2\leqslant r^2} dx\,dy = \iint\limits_{\xi^2+\eta^2\leqslant r^2} \left|\frac{\partial(x,y)}{\partial(\xi,\eta)}\right| d\xi\,d\eta = \frac{1}{|\Delta|} \iint\limits_{\xi^2+\eta^2\leqslant r^2} d\xi\,d\eta = \frac{\pi r^2}{|\Delta|}.$$

If $\pi r^2 \geqslant 4|\Delta|$, then there is a non-zero lattice point (x, y) satisfying (2). Now any ellipse centre at the origin has the equation

$$ax^2 + bxy + cy^2 = r^2. \tag{3}$$

Let

$$\xi = \sqrt{a}\,x + \frac{b}{2\sqrt{a}}y, \qquad \eta = \sqrt{c - \frac{b^2}{4a}}\,y.$$

Then (3) can be written as (2) with

$$\Delta = \sqrt{ac - \left(\frac{b}{2}\right)^2}.$$

Therefore:

Theorem 1.4. *If* $a > 0$, $ac - (\frac{b}{2})^2 > 0$, $\Delta = \sqrt{ac - (\frac{b}{2})^2}$, *then there exists a lattice point* $(x, y) \neq (0, 0)$ *such that*

$$ax^2 + bxy + cy^2 \leqslant \frac{4}{\pi}\Delta. \quad \square$$

This is not the best possible result; in fact we can replace $4/\pi$ by $2/\sqrt{3}$.

Application 3. Take R to be the region given by the hyperbola

$$|\xi\eta| \leqslant r^2. \tag{4}$$

This region is not convex so that we cannot apply Theorem 1.1 directly. Our method is to make a convex region with area at least 4 from this region. We have

$$|\xi\eta| \leqslant \left(\frac{|\xi| + |\eta|}{2}\right)^2, \tag{5}$$

and

$$|\xi| + |\eta| \leqslant 2r \tag{6}$$

is a convex region whose area is given by

$$\iint\limits_{|\xi|+|\eta|\leqslant 2r} dx\,dy = \iint\limits_{|\xi|+|\eta|\leqslant 2r} \left|\frac{\partial(x,y)}{\partial(\xi,\eta)}\right| d\xi\,d\eta = \frac{1}{|\Delta|}\iint\limits_{|\xi|+|\eta|\leqslant 2r} d\xi\,d\eta = \frac{8r^2}{|\Delta|}.$$

Therefore:

Theorem 1.5. *There must be a non-zero lattice point satisfying*

$$|\xi| + |\eta| \leqslant (2|\Delta|)^{\frac{1}{2}}. \quad \square$$

From (5) we deduce at once:

Theorem 1.6. *There must be a non-zero lattice point satisfying*

$$|\xi\eta| \leqslant \tfrac{1}{2}|\Delta|. \quad \square$$

Again this is not the best possible result; it has been proved that $\frac{1}{2}$ can be replaced by $1/\sqrt{5}$.

20.2 The Fundamental Theorem of Minkowski

Let R be a bounded region in n-dimensional space. Suppose that the mid-point of any two points belonging to R also belongs to R. Then we say that R is a *convex region*, or a *convex body*.

Theorem 2.1. *Let R be a convex body in n-dimensional space which is symmetrical about the origin and has a volume $> 2^n$. Then R must contain a non-zero lattice point.*

It is not difficult to generalize the proof of Theorem 1.1 to n-dimensional space. We now give another method to prove Theorem 2.1 in this section.

Proof (Mordell). Let t be a fixed positive integer, and let q_r run over all the integers. The various planes

$$x_r = \frac{2q_r}{t}, \qquad r = 1, 2, \ldots, n$$

divide the space into cubes, with each cube having volume $(2/t)^n$, and $(2q_1/t, \ldots, 2q_n/t)$ is a corner point. Denote by $N(t)$ the number of corners in the region R, and by A the volume of R. Then, from measure theory, we have

$$\lim_{t\to\infty} \left(\frac{2}{t}\right)^n N(t) = A.$$

If $A > 2^n$, then $N(t) > t^n$ for sufficiently large t.

On the other hand there are at most t^n sets $(q_1, \ldots, q_n)$ whose corresponding coordinates are incongruent numbers mod t, so that there must be two points

$$\left(\frac{2q_1}{t}, \ldots, \frac{2q_n}{t}\right), \quad \left(\frac{2q'_1}{t}, \ldots, \frac{2q'_n}{t}\right)$$

in R satisfying $q_i - q'_i \equiv 0 \pmod t$. Since R is symmetrical about the origin, it contains the point

$$\left(-\frac{2q'_1}{t}, \ldots, -\frac{2q'_n}{t}\right).$$

Since R is also convex, it contains the mid-point of the two points

$$\left(\frac{2q_1}{t}, \ldots, \frac{2q_n}{t}\right) \quad \text{and} \quad \left(-\frac{2q'_1}{t}, \ldots, \frac{2q'_n}{t}\right),$$

that is, the point

$$\left(\frac{q_1 - q'_1}{t}, \ldots, \frac{q_n - q'_n}{t}\right).$$

which is a lattice point. The theorem is proved. □

As before we also have:

Theorem 2.2. *In Theorem* 2.1 *we may replace* "$> 2^n$" *by* "$\geqslant 2^n$" *provided that we also replace* "*R must contain a non-zero lattice point*" *by* "*there must be a non-zero lattice which lies in R or on its boundary*". □

We can make the result sharper in the following sense.

Theorem 2.3. *Denote by Q the mid-point of the line joining the origin O to the point P on the convex body R. As P runs over the points of R, the point Q describes a convex body which we denote by $R_{\frac{1}{2}}$. Under the hypothesis of Theorem 2.2 we may strengthen the conclusion by assuming that the lattice point concerned lies outside $R_{\frac{1}{2}}$.*

Proof. Denote by δ the greatest distance between O and a boundary point of R. Take the integer N satisfying $2^{N-1} \leqslant \delta < 2^N$, so that the distance between O and any boundary point of $R_{2^{-N}}$ is less than 1. Since $R_{2^{-N}}$ has no non-zero lattice point, the lattice point in Theorem 2.2 must lie outside $R_{2^{-N}}$. Therefore there exists an integer m with the property that inside or on the boundary of $R_{2^{-m}}$, but outside $R_{2^{-m-1}}$, there is a lattice point $(x_1, \ldots, x_m)$. Now the lattice point

$$(2^m x_1, \ldots, 2^m x_n)$$

lies inside or on the boundary of R but outside $R_{\frac{1}{2}}$. □

20.3 Linear Forms

Let α_{rs} be real numbers, with the determinant

$$\Delta = \begin{vmatrix} \alpha_{11} & \cdots & \alpha_{1n} \\ \cdots & \cdots & \cdots \\ \alpha_{n1} & \cdots & \alpha_{nn} \end{vmatrix} \neq 0$$

and let

$$\xi_r = \alpha_{r1}x_1 + \cdots + \alpha_{rn}x_n, \qquad r = 1, 2, \ldots, n. \tag{1}$$

Take R to be the region

$$|\xi_1| \leqslant \lambda_1, \quad |\xi_2| \leqslant \lambda_2, \quad \ldots, \quad |\xi_n| \leqslant \lambda_n.$$

This is a convex body symmetrical about the origin, and its volume is given by

$$\int_{|\xi_1| \leqslant \lambda_1, \ldots, |\xi_n| \leqslant \lambda_n} \cdots \int dx_1 \cdot dx_2 \cdots dx_n$$

$$= \int_{|\xi_1| \leqslant \lambda_1, \ldots, |\xi_n| \leqslant \lambda_n} \cdots \int \left| \frac{\partial(x_1, x_2, \ldots, x_n)}{\partial(\xi_1, \xi_2, \ldots, \xi_n)} \right| d\xi_1 \cdot d\xi_2 \cdots d\xi_n$$

$$= \frac{1}{|\Delta|} \int_{|\xi_1| \leqslant \lambda_1, \ldots, |\xi_n| \leqslant \lambda_n} \cdots \int d\xi_1 \cdot d\xi_2 \cdots d\xi_n = \frac{2^n \lambda_1 \lambda_2 \cdots \lambda_n}{|\Delta|}.$$

Therefore if $\lambda_1 \lambda_2 \cdots \lambda_n > |\Delta|$, then R contains a non-zero lattice point, and if $\lambda_1 \lambda_2 \cdots \lambda_n \geqslant |\Delta|$, then there is a non-zero lattice point in R or on its boundary. Therefore:

Theorem 3.1. *Let $\xi_1, \ldots, \xi_n$ be n real linear forms in $x_1, \ldots, x_n$ with determinant Δ. Let $\lambda_1, \ldots, \lambda_n$ be positive numbers satisfying $\lambda_1 \lambda_2 \cdots \lambda_n \geqslant |\Delta|$. Then there exist integers $x_1, x_2, \ldots, x_n$, not all zero, such that*

$$|\xi_1| \leqslant \lambda_1, \quad |\xi_2| \leqslant \lambda_2, \quad \ldots, \quad |\xi_n| \leqslant \lambda_n. \quad \square$$

Theorem 3.2. *The conclusion of Theorem 3.1 can be strengthened to the following: there exist integers $x_1, x_2, \ldots, x_n$, not all zero, such that*

$$|\xi_1| \leqslant \lambda_1, \quad |\xi_2| < \lambda_2, \quad \ldots, \quad |\xi_n| < \lambda_n.$$

Proof. Let $\varepsilon > 0$. By Theorem 3.1 there are integers $x_1, \ldots, x_n$, not all zero such that

$$|\xi_1| \leqslant (1 + \varepsilon)^{n-1}\lambda_1, \quad |\xi_2| \leqslant \frac{\lambda_2}{1 + \varepsilon} < \lambda_2, \quad \ldots, \quad |\xi_n| \leqslant \frac{\lambda_n}{1 + \varepsilon} < \lambda_n.$$

Now let $\varepsilon \to 0$. From the discrete nature of integral points the theorem is proved. $\square$

If we replace n by $n+1$, and take

$$\xi_\nu = x_\nu \quad (1 \leqslant \nu \leqslant n), \qquad \xi_{n+1} = \alpha_1 x_1 + \alpha_2 x_2 + \cdots + \alpha_n x_n + x_{n+1},$$

$$\lambda_\nu = t^{1/n} \quad (1 \leqslant \nu \leqslant n), \qquad \lambda_{n+1} = \frac{1}{t},$$

then, from Theorem 3.2, we have:

Theorem 3.3. *There are always integers $x_1, \ldots, x_n$ and y, not all 0, such that*

$$|\alpha_1 x_1 + \cdots + \alpha_n x_n + y| < \frac{1}{t},$$

and $|x_\nu| \leqslant t^{1/n}$, where t is any positive number. □

Again if we take

$$\xi_1 = x_{n+1}, \qquad \xi_{\nu+1} = x_\nu - \alpha_\nu x_{n+1} \qquad (1 \leqslant \nu \leqslant n),$$

$$\lambda_1 = t^n, \qquad \lambda_{\nu+1} = \frac{1}{t} \qquad (1 \leqslant \nu \leqslant n),$$

then we have:

Theorem 3.4. *Let $\alpha_1, \ldots, \alpha_n$ be real numbers and $t \geqslant 1$. Then there exists a non-zero lattice point $(x, y_1, y_2, \ldots, y_n)$ such that*

$$|\alpha_\nu x - y_\nu| < \frac{1}{t}, \qquad 1 \leqslant x \leqslant t^n.$$

In other words there are n rational numbers $y_1/x, \ldots, y_n/x$, with common denominator x, such that

$$\left|\alpha_\nu - \frac{y_\nu}{x}\right| < \frac{1}{x^{1+1/n}}, \qquad 1 \leqslant \nu \leqslant n. \quad \square$$

Let c_n be the greatest real number with the following property: If $0 < c < c_n$, then

$$\left|\alpha_\nu - \frac{y_\nu}{x}\right| < \frac{1}{cx^{1+1/n}}, \qquad 1 \leqslant \nu \leqslant n$$

has infinitely many sets of solutions. From Theorem 10.4.4 we know that $c_1 = \sqrt{5}$. The determination of c_n for $n \geqslant 2$ is an unsolved problem.

In Theorem 3.2 we cannot replace $|\xi_1| \leqslant \lambda_1$ by $|\xi_1| < \lambda_1$ as well. For example:

$$\xi_1 = x_1, \quad \xi_2 = \alpha_{21}x_1 + x_2, \quad \xi_3 = \alpha_{31}x_1 + \alpha_{32}x_2 + x_3, \quad \ldots,$$
$$\xi_n = \alpha_{n1}x_1 + \alpha_{n2}x_2 + \cdots + \alpha_{nn-1}x_{n-1} + x_n. \tag{2}$$

Then from $|\xi_1| < 1$ we have $x_1 = 0$, and from $|\xi_2| < 1$ we have $x_2 = 0$, and so on, giving only the origin satisfying $|\xi_1| < 1, |\xi_2| < 1, \ldots, |\xi_n| < 1$.

Finally we let

$$x_\nu = \sum_{\mu=1}^{n} a_{\nu\mu} y_\mu \qquad (1 \leqslant \nu \leqslant n)$$

be a unimodular transformation. Substituting this into (2) will give a set of equations with properties similar to those in (2). *Question*: Apart from the situation in the given example, can we have $|\xi_1| < \lambda_1$? This is the famous Minkowski's problem, and for decades the situation was settled only for $n \leqslant 7$, until in 1942 the Hungarian mathematician Hajös settled it generally.

20.4 Positive Definite Quadratic Forms

Consider the ellipsoid R:

$$\xi_1^2 + \cdots + \xi_n^2 \leqslant r^2. \tag{1}$$

In order to prove that this is a convex body it suffices to prove that

$$\left(\frac{\xi_1 + \xi_1'}{2}\right)^2 + \cdots + \left(\frac{\xi_n + \xi_n'}{2}\right)^2 \leqslant \tfrac{1}{2}\{(\xi_1^2 + \cdots + \xi_n^2) + (\xi_1'^2 + \cdots + \xi_n'^2)\}. \tag{2}$$

Since

$$\left(\frac{\xi_i + \xi_i'}{2}\right)^2 \leqslant \frac{\xi_i^2 + \xi_i'^2}{2}, \qquad i = 1, 2, \ldots, n,$$

the inequality (2) clearly follows.

Since the volume of the n-dimensional sphere with radius r is $\pi^{n/2} r^n / \Gamma(\frac{n}{2} + 1)$, it follows that the volume of R is given by

$$\int\cdots\int_{\xi_1^2+\cdots+\xi_n^2 \leqslant r^2} dx_1 \cdots dx_n = \int\cdots\int_{\xi_1^2+\cdots+\xi_n^2 \leqslant r^2} \left|\frac{\partial(x_1, \ldots, x_n)}{\partial(\xi_1, \ldots, \xi_n)}\right| d\xi_1 \cdots d\xi_n$$

$$= \frac{1}{|\Delta|} \int\cdots\int_{\xi_1^2+\cdots+\xi_n^2 \leqslant r^2} d\xi_1 \cdots d\xi_n = \frac{1}{|\Delta|} r^n \frac{\pi^{\frac{1}{2}n}}{\Gamma(\frac{1}{2}n + 1)}.$$

Therefore we have:

Theorem 4.1. *There exist integers $x_1, \ldots, x_n$, not all 0, such that*

$$\xi_1^2 + \cdots + \xi_n^2 \leqslant 4\left(\frac{|\Delta|}{J_n}\right)^{2/n},$$

where

$$J_n = \frac{\pi^{n/2}}{\Gamma\left(\frac{n}{2}+1\right)}. \quad \square$$

We can rewrite Theorem 4.1 differently. A positive definite quadratic form

$$Q(x_1, \ldots, x_n) = \sum_{r=1}^{n} \sum_{s=1}^{n} a_{rs} x_r x_s, \qquad a_{rs} = a_{sr}$$

can be represented by

$$Q = \xi_1^2 + \cdots + \xi_n^2.$$

The determinant Δ of $\xi_1, \ldots, \xi_n$ is equal to the square root of $D = |a_{rs}|$. This is because $A = (a_{rs})$ is a positive definite matrix so that there exists a matrix B such that $A = BB'$, $\Delta = |B| = D^{\frac{1}{2}}$. Therefore Theorem 4.1 can be stated as follows:

Theorem 4.2. *Let $Q(x_1, \ldots, x_n)$ be a positive definite quadratic form with determinant D. Then there exists a non-zero point $(x_1, \ldots, x_n)$ such that*

$$Q(x_1, \ldots, x_n) \leqslant 4J_n^{-2/n} D^{1/n}. \quad \square \tag{3}$$

Let γ_n be the least constant with the following property: There exists a non-zero lattice point such that

$$Q(x_1, \ldots, x_n) \leqslant \gamma_n D^{1/n}.$$

In §1 we already remarked that $\gamma_2 = 2/\sqrt{3}$. Up to the present mathematicians have only determined the values of γ_n for $2 \leqslant n \leqslant 10$:

$$\gamma_3 = \sqrt[8]{2}, \quad \gamma_4 = \sqrt{2}, \quad \gamma_5 = \sqrt[5]{8}, \quad \gamma_6 = \sqrt[6]{\tfrac{64}{3}}, \quad \gamma_7 = \sqrt[7]{64},$$

$$\gamma_8 = 2, \quad \gamma_9 = 2, \quad \gamma_{10} = 2\sqrt[10]{\tfrac{4}{3}}.$$

In general, we know that

$$\gamma_n < \frac{2}{\pi}\left(\Gamma\left(2 + \frac{n}{2}\right)\right)^{2/n} \qquad \left(\sim \frac{n}{\pi e} \quad \text{as} \quad n \to \infty\right).$$

20.5 Products of Linear Forms

We first consider the region R:

$$|\xi_1| + |\xi_2| + \cdots + |\xi_n| \leqslant r. \tag{1}$$

This region is clearly symmetrical about the origin, and from

$$\left|\frac{\xi + \xi'}{2}\right| \leqslant \frac{|\xi| + |\xi'|}{2}$$

we see that R is a convex body whose volume is given by

$$\int \cdots \int_{|\xi_1| + \cdots + |\xi_n| \leqslant r} dx_1 \cdots dx_n = \int \cdots \int_{|\xi_1| + \cdots + |\xi_n| \leqslant r} \left|\frac{\partial(x_1, \ldots, x_n)}{\partial(\xi_1, \ldots, \xi_n)}\right| d\xi_1 \cdots d\xi_n$$

$$= \frac{1}{|\Delta|} \int \cdots \int_{|\xi_1| + \cdots + |\xi_n| \leqslant r} d\xi_1 \cdots d\xi_n$$

$$= \frac{2^n}{|\Delta|} \int \cdots \int_{\substack{\xi_1 + \cdots + \xi_n \leqslant r \\ \xi_i \geqslant 0}} d\xi_1 \cdots d\xi_n = \frac{2^n r^n}{n!|\Delta|}.$$

Therefore:

Theorem 5.1. *There exists a non-zero lattice point* $(x_1, \ldots, x_n)$ *such that*

$$|\xi_1| + \cdots + |\xi_n| \leqslant (n!|\Delta|)^{1/n}. \quad \square \tag{2}$$

When $n = 2$, this is the best possible result. For if we take $\xi_1 = x + y$, $\xi_2 = x - y$, then $|\Delta| = 2$ and (2) becomes $|\xi_1| + |\xi_2| \leqslant 2$. But

$$|\xi_1| + |\xi_2| = \max(|\xi_1 + \xi_2|, |\xi_1 - \xi_2|) = 2\max(|x|, |y|),$$

so that if this is less than 2, then $x = y = 0$.

When $n = 3$, Minkowski proved that there is a non-zero lattice point (x_1, x_2, x_3) such that

$$|\xi_1| + |\xi_2| + |\xi_3| \leqslant \left(\frac{108}{19}|\Delta|\right)^{1/3},$$

and that 108/19 is best possible. The problem for $n > 3$ is unsolved.

We now discuss the product of linear forms. We shall use the following result, known as the arithmetic-geometric means inequality.

Theorem 5.2. *If* $a_1 \geqslant 0, \ldots, a_n \geqslant 0$, *then*

$$(a_1 \cdots a_n)^{1/n} \leqslant \frac{a_1 + \cdots + a_n}{n}.$$

Proof. 1) $n = 2^k$. We use induction on k. Since

$$\sqrt{a_1 a_2} \leqslant \frac{a_1 + a_2}{2}$$

we see that the result holds when $k = 1$. Assume now that the result holds when $n = 2^{k-1}$. Then when $n = 2^k$ we have

$$(a_1 \cdots a_{2^k})^{\frac{1}{2^k}} = \{(a_1 \cdots a_{2^{k-1}})^{\frac{1}{2^{k-1}}}(a_{2^{k-1}+1} \cdots a_{2^k})^{\frac{1}{2^{k-1}}}\}^{\frac{1}{2}}$$

$$\leqslant \left\{\left(\frac{a_1 + \cdots + a_{2^{k-1}}}{2^{k-1}}\right)\left(\frac{a_{2^{k-1}+1} + \cdots + a_{2^k}}{2^{k-1}}\right)\right\}^{\frac{1}{2}}$$

$$\leqslant \frac{a_1 + \cdots + a_{2^k}}{2^k}.$$

2) (Backward induction.) We now show that if the result holds for $n + 1$, then it holds for n. Take

$$a_{n+1} = \frac{1}{n}(a_1 + \cdots + a_n).$$

Then, from our induction hypothesis, we have

$$\left(\frac{1}{n}a_1 \cdots a_n(a_1 + \cdots + a_n)\right)^{\frac{1}{n+1}} = (a_1 \cdots a_{n+1})^{\frac{1}{n+1}} \leqslant \frac{a_1 + \cdots + a_{n+1}}{n+1}$$

$$= \frac{1}{n+1}\left\{a_1 + \cdots + a_n + \frac{1}{n}(a_1 + \cdots + a_n)\right\}$$

$$= \frac{a_1 + \cdots + a_n}{n},$$

which gives

$$(a_1 \cdots a_n)^{\frac{1}{n+1}} \leqslant \left(\frac{a_1 + \cdots + a_n}{n}\right)^{1-\frac{1}{n+1}} = \left(\frac{a_1 + \cdots + a_n}{n}\right)^{\frac{n}{n+1}}.$$

The theorem is proved. □

From Theorem 5.1 and Theorem 5.2 we have at once:

Theorem 5.3. *There exists a non-zero lattice point such that*

$$|\xi_1 \cdots \xi_n| \leqslant \frac{n!}{n^n}|\Delta|. \quad \square$$

Note. We can also deduce from Theorem 3.1 that there is a non-zero lattice point such that

$$|\xi_1 \cdots \xi_n| \leqslant |\Delta|.$$

Since $n! < n^n$ whenever $n > 1$, our Theorem 5.3 here gives a better result. Denote by γ_n the least positive constant such that, whenever $\gamma \geqslant \gamma_n$, there is a non-zero lattice point satisfying

$$|\xi_1 \cdots \xi_n| \leqslant \gamma|\Delta|.$$

Up to the present we only know that $\gamma_2 = 1/\sqrt{5}$ and $\gamma_3 = \frac{1}{7}$ (Davenport).

20.6 Method of Simultaneous Approximations

Theorem 6.1. *Let $\alpha_1, \ldots, \alpha_n$ be real numbers. Then there exist a non-zero lattice point $(x_1, \ldots, x_n)$ and an integer $y \geqslant 1$ such that*

$$\left|\alpha_i - \frac{x_i}{y}\right| \leqslant \frac{n}{(n+1)y^{1+\frac{1}{n}}}, \qquad i = 1, 2, \ldots, n.$$

Proof. We first consider

$$|x_i - \alpha_i y| + \left|\frac{y}{t}\right| \leqslant r, \qquad 1 \leqslant i \leqslant n, \qquad t \neq 0.$$

This is a convex body symmetrical about the origin, and its volume is given by

$$\underset{\substack{|\xi_i|+|\xi_{n+1}|\leqslant r\\ i=1,\ldots,n}}{\int\cdots\int} dx_1 \cdots dx_n\, dy \qquad \begin{pmatrix}\text{here } \xi_i = x_i - \alpha_i y,\ 1 \leqslant i \leqslant n,\\ \xi_{n+1} = y/t\end{pmatrix}$$

$$= \underset{\substack{|\xi_i|+|\xi_{n+1}|\leqslant r\\ i=1,\ldots,n}}{\int\cdots\int} \left|\frac{\partial(x_1, \ldots, x_n, y)}{\partial(\xi_1, \ldots, \xi_n, \xi_{n+1})}\right| d\xi_1 \cdots d\xi_n\, d\xi_{n+1}$$

$$= |t| \underset{\substack{|\xi_i|+|\xi_{n+1}|\leqslant r\\ i=1,\ldots,n}}{\int\cdots\int} d\xi_1 \cdots d\xi_n\, d\xi_{n+1}$$

$$= 2^{n+1}|t| \underset{\substack{\xi_i+\xi_{n+1}\leqslant r\\ i=1,\ldots,n\\ \xi_i\geqslant 0,\ \xi_{n+1}\geqslant 0}}{\int\cdots\int} d\xi_1 \cdots d\xi_n\, d\xi_{n+1}$$

$$= \frac{2^{n+1}|t|}{n+1} r^{n+1}.$$

Therefore there is a non-zero lattice point $(x_1, \ldots, x_n, y)$ such that

$$|x_i - \alpha_i y| + \left|\frac{y}{t}\right| \leqslant \left(\frac{n+1}{|t|}\right)^{\frac{1}{n+1}}.$$

From Theorem 5.2 we have

$$\left|(x_i - \alpha_i y)^n \left(\frac{ny}{t}\right)\right|^{\frac{1}{n+1}} \leqslant \frac{n|x_i - \alpha_i y| + n\left|\frac{y}{t}\right|}{n+1}$$

$$\leqslant \frac{n}{n+1}\left(\frac{n+1}{|t|}\right)^{\frac{1}{n+1}}, \qquad i = 1, \ldots, n.$$

Hence

$$\left|\alpha_i - \frac{x_i}{y}\right| \leqslant \frac{n}{(n+1)y^{1+\frac{1}{n}}}, \qquad i = 1, 2, \ldots, n. \quad \square$$

This theorem is a slight improvement on Theorem 3.4. The best results at the present are:

$$c_3 \geqslant \sqrt{\frac{19}{8}}, \qquad \text{(Minkowski)},$$

$$c_n \geqslant \frac{n+1}{n}\left\{1 + \left(\frac{n-1}{n+1}\right)^{n+3}\right\}^{1/n} \qquad \text{(Blichfeldt)}.$$

Exercise. Let $\alpha_v = \beta_v + i\gamma_v$ ($v = 1, \ldots, n$) be complex numbers. Then there are complex integers $z_1, \ldots, z_n, w$ such that

$$\left|\alpha_v - \frac{z_v}{w}\right| \leqslant \frac{n}{n+1} \cdot \frac{2}{\sqrt{\pi}}\left(\frac{2n+1}{n+1} \cdot \frac{4}{\pi}\right)^{\frac{1}{2n}} \frac{1}{|w|^{1+\frac{1}{n}}}.$$

20.7 Minkowski's Inequality

For $a_i \geqslant 0$ ($i = 1, \ldots, n$), $r > 0$ we define

$$M_r(a) = \left\{\frac{1}{n}(a_1^r + \cdots + a_n^r)\right\}^{1/r}. \tag{1}$$

When $r < 0$, and some $a_i = 0$, then the equation (1) has no meaning. In this case, we define $(a_1^r + \cdots + a_n^r)^{1/r} = 0$. Therefore, when $a_i \geqslant 0$, $r \neq 0$ we can always define

$$M_r(a) = \left\{\frac{1}{n}(a_1^r + \cdots + a_n^r)\right\}^{1/r}.$$

From now on we denote $a_i \geqslant 0$ ($i = 1, \ldots, n$) by (a). We write $(a) > 0$ to mean $a_i > 0$ ($i = 1, \ldots, n$), and $(a) \neq 0$ to mean that not all the a_i are zero. We also denote by $\max a$ and $\min a$ the largest and the smallest numbers in a_i respectively.

If there are non-zero real numbers λ, μ such that $\lambda a_i = \mu b_i$ ($i = 1, \ldots, n$), then we say that (a) and (b) are proportional.

Theorem 7.1. $\lim_{r\to\infty} M_r(a) = \max a$.

Proof. We can suppose that $r > 0$, so that

$$\left\{\frac{1}{n}(\max a)^r\right\}^{1/r} \leqslant M_r(a) \leqslant \left\{(\max a)^r\right\}^{1/r},$$

or

$$\left(\frac{1}{n}\right)^{1/r} \max a \leqslant M_r(a) \leqslant \max a.$$

Since

$$\lim_{r \to +\infty} \left(\frac{1}{n}\right)^{1/r} = \left(\frac{1}{n}\right)^0 = 1,$$

we have $\lim_{r \to +\infty} M_r(a) = \max a$. □

Theorem 7.2. $\lim_{r \to -\infty} M_r(a) = \min a$.

Proof. We can suppose that $r < 0$. We first consider the case $(a) > 0$. We have

$$M_r(a) = \left\{\frac{1}{n}(a_1^r + \cdots + a_n^r)\right\}^{1/r} = \frac{1}{\left\{\frac{1}{n}\left[\left(\frac{1}{a_1}\right)^{-r} + \cdots + \left(\frac{1}{a_n}\right)^{-r}\right]\right\}^{1/-r}}$$
$$= \frac{1}{M_{-r}\left(\frac{1}{a}\right)},$$

so that by Theorem 7.1,

$$\lim_{r \to -\infty} M_r(a) = \frac{1}{\lim\limits_{-r \to +\infty} M_{-r}\left(\frac{1}{a}\right)} = \frac{1}{\max \frac{1}{a}} = \min a.$$

Finally when one $a_i = 0$, and $r < 0$, we see that both $M_r(a)$ and $\min a$ are zero. The theorem is proved. □

We write

$$G(a) = (a_1 \cdots a_n)^{1/n},$$

the geometric mean of a_i.

Theorem 7.3. $\lim_{r \to 0} M_r(a) = G(a)$.

Proof. 1) $r < 0$, and some $a_i = 0$. This case is trivial.

2) $r \neq 0$, $(a) > 0$. From (1) we have

$$M_r(a) = \left\{\frac{1}{n} a_1^r + \cdots + a_n^r\right\}^{1/r}$$
$$= e^{\frac{1}{r}\log\left\{\frac{1}{n}(a_1^r + \cdots + a_n^r)\right\}}.$$

We now let $r \to 0$ and apply L'Hospital's rule, giving

$$\lim_{r \to 0} \frac{1}{r} \log\left\{\frac{1}{n}(a_1^r + \cdots + a_n^r)\right\} = \lim_{r \to 0} \frac{\frac{1}{n}\sum\limits_{i=1}^n a_i^r \log a_i}{\frac{1}{n}(a_1^r + \cdots + a_n^r)} = \frac{1}{n}\sum_{i=1}^n \log a_i.$$

Therefore

$$\lim_{r\to 0} M_r(a) = \lim_{r\to 0} e^{\frac{1}{r}\log\left\{\frac{1}{n}(a_1^r + \cdots + a_n^r)\right\}}$$

$$= e^{\frac{1}{n}\sum_{i=1}^{n}\log a_i} = e^{\log(a_1\cdots a_n)^{1/n}} = (a_1 \cdots a_n)^{1/n} = G(a).$$

3) $r > 0$, and some $a_i = 0$. We can assume that

$a_1 > 0, \ldots, a_s > 0$, $a_{s+1} = a_{s+2} = \cdots = a_n = 0$, $s < n$. Then we have

$$M_r(a) = \left\{\frac{1}{n}(a_1^r + \cdots + a_s^r)\right\}^{1/r} = \left\{\frac{s}{n}\cdot\frac{1}{s}(a_1^r + \cdots + a_s^r)\right\}^{1/r}$$

$$= \left(\frac{s}{n}\right)^{1/r}\left\{\frac{1}{s}(a_1^r + \cdots + a_s^r)\right\}^{1/r}.$$

From our earlier result we have

$$\lim_{r\to 0}\left\{\frac{1}{s}(a_1^r + \cdots + a_s^r)\right\}^{1/r} = (a_1 \cdots a_s)^{1/s},$$

and, since $s < n$,

$$\lim_{r\to 0}\left(\frac{s}{n}\right)^{1/r} = 0.$$

Therefore

$$\lim_{r\to 0} M_r(a) = \lim_{r\to 0}\left\{\left(\frac{s}{n}\right)^{1/r}\left\{\frac{1}{s}(a_1^r + \cdots + a_s^r)\right\}^{1/r}\right\}$$

$$= 0 \cdot (a_1 \cdots a_s)^{1/s} = 0 = (a_1 \cdots a_n)^{1/n}. \quad \square$$

Lemma 1. *Let $\alpha + \beta = 1$, $\alpha > 0, \beta > 0$, Then for $s \geqslant 0$, $t \geqslant 0$, we have*

$$s^\alpha t^\beta \leqslant s\alpha + t\beta$$

with equality only when $s = t$.

Proof. The lemma is trivial if $s = t$ or if one of s, t is 0. We assume therefore that s, t are distinct positive numbers.

If $s > t$, then $s/t > 1$. Also, $0 < \alpha < 1$, $1 - \alpha = \beta$, so that

$$\left(\frac{s}{t}\right)^\alpha - 1 = \alpha\int_1^{s/t} y^{\alpha-1}\,dy \leqslant \alpha\int_1^{s/t} dy = \alpha\left(\frac{s}{t} - 1\right).$$

From

$$\left(\frac{s}{t}\right)^\alpha - 1 \leqslant \alpha\left(\frac{s}{t} - 1\right),$$

we have

$$s^\alpha t^\beta \leqslant s\alpha + t\beta.$$

Finally if $s^\alpha t^\beta = s\alpha + t\beta$, then

$$\alpha \int_1^{s/t} y^{\alpha-1}\, dy = \alpha \int_1^{s/t} dy,$$

or

$$\int_1^{s/t} (y^{\alpha-1} - 1)\, dy = 0,$$

which is impossible unless $s = t$. □

Lemma 2 (Hölder's inequality). *Let $\alpha + \beta = 1$, $\alpha > 0$, $\beta > 0$. When (a) and (b) are not proportional we have*

$$\sum_{i=1}^{n} a_i^\alpha b_i^\beta < \left(\sum_{i=1}^{n} a_i\right)^\alpha \left(\sum_{i=1}^{n} b_i\right)^\beta.$$

Proof. Since (a) and (b) are not proportional, there exists i $(1 \leqslant i \leqslant n)$ such that

$$\frac{a_i}{\sum_{j=1}^{n} a_j} \neq \frac{b_i}{\sum_{j=1}^{n} b_j}.$$

Therefore, by Lemma 1,

$$\frac{\sum_{i=1}^{n} a_i^\alpha b_i^\beta}{\left(\sum_{j=1}^{n} a_j\right)^\alpha \left(\sum_{j=1}^{n} b_j\right)^\beta} = \sum_{i=1}^{n} \left(\frac{a_i}{\sum_{j=1}^{n} a_j}\right)^\alpha \left(\frac{b_i}{\sum_{j=1}^{n} b_j}\right)^\beta$$

$$< \sum_{i=1}^{n} \left\{\left(\frac{a_i}{\sum_{j=1}^{n} a_j}\right)\alpha + \left(\frac{b_i}{\sum_{j=1}^{n} b_j}\right)\beta\right\} = \alpha + \beta = 1,$$

or

$$\sum_{i=1}^{n} a_i^\alpha b_i^\beta < \left(\sum_{i=1}^{n} a_i\right)^\alpha \left(\sum_{i=1}^{n} b_i\right)^\beta. \quad \square$$

Lemma 3 (Hölder's inequality). *Let $k > 0$, $k \neq 1$, $(1/k) + (1/k') = 1$. Suppose that (a^k) and $(b^{k'})$ are not proportional, and that $(ab) \neq 0$. Then*

$$\sum_{i=1}^{n} a_i b_i < \left(\sum_{i=1}^{n} a_i^k\right)^{1/k} \left(\sum_{i=1}^{n} b_i^{k'}\right)^{1/k'} \qquad (k > 1), \tag{2}$$

$$\sum_{i=1}^{n} a_i b_i > \left(\sum_{i=1}^{n} a_i^k\right)^{1/k} \left(\sum_{i=1}^{n} b_i^{k'}\right)^{1/k'} \qquad (k < 1). \tag{3}$$

Proof. 1) $k > 1$. Here $k' = k/(k-1) > 1$, $0 < 1/k < 1$, $0 < 1/k' < 1$, $1/k + 1/k' = 1$. By Lemma 2 we have

$$\sum_{i=1}^{n} a_i b_i = \sum_{i=1}^{n} (a_i^k)^{1/k}(b_i^{k'})^{1/k'} < \left(\sum_{i=1}^{n} a_i^k\right)^{1/k} \left(\sum_{i=1}^{n} b_i^{k'}\right)^{1/k'}.$$

2) $0 < k < 1$. Here $k' = k/(k-1) < 0$. If some $b_i = 0$, then by the definition in the beginning of the section we have

$$\left(\sum_{i=1}^{n} b_i^{k'}\right)^{1/k'} = 0.$$

Therefore

$$\sum_{i=1}^{n} a_i b_i > 0 = \left(\sum_{i=1}^{n} a_i^k\right)^{1/k} \left(\sum_{i=1}^{n} b_i^{k'}\right)^{1/k'}.$$

When $(b) > 0$, from $0 < k < 1$, we have

$$0 < \frac{1}{\left(\frac{1}{k}\right)} < 1, \qquad 0 < \frac{1}{\left(-\frac{k'}{k}\right)} = 1 - k < 1, \qquad \frac{1}{\left(\frac{1}{k}\right)} + \frac{1}{\left(-\frac{k'}{k}\right)} = 1.$$

From Lemma 2 we have

$$\begin{aligned}
\sum_{i=1}^{n} a_i^k &= \sum_{i=1}^{n} (a_i b_i)^k b_i^{-k} = \sum_{i=1}^{n} (a_i b_i)^{1/\left(\frac{1}{k}\right)} (b_i^{k'})^{1/\left(-\frac{k'}{k}\right)} \\
&< \left(\sum_{i=1}^{n} a_i b_i\right)^{1/\left(\frac{1}{k}\right)} \left(\sum_{i=1}^{n} b_i^{k'}\right)^{1/\left(-\frac{k'}{k}\right)} \\
&= \left(\sum_{i=1}^{n} a_i b_i\right)^{k} \left(\sum_{i=1}^{n} b_i^{k'}\right)^{-\frac{k}{k'}}.
\end{aligned}$$

From

$$\sum_{i=1}^{n} a_i^k < \left(\sum_{i=1}^{n} a_i b_i\right)^{k} \left(\sum_{i=1}^{n} b_i^{k'}\right)^{-\frac{k}{k'}},$$

we have

$$\sum_{i=1}^{n} a_i b_i > \left(\sum_{i=1}^{n} a_i^k\right)^{\frac{1}{k}} \left(\sum_{i=1}^{n} b_i^{k'}\right)^{\frac{1}{k'}} \qquad (k < 1). \quad \square$$

Theorem 7.4. *Let* $0 < r < s$. *Then*

$$M_r(a) < M_s(a),$$

unless $a_1 = a_2 = \cdots = a_n$.

Proof. Let $r = s\alpha$, $0 < \alpha < 1$. Then

$$M_r(a) = \left\{\frac{1}{n}(a_1^r + \cdots + a_n^r)\right\}^{1/r} = \left\{\frac{1}{n}(a_1^{s\alpha} + \cdots + a_n^{s\alpha})\right\}^{1/s\alpha}$$

$$= \left(\frac{1}{n}\left\{\sum_{i=1}^{n}(a_i^s)^\alpha \cdot 1\right\}\right)^{1/s\alpha}.$$

By Lemma 2 we have

$$M_r(a) = \left(\frac{1}{n}\left\{\sum_{i=1}^{n}(a_i^s)^\alpha \cdot 1^{1-\alpha}\right\}\right)^{1/s\alpha} < \left\{\frac{1}{n}\left(\sum_{i=1}^{n} a_i^s\right)^\alpha\left(\sum_{i=1}^{n} 1\right)^{1-\alpha}\right\}^{1/s\alpha}$$

$$= \left\{\frac{1}{n}\left(\sum_{i=1}^{n} a_i^s\right)^\alpha n^{1-\alpha}\right\}^{1/s\alpha}$$

$$= \left(\frac{(a_1^s + \cdots + a_n^s)^\alpha}{n^\alpha}\right)^{1/s\alpha}$$

$$= \left(\frac{a_1^s + \cdots + a_n^s}{n}\right)^{1/s}$$

$$= M_s(a). \quad \square$$

Theorem 7.5. *Let $r > 0, r \neq 1$. If (a) and (b) are not proportional, then*

$$\left\{\sum_{i=1}^{n}(a_i + b_i)^r\right\}^{1/r} < \left(\sum_{i=1}^{n} a_i^r\right)^{1/r} + \left(\sum_{i=1}^{n} b_i^r\right)^{1/r} \qquad (r > 1)$$

and

$$\left\{\sum_{i=1}^{n}(a_i + b_i)^r\right\}^{1/r} > \left(\sum_{i=1}^{n} a_i^r\right)^{1/r} + \left(\sum_{i=1}^{n} b_i^r\right)^{1/r} \qquad (r < 1).$$

Proof. 1) $r > 1$. Let $r' = r/(r-1)$. Then $r' > 1$ and $1/r + 1/r' = 1$. By (2) in Lemma 3, we have

$$\sum_{i=1}^{n}(a_i + b_i)^r = \sum_{i=1}^{n} a_i(a_i + b_i)^{r-1} + \sum_{i=1}^{n} b_i(a_i + b_i)^{r-1}$$

$$< \left(\sum_{i=1}^{n} a_i^r\right)^{1/r}\left\{\sum_{i=1}^{n}((a_i + b_i)^{r-1})^{r'}\right\}^{1/r'}$$

$$+ \left(\sum_{i=1}^{n} b_i^r\right)^{1/r}\left\{\sum_{i=1}^{n}((a_i + b_i)^{r-1})^{r'}\right\}^{1/r'}$$

$$= \left(\sum_{i=1}^{n} a_i^r\right)^{1/r}\left\{\sum_{i=1}^{n}(a_i + b_i)^r\right\}^{\frac{r-1}{r}}$$

$$+ \left(\sum_{i=1}^{n} b_i^r\right)^{1/r}\left\{\sum_{i=1}^{n}(a_i + b_i)^r\right\}^{\frac{r-1}{r}}$$

$$= \left\{ \left(\sum_{i=1}^{n} a_i^r \right)^{1/r} + \left(\sum_{i=1}^{n} b_i^r \right)^{1/r} \right\} \left\{ \sum_{i=1}^{n} (a_i + b_i)^r \right\}^{\frac{r-1}{r}}.$$

Multiplying through by

$$\left\{ \sum_{i=1}^{n} (a_i + b_i)^r \right\}^{-\frac{r-1}{r}},$$

we have

$$\left\{ \sum_{i=1}^{n} (a_i + b_i)^r \right\}^{1/r} < \left(\sum_{i=1}^{n} a_i^r \right)^{1/r} + \left(\sum_{i=1}^{n} b_i^r \right)^{1/r}.$$

2) $0 < r < 1$. If $a_i + b_i = 0$ for $i = 1, \ldots, n$, then we deduce, from $a_i \geqslant 0, b_i \geqslant 0$, that $a_i = b_i = 0$ for $i = 1, \ldots, n$. This means that $(a) = (b) = 0$ which implies that (a) and (b) are proportional. Therefore, under the hypothesis of the theorem, we can assume that there is some i, $1 \leqslant i \leqslant n$, such that $a_i + b_i > 0$. In fact it is clear that we can assume without loss that $a_i + b_i > 0$ for $i = 1, \ldots, n$.

Now let $r' = r/(r - 1)$. Then, from (3) in Lemma 3, we have

$$\sum_{i=1}^{n} (a_i + b_i)^r = \sum_{i=1}^{n} a_i(a_i + b_i)^{r-1} + \sum_{i=1}^{n} b_i(a_i + b_i)^{r-1}$$

$$> \left(\sum_{i=1}^{n} a_i^r \right)^{1/r} \left\{ \sum_{i=1}^{n} ((a_i + b_i)^{r-1})^{r'} \right\}^{1/r'}$$

$$+ \left(\sum_{i=1}^{n} b_i^r \right)^{1/r} \left\{ \sum_{i=1}^{n} ((a_i + b_i)^{r-1})^{r'} \right\}^{1/r'}$$

$$= \left(\sum_{i=1}^{n} a_i^r \right)^{1/r} \left\{ \sum_{i=1}^{n} (a_i + b_i)^r \right\}^{\frac{r-1}{r}} + \left(\sum_{i=1}^{n} b_i^r \right)^{1/r} \left\{ \sum_{i=1}^{n} (a_i + b_i)^r \right\}^{\frac{r-1}{r}}$$

$$= \left\{ \left(\sum_{i=1}^{n} a_i^r \right)^{1/r} + \left(\sum_{i=1}^{n} b_i^r \right)^{1/r} \right\} \left\{ \sum_{i=1}^{n} (a_i + b_i)^r \right\}^{\frac{r-1}{r}}.$$

Multiplying through by

$$\left\{ \sum_{i=1}^{n} (a_i + b_i)^r \right\}^{-\frac{r-1}{r}},$$

we have

$$\left\{ \sum_{i=1}^{n} (a_i + b_i)^r \right\}^{1/r} > \left(\sum_{i=1}^{n} a_i^r \right)^{1/r} + \left(\sum_{i=1}^{n} b_i^r \right)^{1/r}. \quad \square$$

This theorem is commonly called Minkowski's inequality.

20.8 The Average Value of the Product of Linear Forms

Theorem 8.1. *Let $n \geqslant 2$, and $\xi_1, \ldots, \xi_n$ be linear forms in $x_1, \ldots, x_n$ with determinant $\Delta \neq 0$. Suppose that there are s pairs of forms with complex conjugate coefficients and r forms with real coefficients, where $r + 2s = n$. If $\sigma \geqslant 1$, then there is a non-zero lattice point such that*

$$\left(\frac{|\xi_1|^\sigma + \cdots + |\xi_n|^\sigma}{n}\right)^{1/\sigma} \leqslant \left(\frac{\left(\frac{2}{\pi}\right)^s n^{-\frac{n}{\sigma}} \Gamma\left(1 + \frac{n}{\sigma}\right)|\Delta|}{2^{-\frac{2s}{\sigma}} \Gamma^r\left(1 + \frac{1}{\sigma}\right)\Gamma^s\left(1 + \frac{2}{\sigma}\right)}\right)^{1/n}.$$

Proof. By Theorem 7.5,

$$\left(\frac{|\xi_1|^\sigma + \cdots + |\xi_n|^\sigma}{n}\right)^{1/\sigma} \leqslant T \tag{1}$$

represents a convex body symmetrical about the origin. We first evaluate the integral

$$A = \underset{|\xi_1|^\sigma + \cdots + |\xi_n|^\sigma \leqslant nT^\sigma}{\int \cdots \int} dx_1 \cdots dx_n.$$

Let $\xi_{r+j} = \eta_{r+j} + i\eta_{r+s+j}$, $\xi_{r+s+j} = \bar{\xi}_{r+j}$ $(j = 1, 2, \ldots, s)$ be the s pairs of linear forms with complex conjugate coefficients. Then

$$A = \underset{|\xi_1|^\sigma + \cdots + |\xi_r|^\sigma + 2\sum_{j=r+1}^{r+s} (\eta_j^2 + \eta_{s+j}^2)^{\sigma/2} \leqslant nT^\sigma}{\int \cdots \int} \left|\frac{\partial(x_1, \ldots, x_n)}{\partial(\xi_1, \ldots, \xi_r, \eta_{r+1}, \ldots, \eta_{r+2s})}\right|$$

$$\times\, d\xi_1 \cdots d\xi_r\, d\eta_{r+1} \cdots d\eta_{r+2s}$$

$$= \frac{2^s}{|\Delta|} \underset{|\xi_1|^\sigma + \cdots + |\xi_r|^\sigma + 2\sum_{j=r+1}^{r+s} (\eta_j^2 + \eta_{s+j}^2)^{\sigma/2} \leqslant nT^\sigma}{\int \cdots \int} d\xi_1 \cdots d\xi_r\, d\eta_{r+1} \cdots d\eta_{r+2s}.$$

We make the following substitutions:

$$\xi_1 = \rho_1, \ldots, \xi_r = \rho_r,$$

$$\eta_{r+j} = (\tfrac{1}{2})^{1/\sigma} \rho_{r+j} \cos \vartheta_{r+j}, \qquad \eta_{r+s+j} = (\tfrac{1}{2})^{1/\sigma} \rho_{r+j} \sin \vartheta_{r+j}, \qquad 1 \leqslant j \leqslant s.$$

This then gives

$$A = \frac{2^s \cdot 2^r(\frac{1}{2})^{2s/\sigma}}{|\Delta|} \int\limits_{\substack{\rho_1^\sigma + \cdots + \rho_{r+s}^\sigma \leqslant nT^\sigma \\ \rho_\nu \geqslant 0}} \cdots \int \left(\prod_{\nu=r+1}^{r+s} \rho_\nu\right) d\rho_1 \cdots d\rho_{r+s}$$

$$\times \int_0^{2\pi} \cdots \int_0^{2\pi} d\vartheta_{r+1} \cdots d\vartheta_{r+s}$$

$$= \frac{2^{n-\frac{2s}{\sigma}}\pi^s}{|\Delta|} \int\limits_{\substack{\rho_1^\sigma + \cdots + \rho_{r+s}^\sigma \leqslant nT^\sigma \\ \rho_\nu \geqslant 0}} \cdots \int \left(\prod_{\nu=r+1}^{r+s} \rho_\nu\right) d\rho_1 \cdots d\rho_{r+s}.$$

Let $\rho_\nu^\sigma = nT^\sigma \tau_\nu$, $\nu = 1, 2, \ldots, r+s$, so that

$$A = \frac{2^{n-\frac{2s}{\sigma}}\pi^s}{|\Delta|}(n^{1/\sigma}T)^n \left(\frac{1}{\sigma}\right)^{r+s} \int\limits_{\substack{\tau_1 + \cdots + \tau_{r+s} \leqslant 1 \\ \tau_\nu \geqslant 0}} \cdots \int \tau_1^{\frac{1}{\sigma}-1} \cdots \tau_r^{\frac{1}{\sigma}-1} \tau_{r+1}^{\frac{2}{\sigma}-1} \cdots \tau_{r+s}^{\frac{2}{\sigma}-1} d\tau_1 \cdots d\tau_{r+s}$$

$$= \frac{1}{|\Delta|} 2^{n-\frac{2s}{\sigma}} \pi^s (n^{1/\sigma}T)^n \left(\frac{1}{\sigma}\right)^{r+1} \frac{\Gamma^r\left(\frac{1}{\sigma}\right)\Gamma^s\left(\frac{2}{\sigma}\right)}{\Gamma\left(1 + \frac{n}{\sigma}\right)}$$

$$= (n^{1/\sigma}T)^n 2^{n-\frac{2s}{\sigma}} \left(\frac{\pi}{2}\right)^s \frac{\Gamma^r\left(1 + \frac{1}{\sigma}\right)\Gamma^s\left(1 + \frac{2}{\sigma}\right)}{|\Delta|\Gamma\left(1 + \frac{n}{\sigma}\right)}.$$

When

$$A \geqslant 2^n,$$

that is when

$$T \geqslant \left(\frac{\left(\frac{2}{\pi}\right)^s n^{-\frac{n}{\sigma}} \Gamma\left(1 + \frac{n}{\sigma}\right)|\Delta|}{2^{-\frac{2s}{\sigma}} \Gamma^r\left(1 + \frac{1}{\sigma}\right)\Gamma^s\left(1 + \frac{2}{\sigma}\right)}\right)^{1/n},$$

there is a non-zero lattice point satisfying (1). The theorem is proved. □

Theorem 8.2. *Let the hypothesis in Theorem* 8.1 *be satisfied. Suppose that* $\lambda_1, \ldots, \lambda_n$ *are positive numbers, with* $\lambda_{r+t} = \lambda_{r+s+t}$ $(t = 1, \ldots, s)$ *and* $(\lambda_1 \cdots \lambda_{r+2s}) \geqslant (\frac{2}{\pi})^s|\Delta|$. *Then there is a non-zero lattice point such that*

$$|\xi_1| \leqslant \lambda_1, \ldots, |\xi_n| \leqslant \lambda_n. \quad \square$$

The reader can supply the proof for this.

Theorem 8.3. *Let the hypothesis in Theorem* 8.1 *be satisfied, and let*

$$\xi_v = \eta_v \qquad (1 \leqslant v \leqslant r),$$

$$\xi_{r+v} = \eta_{r+v} + i\eta_{r+s+v}, \qquad \xi_{r+s+v} = \bar{\xi}_{r+v} \qquad (1 \leqslant v \leqslant s).$$

Suppose that $\lambda_1 \cdots \lambda_n \geqslant |\Delta|/2^s$. *Then there is a non-zero lattice point such that*

$$|\eta_v| \leqslant \lambda_v, \qquad 1 \leqslant v \leqslant n.$$

Proof. The absolute value of the determinant for $\eta_1, \eta_2, \ldots, \eta_n$ is $|\Delta|/2^s$, and therefore the theorem follows at once from Theorem 3.1. □

20.9 Tchebotaref's Theorem

Let

$$\xi_i = \sum_{j=1}^{n} \alpha_{ij} x_j \qquad (i = 1, \ldots, n),$$

where α_{ij} are real numbers, and the determinant for the coefficient is

$$\Delta = \begin{vmatrix} \alpha_{11} & \cdots & \alpha_{1n} \\ \cdots & \cdots & \cdots \\ \alpha_{n1} & \cdots & \alpha_{nn} \end{vmatrix} \neq 0.$$

A famous conjecture of Minkowski is as follows: Corresponding to any set of real numbers $\rho_1, \ldots, \rho_n$ there is a set of integers $x_1, \ldots, x_n$ (possibly all zero) such that

$$|(\xi_1 - \rho_1) \cdots (\xi_n - \rho_n)| \leqslant \frac{1}{2^n} |\Delta|.$$

Minkowski himself proved this for $n = 2$, and it has also been settled for $n = 3, 4, 5$. As for the general case we only have the following theorem.

Theorem 9.1 (Tchebotaref). *Let m be the lower bound for* $|(\xi_1 - \rho_1) \cdots (\xi_n - \rho_n)|$ *when* $x_1, \ldots, x_n$ *take integer values. Then we have*

$$m \leqslant 2^{-n/2} |\Delta|.$$

Proof. We can assume, without loss of generality, that $\Delta = 1$ and $m > 0$. Then, given $\varepsilon > 0$, there are integers $x_1^*, \ldots, x_n^*$ such that

$$\prod_{i=1}^{n} |\xi_i^* - \rho_i| = |(\xi_1^* - \rho_1) \cdots (\xi_n^* - \rho_n)| = \frac{m}{1 - \vartheta}, \qquad 0 \leqslant \vartheta < \varepsilon.$$

Let

$$\xi_i' = \frac{\xi_i - \xi_i^*}{\xi_i^* - \rho_i} \qquad (i = 1, \ldots, n),$$

so that

$$\xi_i' = \sum_{j=1}^{n} \beta_{ij}(x_j - x_j^*) \qquad (i = 1, \ldots, n),$$

and the absolute value of the determinant D for the forms is

$$|D| = \left(\prod_{j=1}^{n} |\xi_i^* - \rho_i| \right)^{-1} = \frac{1 - \vartheta}{m}.$$

Since $\prod_{j=1}^{n} |\xi_i - \rho_i| \geqslant m$, it follows that

$$\prod_{i=1}^{n} |\xi_i' + 1| = \prod_{i=1}^{n} \left| \frac{\xi_i - \rho_i}{\xi_i^* - \rho_i} \right| \geqslant 1 - \vartheta.$$

Similarly

$$\prod_{i=1}^{n} |\xi_i' - 1| \geqslant 1 - \vartheta.$$

Therefore

$$\prod_{i=1}^{n} |\xi_i'^2 - 1| \geqslant (1 - \vartheta)^2.$$

Define the convex region C':

$$|\xi_i'| < \sqrt{1 + (1 - \vartheta)^2} \qquad (i = 1, \ldots, n).$$

We now prove that C' has no non-zero lattice point.

Suppose that C' has a non-zero lattice point. Then the corresponding $\xi_1', \ldots, \xi_n'$ must satisfy

$$-1 \leqslant \xi_i'^2 - 1 < (1 - \vartheta)^2 \leqslant 1, \qquad |\xi_i'^2 - 1| \leqslant 1 \qquad (i = 1, \ldots, n).$$

If $\xi_i'^2 - 1 > -(1 - \vartheta)^2$ for some i, then $|\xi_i'^2 - 1| < (1 - \vartheta)^2$ for that i, and therefore

$$\prod_{i=1}^{n} |\xi_i'^2 - 1| < (1 - \vartheta)^2$$

which is impossible. Hence

$$-1 \leqslant \xi_i'^2 - 1 \leqslant -(1 - \vartheta)^2 \qquad (i = 1, \ldots, n),$$

and so

$$|\xi_i'| \leqslant \sqrt{1 - (1 - \vartheta)^2} \leqslant \sqrt{2\vartheta} \qquad (i = 1, \ldots, n).$$

Thus, when ϑ is very small, this lattice point in C' must be very close to the origin, and this leads at once to a contradiction. For, according to Theorem 2.3, any non-zero lattice point must lie outside $C'_{\frac{1}{2}}$, and this clearly contradicts $|\xi'_i| \leqslant \sqrt{2\vartheta}$ $(i = 1, \ldots, n)$. There is therefore no lattice point in C' besides the origin.

Now, by Theorem 2.1, we have

$$\frac{2^n\{1 + (1 - \vartheta)^2\}^{n/2}}{|D|} \leqslant 2^n,$$

or

$$\{1 + (1 - \vartheta)^2\}^{n/2} \leqslant \frac{1 - \vartheta}{m}.$$

As $\varepsilon \to 0$, so that $\vartheta \to 0$, we have $m \leqslant 2^{-n/2}$. □

20.10 Applications to Algebraic Number Theory

Let $\omega_1, \ldots, \omega_n$ be an integral basis for the algebraic number field $R(\vartheta)$ of degree n. Suppose that there are r_1 real numbers, and r_2 pairs of complex conjugate numbers $(r_1 + 2r_2 = n)$ in $\vartheta^{(1)}, \ldots, \vartheta^{(n)}$. Then in the n linear forms

$$\alpha^{(i)} = \omega_1^{(i)}x_1 + \cdots + \omega_n^{(i)}x_n \qquad (i = 1, 2, \ldots, n),$$

there are r_1 forms with real coefficients, and r_2 pairs of forms having complex conjugate coefficients. It is also clear that the absolute value of the determinant of this set of forms is $\sqrt{|\Delta|}$ where Δ is discriminant of the field $R(\vartheta)$. Let $\alpha = \alpha^{(1)}$, and take $\sigma = 1$ in Theorem 8.1, so that there is a set of rational integers $x_1, \ldots, x_n$, not all zero, such that

$$|N(\alpha)|^{1/n} \leqslant \frac{1}{n}\sum_{i=1}^{n} |\alpha^{(i)}| \leqslant \left(\left(\frac{4}{\pi}\right)^{r_2}\frac{n!}{n^n}\sqrt{|\Delta|}\right)^{1/n},$$

and this means that there is a non-zero algebraic integral α satisfying

$$|N(\alpha)| \leqslant \left(\frac{4}{\pi}\right)^{r_2}\frac{n!}{n^n}\sqrt{|\Delta|}. \tag{1}$$

Since $|N(\alpha)|$ is a natural number, and $2r_2 \leqslant n$, we deduce that

$$\sqrt{|\Delta|} \geqslant \left(\frac{\pi}{4}\right)^{r_2}\frac{n^n}{n!} \geqslant \left(\frac{\pi}{4}\right)^{n/2}\frac{n^n}{n!}. \tag{2}$$

Let

$$v_n = \left(\frac{\pi}{4}\right)^{\frac{n}{2}}\frac{n^n}{n!},$$

so that

$$\frac{v_{n+1}}{v_n} = \frac{\sqrt{\pi}}{2}\left(1+\frac{1}{n}\right)^n \geqslant \pi^{\frac{1}{2}} > 1.$$

Thus $\{v_n\}$ is an increasing and unbounded sequence. Also, when $n = 2$,

$$\sqrt{|\Delta|} \geqslant v_2 = \frac{\pi}{2} > 1.$$

We have therefore the following two theorems:

Theorem 10.1. *The only field with discriminant* 1 *is the rational number field.* □

Theorem 10.2. *If Δ is a rational integer, then there is a finite number $n(\Delta)$ such that any algebraic number field with discriminant Δ has degree at most $n(\Delta)$.* □

Actually we can take one step further and prove:

Theorem 10.3. *Corresponding to each fixed rational integer Δ, there are at most a finite number of algebraic number fields with discriminant Δ.*

Proof. By Theorem 10.2 it suffices to prove that, given any natural number n, the number of algebraic number fields with degree n and discriminant Δ is finite.

Let $R(\vartheta)$ be a field with discriminant Δ and of degree n, and let $\omega_1, \ldots, \omega_n$ be its integral basis. Let

$$\alpha^{(i)} = \omega_1^{(i)}x_1 + \cdots + \omega_n^{(i)}x_n \qquad (i = 1, \ldots, n),$$

and define r_1, r_2 as before. We can assume, without loss of generality, that $\alpha^{(1)} = \alpha, \alpha^{(2)}, \ldots, \alpha^{(r_1)}$ have real coefficients, and $\alpha^{(r_1+1)}, \ldots, \alpha^{(n)}$ have complex coefficients, and that

$$\overline{\alpha^{(r_1+v)}} = \alpha^{(r_1+r_2+v)}, \qquad 1 \leqslant v \leqslant r_2.$$

Let

$$\alpha^{(v)} = \eta_v \qquad (1 \leqslant v \leqslant r_1),$$

$$\alpha^{(r_1+v)} = \eta_{r_1+v} + i\eta_{r_1+r_2+v} \qquad (1 \leqslant v \leqslant r_2),$$

so that, by Theorem 8.3, there are rational integers $x_1^*, \ldots, x_n^*$, not all zero, such that

$$|\eta_1^*| \leqslant \tfrac{1}{2}, \quad \ldots, \quad |\eta_{n-1}^*| \leqslant \tfrac{1}{2}, \quad |\eta_n^*| \leqslant 2^{n-1}\sqrt{|\Delta|}. \tag{3}$$

Therefore there exists a constant c, depending only on n, such that

$$|\alpha^{*(i)}| < c\sqrt{|\Delta|} \qquad (i = 1, 2, \ldots, n). \tag{4}$$

If we can prove that

$$\alpha^{*(n)} \neq \alpha^{*(i)} \qquad (i = 1, \ldots, n-1),$$

then, from Theorem 16.3.1, we see that α^* is an algebraic number of degree n, and it is easy to prove that $R(\vartheta) = R(\alpha^*)$. Let the irreducible equation satisfied by α^* be

$$f(x) = x^n + a_1 x^{n-1} + \cdots + a_n = 0, \tag{5}$$

so that each a_k must satisfy

$$|a_k| \leqslant \binom{n}{k}(c\sqrt{|\Delta|})^k \qquad (k = 1, \ldots, n). \tag{6}$$

Thus any field $R(\vartheta)$ of degree n and with discriminant Δ must be the same as a certain $R(\alpha^*)$, where α^* is a root of a certain irreducible equation (5) which satisfies the condition (6). Since the number of such irreducible equations is finite, the theorem follows. Therefore it remains to prove that

$$\alpha^{*(n)} \neq \alpha^{*(i)} \qquad (i = 1, \ldots, n-1). \tag{7}$$

If $r_2 = 0$, then $\alpha^{*(v)} = \eta_v^*$ $(v = 1, \ldots, n)$. From (3) we have

$$1 \leqslant |N(\alpha^*)| \leqslant \frac{1}{2^{n-1}} |\alpha^{*(n)}|.$$

But

$$|\alpha^{*(i)}| \leqslant \tfrac{1}{2} < 2^{n-1} \leqslant |\alpha^{*(n)}| \qquad (i = 1, \ldots, n-1),$$

so that (7) is established.

If $r_2 > 0$, then, for $1 \leqslant v \leqslant r_2 - 1$,

$$|\alpha^{*(r_1+v)}| = |\eta^*_{r_1+v} + i\eta^*_{r_1+r_2+v}| \leqslant \frac{1}{\sqrt{2}},$$

$$|\alpha^{*(r_1+r_2+v)}| = |\eta^*_{r_1+v} - i\eta^*_{r_1+r_2+v}| \leqslant \frac{1}{\sqrt{2}}.$$

Thus

$$1 \leqslant |N(\alpha^*)| \leqslant \frac{1}{(\sqrt{2})^{n-2}} |\alpha^{*(n)}|^2.$$

But

$$|\alpha^{*(i)}| \leqslant \frac{1}{\sqrt{2}} < 2^{\frac{1}{4}(n-2)} \leqslant |\alpha^{*(n)}|, \qquad i \neq n, \qquad i \neq r_1 + r_2,$$

and $\alpha^{*(r_1+r_2)} \neq \alpha^{*(n)}$, since otherwise $\eta_n^* = 0$, so that $|\alpha^{*(n)}| \leqslant \frac{1}{2}$, giving

$$1 \leqslant |N(\alpha^*)| \leqslant \frac{1}{(\sqrt{2})^{n+2}},$$

which is impossible. Therefore (7) also holds when $r_2 > 0$. The theorem is proved. □

Exercise 1. Prove that we can always select an integer α from the ideal $\mathfrak{a}$ such that

$$|N(\alpha)| \leqslant \sqrt{|\Delta|}\, N(\mathfrak{a}).$$

Exercise 2. Prove that, given any ideal class, there is an ideal $\mathfrak{a}$ satisfying

$$N(\mathfrak{a}) \leqslant \sqrt{|\Delta|}.$$

20.11 The Least Value for $|\Delta|$

We saw in the previous section that the discriminant Δ of an algebraic number field of degree n satisfies

$$|\Delta| \geqslant \left(\frac{\pi}{4}\right)^{2r_2} \left(\frac{n^n}{n!}\right)^2.$$

Moreover, from $\Delta \equiv 0$ or $1 \pmod 4$, and $(-1)^{r_2}\Delta > 0$, we can construct the following table:

	$r_2 = 0$	$r_2 = 1$	$r_2 = 2$	
$n = 2$	$\Delta \geqslant 4$	$\Delta \leqslant -3$	–	(I)
$n = 3$	$\Delta \geqslant 21$	$\Delta \leqslant -15$	–	
$n = 4$	$\Delta \geqslant 116$	$\Delta \leqslant -71$	$\Delta \geqslant 44$	
$n = 5$	$\Delta \geqslant 680$	$\Delta \leqslant -419$	$\Delta \geqslant 260$	

But actually the least value for $|\Delta|$ can be calculated to give

	$r_2 = 0$	$r_2 = 1$	$r_2 = 2$	
$n = 2$	$\Delta = 5$	$\Delta = -3$	–	(II)
$n = 3$	$\Delta = 49$	$\Delta = -23$	–	
$n = 4$	$\Delta = 725$	$\Delta = -275$	$\Delta = 117$	

The case $n = 2$ in Table (II) follows at once from considering the quadratic fields $R(\sqrt{5})$ and $R(\sqrt{-3})$.

When $n = 3$, if ϑ satisfies $x^3 + x^2 - 2x - 1 = 0$, then the discriminant of $R(\vartheta)$ is 49, and if ϑ satisfies $x^3 - x - 1 = 0$, then the discriminant of $R(\vartheta)$ is -23.

When $n = 4$, we let ϑ be a root of

$$x^4 - 2ax^2 + (-1)^{\frac{1}{2}(p-1)}p = 0.$$

The following can then be proved:

1) When $a = 7$, $p = 29$, we have $r_2 = 0$, $\Delta = 725$;
2) When $a = 3$, $p = 11$, we have $r_2 = 1$, $\Delta = -275$;
3) When $a = -1$, $p = 13$, we have $r_2 = 2$, $\Delta = 117$.

The actual construction of Table (II) presents a problem. The case $n = 2$ in the table is very easily settled. When $n \geqslant 3$, the proof of Theorem 10.3 gives us a method whereby after a "finite number" of calculations we can arrive at the results given in Table (II). However, in actual practice, this method requires the calculations of the roots of about one thousand polynomial equations and the determination of the discriminants of the corresponding algebraic number fields. In order to solve this concrete problem we need a practical method. We now examine the situation when $n = 3$.

Suppose that the cubic field $R(\vartheta)$ in our discussion has discriminant Δ which satisfies $0 < \Delta \leqslant 49$ $(r_2 = 0)$, or $-23 \leqslant \Delta < 0$ $(r_2 = 1)$. From §10 we see that there is a non-zero integer α in this field such that

$$|\alpha^{(1)}| + |\alpha^{(2)}| + |\alpha^{(3)}| \leqslant \tau, \tag{1}$$

and

$$3 < \tau = \begin{cases} 42^{\frac{1}{3}} \\ 2\left(\dfrac{3}{\pi}\right)^{\frac{1}{3}} 23^{\frac{1}{6}}. \end{cases}$$

The degree of α is either 3 or 1. Suppose that the degree of α is definitely 3 so that α cannot be a rational integer and hence $R(\vartheta) = R(\alpha)$. From the inequality (1) we can determine a bound for the coefficients for the equations satisfied by α, and the eventual result can be obtained after a finite number of calculations. Unfortunately we have no way of ensuring that α is not a rational integer. On the contrary, from $\tau > 3$, we see that $\alpha = \pm 1$ do satisfy (1) and ± 1 belong to $R(\vartheta)$; therefore this method is not applicable.

Let $\rho > 3$ and consider the convex body B:

$$|\xi_1| + |\xi_2| + |\xi_3| \leqslant \rho,$$

$$|\xi_1 + \xi_2 + \xi_3| < 3 \ (< \rho),$$

where

$$\xi_i = \omega_1^{(i)} x_1 + \omega_2^{(i)} x_2 + \omega_3^{(i)} x_3,$$

and $\omega_1, \omega_2, \omega_3$ is an integral basis for $R(\vartheta)$. It is easy to see that B is a convex body symmetrical about the origin.

Denote by $F(t)$ the area of the intersection between the convex body A:

$$|\xi_1| + |\xi_2| + |\xi_3| \leqslant \rho$$

and the plane $\xi_1 + \xi_2 + \xi_3 = t$. Then $F(t) = F(-t)$, and when $t \geqslant 0$, $F(t)$ is decreasing. Therefore

$$\text{Volume of } B = 2\int_0^3 F(t)\,dt = 2\frac{3}{\rho}\int_0^\rho F\left(\frac{3}{\rho}u\right)du$$

$$\geqslant 2\frac{3}{\rho}\int_0^\rho F(u)\,du = \frac{3}{\rho} \times \text{Volume of } A.$$

But

$$\text{Volume of } A = \begin{cases} 2^3 \dfrac{\rho^3}{3!\sqrt{49}}, & \text{when } r_2 = 0; \\ 2^3 \left(\dfrac{\pi}{4}\right)\dfrac{\rho^3}{3!}\dfrac{1}{\sqrt{23}}, & \text{when } r_2 = 1. \end{cases}$$

Therefore, by Minkowski's theorem, there is a non-zero integer α in $R(\vartheta)$ satisfying

$$|\alpha^{(1)}| + |\alpha^{(2)}| + |\alpha^{(3)}| \leqslant \tau' = \begin{cases} \sqrt{14}, & \text{when } r_2 = 0; \\ \sqrt{\dfrac{8}{\pi}\sqrt{23}}, & \text{when } r_2 = 1 \end{cases} \tag{2}$$

and

$$|\alpha^{(1)} + \alpha^{(2)} + \alpha^{(3)}| < 3. \tag{3}$$

Now we see from (3) that α certainly cannot be a rational integer. Therefore α has degree 3 and $R(\vartheta) = R(\alpha)$. Let the irreducible equation satisfied by α be

$$f(x) = x^3 - g_1x^2 + g_2x - g_3 = 0. \tag{4}$$

Then $g_3 \neq 0$, and we can assume that $g_3 > 0$. For, if otherwise, from $-\alpha$ satisfying the equation

$$g(x) = x^3 - (-g_1)x^2 + g_2x - (-g_3) = 0,$$

and $R(\vartheta) = R(\alpha) = R(-\alpha)$, and $-\alpha$ also satisfying (2) and (3), we can replace g_3 by $-g_3$.

From the relationship between the roots and the coefficients we have

$$|g_1| = |\alpha^{(1)} + \alpha^{(2)} + \alpha^{(3)}| < 3,$$

$$g_3 = \alpha^{(1)}\alpha^{(2)}\alpha^{(3)} \leqslant \left(\frac{|\alpha^{(1)}| + |\alpha^{(2)}| + |\alpha^{(3)}|}{3}\right)^3 < 2,$$

so that $|g_1| \leqslant 2$ and $g_3 = 1$. Finally we find a bound for g_2 by

$$\begin{aligned}|g_2| &= |\alpha^{(1)}\alpha^{(2)} + \alpha^{(1)}\alpha^{(3)} + \alpha^{(2)}\alpha^{(3)}| \\ &\leqslant |\alpha^{(1)}\alpha^{(2)}| + |\alpha^{(1)}\alpha^{(3)}| + |\alpha^{(2)}\alpha^{(3)}| \\ &\leqslant \frac{(|\alpha^{(1)}| + |\alpha^{(2)}| + |\alpha^{(3)}|)^2}{3} \leqslant \frac{\tau'^2}{3} < 5,\end{aligned}$$

so that $|g_2| \leqslant 4$. But when $r_2 = 0$ we actually have $|g_2| \leqslant 3$. For we now have $\alpha^{(i)}$ $(i = 1, 2, 3)$ being all real, so that either they all have the same sign or there are exactly two of them having the same sign. In the former situation we have

$$\begin{aligned}|g_2| &\leqslant |\alpha^{(1)}\alpha^{(2)}| + |\alpha^{(1)}\alpha^{(3)}| + |\alpha^{(2)}\alpha^{(3)}| \\ &\leqslant \frac{(|\alpha^{(1)}| + |\alpha^{(2)}| + |\alpha^{(3)}|)^2}{3} = \frac{(\alpha^{(1)} + \alpha^{(2)} + \alpha^{(3)})^2}{3} < 3,\end{aligned}$$

and in the latter situation, we can assume that $\alpha^{(1)}\alpha^{(2)} > 0$ and $\alpha^{(1)}\alpha^{(3)} < 0$, so that

$$\begin{aligned}|g_2| &\leqslant |\alpha^{(1)}\alpha^{(2)} + \alpha^{(1)}\alpha^{(3)} + \alpha^{(2)}\alpha^{(3)}| \\ &\leqslant \max(\alpha^{(1)}\alpha^{(2)}, -\alpha^{(3)}(\alpha^{(1)} + \alpha^{(2)})) \\ &\leqslant \left(\frac{\alpha^{(1)} + \alpha^{(2)} - \alpha^{(3)}}{2}\right)^2 \leqslant \frac{14}{4} < 4,\end{aligned}$$

that is $|g_2| \leqslant 3$.

Summarizing the above, in any cubic field $R(\vartheta)$ with discriminant Δ satisfying $0 < \Delta \leqslant 49$ $(r_2 = 0)$ or $-23 \leqslant \Delta < 0$ $(r_2 = 1)$ there is an integer α such that $R(\vartheta) = R(\alpha)$, and α satisfies an irreducible equation

$$x^3 - g_1x^2 + g_2x - 1 = 0,$$

with $|g_1| \leqslant 2$, $|g_2| \leqslant 4$ (when $r_2 = 0$, $|g_2| \leqslant 3$). Therefore, in order to determine cubic fields $R(\vartheta)$ with discriminant Δ satisfying $0 < \Delta \leqslant 49$ $(r_2 = 0)$ or $-23 \leqslant \Delta < 0$ $(r_2 = 1)$, we need only examine these irreducible equations. But the number of such equations is at most 45 (at most 35 when $r_2 = 0$). Moreover, when $g_1 = g_2$ the equation has the root 1, and when $g_1 + g_2 + 2 = 0$ the equation has the root -1, so that we have no need to examine these reducible equations. Finally since the roots of $x^3 - g_2x^2 + g_1x - 1 = 0$ are the reciprocals of the roots of $x^3 - g_1x^2 + g_2x - 1 = 0$, and $R(\vartheta) = R(1/\vartheta)$, the reciprocal equation to (4) need not be examined either. We are then left with 27 (18 when $r_2 = 0$) equations to be considered. We then calculate the roots ϑ of these 27 (or 18) equations and then determine the discriminants for $R(\vartheta)$ to arrive at the results for $n = 3$ in Table (II).

Bibliography

1. Baker, A.: Linear forms in the logarithm of algebraic numbers. Mathematatika *13* (1966) 204–216. (II) Mathematika *14* (1967) 102–107. (III) Mathematika *14* (1967) 220–228. (IV) Mathematika *15* (1968) 204–216
2. Baker, A.: Contribution to the theory of Diophantine equations I: On the representation of integers by binary forms. Phil. Tran. Roy. Soc. London, *A 263* (1967) 273–291
3. Baker, A.: On the class number of quadratic fields. Bull. London Math. Soc. *1* (1969) 98–102
4. Baker, A.: Transcendental number theory. Cambridge University Press (1975)
5. Balasubramanian, R.: On Waring's problem: $g(4) \leqslant 21$. Hardy-Ramanujan Journal *2* (1979) 1–32
6. Barban, M. B.: Arithmetic functions on thin sets. [Russian]. Dokl. UzSSR *8* (1961) 9–11
7. Barban, M. B.: The density of the zeros of Dirichlet L-series and the problem of the sum of primes and almost primes. [Russian]. Mat. Sbornik (N. S.) *61* (103) (1963) 418–425
8. Bombieri, E.: Sulle formula di A. Selberg generalizzate per classi di funzioni aritmetiche e le applicazioni al problema del resto nel "Primzahlsatz". Riv. Mat. Univ. Parma *2*; *3* (1962) 393–440
9. Bombieri, E.: On the large sieve. Mathematika *12* (1965) 201–225
10. Bombieri, E.: Le grand crible dans la théorie analytique des nombres. Sociéte Mathématique de France *18* (1974)
11. Bombieri, E., and Davenport, H.: Small differences between prime numbers. Proc. Roy. Soc. Ser. *A*, *293* (1966) 1–18
12. Burgess, D. A.: The distribution of quadratic residues and non-residues. Mathematika *4* (1957) 106–112
13. Burgess, D. A.: On character sums and primitive roots. Proc. London Math. Soc. *12* (1962) 179–192
14. Buchstab, A. A.: New results in the investigation of the Goldbach-Euler problem and the problem of prime pairs. [Russian]. Dokl. Akad. Nauk SSSR *162* (1965) 735–738 = Soviet Math. Dokl. *6* (1965) 729–732
15. Cassels, J. W. S.: An introduction to Diophantine approximation. Cambridge Tracts in Mathematics *45* (1957)
16. Chao, K.: On the diophantine equation $x^2 = y^n + 1$, $xy \neq 0$. Sci. Sin. *14*, 3 (1965) 457–460
17. Chen, J. R.: On the circle problem. [Chinese]. Acta Math. Sinica *13* (1963) 299–313
18. Chen, J. R.: On Waring's problem: $g(5) = 37$. [Chinese]. Acta Math. Sinica *14* (1964) 715–734
19. Chen, J. R.: On the representation of a large even integer as the sum of a prime and the product of at most two primes. [Chinese]. Kexue Tongbao *17* (1966) 385–386
20. Chen, J. R.: On the representation of a large even integer as the sum a prime and the product of at most two primes. Sci. Sinica *16* (1973) 157–176
21. Diamond, H., and Steinig, G. J.: An elementary proof of the prime number theorem with a remainder term. Inventiones Math. *11* (1970) 199–258
22. Dickson, L. E.: History of the theory of numbers. (Three volumes). Carnegie Institute, Washington (1919, 1920, 1923)
23. Elliot, P. D. T. A., and Halberstam, H.: Some applications of Bombieri's theorem. Mathematika *13* (1966) 196–203
24. Estermann, T.: Introduction to modern prime number theory. Cambridge Tracts in Mathematics *41* (1952)
25. Gauss, C. F.: Disquisitiones arithmeticae. Leipzig, Fleisher, (1801). English translation: A. A. Clarke, Yale University Press (1966)

26. Hagis, Jr., P.: A lower bound for the set of odd perfect numbers. Math. Comp. *27*; 124 (1973) 951–953
27. Hagis, Jr., P., and McDaniel, W. L.: On the largest prime divisor of an odd perfect number, II. Math. Comp., *29* (1975) 922–924
28. Halberstam, H., and Richert, H.-E.: Sieve methods. Academic Press, London (1974)
29. Hardy, G. H., and Wright, E. M.: An introduction to the theory of numbers. 4th ed. Oxford (1960)
30. Hua, L. K.: Die Abschätzungen von Exponentialsummen und ihre Anwendung in der Zahlentheorie. Enzykl. Math. Wiss., *I, 2*, Heft 13. Teil I. Leipzig (1959)
31. Huxley, M. N.: On the difference between consecutive primes. Inventiones Math. *16* (1972) 191–201
32. Huxley, M. N.: Small differences between consecutive primes. Mathematika, *20*; 2 (1973) 229–232
33. Ingham, A. E.: The distribution of prime numbers. Cambridge Tracts in Mathematics *30* (1932)
34. Kolesnik, G. A.: The refined error term of the divisor problem. [Russian]. "Mat. Zametki" *6* (1969) 545–554
35. Korobov, N. M.: On the estimation of trigonometric sums and its applications. [Russian]. Uspeki Math. Nauk SSSR *13* (1958) 185–192
36. Landau, E.: Handbuch der Lehre von der Verteilung der Primzahlen. (2 Bände). Leipzig, Teubner (1909)
37. Landau, E.: Vorlesungen über Zahlentheorie. (3 Bände). Leipzig, Hirzel (1927)
38. Landau, E.: Über einige neuere Fortschritte der additiven Zahlentheorie. Cambridge Tracts in Mathematics *35* (1937)
39. Lavrik, A. V., and Soberov, A. S.: On the error term of the elementary proof of the prime number theorem. [Russian]. Dokl. Adad. Nauk SSSR *211* (1973) 534–536
40. Linnik, Yu. V.: The dispersion method in binary additive problems. Leningrad, (1961). = Providence, R.I. (1963)
41. Mahler, K.: On the fractional parts of the powers of a rational number, II. Mathematika *4* (1957) 122–124
42. Minkowski, H.: Geometrie der Zahlen. Leipzig, Teubner (1910)
43. Minkowski, H.: Diophantine Approximation. Leipzig, Teubner (1927)
44. Montgomery, H. L.: Topics in Multiplicative Number Theory. Springer Lecture Notes *227* (1971)
45. Pan, C. T.: On the least prime in an arithmetic progression. [Chinese]. Sci Rec., New Ser. *1* (1957) 283–286
46. Pan, C. T.: On the representation of an even integer as a sum of a prime and an almost prime. [Chinese]. Acta Math. Sinica *12* (1962) 95–106 = Chinese Math.-Acta *3* (1963) 101–112
47. Pan, C. T.: On the representation of even numbers as the sum of a prime and a product of at most 4 primes. [Chinese]. Acta Sci. Natur. Univ. Shangtung *2* (1962) 40–62 = Sci. Sinica *12* (1963) 455–474. [Russian]
48. Pan, C. T., Ding, X. X., and Wang, Y.: On the representation of a large even integer as a sum of a prime and an almost prime. Kexu Tongbao *8* (1975) 358–360
49. Richert, H.-E.: Zur multiplikativen Zahlentheorie. J. reine angew. Math. *206* (1961) 31–38
50. Roth, K. F.: On the large sieves of Linnik and Rényi. Mathematika *12* (1965) 1–9
51. Schmidt, W. M.: Simultaneous approximations to algebraic numbers by rationals. Acta Math. *125* (1970) 189–201
52. Schmidt, W. M.: Diophantine Approximations. Springer Lecture Notes *785* (1980)
53. Sierpiński, W.: Elementary theory of numbers. Warszawa (1964)
54. Slowinski, D.: Searching for the 27th Mersenne prime. J. Recreational Mathematics *11* (1979) 258–261
55. Stark, H. M.: A complete determination of the complex quadratic fields of class number 1. Michigan Math. J. *14* (1967) 1–27
56. Stepanov, S. A.: On the estimation of Weyl's sums with prime denominators. [Russian]. Uzv. Akad. Nauk. SSSR, Ser. Mat. (1970) 1015–1037
57. Titchmarsh, E. C.: The theory of the Riemann zeta-function. Oxford (1951)
58. Vaughan, R. C.: A note on Šnirel'man's approach to Goldbach's problem. Bull. London Math. Soc. *8* (1976) 245–250
59. Vinogradov, A. I.: The density hypothesis for Dirichlet *L*-series. [Russian]. Izv. Akad. Nauk SSSR Ser. Mat. *29* (1965) 903–934. Corrigendum: ibid. *30* (1966) 719–720

60. Vinogradov, I. M.: On a new estimation of a function $\zeta(1 + it)$. [Russian]. Izv. Akad. Nauk SSSR, Ser. Mat. *22* (1958) 161 – 164
61. Vinogradov, I. M.: On the problem of the upper estimation for $G(m)$. [Russian]. Izv. Akad. Nauk SSSR, Ser. Mat. *23* (1959) 637 – 642
62. Wang, Y.: On the least primitive root. [Chinese]. Acta Math. Sinica *9* (1959) 432 – 441
63. Wang, Y.: On the estimation of character sums and its applications. [Chinese]. Sci. Record (N. S.) *7* (1964) 78 – 83
64. Wirsing, E.: Elementare Beweise des Primzahlsatzes mit Restglied, II. J. Reine Angew. Math. *214*/215 (1964) 1 – 18
65. Yin, W. L.: On Dirichlet's divisor problem. Sci. Rec., New Ser. *3* (1959) 131 – 134

Index

L.-K. Hua, Y. Wang

Applications of Number Theory to Numerical Analysis

1981. IX, 241 pages. ISBN 3-540-10382-1
Distribution rights for The People's Republic of China: Science Press, Beijing

Number theoretic methods are used in numerical analysis to construct a series of uniformly distributed sets in the s-dimensional unit cube $G_s(s \geq 2)$; these sets are then used to calculate an approximation of a definite integral over G_s with the best possible order of error, significantly improving existing methods of approximation. The methods can also be used to construct an approximating polynomial for a periodic function of s variables and in the numerical solution of some integral equations and PDEs.

Many important methods and results in number theory, especially those concerning the estimation of trigonometrical sums and simultaneous Diophantine approximations as well as those of classical algebraic number theory may be used to construct the uniformly distributed sequence in G_s. This monography, by authors who have contributed significantly to the field, describes methods using a set of independent units of the cyclotomic field and by using the recurrence formula defined by a Pisot-Vijayaraghavan number. Error estimates and applications to numerical analysis are given; the appendix contains a table of *glp* (good lattice point) sets.

The volume is accessible to readers with a knowledge of elementary number theory.

Contents: Algebraic Number Fields and Rational Approximation. – Recurrence Relations and Rational Approximation. – Uniform Distribution. – Estimation of Discrepancy. – Uniform Distribution and Numerical Integration. – Periodic Functions. – Numerical Integration of Periodic Functions. – Numerical Error for Quadrature Formula. – Interpolation. – Approximate Solution of Integral Equations and Differential Equations. – Appendix: Tables. – Bibliography.

Springer-Verlag
Berlin
Heidelberg
New York

R. K. Guy

Unsolved Problems in Number Theory

1981. 17 figures. XVIII, 161 pages
(Problem Books in Mathematics, Volume 1)
ISBN 3-540-90593-6

Number theory has intrigued amateurs and professionals for a longer time than any other branch of mathematics. While it has developed to a level of considerable complexity, there are now more unsolved problems than ever before, challenging mathematicians from one generation to the next.

Unsolved Problems in Number Theory brings together 178 open questions organized into six categories: prime numbers, divisibility, additive number theory, diophantine equations, sequences of integers, and miscellaneous. Extensive and up-to-date references are provided to prevent the repetition of earlier efforts, or the duplication of previously known results, on the part of the research mathematician. The book will also stimulate the interest of the student, giving him rapid access to numerous aspects of number theory. In addition to being the first release in Springer's new *Problem Book* series, this publication is the initial installment of a multi-volume project by H. T. Croft (Cambridge University), and Richard K. Guy (the University of Calgary), entitled *Unsolved Problems in Intuitive Mathematics.*

Springer-Verlag
Berlin
Heidelberg
New York

DATE DUE

FEB 2 0 1984			
JAN 3 1 1989			
DEC 1 6 1989			
JAN 3 1 [illegible]			

DEMCO 38-297